Knowledge Preservation in Nuclear Safety

Bengt Lydell

Knowledge Preservation in Nuclear Safety

Advancements in Probabilistic Assessments of Piping Pressure Boundary Integrity

Springer

Bengt Lydell
Sigma-Phase Inc.
Vero Beach, FL, USA

ISBN 978-3-032-25462-7 ISBN 978-3-032-25463-4 (eBook)
https://doi.org/10.1007/978-3-032-25463-4

This Springer imprint is published by the registered company Springer Nature Switzerland AG
The registered company address is: Gewerbestrasse 11, 6330 Cham, Switzerland

Competing Interests The author has no competing interests to declare that are relevant to the content of this manuscript.

Contents

Nomenclature

Roman

A	Area of hole in pipe
$^{a}\!/_{t}$	Aspect ratio; ratio of a flaw's depth to its length
B	Learning curve factor
C	Number of ruptures, also consequence
C_x	Pipe wall thickness factor
C_z	Design code correction factor
CFP	Conditional failure probability
D	Pipe diameter
d	Diameter of the orifice and also hydraulic diameter
E	Engineering knowledge
$E(p^t)$	Moment generation function of the beta distribution
F	Age factor
F_C	Frequency of a catastrophic rupture
f	Fraction of piping components susceptible to a certain degradation mechanism
g	Gravity
h_l	Loss of static pressure head
$h(t)$	Hazard function
IF	Initiating event frequency
I_{ik}	Integrity management factor, which is a function of the reliability of leak detection and non-destructive examination. Ratio of hazard rates for weld type i and material degradation mechanism k, subject to epistemic uncertainty
ksi	Kilo-pound per square inch
L	Number of leaks
L_D	Number of susceptible locations
l	Leg length of fillet weld
MPa	Megapascal
N	Number of welds, bends, elbows, or linear meters of piping

n_{ik}	Number of failures in pipe component (e.g., weld) type i due to degradation mechanism k; very little epistemic uncertainty. The component boundary used in defining exposure terms is a function of susceptibility to certain material degradation mechanism
P	Pressure (in Pascal)
P_i	Probability of pipe pressure boundary failure
P_{Leak}	Probability of leak
$P_{Severance}$	Probability of a severance
$P(R_x \mid DC_{ik})$	Conditional probability of pipe failure with consequence x in terms of an equivalent break size or through-wall flow rate, given a degraded pipe material condition (DC) for weld of type i susceptible to degradation mechanism k
Q	Volumetric flow rate (liter per second)
Q_E	Influence on integrity by pipe size and "shape differences"
q	Volumetric flow rate (m^3/s)
R	Number of significant structural failures
RF	Range factor
S	Material degradation scenario
SSF	Significant structural failure
T	Total exposure time in reactor operating years, effective full power years (EFFY), or effective shutdown years (ESDY)
T_R	Time to repair; also, MTTR
t_D	Pipe wall thickness
$\sqrt{\frac{x_{95}}{x_5}}$	Range factor

Greek

α	Parameter of a beta or gamma distribution which influences the "peakedness" relative to the mean value
β	Parameter of a beta or gamma distribution which influences the "spread" of the distribution
γ	Cohort effects coefficient
λ_B	Frequency of a break
λ_L	Frequency of a leak
μ	Repair rate
σ	Standard deviation
σ^2	Variance
ω	Markov model transition rate representing inspect-and-repair
$\Gamma(\alpha)$	Gamma function evaluated at α

Abbreviations and Acronyms

ABB	ASEA Brown Boveri
ACRS	Advisory Committee on Reactor Safeguards
AEC	Atomic Energy Commission (of the United States)

AECL	Atomic Energy of Canada Limited
AFCEN	Association Française pour les règles de conception et de construction des matériels des Chaudières Électro Nucléaires (French Association for Design, Construction and Surveillance Rules of Nuclear Power Plant Components)
AFNOR	Association Française de Normalisation
AFW	Auxiliary feedwater
AI	Artificial intelligence
AIF	Atomic Industrial Forum
AIPA	Accident Initiation and Progression Analysis
AIV	Acoustically induced vibration
AKS	Anlagenkennzeichnungssystem
AMP	Ageing Management Program
AMR	Ageing Management Review
ANS	American Nuclear Society
AO	Abnormal occurrence
AOO	Anticipated operational occurrence
AOT	Allowed outage time
APEX	Application Express (of Oracle)
ASAR	As-Operated Safety Analysis Report
ASME	American Society of Mechanical Engineers
ASNR	Autorité de Sûreté Nucléaire et de Radioprotection
ASNT	American Society for Nondestructive Testing
ASP	Accident Sequence Precursor
ASTM	American Society for Testing and Materials
AtSMV	Atomrechtliche Sicherheitsbeauftragten- und Meldeverordnung (Ordinance on the Nuclear Safety Officer and the Reporting of Incidents and other Events)
AVR	Arbeitsgemeinschaft Versuchsreaktor
AWS	American Welding Society
BAM	Bundesanstalt für Materialforschung und -Prüfung (Federal Institute for Materials Research and Testing)
BCCN	Bureau de Contrôle des Chaudières Nucléaires (Nuclear Steam Supply Systems Control Office)
BDBA	Beyond Design Basis Accident
BMFT	Bundesministerium für Forschung und Technologie
BNCS	Board on Nuclear Codes and Standards (of ASME)
BP	Break preclusion
BPC	Break preclusion concept
BPVC	Boiler and Pressure Vessel Code (of ASME)
BTP	Branch Technical Position
BWR	Boiling water reactor
CAD	Computer-aided design
CANDU	Canada Deuterium Uranium

CANES	Center for Advanced Nuclear Energy Systems (of MIT)
CC	Component cooling (water system)
CCDP	Conditional Core Damage Probability
CCWS	Component cooling water system
CDF	Core damage frequency
CDP	Core damage probability
CEA	Commissariat à l'Énergie Atomique
CFD	Computational fluid dynamics
CFP	Conditional failure probability
CFR	Code of Federal Regulations
CFRP	Carbon-fiber reinforced polymer
CGE	Canadian General Electric
CGR	Crack growth rate
CHUG	CHECWORKS™ Users Group (of EPRI)
CIPP	Cured-in-place pipe
CMTR	Certified Material Test Report
CNID	Constrained noninformative distribution
CNNC	China National Nuclear Corporation
CNSC	Canadian Nuclear Safety Commission
COA	Crack opening area
CODAP	Component Operational Experience, Degradation and Ageing Program (of the OECD Nuclear Energy Agency)
CORDEL	Cooperation in Reactor Design Evaluation and Licensing
COSYMA	Code System from MARIA
CRP	Coordinated Research Project
CS	Carbon steel
CSA	Canadian Standards Association
CSD	Cold shutdown
CSN	Consejo de Seguridad Nuclear (Nuclear Safety Council of Spain)
CSNI	Committee on the Safety of Nuclear Installations
CTWC	Circumferential Through-Wall Crack
CVC	Chemical and volume control
CW	Crack width
DBA	Design basis accident
DBMS	Database Management System
DEGB	Double-ended guillotine break
DID	Defense-in-depth
DM	Degradation mechanism
DN	Nominal diameter (mm)
DOE	United States Department of Energy
DSAW	Double submerged arc welding
EAW	European-American Workshop
EBS	Equivalent break size
ECCS	Emergency core cooling system

EDF	Électricité de France SA
EE	Expert elicitation
EFPY	Equivalent full power year
EFW	Emergency feedwater
EHC	Electro-hydraulic control
EIIS	Energy Industry Identification Systems
EJE	Expert judgment elicitation
ELL	Large external leakage
ELS	Small external leakage
EMDA	Expanded Materials Degradation Assessment
ENIQ	European Network for Inspection Qualification
EPIX	Equipment Performance Exchange System (of INPO)
EPR	European Pressurized Water Reactor
EPRI	Electric Power Research Institute
ERDA	United States Energy Research and Development Agency
ESC	Environmental stress cracking
ESCR	Environmental stress cracking resistance
ESDY	Equivalent shutdown years
ESFAS	Engineered safety features actuation system
ESREDA	European Safety, Reliability and Data Association
ESW	Essential service water
EXT	Extraction steam system
FAC	Flow-accelerated corrosion
FBR	Fast breeder reactor
FCAW	Flux core arc welding
FDF	Flaw detection frequency
FEM	Finite element analysis
FFS	Fitness-for-service
FIK	Fits-in-kind
FIV	Flow-induced vibration
FN	False negative
FOI	Factor of improvement
FRAMACEC	FRAMatome ACEC (Ateliers de Constructions Électriques de Charleroi - Belgium) Cockerill (Cockerill-Ougrée-Providence - Belgium)
FSAR	Final Safety Analysis Report
FSWOL	Full structural weld overlay
FW	Feedwater
GALL	Generic Ageing Lessons Learned
GALL-SLR	Generic Aging Lessons Learned for Subsequent License Renewal
GAO	United States General Accounting Office
GCR	Gas-cooled reactor
GDC	General design criterion
GSI	Generic safety issue

GTAW	Gas tungsten arc welding
GUI	Graphic user interface
HAZ	Heat-affected zone
HBD	Heat balance diagram
HCF	High-cycle fatigue
HDPE	High-density polyethylene
HELB	High-energy line break
HF	Human factors
HPB	Helium pressure boundary
HSS	High-safety-significant
HTGR	High-temperature gas-cooled reactor
HTR	High-temperature reactor
HWR	Heavy water reactor
IA	Instrument air
IAEA	International Atomic Energy Agency
IBR	Incorporation by reference
ICCDP	Incremental cumulative core damage probability
ICES	INPO Consolidated Events Database
ID	Inside diameter
IEEE	Institute of Electrical and Electronics Engineers
IF-PSA	Internal Flooding Probabilistic Safety Assessment
I-GALL	International Generic Ageing Lessons Learned
IGSCC	Intergranular stress corrosion cracking
IHSI	Induction heat stress improvement
INPO	Institute of Nuclear Power Operations
INSAG	International Nuclear Safety Group (of the IAEA)
IPE	Individual Plant Examination
IPEEE	Individual Plant Examinations for External Events
IPIRG	International Piping Integrity Research Group
IPPFM	Integrated probabilistic physics of failure methods
IPRG	International Piping Integrity Research Group
IREP	Interim Reliability Evaluation Program (of the USNRC)
IRS	Incident Reporting System (of the IAEA)
ISF	Institut für Sicherheitsforschung (Institute for Safety Research)
ISI	In-Service Inspection
IWB	In-Service Inspection, Welding, and Brazing (in ASME Codes)
IWG-RPPC	International Working Group on Reliability of Reactor Pressure Components
JCNRM	Joint Committee on Nuclear Risk Management (of ASME)
JIC	Just-in-Case (Database)
JIT	Just-in-Time (Database)
J-R	Crack growth resistance
JSME	Japan Society of Mechanical Engineers
KEPCO	Korea Electric Power Corporation

KFA	Kernforschungsanlage Jülich GmbH
KKS	Kraftwerk-Kennzeichensystem
KTA	Kerntechnische Ausschuss (Nuclear Safety Standards Commission of Germany)
LANL	Los Alamos National Laboratory
LAS	Low alloy steel
LBB	Leak-before-break
LBE	Licensing basis event
LCF	Low-cycle fatigue
LCO	Limiting condition for operation
LDS	Leak detection system
LER	Licensee Event Report
LLNL	Lawrence Livermore National Laboratory
LOCA	Loss of coolant accident
LPSD	Low power and shutdown
LPT	Liquid penetrant testing
LR	License renewal
LRFD	Load and resistance factor design
LSS	Low-safety-significant
LTA	Less-than-adequate
LTO	Long-term operation
LTOP	Low temperature overpressure protection
LWR	Light water reactor
MARIA	Methods for Assessing the Radiological Impact of Accidents
MBLOCA	Medium-diameter pipe break LOCA
MCM	Markov chain model
MHA	Maximum hypothetical accident
MIC	Microbiologically influenced corrosion
MIT	Massachusetts Institute of Technology
MIV	Mechanically induced vibration
MR	Median rank
MR	Median rank (in hazard plotting)
MRP	Materials Reliability Program (of EPRI)
MSPM	Multi-state physics model
MSR	Moisture separator reheater
NASA	National Aeronautics and Space Administration
NCD	Nuclear Components, Division 1, Class 2 and 3 (of ASME BPVC Section III)
NDE	Non-destructive examination
NEA	Nuclear Energy Agency (of the OECD)
NES	Nuclear Energy Series (of the IAEA)
NESCC	Nuclear Energy Standards Coordination Collaborative
NHPP	Non-homogeneous Poisson process
NIFG	NDE Improvement Focus Group

NIKIET	Nauchno-Issledovatel'skii Konstruktorskii Institut Energotekhniki (Scientific Research and Design Institute of Energy Technologies)
NIST	National Institute of Standards and Technology
NOS	Normal operating stresses
NPCIL	Nuclear Power Corporation of India Limited
NPE	Nuclear power experience (by Stoller Corporation)
NPS	Nominal pipe size
NPSAG	Nordic Probabilistic Safety Assessment Group
NRMCC	Nuclear Risk Management Coordinating Committee
NROD	NRC Reactor Operating Experience Data
NSRT	Non-Safety-Related with Special Treatment
NUCIA	Nuclear Information Archives (of the Japan Nuclear Safety Institute, JANSI)
NURBIM	Nuclear Risk-Based Inspection Methodology
OAR	Owner Activity Report
OD	Outside diameter
OE	Operating experience
OECD	Organisation for Economic Co-operation and Development
OPDE	OECD Pipe Failure Data Exchange Project
OPG	Ontario Power Generation
OSART	Operational Safety Review Team (of the IAEA)
OWOL	Optimized weld overlay
P&ID	Piping and instrumentation diagram
PARS	Publicly Available Records System (of the USNRC)
PAUT	Phased array ultrasonic testing
PB	Pressure boundary
PBF	Pressure boundary failure
PBMR	Pebble bed modular reactor
PCSG	Pipe Crack Study Group
PCTG	Pipe Crack Task Group
PDA	Performance Demonstration Administrator
PDI	Performance Demonstration Initiative
PFB	Pipe failure probability
PFM	Probabilistic fracture mechanics
PIRT	Phenomena Identification and Ranking Table
PISC	Program for Inspection of Steel Components
PMDA	Proactive Materials Degradation Assessment
PNNL	Pacific Northwest National Laboratory
PNO	Preliminary Notification of Event or Unusual Occurrence
POD	Probability of detection
PPF	Probabilistic physics of failure
PPR	Probabilistic piping reliability
PRA	Probabilistic risk assessment
PRAISE	Piping Reliability Analysis Including Seismic Events

PRD	Pressure relief device
PROSPER	Peer Review of Operational Safety Performance Experience
PSA	Probabilistic Safety Assessment
PSAM	Probabilistic Safety Assessment and Management
PSHA	Probabilistic Seismic Hazard Assessment
PSR	Periodic Safety Review
PVRC	Pressure Vessel Research Council
PWR	Pressurized water reactor
PWROG	PWR Owners Group
PWSCC	Primary water stress corrosion cracking
QOI	Quantity of interest
RADS	Reliability and Availability Data System
RAI	Request for additional information
RAS	Reactor auxiliary system
RBMK	Reaktor Bolshoy Moshchnosty Kanalny ("high-power channel" reactor)
RCC-M	Règles de Conception et de Construction des Matériels Mécaniques des Ilôts Nucleaires PWR (Code for the Design and Construction of Mechanical Components for Pressurized Water Reactor Nuclear Islands)
RCP	Rapid crack growth
RCPB	Reactor coolant pressure boundary
RG	Regulatory Guide
RI-ISI	Risk-Informed In-Service Inspection
RISMET	Risk-Informed In-Service Inspection Methodologies
RMS	Root mean square
RO	Reportable occurrence
ROY	Reactor operating year
RR	Reactor Recirculation System (of a BWR)
RR	Round Robin (in the context of testing for NDE effectiveness)
RRT	Round-Robin Test
RSE-M	Règles de Surveillance en Exploitation des Matériels Mécaniques des Îlots Nucléaires REP (In-Service Inspection Rules for the Mechanical Components of PWR Nuclear Islands)
RSSMAP	Reactor Safety Study Methodology Applications Program (of the USNRC)
SAPHIR	Système d'Analyse de la Physique des Incidents dans les Réacteurs (System for the Physical Analysis of Incidents in Reactors)
SAPIDE	Système d'Analyse des Problèmes d'Intégrité et des Défaillances Événementielles (Event-Based Failure Analysis System for the Integrity of Systems, Structures and Components)
SBLOCA	Small-diameter pipe break LOCA
SCG	Slow crack growth
SCH	Schedule

SDP	Significance determination process
SDR	Standard dimension ratio
SKI	Statens Kärnkraftinspektion (Swedish Nuclear Power Inspectorate)
SKIFS	Kärnkraftinspektionens Föreskrifter (Regulations of SKI)
SLR	Subsequent license renewal
SMAW	Shielded metal arc welding
SME	Subject matter expert
SMiRT	Structural mechanics in reactor technology
SN	Schedule Number (also SCH)
SOK	State-of-knowledge
SPAR	Standardized plant analysis risk (Model)
SQL	Structured Query Language
SRM	Structural reliability model
SS	Stainless steel
SSC	Systems, structures, and components
SSM	Strålsäkerhetsmyndigheten (Swedish Radiation Safety Authority)
STD	Standard pipe
STRYK	Skador i Tryckbärande System i Kärnkraftanläggningar (Failures in Mechanical Devices of Nuclear Power Plants)
SW	Service water
TBS	Transition break size
TF	Thermal fatigue
TFI	Technical facilitator integrator
THTR	Thorium-Hochtemperatur-Reaktor
TMI	Three Mile Island
TP	True positive
TS	Technical specifications
TÜV	Technischer Überwachungs-Verein (of Germany)
TWFR	Through-wall flow rate
TWL	Through-wall leak
UCLA	University of California Los Angeles
UDRI	University of Dayton Research Institute
UKAEA	United Kingdom Atomic Energy Authority
ULP	Underwater laser peening
USNRC	United States Nuclear Regulatory Commission
USP	Ultrasonic shot peening
UT	Ultrasonic testing
UTT	Ultrasonic thickness test
VBWR	Vallecitos Boiling Water Reactor (also referred to as Vallecitos Experimental Boiling Water Reactor)
VT	Visual test
VVER	Vodo-Vodyannoy Energeticheskiy Reaktor; Water-Water Energy Reactor (WWER, water-moderated and water-cooled pressurized water reactor)

WANO	World Association of Nuclear Operators
WBIRS	Web-Based IRS (of the IAEA)
WCR	Water-cooled reactor
WJP	Water jet peening
WOR	Weld overlay repair
WPQ	Welder performance qualification
WPS	Welding process specification
xLPR	Extremely Low Probability of Rupture (Computer Code)
XML	Extensible Markup Language
XS	Extra strong pipe
XXS	Double extra strong pipe
ZOI	Zone-of-influence

Chapter 1
Introduction

1.1 Statement of Intent and General Historical Background

Building upon six decades of accumulated operating experience with nuclear plant piping, this book offers a comprehensive coverage of relevant terminology, analytical frameworks, methods, and techniques. These elements are indispensable for conducting robust probabilistic piping reliability analyses, particularly in the context of risk-informed applications for nuclear facilities. Since the formative years of nuclear safety and reliability (1955–1975), the field has experienced significant advancements (Okrent 1978). Drawing upon historical data, research and development, and operational expertise, new safety frameworks are being developed for future generations of nuclear power reactors. Maintaining and ensuring the accessibility of this accumulated body of knowledge are vital for supporting future engineers and scientists, highlighting the essential role of knowledge preservation in the sector. What are the associated implications and requirements of this process?

This book has a specific, technical focus on explicit and tacit knowledge preservation concerning piping operating experience and its role in probabilistic piping reliability analysis and probabilistic safety assessment (PSA). It discusses its application in risk-informed decision-making and PSA. Effective knowledge preservation can be evaluated based on how well data, methods, and techniques are accessible to future analysts dedicated to long-term careers in nuclear safety. It is a matter of raising awareness of what is available in terms of data, methods, and lessons learned. The subsequent chapters provide summaries of the current state of knowledge in probabilistic piping reliability and address the question of where the necessary data for successful implementation of these methods, techniques, and computational tools can be found.

Research into piping structural integrity dates back to the earliest steam plants. Today, R&D focuses on design practices, mitigating material degradation, developing and modernizing codes and standards, conducting experiments, and advancing computational assessment methods. In the absence of operational data, quantitative

B. Lydell, *Knowledge Preservation in Nuclear Safety*,
https://doi.org/10.1007/978-3-032-25463-4_1

reliability assessments relied on principles of fracture mechanics (Wilkowski et al. 1998), probabilistic structural mechanics (Sundararajan 1995), and probabilistic mechanical design (Haugen 1980).

Consideration of piping pressure boundary integrity has been an integral technical element of the evolution of the fundamental principles of nuclear power plant reliability and safety. The United States Atomic Energy Commission (AEC) published a report in 1957, which was referred to as WASH-740 and with the title "*Theoretical Possibilities and Consequences of Major Accidents in Large Nuclear Power Plants*" (USAEC 1957). Appendix A of the report, entitled "*The Nature and Extent of a Fission Product Release from a Power Reactor*," was concerned with loss-of-coolant accidents (LOCAs) resulting from "*a break in the primary coolant circulating system or from a rupture of the reactor vessel itself.*" Such an event was deemed a maximum hypothetical accident (MHA) (Camp and Haskin 1994). Moreover, in 1963, the AEC initiated a multi-year program to investigate piping reliability in water-cooled reactor systems. The objective of that program was to establish a technical basis for a "*Primary Coolant Pipe Rupture Study*." Major tasks in the study involved an assessment of the range of parameters to be considered in the analysis of piping reliability, a survey of the piping failure experience, a review of pipe fracture modes, a study on the adequacy of stress analysis, and a review of relevant piping design criteria. A conclusion of the early work was that "*piping systems which have been designed and constructed within established code criteria will exhibit high reliability, and catastrophic, complete severance rupture appears unlikely to occur.*"

At the *International Symposium on Fission Product Release and Transport under Accident Conditions* (held in Oak Ridge, Tennessee, April 5–7, 1965), reference was made to "incipient failures in the form of significant cracks having been discovered in primary coolant system piping in two reactors"[1] (Vol. 2, page 914, of the conference proceedings) (USAEC 1965). Operating experience became an important consideration in the early efforts to lay a foundation for how to quantitatively assess the integrity and reliability of piping systems.

About 10 years later, the *Reactor Safety Study* (usually referred to as either "WASH-1400" or the "Rasmussen Study"[2] and performed in the period 1972–1975) included an assessment of nuclear power plant piping reliability which was based on insights from the *Primary Coolant Pipe Rupture Study* and results of evaluations of operating experience from a small number of nuclear power plants combined with nonnuclear industry pipe failure experience. This assessment was aimed at estimating the frequency of the so-called loss-of-coolant accident

[1]Reference was made to pipe cracking events at the Dresden Unit 1 Boiling Water Reactor and Vallecitos Experimental Boiling Water Reactor in the United States.

[2]N. Rasmussen (1927–2003), Professor of Nuclear Engineering at Massachusetts Institute of Technology (MIT) from 1958 until 1994 and co-director of the 1975 Reactor Safety Study (WASH-1400). As stated by Keller and Modarres (Reliability Engineering and System Safety, 89 (2005), 271–285), "... this landmark study introduced a novel probabilistic, systematic and holistic approach to the assessment of safety, which ultimately resulted in a sweeping paradigm shift in safety design and regulation of nuclear power in the United States ... ."

caused by a complete 360-degree circumferential break of the largest pipe connected to the reactor pressure vessel. Such a pressure boundary failure was representative of a design-basis accident (DBA), a postulated accident that a nuclear facility must be designed and built to withstand without loss to the systems, structures, and components necessary to ensure public health and safety. One technical insight of WASH-1400 was that the accident sequences that were found to make the most important contributions to the risk of a severe accident were found not to be correlated to a DBA. Instead, small- and medium-diameter pipe break LOCAs were found to be considerably more important than those initiated by the large pipe breaks that had been considered in earlier analyses. Operating experience insights, experimental work, and theoretical studies performed since 1975 have validated this observation.

After WASH-1400, the German Federal Ministry for Science and Technology (BMFT)[3] awarded a contract for a probabilistic safety assessment (PSA) of a German reference plant.[4] Initial results were published as the *German Risk Study Phase "A"* (1979) and colloquially referred to as the "Birkhofer Study" (GRS 1979)[5] and as an expanded and updated Phase "B" study (GRS 1990). Phase B of this work referred to the use of probabilistic fracture mechanics to estimate the frequency of a complete circumferential break of a large-diameter pipe connected to the reactor pressure vessel.[6]

[3]Initially, a Ministry for Atomic Issues was established in 1955, focusing on nuclear energy research. This evolved into the Federal Ministry of Scientific Research in 1962 and then the Federal Ministry of Education and Science in 1969. A separate Federal Ministry for Research and Technology (BMFT) was created in 1972, with a focus on basic and applied research and technological development. These two ministries merged in 1994, forming the Federal Ministry for Education, Science, Research and Technology, and since 1998 shortened to BMBF.

[4]The reference plant was Unit B of the Biblis Nuclear Power Plant, a 1300 MWe pressurized water reactor designed by KWU/Siemens. First reactor criticality was achieved in March 1972. The two-unit Biblis plant was permanently shut down in August 2011 in response to a political decision. Copies of Phase A and Phase B PSA study documentation can be downloaded from https://www.grs.de/

[5]A. Birkhofer (1934–2019), in 1971, was appointed Chair of the Laboratory for Reactor Control and Reactor Safety of the Technical University of Munich, and from 1977 until 2002, he was also the managing director of GRS (Gesellschaft für Anlagen- und Reaktorsicherheit). He was a contributor to the German approach to the evolution of nuclear reactor safety. He directed the German Risk Study, Phase A and Phase B.

[6]Specifically, Chap. 5 in the main report refers to IRB-Forschungsbericht no. T 1800, "Probabilistische Untersuchung des Rissfortschrittsverhaltens von Reaktorkomponenten" (Probabilistic Study on Crack Propagation Behavior of Reactor Components). This study was performed on behalf of the Bundesministerium des Innern (BMI) by Industrieanlagen-Betriebsgesellschaft mbH (Germany). The abstract of the report reads as follows: "A reliability analysis was carried out on a circumferential weld in the main coolant loop of a PWR—Biblis-B was used as the reference—with the aim of estimating the probability of a leak or break occurring in the planned life cycle of the plant. A computer program was developed for the actual reliability analysis which took the individual input data, in accordance with their statistical parameter, into account in a simulation calculation in line with the Monte Carlo method. Taking a corresponding initial crack magnitude as the starting point, the fatigue crack growth caused by the sequence of load events is calculated, using the Forman crack growth rate equation." Proposed in 1967, the Forman equation is a fatigue

In 1992, members of an American Society of Mechanical Engineers (ASME) task force on risk-based inspection of pressure boundary components made the following statement: "*… the task of estimating piping reliability is complex, uncertain and costly …*" (Balkey et al. 1992).[7] Significant progress has been made over the past 30 years, and this progress is presented in this book.

Recent research and development have examined the application of structural reliability models and statistical assessments of operating experience data as valid inputs for probabilistic piping reliability analysis. Structural reliability models employ principles of reliability engineering with input parameters defined as probabilistic variables. Extensive pipe failure databases have been established to support exploratory studies on structural reliability and serve as resources for evaluating the effectiveness of various analytical methods and modeling approaches. Building upon a historical perspective of previous initiatives, this text evaluates how operating experience data can be most effectively integrated into probabilistic piping reliability analyses and risk-informed decision-making.

1.2 Contents

An outline of the contents of this book is given below, providing reasoning behind its composition. The origin of this text is found in a series of reports that were published in the period 1995–1998.[8] These reports were the result of a research program funded by the Swedish Nuclear Power Inspectorate (SKI). The available methods, techniques, and data resources that had been used to solve many piping reliability analysis problems were reviewed and assessed. Based on those reviews, an analysis framework intended for PSA analysts was developed. Towards the end of the project, SKI organized the "*International Seminar on Piping Reliability*" (SKI 1997), with participants from academia, technical support organizations, and regulators. Some of the conclusions of the seminar are reproduced in Fig. 1.1.

The primary outcome of SKI's research project was the creation of an international collaboration for the exchange and evaluation of piping operating experience data, a collaboration that continues to this day in the form of an OECD Nuclear

crack growth model that modifies the Paris-Erdogan equation by adding a term that accounts for accelerated crack growth as the stress intensity factor approaches fracture toughness.

[7]ASME formed a Risk Analysis Task Force in 1985 to explore the application of risk-based methods in codes and standards development. Following the recommendations of this task force, the ASME Center for Research and Technology Development (CRTD) launched a research program in 1988 to determine how RBI methods could be utilized in setting inspection requirements.

[8]All reports are available as PDF files from https://www.stralsakerhetsmyndigheten.se/ as well as from https://www.iaea.org/resources/databases/inis. The report numbers are as follows: SKI Report 1995-58, 1995-59, 1995-60, 1995-61, 1997-26, and 1998-30.

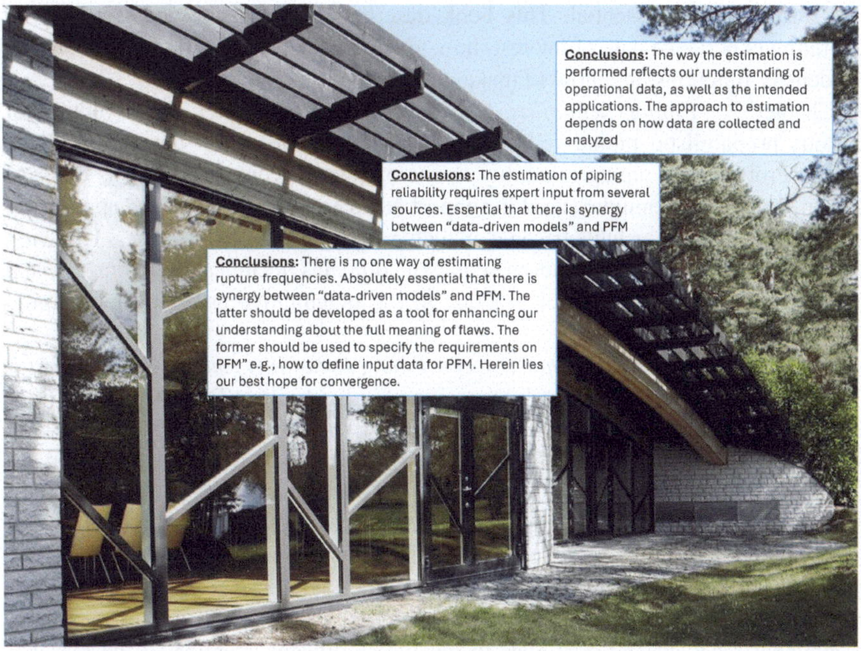

Fig. 1.1 The 1997 International Seminar on Piping Reliability in Sweden

Energy Agency database project.[9] This project gives mechanical engineers, nuclear engineers, PSA engineers, and material scientists access to quality-assured data on the performance of piping systems in different operating environments.

This book highlights how operating experience is crucial for assessing the pressure boundary integrity of piping systems in nuclear power plants. Such evaluations are relevant for both older facilities operating over the long term and new advanced reactor designs. Multiple models of piping reliability have been developed to address gaps created by insufficient operational data for statistically analyzing pipe leak or rupture rates. Some of these models also integrate failure physics, such as crack initiation and growth in thick-walled pipes. The document discusses analytical uses including risk-informed decision-making, operability assessments, creation of probabilistic failure metrics for probabilistic safety assessment (PSA), and support for developing piping reliability and integrity management programs for advanced reactors. All of these applications rely on readily available data and strong knowledge retention.

With hindsight, today's analysts are in the enviable position of having relatively easy access to an enormous volume of experimental data and operating experience data on which to base or inform an analysis of piping reliability. The preservation of

[9] Details on the Component Operational Experience, Degradation and Ageing Programme (CODAP) project are found at https://www.oecd-nea.org/.

this information is essential. This book describes how to gain access to piping operating experience data of interest, how to convert this information into probability density functions, and how to make good use of the conversions.

Chapter 2, "*Piping Reliability and Nuclear Safety*," traces the evolution of various probabilistic piping reliability analysis concepts, from the 1960s to the 2000s. Probabilistic piping reliability is one of several technical elements of PSA. The methods and techniques used in support of PSA model building and application evolved as new risk insights became available from existing PSA studies. Knowing the history of PSA helps explain the contexts, domains, technical requirements, and progression of today's state of practice in the quantitative assessment of piping pressure boundary reliability.

Chapter 3, "*Historical Perspectives on Piping Reliability Analysis*," summarizes the different phases of the many decades of research activities. This chapter identifies five distinct phases of research: Phase 1 (1960–1974), Phase 2 (1975–1984), Phase 3 (1985–1994), Phase 4 (1995–2004), and Phase 5 (2005 to the present day). Consideration of various types of rare events, from common cause failures of redundant active or passive components to "catastrophic failures" of highly reliable piping components, has been an intrinsic aspect of the evolution of nuclear safety principles and risk and reliability analysis methodologies. The question arises on how to quantitatively demonstrate that the structural integrity of a pressurized piping component received and continues to receive significant attention in the research arena as well as in applied risk and reliability analysis. The published works on piping reliability analysis include detailed evaluations of operating experience data and reviews of applications of probabilistic modeling, fracture mechanics, and expert judgment. The chapter includes a summary of significant pipe failures that not only influenced reactor regulations and codes and standards but also influenced probabilistic piping reliability method development.

Chapter 4, "*Material Degradation Mechanisms: Learning from Operating Experience*," gives a high-level summary of what is known about piping material degradation and failure. Design and construction of the first-in-kind nuclear power plants began in the early 1950s and continued into the early 1960s. Examples of such reactors were Calder Hall Unit 1 (a gas-cooled reactor which was commissioned in 1956, the world's first commercial power reactor), Shippingport Atomic Power Station (a PWR commissioned in 1957), Vallecitos Experimental Boiling Water Reactor (1957), and Marcoule G2 (a French gas-cooled, graphite-moderated reactor commissioned in 1958). Dresden-1 (commissioned in 1960) was the first commercial boiling water reactor (BWR) in the United States. It was a dual-cycle plant that incorporated an external steam drum as well as steam generators. Indian Point Unit 1 (1962), the first commercial pressurized water reactor (PWR), was a two-loop plant with a supplemental oil-fired steam superheater. The commissioning and early operational phases of these reactors produced valuable insights into material performance.

This chapter presents the concept of first-in-kind (FIK) pipe failures, highlighting their relevance both historically and in contemporary contexts. These types of failures remain relatively frequent, underscoring the necessity of maintaining a comprehensive pipe failure database that is accessible to engineers and scientists. The piping operating experience has been detailed extensively, offering valuable insights into material behavior in various reactor environments. This chapter covers these findings, including data from newer commercial nuclear power plants. Pipe degradation and failure result from off-normal conditions, extreme loads, less-than-adequate fabrication methods, less-than adequate welding, and weld defects. Structural material degradation is represented by a "triplet" indicating that changes to physical parameters can significantly impact material degradation; <***material; envronment;loading***>. These interactions are often illustrated using a Venn diagram: corrosion mechanisms (material and environment overlap), fatigue mechanisms (material and stress/strain overlap), corrosion-fatigue (environment and stress/strain overlap), and stress corrosion cracking (all three factors overlap).

Chapter 5, "*Pressure Boundary Integrity*," presents a synthesis of the factors that control the integrity of a piping pressure boundary, including the chemical and mechanical properties of materials, the methods of manufacturing and fabrication, and the methods and techniques that are used to prevent failures. In analyzing the direct and indirect causes of pipe failure, it has been shown that often very basic aspects of how pipes are manufactured and fabricated into a piping system determine the ultimate in-service structural integrity. It can be several decades before a design and construction flaw manifests itself as a significant part through-wall flaw or through-pipe-wall leakage that requires repair or replacement actions. The evolution of relevant codes and standards is covered, including piping design, in-service inspection principles, risk-informed in-service inspection, and piping pressure boundary integrity management. Codes and standards are being modernized to incorporate probabilistic considerations for piping pressure boundary integrity.

Chapter 6, "*Analytical Contexts and Analysis Guidelines*," discusses how the performance of probabilistic piping reliability analysis is conditioned by the objectives of an analysis and the domain in which the outcomes of an analysis are to be applied. Each analysis domain comes with different requirements for usability. It is a three-tiered selection and application process that acknowledges the general context of an analysis, end-user expectations and requirements for validation, and relevant national codes and standards. This chapter offers insights into six analysis guidelines. Four guidelines pertain to probabilistic fracture mechanics applications, while the fifth guideline encompasses a broader perspective by recognizing the applicability of different methodologies to a common problem. A sixth guideline addresses probabilistic piping reliability analysis in the context of probabilistic safety assessment of accident scenarios induced by internal flooding caused by loss of piping pressure boundary integrity.

Chapter 7, "*Probabilistic Piping Reliability Analysis*," combines piping fundamentals, operating experience, and modeling to realistically represent failure through validated probabilistic metrics. The analysis starts by identifying and quantifying material degradation scenarios and then evaluates the chances of detecting defects before they compromise pressure boundary integrity. It assesses the probability of such degradation leading to failure, with methods like expert elicitation and Bayesian reliability analysis frequently employed. Examples are provided.

Chapter 8, "*Reliability of Leak Detection and Nondestructive Examination (NDE)*," is concerned with the "line of defense" which is embedded in what is referred to here as the structural integrity risk triplet. All piping in a nuclear power plant is subject to some sort of monitoring for material degradation and leakage. The monitoring is in the form of leak detection and visual inspections during normal plant operation and scheduled maintenance outages and nondestructive examinations that are an integral part of piping integrity management, the effectiveness of which is characterized by the reliability of leak detection and nondestructive examination program implementation. The latter factor is defined by the probability of detecting a subsurface flaw of certain dimensions (e.g., length and depth). It is not an easily measured quantity. It is usually inferred or estimated from experiments and qualification of NDE technologies and processes and based on performance demonstration data. Operating experience offers valuable insights into NDE reliability.

Chapter 9, "*Knowledge Preservation Through Data Curation*," presents the different types of input data, including piping design information (e.g., dimensional data), chemical and mechanical properties of materials (yield stress and fracture toughness), crack initiation frequency, dimensions of initial crack, fluid flow conditions, etc. This chapter is concerned with one aspect of the data requirements, that is, how to utilize operating experience to the fullest to inform model selections and their implementation. The premise on which this chapter is based is that operating experience data, if of sufficient quality and technical detail, is an essential element of all types of probabilistic piping reliability analyses, including the validation of results, and in support of uncertainty and sensitivity analyses. The philosophies behind pipe failure database development, data quality, and data curation are explained.

Chapter 10, "*Operating Experience Data Analytics*," offers technical insights on the many different piping material degradation scenarios, and locations susceptible to material degradation, different material degradation mitigation processes, and role of in-service inspection are drawn from an extensive database, which was established more than 30 years ago and which has been maintained and updated ever since. This chapter includes a high-level summary of the content of a piping operating experience database on which this text is based. More than 14,800 pipe failures of varying severity levels have been recorded. The content of the database has been collected, curated, and continuously updated since 1993, representing a significant effort in terms of the time required to extract, interpret, and classify pipe failure reports. The objective of this chapter is to instill in the readership the immense volume of information that is available for correlating models of structural integrity of piping with service experience insights. Piping operating experience data and its practical use are central to this text, which is concerned with probabilistic piping reliability analysis, including development of models and associated input parameters.

Chapter 11, titled "*Conclusions: Reflecting on the Past and Planning for the Future*," presents final observations on leveraging historical operating experience to inform probabilistic safety assessments for emerging advanced reactor designs. Consistent with earlier chapters, it is directed towards future nuclear safety professionals and researchers. The chapter emphasizes the significance of effective knowledge management, preservation, and ongoing improvement through learning from previous analyses. It also provides recommendations for sustaining knowledge retention within the field.

1.3 Comments

To paraphrase recent publications of the Nuclear Energy Agency,[10] knowledge preservation in nuclear safety is crucial for maintaining the long-term safety and reliability of nuclear power plants. It involves safeguarding knowledge to ensure that entry-level and early career nuclear engineers and scientists have a comprehensive and robust platform on which to base probabilistic assessments of the structural integrity of piping systems. Knowledge preservation includes preserving institutional and "tribal" knowledge concerning the design, operation, structural integrity management, operating experience, and lessons learned from practical applications of analytical tools. As data-driven methodologies and machine learning techniques continue to expand in scope and application, ensuring data quality, integrity, and effective curation is becoming increasingly important.

Probabilistic piping reliability analysis should rely on relevant operating experience and experimental data, provided that these sources meet quality and completeness standards. This approach is widely accepted among PSA engineers and nuclear safety analysts but requires careful consideration of uncertainties. The analysis assesses failure frequency and probability based on material degradation, failure modes, consequences, and associated uncertainties. While abundant data exists, effectively utilizing it can be complex. This book focuses on practical methods for interpreting and applying both historical and current operating experience data. It explains how an analysis can be organized and executed, from model development, input data preparation, and development and execution of computational tools to result interpretation. The scope of the text is narrow. Its focus is on piping systems in commercial nuclear power plants and the unique analytical contexts of nuclear plant safety and reliability.

"Piping" is the term used to describe a connected assembly of straight sections of pipe, fittings such as bends, elbows, reducers and tees, welds, pumps, and valves forming a system for transferring steam, water, helium gas, or liquid metal. A commercial nuclear power plant has on the order of 30,000 linear meters of piping of different diameters and materials and tens of thousands of welds of different types. Since the first commercial nuclear power plants came online in the early 1960s, the rules and regulations for piping design and construction have evolved in a significant way, in part as the result of operating experience insights and in part as the result of very extensive experimental and analytical work to better understand the

[10] NEA Publications No. 7423 (2019) and 7738 (2025); https://www.oecd-nea.org/.

relationships between material degradation and modes of pipe failure. An extensive amount of work has also been performed to determine the conditions which must be met to prevent piping failures. This work remains ongoing, and "surprises" will be forthcoming!

Codes and standards (C&S) related to piping design are referenced in this book. Keeping current with the evolving C&S environment requires effort and attention. Chapter 5 outlines relevant C&S frameworks. Readers should refer to the websites of respective international and national C&S organization and regulatory authority for updated revisions and interpretations. As presented in this book, the interpretations of these frameworks are based on the author's experience in the field.

References

Balkey KR et al (1992) Risk-based inspection—development of guidelines, volume 2—part 1: light water reactor (LWR) nuclear power plant components, CRTD, vol 20-2. American Society of Mechanical Engineers, New York

Camp AL, Haskin FE (1994) Perspectives on reactor safety, NUREG/CR-6042. USNRC, Washington, DC

GRS (1979) Deutsche Risikostudie Kernkraftwerke Eine Untersuchung zu dem durch Störfälle in Kernkraftwerken verursachten Risiko (German risk study nuclear power plants: an investigation of the risk caused by accidents in nuclear power plants). Gesellschaft für Anlagen- und Reaktorsicherheit (GRS) mbH, Cologne

GRS (1990) Deutsche Risikostudie Kernkraftwerke, Phase B (German Risk Study Phase B). Gesellschaft für Anlagen- und Reaktorsicherheit (GRS) mbH, Cologne. ISBN 3-88585-809-6

Haugen EB (1980) Probabilistic mechanical design. John Wiley & Sons, New York. ISBN: 0-471-05847-5

Okrent D (1978) On the history of the evolution of light water reactor safety in the United States, chemical, nuclear, and thermal engineering department. School of Engineering and Applied Science, University of California, Los Angeles

SKI (1997) Proceedings of a seminar on piping reliability, SKI report 97:32. Swedish Nuclear Power Inspectorate, Stockholm

Sundararajan C (ed) (1995) Probabilistic structural mechanics handbook. Chapman & Hall, New York. ISBN: 0-412-05481-7

USAEC (1957) Theoretical possibilities and consequences of major accidents in large nuclear power plants: a study of possible consequences if certain assumed accidents, theoretically possible but highly improbable, were to occur in large nuclear power plants, WASH-740. United States Atomic Energy Commission, Washington, DC

USAEC (1965) International symposium on fission product release and transport under accident condition, CONF-650407. United States Atomic Energy Commission, Washington, DC

Wilkowski GM, Olson RJ, Scott PM (1998) State-of-the-art report on piping fracture mechanics, NUREG/CR-6540. USNRC, Washington, DC

Chapter 2
Piping Reliability and Nuclear Safety

2.1 Fundamental Principles

A synthesis of probabilistic piping reliability analysis is shown in Fig. 2.1; it incorporates piping fundamentals, operational experience, and advanced modeling methodologies. The effectiveness of this analytical approach relies on a thorough grasp of piping design principles in conjunction with risk science. Establishing a comprehensive database of operational experience and conducting methodical statistical evaluations are critical components in the development and application of a robust piping reliability model.

Models of piping reliability are used to predict the frequency and probability of failure. These models can be broadly categorized into elementary statistical models, probabilistic models, and physical models; see Fig. 2.2:

- Basic point process models treat failures as a sequence of random events occurring over time. Counts of pipe failures are obtained from a structured query language (SQL) database, and high-level conclusions about piping integrity are obtained through statistical inference. It is one way of organizing operating experience relative to the evolution of failures over time. The pipe failures are treated as a nonhomogeneous Poisson process (NHPP).
- Probabilistic models decompose pipe failure into a multistage process: first, an analysis of stochastic occurrences of precursors to more significant pipe failures, followed by an analysis of the "line of defense" offered by leak detection and nondestructive examination, and thirdly, analysis of stochastic occurrences of significant structural failures given undetected pipe flaws.
- Physical models describe the mechanisms contributing to failure by analyzing the stresses acting on a flawed piping component. Fracture mechanics theory forms the basis for the study of cracked structures under certain loads and material degradation mechanisms.

B. Lydell, *Knowledge Preservation in Nuclear Safety*,
https://doi.org/10.1007/978-3-032-25463-4_2

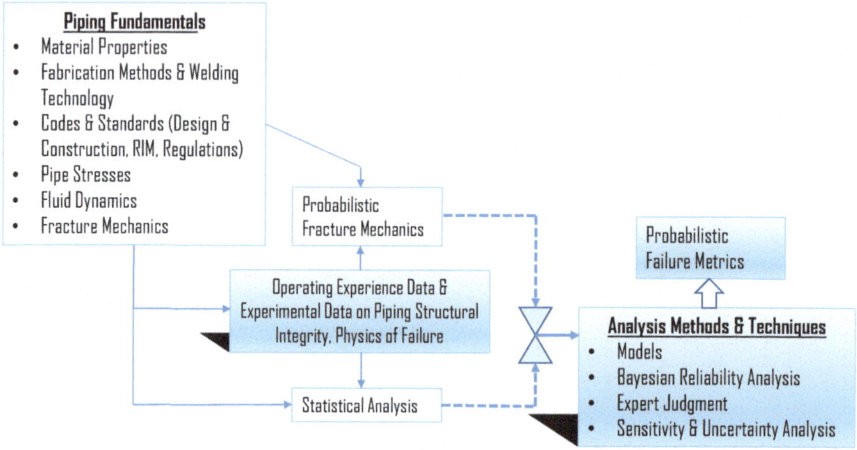

Fig. 2.1 Synthesis of probabilistic piping reliability analysis

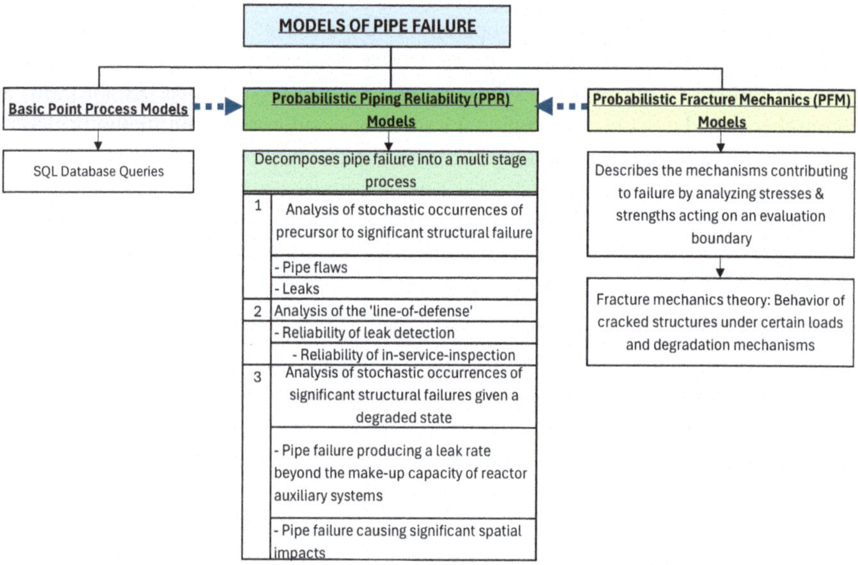

Fig. 2.2 Models of pipe failure

Probabilistic failure metrics, such as "precursor frequency," "frequency of failure," and "conditional probability of failure," are determined using a combination of methods and modeling techniques. These methods ensure that the physics of failure are accurately represented by validating input parameters and results against experimental data and operating experience data. It is a multidisciplinary analysis effort. As part of an IAEA Coordinated Research Project (IAEA 2021, 2023), a benchmark study conducted from 2018 to 2021 compared different methodologies used to assess probabilistic failure metrics for well-defined evaluation boundaries. The study provided insights into the selection and implementation of technical approaches for probabilistic piping reliability analysis:

(a) All implementations required extensive input data preparation, and some form of computer code implementation to facilitate the quantification processes, including the uncertainty analyses.
(b) Implementations of the methods tended to be computationally intense, necessitating pre-planning of the computations and post-processing of results.
(c) Regardless of the selected methodology, a successful implementation depended on the experiences that had been obtained by the analysts from previously performed practical applications.
(d) The value and importance of operating experience data cannot be understated. All methods benefited from and required information on material performance and pipe failures.

Probabilistic piping reliability analysis utilizes statistical and probabilistic models of piping integrity. Statistical models rely on experimental or empirical data and utilize mathematical and statistical concepts to identify trends and relationships between independent covariates such as the influence of pipe size, material and age, and failure parameters such as number of failures, failure rate, time to failure, and probability of failure. Probabilistic models rely on the output from statistical models to predict the outcome of degraded piping when left unattended and exposed to operational loads and/or off-normal pressure transients.

2.2 Probabilistic Piping Reliability in PSA

Probabilistic piping reliability is an integral part of probabilistic safety assessment (PSA). The methodologies and techniques used in PSA have evolved as new risk insights emerged from plant-specific PSA studies. Understanding the history of PSA provides a valuable context regarding the domains, technical requirements, and evolution of the current practices in quantitatively assessing piping pressure boundary integrity and reliability. While related, piping integrity and piping reliability focus on different aspects. Piping integrity refers to the overall condition and safety of piping systems, emphasizing the prevention of flaws, leaks, and other incidents that could affect plant safety. Conversely, piping reliability focuses on the statistical probability of a system or component functioning correctly over a specified period, considering potential failures and their impact on the system's overall operation.

The period from 1945 to 1970 saw significant advancements in the theoretical foundation for reliability and risk analysis. Significant contributions were made by

Gnedenko (1969), Farmer (1967), Green and Bourne (1972),[1] Finetti (1974), Barlow and Proschan (1974), Kaplan (Kaplan and Garrick 1979; Kaplan et al. 1981), and Garrick (1970, 1984). Apostolakis (1974) and Fussell et al. (1976) give good overviews of the early developments in "probabilistic thinking" and "risk-informed" decision-making. The Reactor Safety Study, WASH-1400 (USNRC 1975), continues to be regarded as the inception of PSA, with many methods and risk insights from the study remaining relevant.[2]

WASH-1400 questioned the assumption that conservative safety analyses of design-basis accidents (DBAs) alone could establish an upper boundary for public health and safety risks. A DBA represents accident scenarios that must be addressed through design features and validated during the licensing phase of a specific facility. For instance, each nuclear power plant is required to be engineered to withstand a "double-ended guillotine break" (DEGB) of the largest pipe attached to the reactor pressure vessel. Notably, WASH-740 (1957) employed the term "major hypothetical accident" (MHA) instead of DBA.[3]

WASH-1400 concluded that accidents other than traditional design-basis accidents, specifically smaller and more frequent sequences, were the primary contributors to public risk (Kadak and Matsuo 2007). While it was evident that the consequences of a severe core damage event would surpass those of a DBA, a significant insight revealed that the frequency of severe core damage events was considerably higher than anticipated with traditional defense-in-depth thinking. This approach assumed that accidents necessitate the failure of multiple independent

[1] Both Green and Bourne worked for the Safety and Reliability Directorate (SRD) of the UK Atomic Energy Authority (UKAEA), which was a key safety organization established to manage risks within the United Kingdom's nuclear program.

[2] On January 4, 1974, the AEC issued 10 CFR Part 50, "*Licensing of Production and Utilization Facilities – Acceptance Criteria for Emergency Core Cooling Systems for Light Water-Cooled Nuclear Power Reactors*." The rule required that in the analysis of loss-of-coolant accidents, a spectrum of possible pipe breaks should be considered. The spectrum included instantaneous double-ended breaks ranging in cross-sectional area up to and including that of the largest pipe in the primary coolant system. The analysis should also consider the effects of longitudinal splits in the largest pipes. The intent of the rule was to cover the analysis of the effects of the large and small pipe breaks. This was explained in memorandum from May 1975, "*Note to ESB Personnel: ECCS Break Spectrum Guideline*," details of which can be found in a subsequent "*Request for Additional Information on the ECCS Reanalysis*" dated June 13, 1975 (NRC-ADAMS Accession No. ML15210A060). These documents define the minimum number of breaks required for an acceptable complete pipe break spectrum analysis.

[3] U.S. Atomic Energy Commission (1957), Theoretical Possibilities and Consequences of Major Accidents in Large Nuclear Power Plants, WASH-740, Washington, DC: The term "double-ended guillotine break" (DEGB) refers to the postulated instantaneous severance of a major pipe in a nuclear reactor coolant system, a scenario that has been a design basis for safety regulations like 10 CFR 50.46, *Acceptance Criteria for Emergency Core Cooling Systems for Light-Water Nuclear Power Reactors*. However, this requirement is being revised to a risk-informed approach that uses a transition break size (TBS), recognizing that DEGB is extremely unlikely. This shift is supported by the leak-before-break (LBB) principle, which shows that flaws in pipes would likely be detected by a small leak before a catastrophic break could occur and probabilistic risk assessments that show that the probability of DEGB is very low.

systems and fission product barriers following an initiating event. Additionally, this approach upheld the qualitative assessments made in the deterministic framework of safety analysis, suggesting that accidents more severe than design-basis accidents were so unlikely as to be negligible in defining the design basis and general design criteria.

As the work on WASH-1400 ended, the United States Energy Research and Development Agency (ERDA) initiated a major PSA study for a high-temperature gas-cooled reactor (HTGR) design. The study became known as the "*Accident Initiation and Progression Analysis*" (AIPA)[4] and was conducted in the period 1974–1977. It utilized methods like that of WASH-1400. Extensive work to expand and apply the AIPA methodology to other high-temperature reactor (HTR) design concepts was performed at what was the Institute for Nuclear Safety Research (ISF) of the Kernforschungsanlage Jülich GmbH (KFA)[5] in Germany in the 1975–1990 timeframe (Kröger and Wolters 1986).

WASH-1400 was performed to inform a congressional debate in the United States on the renewal of the Price-Anderson Act of 1957, which limited the liability from reactor accidents.[6,7] It was also motivated by public and political concerns about the safety of nuclear power plants.[8,9] It took 20 years before any significant attempt was made to incorporate risk information into the regulatory process.

---

[4]The AIPA methodology, modeling, and results were documented in eight volumes, many of which are readily accessible at https://www.osti.gov/.

[5]The Nuclear Research Center (KFA) is now the Research Center Jülich (Forschungszentrum Jülich GmbH).

[6]Details of this act can be found at https://crsreports.congress.gov using the search string "Price-Anderson Act of 1957."

[7]In a technical paper ("*Probabilistic Risk Assessment Practices in the USA for Nuclear Power Plants*," Safety Science, 40:177–201), B.J. Garrick and R.F. Christie state that "*... while the details are not completely clear, it is generally believed that WASH-1400 was triggered by a letter from a United States Senator, Senator Pastore, to James Schlesinger, the then Chairman of the US Atomic, Energy Commission ....*" In a November 3, 1972, letter to James Schlesinger, then Chairman of the Atomic Energy Commission (AEC), Senator John O. Pastore primarily addressed US nuclear cooperation with the United Kingdom, not domestic nuclear risk. At the time, Pastore was the chairman of the congressional Joint Committee on Atomic Energy (JCAE) and was being formally notified by Schlesinger of a decision approved by President Nixon. The declassified letter focused on the United States assisting the United Kingdom with underground nuclear tests at the Nevada Test Site for their "Super Antelope missile improvement program." The letter, along with a preceding communication from an AEC Commissioner, emphasized meeting safety standards and minimizing the project's "public visibility." While this particular letter did not address commercial nuclear power risk, 1972 was a period of significant public concern about reactor safety. The AEC initiated the Reactor Safety Study (WASH-1400) around this time in response to pressure regarding the potential for accidents.

[8]https://www.ans.org/news/article-3216/the-origins-of-the-reactor-safety-study/

[9]For additional perspectives, the nuclear safety scholar is directed to the Swedish "*Urban Siting Study of Nuclear Power Plants*." Conducted in 1971–1974, it was a comprehensive study on risks associated with the siting of nuclear power plants below ground, in very stable bedrock formations and in close proximity of city centers. It was a "groundbreaking" study. The reference is Carlbom, L., Hedgran, A., Larsson, Y., Ryman, J-E.-E., Svensson, K., Wivstad, I. (1974), Urban Siting of

Political debates sparked controversy over the value of PSA in decision-making, which prevented the full appreciation of its potential benefits. This resulted in the United States Nuclear Regulatory Commission (USNRC) organizing the *Risk Assessment Review Group* in 1977 with four elements to its charter:

1. Clarify the achievements and limitations of WASH-1400.
2. Assess the peer comments thereon and responses to those comments.
3. Study the present state of risk assessment methodology.
4. Recommend how (and whether) such a methodology can be used in the regulatory and licensing process.

Professor H.W. Lewis of the University of California, Santa Barbara,[10] led the group and in September 1978, the final results were published as NUREG/CR-0400 (Lewis et al. 1978). Among its recommendations, the group introduced the notion of defining "precursor events" and subjecting them to probabilistic safety analysis formalism in order to determine their risk significance and to establish a technical basis for proactive risk management. The members of the review group acknowledged that the definition of a precursor may vary from highly specific criteria such as exceeding a specific quantitative threshold to broad definitions that encompass a wide range of events and circumstances.[11]

The recommendations made by the Risk Assessment Review Group led the USNRC to issue a policy statement in January 1979 that in effect discouraged the use of PSA as a tool to resolve safety and licensing issues in the regulatory process.[12]

Nuclear Power Plants: A Report of the Commission on Urban Siting, SOU 1974:56, ISBN: 91-38-01579, Department of Industry (now the Ministry of Climate and Enterprise), Stockholm, Sweden.

[10]Harold Lewis (1923–2011) was an American Emeritus Professor of Physics and former Department Chairman at the University of California, Santa Barbara (UCSB). In his 1990 book Technological Risk (ISBN: 978-0393308297), he warned of the risks of global warming.

[11]For a current perspective: Management Directive (MD) 8.3, "NRC Incident Investigation Program," DT-23-06, https://www.nrc.gov/reading-rm/doc-collections/management-directives/volumes/vol-8.

[12]NRC Statement on Risk Assessment and the Reactor Safety Study Report (WASH-1400) in Light of the Risk Assessment Review Group Report, January 18, 1979: This statement concluded that "*The Commission accepts the Review Group Report's conclusion that absolute values of the risks presented by WASH-1400 should not be used uncritically either in the regulatory process or for public policy purposes and has taken and will continue to take steps to assure that any such use in the past will be corrected as appropriate. In particular, in light of the Review Group conclusions on accident probabilities, the Commission does not regard as reliable the Reactor Safety Study's numerical estimate of the overall risk of reactor accident. .... Commission correspondence and statements involving WASH-1400 are being reviewed and corrective action as necessary will be taken. With respect to the component parts of the Study, the Commission expects the staff to make use of them as appropriate, that is, where the data base is adequate and analytical techniques permit. Taking due account of the reservations expressed in the Review Group Report and in its presentation to the Commission, the Commission supports the extended use of probabilistic risk assessment in regulatory decision making. The Commission has provided additional detailed instructions to the NRC staff concerning continued use of risk assessment techniques and results in response to specific criticisms raised by the Risk Assessment Review Group.*"

Subsequently, the Chairman of the Subcommittee on Energy Conservation and Power, Committee on Energy and Commerce, United States House of Representatives, requested of the United States General Accounting Office (GAO)[13] to determine whether the USNRC adequately considered the potential problems and disadvantages of PSA. In its conclusions, the GAO made the following statements:[14]

> *Probabilistic risk assessment (PRA) is a method of quantifying the probabilities of potential accidents and their consequences at nuclear power plants. PRA analysts use complex computer models to help them predict which risks appear to be the greatest and to identify corrective actions that address factors contributing to those risks. Nuclear utilities and the Nuclear Regulatory Commission (NRC) have used PRA since 1975 to help improve plant safety and thereby reduce risks to human health and the environment.*
>
> *GAO believes that NRC is making reasonable use of PRA. However, the uses and effectiveness of PRA are still evolving. Since relatively few empirical data on actual plant accidents are available, PRA analyses are and will continue to be affected by many unknowns and uncertainties about internal plant behavior, as well as external events, such as floods. GAO therefore cautions that NRC should not use PRA risk estimates as the sole or primary basis for regulatory decisions. Rather, NRC should use PRA to supplement its more traditional analytical and engineering methods.*

Drawing on technical insights from applying PSA methodology and learning from incidents like the Three Mile Island Unit 2 (TMI) accident, PSA experts realized that severe core damage events are actually more probable than design-basis accidents (DBAs). Additionally, they discovered that nuclear power plant risk profiles vary significantly based on both design and the specific plant. If one were to reassess the frequency of DBA-initiating events using real-world probabilities for success and failure scenarios in safety analysis reports, it becomes evident that these frequencies would be much lower than current core damage frequency calculations—meaning DBAs occur less often than core damage events.

Two of the major investigations of the 1979 TMI accident, the "Kemeny Report" (1979) and the "Rogovin Report" (1980), strongly supported the increased use of PSA in regulatory processes. The Reactor Safety Study Methodology Applications Program (RSSMAP) was a program funded by the USNRC Office of Research that applied the methodology of WASH-1400 to different reactor designs. Its main objectives were to identify risk-dominating accident sequences for significantly different reactor designs, compare them with those in WASH-1400, and identify significant design differences impacting risk. The RSSMAP studies were published in the 1979–1983 timeframe and produced risk insights applicable to Calvert Cliffs

---

[13] The U.S. General Accounting Office (1985), Probabilistic Risk Assessment: An Emerging Aid to Nuclear Power Plant Safety Regulation, GAO/RCED-85-11, Washington, DC: The GAO was established as the General Accounting Office by the Budget and Accounting Act of 1921. The act required the head of the GAO to investigate, at the seat of government or elsewhere, all matters relating to the receipt, disbursement, and application of public funds and shall make to the President ... and to Congress ... reports [and] recommendations looking to greater economy or efficiency in public expenditures. The name was changed in 2004 to the Government Accountability Office.

[14] The Chairperson was the United States Representative Mr. Edward J. Markey from Massachusetts and, since 2013, Senator from the same state.

Unit 1 (Combustion Engineering PWR), Grand Gulf (General Electric BWR), Oconee Unit 3 (Babcock & Wilcox PWR), and Sequoyah Unit 1 (Westinghouse PWR) nuclear power plants. This program helped validate and refine PSA techniques for broader future applications in reactor safety and licensing.

More or less in parallel, the Interim Reliability Evaluation Program (IREP) produced five plant-specific PSA studies by mid-1983: Crystal River-3, Arkansas Nuclear One Unit 1, Browns Ferry-1, Millstone-1, and Calvert Cliffs-1. Objectives of IREP were to identify those accident sequences that were dominant contributors to core damage frequency and to expand the cadre of experienced PSA practitioners. Like RSSMAP, the IREP evaluations were limited in scope in that they were carried forward only to the point of estimating core damage frequency. They did not consider external events, their containment analysis was limited, and risk estimates were not included in the calculations. The IREP studies were also completed in the 1979–1983 time period.

Similar to other countries' nuclear safety programs, the use of PSA in the United States evolved in three phases. The evolution of PSA can be broadly categorized into the following phases (Fleming 2003; Jackson 2017):

Phase 1: Initial development (1970s and early 1980s). This phase was characterized by the groundbreaking WASH-1400 and numerous follow-up studies.

Phase 2: Refinements and addressing uncertainties (1980s and 1990s). Experience with the early PSAs led to significant improvements in methodology and data analysis. This phase focused on refining models, better characterizing uncertainties, and enhancing the understanding of system interactions and potential failure modes. The development of industry standards and guides helped ensure the quality and consistency of PSA applications.

Phase 3: Integration into risk-informed regulation and decision-making (2000s and beyond) (Apostolakis et al. 2002; Kadak and Matsuo 2007). This phase emphasized incorporating PSA insights into decisions and risk management processes. PSA became a vital tool for evaluating plant design and operation, prioritizing in-service inspections, assessing the significance of safety issues, and enhancing codes and standards.

Before the Individual Plant Examination (IPE) phase began in 1988, a few plant-specific PSA studies addressed issues such as challenges to regulatory decisions to license plants near large population centers, emergency planning issues, and seismic design controversies[15] and the issue of whether to restart Three Mile Island Unit 1 after the Unit 2 accident on March 28, 1979. During this first phase, several plant owners voluntarily conducted plant-specific PSAs to support design and safety improvements of their facilities. An insight from these early studies was that leaks

[15]Noteworthy examples were the Zion and Indian Point probabilistic safety assessments. These were the first industry-sponsored studies in the United States. The studies were notable from a technical perspective for their systematic treatment of uncertainties, internal hazards, and external events. For more details, see Garrick, B.J. (1984), "Recent Case Studies and Advancements in Probabilistic Risk Assessment," Risk Analysis, 4: 267–279.

in large-diameter primary system piping were found to have a small risk impact and that a significant structural failure of a small-diameter pipe could have a more risk impact.

A comprehensive description of the state-of-the-art PSA in the United States as it was practiced in the middle of this phase is found in NUREG/CR-2300, the "*PRA Procedures Guide*" (Hickman et al. 1983). The two-volume guide was published in 1983, and responsible for the work was a steering committee with representatives from the American Nuclear Society (ANS), the Institute of Electrical and Electronics Engineers (IEEE), the USNRC, the United States Department of Energy (DOE), the Atomic Industrial Forum (AIF),[16] the Electric Power Research Institute (EPRI), and utilities. Later in this period, the USNRC performed a major update of their version of full-scope PSA covering a representative set of reactor and containment types, NUREG-1150 (USNRC 1990). These PSAs included full treatment of internal and external events and went further to incorporate both epistemic and aleatory uncertainties than had been done previously. This uncertainty analysis was supported by an extensive research program.

Efforts to conduct PSAs on most of the remaining plants in the United States did not materialize until the regulator issued a requirement to perform the Individual Plant Examinations (IPEs) (USNRC 1988) and Individual Plant Examinations for External Events (IPEEEs) (USNRC 1991), which occurred in the late 1980s and were completed by the mid-1990s. The IPEs included internal flooding accident sequences caused by pipe failure, and, in some cases, these accident sequences were found to be major contributors to a PSA risk metric such as core damage frequency.

Prior to the issuance of Generic Letter 88-20[17] in November 1988, which requested a systematic examination for severe accident vulnerabilities, the USNRC had tried unsuccessfully to initiate requirements for having each of the plants perform a full-scope plant-specific PSA that exceeded the IPE requirements but was unable to define a legal basis because PSAs were not needed to meet existing licensing requirements. PSA methods and modeling techniques were developed and applied globally, not just in the United States. Important contributions were made elsewhere, and simultaneous plant design and PSA received considerable attention in the 1970s. The German contributions to probabilistic piping reliability are significant. As an example, an eight-step procedure for how to perform an analysis was presented in 1992 (Beliczey 1992; GRS 1995):

[16]The Atomic Industrial Forum (AIF) was established in 1953 as an industrial policy organization for the commercial development of nuclear power and energy. In 1987, the AIF was reconfigured into the Nuclear Utility Management and Resources Council (NUMARC), which addressed generic regulatory and technical issues, and the United States Council for Energy Awareness (USCEA). In 2011, the Nuclear Energy Institute became the leading organization representing the nuclear industry in the United States.

[17]Published on November 23, 1988, Generic Letter 88-20 (Individual Plant Examination for Severe Accident Vulnerabilities - 10 CFR 50.54(f)) requested each licensee to perform an individual plant examination of its plant(s) for severe accident vulnerabilities. Licensees were expected to submit the IPE results within 3 years.

1. Definition and classification of safety-related leakages (Beliczey and Schultz 1987).
2. Determination of relevant material degradation mechanisms.
3. Definition of equivalent break sizes and corresponding through-wall leak/flow rates (Schultz 1988).
4. Definition of evaluation boundaries, organized by system, material, pipe diameter, and isometric drawings.
5. Identification of locations within a piping system that are susceptible to material degradation and failure. This step involves a formalized degradation mechanism analysis.
6. Critical review of operating experience with respect to material, in-service inspection, quality of reporting of degraded/failed conditions, process media and chemistry control, and history of piping design changes.
7. Expanded review of events involving through-wall leaks and the steps taken to prevent failures from plant-specific perspectives and generically.
8. Assess the frequency of failure based on the results of Steps 1 through 7.

PSA method development and implementation in the Nordic countries followed a three-phase process. The first phase coincided with the design, construction, and commissioning of commercial nuclear power plants in the 1960s and 1970s (Pershagen 1993). In the early 1980s, the Swedish regulatory authority mandated PSA to be an element of the periodic safety review (PSR) process every 8–10 years for the operating reactors (Carlsson and Hammar 1991). Following the recommendations of the "*Commission on Safe Nuclear Power*" in 1979, the Swedish Parliament ruled that every nuclear power plant should be made subject to at least three complete safety reviews during its useful life.[18] Reports were to be submitted every 8–10 years to the government by the regulatory authority. These reports were to be compiled on the basis of analyses carried out by the owners of the plants. The intention was that the depth of the reviews should correspond to the Final Safety Analysis Reports (FSAR) required for the original licensing of the plants.[19] For this reason, the acronym ASAR (As-operated Safety Analysis Report) was chosen for these periodical safety reviews.

The principal emphasis of the ASAR was to be varied from one periodic safety review to the next periodic safety review, but with key elements of these reviews consisting of plant-specific PSAs. Only Oskarshamn Unit 3 had the performance of a

[18] The Commission was formed on April 26, 1979, in response to the TMI-2 accident, with the results published on November 19, 1979. The final report was entitled "Säker Kärnkraft?: Betänkande av Reaktorsäkerhetsutredningen, SOU 1979:86" (Safe Nuclear Power? Report of the Reactor Safety Commission).

[19] A Final Safety Analysis Report (FSAR) is a comprehensive document submitted by a nuclear facility applicant that provides a detailed analysis of the facility's design, operation, and safety to demonstrate that it can be operated safely and protect public health. It serves as the basis for an operating license, outlining safety limits, controls, accident scenarios, and risk assessments and is used by regulatory bodies for oversight and for licensees to manage changes.

plant-specific PSA as a licensing requirement.[20] This reactor unit entered into commercial operation in August 1985. In 1986, the Swedish regulatory authority initiated a comparative review of all PSAs, the SUPER-ASAR project. The aim of the project was to review and compare the existing Swedish PSA studies. For that purpose, there were five studies available in 1987: Barsebäck 1&2, Forsmark 3, Oskarshamn 1, Oskarshamn 3, and Ringhals 1. These studies were made for BWR units. The PSA made for Ringhals 2, which is a PWR, was also partly reviewed in the project. LOCA frequencies were based on WASH-1400, with top-down allocation of those frequencies to individual systems and locations within the reactor coolant pressure boundary.

A second phase involved a Nordic research program which was coordinated through the NKS Nordic Nuclear Safety Research organization[21] (https://www.nks.org/): NKA/SÄK-1 "*PRA Uses and Techniques, a Nordic Perspective*" (Dinsmore 1985) and NKA/RAS-400 "*Risk Analysis and Safety Rationale*" (Bengtsson 1989). In the 1990s, an objective of this R&D was to define and demonstrate the practical use of living PSA and operational safety indicators for safety evaluation and for identification of possible improvements in operational safety.

In 1989, the OECD/NEA Committee on the Safety of Nuclear Installations' Principal Working Group on Risk Assessment (PWG5) began the *State-of-the-Art Level 1 PSA Methodology* task. The group observed that core damage frequencies were mainly influenced by small loss-of-coolant accident, which depended on analysis factors like database selection and reactor operating years. The French, German, and US studies relied on generic data derived from respective national operating experience. The Japanese and Swedish studies pooled the national operating experience with the operating experience of the United States. Table 2.1 shows the pipe failure-induced initiating event frequencies for the relevant plants (OECD/NEA 1994a, b).

---

[20] Construction of Oskarshamn Unit 3 began on May 1, 1980.

[21] NKS is a Nordic forum for research, competence building, experience exchange, and networking in nuclear and radiation safety with a focus on reactor safety, including decommissioning, and emergency preparedness. This is achieved by joint activities of interest to the financing organizations and other end users resulting in scientific articles and technical reports. The results are used by the participating organizations in their competence building, decision-making processes, and information activities. NKS was created in 1975 as an ad hoc committee under NKA (Nordic Liaison Committee for Atomic Energy) to assure the safety of the growing Nordic nuclear power program and secure funds to that end. NKA in turn was an organization under the Nordic Council, with joint Nordic funding. NKS was instructed to prepare a research program which would take up current safety issues.

Table 2.1 Pipe failure-induced initiating event frequencies in selected PSA studies

Initiating event		**BWR initiating event frequency (1/year)**				
		Grand Gulf	**Forsmark 1/2**	**Gundremmingen**	**Japan 1100 MWe**	**Grand Gulf**
Large steam line break inside containment		1.0E−04	1.0E−04	<1.0E−7	8.0E−05	1.0E−04
Intermediate steam line break inside containment		3.0E−04	3.0E−04	–	2.7E−04	3.0E−04
Small steam line break inside containment		3.0E−03	1.0E−03	4.3E−03	8.0E−04	3.0E−03
Initiating event	**PWR initiating event frequency (1/year)**					
	Surry and Sequoyah (USA)	**EDF/REP-900 (France)**	**EDF/REP-1300 (France)**	**Biblis "B" (Germany)**	**Japanese PWR (1100 MWe)**	**Ringhals-3/4 (Sweden)**
Large LOCA	5.0E−04	1.0E−04	1.0E−04	$<10^{-7}$	4.8E−05	4.0E−04
Medium LOCA	1.0E−03	3.0E−04	3.0E−04	$<10^{-7}$	1.5E−04	8.1E−04
Small LOCA	1.0E−03	2.0E−03	2.0E−03	3.0E−03	4.8E−04	1.1E−02

2.3 Piping Integrity Research

Nuclear piping integrity research has been ongoing since the 1960s, e.g., the 1963–1979 "*Primary Coolant Pipe Rupture Study*." In 1981, the USNRC started work at Battelle Columbus Laboratories to plan for a comprehensive experimental program (Kanninen et al. 1983). The results of this program were referred to as the "*Phase I Degraded Piping Program*."[22] It included development of new elastic-plastic fracture analysis solutions, an in-depth review and evaluation of related prior and ongoing experimental work; identification of field-degraded and surplus nuclear plant piping available for testing; development of a comprehensive test matrix; and conceptual design and feasibility studies of test configurations. The program was motivated by the then widespread nature of intergranular stress corrosion cracking in boiling water reactors. One of the technical tasks was to collect material property data and to review the in-service pipe cracking experience. The results of the Phase I study formed the basis for the Phase II experimental work (1984–1989) (Schmidt et al. 1992).

As a second example, the International Piping Integrity Research Group (IPIRG) program was established in 1986, which was managed by the USNRC and funded by a consortium of organizations from nine countries. The IPIRG program was conceived as a means to develop data for the verification of fracture mechanics analyses for a cracked pipe under dynamic loading. The IPIRG task structure was developed at a meeting held in conjunction with a USNRC seminar entitled, "*Leak-Before-Break: International Policies and Supporting Research*." This meeting was held in October 1985, in Columbus, Ohio (USNRC 1986). The work within the IPIRG program was performed in two phases, IPRG-1 and IPRG-2, and consisted of five tasks (Scott et al. 1997; Hopper et al. 1997; Wilkowski et al. 1998):

- Task 1: Leak-before-break under simulated seismic/dynamic stresses. An experimental facility was designed and constructed; see Fig. 2.3 (Scott et al. 1997).
- Task 2: Experimental pipe fracture and pipe material property database development.
- Task 3: Fracture of piping containing high-energy fluids.
- Task 4: Resolution of unresolved issues from the USNRC Degraded Piping Program and IPIRG Research Tasks.
- Task 5: Information exchange seminars and program administration.

The piping system of the pipe loop test facility was fabricated as an expansion loop with over 30 linear meters of DN400 thick-walled pipe and five long-radius elbows (Scott et al. 1997). Cracked pipe test specimens were welded into the pipe loop at a location of high bending stress as identified from engineering calculations. Each experiment was conducted by simply replacing the failed test section with a

[22] Phase I was documented in two reports, NUREG/CR-3142, Volume 1 and Volume 2. The latter consists of ten appendices with supporting technical details. Appendix I, Review of Experimental Techniques, describes the design of different types of experimental apparatuses.

Fig. 2.3 Photograph of the pipe loop test facility at Battelle. (Reproduced from Appendix A of NUREG/CR-6389, IPIRG-2 Task 1 Pipe System Experiments with Circumferential Cracks in Straight-Pipe Locations provides the geometric details of the pipe loop facility used in the experiments. https://www.osti.gov/servlets/purl/453762)

new one. The pipe loop was supported and constrained at several locations using various specially designed devices. These supports were designed to produce specific well-defined boundary conditions that could be accurately modeled in stress analyses. A large 1950 kg mass, which simulated a swing check valve, was located in one leg of the pipe loop. System loads were applied by a 1560 kN hydraulic actuator, a water circulation pump, and an expansion tank that completed the basic pipe system. All cracked pipe experiments contained subcooled water at 288 °C and 15.5 MPa.

The emphasis of the experiments was to gather data to test analytical capabilities of probabilistic fracture mechanics codes. Experiments were also conducted on DN150 thick-walled pipe samples containing circumferential through-wall cracks and loaded in four-point bending under displacement control. Additional experiments were conducted with internal surface cracks. Pipe specimens were either wrought ASTM A312 TP304 stainless steel or ASTM A106 Grade B carbon steel. The results of IPRG-1 were documented in three reports:

1. Evaluation and Refinement of Leak-Rate Estimation Models, NUREG/CR-5128 (Paul et al. 1994).
2. Loading Rate Effects on Strength and Fracture Toughness of Pipe Steels Used in Task 1 of the IPIRG Program, NUREG/CR-6098 (Marschall et al. 1993).
3. Stability of Cracked Pipe Under Inertial Stresses, NUREG/CR-6233 (Scott et al. 1994).

A goal of the IPIRG-2 program (1991–1996) was the development of data for the verification of fracture mechanics analyses for cracked pipes and fittings subjected to dynamic and/or cyclic load histories. The program explored different crack geometries/locations than had been considered in the IPIRG-1 experiments, and laboratory specimen and separate effect pipe experiments were conducted to provide better insight into the effects of dynamic and cyclic load histories. The following tasks were performed (Hopper et al. 1997):

- Task 1: Pipe system experiments with flaws in straight pipe and welds
- Task 2: Fracture of flawed fittings such as elbows
- Task 3: Cyclic and dynamic load effects on fracture toughness
- Task 4: Resolution of issues from IPIRG-1 and related programs
- Task 5: Seminars, workshops, and program management

2.4 Operating Experience and LOCA Frequency

A breach in the primary pressure boundary in 1992 at the Barsebäck nuclear power plant in Sweden,[23] (Figs. 2.4 and 2.5) was to have a major influence on how to model loss-of-coolant accident scenarios, including the estimation of LOCA frequencies and the development of pipe failure databases. On July 28, 1992, a breach in the steam line pressure boundary occurred when a safety relief valve inadvertently opened in the drywell of reactor unit 2 (OECD/NEA 1994a, b; Löfstedt 2005).

With the reactor at about 2% power and 3.2 MPa pressure, a mis-calibrated and leaking pilot valve caused a depressurization of the main safety relief valve, which then opened. When the main valve opened, a rupture disc, with a design pressure of 3 MPa, broke causing an opening into the containment drywell. The resulting steam jet stripped fibrous insulation from adjacent pipework. Part of that insulation debris was transported to the suppression pool and subsequently clogged the intake strainers for the containment vessel spray system about 1 h into the event sequence. The rupture disc had not been properly installed. What was so significant about the so-called strainer event?

The event confirmed earlier findings on location-specific spatial effects of a piping pressure boundary breach inside the containment. This highlighted the need for a more accurate LOCA frequency assessment. Research from international strainer event response programs (1992–1996) built a substantial knowledge base on ECCS suction strainer clogging phenomena (Natesan et al. 2014). The findings from the research to resolve BWR strainer clogging issues raised questions about the adequacy of PWR containment sump designs. In the United States, this reinvestigation of PWR suction strainer issues was labeled as Generic Safety Issue

[23] The two-unit Barsebäck Nuclear Power Plant was operated by Sydkraft AB, and the reactors were designed by Asea-Atom (AA) of Sweden. The two, third-design-generation reactors were commissioned in 1975 and 1977, respectively, and were permanently shut down in 1999 and 2005, respectively. These shutdowns were politically motivated; for more details on this, the reader is encouraged to read the paper by Ragnar E. Löfstedt, "Fairness Across Borders: The Barsebäck Nuclear Power Plant," 7 RISK 135 (1996).

Fig. 2.4 Aerial view of the Barsebäck plant ca. 2005. (Reproduced from https://malmo.se/Uppleva-och-gora/Konst-och-museer/Malmo-museum/Om-oss/Vara-samlingar/Aktuella-insamlingar/Vad-minns-du-av-karnkraftverket-i-Barseback.html)

The Strainer Event at Barsebäck 2, July 1992

A containment isolation event was signalled in the control room during the first heat up and pressurisation of the reactor system, following the completion of the annual refuelling of the reactor. At that time the reactor was at less than 2% of its nominal power, and the pressure in the reactor vessel was only 3 MPa, hardly half of its normal operating value.

As an automatic response, the reactor tripped and all safety functions worked as intended. Within a minute it was seen that one of the safety valves in the automatic depressurisation system had opened spuriously – due to the action of an incorrectly adjusted pilot-valve – and a vigorous blowdown could be seen on TV-cameras. The operators returned the plant to a cold shutdown condition, and the open valve was closed after about 25 minutes. By then the reactor pressure had fallen to around 1.1 MPa.

After an hour's operation of the containment cooling system an alarm indicating a high pressure drop over one of the intake strainers was received. This partial blockage occurred sooner than the 10 hours that previous calculations had predicted, for the case of a large pipe rupture at full power conditions.

During the subsequent inspection of the reactor containment it was found that about 200 kg of insulation material (mineral wool) had blown away in the region of the activated safety valve, and become distributed around the containment. It was estimated that about half of the finely-divided material had followed the vented steam down into the condensation pool, where the two intake strainers for the containment cooling system were placed.

From the view point of risk for core damage, the occurrence was of minor importance. It did, however result in many far reaching and valuable lessons.

Fig. 2.5 "Strainer event" narrative. (Reproduced from NUCLEUS-02-03, the English edition of the R&D periodical published by the Swedish Nuclear Power Inspectorate)

(GSI) 191, "*Assessment of Debris Accumulation on PWR Sump Performance*" (Rao et al. 2002). The updated research investigated all aspects of PWR emergency core cooling system (ECCS) suction strainer performance following a LOCA: debris generation, debris fragmentation, protective coating performance, debris transport, chemical effects, suction strainer prototype testing, downstream effects, and risk characterization using PSA models. As stated in SECY-10-0113 ("*Closure Options for Generic Safety Issue—191, Assessment of Debris Accumulation on Pressurized Water Reactor Sump Performance*"),[24] resolution of GSI-191 involved two distinct but related safety concerns: (1) potential clogging of the sump strainers that results in ECCS pump failure and (2) potential clogging of flow channels within the reactor vessel because of debris bypass of the sump strainer, so-called in-vessel effects.

Clogging at either the strainer or in-vessel channels can result in loss of the long-term cooling safety function. SECY-10-0113 presented three options for bringing GSI-191 to closure:

Option 1: Maintain an integrated resolution process, including evaluating new methods for testing to justify that assumptions that the NRC has determined have not been technically justified in the past.

Option 2: Develop new risk-informed implementation using either the existing regulatory framework or the proposed risk-informed rulemaking, "*Risk-Informed Redefinition of Large Break LOCA ECCS Requirements*" at Title 10 of the Code of Federal Regulations (10 CFR) 50.46a.

Option 3: Allow General Design Criterion (GDC) 4, "*Environmental and Dynamic Effects Design Bases*," leak-before-break (LBB) credit for GSI-191.

Regulatory Guide 1.229 ("*Alternative Risk-Informed Approach for Addressing the Effects of Debris on Post-Accident Long-Term Core Cooling*") (USNRC 2017) states that determining the LOCA frequency is one of several requirements in performing a risk-informed resolution of GSI-191, Option 2. This is a critical task because the quantity of debris that can result from LOCA has been shown to be highly dependent on the location of the pipe break due to large variations in piping insulation inventory around the reactor coolant system piping circuit. Hence, knowing the location dependency of the LOCA frequency is critical, and simply assuming that the LOCA frequency is uniformly distributed around the circuit is likely to yield optimistic results.

In support of an effort to develop a risk-informed revision of the emergency core cooling system (ECCS) requirements for commercial nuclear power plants, estimates of LOCA frequencies have been developed which enable redefinition of the design-basis break size for these requirements. These LOCA frequency estimates have been developed using an expert elicitation process by consolidating service history data and insights from probabilistic fracture mechanics studies with knowledge of plant design, operation, and material performance. This expert elicitation to develop LOCA frequency estimates is described in NUREG-1829, "*Estimating*

[24] https://www.nrc.gov/reading-rm/doc-collections/commission/secys/2010/index.html

Loss-of-Coolant Accident (LOCA) Frequencies Through the Elicitation Process" (USNRC 2008).

The ECCS requirements in the United States are contained in 10 CFR 50.46, Appendix K to Part 50, and General Design Criterion (GDC) 35. Specifically, ECCS design, reliability, and operating requirements exist to ensure that the system can successfully mitigate postulated LOCAs. Consideration of an instantaneous break with a flow rate equivalent to a double-ended guillotine break (DEGB) of the largest primary piping system in the plant generally provides the limiting condition in the required 10 CFR Part 50, Appendix K, analysis. Given the unlikelihood of a DEGB, a risk-informed revision of the design-basis break size requirements is in progress. A central consideration in selecting a risk-informed design-basis break size is an evaluation of the LOCA frequency as a function of break size. Prior to NUREG-1829, the USNRC sponsored a study of pipe break frequencies, which was published in 1999 as NUREG/CR-5750 (Poloski et al. 1999). Those estimates were determined not to be for design-basis break size selection because they did not discriminate among breaks having diameters greater than 150 mm effective break size.

NUREG-1829 noted that there had been two approaches traditionally used to estimate LOCA frequencies and their relationship to pipe size: (1) estimates based on statistical analysis of operating experience data and (2) PFM analysis of specific postulated failure mechanisms. Neither approach was deemed fully suitable for evaluating LOCA event frequencies due to the rarity of these events and the modeling complexity. NUREG-1829 used an expert elicitation process, with elicitation responses from a panel of 12 experts determining individual LOCA frequency estimates for the 5th percentile, median, mean, and 95th percentile of the frequency distribution for each of the six LOCA categories. Group estimates were determined by aggregating the individual estimates using the geometric mean of the individual estimates for each frequency parameter (i.e., median, mean, 5th and 95th percentiles). Group variability was estimated by calculating 95% confidence bounds for each of the group frequency parameters. A number of sensitivity analyses were conducted to examine the effects on the quantitative results from variation of the assumptions, structure, and techniques of the baseline analysis procedure.

Periodic LOCA frequency reevaluations and updates were to continue. The evolution of quantitative estimates of large pipe break LOCA frequencies developed for PWRs is shown in Fig. 2.6. To a large extent, the reduction in LOCA frequency estimates is attributed to operating experience insights, effects of improvements in piping reliability and integrity management, as well as improvements in LOCA frequency assessment methods. Additional perspectives on the frequency of a "large" LOCA are captured in Figs. 2.7 and 2.8.

- BWR large (≥Cat3) LOCA frequency estimates, 1982–2004: see Fig. 2.7.
  - NUREG-1829 (USNRC 2008) large LOCA frequency based on PWR expert elicitation estimates.
  - PSAM-0282 (Galyean 2006): The total LOCA frequency is calculated using US operating experience of zero LOCAs in 2924 LWR-calendar-years of operation (as of 12/31/2005). A Bayesian update of a non-informative prior

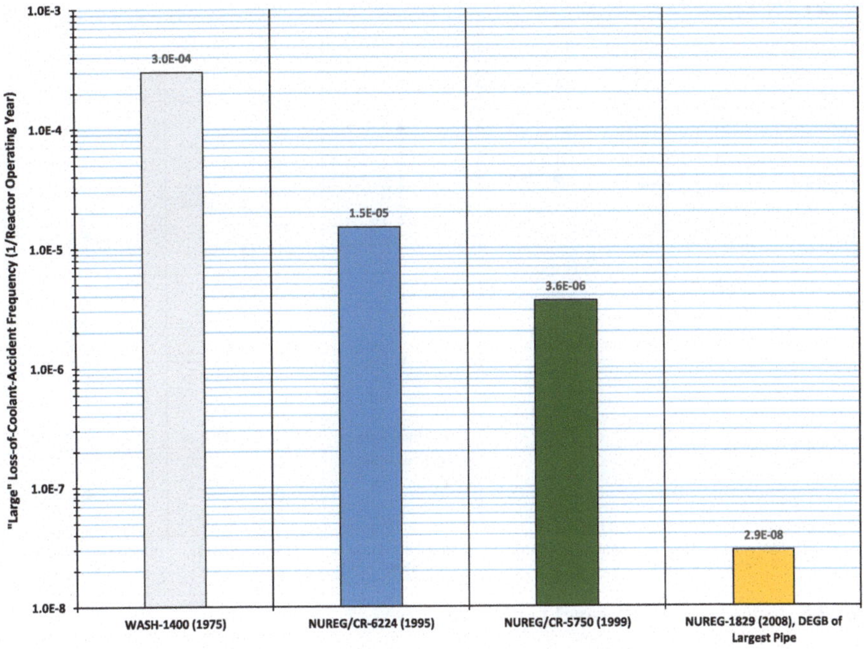

Fig. 2.6 Evolution of large LOCA frequency estimates 1975–2008

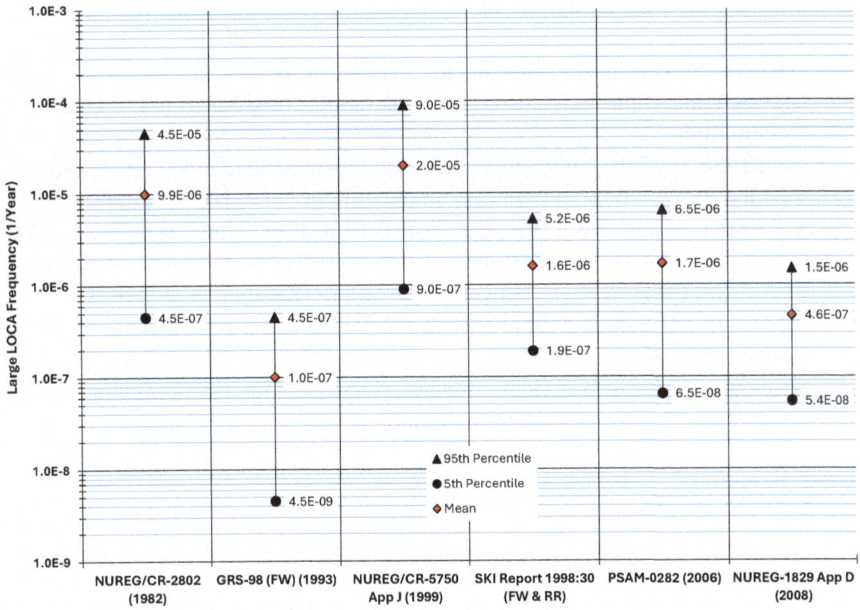

Fig. 2.7 Estimates of BWR "large" LOCA frequency

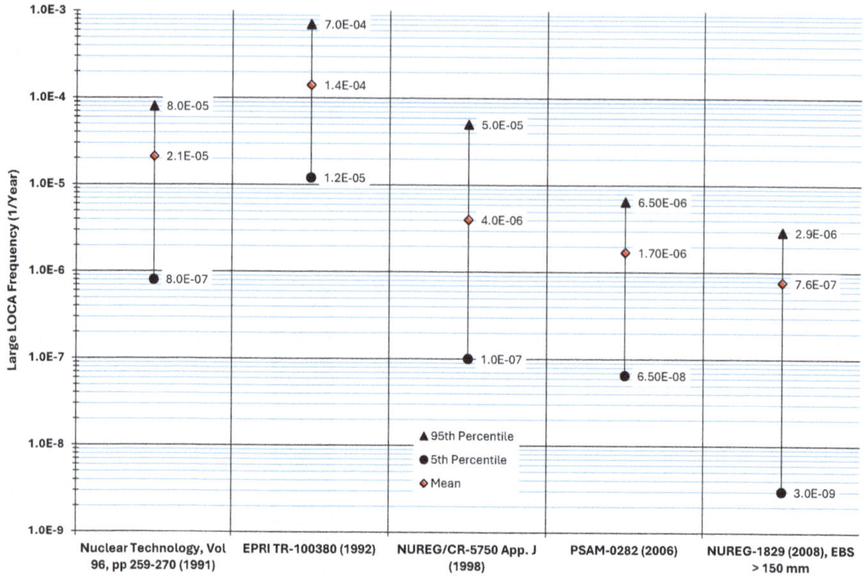

Fig. 2.8 Estimates of PWR "large" LOCA frequency

distribution was performed to produce a total LOCA frequency of 1.7×10^{-4} per LWR-year. This LOCA frequency was allocated among the small, medium, and large LOCA categories by the ratios of 89%, 10%, and 1%, respectively.

- SKI 98:30 (Feedwater/Reactor Recirculation) (Lydell 1998): The report assessed the contribution from thermal fatigue in Code Class 1 feedwater system piping.
- NUREG/CR-5750 (Appendix J) provided recommended pipe LOCA frequencies for PSA.
- GRS-98 was a probabilistic safety assessment of the Gundremmingen-BWR plant. The given LOCA frequency value range accounted only for contributions from Code Class 1 feedwater pipe failure (Kersting et al. 1993).
- BFN-1 (NUREG/CR-2802) (Mays et al. 1982) was a PSA of Browns Ferry Unit 1 performed as part of the NRC-sponsored Interim Reliability Evaluation Program. The given LOCA frequency value range accounts for reactor recirculation pump suction piping failure.

- PWR large (≥Cat3) LOCA frequency estimates: see Fig. 2.8.
  - NUREG-1829 large LOCA frequency based on PWR expert elicitation estimates.
  - PSAM-0282 did not differentiate the PWR estimate from the BWR estimate.
  - NUREG/CR-5750 (Appendix J) provided recommended pipe LOCA frequencies for PSA.

– Nuclear Technology, Volume 96 (1991), pp. 259–270 (Vo et al. 1991).
– EPRI TR-100380 (Jamali 1992) offered recommended BWR and PWR LOCA frequencies.

2.5 Expanding Perspectives

A new PSA era began in the United States when the nuclear regulator issued a revised policy statement in 1995 on the use of PSA in decision-making (USNRC 1995), a policy statement that included a more positive view on the role that risk information should have in supporting regulatory decisions. With the publication of Regulatory Guide (RG) 1.174 of July 1998 (*An Approach for Using Probabilistic Risk Assessment in Risk-Informed Decisions on Plant-Specific Changes to the Licensing Basis*) and associated application-specific regulatory guides such as 1.178 (*Risk-Informed Decision Making for In-Service Inspection of Piping*) later that year, USNRC provided clear criteria for the review of risk-informed changes to the licensing basis. These regulatory guides included quantitative risk acceptance guidelines for judging whether a calculated change in core damage frequency or large early release frequency would be considered large enough to impact a decision.

Prior to these guides, industry attempts to present risk-informed arguments to get relief from a regulation were difficult to develop and for the USNRC staff to review, due to lack of criteria for judging "how safe is safe enough" and lack of standards for PSA quality. It is noteworthy that the concept and principles of PSA quality had already received considerable attention in the mid-1970s. As an example, Sandia National Laboratories supported a research project at the University of California Los Angeles (UCLA) School of Engineering and Applied Science, results of which were published as "*On PRA Quality and Use*," UCLA-ENG-8269 (Okrent et al. 1982). This report dealt with several interrelated aspects of probabilistic safety assessment. Some prior opinions regarding quality assurance, methodology, and questions of peer review were examined, followed by comments by the authors on these and related subjects. Problems arising in decision-making by different groups concerning the meaning and validity of a PSA were examined, and the role of performance criteria in helping to achieve consensus was treated. Finally, a general approach to the development of performance criteria for systems and functions by the retrospective comparison of existing probabilistic safety assessments was proposed and examined in a preliminary fashion.

In response to regulatory guides RG 1.174 and 1.178, significant initiatives were made towards development of industry standards for the development, review, and maintenance of PSA models. In 2004, the Board on Nuclear Codes and Standards (BNCS) of the American Society of Mechanical Engineers (ASME) and the American Nuclear Society (ANS) Standards Board mutually agreed to form a Nuclear Risk Management Coordinating Committee (NRMCC), with participation from reactor owners' groups,[25] the U.S. Department of Energy (DOE), the Electric Power Research Institute (EPRI), the National Aeronautics and Space Administration (NASA), the USNRC, and other industry groups. This committee was chartered to coordinate and harmonize standards activities related to PSA for nuclear power plants and other nuclear installations among all interested standards development organizations and other interested parties. The NRMCC, ASME, and ANS subsequently formed a Joint Committee on Nuclear Risk Management (JCNRM) to develop and maintain PSA standards.[26]

Current PSA standards for light water reactors and advanced non-light water reactors are identified as ASME/ANS RA-S-1.1-2024 and ASME/ANS RA-S-1.4-2021, respectively. These standards include requirements for the estimation of initiating event frequencies from pipe leaks and ruptures. PSA results and operating experience have greatly influenced methods development, from a cursory analysis of piping pressure boundary integrity in the early PSA studies to today's state of practice. Internal flooding events caused by piping pressure boundary failure are now subjected to detailed analyses that follow industry guidelines. Considerations of pipe failure in PSA development and applications evolved significantly during the period 1960–2000 and beyond; this is shown in Fig. 2.9.

Each technical aspect of PSA has been subjected to extensive research and development, including the development of computational tools and databases that support the quantification of probabilistic failure metrics. Listed in chronological order are selected, post-year 2000 examples of collaborative programs that were established to support probabilistic piping reliability method development and applications:

- Nuclear Risk-Based Inspection Methodology (NURBIM; 2001–2004) (Brickstad et al. 2004): Funded by the European Community and a consortium of nuclear utility organizations and technical support organizations, the project established technical requirements for probabilistic fracture mechanics (PFM) analyses. Different PFM computational tools were evaluated in a series of benchmarks. The project produced several work packages (WPs), including WP-4, a review of PFM computer codes, results of benchmarks, and requirement and recommendations with respect to PFM applications, and WP-9, a handbook of best practices for risk-informed in-service inspection.
- Transition break size (TBS) assessment: Based on results of an expert elicitation in 2003–2004, the USNRC issued a draft Regulatory Guide DG-1216, "*Plant-Specific Applicability of Transition Break Size Specified in 10 CFR 50.46a*," which would have risk-informed the US regulatory requirements for emergency

[25] The reactor owners' groups, such as the BWROG, PWROG, and the European BWR Club, bring together plant owners and operators to address plant operability and reliability, to exchange operating experience, and to share R&D insights.

[26] https://cstools.asme.org/csconnect/CommitteePages.cfm?Committee=100186782

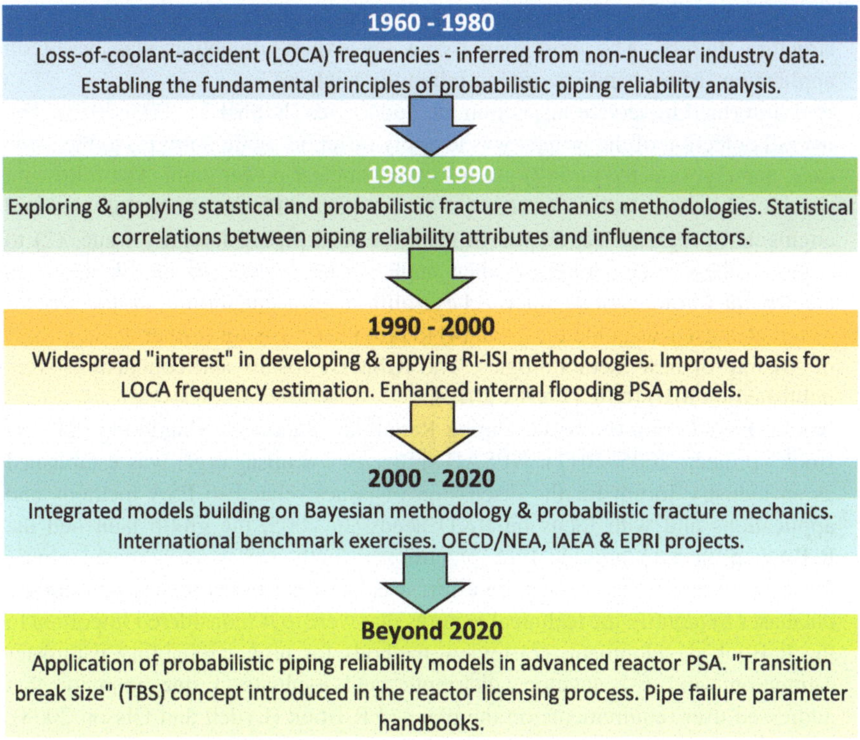

Fig. 2.9 Evolution of probabilistic piping reliability

core cooling systems.[27] TBS is a break of area equal to the cross-sectional flow area of the inside diameter of specified piping for a specific reactor. The specified piping for a PWR reactor is the largest piping attached to the reactor coolant system. The specified piping for a BWR is the larger of the feedwater line inside containment or the residual heat removal line inside containment. The proposed regulatory rule would allow plant changes such as optimization of safety system design and set points for the more likely (smaller) pipe breaks. In essence, the need for analyzing DEGB of the largest pipe connected to the reactor pressure

[27] Draft Regulatory Guide DG-1216 was a proposed new regulatory guide to provide implementation guidance for a proposed ECCS rulemaking, which would have provided a voluntary, risk-informed alternative to the existing, deterministic requirements for evaluating ECCS performance. The proposed ECCS rule was published in the Federal Register on November 7, 2005 (70 FR 67597), with a supplemental proposed rule published on August 10, 2009 (74 FR 40006). However, in SECY-16-0009, "Recommendations Resulting from the Integrated Prioritization and Re-Baselining of Agency Activities," dated January 31, 2016 (ADAMS Accession No. ML16028A189), the USNRC staff recommended that the ECCS rulemaking be discontinued. In 2024, steps were taken to develop a revised ECCS rulemaking and to perform a new expert elicitation to take advantage of piping operating experience since 2005 as well as advances in probabilistic piping reliability analysis.

vessel would be precluded since leaks would be detected well before major structural failure. The work underlying the expert elicitation was based on applications of probabilistic piping reliability analyses.

- Risk-informed in-service inspection methodologies (RISMET, 2005–2010): The overall objective of the project was to apply different methodologies to the same case, namely, selected piping systems in one nuclear power plant. The following specific objectives were defined: (1) to compare different methodologies and augmented programs developed in response to a piping integrity issue, (2) to compare these methodologies when applied to a benchmark problem, (3) to identify how the various approaches fulfil recommendations put forward in different consensus guidelines, and (4) to provide a basis for further development of existing or new methodologies. The RISMET results and conclusions were published in 2010 (OECD/NEA 2010).
- Nordic PSA Group (NPSAG) Piping Reliability Parameter Handbook (the "R-Book" project, 2005–2011): NPSAG (https://www.npsag.org/) was established as an industry forum for the discussion of issues related to PSA methods and applications and with focus on R&D needs. In 2005, the group launched the R-Book project to provide PSA practitioners with access to validated piping reliability parameters. A planning effort was launched to (1) review pipe failure databases to identify the technical features that were to be considered important to the R-Book development; (2) review methods for piping reliability parameter estimation; and (3) develop, distribute, and evaluate a questionnaire that addressed user requirements on the planned R-Book (Lydell and Olsson 2008). A final electronic version of the R-Book was issued in 2011.
- U.S. Nuclear Regulatory Commission and Electric Power Research Institute (EPRI) Extremely Low Probability of Rupture Code (xLPR, 2011–2021): A state-of-the-art probabilistic fracture mechanics code for piping applications was jointly developed by the USNRC Office of Nuclear Regulatory Research and EPRI. The code models' failure probabilities are associated with piping system components subject to active degradation mechanisms. Its core capabilities include modeling thermal fatigue, stress corrosion cracking, in-service inspection, chemical and mechanical mitigation, leakage rates, and seismic effects. The associated technical reports are available at https://www.nrc.gov/docs/ML1634/ML16341B038.html. The experiences gained from developing and applying this computer code are documented as "Technical Basis for the Use of Probabilistic Fracture Mechanics in Regulatory Applications" and in a Regulatory Guide RG 1.245 "*Preparing Probabilistic Fracture Mechanics Submittals.*"[28]
- IAEA *Methodology for Assessing Pipe Failure Rates in Advanced Water-Cooled Reactors* (WCRs) (2017–2021): To assist member states in applying adequate methodologies to pipe failure rate analysis in advanced WCRs, the Agency

[28] Also, USNRC (2022), Preparing Probabilistic Fracture Mechanics Submittals, Regulatory Guide 1.245 Revision 0, Washington, DC.

organized a 3-year coordinated research project (CRP I31030) which took place from 2018 to 2021. There were two main activities: (1) a contrast-and-comparison of different probabilistic piping reliability methodologies in the form of a benchmark study and (2) development of a good practice framework for how to derive piping probabilistic failure metrics. The results of the former activity were published as TECDOC-1988 on "*Technical Insights from Benchmarking Different Methods for Predicting Pipe Failure Rates in Water Cooled Reactors*" (IAEA 2021). The results of the latter activity were published as a Nuclear Energy Series (NES) Report NR-T-2.16 titled "*Methodologies for Assessing Pipe Failure Rates in Advanced Water-Cooled Reactors*" (IAEA 2023).

- EPRI *Pipe Rupture Frequencies for Internal Flooding Probabilistic Risk Assessments* (2005–2023) (Fleming et al. 2023):[29] Influenced by lessons learned from the Byron and Braidwood Internal Flooding PSA (2000–2001) and a subsequent pilot study on pipe failure rates and rupture frequencies for use in internal flooding PSA (IF-PSA),[30] this EPRI project developed estimates of piping system failure rates for use in PSA of accident sequences induced by internal flooding and high-energy line breaks (HELBs). The latest edition of this handbook was published in 2023. This edition offered advancements in the methodology for passive component reliability, including the quantification of aging effects on pipe rupture frequencies and the capability to adjust these frequencies to account for enhancements to piping pressure boundary integrity management strategies associated with leak inspections and nondestructive examinations. It relied on US operating experience and a methodology originally developed for risk-informed in-service inspection program development.
- OECD/Nuclear Energy Agency Component Operational Experience, Degradation and Ageing Program (CODAP, 2002–2026) (Lydell et al. 2007, 2021): At this point, 14 nations are exchanging operating experience data. CODAP periodically prepares topical reports on the international piping operating experience. These reports aim to promote technical cooperation, understand national differences in plant aging management, and preserve knowledge on material degradation. The topical reports present results of systematic evaluations of the international operating experience with piping systems. An overriding objective of these reports is to foster technical cooperation and to preserve institutional knowledge with respect to material degradation phenomena. To date, the project has prepared nine such reports which are available from https://www.oecd-nea.

[29] A summary of this work is found in Fleming, K., Lydell, B., Presley, M., Mosleh, A., Chalgham, W. (2025), "Initiating Event Frequencies for Internal Flooding and High Energy Line Break PRAs," Journal of Nuclear Engineering, 6, 37. https://doi.org/10.3390/jne6030037.

[30] Results were presented at the NPSAG Castle Meeting in June 2003. It is unclear if the proceedings of this meeting were published. A copy of the referenced paper, including the PowerPoint presentation, is available on request. The conference paper summarized the results of an unpublished pilot study ("*A Risk-Informed Perspective on the Reliability of Raw Water Piping Systems in Commercial Nuclear Power Plants: A Resource Document for Internal Flooding PSA Initiating Event Frequency Estimation and Risk-Informed In-Service (RI-ISI) Applications*").

org/. A Web-based piping operating experience database, CODAP collects and analyzes information on passive metallic and high-density polyethylene (HDPE) component degradation and failures to promote a better understanding of underlying causes, impact on operations, safety, and prevention.

References

Apostolakis GE (1974) Mathematical methods of probabilistic safety analysis, UCLA-ENG-7464. UCLA School of Engineering and Applied Science, Los Angeles

Apostolakis GE, Golay MW, Koser JP, Knight A, Sato G, Silsdorf ME (2002) Enhancing the competitiveness of nuclear power plants through risk informed decision making, MIT-NSP-TR-016. MIT Center for Advanced Nuclear Energy Systems (CANES), Cambridge

Barlow RE, Proschan F (1974) Statistical theory of reliability and life testing: probability models. Holt, Rinehart and Winston, New York

Beliczey S (1992) Steps of the reliability analysis of NPP-piping. In: CSNI international workshop on the reactor coolant system leakage and failure probabilities Cologne, Germany, December 9–11, 1992, NEA/CSNI/R(1995)6, Boulogne-Billancourt, France

Beliczey S, Schultz H (1987) The probability of leakage in piping of pressurized water reactors on the basis of fracture mechanics and operating experience. Nucl Eng Des 102:432–438

Bengtsson G (ed) (1989) Risk analysis and safety rationale, NKA/RAS-400. NKS Secretariat, Roskilde

Brickstad B et al (2004) Project NURBIM (nuclear RI-ISI methodology for passive components): benchmarking of structural reliability models and associated software. In: PVP2004-2551, proceedings of the ASME/JSME pressure vessels and piping conference. American Society of Mechanical Engineers, New York

Carlsson L, Hammar L (1991) Probabilistic safety analysis as a component of periodic safety reassessment of Swedish NPPs—experience from the eighties and prospects for the nineties. In: IAEA-SM-321/1, proceedings use of probabilistic safety assessment for operational safety. IAEA, Vienna. ISBN: 92-0-000292-7

de Finetti B (1974) Theory of probability: a critical introductory statement. John Wiley & Sons, New York. ISBN: 0-471-20141-3

Dinsmore S (ed) (1985) PRA uses and techniques—a Nordic perspective, SAK-1. NKS Secretariat, Roskilde

Farmer FR (1967) Reactor safety and siting: a proposed risk criterion. Nucl Saf 8:539–548

Fleming KN (2003) Issues and recommendations for advancement of PRA technology in risk-informed decision making, NUREG/CR-6813. USNRC, Washington, DC

Fleming K, Lydell B, Presley M, Mosleh A, Chalgham W (2023) Pipe rupture frequencies for internal flooding probabilistic risk assessments revision 5, 3002024904. EPRI, Palo Alto

Fussell JB, Rasmuson DM, Wilson JR, Burdick GR, Zipperer JC (1976) A collection of methods for reliability and safety engineering, ANCR-1273. Idaho National Engineering Laboratory, Idaho Falls

Galyean WJ (2006) LOCA frequencies estimated from operating experience. In: PSAM-0282, proceedings of the 8th international conference on probabilistic safety assessment and management, May 14–18, 2006, New Orleans, Louisiana

Garrick BJ (1970) Principles of unified systems safety analysis. Nucl Eng Des 13(2):245–321

Garrick BJ (1984) Recent case studies and advancements in probabilistic risk assessment. Risk Anal 4(4):267–279

Gnedenko BV (1969) The Theory of Probability, First Edition, MIR Publishers, Moscow, Russia.

Green AE, Bourne AJ (1972) Reliability technology. John Wiley & Sons Ltd, London

GRS (1995) Fachseminar: Ermittlung der Häufigkeiten von Lecks und Brüchen in druckführenden Systemen für probabilistische Sicherheitsanalysen (Topical seminar: estimation of frequencies of leaks and breaks in pressurized systems for use in PSA). Gesellschaft für Anlagen- und Reaktorsicherheit (GRS) mbH, Cologne

Hickman JW et al (1983) PRA procedures guide: a guide to the performance of probabilistic risk assessments for nuclear power plants, NUREG/CR-2300, vols 1 & 2. USNRC, Washington, DC

Hopper A, Wilkowski G, Scott P, Olson R, Rudland D, Kilinski T, Mohan R, Ghadiali N, Paul D (1997) The second international piping integrity research group (IPIRG-2) program final report, NUREG/CR-6452. USNRC, Washington, DC

IAEA (2021) Technical insights from benchmarking different methods for predicting pipe failure rates in water cooled reactors. Final report of a coordinated research project, IAEA-TECDOC-1988, Vienna, Austria

IAEA (2023) Methodologies for assessing pipe failure rates in advanced water-cooled reactors. Nuclear energy series publication NR-T-2.16, Vienna, Austria. ISSN 1995-7807

Jackson C (2017) Risk informed decision making: a survey of United States experience. Report prepared by The B. John Garrick Institute for the Risk Sciences, University of California, Los Angeles and The Nuclear Risk Research Center, Central Research Institute of the Electric Power Industry, Tokyo

Jamali K (1992) Pipe failures in U.S. Commercial Nuclear Power Plants, TR-100380. EPRI, Palo Alto

Kadak AC, Matsuo T (2007) The nuclear industry's transition to risk-informed regulation and operation in the United States. Reliab Eng Syst Saf 92:609–618

Kanninen MF et al (1983) The development of a plan for the assessment of degraded nuclear piping by experimentation and tearing instability fracture mechanics analysis, NUREG/CR-3142, volumes 1 & 2. USNRC, Washington, DC

Kaplan S, Garrick BJ (1979) On the use of a Bayesian reasoning in safety and reliability decisions three examples. Nucl Technol 44(2):231–245

Kaplan S, Apostolakis G, Garrick BJ, Bley DC, Woodward K (1981) Methodology for probabilistic risk assessment of nuclear power plants, PLG-0209. Pickard, Lowe and Garrick, Inc., Irvine

Kemeny JG et al (1979) Report of the President's Commission on the accident at Three Mile Island, the need for change: the legacy of TMI. U.S. Government Publishing Office, Washington, DC

Kersting E, von Linden J, Müller-Ecker D, Werner W (1993) Safety analysis for boiling water reactors. A summary, GRS-98. Gesellschaft für Anlagen und Reaktorsicherheit (GRS) mbH, Cologne

Kröger W, Wolters JP (1986) Experience in the use of probabilistic safety analysis for the development of safety concepts for commercial high-temperature reactors. Nucl Technol 74(1):53–64

Lewis HL, Budnitz RJ, Kouts HJC, Rowe WD, von Hippel F, Loewenstein WB, Zachariasen F et al (1978) Risk assessment review group report to the U.S. Nuclear Regulatory Commission, NUREG/CR-0400, Washington, DC

Löfstedt RE (2005) Sweden: Barsebäck, risk management and trust. In: Risk management in post-trust societies. Palgrave Macmillan, London

Lydell B (1998) Failure rates in Barsebäck-1 reactor coolant pressure boundary piping. An application of a piping failure database, SKI report 98-30. Swedish Nuclear Power Inspectorate, Stockholm, Sweden

Lydell BOY, Olsson A (2008) Reliability data for piping components in Nordic nuclear power plants "R-book" project phase I, SKI report 2008:1 revision 1. Swedish Nuclear Power Inspectorate, Stockholm

Lydell B, Huerta A, Gott K (2007) Progress with the international pipe failure data exchange project, PVP2007-26278. In: Proceedings of the 2007 ASME pressure vessel and piping division conference. ASME, New York

Lydell B, Escrig Forano D, Riznic J (2021) OECD nuclear energy agency CODAP database project on passive component operating experience. An international collaboration in materials research. Nucl Eng Des. https://doi.org/10.1016/j.nucengdes.2021.111280

Marschall CW, Landow MP, Wilkowski GM (1993) Loading rate effects on strength and fracture toughness of pipe steels used in task 1 of the IPIRG program, NUREG/CR-6098. UNRC, Washington, DC

Mays SE, Poloski JP, Sullivan WH, Trainer JE, Bertucio RC, Leahy TJ (1982) Interim reliability evaluation program; analysis of the Browns Ferry unit 1 nuclear power plant, NUREG/CR-2802. USNRC, Washington, DC

Natesan K, Moisseytsev A, Bahn C, Diercks DR, Shaffer C (2014) Knowledge base report on emergency core cooling sump performance in operating light water reactors, NUREG/CR-7172. USNRC, Washington, DC

OECD/NEA (1994a) Proceedings of the workshop on the Barsebäck strainer event, NEA/CSNI/R (1994)14, Boulogne-Billancourt, France

OECD/NEA (1994b) Results of recent studies in France, Germany, Japan, Sweden and the United States, a supplement report of NEA/CSNI/R(92) "state of the art of level 1 PSA methodology," NEA/CSNI/R(94)10, Boulogne-Billancourt, France

OECD/NEA (2010) EC-JRC/OECD-NEA benchmark study on risk informed in service inspection methodologies (RISMET), NEA/CSNI/R(2010)13, Boulogne-Billancourt, France

Okrent D, Apostolakis G, Whitley R, Garrick BJ (1982) On PRA quality and use, UCLA-ENG-8269. UCLA School of Engineering and Applied Science, Los Angeles

Paul DD, Ahmad J, Scott PM, Flanigan LE, Wilkowski GM (1994) Evaluation and refinement of leak-rate estimation models, NUREG/CR-5128. USNRC, Washington, DC

Pershagen B (ed) (1993) Nordic studies in reactor safety, final report of the Nordic nuclear safety research program SIK, TemaNord-1994:544. NKS, Roskilde

Poloski JP, Marksberry DG, Atwood CL, Galyean WJ (1999) Rates of initiating events at U.-S. Nuclear Power Plants: 1987–1995, NUREG/CR-5750. USNRC, Washington, DC

Rao DV, Letellier BC, Schaffer C, Ashbaugh S, Bartlein LS (2002) GSI-191 technical assessment: parametric evaluations for pressurized water reactor recirculation sump performance, NUREG/CR-6762, vol 1. USNRC, Washington, DC

Rogovin M, Frampton GT (1980) Three Mile Island; a report to the commissioners and to the public, NUREG/CR-1250. USNRC, Washington, DC

Schmidt R, Wilkowski G, Scott P, Olson R, Marschall C, Vieth P, Paul D (1992) International piping integrity research group (IPIRG) program final report, AECB project no. 2.111.1. Atomic Energy Control Board, Ottawa

Schultz H (1988) Comments on the probability of leakage in piping systems as used in the PRAs. Nucl Eng Des 110:229–232

Scott P, Wilson M, Olson R, Marshall C, Schmidt R, Wilkowski G (1994) Stability of cracked pipe under inertial stresses, NUREG/CR-6233. USNRC, Washington, DC

Scott P, Wilson M, Olson R, Marshall C, Rudland D, Francini R, Wolterman R, Hopper A, Wilkowski G (1997) IPIRG-2—pipe system experiments with circumferential cracks in straight-pipe locations, NUREG/CR-6389. USNRC, Washington, DC

USNRC (1975) Reactor safety study: an assessment of accident risks in U.S. Commercial Nuclear Power Plants, NUREG-75/014 (WASH-1400), Washington, DC

USNRC (1986) Leak before break: international policies and supporting research. In: Proceedings of a seminar held in Columbus, Ohio on October 28–30, 1985, NUREG/CP-0077, Washington, DC

USNRC (1988) Individual plant examination for severe accident vulnerabilities—10 CFR 50.54(f), Generic Letter 88-20, Washington, DC

USNRC (1990) Severe accident risks: an assessment for five U.S. Nuclear Power Plants, NUREG-1150, Appendix C, Section C.15, Washington, DC

USNRC (1991) Individual plant examination of external events (IPEEE) for severe accident vulnerabilities—10CFR 50.54(f), Generic Letter 88-20 Supplement 4, Washington, DC

USNRC (1995) Use of probabilistic risk assessment methods in nuclear regulatory activities; final policy statement, 60 FR 42622. Federal Register, Washington, DC

USNRC (2008) Estimating loss-of-coolant accident (LOCA) frequencies through the elicitation process, NUREG-1829, vols 1 & 2. USNRC, Washington, DC

USNRC (2017) Risk-informed approach for addressing the effects of debris on post-accident long-term core cooling. Draft regulatory guide 1.229, Washington, DC. NRC-ADAMS Accession No. ML17025A060

Vo TV, Heasler PG, Doctor SR, Simonen FA, Gore BF (1991) Estimates of rupture probabilities for nuclear power plant components: expert judgment elicitation. Nucl Technol 96(3):259–271

Wilkowski GM, Olson RJ, Scott PM (1998) State-of-the-art report on piping fracture mechanics, NUREG/CR-6540. USNRC, Washington, DC

Chapter 3
Historical Perspectives

3.1 Timeline

Much of the initial work to develop quantitative estimates of piping reliability was focused on the double-ended guillotine break of the largest primary coolant pipe connected to the reactor pressure vessel. This work was conducted without the benefit of operating experience and before the advent of probabilistic fracture mechanics tools. As operational data became available, basic statistical characterizations of the operating experience were performed in conjunction with informed engineering judgment. Several significant contributions were made by researchers and risk science practitioners during the five periods covered in this chapter, from 1964 to 2025 (Fig. 3.1), contributions that should be recognized by current and future analysts. Especially noteworthy is the work by the investigators that worked on the *Reactor Primary Coolant System Pipe Rupture Study* during 1964–1974. This chapter covers the following:

- **Early work**:
  - 1964–1974: A study of reactor primary coolant system pipe rupture consisted of multiple tasks to develop methods for estimating the reliability of a reactor primary piping system in terms of leak or severance modes of failure in a 40-year reactor service period and for identifying the principal sources of uncertainty in the estimated probabilistic failure metrics.
  - 1975–1984: Work performed built or expanded on the insights generated by the Reactor Safety Study. It involved assessing the applicability of piping operating experience and exploring the applicability of probabilistic fracture mechanics.
  - 1985–1994: Increased interest in piping operating experience and the statistical characterizations thereof. Initial steps towards development of a methodology for risk-informed in-service inspection of piping that included pipe failure potential assessments.

B. Lydell, *Knowledge Preservation in Nuclear Safety*,
https://doi.org/10.1007/978-3-032-25463-4_3

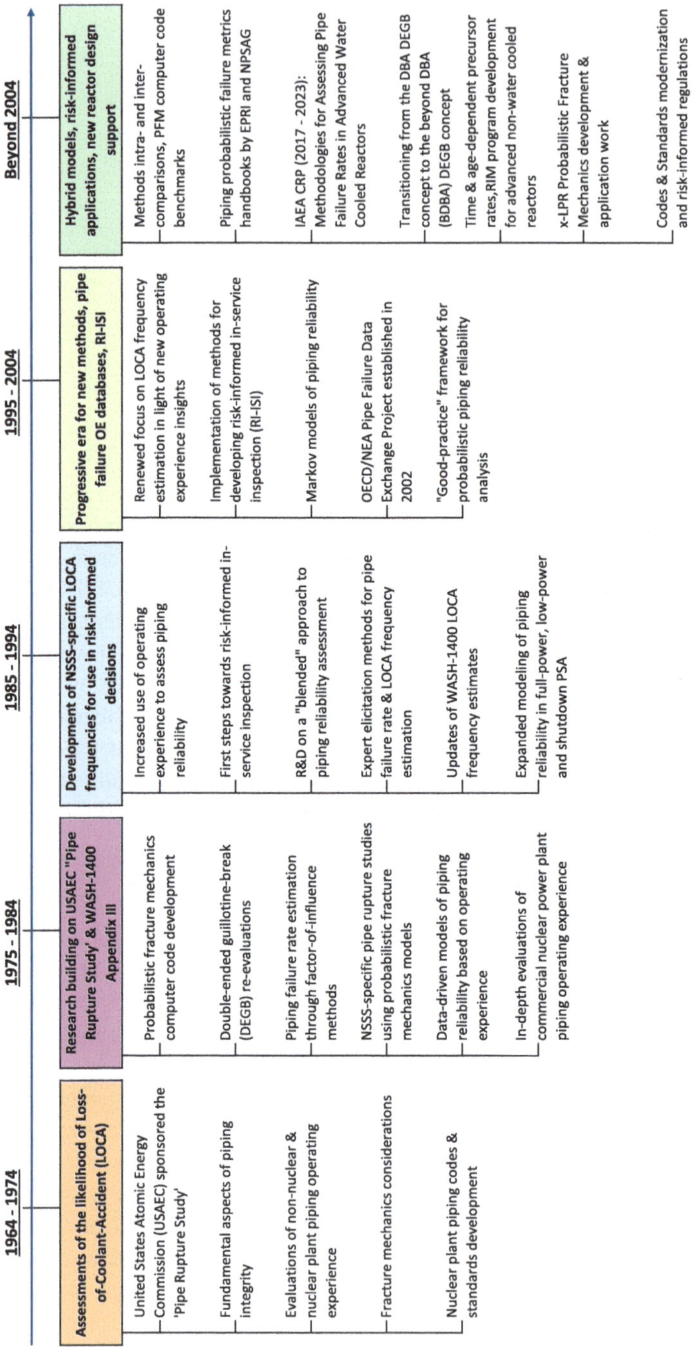

Fig. 3.1 Probabilistic piping reliability method development timeline

- **Progressive work**:
  - 1995–2004: Significant interest in probabilistic piping reliability for use in risk-informed in-service inspection program development. The first steps were made towards the creation of an international pipe failure data exchange project, a project that remains active today.

- **Current progressive work**:
  - 2005–2024: Knowledge preservation through international efforts in areas of piping operating experience exchange, methodology intra- and intercomparisons, good practice analysis frameworks, and pipe failure parameter handbooks. Transitioning from the double-ended guillotine break of the largest primary coolant pipe being seen as to a design-basis accident to being a beyond-design-basis accident per today's state of knowledge.

3.2 Early Work I: 1964–1974

Summaries of pioneering work are provided for reference and to assist in placing in perspective the significant progress that has been made in the collection and analysis of piping operating experience data as well as method development. Table 3.1 lists results of some early assessments of plant-level primary system pipe failure frequencies in terms of point estimates and interval estimates (Gibbons and Hackney 1964; Kilsby 1964). The quantitative estimates were primarily inferred from nonnuclear industry data sources.[1]

The Phillips and Warwick report (AHSB(S) R162) mainly analyzed pressure vessel failures, but some pipe failure data were considered and evaluated as well. Their study was subsequently updated, and the results were presented in the *International Journal of Pressure Vessels and Piping* in 1974 (Smith and Warwick 1974). The nonnuclear operating experience covered the period 1962–1972, and it was comprised of a total of 139 pressure vessel failures of different severity levels: all itemized in the journal paper. To better correlate with nuclear applications, vessel ages were restricted to be less than 30 years, had associated working pressures above 1 MPa, and were built to the English "Class 1 standards."[2] Because the reports on

[1] Page III-93 of WASH-1400 Appendix III lists the information used in compiling these results. It is stated that the Nuclear Safety Information Service (NISC), Oak Ridge National Laboratory, performed a literature search on "pipe failure" which yielded 317 references that documented pipe failures in industrial and nuclear installations. The author of this book has not been able to verify the Wells-Knecht (1965) pipe frequency estimate.

[2] Class 1 welded pressure vessels are those of a "sufficiently high standard of manufacture" to comply with the requirements of certain recognized specifications. Heeley E.J. (1946). Some

Table 3.1 Examples of early assessments of primary system pipe failure frequency

Source	Plant-level pipe failure frequency (1/year)	Type of pressure boundary failure
Derived from GEAP-5474 (Gibbons and Hackney 1964)	4.0×10^{-2}	All failures
	2.0×10^{-3}	Severed pipe
Wells-Knecht (1965) German nonnuclear pressure vessel operating experience	1.0×10^{-7}	Rupture of primary coolant system piping event.[a] Believed to be an assumed value
Phillips and Warwick (1967) Risø Report No. 214, pp 323–334, and UKAEA AHSB(S) R162	1.5×10^{-2}	Potentially dangerous event
	2.5×10^{-3}	Catastrophic event
A. J. Bourne, G. Hensley, A. R. Eames, A. Aitken, "Reliability Assessment of the SGHWR Liquid Shutdown System," AHSB(S)R 144 (March 1968)	3.0×10^{-6}–2.0×10^{-3}	Large-scale rupture of a primary system
R. Salvatori, Westinghouse Electric Co. (1971); IEEE Transactions on Nuclear Science, NS-18, Issue 1, pp 495–504	1.0×10^{-4}	

[a] It remains unclear how this estimate was determined. The "Wells-Knecht" pressure vessel operating experience data refers to a statistical analysis of German steam boiler and pressure vessel failures in the 1960s. The study was significant for providing early, comprehensive data on the reliability of passive components in reactor systems. The research was conducted by the Institut für Reaktorsicherheit (Institute for Reactor Safety) of the German Technischer Überwachungs-Verein (TÜV). An updated analysis was published in 1978, results of which are sometimes referred to as the "Wells-Knecht/Oberender" data

pipe failures were not as detailed as the reports on pressure vessel failures and because of extrapolation uncertainties, the results were viewed as having larger uncertainties.

A subsequent paper by Kellerman et al. (1974) also included results of a survey of nuclear primary system pipe failures; a total of 40 such failures were identified, of which 22 occurred in the United Kingdom. Short descriptions of the failures were included in this reference.

The Atomic Energy Commission (AEC) of the United States, and following the Energy Reorganization Act of 1974,[3] the Nuclear Regulatory Commission (NRC)

Considerations in the Design of Class 1 Pressure Vessels, Proceedings of the Institution of Mechanical Engineers, 154(1):22–31. https://doi.org/10.1243/PIME_PROC_1946_154_013_02.

[3] On October 11, 1974, President Gerald R. Ford signed the Energy Reorganization Act of 1974. The Act abolished the Atomic Energy Commission and created three new federal entities: the Energy Research and Development Administration (ERDA), the Nuclear Regulatory Commission (NRC), and an Energy Resources Council composed of the secretaries of state and interior, the administrators of ERDA and FEA, and the director of the Office of Management and Budget. Three years after the creation of ERDA, President Carter signed the Department of Energy Organization Act, whereby ERDA was merged with the Federal Energy Administration to form the Department of Energy (DOE).

sponsored the "*Reactor Primary Coolant System Pipe Rupture Study*" which was performed over a 15-year period, 1964–1979. The research program consisted of multiple tasks to develop methods for estimating the reliability of a reactor primary piping system in terms of leak or severance modes of failure in a 40-year reactor service period[4] and for identifying the principal sources of uncertainty in the estimated probabilistic failure metrics. Early outcomes of the research included the following technical reports, reports which remain readily available for a detailed review at https://www.osti.gov/:

- GEAP-4445 (1964), A Survey of Water Reactor Primary System Conditions Pertinent to the Study of Pipe Rupture (Kilsby 1964)
- GEAP-4574 (1964), Survey of Piping Failures (Gibbons and Hackney 1964)
- GEAP-4578 (1964), Mechanical Design Considerations in Primary Nuclear Piping (Tagert 1964)
- GEAP-10452 (1972), Estimating Pipe Reliability by the Distribution of Time to Damage Method (Wilson 1972)

3.2.1 Kilsby (1964)

Under the terms of the project agreement, the investigators developed a questionnaire to perform a survey to establish the range of parameters to be considered in subsequent tasks in the *Reactor Primary Coolant System Pipe Rupture Study.* Based on the data received from the survey participants, a partial list of parameters, conditions, and criteria for consideration in the study of "pipe rupture" are reproduced below:

3.2.1.1 Materials

- For future applications, Inconel-600 piping and carbon steel piping clad with austenitic stainless steel or Inconel should also be considered.
- Dissimilar metal joints such as reactor vessel nozzle joints and pipe flange joints should receive close attention in the study of pipe rupture.

[4]The 40-year lifespan for early commercial nuclear reactors was set mainly for economic and accounting reasons, matching utility depreciation schedules rather than establishing a safety limit. Regulators and industry expected technology to advance and eventually replace older plants, even though many parts were designed for replacement and extensions became possible as reliability was proven. The main reasons for the initial 40-year limit were (1) economic depreciation (utilities planned investments over 40 years), (2) expected technological advancements, and (3) regulatory standards that set 40 years as the default license period.

- Pipe sizes of interest range from ½-inch diameter to 30-inch diameter. Wall thicknesses associated with each of the diameters should be those calculated by appropriate design code formulas.

3.2.1.2 Loadings

- Major shock loadings to be considered were those associated with reactor scram. Temperature variations were at rates of 50–360 °F per minute with magnitudes of 300–400 °F.
- Mechanical loadings of all types exist and should be considered. The most critical geometric problem was considered to be that of a constrained branch connection used to introduce cold water into a high-temperature mainstream.
- Critical piping design problems requiring investigation were the dissimilar metal joint, the nozzle connection, and the branch connection, under various combinations of pressure and temperature and with imposed restraints.

3.2.1.3 Fabrication and Quality Control

- Survey replies generally indicated that welding procedures and weld tests are considered to be adequate. Attention should be given to possible improvements in techniques and interpreting test data.
- In-service inspection of nuclear plant piping is quite limited and is not regularly scheduled. Leak detection provisions are also inadequate. The development of tools and procedures for surveillance of piping systems in nuclear plants is an area requiring attention in the study of piping reliability.

3.2.1.4 Most Significant Problems

- The most important goal of the pipe rupture study was the determination of a maximum credible pipe rupture and the development of necessary tools for assessing the probability of such a failure occurring.
- In connection with the above goal, recommendations for additions, changes, and/or improvements to standards for design, fabrication, and inspection of piping systems should be made whenever uncertainties or over-conservatisms were found to exist.

3.2.2 Gibbons and Hackney (1964)

Gibbons and Hackney (1964) performed a survey of pipe failures in fossil power plants, petroleum refineries, chemical processing facilities, and marine applications.

The survey produced 399 failure case histories of interest. The term "failure" was defined as any defective condition encountered during start-up, testing, and/or service that required repair or replacement corrective action. Qualitative conclusions were derived from these, mostly nonnuclear systems, as to the nature of possible failures in nuclear primary piping systems. As an example, piping systems which had been designed and constructed within established code criteria would be expected to exhibit high reliability, and a catastrophic, complete severance rupture would appear to be unlikely to occur. The authors also noted that "*less severe failures might be expected*." Key conclusions were as follows:

> *Cracking caused by low-cycle fatigue was determined to be the most probable cause of failure and attributed to the inability of the designer to predict environmental and service conditions on a local basis. The most likely location for failure to occur is in the pipe or pipe-weld heat-affected zone (HAZ) particularly in association with a structural discontinuity. The growth of cracking-type failures can generally be expected to result in leakage prior to gross or major failure. The detection of leakage associated with relatively small cracks should permit repair and prevention of further cracking and aid the study of design inadequacies which allowed cracking to occur. The importance of adequate monitoring and surveillance must therefore be emphasized. The type and amount of monitoring and surveillance conducted must be considered in terms of possible nuclear hazards and the ability of monitoring and surveillance procedures to detect incipient failures.*

Appendix A of the technical report (GEAP-4574) consists of failure data tabulations and illustrations of failed piping. It covers events in nonnuclear industries that occurred in the period 1929–1962. The summary and conclusions of the study noted the following:

> **Piping Reliability**: *It is apparent that in simple numerical terms, many piping failures have occurred. Just on the basis of the limited survey conducted for this study, a total of 399 failure case histories of interest were collected. 701 contacts returned 94 replies describing failures known to them. Of these 94 replies, 57 were pertinent to the piping area of interest to this study. These 57 replies in combination with what failure data had been published provided the total of 381 failure case histories—or an average of 7 failure cases per failure reply of interest. Viewed in this manner, it could be concluded that the piping failure rate must be a significant figure. It does seem, however, that with respect to all of the piping in operation since the mid-1930s, with the accumulation of all of the operating hours for all of the piping in this country and some in Europe, the total yield of 399 failure cases of interest represents what must be a quite low failure rate. Even if this rate is low by a factor of 2 or 3, this is still a very small number when compared to all of the piping in use. Many of the sources contacted believed this to be so—including some of those whose business is to insure against piping-failure accidents. The average failure-free service life of 9.8 years also reinforces a low rate of failure value. This is not to say that failures will not occur, but rather to say that certain types of failures resulting from certain causes of failure are highly improbable in the systems of interest. On the basis of this review, it appears to be a reasonable, conservative conclusion that if a primary piping system has been designed, fabricated, and erected within the established code-allowable stresses, defect acceptability criteria, etc. and is operated within the code-allowable stress limits, the probability of a complete severance type of failure is very low and the probability of any other type of failure is a reasonably low value. The real problem appears to be ensuring that what is designed does in fact represent what will be required in reality, i.e., the design predicts all of the conditions that will be present during operation.*

Pipe Failure Expectation: *On the basis of this review of tabulated failure cases, if a failure should occur in a nuclear reactor primary piping system, the hypothetical nature of such a failure would be expected to be as follows. It would be located in either the pipe material or the pipe material weld structural discontinuity, a transition, or a surface defect. Fracture initiation and propagation would result from low-cycle (plastic) fatigue, high-cycle (elastic) fatigue, or a combination of both. Gradual crack growth would proceed to traverse the component wall thickness until a localized (with respect to the pipe circumference) surface opening was developed. Leakage of the contained medium would ensue and increase with continued extension of the crack. Either crack or leak detection would be expected to be effected prior to the development of any critical crack size at which rapid crack propagation could take place. The prediction of the hypothetical failure discussed above is predicated on the following three assumptions:*

i. Defects in excess of the established code defect acceptability criteria are not present. This implies ultrasonic nondestructive testing inspection of piping base materials, components, and radiographic and liquid penetrant inspection of cast components in addition to weld joint inspection.

ii. Fabrication and erection practices have not introduced metallurgical defects not detectable by nondestructive testing inspection(s). That is, adequate process procedures and controls have been effected.

iii. The inadvertent substitution of a wrong material has not occurred. This implies either adequate process control all along the line and/or a final material check after installation with a portable spectrograph.

The degree to which the above assumptions are satisfied by nondestructive testing, inspection, and process control determines the magnitude of failure probability. If economic considerations dictate that less nondestructive testing, inspection, and process control be imposed, then the probability of piping failure must necessarily increase, and the hypothetical failure discussed previously will not necessarily be the most likely to occur.

Surveillance: *The implication of the preceding hypothetical failure manner is clear. The crack or leak detection time period must be less than the crack growth to a critical crack size for rapid propagation time period. Assurance that present surveillance methods will satisfy this requirement has not been established. The VBWR recirculation piping failure did fall within the hypothetical case, but this one example does not conclusively indicate that present surveillance procedures and methods are sufficient.*[5] *To the contrary, with respect to the amount of operating experience accumulated, it is presently untenable to completely support the case that present procedures will adequately ensure detection prior to development of a more serious failure mode.*

Role of Operating Experience Exchange: *It seems apparent that there is an urgent need for complete documentation of failure cases—especially for those cases that occur in nuclear power plant piping systems. Only by collecting and disseminating as much of the detailed information as is available for each such failure will conclusive and reliable statistics be generated and a firmer safety basis established.*

[5]The authors were referring to a primary system pipe failure that occurred in 1962 at the Vallecitos Experimental Boiling Water Reactor. This event is described in Chap. 4.

3.2.3 *Kilsby and Tagart (1964)*

Based on an industry survey, Kilsby (1964) and Tagart (1964) evaluated the range of parameters, conditions, and criteria for consideration in studying pipe ruptures. The survey yielded preliminary piping design information, and the survey participants identified the following design parameters of interest for current and future applications:

- Materials
  - Austenitic stainless steel, type 304 and 316.
  - Low-alloy steels, $2\frac{1}{4}$ Cr–1 Mo and $1\frac{1}{4}$ Cr–$\frac{1}{2}$ Mo.
  - Carbon steels.
  - For future applications, Alloy 600 piping and carbon steel piping clad with stainless steel or nickel-base material should also be considered.
  - Dissimilar metal welds such as reactor pressure vessel nozzle to primary coolant piping should receive close attention in the study of pipe rupture.
  - Material properties of particular interest are those associated with fatigue, corrosion, creep, and stress rupture.
- Loadings
  - For immediate applicability to present water-cooled reactor plants, the pipe rupture study should concentrate on the pressure range from 7 to 17 MPa with associated temperatures ranging from 10 to 340 °C.
  - Future efforts should extend to supercritical pressures and temperatures up to 650 °C.
  - Principal cyclic variations are those resulting from startup and shutdown. The small number of such cycles coupled with conservative design assumptions indicates that gross cyclic loading is not of major significance to the pipe rupture study. No similar conclusion can be reached regarding localized cyclic loadings. Local cycling and component fatigue will require further investigation in this study.
  - Major shock loadings are those associated with reactor scram. Temperature variations at rates of 10–180 °C per minute with magnitudes of 150–205 °C are indicated.

    Mechanical loadings of all types can exist and should be considered. The most critical geometric problem is that of a constrained branch connection used to introduce cold water into a high-temperature mainstream:
  - The dissimilar metal joint, the nozzle connection, and the branch connection, under various combinations of pressure and temperature and with imposed restraints, are critical piping design problems requiring investigation:

    Present design precautions are considered adequate in most cases. Improvements recommended include (a) development of criteria to assist in early identification of sensitive locations in piping systems, (b) determination of accurate design safety factors, (c) improvement of analytical tools, and (d) improvement of plant layout to provide more piping system flexibility.

- Water Conditions
  - General corrosion is not considered to be a problem. Local corrosion phenomena and particularly chloride stress corrosion should receive closer scrutiny in this study.
- Radiation
  - Radiation is not considered to be a problem in the study of pipe rupture.

3.2.4 Wilson (1972)

Wilson (1972) developed a method referred to as the "distribution of time to damage" (D-T-D) to assess probabilistic failure metrics. An outline of the method was described as follows:

(a) Draw a probability diagram. The diagram shows a block for each stage of damage, such as crack initiation, leakage, and severance. The diagram can be likened to an event sequence diagram. Lines connect the blocks to show the logic of how one stage of damage can lead to another. Three ideas relate to each line: (1) a damage mechanism, such as low-cycle fatigue; (2) a time-to-damage prediction equation, such as for the number of cycles to grow a crack from initial depth through to leakage; and (3) the probability of this event in a chosen time such as 40 years. Thus, many lines connect each block, one for each damage mechanism.
(b) Develop a damage prediction equation for each line, i.e., for each mechanism for each stage of damage. This is an application of fracture mechanics cycles-to-leak equation as proposed by Paris and Erdogan.[6]
(c) Select a piping segment, and assign a nominal value and an uncertainty distribution to each variable in the equation. The nominal values include dimensions (pipe diameter and wall thickness), operating pressure, and thermal expansion stress.

[6]Reproduced from Wikipedia (2023): Paris law (or the Paris-Erdogan equation) is a crack growth equation that gives the rate of growth of a fatigue crack. The stress intensity factor K characterizes the load around a crack tip, and the rate of crack growth is experimentally shown to be a function of the range of stress intensity ΔK seen in a loading cycle. The Paris equation is $\frac{da}{dN} = C \times \Delta K^m$ where a is the crack length and da/dN is the fatigue crack growth for a load cycle N. The material coefficients C and m are obtained experimentally and depend on environment, frequency, temperature, and stress ratio.

(d) Select only those variables making the major contributions to the variability of the equation, e.g., initial crack size or number of fatigue cycles. This was done by a propagation of errors method.
(e) Form the distribution of the equation from the nominal values and distributions of its variables. Or only estimate the fraction of the equation distribution which lies less than 40 years. This fraction is the desired probability and was estimated by Monte Carlo simulation using importance sampling.

Wilson estimated the frequency of pipe leakage and the conditional probability of severance over a 40-year service life due to low-cycle fatigue. It was an early probabilistic fracture assessment with the Paris-Erdogan formula for fatigue crack growth as a starting point:

$$\frac{d(a)}{d(N)} = A\Delta K^B \tag{3.1}$$

$$n_t = A^{-1} \int_{a(0)}^{t} \Delta K^{-B} da \tag{3.2}$$

In this equation, a is the crack length, $d(a)/dN$ is the crack growth rate for a load cycle N, K is the stress intensity factor, ΔK is the stress intensity range, and material coefficients A and B are obtained experimentally and are characteristic of material at temperature. The crack of initial size, a_i, grows incrementally in each fatigue cycle, depending on the stress level and its actual size. This process will terminate in one of the three possible ways:

1. The pipe will be removed from service before the crack has grown through the wall or initiated a gross failure.
2. The crack grows through the wall and becomes a leaker before the applied load overpowers the resistance.
3. The crack grows to a size where its associated stress intensity factor exceeds the resistance capacity of the materials, resulting in rupture.

The estimate for the conditional probability of leakage in a 40-year plant life due to low-cycle fatigue for the selected evaluation boundary and under an assumption of "no checking or inspection" during design, fabrication, or operation was calculated to be 7.1×10^{-3}. In lieu of evaluation boundary, Wilson used the term "piping study surface," defined as surface area of a section of DN150 primary loop piping over which the value of each variable was assumed to be constant, specifically chosen to have an area of 25 mm length and 6 mm circumference of the DN150 stainless steel pipe of grade AISI Type 304.

Table 3.2 WASH-1400 loss-of-coolant accident frequencies

LOCA category	LOCA initiating event frequency [1/ROY]	
	Median	Lognormal range factor
Small	1.0×10^{-3}	10
Medium	3.0×10^{-4}	10
Large	1.0×10^{-4}	10

3.2.5 The Reactor Safety Study (1974)

A limited evaluation of nuclear piping reliability was performed for the Reactor Safety Study (WASH-1400) (USNRC 1975a, b). The objective of the study was to derive "order-of-magnitude" plant-level loss-of-coolant accident (LOCA) frequencies (Table 3.2) for input to event tree analysis and pipe failure rates for input to system fault trees. WASH-1400 examined several different sources to obtain failure rates for small-diameter and large-diameter pipes. The reason for using several data sources was the interest in pipe ruptures resulting in reactor coolant loss, and none had occurred in the United Sates in the 150 commercial nuclear reactor operating years (ROYs) considered by the analysts. Therefore, other pipe failure data sources were sought for extrapolating pipe failure rates for use in PSA. A criticism of the approach of WASH-1400 was that the database on significant pipe failures was limited to 11 "significant events."

3.3 Early Work II: 1975–1984

Work performed in the period 1975–1984 built or expanded on the insights generated by Reactor Safety Study (WASH-1400). It involved assessments of piping operating experience and explored the applicability of probabilistic fracture mechanics. The "Thomas model" of piping reliability was presented as an extended paper published in the *Reliability Engineering* journal[7] in 1981.

3.3.1 Pacific Northwest National Laboratory (1975)

The Pacific Northwest National Laboratory (PNNL) performed an independent assessment of piping reliability based on light water reactor and nonnuclear operating experience (Bush 1976). Differences between WASH-1400 and PNNL results were attributed to how the limited failure data was interpreted. The study by PNNL

[7] Reliability Engineering was published from 1980 until 1987 (Volume 19). Effective from 1988 (Volume 20), the journal continued as Reliability Engineering and System Safety (RESS).

addressed the role of periodic inspection and addressed failures due to intergranular stress corrosion cracking. Among the conclusions were the following:

- The failure frequencies for larger sizes of piping are in the range of 1.0×10^{-4} to 1.0×10^{-6} per reactor year, exclusive of considerations of intergranular stress corrosion cracking.
- Smaller pipe sizes, of lesser safety significance, have much higher failure rates.
- In BWRs, intergranular SCC can cause failure rates much higher than 1.0×10^{-4} in piping of size DN100 to DN 250.
- "Catastrophic failures" would appear more likely from operator error or design and construction errors, water hammer, improper handling of dynamic loads, and undetected fabrication defects rather than flaw initiation and growth by fatigue.

3.3.2 *Basins and Burns (1977)*

The work of Basins and Burns (EPRI 1977) documents a statistical characterization of pipe system failures derived from reports that had been submitted by plant operators to the USNRC. A total of 237 pipe failures were identified. The data was mainly obtained from the Nuclear Safety Information Center (NSIC) which, at the time, acted as a clearinghouse for reportable events submitted to the regulator.[8] Pipe system failures were compared from the standpoint of frequency of occurrence within plant types, i.e., BWR versus PWR, frequency of occurrence within plant subsystems, failure modes, pipe size categories, and time to failure. Hypotheses regarding differences between frequency distributions within BWRs versus PWRs were tested via standard nonparametric statistical methods. The locations of pipe failure were distributed as follows:

- Welds and heat-affected zones (54%)
- Pipe wall (base metal) (40%)
- Threaded, small-diameter joints (6%)

Figure 3.2 contrasts the above pipe failure distributions against the 2025 state of knowledge regarding the locations within a piping system most vulnerable to degradation. The operating experience data that are represented in this figure are for primary system piping and organized according to nominal pipe size (NPS) in mm, a total of 2800 pipe failures of varying severity levels as of 2025.

Based on pipe size and mode of failure, differences between the time to failure for BWR and PWR piping were assessed. It was determined that PWR piping exhibited exponentially distributed failure times. The average time to failure of a PWR pipe or

[8]The Atomic Energy Commission (AEC) established the Nuclear Safety Information Center (NSIC) at the Oak Ridge National Laboratory in March 1963. Its mission was nuclear community by collecting, storing, evaluating, and disseminating safety information relevant to the design and operation of nuclear facilities.

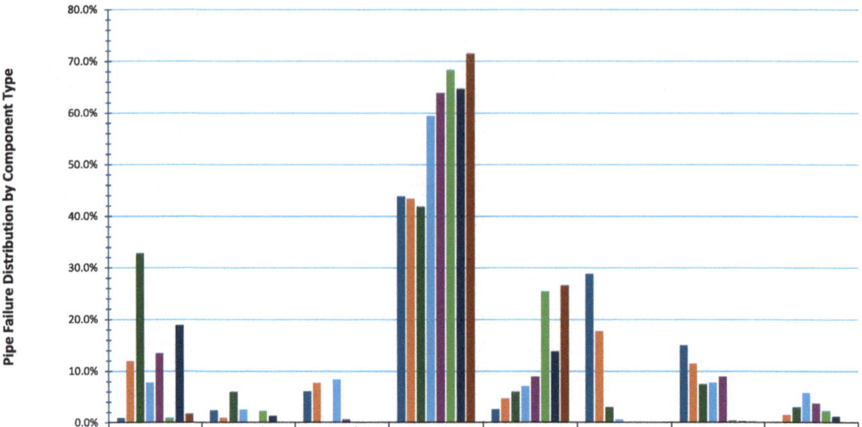

	Bend / Elbow	Branch Connection / Butt Welded	Branch Connection / Socket Welded	Butt Weld / Field Welded	Butt Weld / Shop Welded	Socket Weld	Pipe / Base Metal	Tee / Mixing Tee
≤ DN25	0.9%	2.4%	6.1%	43.9%	2.6%	28.8%	15.1%	0.2%
25 < DN ≤ 50	12.0%	1.0%	7.9%	43.5%	4.7%	17.8%	11.5%	1.6%
50 < DN ≤ 75	32.8%	6.0%	0.0%	41.8%	6.0%	3.0%	7.5%	3.0%
75 < DN ≤ 100	7.8%	2.6%	8.5%	59.5%	7.2%	0.7%	7.8%	5.9%
100 < DN ≤ 150	13.5%	0.0%	0.8%	63.9%	9.0%	0.0%	9.0%	3.8%
150 < DN ≤ 250	0.9%	2.4%	0.0%	68.4%	25.5%	0.0%	0.5%	2.4%
250 < DN ≤ 550	18.9%	1.3%	0.0%	64.6%	13.8%	0.0%	0.2%	1.2%
> DN550	1.7%	0.0%	0.0%	71.5%	26.6%	0.0%	0.2%	0.0%

Fig. 3.2 Distribution of pipe failures by component type as of 2025

pipe component was assessed to be 35.8 months with a standard deviation of 49.6 months. The BWR pipe failure times appeared to have a unimodal distribution with the mode between 50 and 60 months from initial criticality. Other than that, the BWR distribution possessed a relatively long right-hand tail as did the PWR distribution. The average time to failure for a BWR pipe was 56.3 months with a standard deviation of 42.6 months.

Since the failure times of BWR and PWR piping appeared to have been drawn from different distributions, the failure times were separated into distinct groups. Frequency histograms of the failure times associated with each failure mode were constructed. One of the most conspicuous differences between the two sets of histograms was that corrosion was not observed or at least not reported to the same extent in PWRs as it was in the case of BWR plants. In the case of BWR piping failures, it appeared that several failure modes contributed to the unimodal character of the BWR distribution. Another general observation, relative to the differences between the PWR and BWR time distributions, was that the PWR failures, regardless of failure mode, tended to occur early in the life of a plant, whereas the onset of BWR pipe failures occurred later. Twenty-two of the 237 reported pipe failures (i.e., approximately 9.3%) involved pipe ruptures, breaks, or failures of such a nature so one would conservatively classify them as ruptures or breaks. A "rupture" was defined as a failure which causes a "rapid loss of fluid from the pipe." It was noted that these failures were induced by mechanisms that were

Table 3.3 Estimated occurrence rates for pipe rupture

Pressure boundary failure	Failure frequency (1/reactor operating year)
Pipe rupture	8.0×10^{-2} (One pipe rupture per plant in 12 years)
Pipe rupture causing a reactor trip	2.0×10^{-2} (One pipe rupture-induced reactor trip every 50 years)
Large-diameter (>DN250) pipe rupture	1.2×10^{-2} (One large-diameter pipe rupture every 83 years)

difficult to predict. Vibration and pressure surges appeared to be principal modes of these failures. The estimated pipe rupture occurrence frequencies are reproduced in Table 3.3.

3.3.3 GRS (1980)

The Federal Minister for Research and Technology (BMFT) in 1976 commissioned the Gesellschaft für Reaktorsicherheit (GRS) mbH to perform the "German Risk Study" (*Deutsche Risikostudie Kernkraftwerke: Eine Untersuchung zu dem durch Störfälle in Kernkraftwerken verursachten Risiko*). An overriding objective was to apply the WASH-1400 methodology to a German reference plant. Under the scientific direction of the managing director of GRS, Prof. Dr. A. Birkhofer, the associated work was carried out together with other German technical and scientific organizations, and a summary of the results were published in August 1979. The supporting technical documentation was published in 1980 and consisted of eight volumes (BMFT 1980).

Volume 3, "Reliability Parameters & Operating Experience," includes an assessment of pressure vessel and piping reliability. Pages 13–22 and 84–85 document the assessment of pipe failure-induced LOCA initiating event frequencies (Albertz et al. 1980). The assessment was built on the related work done in support of WASH-1400. LOCA frequencies were derived based on "engineering analyses," "engineering judgment," and literature reviews (Fig. 3.3).

It was noted that because of the relative risk insignificance of pipe failures, an in-depth quantitative evaluation of piping reliability was not pursued. Instead, order-of-magnitude pipe failure rate estimates were obtained using literature data. In support of the LOCA frequency determination, the authors used a pipe failure rate of 1.0×10^{-9} per linear meter of piping and year. No uncertainty analysis was pursued.

The German Risk Study used Biblis-B Pressurized Water Reactor (PWR) as its technical reference plant. It was a KWU-type plant with a thermal output of 3750 MW that began commercial operation in March 1976. Biblis-B was the first 1300 MWe nuclear power plant in the world. This reactor unit was chosen for the technical analysis, and it was assumed that all the other reactors considered in the

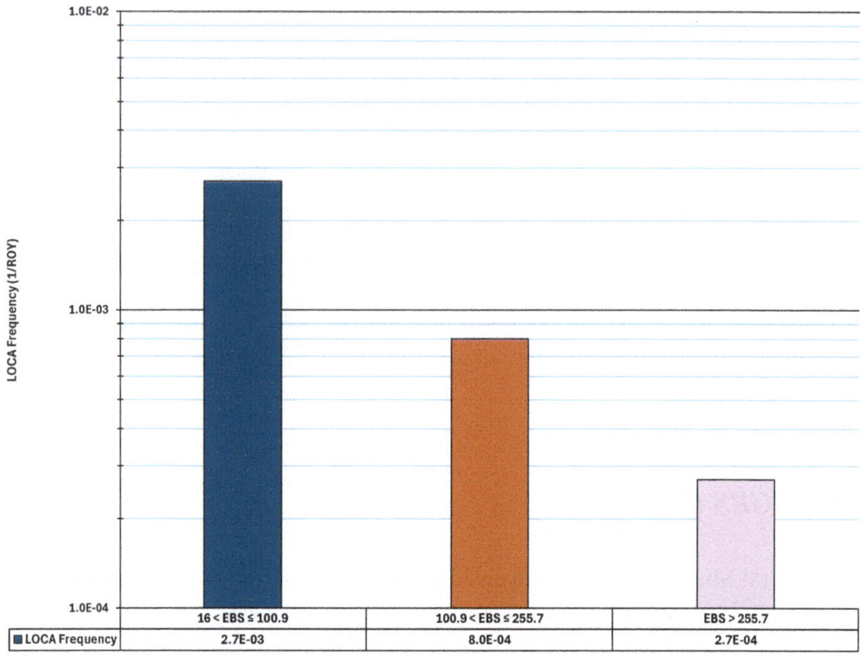

Fig. 3.3 LOCA frequencies in "German Risk Study Phase A"

study, which included a projected set of 20 plants with 600 MW or more electrical output, were technically identical to this reference plant. Following the 2011 Fukushima nuclear accident and for political reasons, Biblis Units A and B were ordered to be permanently shut down in August 2011.

3.3.4 *Janzen (1981, 1984)*

Because of the limited operating experience of CANDU reactors and the availability of documentation of the much longer history with light water reactors in the United States, the latter was chosen as the initial subject of an analysis of the expected performance of primary heat transport system (PHTS) piping in CANDU reactors. P. Janzen of Chalk River Nuclear Laboratories (Canada) studied the US light water reactor pipe failure data for the period 1959 through 1978, representing 409 reactor operating years of experience (Janzen 1981). The data sources were obtained from the public document room of the USNRC. Another objective was to determine whether the additional operating experience that had accumulated since publication of WASH-1400 in 1974 warranted new pipe failure rates to be used in probabilistic safety assessment.

Table 3.4 LWR primary system pipe failure rates (Janzen 1981)

Nominal pipe diameter (DN) (mm)	Failure rate [1/ROY] Upper boundary @ 95% confidence
Small-diameter piping ≤ 25	3.9×10^{-2}
Intermediate-diameter piping 25 < DN < 150	7.3×10^{-3}
Large-diameter piping DN ≥ 150	7.3×10^{-3}

The pipe failure events were classified according to (1) severance, (2) leak, and (3) "defect." A total of 87 pipe failures were determined to be of relevance to the study. Abstracts of the corresponding failure reports were included in appendices to the report: Appendix A (large-diameter piping, 7 events), Appendix B (intermediate-diameter piping, 33 events), and Appendix C (small-diameter piping 47 events). None of the recorded failures involved piping connected to the reactor pressure vessel, however. A statistical analysis was limited to an estimation of confidence limits for failure rates using the Chi-square distribution. Because of uncertainties in the pipe failure database and assumptions in interpretation of the data, the order-of-magnitude failure rate estimates by WASH-1400 were viewed by the Janzen as representative of "true" pipe failure rates. Table 3.4 shows failure frequency estimates for primary system pipe "breaks." In retrospect, a problem with these estimates is the lack of specificity with respect to how they relate to safety class, system, operating environment, degradation mechanisms, and reliability and integrity management.

A follow-up study was published in 1984 as AECL-MISC-252 (Janzen 1984). This time, the author accessed pipe failure data from Ontario Hydro (now Ontario Power Generation, OPG) and its CANDU reactors which had been in operation for about 15 years. Janzen collected 78 pipe failure records that were relevant to the primary heat transport system. These raw datasets were subjected to detailed screening and analysis along with engineering data (pipe lengths) supplied by Ontario Hydro (Table 3.5).

According to the follow-up study, 8%, 39%, and 55% of total plant piping in the large, intermediate, and small pipe size ranges, respectively, were classified as "nuclear piping." The pipe failure rate results were organized in three pipe size categories: Table 3.6. At the time, the work of P. Janzen represented an important contribution to the study of piping reliability, especially with respect to the general format for systematic evaluation and assessment of operating experience data. The author made the following observations:

- Results of analyses of pipe failure data are credible and accurate only to the extent that underlying information is complete and precise.
- Many event reports are being used for purposes not foreseen at the time the information was gathered. This has meant that many factors of current interest which characterize pipe failure events are stated imprecisely or not at all. Therefore, inevitably, an element of interpretation was introduced.
- Failure rate calculations in units per component-hour of service require engineering data, in this instance pipe length. Accurate estimates of pipe length appeared

Table 3.5 CANDU reactor pipe lengths[a]

Class of pipe lengths (linear kilometers)	Pipe size range		
	Small diameter DN ≤ 25	Intermediate diameter 25 < DN < 150	Large diameter DN ≥ 150
Bruce Nuclear Generating Station A			
Nuclear: all	58	81	13
Nuclear: feeder piping	–	65	–
Nonnuclear	47	118	60
Total length	105	199 (incl. feeders)	73
Pickering Nuclear Generating Station A			
Nuclear	40	49	9
Nuclear: feeder piping	–	37	–
Nonnuclear	33	82	41
Total length	73	131 (incl. feeders)	50

[a] At the time of the study, Bruce NGS consisted of four reactor units. Unit A-1 achieved first criticality on December 17, 1976. Pickering NGS consisted of four reactor units, two of which since have been permanently shut down. Unit A-1 achieved first criticality on February 25, 1971

Table 3.6 CANDU pipe severance rates for the primary heat transport system

Class of PHTS pipe	Frequency of pipe severance (1/m and reactor operating year)
Small pipe	$8.5 \times 10^{-8} \leq \lambda_{90} \leq 7.9 \times 10^{-6}$
Intermediate pipe	$\lambda_{95} \leq 3.9 \times 10^{-6}$
Large pipe	$\lambda_{95} \leq 2.2 \times 10^{-5}$

difficult to determine, and furthermore, the delineation between nuclear piping and conventional (nonnuclear) piping was at times vulnerable to interpretation.

3.3.5 Thomas (1981)

In 1981, H.M. Thomas of Rolls-Royce & Associates Ltd. published an extended technical paper which appeared in Volume 2 of the Reliability Engineering journal (the predecessor of Reliability Engineering and System Safety) (Thomas 1979, 1981). The technical paper presented a generalized approach to the interpretation of piping operating experience data with recommendations on how to adjust generic industry data to plant-specific pipe failure metrics. Interpreting data from WASH-1400 (1975), EPRI-sponsored work, and other industry data sources, Thomas provided statistical correlations among piping reliability attributes and influence factors. The frequency of "catastrophic pipe rupture" F_C was expressed as the frequency of leakage λ_L times a conditional rupture probability. It was referred to as the "broad strategy":

$$F_C = \lambda_L \times P\left(\frac{C}{L}\right) \tag{3.3}$$

The Thomas correlation was seen by many as the only practical approach to estimate pipe leak and rupture frequency. From time to time, it has been proposed that the single most important advancement in piping reliability would be to simply validate the original correlation using today's failure statistics. The following correlation between piping design and operational parameters and the frequency of through-pipe wall leakage was proposed:

$$\lambda_{F-Tot} = \lambda_{Base} \times Q_E \times F \times B \tag{3.4}$$

The term $\lambda_{F\ -\ Tot}$ is the estimated total leakage frequency, and λ_{Base} is a "baseline" or generic leak frequency suggested to be in the range of 1.0×10^{-9} to 1.0×10^{-7} per year, with a recommended value of 1.0×10^{-8} per. The multiplier Q_E represents the influence on integrity by pipe size and "shape differences." Parameter F is a "plant age factor." As suggested by Thomas, the frequency of leakage declines with plant age. Parameter *B is a* "design learning curve factor." According to Thomas, new piping designs have higher-than-average failure frequency. The factor Q_E accounts for the piping dimensions and the type and number of fittings included in a piping system, for which the following relationship was suggested:

$$Q_E = Q_P + A \times Q_W \tag{3.5}$$

$$Q_P = L \times \left(\frac{D}{t^2}\right) \tag{3.6}$$

$$Q_W = N \times 1.75 \times \left(\frac{D}{t}\right) \tag{3.7}$$

"*A*" is a "weld penalty factor," expressing that a weld is considered more vulnerable to leakage than the pipe base metal.[9] According to Thomas, data analysis suggests that a value of 50. L is the length of the pipe, D is the pipe diameter, t is the pipe wall thickness, and N is the number of circumferential welds in the piping system. The factor of 1.75 was applied to account for the fact that the length of the pipe that is influenced by the weld is assumed to be approximately 1.75 times the width of the weld. In other words, the heat-affected zone of a weld would extend 1.75 times the width of the weld into the base metal. Perhaps the most interesting aspect of the work of Thomas is the "pipe wall thickness scaling factor." According

[9]For an additional perspective, the reader is encouraged to review ASME B31.1-Power Piping, which prescribes minimum requirements for the design, materials, fabrication, erection, test, examination, inspection, operation, and maintenance of piping systems. This piping design code defines weld quality factors, denoted as E_j and with values ranging from 0.6 to 1.0 that reduce the allowable stress of pipe based on welding method and nondestructive examination method.

to the Thomas method, the pipe failure probability is proportional to length L, diameter D, and wall thickness t:

$$\lambda \propto (L \times D)/t^2 \tag{3.8}$$

Thomas stated that for the prevailing pipe fabrication technology, a pipe has fewer but larger size flaws as the pipe wall thickness increases. In general, the pipe failure rate decreases as the wall thickness increases. If the pipe length and nominal diameter are held constant and the change in the pipe failure rate is sought for two different pipe wall thicknesses, then the ratio of failure rate for thickness "1" (λ_1) to thickness "2" (λ_2) is

$$\frac{\lambda_1}{\lambda_2} = \left[(L_1 \times D_1)/t_1^2)/(L_2 \times D_2)/t_2^2)\right] = t_2^2/t_1^2 \tag{3.9}$$

Thus, the leak probability quotient is found to be a ratio of the thickness of the two pipes. The thicker part carries stresses better and the failure rate lower for thicker pipe. The failure rate, say for a Schedule 160 pipe versus the failure rate for a Schedule 80 pipe, becomes

$$\lambda_{Sch160} = \lambda_{Sch80} \times \left(t_{Sch80}^2/t_{Sch160}^2\right) \tag{3.10}$$

Thomas made compelling technical arguments supporting his piping reliability correlations. The various attributes and influence factors in the equations above were each discussed in detail. His work has found relatively widespread use, and in some cases, model extensions have been made. The paper by Medhekar et al. (1993) is an example of such an extension. The authors developed scaling factors to account for the influences of corrosion, fatigue, and in-service inspection. To account for uncertainty in $P\left(\frac{C}{L}\right)$, the authors assigned a subjective probability distribution. The bulk of the probability was assigned values less than 5×10^{-2}.

The validity of the relative importance of the attributes of the Thomas modeling system was reviewed 20 years later and against the then available operating experience (Lydell 2000). In that review, the validity of several of the attributes was evaluated, including the weld penalty factor, the age factor, the learning curve factor, the proposed base line failure frequency, and the conditional failure probability. The review paper offered the following conclusion:

> *The Thomas correlation is seen by many as the only practical approach to estimate pipe leak and rupture frequency. From time to time, it has been proposed that the single most important advancement in piping reliability would be to simply validate the original correlation using today's failure statistics. The Thomas paper was well written, and the author elaborated on the technical justifications for his reliability correlation. He also explained at length the many attached analytical caveats enabling a careful user to generate leak and rupture frequency estimates. The demands for quality risk and reliability analysis have become a major challenge for today's practitioners. Piping reliability analysis is now integrated with PSA applications to support the development of risk-informed in-service*

Table 3.7 PWR pipe leak and rupture frequencies based on PFM analysis

Location of pipe failure	Frequency of a leak @ 40 years (1/(ROY and weld))	Frequency of a DEGB @40 years (1/(ROY and weld))
PWR reactor coolant loop (cold leg, hot leg, and crossover leg)	2.5×10^{-8}	2.5×10^{-13}
PWR pressurizer surge line	2.5×10^{-7}	2.5×10^{-10}
PWR main steam line	2.5×10^{-8}	2.5×10^{-14}

inspection programs for piping. In such contexts, the entire analytical process behind derived pipe leak and rupture frequencies must be certifiable. The existing insights from working with a large, current piping failure database points to major limitations of the Thomas correlation. It does not have sufficient discriminating power to develop realistic insights about piping reliability. It should be used with caution and only if actual and current failure statistics cannot be accessed.

3.3.6 Woo and Chou (1983)

Based on a survey of piping failure data collected for the US light water reactors (LWRs), the researchers concluded that about 57% of pipe failures occurred in connections (welds, heat-affected zones, areas adjacent to welds, mechanical joints, fittings, etc.) (Woo and Chou 1983; Woo 1983). This suggested that pipe connections are the locations most likely to fail. Probabilistic fracture mechanics assessments used a new computer program PRAISE, which stood for *Piping Reliability Analysis Including Seismic Events* (Harris et al. 1982). Part-circumferential semielliptical surface cracks were considered. Results indicated that the frequency of leak probabilities per plant life (40 years) for these welds was on the order of 10^{-10} to 10^{-8} for the reactor coolant loops, 10^{-8} to 10^{-7} for the pressurizer surge line, and 10^{-8} for the main steam line. The frequency of double-ended guillotine break (DEGB) per plant life for these locations ranged from 10^{-17} to 10^{-13} for the reactor coolant loops, from 10^{-13} to 10^{-10} for the pressurizer surge line, and from 10^{-16} to 10^{-14} for the main steam line. It was noted that DEGB probabilities were several orders of magnitude (10^3 to 10^8) smaller than leak probabilities. Translating these results into pipe failure frequencies, we get Table 3.7.

3.3.7 Eason and Shusto (1984)

Eason and Shusto analyzed operating experience with small-diameter primary system piping in BWR plants in the United States (EPRI 1984). The objective of this work was to estimate the frequency of "principal" failure modes, analyze safety consequences and countermeasures, and estimate the expected cost of repairs. The

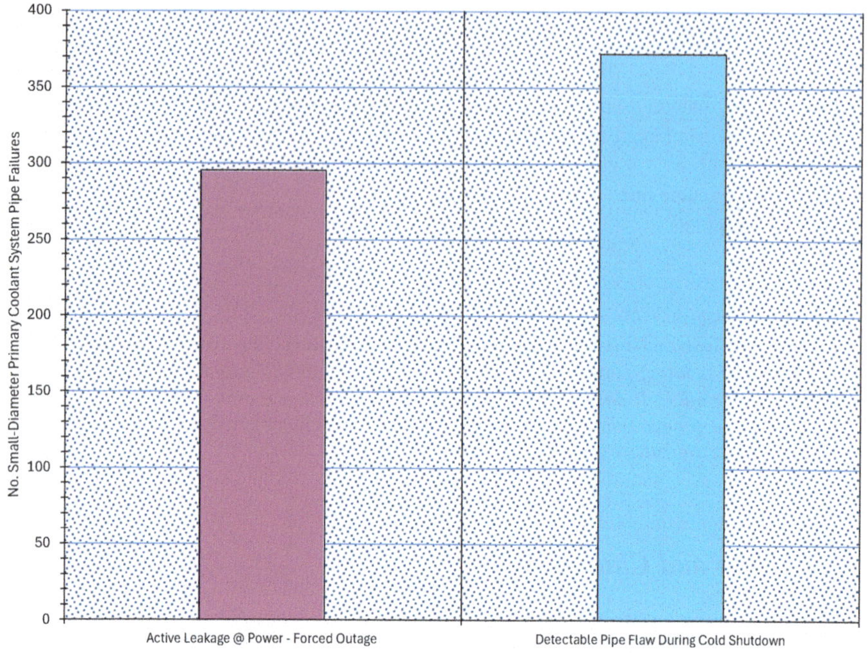

Fig. 3.4 Weld repairs during planned outage vs. forced outage

initial emphasis was on intergranular stress corrosion cracking (IGSCC), but other failure modes, including low-cycle fatigue, high-cycle fatigue, poor welding, and erosive wear, were found to be more significant causes of small-diameter pipe failure than IGSCC. Small diameter was defined as piping having an outside diameter (OD) equal to or greater than 26.7 mm (DN20) and less than 114.3 mm (DN100). The researchers recognized that the appropriate definition of failure depends on the purpose of an analysis. Three failure definitions were used in the analysis:

- Active through-wall leak discovered during routine plant operation or during pressure testing prior to returning to power operation following a planned maintenance or refueling outage.
- Repair needed to prevent further crack growth. Once a reactor is shut down to repair one crack, other cracks discovered in the same piping may be repaired at a little extra cost.
- Detectable crack discovered during performance of visual inspection and/or nondestructive examination using ultrasonic techniques.

The failure rate calculations were performed using the Weibull curve fit approach to obtain the shape, scale, and location parameters (Figs. 3.4 and 3.5). Underlying the calculation was a review of the small-diameter weld populations that produced pipe failures. An interesting aspect of this work is the information on weld populations, which was organized by system and nominal pipe size (Fig. 3.6).

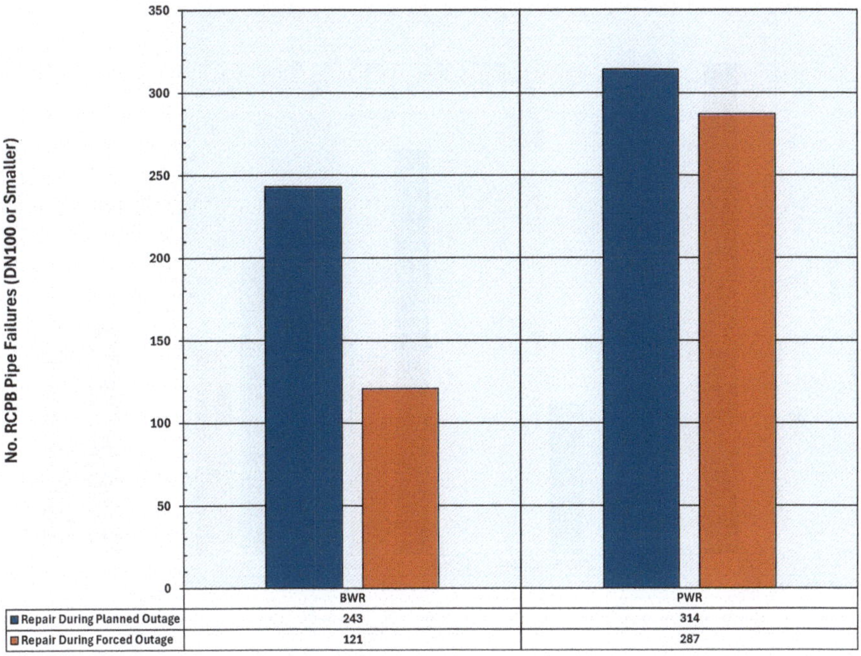

Fig. 3.5 Small-diameter RCPB pipe repairs during forced vs. planned outage

Using Weibull statistics, the study estimated the expected number of IGSCC repairs, leaks, and crack detections by the time a BWR/4 plant[10] reached 10 and 30 years of age. How do the predicted number of failures compare to the observed number of failures? Today's state of knowledge is illustrated in Fig. 3.7.

3.3.8 Lawrence Livermore Laboratory (1984)

The Lawrence Livermore National Laboratory (LLNL) performed probabilistic piping reliability analyses of PWR and BWR reactor coolant piping. LLNL estimated the probability of a double-ended guillotine break (DEGB) in the reactor coolant loop piping in PWR plants and in the main steam, feedwater, and recirculation piping of BWR plants. For these piping systems, the intent of the study was to provide USNRC with one technical basis on which to:

[10] Seven General Electric NSSS designs, BWR/1, 2, 3, 4, 5, and 6 and the ABWR, have been commercialized. In 2024, there were 17 BWR/4 reactors in commercial operation. Technical details are found in Kevin Theriault. 21 Oct 2016, Boiling Water Reactors from: Nuclear Engineering Handbook CRC Press Accessed on 16 Jan 2024 https://www.routledgehandbooks.com/doi/10.1201/9781315373829-5.

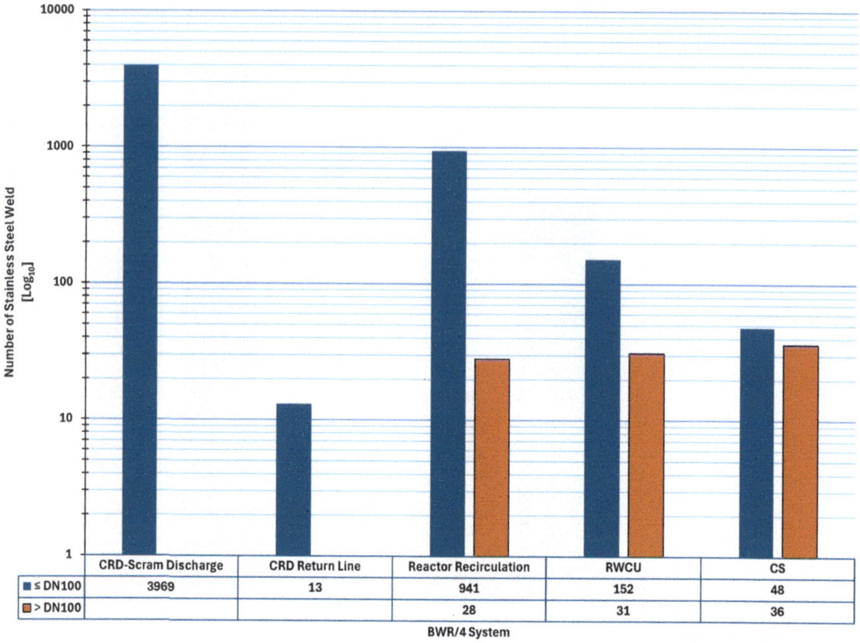

Fig. 3.6 Average number of ASME III Class 1 and 2 welds by system for a BWR/4

– Reevaluate the then current general design requirement that double-ended guillotine break (DEGB) be assumed in the design of nuclear power plant structures, systems, and components against the effects of a postulated pipe break.
– Determine if an earthquake could induce a DEGB and thus reevaluate the current design requirement that pipe break loads be combined with loads resulting from a safe shutdown earthquake (SSE).
– Make licensing decisions concerning the replacement, upgrading, or redesign of piping systems or addressing such issues as the need for pipe whip restraints on reactor coolant piping.

To estimate the probability of DEGB, two causes of pipe break were considered: (1) pipe fracture due to the growth of cracks at welded joints ("direct DEGB") and (2) pipe rupture indirectly caused by a seismically induced failure of critical supports or equipment ("indirect DEGB"). For each weld, the leak or break probability was estimated using a Monte Carlo simulation technique. Each replication of the simulation started with a preexisting flaw having initial length and depth randomly selected from appropriate distributions. These distributions in turn related the conditional probability of crack existence. Fatigue crack growth was then calculated using a Paris-Erdogan crack growth equation, which was applied to stresses associated with normal operating conditions and postulated seismic events. Selected results of the study are presented in Table 3.8 (Lo et al. 1984; Ravindra et al. 1985; Holman and Chou 1988).

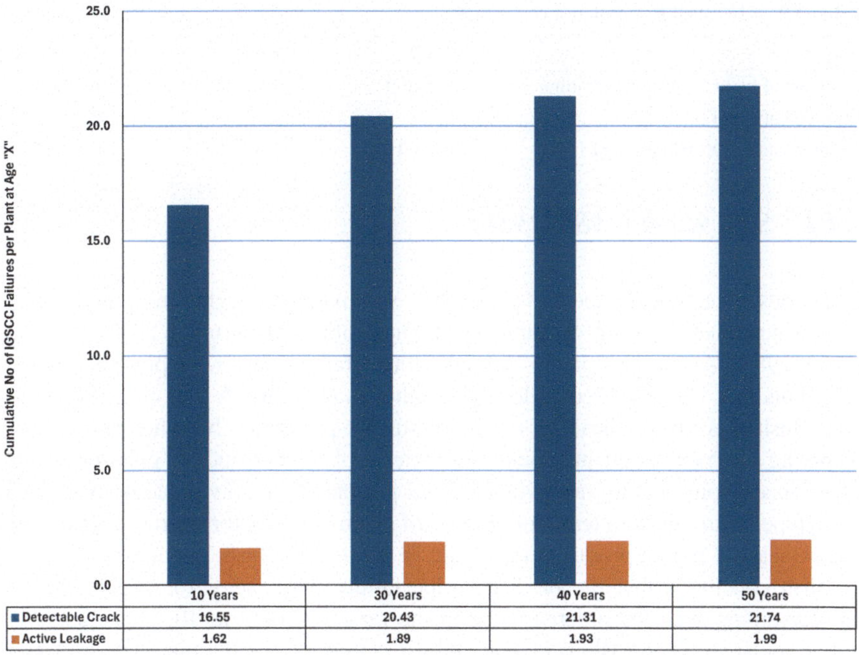

Fig. 3.7 Cumulative number of IGSCC leaks and detections per 2025 state of knowledge

Table 3.8 Frequencies of leak and direct DEGB for PWR reactor coolant loops

Analysis case	Failure mode	Confidence limit		
		10%-tile	50%-tile	90%-tile
Plants east of the Rocky Mountains—generic seismic hazard curves	Leak	5.6×10^{-10}	1.1×10^{-7}	2.4×10^{-7}
	Direct DEGB	5.0×10^{-17}	4.4×10^{-12}	7.5×10^{-12}
West Coast plants—generic seismic hazard curves	Leak	2.0×10^{-8}	5.5×10^{-8}	1.5×10^{-7}
	Direct DEGB	2.6×10^{-17}	2.2×10^{-13}	1.0×10^{-9}

3.4 Early Work III: 1985–1994

There were increased interest in the probabilistic assessment of pipe failure rates, including the development of probabilistic fracture mechanics applications; continued exploration of the expanding operating experience data; and hybrid piping reliability models built on data-driven models and probabilistic fracture mechanics. Expert elicitations were conducted to assess BWR- and PWR-specific pipe failure rates. There was continued skepticism regarding the viability of using operating experience data in probabilistic piping reliability analysis.

Table 3.9 S.H. Bush pipe failure rate synthesis

Data	BWR	PWR	Combined
Number of plants × length of piping	315,000	280,000	595,000
Number of failures	4	4	8
Failure rate (>DN100 piping)	1.3×10^{-5}	1.4×10^{-5}	1.3×10^{-5}

3.4.1 S.H. Bush (1985, 1988)

In a series of technical papers, S.H. Bush/[11] summarized insights from analyses of light water reactor piping operating experience obtained during a 20-year period (1965–1985). As an example, in WASH-1400,[12] failure rates were presented based on events that had occurred in locations other than welds. A key contribution of S.H. Bush was his synthesis of results of studies performed by other researchers. Applying his own expert judgment and reviews of the US nuclear plant operating experience, a pipe failure rate synthesis was presented. A convoluted derivation of these pipe failure rates in terms of number of events per reactor operating years and linear meter of piping produced the results in Table 3.9 (Bush 1985a, b).

Bush concluded that the predicted pipe failure frequencies for larger sizes of nuclear piping were considered to be in the range of 10^{-4} to 10^{-6} (exclusive of intergranular stress corrosion cracking, IGSCC). Further, smaller pipe sizes, of lesser safety significance, were predicted to have much higher failure rates. In his analyses, Bush differentiated "failure" from "catastrophic failure," the former corresponding to a pressure boundary breach resulting in a controlled, well-managed shutdown of a reactor. He went on to state that a "catastrophic failure" would appear more likely from operator error or design and construction errors (e.g., water hammer, improper handling of dynamic loads, undetected fabrication defects) rather than conventional flaw initiation and growth or metal fatigue. Finally, and prophetically, Bush made the following statement:[13]

---

[11] S.H. Bush (1920–2005), an international authority, sometimes referred to as the "walking database," on metallurgy and piping integrity. He was the driving force in the development of ASME Section XI. For more information, see for example "Recent Advances in Non-Destructive Examination and ASME B&PVC Section XI – a Memorial Symposium in Honor of Spencer H. Bush, July 22–26, 2007." He was a member of the USNRC Advisory Committee on Reactor Safeguards (ACRS) from 1966 to 1977.

[12] USNRC (1975), Reactor Safety Study: An Assessment of Accident Risks in U.S. Commercial Nuclear Power Plants, Appendix III, Failure Data, NUREG-75/014 (WASH-1400), Washington, DC. NRC-ADAMS Accession No. ML070610293.

[13] The author of this text had the good fortune to personally interact with S.H. Bush, the first time at an ASME subcommittee meeting on risk-based inspections in 1994. We communicated from time to time regarding piping operating experience; he had a very deep interest in the topic. In late August 2005, I received an e-mail from him in which he mentioned that he was in the process of clearing out a storage unit near his home. Would I be interested in his storage boxes with pipe failure information? A week or two later, those boxes were delivered to my home office. I remain grateful. His statement on international collaboration comes from Bush, S.H. (1985); see above.

Table 3.10 Updated plant-level LOCA frequencies before expert elicitation

Leak rate (gpm)	Failures	Operating years	Lower bound	Point estimate	Upper bound
BWR					
500–5000	0	313.36	0	5.0×10^{-4}	6.2×10^{-3}
>5000	0	313.36	0	5.0×10^{-4}	6.2×10^{-3}
PWR					
50–5000	0	484.73	0	7.0×10^{-4}	9.6×10^{-3}
>5000	0	484.73	0	7.0×10^{-4}	9.6×10^{-3}
LWR					
>50	0	798.09	0	3.0×10^{-4}	3.8×10^{-3}

Table 3.11 Updated plant-level LOCA frequencies after expert elicitation

	Pipe rupture frequency (1/ROY)		
EBS (mm)	5%-tile	Median	95%-tile
BWR			
12–50	2.6×10^{-3}	9.6×10^{-3}	2.5×10^{-2}
50–150	1.1×10^{-3}	6.4×10^{-3}	2.0×10^{-2}
>150	6.4×10^{-3}	1.6×10^{-2}	3.3×10^{-2}
PWR			
12–50	7.0×10^{-4}	4.1×10^{-3}	1.3×10^{-2}
50–150	2.8×10^{-3}	8.3×10^{-3}	1.9×10^{-2}
>150	1.7×10^{-3}	6.2×10^{-3}	1.6×10^{-2}

There should be an international body to which all failures from an agreed type of plant could be reported. Alternatively, some independent national body could be made responsible for collecting data from plant within agreed selective guidelines. Either type of organization should pool or report data periodically at a suitable forum.

3.4.2 Idaho National Laboratory (1987)

Published in 1987, the objective of a study performed by the Idaho National Laboratory (INL) was to update the pipe failure rate estimates of WASH-1400 by utilizing the available, accumulated US nuclear operating experience as of December 1984 (Wright et al. 1987). It is a well-documented study which utilized an elaborate expert elicitation approach to estimate pipe failure rates. The updated, plant-level LOCA frequencies are summarized in Tables 3.10 and 3.11. The former shows the results when using a chi-squared distribution on zero failures on the indicated reactor years of operation and with the as-expected results. The latter table shows the results of an expert elicitation process. Selected pipe break frequency estimates categorized by system are reproduced in Table 3.12.

The initial approach was to collect all applicable failure and population data from the US commercial nuclear power plants and military reactors, foreign nuclear power plants, and fossil electrical generating and steam processing plants. Then, a

Table 3.12 Frequencies of pipe break categorized by system

System	Numerator (*N* failures)	Denominator (*T* years)	Lower bound $\chi^2_{(0.05,2N)}/2T$	Point estimate	Lower bound $\chi^2_{(0.95,2N+2)}/2T$
BWR					
High-pressure coolant injection	2	313.36	1.1×10^{-3}	6.4×10^{-3}	2.0×10^{-2}
Condensate	4		4.5×10^{-3}	1.3×10^{-2}	2.9×10^{-2}
Main feedwater	3		2.6×10^{-3}	9.6×10^{-3}	2.5×10^{-2}
Service water	1		2.0×10^{-4}	3.2×10^{-3}	1.5×10^{-2}
All other systems	0		0	7.0×10^{-4}	9.6×10^{-3}
PWR					
Residual heat removal	2	484.7	7.0×10^{-4}	4.1×10^{-3}	1.3×10^{-2}
Chemical and volume control	5		4.1×10^{-3}	1.0×10^{-2}	2.2×10^{-2}
Main feedwater	2		7.0×10^{-4}	4.1×10^{-3}	1.3×10^{-2}
All other systems	0		0	5.0×10^{-4}	6.2×10^{-3}

Point estimate $\lambda = N/T$ if $N > 0$; $\lambda = \chi^2_{(0.50,2N+T)}/2T$

statistical analysis would be used on the collected data to determine the overall pipe break frequency by pipe size, as was done in WASH-1400, and estimate the conditional pipe break failure frequencies that would meet the objectives of this study.

Once failure data, operating experience times, and subject data had been collected, the final process to determine piping failure rates combined the different sources. Bayesian techniques were used to perform this integration of data sources. To develop the prior pipe failure rate distribution, an expert elicitation was performed. Piping experts responded to a questionnaire that elicited failure rates for pipes under certain conditions. The resulting subjective probabilities formed the prior probability density function expressing the degree of belief about the value of λ prior to observing a sample of the random variable x, the number of failures. Piping experts responded to a questionnaire that elicited failure rates for pipes under certain conditions. The prior distribution of parameter *A* was defined as a probability density function expressing the degree of belief about the value of *A* prior to observing a sample of the random variable x: the number of failures.

A minor adjustment to the prior distribution was made when the confidence levels submitted by the respondents were taken into consideration. The levels were coded 1–3 with 1 indicating a low confidence and 3 indicating high confidence. In the example below (Table 3.13), while two respondents submitted rates of 1×10^{-2} and

Table 3.13 Adjusted prior probabilities

Respondent (i)	λ_i	Confidence	Weight	Adjusted λ	Prior subjective probability
1	1.0×10^{-2}	1	1/6	–	
2	1.0×10^{-3}	2	2/6	1.0×10^{-2}	2/6
3	1.0×10^{-2}	1	1/6	1.0×10^{-3}	3/6
4	1.0×10^{-4}	1	1/6	1.0×10^{-4}	1/6
5	1.0×10^{-3}	1	1/6	–	

two other respondents submitted rates of 1×10^{-3}, one submitting the latter did so with a higher confidence of 2. The $\lambda = 1 \times 10^{-3}$ was thus weighted to give it a slightly higher prior probability than that for $A = 1 \times 10^{-2}$. Similarly, a response with a confidence of 3 would be weighted the same as three responses with a confidence of 1.

These adjusted prior probabilities were compared at this point with the ratio of number of failures (n) to time (t). For the actual data to have the opportunity to influence the resulting posterior probabilities, the ratio of actual failures to time had to lie within the range of prior probabilities for the given set of conditions. If it did not, a new A was created and assigned the value of n/T. The prior probability associated with it was initially set equal to the minimum existing prior probability for that set of conditions, i.e., system, pipe size, and operational mode; then all the probabilities were again normalized (as in the example above), so their sum was 1.

3.4.3 L'Institut de Radioprotection et de Sûreté Nucléaire (IRSN) (1982–1990)

The basis for the LOCA frequencies applied in the PSA of the standardized French 900 MWe PWR is described in Section D.2.2 of "*A Probabilistic Safety Assessment of the Standardized French 900 MWe Pressurized Water Reactor*" (IRSN 1990). The LOCA frequency assessment considered primary system pipe failures during four plant operating states (Table 3.14):

- State 1: At power, with reactor pressure with all safety injection signals operational
- State 2: Power reduction and initiation of cooldown, from about 13 MPa primary pressure, with safety injection signal blocked, to 2.9 MPa at which point the residual heat removal (RHR) suction valves are opened
- State 3: Cold shutdown with the residual heat removal system (RHRS) in operation and with bubble in the pressurizer or solid reactor coolant system
- State 4: Cold shutdown with the RHRS in operation and an open reactor coolant system

Table 3.14 LOCA frequencies in the French EPS900 PSA

Initiating event	Frequency (1/ROY)	Technical basis
Large pipe break: $>$DN150	1.0×10^{-4}	Worldwide operating experience (OE): 0 failures
Intermediate break: $50 \le DN \le 150$	3.0×10^{-4}	
Small break: $10 < DN < 50$	2.0×10^{-3}	Worldwide OE: 2 failures
Break at pressurizer safety valve	5.0×10^{-5}	Pressurizer breaks were evaluated on the basis of the probabilities of demand and non-reclosure or spurious opening of valves
Very small break: $\le$DN10	3.0×10^{-1}	EDF operating experience
Large pipe break: $>$DN150	5.0×10^{-7}	Based on the first four line items in this table and with adjustments for the time in respective plant operating state. Pressurizer breaks were evaluated based on the probabilities of demand and non-reclosure or spurious opening of valves. RHRS = residual heat removal system
Intermediate break: $50 \le DN \le 150$	1.5×10^{-6}	
Small break: $10 < DN < 50$	1.0×10^{-5}	
Break at pressurizer safety valve	2.4×10^{-4}	
Large RHRS break: $\varnothing \ge 115$ mm	3.5×10^{-6}	
Intermediate RHRS break: $50 \le \varnothing < 115$ mm	1.1×10^{-5}	
Small RHRS break: $\varnothing < 50$ mm	7.1×10^{-5}	
Break at RHRS safety valves	6.7×10^{-6}	
RHR system break: $\le$DN50	1.2×10^{-4}	

3.4.4 *GRS (1987–1990)*

In support of the Phase B of the German Risk Study (1981–1989), GRS performed piping reliability studies in recognition of the significant limitations of the then available piping reliability estimation techniques. GRS proposed two general approaches: (1) statistical evaluation of relevant operating experience data and (2) probabilistic fracture mechanics in the absence of any relevant operating experience data. The former approach was intended for small-diameter piping susceptible to fatigue, with latter approach to be used for piping susceptible to degradation mechanisms that produce long surface flaws without leaking. The LOCA frequencies developed for Phase B of the German Risk Study are shown in Fig. 3.8. It was the papers by S. Beliczey and S. Schulz that introduced the concept of the conditional probability of failure as an inverse function of the nominal pipe diameter in millimeters (Beliczey and Schulz 1987; Schulz 1988). The authors were careful in

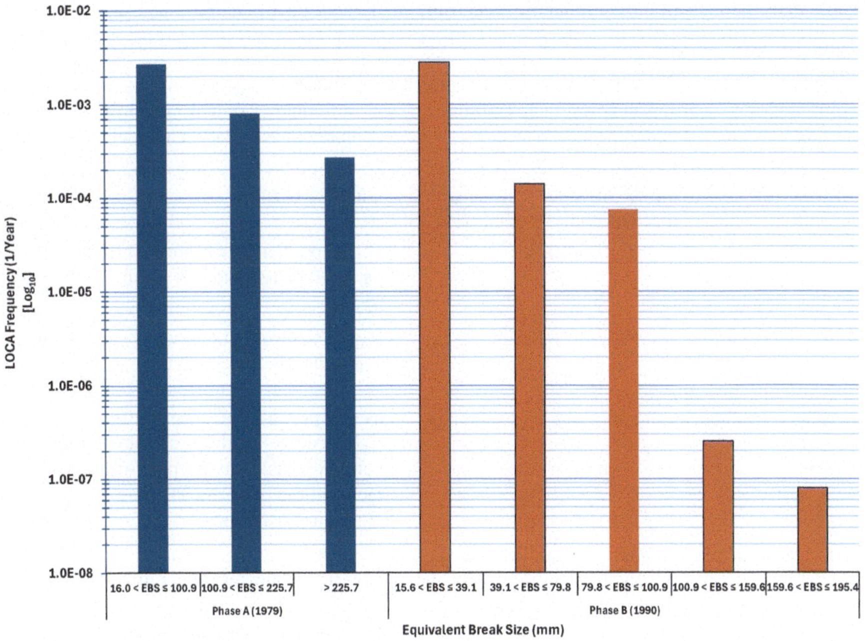

Fig. 3.8 LOCA frequencies in German Risk Study, Phase B

stating that such a relationship would only apply to small-diameter piping susceptible to high-cycle fatigue.

For piping of nominal diameter less than DN25, Beliczey and Schulz provided technical arguments for using operating experience data to directly estimate the desired probabilistic failure metrics. For piping of nominal diameter (DN) $25 < \mathrm{DN} \leq 250$, the authors used insights from experimental data and statistical investigations of pressure boundary components (piping and pressure vessels) to assess the conditional probability of "severance" given through-wall leakage. The following polynomial equation was proposed:

$$\frac{P_{Severance}}{P_{Leak}} = \left(9.6 \times \frac{\mathrm{DN}}{25} + 0.4 \times \frac{\mathrm{DN}^2}{25}\right)^{-1} \tag{3.11}$$

About 20 years after the original work, an updated evaluation of the German operating experience was performed and extended to incorporate new methodical aspects and data. A major point was the inclusion of structural reliability models based on fracture mechanics calculation procedures. The method used to produce statistical estimates for leak and break frequencies of piping systems was updated. Moreover, the evaluation of the operational experience was extended to 341 years with respect to cracks, leaks, and breaks in the volume control system of German pressurized water reactors. Table 3.15 is a summary of the German approach which

Table 3.15 Models of piping reliability according to German PSA practice

Nominal pipe diameter (mm)	Method	Comment
≤50	λ_L can be estimated from operating experience	λ_L = frequency of leak [1/CB.ROY]. CB is component boundary, e.g., weld
	$CFP = \lambda_B/\lambda_L = 2.5/DN$ for piping susceptible to fatigue mechanisms	λ_B = pipe break frequency. D = pipe diameter in mm
50 < DN ≤ 150	$\lambda_L = C_x(L_D D)/t_D^x$	D = pipe diameter in mm. The other parameters are obtained using the Thomas model
150 < DN ≤ 250	$\lambda_L = C_x(L_D D)/t_D^x$ $C_x = \frac{\sum_{50}^{150} N_{L,D}}{\sum_{50}^{150} L \times D/t_D^x \times T}$	$D = 150$, the other parameters are obtained using the Thomas model L_D = number of susceptible locations t_D = pipe wall thickness [mm] C_x = a dimensionless parameter, which is being referred to as the pipe wall thickness factor in the paper by Thomas N = number of leaks T = ROYs
>250	Assumption: $\lambda_L < 1 \times 10^{-7}$ per reactor operating year	For large-diameter piping, it is recommended that estimation of the CFP be based on probabilistic fracture mechanics

formed a part of the German PSA procedure guide (Beliczey and Schulz 1990; Beliczey 1991; GRS 1995; Berg et al. 2010; BMFT 1980; BfS 2016).

3.4.5 NUREG-1150 (1990)

WASH-1400 was published in 1975, and the Risk Assessment Review Group was organized in 1977 with four elements to its charter: (1) clarify the achievements and limitations of WASH-1400, (2) assess the peer comments thereon and responses to those comments, (3) study the present state of probabilistic safety assessment methodology, and (4) recommend to the USNRC how (and whether) such methodology can be used in the regulatory and licensing process. The reviewers identified a number of shortcomings in the methodology. The USNRC subsequently sponsored a major update of the original work and expanded the analysis to include five nuclear power plants covering a representative set of reactor designs. The results were published in 1990 as NUREG-1150, *Severe Accident Risks: An Assessment for*

Five U.S. Nuclear Power Plants (USNRC 1990). One of the reference plants was the Zion Nuclear Power Plant, consisting of two PWR units.[14]

The design of the component cooling water system (CCWS) at Zion was such that a potential existed for a pipe rupture in certain parts of the CCWS that would fail the entire system. Such an event would constitute an initiating event that could lead to core damage. Emergency core cooling would be lost, and the reactor coolant pump seals would be without cooling as well. This could lead to a potential LOCA through the seals with no capability to replace the lost inventory.

The issue of concern was the frequency with which, under normal operating conditions, the CCWS serving the Zion units would fail due to a pipe rupture event. It was selected as an expert elicitation issue due to the potentially significant impact of the CCWS pipe rupture rate upon the overall core damage frequency at Zion Unit 1. This issue had been raised in an earlier review of the Zion Probabilistic Safety Study (ZPSS).[15] At the time of the review, no data on low-temperature/low-pressure (65 °C and 1.0 MPa design temperature and design pressure, respectively) pipe rupture were available, so the CCWS pipe rupture sequence was quantified using pipe rupture frequencies associated with high-energy piping systems. The owner/operator of the nuclear power plant Zion agreed about the potential severity of the CCWS pipe rupture scenario but felt that the design and construction of the piping were of a "superior nature" and that the application of high-energy pipe rupture frequencies to moderate-energy piping was inaccurate.

Therefore, the NUREG-1150 study team formed an expert panel to assess the Zion CCWS pipe rupture frequency. There was no suggested decomposition or specific structure given to the panel. Each expert developed his own model to describe the CCWS pipe rupture. The models were very different between experts, but the individual expert results were characterized in a similar format. Because of the scarcity of data on CCWS pipe rupture in particular and cold water piping in general, the experts considered data on auxiliary building flooding and non-primary system leaks. They considered factors such as the location of the Zion CCWS (for detection of leaks before they propagate), quality of CCWS piping, and applicability of the data to CCWS piping to derive rupture frequencies for the CCWS. Each expert developed a curve to express his beliefs regarding the uncertainty of the estimate for pipe rupture frequency; the three curves were aggregated by taking the arithmetic average. The aggregated pipe rupture frequency results were as follows (Table 3.16).

Moderate-energy piping has a maximum operating temperature less than 94 °C or maximum operating pressure less than 1.9 MPa. A typical CCWS operates at around 60 °C and 0.5 MPa. Chapter 8 of this book offers additional details on this expert elicitation.

[14]The Zion Power Station consisted of two four-loop PWRs designed by Westinghouse. Unit 1 reached first criticality on June 19, 1973, and Unit 2 on December 24, 1973. The units were permanently shut down in 1998.

[15]ZPSS, a full-scope PSA, was completed in 1981. It was prepared by Pickard Lowe & Garrick Inc., Westinghouse Electric Co., and Fauske & Associates LLC.

Table 3.16 Zion component cooling pipe rupture frequency

Quantile	Aggregated frequency (year^{-1})
5th	8.1×10^{-8}
50th	3.0×10^{-6}
Mean	1.1×10^{-4}
95th	3.5×10^{-4}

3.4.6 Idaho National Laboratory (1991)

To establish a source of pipe leak and rupture frequencies for use in internal flooding probabilistic safety assessment (IF-PSA), the Idaho National Laboratory prepared the report "*Component External Leakage and Rupture Frequency Estimates*" (Eide et al. 1991). Based on a search of Licensee Event Reports (LERs) contained in the S.M. Stoller Corporation's Nuclear Power Experience (NPE) database and based on estimates of component populations and exposure times, component external leakage and rupture frequencies were generated.[16]

Probability of ruptures, *R*, given an external leakage, *L*, or rupture event, was calculated using a Bayesian update of a non-informative prior:

$$P_R = \frac{2R + 1}{(2L + 2)} \tag{3.12}$$

This methodology predicts mean values even if no pipe ruptures have occurred. In contrast, the Chi-square approach only predicts a 95% value (or upper confidence bound) if no events have occurred. A non-informative prior was used because the NPE search covered all commercial reactors in the United States, and no other information was believed to be applicable. For piping (including elbows), the external rupture probability given that an external leakage or rupture has occurred was estimated to be 4.0×10^{-2} for non-primary coolant systems and 1.0×10^{-2} for the primary coolant system (PCS) (Table 3.17).

A "rupture" was referred to as a pressure boundary failure that results in a pipe-wall leak with a flow rate exceeding 3.2 kg/s or an equivalent break size larger than 15 mm. Appendix A of the study provides a list of all recorded pipe failure incidents used to determine the frequencies of leaks and ruptures.

[16]Produced by S.M. Stoller Corporation of Boulder, Colorado, the Nuclear Power Experience "database" (NPE) was an information service available by subscription. It was a compilation of U.S. abnormal occurrence reports and licensee event reports. During a circa 15-year period, subscribers received periodic updates in the form of hole-punched paper copies to be inserted in loose-leaf binders: one for BWR ("Red Book") and another for PWR ("Blue Book"). Each "book" consisted of multiple volumes and was organized by subject matter, e.g., valves, pipes, and pumps.

Table 3.17 Plant-level pipe rupture frequencies for internal flooding PSA

System/component	Operating experience from NPE (1960–1991) No. of ruptures	No. of leaks and ruptures	Estimated conditional rupture probability	Recommended rupture probability
Non-PCS piping including elbows	17	531	3.3×10^{-2}	4.0×10^{-2}
PCS piping including elbows	0	60	8.0×10^{-3}	1.0×10^{-2}

3.4.7 Electric Power Research Institute (1992–1993)

The Electric Power Research Institute (EPRI) commissioned a study to create a US nuclear plant pipe failure rate database reflecting the additional experience generated since the publication of WASH-1400 in 1975. The sources of data and methodology used to estimate pipe failure rates are summarized as follows; it is an adaptation of the Executive Summary (EPRI 1992):

- The principal sources of nuclear plant component failure data used in the study were Licensee Event Report (LER) abstracts, Nuclear Power Experience (NPE), and Nuclear Plant Reliability Data System (NPRDS)[17] operated by the Institute of Nuclear Power Operations (INPO). The methodology used "auxiliary parameters" to account for equivalent break areas in larger pipe sections. This was deemed necessary since a significant fraction of failures in larger pipes were found to result in break areas that were equivalent to complete severance of smaller pipes.
- The methodology also accounted for factors that were postulated to significantly affect the values of the failure rates and were also quantifiable from the database. These include reactor vendor, system type, pipe size, and plant age. Techniques of analysis of variance were used to estimate the effect of system types on leakage failure rates. These results were then used, in part, in calibration (within a generic system type, and across reactor vendors) of some of the rupture failure rates. The rupture (or break) failure mode was defined as those events involving "severe leakage" (>3.3 kg/s) or catastrophic failure of the pipe pressure boundary.

The results of the assessment of the time-independent pipe failure rates and their associated parameters are summarized in Table 3.18. In addition to the average size-independent failure rates, the failure rate for each of the three pipe size groups was

[17] The Nuclear Plant Reliability Data System (NPRDS) was established in 1979 by the US nuclear power industry, following the Three Mile Island incident and the recommendations of the Kemeny Commission. The Institute of Nuclear Power Operations (INPO) assumed management of NPRDS in January 1982. NPRDS served as a central repository for operating experience data from nuclear power plants, aiming to improve safety and reliability. Today, the database is referred to as INPO Consolidated Events Database (ICES), previously known as the Equipment Performance Exchange System (EPIX).

Table 3.18 Piping reliability parameters

Pipe system	**Failure rate (per pipe segment and reactor operating year)**		
Reactor coolant system piping			
– BWR (General Electric)	4.7×10^{-6}		
– PWR (Westinghouse)	2.0×10^{-6}		
Safety injection and recirculation system piping			
– BWR (General Electric)	8.4×10^{-6}		
– PWR (Westinghouse)	2.3×10^{-6}		
Feedwater and condensate system piping			
– BWR (General Electric)	9.1×10^{-6}		
– PWR (Westinghouse)	5.4×10^{-6}		
"Other" safety-related piping (except feedwater and condensate)			
– BWR (General Electric)	2.1×10^{-6}		
– PWR (Westinghouse)	9.8×10^{-7}		
Conditional pipe failure probabilities			
$CFP_{1 \mid 2}$	1.0×10^{-1}		
$CFP_{2 \mid 2}$	9.0×10^{-1}		
$CFP_{3 \mid 3}$	4.7×10^{-1}		
$CFP_{2 \mid 3}$	3.3×10^{-1}		
$CFP_{1 \mid 3}$	3.3×10^{-1}		
Pipe size attribute values (Ci)			
System	**C1**	**C2**	**C3**
	$12.7 \le ID < 50$	**$50 \le ID < 150$**	**$ID > 150$**
BWR—feedwater and condensate piping	0.93	1.4	0.78
PWR—feedwater and condensate piping	0.77	0.53	2.5
All other systems	1.2	0.6	1.4

obtained by using the "pipe size attribute" value, which was defined as the ratio of the failure rate in each size group to the average failure rate. Conditional failure probabilities in this table represent the likelihood that a break in size group j will have the equivalent complete rupture area in size group i, each size group defined by the inside diameter (ID) of a pipe.

A "pipe segment" was defined as a section of piping between major discontinuities and for a system or a subsystem comprised of a set of pipe sections, with leakage rates equivalent to one of the three pipe size categories. This definition was introduced to simplify the derivation of system-level piping reliability metrics. The underlying assumption was that an analyst would base an analysis on information from piping and instrumentation diagrams (P&IDs) rather than isometric drawings. As an alternative to the weld or pipe length definition, the EPRI report defined a pipe section as "*a segment of piping, between major discontinuities such as valves, pumps, reducers, and tees.*" The report went on to state that a "*segment ranges from 10 feet* (3 m) *to 100 feet* (30 m) *in length.*" Furthermore, a pipe section was assumed to contain 4–8 welds and possibly several elbows and flanges.

Table 3.19 Comparison of isometric dimensional information vs. "P&ID pipe segments"[a]

	PWR ASME III Class 3 service water system piping isometric information		Pipe segment definition "A"		Pipe segment definition "B"	
			Linear ft	No. of welds per segment	Linear ft	No. of welds per segment
			10	4	100	8
Nominal pipe diameter (mm)	Length (ft)	No. of welds	No. of segments per pipe size class	No. of welds per pipe size class	No. of segments per pipe size class	No. of welds per pipe size class
DN ≤ 25	266	221	27	106	3	21
25 < DN ≤ 50	1133.80	872	113	454	11	91
50 < DN ≤ 100	620.7	269	62	248	6	50
100 < DN ≤ 150	779.6	227	78	312	8	62
150 < DN ≤ 250	1251.20	249	125	500	13	100
DN > 250	9459.40	1906	946	3784	95	757
Total	13,510.70	3744	1351	5404	135	1081

[a] This table is based on an extensive review performed by the author and with access to a complete set of isometric drawings

Table 3.20 Comparison of pipe failure rates from EPRI (1992) and WASH-1400

	Pipe failure rate (1/pipe segment and hour)			
Data source	12 ≤ ID < 50	50 ≤ ID < 75	75 ≤ ID < 150	ID ≥ 150
EPRI (1992)—point estimate	6.0×10^{-10}	3.0×10^{-10}	3.0×10^{-10}	7.0×10^{-10}
WASH-1400—median	1.0×10^{-9}	1.0×10^{-9}	1.0×10^{-10}	1.0×10^{-10}
WASH-1400—mean	9.0×10^{-9}	9.0×10^{-9}	9.0×10^{-10}	9.0×10^{-10}

Instrumentation connections were not considered to be major discontinuities. Engineering judgment would be required to apply this definition (Table 3.19).

Table 3.20 compares the generic rupture rates from this study with the mean and median rupture rates from WASH-1400. The derived point estimates were intentionally not equated to any fixed parameter, such as the mean or the median, of the underlying uncertainty distributions. The point estimates from the study were smaller than WASH-1400 means in all categories. The largest difference was for the pipe diameter range of 50–75 mm, where the WASH-1400 mean is larger by a factor of 30. The overall disagreement is in the way the failure rate varies with size.

Generic LOCA frequencies, shown in Table 3.21, were calculated by using the average number of pipe sections per reactor type in each of the three size groups. The estimates were averaged over all plant ages. It was noted that, in general, plant-specific frequencies could be significantly different from generic values, due to the number of pipe sections in each size group, nuclear steam supply system (NSSS) vendor, and correction factors for plant age.

Unsurprisingly, the generic point estimate BWR and PWR LOCA frequencies reported in the EPRI study agreed closely with the corresponding WASH-1400 point

Table 3.21 Comparison of LOCA frequencies from EPRI (1992) and WASH-1400

LOCA frequency (1/ROY)			
LOCA category	Point estimate	Upper bound	Lower bound
BWR LOCA frequencies (EPRI 1992)			
Large	3.0×10^{-4}	1.5×10^{-3}	2.5×10^{-5}
Medium	2.8×10^{-4}	1.4×10^{-3}	2.3×10^{-5}
Small	1.8×10^{-3}	9.0×10^{-3}	1.5×10^{-4}
PWR LOCA frequencies (EPRI 1992)			
Large	1.4×10^{-4}	7.0×10^{-4}	1.2×10^{-5}
Medium	3.2×10^{-4}	1.6×10^{-3}	2.8×10^{-5}
Small	1.0×10^{-3}	5.0×10^{-3}	8.3×10^{-5}
WASH-1400 BWR/PWR			
Large	2.7×10^{-4}	1.0×10^{-3}	1.0×10^{-5}
Medium	8.0×10^{-4}	3.0×10^{-3}	3.0×10^{-5}
Small	2.7×10^{-3}	1.0×10^{-2}	1.0×10^{-4}

estimates. The significant differences are in the uncertainty bands, which became smaller in the EPRI study. An updated pipe failure study was published in 1993 as EPRI Report TR-102266 (EPRI 1993). It benefited from an expanded review of operating experience data. One important departure from the initial work was the introduction of actual weld counts and pipe lengths organized by system and reactor type.

3.4.8 Low Power and Shutdown PSA (1989–1994)

During 1989, the USNRC initiated an extensive program to examine the potential risks during low-power and shutdown operations. Two plants, Grand Gulf Nuclear Station (BWR) and Surry Power Station (PWR), were selected as the reference plants to be studied by Sandia National Laboratories (Grand Gulf) and Brookhaven National Laboratory (Surry). A phased approach was used for the overall study. In Phase 1, the objectives were to identify potential vulnerable plant configurations, to characterize (on a high, medium, or low basis) the potential core damage accident scenario frequencies, and to provide a foundation for a detailed Phase 2 analysis. Completed in 1994, the two studies were published as NUREG/CR-6143 (Grand Gulf) and NUREG/CR-6144, respectively (Chu et al. 1994; Dandini et al. 1994; Kohut 1994; Whitehead et al. 1994; Chu and Pratt 1995; Whitehead 1995). Pipe rupture frequencies were developed for loss-of-coolant accidents during periods with the primary system pressurized and for internal flooding events caused by pipe failure.

In the BWR low-power and shutdown analysis, the pipe failure-induced internal flooding initiating event frequencies were based on WASH-1400. The pipe failure-induced LOCA frequencies were based on NUREG/CR-4407 (Wright et al. 1987), with adjustments made for the operational state in which a pipe failure could occur: starting up, normal operation, shutting down, shutdown, and "transients" referring to plant conditions that result in a reactor trip. NUREG/CR-4407 provided pipe failure point estimates for each of the five operational states, from which weighting factors were derived to normalize the point estimates: in hindsight, a rather convoluted process generating questionable results. Table 3.25 summarizes the pipe failure frequencies derived for the Grand Gulf low-power and shutdown (LPSD) PSA. Of interest are the LOCA frequencies during hydrotesting as mandated by ASME Section XI at the completion of a refueling outage. It is unclear how these LOCA frequencies were obtained, however.[18]

In the PWR low-power and shutdown analysis, the pipe failure-induced internal flooding initiating event frequencies were calculated using the "Thomas model." A generic pipe failure rate was developed for moderate-energy piping, e.g., circulating water and raw water cooling piping. No details are given for the derivation of pipe failure frequencies. The pipe failure-induced LOCA frequency analysis assessed the potential loss-of-coolant accident scenarios that are unique to shutdown states.

Full-power LOCA frequencies were taken from NUREG-1150 and adjusted to account for the unique low-power/shutdown system configurations (Table 3.22). For plant operational states (POS) 1 and 15, full-power frequencies were used. For states 2 and 14, the full-power frequencies were divided by a factor of 20. For states 3, 4, 12, and 13, the full-power frequencies were divided by a factor of 30. As a justification, these ratios were derived from "*statistical considerations of structural mechanics in the case of the pressurizer surge line (large LOCA)*" (Vo et al. 1991).[19] Table 3.23 summarizes the full-power LOCA frequency adjustment factors developed for the Surry LPSD PSA.

[18] ASME Boiler & Pressure Vessel Code Section XI originally stipulated that a hydrostatic pressure test be performed once every 10 years. Hydrostatic testing only subjects the piping components to a small increase in pressure over the design pressure: therefore, piping dead weight, thermal expansion, and seismic loads present far greater challenges to the structural integrity of a system. Consequently, Section XI hydrostatic pressure test is primarily regarded as a means to enhance leak detection during the examination of components under pressure, rather than as a method to determine the structural integrity of the components. Industry experience indicates that leaks are not being discovered as a result of hydrostatic test pressures causing a preexisting flaw to propagate through the wall; in most cases, leaks are being found when the system is at normal operating pressure.

[19] This quote is from pages 4–76 of NUREG/CR-6144, Volume 2, Part 1A (Evaluation of Potential Severe Accidents During Low Power and Shutdown Operations at Surry, Unit 1, Analysis of Core Damage Frequency from Internal Events During Mid-Loop Operations Main Report (Chapters 1–6)). The technical basis is likely to be based on NUREG/CR-4469, Nondestructive Examination (NDE) Reliability for Inservice Inspection of Light Water Reactors, in which probabilistic fracture mechanics analyses were performed to establish pipe rupture frequencies. Additional relevant information is to be found in Vo, T.V., Heasler, P.G., Doctor, S.R., Simonen, F.A., Gore, B.F. (1991), "Estimates of Rupture Probabilities for Nuclear Power Plant Components: Expert Judgment Elicitation," Nuclear Technology, 96(3);259–271.

Table 3.22 Internal flooding and LOCA frequencies during low-power and shutdown operations

Initiating event	Mean frequency per year for refueling outage
Grand Gulf LPSD-PSA-NUREG/CR-6143—BWR	
Generic pipe rupture frequency for internal flooding scenarios	6.1E−08 (per linear meter and year)
Large LOCA at low pressure	3.6E−05
Large LOCA during hydrotest at high pressure	1.3E−04
Intermediate LOCA at low pressure	3.6E−05
Intermediate LOCA during hydrotest at high pressure	1.3E−04
Small LOCA at low pressure	3.6E−05
Small LOCA during hydrotest at high pressure	1.3E−04
Small-small LOCA at low pressure	3.6E−05
Small-small LOCA during hydrotest at high pressure	1.3E−04
Surry LPSD-PSA-NUREG/CR-6144—PWR	
Auxiliary building flooding caused by pipe break, $>6.3 \times 10^{-2}$ m^3/s	2.4E−04
Auxiliary building flooding caused by pipe break, $>1.9 \times 10^{-1}$ m^3/s	3.0E−04
Auxiliary building flooding caused by pipe break, $>6.3 \times 10^{-1}$ m^3/s	7.5E−06
Large LOCA during RCS heat-up and reactor startup	6.6E−06
Intermediate LOCA during RCS heat-up and reactor startup	1.3E−05
Small LOCA during RCS heat-up and reactor startup	1.1E−05
Small-small LOCA during load reduction and cooldown	6.8E−06
Small-small LOCA during RCS heat-up and reactor startup	1.7E−04

3.5 "Progressive Work" (1995–2004)

The period 1995–2000 saw significant, renewed interest in probabilistic piping reliability and in the contexts of risk-informed in-service inspection methodology, development, and more refined LOCA frequency and high-energy line break frequency assessments to address spatial impacts of pipe failure within containment structures and turbine buildings, respectively. The first steps were made towards the creation of an international pipe failure database in support of piping reliability method development, validation of probabilistic fracture mechanics results, and plant-specific applications in support of PSA.

Table 3.23 LOCA frequency adjustment factors for LPSD PSA

Plant operational state	Definition—NUREG/CR-6144	Full-power LOCA frequency multiplier
1	Low-power operation and reactor shutdown	1
2	Cooldown with steam generators to 175 °C	5.0E−02
3	Cooldown with RHR to 95 °C	3.3E−02
4	Cooldown to ambient temperatures (with RHR)	3.3E−02
5	Draining the reactor coolant system (RCS) to mid-loop	Not assessed (n/a)
6	Mid-loop operations	n/a
7	Fill for refueling	n/a
8	Refueling	n/a
9	Draining RCS to mid-loop after refueling	n/a
10	Mid-loop operations after refueling	n/a
11	Refill RCS completely after mid-loop operation	n/a
12	Preparing to start one reactor coolant pump with RCS solid and pressurizer heater energized to draw bubble in pressurizer	3.3E−02
13	RCS heat up to 175 °C	3.3E−02
14	Startup with steam generators	5.0E−02
15	Reactor startup and low-power operation	1

3.5.1 Appendix A of NUREG/CR-6224 (1995)

NUREG/CR-6224 (*Parametric Study of the Potential for BWR ECCS Strainer Blockage Due to LOCA Generated Debris*) (Zigler et al. 1995) documents a plant-specific study on a BWR/4 with a Mark I containment[20] that evaluated the potential for LOCA-generated debris and the probability of losing long-term recirculation capability due to emergency core cooling pump suction strainer blockage. The study consisted of the following elements:

(a) Acquisition of detailed piping layouts and installed insulation details for a reference BWR plant
(b) Analysis of plant-specific piping weld failure probabilities to estimate the LOCA frequency
(c) Development of insulation and other debris generation and drywell transport models
(d) Modeling of debris transport in the suppression pool

[20] A Mark I containment design is a drywell in the shape of an inverted incandescent light bulb containing the reactor vessel and primary piping attached with several large vent pipes to a torus-shaped suppression chamber located below the drywell. The steam escaping from the break in the reactor coolant piping would vent, along with the drywell atmosphere, down into the suppression chamber. It would be distributed through a header to many downcomer pipes whose open ends were submerged in the suppression pool, which fills about half the suppression chamber.

(e) Development of strainer blockage head loss models for estimating loss of net positive suction head margin
(f) Estimation of core damage frequency attributable to loss of emergency core cooling recirculation capability following LOCA

A point estimate of overall DEGB pipe break frequency of 1.6×10^{-4} per reactor year was calculated, with a corresponding overall ECCS loss of net positive suction head (NPSH) frequency of 1.6×10^{-4} per reactor year. The calculated point estimate of core damage frequency (per reactor year) due to blockage-related accident sequences for the reference BWR ranged from 4.2×10^{-6} to 2.5×10^{-5}. The results showed that unacceptable strainer blockage and loss of NPSH margin could occur within the first few minutes after ECCS pumps achieve maximum flows when the ECCS strainers are exposed to LOCA-generated fibrous debris in the presence of particulates such as sludge, paint chips, and concrete dust. Appendix A of the study offered details on the analysis, which was based on an application of the PRAISE probabilistic fracture mechanics computer code. DEGB frequencies were estimated for primary piping made of Type 304 stainless steel and Type 316NG stainless steel. Table 3.24 reproduces the estimated DEGB frequencies for all welds within the primary pressure boundary of the reference plant.

The study acknowledged that there were limitations associated with the analysis. Because of the overwhelming contribution of IGSCC to the predicted weld break frequencies, efforts were focused on identifying the most significant limitations associated with the IGSCC portion of the analysis. Some of the limitations of the IGSCC analysis that were identified in this study included the following:

- Certain local phenomena were not considered in the analysis, for example the effect of coolant flow velocity on possible flushing of impurities that otherwise could aggravate susceptibility to IGSCC.
- The model used "harsh" laboratory conditions to predict growth rates and time to initiation. It was noted that extrapolating the laboratory data to the relatively benign conditions that exist in reactor facilities was conservative.
- The failure probability is very sensitive to the type of residual stress assumed in the analysis. Consequently, plant-to-plant experiences could vary significantly, depending on residual stresses that remain following pipe assembly welding and "fit-up." Worst-case stress assumptions were used in the analysis.
- The analysis did not give credit for actions to mitigate the effects of IGSCC, specifically in-service inspections, weld overlay, or induction heat stress improvement (IHSI). In addition, the analysis did not address the mitigating effects of corrosion control programs.

3.5.2 SKI Research (1995–1998)

The Swedish Nuclear Power Inspectorate (SKI) started a 5-year research project on piping reliability in 1994. The research objectives were to:

Table 3.24 Summary of estimated DEGB frequencies

			Pipe DEGB frequency (1/ROY)	
Pipe diameter	Total no. of welds	Per-weld DEGB frequency	Individual pipe size category	Total
Reactor recirculation system				
25	25	1.0E−06	2.5E−05	
30	2	1.0E−06	2.0E−06	
50	2	1.0E−06	2.0E−06	
100	26	1.0E−06	2.6E−05	
250	40	1.0E−06	4.0E−05	
400	6	2.0E−06	1.6E−05	
550	37	2.0E−07	7.4E−06	
Subtotal	140	–	–	1.2E−04
Main steam system				
25	16	2.4E−06	3.2E−06	
50	12	2.0E−07	2.4E−06	
150	24	2.0E−07	4.8E−06	
500	63	2.0E−07	1.3E−05	
Subtotal	115	–	–	2.3E−05
Feedwater system				
250	58	2.0E−07	1.2E−05	
400	10	2.0E−07	2.0E−06	
Subtotal	68	–	–	1.4E−05
High-pressure coolant injection system				
250	6	2.0E−07	1.2E−06	
Subtotal	6	–	–	1.2E−06
Residual heat removal system				
450	6	2.0E−07	1.2E−06	
500	10	2.0E−07	2.0E−06	
Subtotal	16			3.2E−06
Total no. of welds	345	Total		1.6E−04

1. Define the requirements for appropriate and sufficient service data and analysis techniques for parameter estimation in support of PSA applications and PSA-based evaluations of licensee submittals involving piping system modifications.
2. Investigate the need for improved treatment of piping reliability in PSA studies.
3. Develop a taxonomy and process for piping operating experience data collection and evaluations.
4. Develop a "good practice" framework for piping reliability analysis.

5. Develop a piping operating experience database.
6. Organize an international piping reliability seminar to present the results of the R&D and to seek expert input on the roles of probabilistic fracture mechanics and data-driven models in PSA.
7. Utilize the output from steps 1 through 6 to perform a piping reliability analysis pilot study.

The results of the R&D are documented in a series of SKI reports, as well as in the proceedings of the international piping reliability seminar (Lydell et al. 1995a, b, 1997; Tomic et al. 1995a, b; SKI 1997; Lydell 1999). Perhaps the most important outcome of this R&D was the decision by SKI to transfer the pipe failure database including the lessons learned to an international collaborative effort under the auspices of the Committee on the Safety of Nuclear Installations (CSNI) of the OECD Nuclear Energy Agency (NEA).

3.5.3 Markov Models (1998–2002)

More than 40 years ago, Markov chain models were used to describe the interaction of crack growth and in-service inspection effectiveness in the US research program on "*Integration of NDE Reliability and Fracture Mechanics*" (Heasler and Simonen 1983). At a later stage, an application of Markov[21] chain theory to the assessment of the effectiveness of leak detection and nondestructive examination (NDE) in the context of risk-informed in-service inspection was published in a series of conference papers and technical reports (Fleming and Gosselin 1997; EPRI 1998; Fleming and Mikschl 1998; Fleming 2004). Markov models were subsequently used in the development of multiple plant-specific risk-informed in-service inspection programs.[22]

Overviews of the different risk-informed in-service inspection (R-ISI) methodologies are presented in Chap. 5 of this book. The effect of implementing a risk-informed inspection program is to alter the number of locations inspected and effectiveness of inspections in relation to an existing, traditional prescriptive inspection program. The impact of an RI-ISI strategy on a specific location could include adding locations to the inspection program, removing locations, or changing the effectiveness of the program. When an element is selected for inspection, the effectiveness of the program is expected to increase because of the knowledge gained from the RI-ISI evaluation to determine which degradation mechanisms are

---

[21] Andrey Andreyevich Markov (1856–1922) was a Russian mathematician best known for his work on stochastic processes. A primary subject of his research later became known as the Markov chain.

[22] RI-ISI programs that utilized Markov chain theory were prepared for the following nuclear power plants: Braidwood, Byron, Clinton, Dresden, Koeberg, LaSalle, Limerick, Peach Bottom, Quad Cities, and Three Mile Island.

most likely to be present at that location. Therefore, the inspection strategy is switched from "traditional" (or deterministic) strategy to risk-informed strategy; there are three possibilities for each location in a piping system to change its failure potential:

1. If the location was inspected in a traditional program and retained in a risk-informed program, the change in the pipe rupture frequency, if any, would be to reduce the pipe rupture frequency due to the capability to inspect for the most probable degradation mechanism.
2. If the location was not inspected in an existing program and added to the risk-informed program, the pipe failure potential could decrease.
3. If the location was inspected in a traditional program and not retained in the risk-informed program, there may be increases in pipe failure potential.

Based on the results of early implementations of risk-informed in-service inspection program, EPRI initiated a research program to develop models of the interactions between material degradation mechanisms, inspections, and conditions that may produce a pipe rupture. Intended for piping that meets leak-before-break criteria, these reliability models use Markov chain theory processes to model the interactions between material degradation mechanisms, pipe inspection, and leak detection processes that cause and mitigate pipe cracks, leaks, and ruptures. The models support the calculation of time-dependent probabilities of various pipe states such as the normal state, presence of detectable flaw, presence of detectable leak, and rupture. The model's primary output is a time-dependent hazard rate, which measures the time-dependent growth in pipe rupture frequencies due to degradation mechanism. The Markov model explicitly models the role of the inspection and leak detection programs in reducing the likelihood of pipe ruptures.

An implicit consequence of relying on operating experience data is that the derived reliability parameters account for different types of leak detection technologies and in-service inspection processes. The influence of reliability and integrity management (RIM) such as leak detection and NDE on pipe failure rates is accounted for by a structural integrity management factor, I_{ik}, in the equation for the pipe rupture frequency ρ_{ix}. A Markov model is used to quantify I_{ik} in the equation below:

$$\rho_{ix} = \lambda_i \times CFP_{ix} \times I_{ik} \tag{3.13}$$

Markov process models the interactions between piping material degradation, in-service inspection, flaw detection, and repair. The model starts with a representation of the integrity of a piping component in a set of discrete and mutually exclusive states. At any instant of time, the piping component is permitted to change state in accordance with whatever competing processes are appropriate for that plant state. A Markov state refers to the existence of flaws, leaks, or ruptures. The processes that can create a state change are degradation mechanisms acting on the pipe and the process of inspecting or detecting flaws and leaks and repair of damage before progressing to a complete structural failure. The degradation mechanisms that act on a piping component are represented by precursor rates.

3.5.4 *Appendix J of NUREG/CR-5750 (1999)*

Prepared by the Idaho National Laboratory for the USNRC, (Poloski et al, 1999) is an initiating event frequency handbook for probabilistic safety assessments. Appendix J of the report documents an effort to estimate frequencies of loss-of-coolant accidents (LOCAs) caused by failure of piping connected to the reactor pressure vessel. Estimates were made for both BWRs and PWRs using available operating experience data, information on corrosion mechanisms acting on primary pressure boundary piping, and information from fracture mechanics analysis on crack initiation and crack propagation mechanisms. The authors of the report noted that no definitive LOCA frequency estimates had been made since NUREG-1150 in 1990, which used WASH-1400 values in many cases. In view of the progress with operating experience data and engineering understanding of pipe failures, the estimates presented in the report were considered to represent a reasonable but conservative adjustment to the understanding of the probability of pipe ruptures and LOCA frequencies. Considering this experience, a more complete analysis using data, fracture mechanics analyses, and results from pipe fracture experiments would likely produce more definitive estimates and uncertainties.

The small pipe break LOCA (SBLOCA) frequency was estimated from available operating experience data in a simple Bayes update of the SBLOCA frequency from WASH-1400. Because no difference could be discerned between the PWR and BWR operating experience data and the dominant failure mechanisms (compression fitting failure and failure of socket weld from high-cycle fatigue), the data were combined into a single dataset. This combined data was used to update a prior distribution based on the WASH-1400 estimate to produce a single SBLOCA frequency estimate appropriate for both PWRs and BWRs. The frequency estimates for medium- and large-break LBLOCAs relied on a "precursor type" of analysis of the data (i.e., through-wall crack and leak events), which were then combined with a conditional probability of the precursor event transitioning into a rupture. This conditional probability of a break given a through-wall crack was based on a technical review of readily available information on fracture mechanics, data on high-energy pipe failures and cracks, and an assessment of pipe break frequencies estimated by other entities since WASH-1400. Due to differences observed in both operating experience and engineering characteristics, separate frequency estimates were given for PWRs and BWRs. Also, wherever possible, the LOCA frequencies and the parameters used to calculate them were compared to similar values derived from or presented in the available literature. The results are given in Table 3.25.

Table 3.25 Estimated LOCA frequencies in NUREG/CR-5750

		LOCA frequency (1/ROY)		
Source	Pipe break size	Lower bound	Mean	Upper bound
NUREG/CR-5750	Small—ID ≤ 50 mm	1.0×10^{-4}	4.0×10^{-4}	1.0×10^{-3}
WASH-1400		1.0×10^{-4}	3.0×10^{-3}	1.0×10^{-2}
NUREG/CR-5750	BWR medium—50 < ID ≤ 125 mm	9.7×10^{-7}	3.0×10^{-5}	9.0×10^{-5}
WASH-1400		3.0×10^{-5}	8.0×10^{-4}	3.0×10^{-3}
NUREG/CR-5750	PWR medium—50 < ID ≤ 150 mm	1.0×10^{-6}	3.0×10^{-5}	1.0×10^{-4}
WASH-1400		3.0×10^{-5}	8.0×10^{-4}	3.0×10^{-3}
NUREG/CR-5750	BWR large—ID > 125 mm	9.7×10^{-7}	2.0×10^{-5}	9.0×10^{-5}
WASH-1400		1.0×10^{-5}	3.0×10^{-4}	1.0×10^{-3}
NUREG/CR-5750	PWR large—ID > 150 mm	1.0×10^{-7}	4.0×10^{-6}	1.0×10^{-5}
WASH-1400		3.0×10^{-5}	8.0×10^{-4}	3.0×10^{-3}

3.5.5 Appendix A of NUREG/CR-6928

NUREG/CR-6928, *Industry-Average Performance for Components and Initiating Events at U.S. Commercial Nuclear Power Plan*ts (Eide et al, 2007), provides industry-average component and initiating event parameter estimates. It outlines methods for calculating failure probabilities, failure rates, maintenance unavailabilities, and initiating event frequencies for the level 1 standardized plant risk analysis (SPAR) models.[23] The results were obtained by applying standard estimation methods that are described in NUREG/CR-6823 (Atwood et al. 2003).

Appendix A includes the results for piping. Two failure modes were considered: small external leakage (ELS) and large external leakage (ELL). ELS was defined as an external leak in the interval 6.3×10^{-2} to 3.2 kg/s, and ELL was defined as >3.2 kg/s external leakage. Pipe failure data for the period 1997–2004 and for 12 different BWR and PWR piping systems was collected from the Equipment Performance and Information Exchange (EPIX) database. A gamma distribution was used to estimate the failure rates. For essential service water (ESW) piping, the ELL mean was taken to be the ELS mean multiplied by a gamma distribution α parameter

[23] Standardized plant risk analysis (SPRA) refers to a methodology developed by the USNRC for assessing the risk of nuclear power plants.

Table 3.26 NUREG/CR-6928 pipe failure rates[a]

System	Failure mode	Failure rate, λ (1/linear meter and year)				Distribution (1/linear foot and hour)		
		5%	Median	Mean	95%	Type	α	β
ESW	ELS	7.18E–08	8.62E–06	2.01E–05	7.18E–05	Gamma	0.5	7.14E+08
	ELL	4.31E–10	8.62E–07	4.31E–06	1.73E–05		0.5	2.00E+09
Non-ESW	ELS	2.87E–08	2.87E–08	3.45E–06	7.18E–06		0.5	2.00E+09
	ELL	7.18E–11	1.73E–07	7.18E–07	3.45E–06		0.5	1.20E+10

[a] This table corresponds to Table A.2.33-5, "Selected industry distributions of λ for PIPEs." The failure rates in Table 3.29 have been converted to units of 1/linear meters and year

of 0.3. Results were generated for ESW piping and non-ESW piping as shown in Table 3.26.

3.6 Current "Progressive Work" (2005–2024)

During this period, significant advancements were made in the development of piping operating experience databases. Concurrently, comprehensive analysis frameworks were established, leveraging extensive service data, fracture mechanics insights, and expert elicitation to create an integrated, risk-informed methodology for estimating piping reliability parameters while thoroughly accounting for inherent uncertainties.

3.6.1 NPSAG R-Book

Established in the year 2000, the Nordic PSA Group (NPSAG) is a common forum for discussion of issues related to PSA, but with a focus on R&D needs. The group follows and discusses current issues related to PSA nationally and internationally, as well as PSA activities within the participating utilities. During the transition from the Swedish piping reliability R&D program (1994–1998) to the initiation of the OPDE international collaborative project (Lydell et al. 2002) in 2002, the NPSAG was kept abreast about proposed objectives for the new international pipe failure data collaboration. Based on technical discussions and seminars within the framework of planned activities for R&D, a formal decision to launch the "R-Book" project was made in 2005. Phase I of the project consisted of five elements:

1. Review of pipe failure databases and identification of technical features that are considered important to the statistical estimation processes that are considered for use in the development of the handbook.
2. Review of methods for piping reliability parameter estimation.
3. Development, distribution, and evaluation of a questionnaire that addresses user requirements on the planned R-Book (content, including level of detail, and updating philosophy). Input from the international PSA community was sought.
4. Development, distribution, and evaluation of a questionnaire addressing availability and access to piping system design information including weld counts and pipe length information organized by system, diameter, material, process medium, and safety classification.
5. Detailed work plan for the R-Book development (Phase II). All calculations were to be performed in an Excel workbook format with R-DAT and Oracle Crystal Ball add-in programs for Bayesian update and uncertainty analysis, respectively.[24]

Elements 1 through 5 were completed in 2007, with the results published as SKI Report 2008:01 (Lydell and Olsson 2008). Phase II of the project was launched through a workshop in mid-2008, and the first edition of the R-Book was issued in 2011. Only an electronic version of the R-Book has been issued (Fig. 3.9). It consists of a password-protected CD with the project files structured as follows:

- Summary report (Word file and as a PDF), NPSAG Report 04-007:01: This report documents the data processing routines: from extracting event population data from the OPDE database, via definition of relevant exposure term data (i.e., the "number of reactor operating year that produced a certain event population" times the number of welds susceptible to a certain degradation mechanism), to the definition of calculation case and to the execution of a certain calculation.
- Theory manual: This document contains an overview of the general calculation format, including a technical basis for how to define exposure terms specific to a certain calculation case.
- For each of the 26 systems, a file folder with the following system-specific computer files:
- Text file (Word and PDF formats) summarizing the underlying degradation mechanism analysis results, input data, result summary, and a discussion on how to use the R-Data results.
- Data processing and result summary, Microsoft® Excel files.
- Oracle Crystal Ball® reports (Microsoft® Excel workbook files) for each calculation case: These files enable a user to not only reproduce a certain application but also perform different types of sensitivity analyses.

[24] Reliability data analysis tool (R-DAT) is a specialized software platform developed by Prediction Technologies Inc. for risk analysts and reliability engineers to perform Bayesian updates on reliability parameters. It enables users to update system-specific models using Poisson and binomial likelihoods or incorporate expert judgment via lognormal/normal error models: https://www.prediction-technologies.com/rdat and https://www.riskspectrum.com/r-dat.

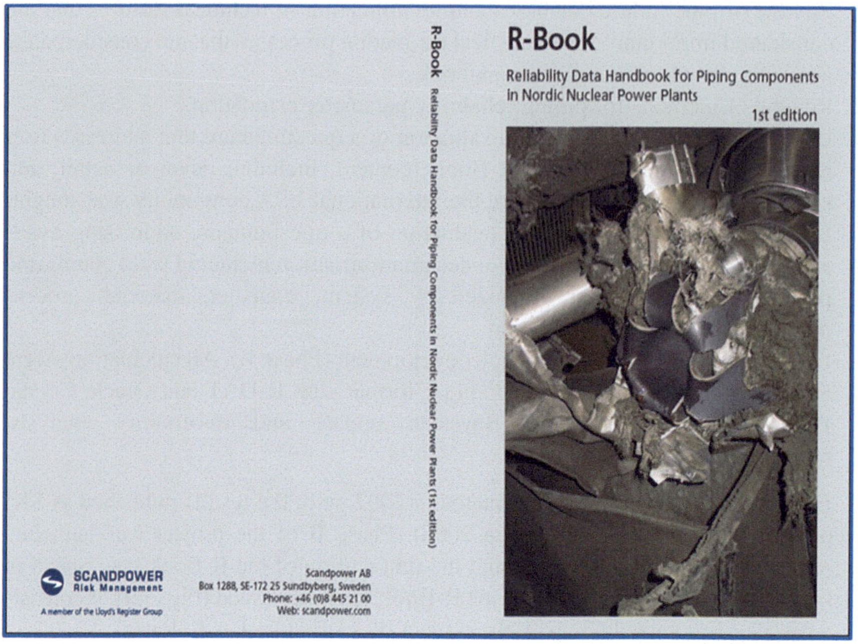

Fig. 3.9 "R-Book" CD cover sheet

3.6.2 EPRI Pipe Failure Rates and Rupture Frequencies

Since 2006, the Electric Power Research Institute (EPRI) has published reports on piping system failure rates for use in PSA studies of internal plant flooding and high-energy line breaks (HELBs). The most recent edition of the EPRI report was published in 2023 (EPRI 2023). These estimates are intended to satisfy requirements in the ASME and ANS PSA Standards RA-S-1.1-2022 (ASME 2022), RA-S-1.4-2021 (ASME 2021), and RA-S-1.5-2023 (ASME 2023). The objectives of this EPRI activity remain as follows:

- To develop generic estimates of piping system failure rates and rupture frequencies and provide additional information on the frequency of pipe breaks as a function of pipe diameter, equivalent break size, and flood rates.
- To capture piping system failure modes, including events involving water sprays and those involving flooding in different flood rate intervals.
- To provide piping system failure rates, flood frequencies, and plant-level flood events that account for the operating experience in US operating BWR and PWR plants, as well as all known and experienced pipe failure mechanisms.
- To provide a method and associated data that enables the estimation of pipe system flood frequencies as a function of piping system age and accounts for

increasing trends in pipe failure rates with system age evident in the plant service data.

- To the extent practical, quantify the uncertainties inherent in pipe failure rate estimation, including those resulting from scarcity of data, plant-to-plant variability, exposure data uncertainty, and modeling assumptions. Uncertainty distributions are developed and fitted to lognormal distributions defined by mean values and range factors.
- To provide information that will help PSAs meet the ASME/ANS PSA standard associated with pipe failure-induced internal flooding events for Capability Category II.[25]
- The more recent editions of the report (Revision 3 and later) are available to non-funding individuals or entities for a fee. Revision 5 is a 288-page document plus data tables in the form of Excel files with user instructions.

The methodology used in EPRI's work combines Bayesian updating with data specialization and periodic updates to refine pipe failure rate predictions and adapting to new information. This approach allows models to learn incrementally, moving from a general prior belief (as represented in the first edition of the pipe failure report) to a more specific posterior belief as new, specialized data has been incorporated in subsequent editions. Periodic updates have been performed to ensure that the models used to assess piping reliability have remained current and relevant.[26]

3.6.3 Estimation of PWR Loss-of-Coolant Accident Frequencies (MRP-480)

The Extremely Low Probability of Rupture (xLPR) probabilistic fracture mechanics code was used to perform a PFM evaluation of all PWR pipeline sizes represented in NUREG-1829 (USNRC 2008) within the scope of interest. For each PWR pipeline, a base case and several sensitivity cases were developed. Sensitivity cases were defined considering inputs known to have an influence on xLPR results, as well as modeling decisions made during input development. Prior xLPR analysis case results were leveraged where possible and supplemented with additional xLPR analysis cases to cover the remaining piping systems of interest. Key findings of this work were as follows (EPRI 2024):

[25] The ASME PSA standard defines supporting technical requirements for three "capability categories." The distinction between respective categories is based on the treatment of significant contributors to plant risk. As an example, for an internal flooding PSA, to satisfy capability category II, the analysts must collect plant-specific information on plant design, operating practices, and conditions that may impact flood-induced initiating event frequency.

[26] The first edition was published in 2006, with subsequent editions published in 2010, 2015, 2018, and 2023.

- Results from xLPR considering primary water stress corrosion cracking (PWSCC) and fatigue provided valuable information regarding conservatism or non-conservatism of the NUREG-1829 LOCA frequencies in the context of the material degradation mechanisms considered in xLPR.
- Ruptures did not occur when evaluating in-service inspection or leak rate detection in all base cases and all but three sensitivity cases. When interpreted as LOCA frequencies, these 80-year results are on a similar or lower order of magnitude than NUREG-1829. Notably, the three sensitivity cases that led to isolated realizations ending in ruptures with in-service inspection or leak detection included modeling assumptions which are not representative of current plant conditions and operations.
- The time between detectable leakage and LOCA was calculated for the xLPR analysis cases considered. This output provided an important understanding of the likelihood for leakage to be detected in sufficient time to shut down the reactor and reduce decay heat generation prior to LOCA occurring. Depending on the analyzed component, this investigation showed that either LOCAs did not occur, substantial time (lower bound 95/95 one-sided tolerance intervals of at least 19 months) existed between detectable leakage and LOCA, or when ISI is credited, any resulting LOCA scenarios are highly unlikely (annual frequency of occurrence on the order of 1.0×10^{-12} per year). A (95/95) lower one-sided tolerance interval signifies a 95% confidence level that at least 95% of the population values fall below the specified lower bound.

3.6.4 NUREG-1829 Loss-of-Coolant Accident Frequency Estimates

NUREG-1829 (USNRC 2008) documents estimated frequencies of loss-of-coolant accidents (LOCAs). To provide the technical basis for a risk-informed definition of the design-basis break size, the study developed LOCA frequency estimates using an expert elicitation process. The results from the expert elicitation provided LOCA frequency estimates for piping and non-piping passive systems, as a function of effective break size and operating time for BWRs and PWRs, respectively (Tables 3.27 and 3.28).

To provide a technical basis for a risk-informed definition of the design-basis break size, NUREG-1829 developed LOCA frequency estimates using an expert elicitation process. The results from the expert elicitation provided LOCA frequency estimates for piping and non-piping passive systems, as a function of effective break size and operating time for both boiling and pressurized water reactors, respectively. The panelists generally agreed on the important technical issues and LOCA-contributing factors. The panelists' estimates exhibited both significant uncertainty and diversity. This uncertainty was reflected in the estimated parameters (mean, median, 5th and 95th percentiles) of the individual LOCA frequency distributions,

Table 3.27 NUREG-1829 LOCA frequencies

LOCA category	Non-pipe to pipe ratio	Total mean	Non-pipe mean	Pipe			
				Mean	5%-tile	50%-tile	95%-tile
BWR LOCA frequencies @ 25 years of plant age							
1	0.24	5.60E–04	1.08E–04	4.52E–04	1.77E–06	2.05E–04	1.73E–03
2	0.4	1.10E–04	3.14E–05	7.86E–05	3.09E–07	3.58E–05	3.02E–04
3	0.11	2.40E–05	2.38E–06	2.16E–05	8.51E–08	9.85E–06	8.32E–05
4	0.06	6.20E–06	3.51E–07	5.85E–06	2.80E–08	2.66E–06	2.25E–05
5	0.07	1.10E–06	7.20E–08	1.03E–06	4.05E–09	4.68E–07	3.95E–06
6	N/A	3.30E–09	RPV failure				
PWR LOCA frequencies @ 25 years of plant age							
1	32	7.30E–03	7.08E–03	2.21E–04	8.70E–07	1.01E–04	8.50E–04
2	6.7	6.00E–04	5.22E–04	7.79E–05	3.06E–07	3.54E–05	2.99E–04
3	0.36	1.60E–05	4.24E–06	1.18E–05	4.63E–08	5.35E–06	4.52E–05
4	1.1	1.60E–06	8.38E–07	7.62E–07	3.00E–09	3.47E–07	2.93E–06
5	0.5	2.00E–07	6.67E–08	1.33E–07	5.24E–11	6.07E–09	5.12E–08
6	1.4	2.90E–08	1.69E–08	1.21E–08	4.75E–11	5.49E–09	4.64E–08

Table 3.28 NUREG-1829 LOCA categories

Plant type	LOCA category	Equivalent break size (mm)
BWR	1	12.7
	2	47.6
	3	82.5
	4	177.8
	5	457.2
PWR	1	12.7
	2	41.3
	3	76.2
	4	177.8
	5	355.6
	6	787.4

and the diversity was captured by the confidence bounds on the group estimates. In addition, NUREG-1829 considered the sensitivity of the results to various analysis approaches.

The results were found most sensitive to the method used to aggregate the individual panelists' estimates to obtain group estimates. A geometric mean aggregation produced group frequency estimates that approximated the medians of the panelists' estimates deemed generally consistent with both operating experience and prior LOCA frequency estimates except where increases are supported by specific material aging-related concerns. However, arithmetic mean and mixture distribution aggregations were found to be alternative methods that led to significantly higher mean and 95th percentile group estimates.

The expert elicitation process was an adaptation of the processes used to evaluate reactor risk in NUREG-1150 (USNRC 1990). The process consisted of a number of steps. A group of 12 panel members was selected for the formal elicitation. A unified analysis format was developed to ensure consistency in processing the panelists' inputs. The panelists' mid-value, upper bound, and lower bound responses were assumed to represent the median and 95th and 5th percentiles, respectively, of their subjective uncertainty distributions for each elicitation response. The analysis structure assumed that all the responses correspond to percentiles of lognormal distributions. These distributions were then combined using a lognormal framework. The final outputs for each panelist are estimates for the means, medians, and 5th and 95th percentiles of the total BWR and PWR LOCA frequencies. The panelists' estimates were then aggregated to obtain group LOCA frequency estimates, along with measures of panel diversity.

3.6.5 Reexamination of NUREG-1829 LOCA Frequency Estimates

The NUREG-1829 LOCA frequency estimates from 2008 were reassessed in 2024 using probabilistic fracture mechanics and statistical models of piping reliability (Sallaberry et al. 2025). The Extremely Low Probability of Rupture (xLPR), Version 2.3 PFM code, was used to reanalyze four of the NUREG-1829 base cases. The xLPR code was initially developed to model the effects of PWSCC. For the BWR base cases, IGSCC was modeled with the assumption of an initial crack at the beginning of plant operation, utilizing a generic stress corrosion cracking model available in the xLPR code with "some modifications."

The statistical approach involved updating NUREG-1829 LOCA frequency estimates using operating experience data from 2005 to 2024, focusing on LOCA categories 4, 5, and 6. This analysis used US operational experience relevant to the risk-significant base cases outlined in NUREG-1829. Additionally, domestic data was examined for primary system piping branch connections larger than DN150.

Table 3.29 Change in LOCA frequency, 2004–2024

NUREG-1829 LOCA categories	Equivalent break size (mm)	Plant-level LOCA frequency (1/year)			
		BWR—piping		PWR—piping	
		NUREG-1829	2024 update	NUREG-1829	2024 update
Cat4	175	7.60E−07	2.40E−08	7.60E−07	6.00E−08
Cat5	350–450	1.30E−07	4.30E−09	1.30E−07	2.60E−08
Cat6	775			1.20E−08	4.00E−10

Reviews were conducted across two timeframes—1970 to the end of 2004 and 2005 through mid-2024—to evaluate differences and identify key influencing factors.

In the second step, the piping failure precursor frequencies were calculated using a Bayesian analysis approach. A constrained, non-informative distribution method was used to develop prior piping failure precursor frequency distributions representative of the state of knowledge in 2004. The prior distribution was then updated using a Bayesian approach.

In the third step, the BWR and PWR base case conditional probability uncertainty distributions were developed from NUREG-1829 by first obtaining a geometric mean of the expert elicitation participant results for each base case. This value represented a target LOCA frequency. Next, the geometric mean of the precursor failure frequency results of the respective base case analyses was fit to a lognormal distribution. Finally, the median conditional probability of LOCA was obtained by dividing the median target LOCA frequency by the median system-level precursor failure rate. The resulting conditional probability of LOCA represents uncertainty in the 2004 state of knowledge when the NUREG-1829 expert elicitation was completed. This conditional probability of LOCA uncertainty distributions was subsequently updated with evidence of precursor events after 2004 from the operating experience data.

Lastly, in the fourth step, the uncertainty distributions developed in steps 2 and 3 were convolved in a converged Monte Carlo uncertainty propagation to obtain the respective LOCA frequency uncertainty distributions. The updated LOCA frequency results indicate at least an order of magnitude reduction for LOCA categories 4, 5, and 6 (greater than 150 mm equivalent break size) (Table 3.29).

3.7 Methods Intercomparisons

Many evaluations of the capabilities and potential limitations of different methodologies have been performed. Some of these evaluations have been based on intercomparisons of a methodology when applied by different teams of analysts. Such comparisons usually involve different implementations of a methodology, such as probabilistic fracture mechanics. Other evaluations have been intra-comparisons of different methodologies when applied to a single, well-specified evaluation boundary in terms of material properties, degradation mechanism, and location within a piping system. Three examples of intra-comparisons are highlighted in this section, including a recent benchmark exercise that was performed as part of an IAEA Coordinated Research Project.

3.7.1 VTT (2002)

A report prepared by the VTT Technical Research Centre of Finland describes a comparative study of two approaches to estimate pipe leak and rupture frequencies for piping (Simola et al. 2002). One method was based on a probabilistic fracture mechanistic (PFM) model, while the other was based on statistical estimation of rupture frequencies from a large database. To be able to compare the approaches and their results, the rupture frequencies of some selected welds were estimated using both methods. The report highlighted the differences in methods, input data, need, and use of plant-specific information and need of expert judgment. The study focused on one specific degradation mechanism, namely the intergranular stress corrosion cracking (IGSCC). This is the major degradation mechanism in old stainless steel piping in BWR environment, and its growth is influenced by material properties, stresses, and water chemistry.

The work by VTT had been motivated by two independent studies to estimate pipe leak and rupture frequencies for piping in Swedish BWRs. A pilot study to apply a piping failure database to estimate the leak and rupture frequency in reactor coolant pressure boundary (RCPB) piping was conducted in 1998 (Lydell 1999). The LOCA frequencies of the Barsebäck 1 RCPB piping were estimated statistically on the basis of a large database consisting of nuclear power plant operating experience. At the same time, a method based on a probabilistic fracture mechanistic (PFM) model was developed within the framework of a Nordic nuclear safety R&D project referred to as the NKS/RAK-1 project (Andersson 1997), and the methodology had been applied to establish ISI priorities for piping components at the Swedish Oskarshamn 1 nuclear power plant (Brickstad 2000). In its comparison, the authors of the VTT report concluded as follows:

> *The results of the rupture frequencies obtained by the two alternate approaches were quite different, but one approach does not give systematically higher values than the other one. The rupture frequencies calculated by the PFM codes have larger variation, which is understandable since the models use weld-specific information on stresses that may vary quite significantly.*
>
> *Some of the differences may be explained by different expert judgments made while applying the models. As an example, both approaches take into consideration the material properties of the weld but may have different interpretations depending on whether this information is from the upstream side or downstream side of the welds.*
>
> *Some differences were related to the different possibilities to account for, e.g., weld-specific information. In some cases, the results are not directly comparable, because the*

considered degradation mechanisms are not the same. For instance, the effect of vibrations is accounted for in the PFM modeling, while the statistical approach considered only the IGSCC.

It is worth noticing that there are some common features in the two approaches, such as the use operating experience, i.e., information on observed cracks and leaks. While the PFM model uses Swedish data for estimating crack initiation and aspect ratio, statistical approach uses a larger database for estimating leak and rupture frequencies.

Compared to the pipe failure estimates in WASH-1400, both methods generally give significantly lower frequencies except for some welds with high local stresses, where the PIFRAP code gives very high failure frequencies. Information based on statistical evidence should be taken into consideration, e.g., in updating of LOCA frequencies. For this purpose, the principles of the statistical approach might be sufficiently accurate, especially if it is completed with structured expert judgment. Although we have considered here only one degradation mechanism, corresponding databases are being developed for other mechanisms too.

The PFM approach requires a lot of weld-specific information from the stresses, and it may be argued that the approach may be too laborious for some applications. However, in RI-ISI applications where the procedure requires quite detailed analysis of the piping system, the quantification of leak and break probabilities may provide additional support for decision-making with relatively small additional efforts.

Despite large uncertainties related to the quantification of these probabilities, a PFM approach could be considered as an appropriate decision support in the selection of locations to be subjected to in-service inspection based on their susceptibility to material degradation. As the primary interest in RI-ISI is in the risk ranking of the welds, the absolute quantitative results are of less importance than the relative results, which are not sensitive for eventual conservative assumptions of the model. One advantage of the PFM model is also the explicit treatment of inspection reliability, which enables sensitivity studies with different inspection policies. It should be mentioned that the statistical analysis procedure has also been completed with a Markov model to allow for explicit modeling of inspection reliability.

The further development of both PFM and statistical estimation approaches and especially the discussions between PSA and material experts should be strongly encouraged. As an example, we would recommend the material and structural engineer's expertise to be included in the estimation of ruptures, especially when there is no experience from them. This could be easily incorporated in the statistical model, as it uses a Bayesian approach in the estimation of the probability of a rupture given a leak.

3.7.2 IAEA-CRP (2018–2021)

In 2018, the IAEA launched a coordinated research project (CRP) entitled "*Methodologies for Assessing Pipe Failure Rates in Advanced Water-Cooled Reactors*" (IAEA 2021, 2023). The primary objective of the coordinated research project was to provide member states with a strong technical basis for the estimation of piping reliability parameters consistent with required standards and relevant to deployable advanced water-cooled reactors. To meet this objective, two interrelated activities were carried out. First, a set of benchmark studies were developed to compare different methodologies when applied to a common set of problems. The results of the benchmark case studies were compared, and the reasons for differences were evaluated. The benchmark results provided input to the second activity that developed an analysis framework on piping reliability good practices.

Two benchmarks were developed and specifics of each defined. The first benchmark considered the chemical and volume control system (CVC) letdown branch connection on the reactor coolant system in a PWR. This evaluation boundary was chosen because of the availability of well-documented operating experience, including the associated root cause analyses. Hence, a motivation behind the first benchmark was to address the possibilities for reconciling the results of PFM with operating experience data. The motivation behind the second benchmark was to address the ability of different methods to address an evaluation boundary for which no failure data exists beyond results from laboratory testing and for which knowledge gaps exist regarding crack initiation times and crack propagation phenomena. Specifically, this evaluation boundary consisted of a dissimilar metal weld between the reactor pressure vessel and the reactor coolant system hot leg in a PWR. The assumed dissimilar metal weld consisted of nickel-base alloys 152/52. Seven teams applied three different methodologies, and the results were evaluated as follows:

- Intra-comparison of results: The results obtained within respective methodology grouping were reviewed based on the input and output parameters, result presentation, and treatment of uncertainties.
- Intercomparison of results: The results obtained from the application of the three different methodologies were reviewed based on input and output parameters, result presentation, and treatment of uncertainties.
- Development of a synthesis on the strengths and limitations of respective methodology: This step in the evaluation process also addressed the applicability of respective methodology to specific analytical contexts with clearly defined end-user requirements.

3.8 The "Double-Ended Guillotine Break"

The "design-basis accident" (DBA), "maximum credible accident" (MCA), or "maximum hypothetical accident" (MHA) have been used as terms describing what was generally the double-ended guillotine break (DEGB). The concept was originated by the United States Atomic Energy Commission for the purpose of sizing containments and establishing "accident" radiation doses and, later, the sizing of emergency core cooling systems. The original concept was quite straightforward; namely, an instantaneous DEGB of a major pipe in the primary system of a light water reactor (LWR) would maximize the fluid release and establish an upper bound for the design pressure established for a containment. This optimized the containment volume vis-a-vis a reasonable design accident pressure.

Subsequent changes in regulatory philosophy, particularly concerning seismic design, gradually transformed the DEGB from a theoretical accident to one considered more plausible. This shift led to a growing belief in the likelihood of major pipe breaks. It was a relatively short step from the hypothetical to a belief in major pipe breaks. A natural consequence of an accepted pipe break was the assumption of a terminal end (reactor pressure vessel nozzle) break and the asymmetric loading of the reactor pressure vessel. If one accepts a DEGB, then massive pipe restraints to minimize pipe deflection become a natural consequence, and backfitting requirements follow automatically. A reassessment of the overall probability of a large pipe break, particularly in reactor primary systems, undermined the basic premise that a DEGB was an accepted event. Probabilistic studies, deterministic studies, and an assessment of failure statistics in large pipes and nonnuclear vessels led to the same conclusion: the probability of a DEGB is extremely low; it is a beyond-design-basis event.

A DEGB in the primary coolant piping introduces severe design loading conditions, which include asymmetric slowdown, pipe whip, and a combination of loads due to DEGB and a safe shutdown earthquake (SSE). A "large pipe break" is normally considered to be double ended. This means that a process medium is expelled through both ends of a severed pipe or the equivalent of two pipes that are about 750 mm in diameter. In a relatively recent technical paper by Wilkowski, Uddin, Brust, and Kalyanam, reference was made to results from leak-before-break analyses from which it was concluded that (Wilkowski et al. 2019):

> *[W]hen circumferential through-wall cracks were put in the highest stressed locations, the applied moment dropped for both normal operating and N+SSE loading as the crack length increased. With loading three times higher than SSE loads, the through-wall crack size for causing a double-ended guillotine break (DEGB) was greater than 90% of the circumference. Similar results were also found for a petrochemical pipe system where thermal expansion stresses are much higher when compared to other primary stresses (pressure, gravity). Even with very low toughness materials of petrochemical plant (due to high-temperature hydrogen attack), the critical crack size leading to DEGB was greater than 80%. The implication of this work is that pragmatically a DEGB is not a critical concern for nuclear plant operation, and efforts would be better focused on the potential for a small-break loss-of-coolant accident (SB-LOCA).*

3.9 Pipe Failures That Influenced Structural Integrity Research

Reviews of piping operating experience have been ongoing since the first commercial nuclear power plants came online in the 1960s. For example, in 1975, the United States Nuclear Regulatory Commission (USNRC) established a first *Pipe Crack Study Group* (PCSG) charged with the task of evaluating the significance of stress corrosion cracking in boiling water reactors (BWRs) and pressurized water reactors (PWRs). The Pipe Crack Study Group reviewed intergranular stress corrosion cracking (IGSCC) of piping in boiling water reactors (BWRs) and in 1975 issued its findings in NUREG-75/067 (USNRC 1975b). During 1978, such cracking was reported for the first time in large-diameter piping (DN600) in Gundremmingen Unit

A[27] in Germany (Burghard et al. 1982). This discovery, together with the reported questions concerning the interpretation of ultrasonic inspections, led to the establishment of a new Pipe Crack Study Group (USNRC 1984). The charter of the new group was expanded to include (1) a review of the potential for stress corrosion cracking in pressurized water reactors (PWRs) as well as in BWRs, (2) an examination of the operating experience in foreign reactors relevant to IGSCC, and (3) a response to five specific questions concerning IGSCC. The five questions were as follows:

1. What is the significance of the cracks discovered in large-diameter pipes relative to the conclusions and recommendations set forth in the referenced report (NUREG-75/067) and in its implementation document NUREG-0313?
2. What is the resolution of concerns raised over the ability to use ultrasonic techniques to detect cracks in austenitic stainless steel?
3. What is the significance of cracks found in large-diameter sensitized safe ends and recommendations regarding the current NRC program dealing with the matter?
4. What is the potential for stress corrosion cracking in PWRs?
5. What is the significance of cracking in the Inconel safe ends that has been experienced at the Duane Arnold operating facility, and what are the recommendations regarding NRC actions taken or to be taken?

At the time, intergranular stress corrosion cracking of welded Type 304 stainless steel in the recirculation piping of boiling water reactors affected plant availability and reliability. Investigations of the problem resulted in significant progress in understanding the phenomenon and providing an engineering resolution by developing and qualifying countermeasures. A number of these countermeasures including solution heat treatment, corrosion-resistant clad, alternate pipe materials, induction heating stress improvement, and heat sink welding were implemented by plant owners. A seminar was held in Palo Alto, California, on January 22–24, 1980, to disseminate information on IGSCC countermeasures (EPRI 1980). It was attended by representatives from the utilities, vendors, architect engineers, regulatory groups, universities, steel companies, and pipe fabricators. Some 200 people attended the seminar, including a large contingent from Japan and representatives from Canada, Italy, the Netherlands, Sweden, and the United Kingdom.

The *Pipe Crack Task Group* (PCTG), a subset of the PCSG, concentrated on the SCC problem. The results of their work were published in 1984 as NUREG-1061 (USNRC 1984) and contain nearly 23 pages of conclusions and recommendations. The PCTG recommended the following:

> *Mitigating actions to control IGSCC in BWR piping must be designed to alleviate one or more of the three synergistic factors: sensitized material, convention BWR environment, and high tensile stresses.*

[27] A first-in-kind German boiling water reactor, which achieved initial criticality in August 1966. The reactor unit was permanently shut down in January 1977.

Because mitigating actions addressing each one of these factors may not be fully effective under all anticipated operating conditions, mitigating actions should address two and preferably all three of the causative factors, e.g., material plus some control of water chemistry or stress reversal plus controlled water chemistry.

All Types 304 and 316 austenitic piping systems operating over 93°C should receive augmented in-service inspection, unless they have been treated with effective countermeasures.

The extent and frequency of examinations should depend on the resistance of materials to IGSCC and the effectiveness of any processes used to prevent cracking.

The primary basis for sample welds selected for examination should be field experience not the stress rule index and carbon content. Other factors such as weld prep, excessive grinding, extensive repairs, or high-stress locations should also be considered in the sample selection.

Flaw evaluation criteria should limit the length of the cracks accepted for continued operation without repair. The limitation on acceptable crack length is primarily a result of the lack of confidence in flaw depth sizing capability and is intended to ensure leak-before-break conditions. The maximum allowable through-wall crack length can be determined based on weld joint-specific loads.

The maximum crack length allowable without repair for a specific weld joint should be the minimum of either 1) the through-wall crack length demonstrated by elastic-plastic fracture mechanics analyses to be stable under normal operating plus safe shutdown earthquake (SSE) loading conditions, 2) the through-wall crack length that would still permit the pipe to withstand normal operating plus SSE loading conditions as demonstrated by net section collapse (limit-load) analyses, or 3) the maximum crack length that would result in a leak rate greater than the plant's normal makeup capacity. Shorter cracks can be evaluated using the ASME XI IWB-3640 criteria as modified by the USNRC staff in SECY 83-267C (Staff Requirements for Reinspection of BWR Piping and Repair of Cracked Piping, November 3, 1983).[28] *Calculations indicate that in the majority of cases, the maximum crack length acceptable under the above criteria will be approximately 25–30% of the pipe circumference.*

It is recommended that a formal study and updating of the pipe failure database be undertaken. Some statistics may be only meaningfully derived or assessed analytically. The USNRC has developed a computer code PRAISE (Piping Reliability Analysis Including Seismic Events), which is currently being modified to include a model for IGSCC based on current understanding.

It is recommended that PRAISE be used to a) investigate the impact of IGSCC on primary piping according to pipe size; b) study the failure frequencies of components due to IGSCC, fatigue, etc.; and c) calculate the conditional probability of multiple failures associated with IGSCC to check the point value and to test the assumption of a lognormal distribution.

On January 25, 1988, the USNRC issued Generic Letter (GL) 88-01, "*NRC Position on IGSCC in BWR Austenitic Stainless Steel Piping*."[29] A generic letter is a formal request for licensee action and/or information to address issues regarding emergent or routine matters of safety, security, safeguards, or environmental significance, and it required a written response.

[28] SECY stands for the USNRC's Secretary who submits written "issue papers" to the NRC's Commission. These papers cover various matters related to policy, rulemaking, and adjudication.

[29] https://www.nrc.gov/reading-rm/doc-collections/gen-comm/gen-letters/1988/index.html

Table 3.30 Summary of inspection schedule for BWR piping welds

Description of weld	IGSCC category	Inspection extent and schedule
Resistant materials	A	25% every 10 years (at least 12% in 6 years)
Nonresistant materials with stress improvement (SI) within 2 years of operation	B	50% every 10 years (at least 25% in 6 years)
Nonresistant materials with stress improvement (SI) after 2 years of operation	C	All weld within the next 2 refueling cycles, then all every 10 years (at least 50% in 6 years)
Nonresistant materials—no SI	D	All welds every 2 refueling cycles
Cracked—reinforced by weld overlay or mitigated by SI	E	50% next refueling outage, then all welds every 2 refueling cycles
Cracked—inadequate or no repair	F	All welds every refueling outage
Nonresistant—not inspected	G	All welds every refueling outage

On inspection schedule and qualification, the USNRC stated that nondestructive examination (NDE) performed under the generic letter should comply with the applicable edition and addenda of the ASME Code, Section XI:

- As specified in paragraph (9), "In-service Inspection Requirements" of 10CFR50.55a, codes and standards, or as otherwise approved by the USNRC.
- In addition, the detailed procedure, equipment, and examination personnel shall be qualified by a formal program approved by the USNRC such as that being conducted in accordance with the NDE Coordination Plan agreed upon by USNRC, EPRI, and the Boiling Water Reactor Owners Group for IGSCC Research, being implemented at the EPRI NDE Center in Charlotte, North Carolina.

Table 3.30 is a summary of the inspection schedule that was to be implemented by all BWR operating license holders (USNRC 1988).

After the 1986 flow-accelerated corrosion (FAC)-related pipe rupture at Surry Unit 2 Power Station, the USNRC convened a multidisciplinary panel in 1987 to discuss factors contributing to FAC and ways to predict and prevent such failures. The condensate pipe failure was attributed to FAC-induced thinning, but the complex variables affecting FAC rates were not yet fully understood. The panel reviewed parameters that may have played a role in the feedwater piping system's erosion-corrosion. At the time, the complex interactions of the individual variables which affect the FAC phenomenon and influence the rate at which it proceeds had not been thoroughly established by available research activity. The panel discussion elaborated on the role of those parameters which could have potentially contributed to erosion-corrosion in the feedwater piping system. Key observations from the panel discussion were as follows (USNRC 1987):

The phenomena associated with piping wear due to erosion, FAC, or cavitation had not been recognized as significant problems by piping designers. Some architect-engineers provide guidance in system design specifications to minimize erosion-corrosion by limiting maximum bulk flow velocity in water-filled systems, such as feedwater and service water, but most architect-engineers focus primarily on meeting code minimum pipe wall requirements, with some variable added wall thickness, and on meeting stipulated pressure drop requirements for equipment operability.

The amount of FAC damage and the rate at which it proceeds are complex phenomena depending on several variables. In carbon steel systems, the process is dependent on the dissolution of the metallic oxide (magnetite) protective surface layer. This rate of wear decreases with an increase in oxygen concentration and with an increase in pH above 9.2. Flow velocity and regions of turbulence caused by pipe fittings, particularly, 90° splitting tees, also abet the wear process. Temperature plays a role in the process with a peak in the amount of damage occurring in the range of 250–340 °F.

In the case of Surry and other US plants, the three most important variables influencing the corrosion-erosion process are material, local fluid velocity (turbulence), and water chemistry. The panel could not identify which of these three variables was controlling but believed that the Surry-specific piping configuration with its high bulk flow velocity and turbulence at the splitting tee and close-coupled elbow configuration had a significant role. The role of the past operating water chemistry conditions is uncertain. It is possible that most of the damage could have occurred early in life, with an appreciably lower damage rate following installation of the condensate polishing system.

Key recommendations of the panel were as follows:

The panel believes that adjustments to pH or oxygen content from levels now in use to protect steam generators should not be made without a thorough study of the possible global effects of such changes on the entire system. Often, changes are made to fix a specific problem in a nuclear plant without considering the total ramifications of the fix. The addition of hydrogen to BWR systems was cited as an example of another proposed fix, which requires careful study. Long-term water chemistry needs for nuclear plants, particularly, a better understanding of the effects of pH and oxidizing or reducing environments on materials of system components, were cited as a possible topic for additional research. Another important topic mentioned for additional research was the effect of local fluid dynamics in typical pipe configurations.

The panel believes that the factors influencing the rate at which the erosion-corrosion phenomenon proceeds in single-phase systems cannot be ranked because the available current quantitative data are insufficient. However, the panel believes that the information which is available relative to the identity of the involved parameters and how they influence erosion-corrosion in single-phase systems should be presented to the utilities so that they may formulate programs to ultrasonically measure existing (baseline) pipe wall thickness at susceptible locations in specific systems. Screening criteria, such as that developed by Virginia Power, should be used on the most susceptible systems to identify potential locations of maximum wear. The need for periodic monitoring could be determined from the results of such baseline measurements.

Data resulting from baseline wall thickness measurements taken as a result of recommendation (2) should be correlated with flow velocity, turbulence, water chemistry, temperature, actual material chemical composition, and installed original wall thickness, to the extent that such data are available. The data should be centrally collected and processed to increase overall knowledge and understanding of erosion-corrosion. Quantitative erosion-corrosion data for single-phase systems appear to be more generally known and available in Europe and apparently receive some consideration there in the design process. The Netherlands, France, the Federal Republic of Germany, and the Soviet Union were mentioned as having joint government utility-sponsored research on erosion-corrosion.

In response to thermal fatigue failures, the Committee on the Safety of Nuclear Installations (CSNI) of the Nuclear Energy Agency (NEA) and the World

Organization of Nuclear Operators (WANO) organized the "*Specialists Meeting on Experience with Thermal Fatigue in LWR Piping Caused by Mixing and Stratification*" (OECD/NEA 1999). The meeting was prompted by several events that had resulted in losses of reactor primary coolant. Conclusions of the meeting were as follows:

> *Thermal fatigue due to stratification and mixing of hot and cold water is a recurring phenomenon, but with a relatively low frequency. It can affect safety-related piping of many different systems (ECCS, RHR, AFW, pressurizer surge line, and safety/relief valve discharge lines) at various locations. It may happen that these phenomena are not included in the design conditions. Recently, similar events have occurred in sections of piping in the primary circuit, and therefore, the plant owners and the safety authorities are now alerted over some conditions not being consistent with licensing basis and inspection commitments. Of particular concern are those problems arising from unanticipated thermal fatigue in un-isolable piping connected to the reactor coolant system.*
>
> *There is a need to develop further accepted methods to identify locations with potential risk of thermal fatigue. There are proposals for simplified screening criteria based on semiempirical models to determine the areas where there is a risk of thermal fatigue, but these are not generally accepted. As there are many uncertainties, it is possible to consider probabilistic treatment, although the data for this is also limited. As a first step in this direction, the use of best estimate analysis methods should be considered, so that attention is focused on the areas most at risk for thermal fatigue, although these may be difficult to define. Then, at those locations, monitoring of temperature and pressure, if necessary, should be implemented. In addition, periodic verification of the leak tightness of the nearest valves could further reduce the risk of thermal fatigue.*
>
> *Monitoring of temperature fluctuations can be seen as an important part of the defense, and, at present, it remains the most reliable method to avoid unanticipated incidents. At present, there are different strategies in use and no single method provides defense. There is a need for combining redesign and revised operating practices. A small internal cold water leakage into a hot section of pipe can lead to a quick propagation of cracks by thermal fatigue in some sensitive zones. The manufacturing process probably has a very important impact on the rate of crack initiation. It is not possible to draw up simple criteria for such parameters as allowable valve leakage rates or limiting pipe diameters, as there is a great variety in the systems and operating procedures.*

References

Albertz V et al (1980) Deutsche Risikostudie Kernkraftwerke: Eine Untersuchung zu dem durch Störtälle in Kernkraftwerken verursachten Risiko, Fachband 3: Zuverlässig-keitskenngrößen und Betriebserfahrungen. Der Bundesminister für Forschung und Technologie, Bonn. ISBN: 3-88585-014-1

Andersson K (1997) Strategies for reactor safety. Final report of the Nordic Nuclear Safety Research Project RAK-1, NKS(97)FR1, Roskilde, Denmark

ASME (2021) Probabilistic risk assessment standard for advanced non-light water reactor Nuclear Power Plants, RA-S-1.4-2021, New York. ISBN: 9780791874370

ASME (2022) Standard for level 1/large early release frequency probabilistic risk assessment for nuclear power plant applications, RA-S-1.1–2022, New York, NY. ISBN 9780791875315

ASME (2023) Probabilistic risk assessment standard for advanced light water reactor nuclear power plants during construction and pre-operational stages, RA-S-1.5-2023

Atwood CL, LaChance JL, Martz HF, Anderson DJ, Englehardt M, Whitehead D, Wheeler T (2003) Handbook of parameter estimation for probabilistic risk assessment, NUREG/CR-6823. USNRC, Washington, DC

Beliczey S (1991) Frequencies of leaks and breaks in safety related piping of PW-plants as initiating events for LOCAs. In: Proceedings of the OECD/BMU-workshop on special issues of level 1 PSA, GRS-86, Cologne, Germany, pp 364–380

Beliczey S, Schulz H (1987) The probability of leakage in piping systems of pressurized water reactors on the basis of fracture mechanics and operating experience. Nucl Eng Des 102:431–438

Beliczey S, Schulz H (1990) Comments on probabilities of leaks and breaks of safety-related piping in PWR plant. Int J Press Vessels Pip 43:219–227

Berg H-P, Gersinska R, Sievers J (2010) Procedure for probabilistic safety assessment of leaks and breaks of piping systems. Int J Press Vessels Pip 87:94–99

BfS (Bundesanstalt für Strahlenschutz) (2016) Methods and data for probabilistic safety assessment of nuclear power plants, BfS-SCHR-61/16 (in German), Salzgitter, Germany

BMFT (1980) Deutsche Risikostudie Kernkraftwerke: Eine Untersuchung zu dem durch Störtälle in Kernkraftwerken verursachten Risiko, 2nd edn. Bonn, Germany. ISBN: 3-911059-67-4

Brickstad B (2000) The use of risk based methods for establishing ISI-priorities for piping components at Oskarshamn 1 Nuclear Power Station, SKI Report 00:48. Swedish Nuclear Power Inspectorate, Stockholm

Burghard HC Jr et al (1982) Metallurgical examination and destructive assay of piping samples from the KRB nuclear reactor, report NP-2382-SY-LD. EPRI, Palo Alto

Bush SH (1976) Reliability of piping in light water reactors. Nucl Saf 17:568–557

Bush SH (1985a) Statistics of pressure vessel and piping failures. J Press Vessel Technol 110:225–233

Bush SH (1985b) Statistics of pressure vessel and piping failures. In: Sundararajan C (ed) Pressure vessel and piping technology—1985—a decade of progress. ASME, New York, pp 875–893

Chu TL, Pratt WT (eds) (1995) Evaluation of severe accidents during low power and shutdown operations at surry, unit 1, summary of results, NUREG.CR-6144, vol 1. USNRC, Washington, DC

Chu TL et al (1994) Evaluation of potential severe accidents during low power and shutdown operations at Surry, Unit 1, analysis of core damage frequency from internal events during mid-loop operations, main report (Chapters 1-6), NUREG/CR-6144, vol 2. USNRC, Washington, DC

Dandini V, Forester J, Staple B, Kirk H, Whitehead D (1994) Evaluation of potential severe accidents during low power and shutdown operations at Grand Gulf, Unit 1, analysis of core damage frequency from internally induced flooding events for plant operational state 5 during a refueling outage, NUREG/CR-6143, vol 4. USNRC, Washington, DC

Eason ED, Shusto LM (1984) Analysis of cracking in small-diameter piping, NP-4394. Electric Power Research Institute, Palo Alto

Eide SA, Khericha ST, Calley MB, Johnson DA, Maryteeny ML (1991) Component external leakage and rupture frequency estimates, EGG-SSRE--9639. Idaho National Engineering Laboratory, Idaho Falls

Eide SA, Wierman TE, Gentillon CD, Rasmuson DM, Atwood CL (2007) Industry-average performance for components and initiating events at U.S. Commercial Nuclear Power Plants, NUREG/CR-6928. USNRC, Washington, DC

EPRI (1977) Characteristics of pipe system failures in light water reactors, NP-438. EPRI, Palo Alto

EPRI (1980) Proceedings: seminar on countermeasures for pipe cracking in BWRs, WS 79-114, vols 1, 2, 3 and 4, Palo Alto

EPRI (1984) Analysis of cracking in small-diameter piping, NP-4394, Palo Alto

EPRI (1992) Pipe failures in U.S. Commercial Nuclear Power Plants, TR-100380, Palo Alto

EPRI (1993) Pipe failure study update, TR-102266, Palo Alto

EPRI (1998) Piping system reliability and failure rate estimation for use in risk-informed in-service inspection applications, TR-110161, EPRI, Palo Alto

EPRI (2023) Pipe rupture frequencies for internal flooding probabilistic risk assessments, revision 5, 3002024904, Palo Alto

EPRI (2024) Materials reliability program: xLPR estimation of PWR loss-of-coolant accident frequencies (MRP-480), 3002023895. EPRI, Palo Alto

Fleming KN (2004) Markov models for evaluating risk-informed in-service inspection strategies for nuclear power plant piping systems. Reliab Eng Syst Saf 83:27–45

Fleming KN, Gosselin S (1997) Application of Markovian technique to modeling influences of inspection on pipe rupture frequencies. In: Proceedings of the SKI seminar on piping reliability, SKI report 97:32. Swedish Nuclear Power Inspectorate, Stockholm

Fleming K, Mikschl T (1998) Application of Markovian technique to modeling influences of inspection on pipe rupture frequencies. In: Proceedings of the probabilistic safety assessment and management: proceedings of the 4th international conference on probabilistic safety assessment and management (PSAM), 13-18 September 1998. Springer, New York, pp 463–468. ISBN 978-3540762621

Gibbons WS, Hackney SD (1964) Survey of piping failures for the reactor primary coolant pipe rupture study, GEAP-4574, prepared for the U.S. AEC under Contract AT-04-3-189. Atomic Power Equipment Department, General Electric Company, San Jose

GRS (Gesellschaft für Anlagen- und Reaktorsicherheit) (1995) Workshop compendium on the estimation of leak and rupture frequencies of pressure boundary components for probabilistic safety assessment (in German), September 18-20, Cologne, Germany

Harris DO, Lim EY, Dedhia DD, Woo HH, Chou CK (1982) Fracture mechanics models developed for piping reliability assessment in light water reactors, NUREG/CR-2301. USNRC, Washington, DC

Heasler PG, Simonen FA (1983) Using Markov Chain models to describe crack growth and inspection effectiveness. In: ASME paper no. 83-PVP-88 presented at the 4th national congress on pressure vessel and piping technology, June 19-24, 1983, Portland

Holman GS, Chou CK (1988) Probability of failure in BWR reactor coolant piping, NUREG/CR-4792. U.S. Nuclear Regulatory Commission, Washington, DC

IAEA (2021) Technical insights from benchmarking different methods for predicting pipe failure rates in water cooled reactors. Final report of a coordinated research project. IAEA-TECDOC-1988, Vienna

IAEA (2023) Methodologies for assessing pipe failure rates in advanced water-cooled reactors, NES NR-T-2.16, Vienna, Austria. ISBN: 978-92-0-150322-0

IRSN (Institute for Nuclear Safety and Protection) (1990) A probabilistic safety assessment of the standardized French 900MWe pressurized water reactor, main report, INIS-XA-N-196. Fontenay-aux-Roses, Paris

Janzen P (1981) A study of piping failures in U.S. Nuclear Power Reactor, AECL-Misc-204. Chalk River Nuclear Laboratories, Chalk River

Janzen P (1984) Piping performance in Canadian CANDU Nuclear Generating Station, AECL-MISC-252. Chalk River Nuclear Laboratories, Chalk River

Kellermann O, Costes D, Hazelton WS, O'Neil R, Smidt D, Witt FJ (1974) Structural safety and reliability of primary system pressure boundaries for water cooled power reactors. Nucl Eng Des 28(2):218–238

Kilsby ER (1964) A survey of water reactor primary system conditions pertinent to the study of pipe rupture, GEAP-4445. Atomic Power Equipment Department, General Electric, San Jose

Kohut P (1994) Evaluation of potential severe accidents during low power and shutdown operations at Surry, Unit 1, analysis of core damage frequency from internal floods during mid-loop operations, NUREG/CR-6144, vol 4. USNRC, Washington, DC

Lo TY, Mensing RW, Woo HH, Holman GS (1984) Probability of pipe failure in the reactor coolant loops of combustion engineering PWR plants, vol 2, pipe failure induced by crack growth, NUREG/CR-3663. USNRC, Washington, DC

Lydell B (1999) Failure rates in Barsebäck-1 reactor coolant pressure boundary piping. An application of a piping failure database, SKI report 98:30. Swedish Nuclear Power Inspectorate, Stockholm
Lydell B (2000) Pipe failure probability: the 'Thomas paper' revisited. Reliab Eng Syst Saf 68:207–217
Lydell B, Olsson A (2008) Reliability data for piping components in Nordic Nuclear Power Plants: 'R-Book' project phase I, SKI report 2008:01. Swedish Nuclear Power Inspectorate, Stockholm
Lydell B, Tomic B, Nyman R, Erixson S (1995a) Reliability of piping system components, vol 1: piping reliability—a resource document for PSA applications, SKI report 95:58, Stockholm, Sweden
Lydell B, Tomic B, Nyman R, Erixson S (1995b) Reliability of piping system components, vol 4: the pipe failure event database, SKI report 95:61, Stockholm, Sweden
Lydell B, Hegedus D, Tomic B, Nyman R (1997) Reliability of piping system components: framework for estimating failure parameters from Service Data, SKI report 1997:26, Stockholm (Sweden)
Lydell B, Mathet E, Gott K (2002) OECD pipe failure data exchange (OPDE) project: a framework for international cooperation in piping reliability. In: Proceedings of the international topical meeting on probabilistic safety assessment. American Nuclear Society, LaGrange Park
Medhekar SR, Bley DC, Gekler WC (1993) Prediction of vessel and piping failure rates in chemical process plants using the Thomas model. Process Saf Prog 12(2):123–126
OECD/NEA (1999) Specialists meeting on experience with thermal fatigue in LWR piping caused by mixing and stratification, NEA/CSNI/R(1998)8. Boulogne-Billancourt, Paris
Phillips CAG, Warwick RG (1967) A survey of defects in pressure vessels built to high standards of construction and its relevance to nuclear primary circuits, AHSB(S) R-162. U.K. Atomic Energy Authority, London
Poloski JP, Marksberry DG, Atwood CL, Galyean WJ (1999) Rates of initiating events at U.-S. Nuclear Power Plants: 1987–1995, NUREG.CR-5750 – 1995. U.S. Nuclear Regulatory Commission, Washington, DC
Ravindra MK, Campbell RD, Kennedy RP, Banon H (1985) Probability of pipe failure in the reactor coolant loops of combustion engineering PWR plants. In: Double-ended guillotine break indirectly induced by earthquakes, NUREG/CR-3663, vol 3. USNRC, Washington, DC
Sallaberry CJ, Brust FW, Kurth-Twombly E, Kurth RE, Lydell BOY (2025) Reexamination of NUREG-1829 loss-of-coolant accident (LOCA) frequency estimates, technical letter report TLR-RES/DE/REB-2025-03. USNRC, Washington, DC
Schulz H (1988) Comments on the probability of leakage in piping systems as used in PRAs. Nucl Eng Des 110:229–232
Simola K, Pulkkinen U, Talja H, Saarenheimo A, Karjalainen-Roikonen P (2002) Comparative study of approaches to estimate pipe break frequencies, NKS-79. Nordic Nuclear Safety Research, Roskilde. ISBN 87-7893-135-5
Smith, TA, Warwick, RG (1974) "A Survey of Defects in Pressure Vessels Built to High Standards of Construction and Its Relevance to Nuclear Primary Circuits," International Journal of Pressure Vessels and Piping, 2(4):283–322
SKI (Swedish Nuclear Power Inspectorate) (1997) Seminar on piping reliability, SKI report 1997: 32, Stockholm, Sweden
Tagert SW (1964) Mechanical Design Considerations i Primary Nuclear Piping, GEAP-4578, Atomic Power Equipment Department, General Electric Company, San Jose, CA
Thomas HM (1979) In-service inspection is over-rated. J Inst Nucl Eng 20(2):43–46, 51–56
Thomas HM (1981) Pipe and vessel failure probability. Reliab Eng 2:83–124
Tomic B, Lydell B, Nyman R, Erixson S (1995a) Reliability of piping system components, vol 2: PSA LOCA database; review of methods for LOCA evaluation since WASH-1400, SKI report 95:59, Stockholm, Sweden
Tomic B, Lydell B, Nyman R, Erixson S (1995b) Reliability of piping system components, vol 3: piping reliability—a bibliography, SKI report 95:60, Stockholm, Sweden

USNRC (1975a) Reactor safety study: an assessment of accident risks in U.S. Commercial Nuclear Power Plants, Appendix III, failure data, NUREG-75/014 (WASH-1400), Washington, DC

USNRC (1975b) Investigation and evaluation of cracking in austenitic stainless-steel piping of boiling water reactor plants, NUREG-75/067, Washington (DC)

USNRC (1984) Report of the U.S. Nuclear Regulatory Commission Piping Review Committee. In: Investigation and evaluation of stress corrosion cracking in piping of boiling water reactor plants, NUREG-1061, vol 1. USNRC, Washington, DC

USNRC (1987), Outcome of NRR 870115 technical meeting to consider generic implications of 861209 feedwater line failure, contributing parameters & means to predict or mitigate effects of erosion-corrosion in piping systems, external memorandum, NRC-ADAMS Accession No. ML20207Q695, Washington, DC

USNRC (1988) NRC position on IGSCC in BWR austenitic stainless steel piping, Generic Letter No. 88-01, Washington, DC

USNRC (1990) Severe accident risks: an assessment for five U.S. Nuclear Power Plants, NUREG-1150. Office of Nuclear Regulatory Research, Washington, DC

USNRC (2008) Estimating loss of coolant accident (LOCA) frequencies through the elicitation process, NUREG-1829. USNRC, Washington, DC

Vo TV, Heasler PG, Doctor SR, Simonen FA, Gore BF (1991) Estimates of rupture probabilities for nuclear power plant components: expert judgment elicitation. Nucl Technol 96(3):259–271

Whitehead DW (ed) (1995) Evaluation of severe accidents during low power and shutdown operations at Grand Gulf, Unit 1, summary of results, NUREG.CR-6143, vol 1. USNRC, Washington, DC

Whitehead D et al (1994) Evaluation of potential severe accidents during low power and shutdown operations at Grand Gulf, Unit 1, analysis of core damage frequency from internal events for plant operational state 5 during a refueling outage, main report (sections 1-9), NUREG/CR-6143 vol 2, Part 1A. USNRC, Washington, DC

Wilkowski G, Uddin M, Brust FW, Kalyanam S (2019) The huge missing factor in LBB analysis—how a circumferential through-wall crack in a pipe system changes the flexibility and reduces the applied moments. Transactions, SMiRT-25, Division II (Fracture Mechanics and Structural Integrity). https://www.lib.ncsu.edu/resolver/1840.20/37580

Wilson SA (1972) Estimating pipe reliability by the distribution of time-to-damage method, GEAP-10452. Atomic Power Equipment Department, General Electric Company, San Jose

Woo HH (1983) The impact of inservice inspection on the reliability of nuclear piping, UCID-19725. Lawrence Livermore National Laboratory, Livermore

Woo HH, Chou CK (1983) Piping reliability model validation and potential use for licensing regulation development, NUREG/CR-2801. USNRC, Washington, DC

Wright RE, Steverson JA, Zuroff WF (1987) Pipe break frequency estimation for nuclear power plants, NUREG/CR-4407. USNRC, Washington, DC

Zigler G, Brideau J, Rao DV, Shaffer C, Souto F, Thomas W (1995) Parametric study of the potential for BWR ECCS strainer blockage due to LOCA generated debris, NUREG/CR-6224. USNRC, Washington, DC

Chapter 4
Material Degradation: Learning from Operating Experience

4.1 Conjoint Requirements for Pipe Failure

Pipe material degradation and failure are due to synergistic effects caused by off-normal operating and environmental conditions, unusual or extreme loading conditions, and (sometimes strong) influence of less-than-adequate application of methods used to fabricate and weld piping components to form a piping system. The triplet <***material;envronment;loading***> is used to represent the conjoint requirements for material degradation. Making, sometimes, subtle changes to any of the physical parameters that are embedded in this triplet can have a profound effect on the propensity for pipe material degradation.

Venn diagrams have been used to show how system parameters affect material degradation, but its limitations arise when a quantitative analysis is needed. In its simplest form, the area of intersection of three circles representing material, environment, and stress/strain, respectively, would give some indication of the effect on degradation susceptibility. The overlap between material and environment pertains to corrosion mechanisms (Sect. 4.5). The overlap between material and stress/strain concerns fatigue mechanisms (Sect. 4.6). The overlap between environment and stress/strain relates to corrosion fatigue (Sect. 4.6.5). Lastly, the overlap among material, environment, and stress/strain involves stress corrosion cracking mechanisms (Sect. 4.8). A Venn diagram as it applies to flow-accelerated corrosion is found in Sect. 4.7.3.

4.2 First-in-Kind Material Degradation Mechanisms

Design and construction of first-in-kind nuclear power plants began in the 1950s. Examples of such reactors were Calder Hall Unit 1 (commissioned in 1956, it was the world's first commercial power reactor), Shippingport Atomic Power Station

B. Lydell, *Knowledge Preservation in Nuclear Safety*,
https://doi.org/10.1007/978-3-032-25463-4_4

(a pressurized water reactor, PWR, commissioned in 1957), Vallecitos Experimental Boiling Water Reactor (1957), and Marcoule G2 (a gas-cooled, graphite-moderated reactor commissioned in 1958).[1] Dresden 1 (commissioned in 1960) was the first commercial boiling water reactor (BWR) in the United States. It was a dual-cycle plant that incorporated an external steam drum as well as steam generators. Indian Point Unit 1 (1962), the first commercial PWR, was a two-loop plant with a supplemental oil-fired steam superheater.

As new reactors were commissioned, new forms of material degradation mechanisms were identified every 2–5 years. The industry was learning from these initial first-in-kind (FIK) failures. Some of these failures stemmed from suboptimal piping design, including incorrect material selections and welding practices, as well as inadequate in-service inspection program implementations, such as the use of unvalidated nondestructive examination techniques. In other cases, significant structural failures occurred after less than a year of operation. There have also been instances where four decades would elapse until the discovery of an FIK material failure (Table 4.1). These material degradation mechanisms and associated field experience are described in Sects. 4.4–4.10. There are examples of how the industry failed to proactively incorporate preventive measures in piping system design at an early stage. The emergence of primary water stress corrosion cracking is such an example of how a considerable lag came to be between a first discovery and broader recognition of a problem.

The onset of primary water stress corrosion cracking (PWSCC) in pressurized water reactors (PWRs) during the 1980s serves as an example of a failure in knowledge integration and illustrates how preventive measures could have been incorporated into piping system design at an early stage. In the environment of a PWR primary system, intergranular stress corrosion cracking (SCC) in wrought nickel-base materials is commonly referred to as primary water SCC, or PWSCC. The predominant nickel-base alloys used in PWRs are Alloy 600 and Alloy 690. Alloy 600 was developed in the 1950s, and Alloy 690 followed in the 1960s, with full-scale production commencing in the 1960s and 1970s, respectively.

PWSCC in high-purity water had been extensively studied since its first documented observation in laboratory tests conducted at the Commissariat à l'Énergie Atomique (CEA) in France in 1959. These tests were performed on Alloy 600 material, and the discovery of PWSCC is credited to Dr. Henri Couriou. Despite this discovery, Alloy 600 became a preferred alloy when commercial PWR plants were designed and constructed between the 1960s and 1990s. The initial instances of Alloy 600 PWSCC in operational plants were reported in the early 1980s. This delay between the identification of Alloy 600's susceptibility to SSC in

[1] Construction of Calder Hall began in August 1953. It was the first full-scale nuclear power station to enter operation. Construction of Shippingport began in January 1954. It was light water-moderated, thermal breeder reactor and is considered to be the first full-scale atomic electric power plant devoted exclusively to peacetime uses. Construction of Marcoule G2 started in March 1955.

Table 4.1 Summary of FIK pipe failure and the underlying failure mechanisms

Material damage and degradation mechanism		Time period when first discovered in service												Note
		1965–1969	1970–1974	1975–1979	1980–1984	1985–1989	1990–1994	1995–1999	2000–2004	2005–2009	2010–2014	2015–2019	2020–2024	
Corrosion mechanisms	Corrosion—general	☒	☐	☐	☐	☐	☐	☐	☐	☐	☐	☐	☐	(1)
	Corrosion under insulation	☐	☒	☐	☐	☐	☐	☐	☐	☐	☐	☐	☐	–
	Corrosion—crevice	☐	☐	☒	☐	☐	☐	☐	☐	☐	☐	☐	☐	–
	Corrosion—pitting	☐	☒	☐	☐	☐	☐	☐	☐	☐	☐	☐	☐	–
	Dealloying—selective leaching (Al-bronze)	☐	☐	☐	☐	☒	☐	☐	☐	☐	☐	☐	☐	–
	Galvanic corrosion	☒	☐	☐	☐	☐	☐	☐	☐	☐	☐	☐	☐	–
	Graphitic corrosion	☐	☐	☐	☒	☐	☐	☐	☐	☐	☐	☐	☐	(2)
	Microbiologically influenced corrosion—carbon steel	☐	☒	☐	☐	☐	☐	☐	☐	☐	☐	☐	☐	
	Microbiologically influenced corrosion—stainless steel	☐	☐	☐	☒	☐	☐	☐	☐	☐	☐	☐	☐	
Flow-assisted degradation	Erosion	☐	☒	☐	☐	☐	☐	☐	☐	☐	☐	☐	☐	
	Erosion-cavitation	☐	☒	☐	☐	☐	☐	☐	☐	☐	☐	☐	☐	
	Erosion-corrosion	☐	☒	☐	☐	☐	☐	☐	☐	☐	☐	☐	☐	
	Flow-accelerated corrosion (FAC)—balance of plant	☒	☐	☐	☐	☐	☐	☐	☐	☐	☐	☐	☐	(3)
	Flow-accelerated corrosion (FAC) at low temperatures	☐	☐	☐	☐	☐	☐	☐	☒	☐	☐	☐	☐	(4)
	FAC—BWR RPV drain	☐	☐	☐	☐	☐	☐	☐	☐	☒	☐	☐	☒	(5)

(continued)

Table 4.1 (continued)

Material damage and degradation mechanism		Time period when first discovered in service 1965–1969	1970–1974	1975–1979	1980–1984	1985–1989	1990–1994	1995–1999	2000–2004	2005–2009	2010–2014	2015–2019	2020–2024	Note
	FAC entrance effect (or leading edge effect)	☐	☐	☐	☐	☐	☐	☒	☐	☐	☐	☐	☐	(6)
	Liquid droplet impingement erosion	☐	☐	☐	☒	☐	☐	☐	☐	☐	☐	☐	☐	
Fatigue	Corrosion fatigue	☐	☒	☒	☐	☐	☐	☐	☐	☐	☐	☐	☐	
	High-cycle fatigue	☒	☐	☐	☐	☐	☒	☐	☐	☐	☐	☐	☐	(7)
	Low-cycle fatigue	☐	☒	☐	☐	☐	☐	☐	☒	☐	☐	☐	☐	
	Thermal stratification, cycling, and striping	☒	☐	☐	☐	☒	☐	☐	☐	☐	☐	☐	☐	(8)
Stress corrosion cracking	External chloride-induced SCC/ODSCC	☐	☒	☐	☐	☐	☐	☐	☐	☐	☐	☐	☐	(9)
	Intergranular SCC (IGSCC)—BWR primary: Type 304 stainless steel	☒	☐	☐	☐	☐	☐	☐	☐	☒	☐	☐	☐	(10)
	IGSCC—BWR primary: Type 304L/316L	☒	☐	☐	☒	☐	☐	☐	☐	☐	☐	☐	☐	(11)
	IGSCC—BWR primary: alloy 600/82/182	☐	☐	☒	☐	☐	☐	☐	☐	☐	☐	☐	☒	(12)
	IGSCC—stabilized austenitic stainless steel	☐	☐	☐	☒	☐	☐	☐	☐	☐	☐	☐	☐	(13)
	Intergranular SCC—PWR primary system: austenitic stainless steel	☐	☐	☐	☒	☐	☐	☒	☐	☐	☒	☒	☒	(14)
	Primary water SCC—alloy 600/82/182	☒	☒	☐	☒	☐	☐	☐	☒	☐	☐	☐	☐	(15)

	Strain-induced corrosion cracking	☐	☐	☒	☐	☐	☐	☐	☐	☐	☐	☐	☐	(16)
	Transgranular SCC	☒	☐	☐	☐	☐	☐	☐	☐	☐	☒	☐	☐	(17)
Overstressed	Hydrogen deflagration	☐	☐	☐	☐	☐	☐	☒	☐	☐	☐	☐	☐	
	Hydrogen embrittlement	☐	☐	☐	☐	☒	☐	☐	☐	☐	☐	☐	☐	
	Steam hammer/water hammer	☒	☐	☐	☐	☐	☐	☐	☐	☐	☐	☐	☐	

Notes: (1) A first corrosion under insulation-related (CUI) standard in the United States was issued in 1971: ASTM-C692 (standard test method for evaluating the influence of thermal insulations on external stress corrosion Cracking tendency of austenitic stainless steel). (2) Beaver Valley Unit 1 (PWR) experienced a fire water system pipe break on 8 September 1982. The failure occurred in a section of belowground cast iron (ANSI 21.6) piping. (3) Big Rock Point (BWR) and Yankee Rowe (PWR) experienced multiple extraction steam line FAC failures in the period 1965–1970. (4) Horowitz documents the relevant US and international operating experience. (Horowitz, J.S. (2007), Investigation into Flow-Accelerated Corrosion at Low Temperatures, 1015170, EPRI, Palo Alto, CA.) (5) Horowitz documents a technical background to this issue. (Horowitz, J.S. (2006), An Evaluation of Flow-Accelerated Corrosion in the Bottom Head Drain Lines of Boiling Water Reactors, 1015072, EPRI, Palo Alto, CA.) In 2021, a US plant reported FAC wear in a DN50 reactor pressure vessel drain line. (6) Horowitz describes the underlying phenomenology and relevant operating experience. (Horowitz, J.S. (2007), Flow-Accelerated Corrosion - The Entrance Effect, 1013013, EPRI, Palo Alto, CA.). (7) A significant high-cycle fatigue failure occurred in Pickering 2 in December 1994. An instrument air pipe rupture caused a heat transport system relief valve to fail open and release primary water. A support system piping failure resulted in a primary system leakage (loss-of-coolant accident). (8) A mixing tee failure caused by intermittent mixing of multiple coolant streams at very different temperatures was reported at the Swedish Ågesta heavy water reactor in 1964. Un-isolable primary coolant leaks that were attributed to thermal stratification due to valve leak-through were reported in the period 1987–1988. (9) First reported in 1974 when the Dragon HTGR in the United Kingdom experienced a small helium coolant leak. The leak was caused by ECSCC in stainless steel piping leading to the helium purification plant. The first ECSCC failure of stainless steel piping in an LWR environment was reported in 1975. (10) Dresden 1 (BWR) reported intergranular SCC in DN150 piping in 1965. Several experimental and prototype reactors experienced intergranular SCC prior to 1965. According to a 1962 report (APED 4116) by W.R. Smith, Sr., and J.B. Violette, the failure of austenitic stainless steel recirculation piping at the Vallecitos Boiling Water Reactor (VBWR) was among the first recorded instances of pipe cracking in a nuclear power plant. (11) First reported in 1969 (Type 304L) and in 1984 (Type 316L). (12) First reported in Germany in 1980 (KKI Isar-1). (13) Duane Arnold, June 14, 1978. A discovery of a primary system leak in the spring of 2025 is of particular interest. Oskarshamn Unit 3, commissioned in 1985, reported a through-wall crack (TWC) in a dissimilar metal weld. The particular location had been subjected to in-service inspection every 3 years. A minor flaw was recorded during the annual refueling outage in 2020. A fitness-for-service (FFS) evaluation found the flaw to be acceptable for continued operation of the plant. During the in-service inspection in 2023, the 3 mm deep flaw was again subjected to a formal FFS and again deemed to be acceptable. The dissimilar metal weld connects a 300 mm OD ECCS stainless steel pipe (16 mm wall thickness) to a reactor pressure vessel nozzle. During the 2025 refueling outage, a drywell inspection found a through-wall leak in the location. The IGSCC flaw had grown in size from 3 to 16 mm in about 3 years. This was an FIK event in Sweden. (14) First reported in 2010 in China. In August 2020, Ohi Unit 3 in Japan reported rejectable weld flaws in a section of pressurizer spray piping. IGSCC was discovered in Safety Class 1 LPSI and RHR piping at multiple PWR

(continued)

units in France in 2021–2025. Additional, relevant information can also be found in References, which summarizes the Chinese operating experience. Some "precursor events" were identified already in 1983 (France) and 1996 (USA), however. (15) Primary water SCC in steam generator tubes was first detected in 1972 in the German plant Obrigheim (KWO). Cracking in thick-section wrought Alloy 600 nozzles was first detected in PWR pressurizer nozzles in the 1980s and in large-diameter RCS hot leg nozzles in the year 2000. Stress corrosion cracking of Alloy 600 was reported at the Ågesta Nuclear Power Plant Reactor in 1966 (Grönwall et al. 1966). (16) Strain-induced corrosion cracking events reported in Germany in the period 1975–1985. (17) Several experimental and prototype reactors experienced transgranular SCC in the 1960s. Extensive TGSCC found in the 2010–2014 timeframe in cold-formed safety class 1 bends in PWR low-pressure safety injection and residual heat removal piping

1959 and the broader recognition of the problem many years later is referred to as the "*Couriou effect*" (Féron and Staehle 2016; Féron et al. 2019).[2]

Intergranular stress corrosion cracking (IGSCC) in boiling water reactor (BWR) environments was first observed in the 1960s and became more prevalent in the 1970s. The main factors contributing to IGSCC in BWR non-sensitized stainless steel include cold working, which introduces stresses and strains during pipe fabrication at or near room temperature.

Welding may lead to sensitization of the weld heat-affected zone and produce high residual stresses. Exposure to oxidizing primary water makes both non-sensitized and sensitized materials susceptible to IGSCC. Weld residual stresses refer to internal stresses that persist within a welded structure after the welding and cooling processes are complete, without any external forces applied. These stresses primarily result from the localized heating and cooling cycles associated with welding, which cause differential thermal expansion and contraction of the material, creating internal forces as the weld cools and solidifies. IGSCC was not generally seen as an issue affecting austenitic stainless steel piping in the PWR primary system operating environment, which is nonoxidizing. Initial concerns were raised by French material scientists in the 1980s, however, and confirmed by more recent, 2010–2025, operating experience. Cracking in stainless steel components in PWRs has been attributed to the presence of anionic contaminants such as chloride and sulfate together with trapped air bubbles in reactor coolant system piping dead legs after refueling outages.

4.3 Historical Perspectives on Material Degradation Mechanisms

Extensive piping material performance information exists, and it covers operating experience that has been generated over a period of about six decades. Examples of how the piping operating experience (OE) has evolved are shown in Tables 4.2, 4.3, and 4.4.[3]

[2]The history of the "Coriou effect" is told by R.W. Staehle. The work of Coriou "started with showing that the alloy common at the time, Inconel (Alloy 600), was not suitable for use despite the popular view that this was the right material. He was opposed by the International Nickel Company (INCO) and their colleagues. By 1968, their opposition withered, and the community accepted Coriou's findings, which he had organized a very substantial program to support. INCO organized a program that was weak technically and of little merit. At the same time, Coriou developed Alloy 800, a greatly superior alloy. Despite Coriou's strong and valid opposition, reactor programs of France, the United States, and Japan selected Inconel for its material for tubing in steam generators; too late ..." https://doi.org/10.1016/B978-0-08-100049-6.00001-X.

[3]These compilations were prepared in the first quarter of 2025.

Table 4.2 Pipe failure state of knowledge in 1977 vs. today's knowledge base

Damage/degradation mechanism	Plant type	Basins and burns (EPRI 1977) No. of failures	% of total	2025 OE perspective No. of failures	% of total (%)
Corrosion	BWR	6	3.00%	1128	8.40
	PWR	0	0.00%	2498	18.61
Corrosion fatigue	BWR	–	–	7	0.05
	PWR	–	–	60	0.45
Erosion	BWR	9	4.40%	9	0.07
	PWR	5	2.50%	10	0.07
Erosion-cavitation	BWR	–	–	119	0.89
	PWR	–	–	202	1.50
Erosion-corrosion	BWR	–	–	218	1.62
	PWR	–	–	467	3.48
Fabrication and manufacturing defects	BWR	13	6.40%	112	0.83
	PWR	13	6.40%	212	1.58
Flow-accelerated corrosion	BWR	–	–	672	5.01
	PWR	–	–	1576	11.74
Flow-induced vibration and fretting	BWR	–	–	47	0.35
	PWR	–	–	73	0.54
Jet impingement erosion—water droplets in steam: low alloy steel and stainless steel	BWR	–	–	36	0.27
	PWR	–	–	41	0.31
Low-cycle fatigue	BWR	–	–	100	0.74
	PWR	–	–	201	1.50
Stress corrosion cracking—Ni-base alloys and stainless steel	BWR	20	9.90%	1872	13.95
	PWR	3	1.50%	931	6.94
Thermal fatigue	BWR	9	4.40%	104	0.77
	PWR	2	1.00%	164	1.22
Unknown or not specified	BWR	41	20.20%	0	0.00
	PWR	31	15.30%	0	0.00
Vibratory fatigue—high cycle (acoustically/mechanically induced)	BWR	23	11.30%	827	6.16
	PWR	28	13.80%	1688	12.58
Water hammer—hydraulic transient: leaks and ruptures	BWR	–	–	21	0.16
	PWR	–	–	28	0.21

4.4 Pipe Damage and Degradation Mechanisms

Pipe failures are attributed to material degradation mechanisms and damage mechanisms. Pipe failure occurs due to synergistic effects involving operating environments and loading conditions. The operating experience is synthesized in Fig. 4.1. The pipe failures are of two types:

Table 4.3 Evolution of the pipe failure state of knowledge

	Janzen (1981)		2025 state of knowledge 1960–2025 operating experience			
Nominal pipe diameter (mm)	Number of occurrences	Percent (%)	Number of occurrences (all safety classes)	Percent	Safety Class 1—RCPB: number of occurrences	Percent
ø ≤ 25	75	59.50	3674	24.94%	702	24.29%
25 < ø ≤ 50	5	4.00	1871	12.70%	202	6.99%
50 < ø ≤ 100	9	7.10	2122	14.41%	270	9.34%
100 < ø ≤ 150	2	1.60	1210	8.21%	234	8.10%
150 < ø ≤ 250	1	0.80	1827	12.40%	115	3.98%
250 < ø ≤ 550	9	7.10	2880	19.55%	952	32.94%
>550	5	4.00	1146	7.78%	415	14.36%
Unknown	20	15.90	–	–	–	–
Total	126	100.00	14,730	100.0%	2890	100.0%

1. **Failures attributed to time-dependent environmental degradation**: Defined by the conjoint requirements that include operating environment, material, and loading conditions. These conjoint requirements differ across the different types of piping system designs and are influenced by routing, material, diameter, wall thickness, method of construction/fabrication, etc. Similarly, pipe flaw incubation times and flaw growth rates differ across the many combinations of degradation susceptibility and operating environment. Nondestructive examination techniques usually detect the onset of material degradation before a crack propagates through a pipe wall.
2. **Event-driven pipe failure**: Mechanically stress driven and attributed to conditions involving combinations of equipment failures other than the piping itself, for example, loose/failed pipe support, leaking valve, failed pump bearing, spurious opening of a valve, and stress risers or unanticipated loading conditions (e.g., hydraulic transient or operator error that causes an inadvertent valve operation). Short-term degradation is sometimes used in lieu of event-driven degradation.

A failure refers to any condition that negatively affects operations and leads to pipe repairs or replacement, whether or not it causes a reactor shutdown. Nondestructive examination methods or metallographic analysis is used to identify non-through-wall defects, which are described by how deep a crack is compared to the pipe wall thickness, its orientation, and its length. Through-wall defects are defined by the shape and size of flaws that penetrate the pipe wall, as well as the resulting leak or flow rate. Downward arrows in Fig. 4.1 indicate possible combined effects from different damage and degradation mechanisms. For instance, weld defects like lack of fusion, slag inclusions, or small cracks not connected to the inner diameter of the pipe often serve as starting points for cracks that eventually lead to stress corrosion cracking (SCC) failures. Also, thermal fatigue can initiate cracks, while

Table 4.4 Pipe failure state of knowledge 1985 vs. 2025

Damage/degradation mechanism	Plant type	Bush (1985)	Operating experience as of 2025
Stress corrosion cracking	BWR	47.9%	13.5%
	PWR	6.3%	5.7%
Thermal fatigue	BWR	16.0%	0.7%
	PWR	7.7%	1.3%
High-cycle fatigue	BWR	1.7%	5.9%
	PWR	7.2%	12.8%
Erosion-corrosion	BWR	5.0%	1.6%
	PWR	1.4%	3.3%
Erosion-cavitation	BWR	–	0.8%
	PWR	–	1.6%
Flow-accelerated corrosion	BWR	–	4.9%
	PWR	–	13.0%
Flow-induced vibration/fretting damage	BWR and PWR	–	0.9%
Jet impingement erosion	BWR and PWR	–	0.5%
Corrosion fatigue (environmental degradation)	BWR	1.1%	0.1%
	PWR	1.1%	0.4%
Corrosion (multiple mechanisms)	BWR	–	7.4%
	PWR	–	18.2%
Low-cycle fatigue	BWR and PWR	–	2.4%
Overstressed/water hammer	BWR and PWR	–	2.5%
Construction defects	BWR and PWR	–	2.4%
Unknown cause	BWR	2.8%	–
	PWR	1.9%	–
	Total no. of failures	363	14,730

Bush used the term "erosion-corrosion" in lieu of flow-assisted degradation, which accounts for four different types of wall thinning mechanisms: (1) erosion-corrosion, (2) erosion-cavitation, (3) flow-accelerated corrosion, and (4) flow-induced vibration, which can cause fretting damage to outside pipe wall

an SCC process may cause these cracks to grow through the pipe wall. The color intensity of the horizontal bars reflects the number of events observed: that is, strong colors represent multiple incidents, while weak colors show very few, if any, significant structural failures.

Proneness to pipe degradation and failure is illustrated in Fig. 4.2 in which "proneness" is measured as the ratio of degradation mechanism (DM) i to the number of intergranular stress corrosion (IGSCC) failures prior to 1989 ("unmitigated IGSCC"). Why "prior to 1989"? Section 4.8.1 summarizes regulatory and industry response to IGSCC failures and the processes that were implemented to mitigate IGSCC in BWR operating environments. Major progress had been made by 1989 in regard to proactive material degradation monitoring.

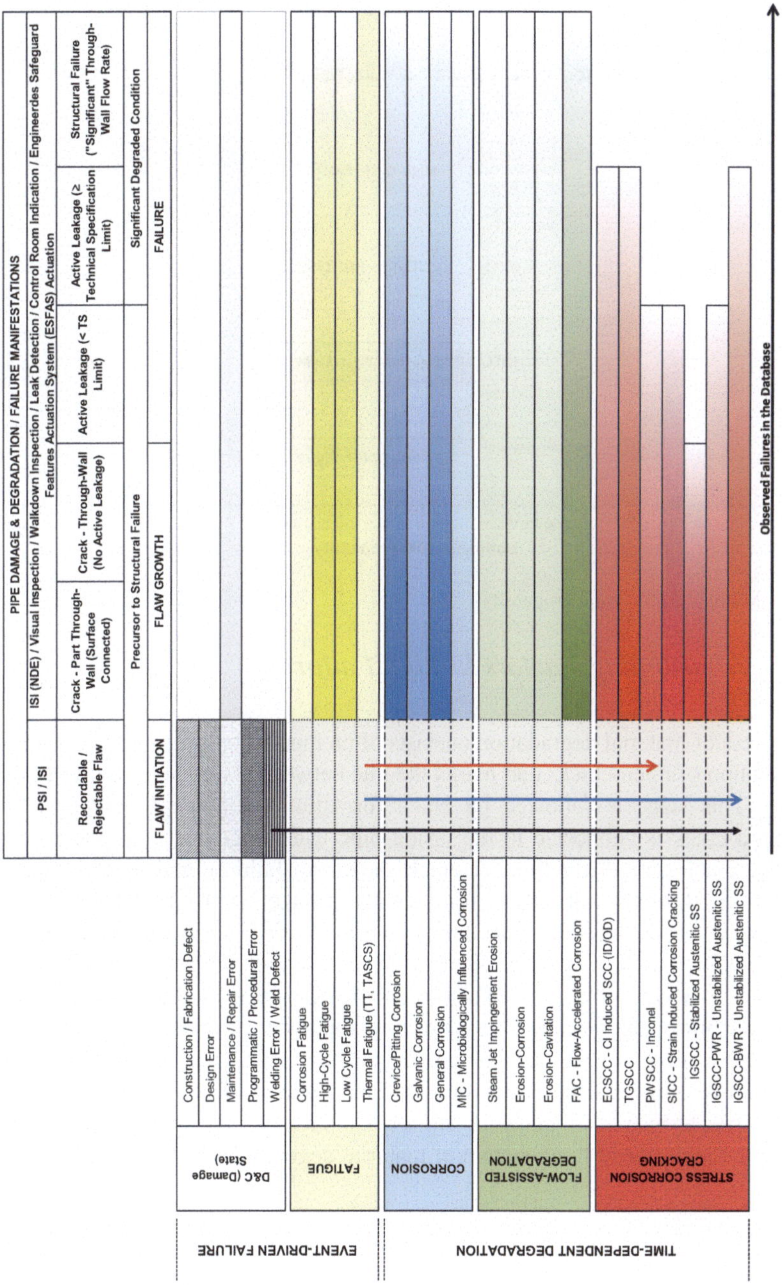

Fig. 4.1 Schematic of pipe failure manifestations

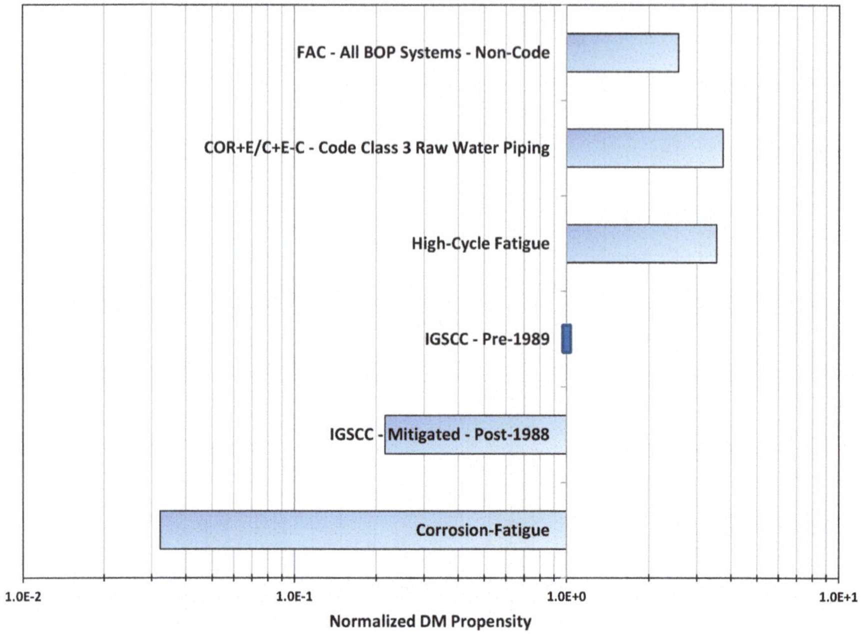

Fig. 4.2 Material degradation propensity

4.4.1 Contributing Factors to Pipe Failure

The process of material degradation consists of an incubation stage, which involves the formation of microcracks and marks the line between reversible and irreversible damage. This stage is followed by crack formation (or Stage 1 crack growth, meaning a crack is connected to the inside pipe wall, in contact with the process medium), crack growth (or Stage 2 crack growth to failure), and, if unmitigated, ultimately a through-wall crack and leakage (Stage 3). CORROSIONPEDIA (https://www.corrosionpedia.com/) defines the incubation period as "*the total time that a certain structure goes through crack nucleation until the time the crack progresses to a critical degree.*" Factors influencing material degradation include cold deformation and forming. Welding, heat treatment, machining, and grinding can introduce residual stresses. The residual stresses set up because welding operations tend to approach the yield strength. The buildup of corrosion products in confined spaces can also generate significant stresses. The following sections describe the many underlying factors of material degradation and failure.

4.4.2 *Design and Construction Defects*

Root cause analyses demonstrate that human error, encompassing organizational factors, is a significant contributor to pipe failures. Industry data reveals that between 20% and 90% of incidents involve some form of human error, which may be latent or active in nature. Latent errors—such as design deficiencies or incomplete in-service inspection (ISI) programs—can remain undetected for extended periods before manifesting as issues. Conversely, active errors, such as improper system restoration following maintenance, tend to have immediate consequences (e.g., water hammer). Research has been conducted to evaluate the extent of human error in pipe failure, identifying its sources in system design, fabrication and welding processes, and suboptimal operating procedures (Taylor 1974; Hattori et al. 1988; Bellamy et al. 1989).

Hurst et al. (1991) analyzed pipe failures in the chemical process industry and showed that "operating error" was the largest immediate contributor to piping failure (30.9% of all known causes). Overpressure (20.5%) and corrosion (15.6%) were the next largest categories of known immediate causes. The other major areas of human contribution to immediate causes were human-initiated impact (5.6%) and incorrect installation of equipment (4.5%). The total human contribution to immediate causes was therefore about 41%. For the underlying causes of pipe failure, maintenance (38.7%) and design (26.7%) were the largest contributors. About 90% of all failures could have been prevented through adequate reliability integrity management processes.

Service-induced material degradation is a result of synergies among material properties, loading (e.g., stress riser), and environmental conditions. Through-wall pipe flaws involve initiation and incubation. That is, a preexisting flaw acting as a stress riser (e.g., an embedded slag inclusion in a weld) or exposed to an adverse environment eventually progresses to a through-wall defect or non-through-wall defect that is connected to the inside surface of a pipe. Most pipe flaws that result in a corrective action are attributed to a readily identifiable active degradation mechanism or an off-normal loading condition. A relatively small subset of all recordable or rejectable pipe failures involves a preexisting defect that grows over a long time and is detected through a visual examination or liquid penetrant testing. A design-and-construction (D&C) event sequence diagram shows an example of how to classify weld flaws for which no active degradation mechanism is present (Fig. 4.3).

Assuming that a preexisting flaw is discovered during an ISI but remains unmitigated (i.e., no repair is performed), then crack growth may occur given that it is subjected to an adverse operating environment (e.g., high temperature, corrosive conditions, subject to high stresses) and the material is susceptible to degradation. The potential of through-wall cracking would be high if the conjoint requirements for degradation are met, indicated as "through-wall cracking Potential-H." In developing an initiating event frequency model, precursor rates and conditional failure probabilities are derived for the evaluation boundaries considered. The estimation process considers all credible damage and degradation mechanisms. For locations

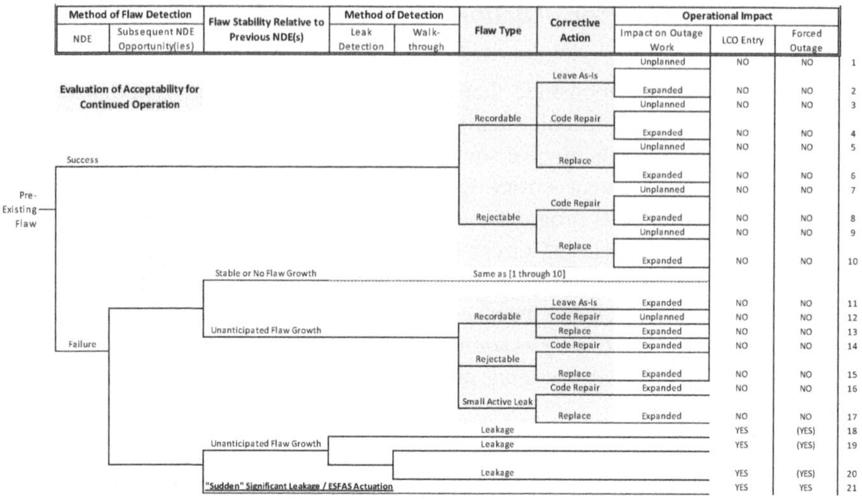

Fig. 4.3 D&C event sequence diagram

without any readily identifiable damage or degradation susceptibility, an assumption is made that a preexisting weld flaw may exist and eventually grow in the through-wall direction. The flaw growth mechanism is termed low-cycle fatigue and pressure loading and accounts for the effects of normal operation including cooldown and heat cycles. Therefore, a unique conditional failure probability model is needed to resolve the low-cycle fatigue analysis cases. Event sequences 16 through 21 represent a "high likelihood" that a preexisting defect eventually grows and penetrates the outside pipe wall.

4.4.3 Welding Processes

Mechanical and metallurgical properties of pipe welds are influenced by welding parameters, including the type and speed of welding employed. Differential thermal expansion and contraction between the weld metal and base material generate residual stresses within the weldment. In highly constrained scenarios—common in most structural applications—these stress levels may approach the yield strength of the material. The presence of tensile weld residual stresses (WRS) is a prerequisite for fatigue and stress corrosion cracking, as crack initiation can occur after sufficient load cycles at elevated tensile stress. Operational experience provides valuable qualitative insights into how welding processes contribute to material degradation.

Operating experience shows a distinction between field and shop weld failures, raising questions about their relative reliability. Assessing the factors impacting welding quality is complex. Shop welds undergo solution heat treatment to minimize sensitization and residual stress, possibly making them more reliable than field

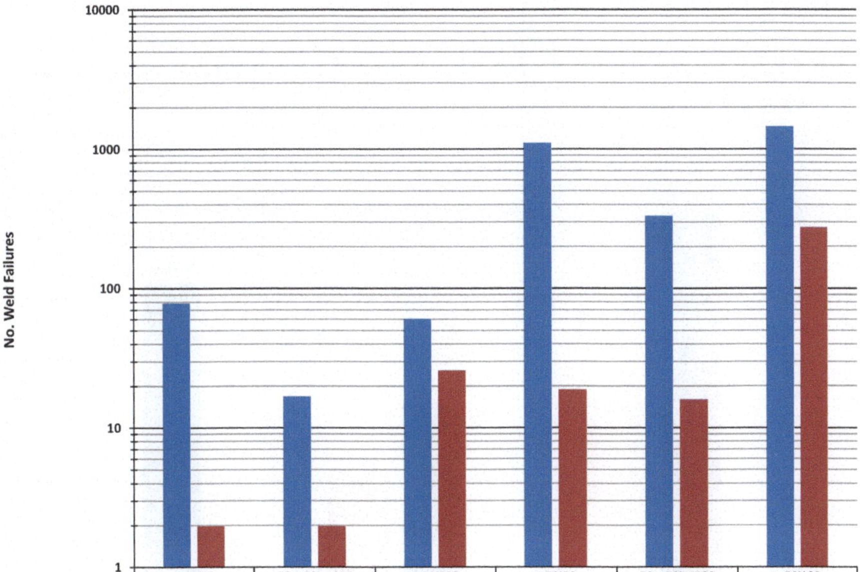

	≤ DN50	50 < DN ≤ 100	> DN100	≤ DN50	50 < DN ≤ 100	> DN100
		No Active DM			Active DM	
Weld - Field Welded	79	17	61	1106	333	1455
Weld - Shop Welded	2	2	26	19	16	277

Fig. 4.4 BWR primary system field and shop weld operating experience

welds. Figure 4.4 summarizes field and shop weld performance; in BWR plants with external recirculation loops, about 40% of welds are shop welded. Studies on IGSCC in RBMK plants indicate that vertical pipe welds have higher failure rates than horizontal ones due to higher heat input during construction, increasing sensitization and residual stress (Nedzinskas and Khmagauskas 2003) (Fig. 4.5).

4.4.4 Sensitization of Austenitic Stainless Steel

Sensitization refers to the susceptibility of materials to intergranular corrosion and comes about by a combination of factors in the materials' composition and treatment. Contributing factors can include heating and cooling through specific temperature ranges, increased C content, or decreased Cr content, among other factors. An alloy is said to be "sensitized" if it is more susceptible to intergranular corrosion than a non-sensitized sample of the same alloy. Atkinson and Van Droffelaar (1982) made the following observations on the role of sensitization:

- Sensitization occurs when the alloys are slowly heated or cooled through the range of approximately 500–800 °C.
- Fast cooling through this temperature range does not result in sensitization.

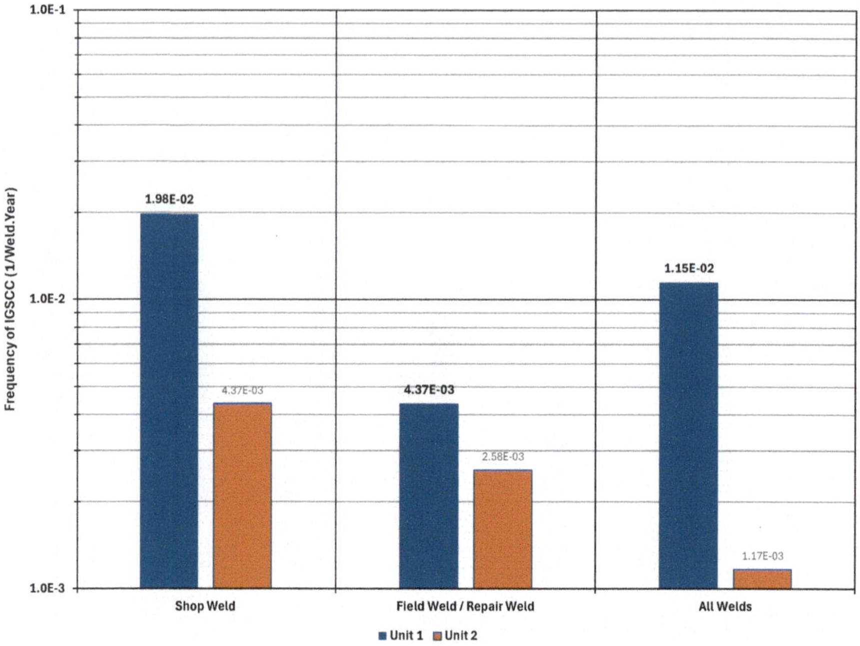

Fig. 4.5 RBMK primary system field and shop weld operating experience

- Quenching from elevated temperatures followed by reheating in the range of 500–800 °C causes the steel to become sensitized.
- The degree of sensitization increases markedly with increasing carbon content and to a lesser extent with decreasing chromium content.
- The degree of sensitization varies within the sensitization temperature range.
- The degree of sensitization also increases with time at the sensitization temperature.
- The prominent feature of sensitized austenitic stainless steel is the presence of precipitates of chromium carbides at the grain boundaries.

A steel which has been sensitized may appear essentially unaffected, yet it will be found to possess reduced strength and corrosion resistance which, if severe enough, will cause the steel to lose its characteristic metallic ring if struck with a metal object, while bending or twisting will produce cracks. The most popular explanation for the sensitization phenomenon is the chromium depletion theory. Intergranular corrosion occurs along the grain boundaries of a metal (Miller and Boulton 1982). Sensitization resulting from isothermal exposure is normally represented by time-temperature-sensitization (TTS) diagrams, which are plots of aging time versus the temperature necessary for sensitization. These are "C"-shaped curves, which separate the sensitized and non-sensitized regions. The TTS diagrams show the time required for isothermal sensitization at various temperatures and can be used to solve problems such as the selection of conditions of annealing, which will not result in sensitization. The nose of this curve specifies the critical temperature at which the minimum time, t_{min}, is required for sensitization.

4.4.5 Hot Cracking

"CORROSIONPEDIA" (https://www.corrosionpedia.com) defines hot cracking (or micro-fissuring) as the formation of shrinkage cracks during the solidification of weld metal. These cracks are known as hot cracks because they occur immediately after welds are completed and sometimes while the welds are in progress, when the weld metal tends to solidify from the corners of the base metal to which it is joined. As the solidification proceeds, the low melting eutectics are concentrated in the center and remain liquid, which are then torn apart by the stress associated with the welding, resulting in a centerline crack which compromises the weld integrity. Hot cracking occurs when the available supply of liquid weld metal is insufficient to fill the spaces between solidifying weld metal, which are opened by shrinkage strains. In summary, hot cracking occurs when:

- The strain on the welding pool is too high.
- Liquid cannot reach the regions where it is needed due to inadequate supply or blockage/narrow channels between solidifying grains.
- The temperature is >1200 °C.
- Impurities such as carbon, sulfur, or phosphorus are present.

To control hot cracking, the following principal factors need to be manipulated:

- Weld metal composition.
- Weld solidification pattern.
- Strain on the solidifying weld metal.
- Adding manganese, chromium, or molybdenum.

4.4.6 Weld Repair and "Repair Welds"

Weld quality can be expressed in terms of the shape, size, location, and frequency of the defects or discontinuities present, as well as in terms of mechanical and metallurgical properties. Discontinuities in welding are normally classified from the fabricator's point of view as one of five major categories:

1. Crack or crack-like.
2. Geometric.
3. Lack of fusion/penetration.
4. Porosity.
5. Slag.

Crack-like fabrication defects normally result from unsuitable materials and/or welding procedure and are exacerbated by poor workmanship. Solidification cracking is caused by high thermally induced strain acting on insufficiently ductile weld metal, but this is very rare in structural steelwork nowadays. Hydrogen cracking can occur in the heat-affected zone (HAZ) or weld metal, the former being the more common. It is caused by hydrogen diffusion from contaminated weld metal, which embrittles the microstructure to such an extent that only a low level of strain results in fracture.

Geometric discontinuity in this context is a sudden change of shape or a surface irregularity. Weld profile is usually considered as a geometric discontinuity, although most are inherent in the design, for example, sudden changes of section at a weld joint, or the provision of a permanent backing bar beneath a single-sided butt weld. Misalignment is a source of geometric discontinuity not directly associated with the weld, as is angular distortion. Weld profile discontinuities can be classified as:

1. Undercut.
2. Concavity or convexity.
3. Excessive (or insufficient) reinforcement.
4. Poor reinforcement angle.
5. Overlap.
6. Burn-through.
7. Shrinkage.
8. Surface irregularity.

Irregularities such as lack of fusion or lack of penetration are usually planar; they differ from cracks in that their extremities are relatively blunt. Their nature and appearance are self-evident, and they are both indicative of incorrect welding procedure, poor workmanship, or poor joint design. Lack of penetration can be deliberate, as in a partial penetration butt weld. Lack of sidewall fusion can be through-thickness, especially in a single-pass weld. Lack of inter-run fusion is associated with multi-pass welds.

Slag inclusions occur predominantly in multiple-pass welds and may be intermittent or continuous. This type of discontinuity is largely process controlled and is influenced particularly by choice of flux/electrode and weld geometry. The former influences the formation of slag, and the latter influences its detection and removal. For example, a well-rounded weld bead in a deep, narrow preparation is much more likely to trap slag along the weld toes than a flatter weld bead in a more open joint. The presence of buried slag is often indicative of poor workmanship, because although the formation is a function of process and consumables, most slag should normally be removed by the operator before the next weld pass is made.

Porosity results from gas in the molten weld metal failing to escape completely to the surface. The formation of the gas usually arises from the presence of contaminants on either the consumable or the metal surfaces and from failure of shielding gas in gas metal arc (GMA) processes or loss of flux (submerged arc or SMA processes).

As such, it is a result of inadequate cleanliness and is indicative of poor welding procedure or workmanship.

In-service-related stress corrosion cracking in welds has occurred in areas where repairs were performed during initial fabrication and construction. Such welds are oftentimes referred to as "repair welds." Weld repair of benign weld flaws often contributes to in-service cracking due to high local weld residual stress and/or sensitization. Welding residual stress develops as the nearby material pulls back to maintain a bond with the shrinking weld material. There are no ASME B31.1 or Section III Code requirements limiting the number of repairs. Some plant owners/ architect-engineers have limited localized repairs to two and consider full cutouts to be new welds.

The United Kingdom Welding Institute has published results of a study of the effects of repair welding on pressure boundary integrity (Slater 1983). It was in the form of an extensive literature survey to establish the knowledge base at the time and included reviews of operating experience from nonnuclear industries. Examples of recent multi-physics studies based on finite element analysis and simulation of weld stresses are documented in three technical papers (Hu et al. 2020; Luo et al. 2020; Lee et al. 2021). These studies addressed the effects of heat input and weld height on SCC susceptibility. The effect of "repair welding" on residual stress is often greater than that of the initial welding, and the smaller the repair length, the greater the transverse residual stress. During welding, localized heat is applied, causing rapid heating and cooling. This leads to uneven expansion and contraction, resulting in plastic deformation and residual stress.

4.4.7 High-Density Polyethylene Piping Material

High-density polyethylene (HDPE) is used in nuclear power plants for belowground applications like safety-related service water systems and firewater systems because of its high resistance to corrosion and erosion. These pipe materials fail by one of the three modes: (1) ductile rupture, (2) slow crack growth failures, or (3) rapid crack propagation failures.

Ductile rupture failures occur due to the presence of high internal pressures. The failure mode is manifested in large, localized plastic permanent deformations of the pipe wall (Krishnaswamy 2005). They occur as a result of increasing pressure to levels greater than 2.7 MPa. Increasing pressures cause the pipe to undergo large diametric expansions, typically resulting in wall thinning and stretching until a point where the remaining wall ligament is not sufficiently large to withstand the induced high circumferential hoop stresses.

Slow crack growth (SCG) failures occur over long periods of time (decades) at relatively low loads below the yield point of the material and are characterized by brittle (slit) fractures, which exhibit very little material flow or deformation (Villabon et al. 2024). Using high-magnification scanning electron microscopy, surface morphology of an SCG fracture can be characterized. Examinations show

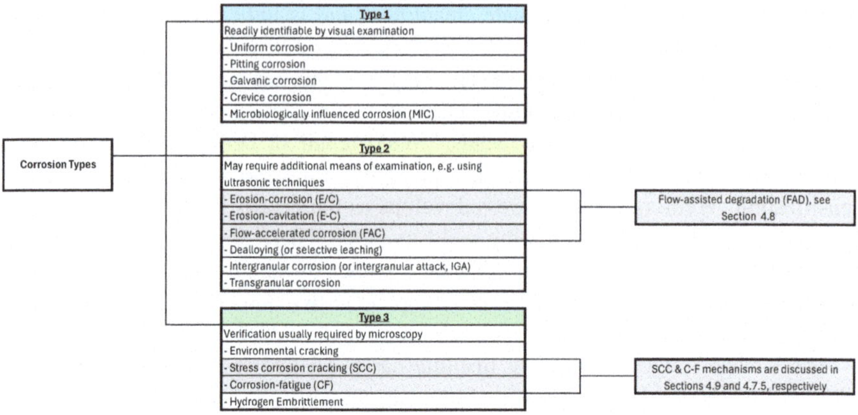

Fig. 4.6 Three types of corrosion mechanisms

that cracks initiate and radiate from an initiation point. Initiation points are stress risers caused by inclusions, contaminants, scratches, defects, cavities, dimples, high stress risers, etc. SCG failures grow stepwise and are associated with the sequential formation and fracture of the craze damage zone formed ahead of the crack tip.

Rapid crack propagation (RCP) failures are manifested in the form of a large-scale brittle crack that propagates at high speeds over a long span of pipe (Krishnaswamy et al. 2006). RCP failures could be catastrophic due to the rapid release of high volumes of gas over long spans. In order for RCP to occur, a "critical" initial axial notch must exist in the pipe wall, and the driving force has to exceed the dynamic fracture resistance of the material. The dynamic fracture resistance decreases with decreasing temperatures making PE materials more susceptible to RCP ruptures at lower field temperatures. The susceptibility to RCP also increases with increasing field service pressure, increasing pipe diameter, increasing standard dimension ratio (SDR), and decreasing modulus of elasticity of the material. The SDR is the ratio of pipe outside diameter to its minimum wall thickness.

4.5 Corrosion Mechanisms

Corrosion is a naturally occurring phenomenon defined as the deterioration of a metal that results from a chemical or electrochemical reaction with its environment. Corrosion takes many different forms and is affected by numerous environmental factors. The Association for Materials Protection and Performance (AMPP)[4] defines three types of corrosion (Fig. 4.6):

[4]NACE and the Society for Protective Coatings merged in 2021 to form the Association for Materials Protection and Performance (AMPP).

- Type 1: Readily identifiable by visual examination. Corrosion mechanisms of type 1 include uniform corrosion, crevice corrosion, and microbiologically influenced corrosion (MIC).
- Type 2: May require additional means of examination, e.g., nondestructive examination. Corrosion mechanisms of type 2 include erosion-corrosion (E/C), which is a form of flow-assisted degradation (FAD).
- Type 3: Verification is usually required by microscopy in the laboratory. Corrosion mechanisms of type 3 include stress corrosion cracking (SCC) and corrosion fatigue (CF).

The modes of corrosion, which can result from microbiologically produced local environments, include general corrosion, pitting, crevice corrosion, dealloying, galvanic corrosion, intergranular corrosion, and corrosion fatigue. Slowly flowing environments tend to favor the growth of microbes by providing a ready supply of nutrients (in particular, periodically flushed systems). Stagnant areas adjacent to flowing areas are particularly susceptible to microbiologically influenced corrosion. Corrosion of carbon and low-alloy steel components by leaking borated water has caused significant corrosion problems for many PWR plants. The mode of corrosion of greatest concern due to leakage of borated water is uniform corrosion of the metal surface, referred to as "wastage."

4.5.1 Crevice Corrosion

Crevice corrosion occurs at locations where oxygen cannot freely circulate, such as tight joints or between two surfaces, at least one of which is metal. The gap or crevice can be formed between two metals or a metal and nonmetallic material. Environmental conditions in a crevice can be significantly different than on adjacent bare metal surfaces. Under many conditions, the environment in the crevice can be more aggressive than outside the crevice, and accelerated local corrosion can take place.

Geometrical crevices can exist in socket welds or between pipe metal and protective liners. Chloride salts, pollutants, and moisture from the environment accumulate in the crevice. The environment inside the crevice becomes depleted of oxygen, enriched in chlorides, and acidified, which promotes the breakdown of the passive film and anodic dissolution. Crevice corrosion is also referred to as localized attack on a metal surface at or immediately adjacent to the gap or crevice between two joining surfaces.

4.5.2 Microbial Corrosion

Process piping in a raw water environment is susceptible to microbial or microbiologically influenced corrosion (MIC) (EPRI 1988a, b). This mode of corrosion is common in low-temperature systems such as firewater, service water, and circulating water systems and typically occurs in two general locations. One is on external surfaces where there is moisture and other materials, such as organic debris buildup, which contains nutrients suitable for bacterial or fungal growth. The second occurs on internal surfaces in low-temperature components, primarily those where water is flowing slowly or is periodically flushed; both situations provide a good supply of nutrients for microbiological activity and growth.

Stagnant or intermittent-flow conditions, as in the case of emergency diesel service water supply headers, are conducive to the growth of microorganisms that can accelerate corrosion rates. Service water supply lines to emergency diesel generators are stagnant because motor-operated isolation valves are normally maintained shut (except during monthly surveillance testing). Crevices such as those in piping welds that lack penetration can enhance microbiologically influenced corrosion attack by giving a place for deposits and, therefore, for the bacteria to collect. Microbial films form when aerobic species, such as iron-oxidizing bacteria, create anaerobic conditions underneath them for microorganisms, such as sulfate-reducing bacteria, to accumulate at the metal surface. Sulfate-reducing bacteria attack the metal surface, produce corrosive chemicals, and cause deep pitting. Microbiologically influenced corrosion on carbon steel will increase general corrosion, through-wall pitting, and formation of tubercles. Tubercles consist of corrosion products, microbes, and debris. Tubercle growth could restrict cooling water flow to equipment.

Stainless steel piping is not immune to microbiologically influenced corrosion because microbes can attack at the weld heat-affected zone (HAZ) in stainless steels when this zone becomes sensitized. Microbiologically influenced corrosion can also damage metals lined with polymeric materials, typically at coating imperfections. Invariably, MIC failures of stainless steel welds result in pinhole-type minor leaks. It may be necessary to replace materials if microbiologically influenced corrosion severely damages them or where mitigation measures cannot bring the system condition under control. Possible alternatives include replacing carbon steel with stainless steel or replacing stainless steel with more resistant materials, such as 6% molybdenum stainless steels, nickel-base alloys, titanium, or nonmetallic materials such as high-density polyethylene (HDPE) (Henriksson 1981; Krishnaswamy 2007).

4.5.3 Selective Leaching

Selective leaching, also known as dealloying, is a corrosion process that selectively removes one or more elements in an alloy, leaving behind a porous structure composed of the remaining constituents or their corrosion products. Piping materials susceptible to this form of degradation mechanism include gray cast iron, copper alloys with greater than 15% zinc, and aluminum bronze when exposed to treated water and raw water. In service water piping of aluminum bronze materials, "weeping" leaks have been observed in small- and large-diameter pipe fittings such as tees and in welds with backing rings (STP 2016). The latter are strips of metals rolled up to fit inside a pipe and span the gap between pipe ends or between a pipe end and a cast valve body or cast tee.

The safety-related (ASME III Class 3) service water system piping at some plants is made of aluminum bronze material. This material has high yield and tensile strength and is resistant to erosive wear. Between the time of commissioning of these reactor units and the present day, the plants have experienced on the order of one hundred "weeping leaks" mainly in smaller-diameter piping. These through-wall defects were initiated by preexisting weld defects and propagated by a dealloying phenomenon. Engineering evaluations performed by the plant operator indicate that this form of material degradation is slow and that rapid flaw propagation is very unlikely. Laboratory failure analysis data has indicated that in cases that leaked, a preexisting crack penetrating into the central core of the weld was present. Poor fit-up may have contributed to such a root pass cracking. Crack growth appears to have occurred by a process of the crack tip dealloying locally and the crack propagating through the dealloyed zone.

4.5.4 Operating Experience Insights

Service water (SW) piping systems carry cooling water to various heat exchangers throughout a power plant. It removes the heat from such auxiliary systems as component cooling heat exchangers, emergency diesel generators, containment coolers, lube oil coolers, room coolers, and chiller condensers. It consists of an intake structure for the cooling water, a distribution system within the plant, and an outlet structure. Suction is taken from the ultimate heat sink (UHS, which can be ocean, river, lake, or reservoir), heat is removed via various heat exchangers, and water is discharged back to the UHS or self-contained UHS (e.g., spray cooling pond or cooling tower). With an infinite supply of water, the SW systems represent an important potential flood source. Relative to the in-plant SW piping design, there are four basic types of SW system designs: (1) open-loop system, (2) open recirculating system with a self-contained ultimate heat sink such as spray cooling pond or a cooling tower, (3) closed-loop system, and (4) closed-loop system with a self-contained UHS. Each of the four system designs has advantages and disadvantages relative to SW pipe aging characteristics. The piping system and equipment such as heat exchangers are designed for design basis accident conditions. Because of diverse and redundant design and varying heat load demands, the piping operating environment is subject to intermittent or low-flow condition. It is a very large piping system consisting of many kilometers of piping of different sizes, and it "permeates" the entire nuclear power plant.

The open-loop system, or "straight-through" system, includes throttle valves to maintain flow velocities according to the various component temperature limits. In the closed-loop system, the component temperature limits are maintained by dedicated SW load heat exchangers. Relative to the open-loop system, the closed-loop system contains a larger number of components. Regardless of system design type, the SW piping extends throughout a plant running from the cooling water intake structure and through the turbine building, emergency diesel generator building, auxiliary building, reactor building, and containment. A pipe break anywhere inside the plant would result in UHS leakage into the buildings and subsequent spraying and/or flooding of adjacent equipment, including potentially vital, safety-related equipment areas. The safety significance associated with SW piping degradation and failures has warranted corrective actions to reduce both the frequency and potential consequences of operating events involving degradation and failures. The utility industry invests significant resources on SW piping reliability management, including water chemistry control, cleaning (e.g., slime removal), nondestructive examination (NDE), and piping replacement using improved materials to minimize or eliminate susceptibilities to degradation. The SW piping operating experience is extensive. The US SW pipe failure data is shown in Fig. 4.7, in which a "failure" corresponds to through-wall pipe leakage of varying magnitudes.[5]

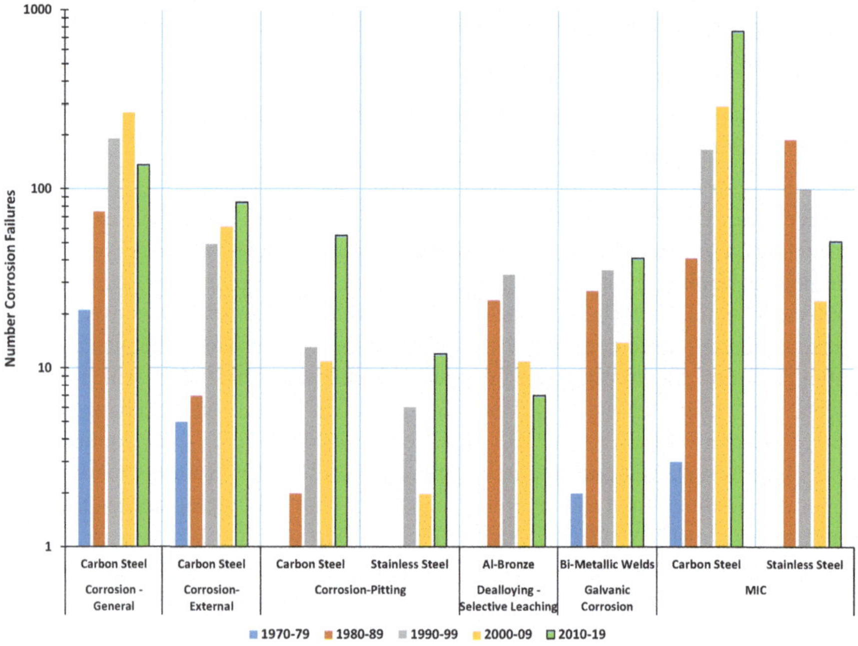

Fig. 4.7 US service water piping operating experience

[5]The failure data is not normalized against the population that produced failures, nor is it normalized against the BWR and PWR piping exposure terms, as in linear meter of piping or number of welds.

Raw water piping systems such as the SW piping system operate under corrosive and erosive service conditions. These systems are prone to degradation involving localized pitting and wall thinning and through-wall leaks across the plant population. Factors influencing the rate of degradation include the following:

- System design: For SW piping, the service conditions tend to be somewhat more aggressive in open or straight-through systems than the closed systems.
- Piping material selections: Examples of piping materials include the following:
  - Carbon steel: a predominant material type is ASTM A-106 Gr. B.
  - Stainless steel (Type 304): mainly used for small- to medium-diameter (≤DN200) piping. The weld heat-affected zones can be as susceptible to MIC as carbon steel.
  - "Super-austenitic" or "high-performance" stainless steels developed especially for saltwater systems (available under different trade names such as AL6XN® and 254 SMO®): these steels are virtually immune to corrosion.
  - Copper-nickel alloys, limited to small- and medium-diameter piping and fittings.
  - Lined carbon steel piping: many plants use cement-, rubber-, or epoxy/PVC-lined carbon steel piping where the service environment is erosive. Lining failure due to thermal cycling and improper welding frequently causes localized pitting and pinhole leaks.

4.6 Fatigue Mechanisms

Fatigue involves cracking of materials due to cyclic stress. It is a process of progressive localized permanent structural change occurring in materials subjected to conditions which produce fluctuating stresses and strains at some point or points and which may culminate in crack or complete fracture after a sufficient number of fluctuations. Four basic types of fatigue mechanisms are (1) vibration-induced high-cycle fatigue, (2) low-cycle fatigue, (3) thermal fatigue, and (4) corrosion fatigue (Fig. 4.8).

High-cycle fatigue (HCF) involves many repeated, low-stress cycles that cause failure primarily through elastic deformation, while low-cycle fatigue (LCF) involves fewer, high-stress cycles that result in significant plastic deformation and eventual fracture. The term LCF is used to characterize crack growth in the pipe through-wall direction through applied stress and normally occurring cooldown/heat-up cycles. An underlying assumption is that of a preexisting weld flaw attributed to original construction, fabrication, or welding defects missed by pre-service inspections and/or subsequent in-service inspections. Thermal fatigue involves temperature fluctuations that result in high-cycle fatigue. The three stages of fatigue failure are (IAEA 2014; Merikoski 2017; OECD/NEA 2023):

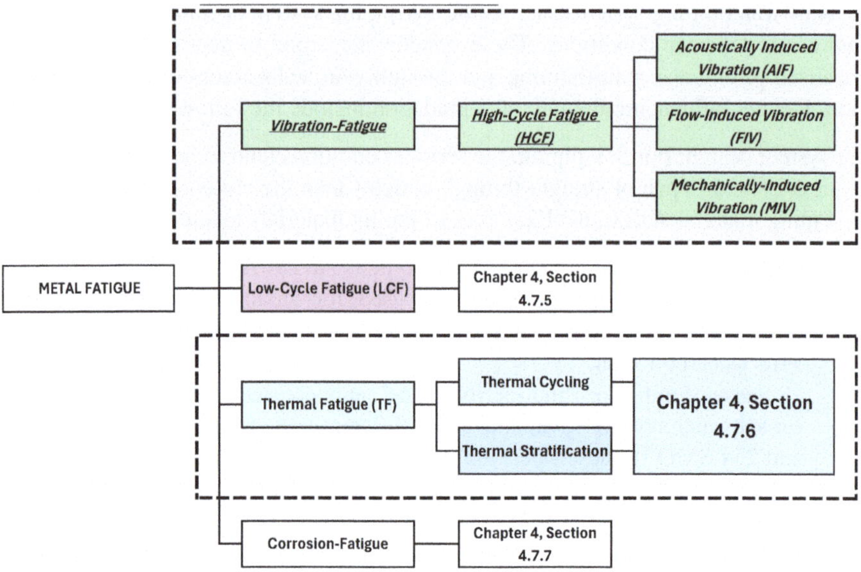

Fig. 4.8 The different forms of metal fatigue

1. **Initiation**: The initial fatigue crack occurs at localized discontinuities in the metal crystal structure. The generation and movement of discontinuities strengthen the metal through plastic deformation. This is referred to as work hardening. As the metal becomes work hardened, it loses its ability to plastically deform in the localized area where cyclic stresses are evident. Once the metal reaches its plastic deformation limit, the discontinuity becomes a small crack.
2. **Propagation**: Once a crack is initiated, continued cyclical stresses repeat the process, slowly growing the microcrack, which becomes a threat to structural integrity.
3. **Failure**: The final failure stage can be ductile or brittle, depending on the material, thickness, temperature, and applied stress. Fatigue-induced fracture surfaces of a broken component are typically smooth and show no evidence of plastic deformation.

Possible sources of piping vibration include compressors, pumps, vortex shedding near a valve, mixing of fluid streams at tees, reducing orifices, acoustic resonance, water hammer, external vibration transmitted through pipe supports, and two-phase process media. Thermal fatigue describes the cyclic loads leading to fatigue caused by temperature fluctuations. Thermal cycling can be the result of operating transients (low-cycle fatigue) or complex flow phenomena, such as swirl penetration thermal cycling (low-cycle fatigue) or thermal mixing (high-cycle fatigue). Thermal stratification refers to the phenomenon where fluids of different temperatures separate within a pipe, creating distinct layers with hotter fluid rising to the top and cooler fluid settling at the bottom.

The following conditions apply for an event to be classified as low-cycle fatigue: (1) no active environmental degradation mechanism can be identified, and (2) the root cause evaluation points to the presence of a weld flaw such as lack of fusion. Thermal fatigue is due to the cyclic stresses that result from changing temperature conditions in a component or in the piping attached to the component. Thermal fatigue may involve a relatively low number of cycles at a higher strain (e.g., plant operational cycles or injection of cold water into a hot nozzle) or a high number of cycles at low stress amplitude. High-cycle fatigue involves a high number of cycles at relatively low stress amplitudes (typically below the material's yield strength but above the fatigue endurance limit of the material). The crack initiation phase is dominant since crack growth is usually rapid. High-cycle fatigue may be due to acoustically or mechanically induced vibration or pressure pulses or due to flow-induced vibration (FIV). More than 2600 high-cycle fatigue failures were recorded between 1970 and 2025.

4.6.1 The Contributing Factors to High-Cycle Fatigue

Susceptibilities to high-cycle fatigue are functions of the adequacy of pipe supports and presence of an excitation source such as pressure pulsations in positive displacement or centrifugal pumps. While other descriptions are found, the following description is taken from IAEA-TECDOC-1736 (2014):

> *Inadequate piping support can induce fatigue in piping. The type and distance of piping supports greatly influence the natural frequency of a piping system. Tightly spaced restraints increase the natural frequency. Widely spaced restraints reduce the piping system's natural frequency. If an excitation source exists in the natural frequency range of the piping system, resonance may occur, and high-frequency fatigue damage may result. There are three main contributing factors to vibration fatigue: 1) mechanical excitation mechanisms, 2) cavitation, and 3) flashing.*

Flow-induced vibration (FIV) and cavitation are closely linked fluid dynamics phenomena where rapid fluid flow creates pressure fluctuations that can cause structural vibration, while cavitation involves vapor bubbles forming from low pressure and violently collapsing, generating significant noise, vibration, and loss of base metal (pitting). Cavitation itself is a type of FIV, but FIV also includes turbulence, vortex shedding, and water hammer, with both mechanisms causing stress, reduced efficiency, and potential failure in pumps, pipes, and hydrofoils, often exhibiting strong correlations in frequency and amplitude.

Flashing occurs when coolant temperature is higher than its saturation temperature at a given pressure and the coolant flashes into steam. This results in broadband pressure pulsations, causing vibration of the piping downstream of the flashing component and results in steam hammer phenomena. This is the reason that control valves regulating the flow of oversaturated coolant into a vessel kept at lower pressure are located as close as possible to the vessel entry nozzle, in order that flashing may occur in the larger volume of the vessel and not in the piping.

4.6.1.1 Acoustically Induced Vibration

Acoustically induced vibration is defined as vibrations induced by acoustic pressure waves generated by a noise source in the reactor, by a high-speed or unsteady gas flow, or by interaction of gas flows and solid objects. These pressure waves propagate in the fluid system with the speed of sound and are strong enough to cause unacceptable vibrations. If pressure pulsations happen to coincide with a structural frequency of the piping system, severe vibratory fatigue damage may take place in the system.

Piping systems handling fluids such as steam through a pressure-reducing device at high-pressure and velocity conditions can produce severe acoustic vibration and metal fatigue in the piping system. This vibration and fatigue are caused by high internal acoustic loading inside the piping system generated by the flow passing through the pressure-reducing device. It has been shown that the structural fatigue of a thin-walled piping system with a diameter-to-thickness ratio of $D/t \geq 20$ is governed directly by the product of the Mach number M downstream of the pressure-reducing device and the pressure drop Δp through the device.

These acoustic energies excite the pipe wall circumferentially due to high-frequency sound waves in the range of 500–2500 Hz, where most of the energy is captured. The circumferential mode of vibration causes the pipe to displace radially, and this leads to fatigue failures where stress concentrations occur downstream, such as at pipe fittings and welded pipe supports. Typical sources of AIV are relief valves, vents, control valves, and blowdown valves. The focus is always on the downstream piping where the acoustic energy is being concentrated. Most vibration concerns are downstream and associated with stress points at small-bore socket-welded connections, drain valves, and welded pipe supports. Resonance seldom occurs where the acoustic frequency matches the mechanical pipe natural frequency, leading to drastic vibration amplitude and stress levels.

4.6.1.2 Flow-Induced Vibration

Flow-induced vibration (FIV) is a type of transient which is characterized by fluid instability, system (fluid-structure interaction) instability, or a resonant condition at the natural frequency of some portion of a system (Siba et al. 2016). In each case, the fluid is alternately accelerated and decelerated with resulting pressure increases and decreases to which the system responds. In flow instability, the fluid motion could be caused by such conditions as erratic flow through a valve due to it being too far shut or due to intermittent cavitation in an orifice. System instability could be caused by insufficient support of piping or components or erratic operation of some component, such as a control valve. If the exciting frequency of the fluid or system happens to be at a natural frequency, very high forces could exist which could cause destructive results in the system as well as nearby systems. FIV may lead to fretting wear (loss of material) or fatigue problems in piping systems.

Flow-induced vibration (FIV) may lead to fretting wear (loss of material) or fatigue problems in piping systems. FIV can be caused by internal fluid flow due to pressure pulsations caused by compressors or pumps or by pressure waves from water hammer or valve chatter/vibrations. An "encyclopedic" text on FIV has been prepared by a technical committee of the Japan Society of Mechanical Engineers (JSME) (Kaneko et al. 2014).

4.6.1.3 Mechanically Induced Vibration

The energy responsible for piping vibration is usually provided by rotating or reciprocating pumps and is transmitted to the piping by either direct mechanical contact, pressure pulsations, or turbulence in the pumped fluid. Mechanically induced vibration occurs when a mechanical natural frequency of the piping system is excited by the driving force. Pump-induced vibrations typically are characterized by low levels of amplitudes and a high number of cycles over the lifetime of plant operation. Pump bearing failures aggravate the piping vibrations.

Mechanically induced piping vibration occurs when rotating/reciprocating equipment (pumps, compressors) shakes a pipe. Cavitation excitation occurs when the fluid pressure approaches its vapor pressure. Under such conditions, small fluctuations may induce the formation of vapor pockets, which rapidly collapse, generating intense shock waves. The resulting pressure pulsation can cause severe piping vibration downstream of the component. In addition to severe vibrations, the collapse of the cavities and the resulting impact on a solid surface remove material by mechanical erosion.

4.6.2 Fatigue-Induced Socket Weld Failure

High-cycle fatigue failures of socket-welded connections in safety-related piping systems are frequent occurrences (Fig. 4.9). Failures predominantly occur at the socket-weld toe when subjected to a high-stress, low-cycle operating environment. When subjected to lower stress amplitude and higher cycles, the failures tend to initiate predominantly at the roots of the welds.

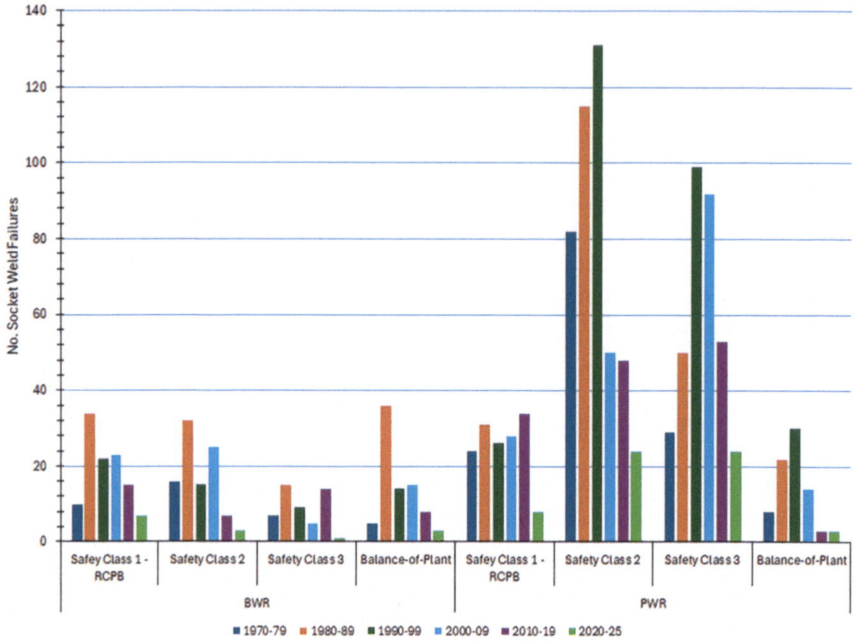

Fig. 4.9 Socket weld failure data

4.6.3 Low-Cycle Fatigue

From a codes and standards perspective, low-cycle fatigue (LCF) refers to the evaluation of how light-water reactor coolant environments affect the fatigue life of reactor pressure vessel nozzles and nozzles welded into primary coolant piping. In this context, a cumulative usage factor (CUF) is calculated for components exposed to combined pressure and thermal stresses. Cyclic loading on reactor pressure boundary components results from variations in mechanical and thermal conditions as the system transitions between load states. The total number of cycles during the design life of these components typically does not exceed 100,000 and is often limited to several thousand, which defines low-cycle fatigue.

Drawing upon operating experience with piping pressure boundaries, low-cycle fatigue is characterized by high stress and strain cycles, typically arising during startups, shutdowns, and thermal transients. This process leads to substantial plastic deformation and repeated crack initiation and propagation. Often described as a "long-term" form of material degradation, low-cycle fatigue failures tend to be periodic, generally occurring after a similar number of reactor startup or shutdown events. It is not limited to primary coolant piping. Most documented cases have been attributed to design configurations, such as interference between piping and support structures, or to construction and manufacturing defects that were not identified during pre-service or subsequent periodic inspections.

4.6.4 *Thermal Fatigue*

Thermal fatigue is due to the cyclic stresses that result from changing temperature conditions in a component or in the piping attached to the component. Thermal fatigue may involve a relatively low number of cycles at a higher strain (e.g., plant operational cycles or injection of cold water into a hot nozzle) or a high number of cycles at low stress amplitude (e.g., local leakage effects or cyclic stratification).

There are several different mechanisms by which thermal fatigue can occur. When a piping system is exposed to a thermal transient, through-wall thermal gradient stresses occur, which is a function of the rate at which the pipe cross section responds to sudden changes in temperature. These stresses are considered in the ASME code design stress analysis and, although they can cause fatigue failure, have been accounted for in the design. Piping thermal expansion, if constrained and occurring with sufficient cycles, can also cause fatigue damage and is also considered. However, there are several fatigue phenomena that are not considered in the piping design code stress analysis, including (1) thermal stratification, (2) thermal cycling, (3) thermal striping, and (4) valve in-leakage/out-leakage.

Thermal fatigue cracks initiate and grow because of alternating thermal loads. The loads cause tensional/compressive stresses due to gradients in thermal expansion. The micro-mechanism for crack initiation and growth by thermal loads is basically the same as for mechanical loads, but the resulting crack morphology, in some respects, deviates from that of mechanical fatigue. A major difference between mechanical and thermal fatigue is the crack pattern on the surface of the component. Unlike mechanical fatigue, with few exceptions, thermal fatigue cracking results in many randomly distributed cracks. This is due to the absence of a well-defined major stress direction under thermal loads. The cracks often occur in a fabric-like pattern known as "cobble stone" pattern. In the through-wall direction, thermal fatigue cracks are characterized by a low tendency for crack branching and transgranular crack growth. The shape in the through-wall direction is in most cases straight, although bent and winding cracks have been observed.

4.6.4.1 Thermal Stratification

Thermal stratification is a condition in which two streams of fluid of different temperatures flow in separate layers without appreciable mixing (Fig. 4.10). When the flow rate is low, turbulence is low and the potential for mixing is minimal. The density of water varies depending on temperature. When streams of water at different temperatures meet, the warmer fluid, which is less dense, tends to seek the upper portion of the pipe, while the cooler fluid remains at the bottom. The two layers can remain in thermal equilibrium without mixing under a wide range of flow conditions. An indicator of whether stratification will occur is the Richardson number, Ri:

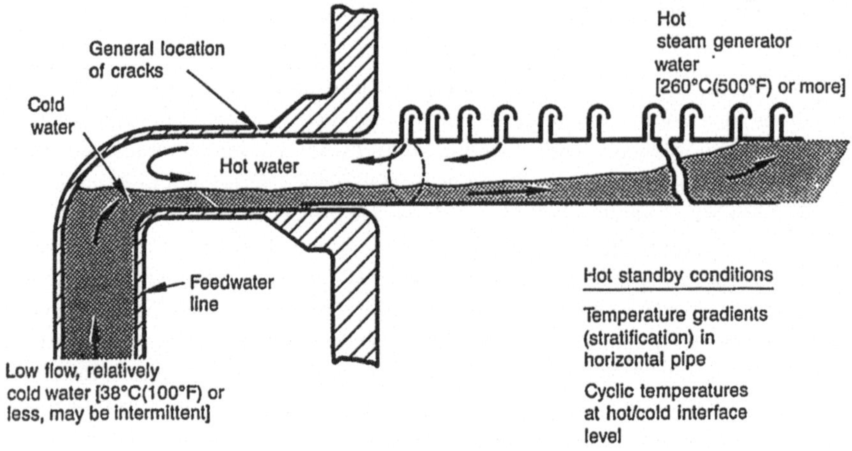

Fig. 4.10 Thermal stratification in a PWR feedwater line. (Reproduced from NUREG-0691 (1980))

$$Ri = \frac{(g \times \Delta\rho \times \mathrm{ID})}{(\rho \times \upsilon^2)}$$

where g is the gravitational constant, ρ is the fluid density, $\Delta\rho$ is the difference in density of hot and cold fluid, υ is the flow velocity, and ID is the pipe inside diameter. The Richardson number does not address the effects of piping configuration such as pipe length and slope or the mixing effects of elbows and tees. Nonetheless, the Richardson number is used as a screening criterion for assessing the potential for thermal stratification. When $Ri > 4.0$, a susceptibility exists for thermal stratification. Locations where there can be leakage past valves separating hot and cold fluids and regions where there might be intermittent mixing of hot and cold fluids caused by fluid injection are susceptible from thermal fatigue. Exceptions are for pipe segments where the diameter is 25 mm or less, or the slope of the segment is 45° or more from the horizontal. When these criteria are exceeded, additional evaluations can be performed to determine if the maximum temperature difference is greater than 10 °C or $Ri > 4.0$.

The difference in temperature between the top and bottom of the pipe causes greater thermal expansion at the top, resulting in bowing of the pipe. If the bowing is restrained, global bending moment stresses result. In addition, the interface between the two fluid layers causes local stress in the pipe due to the thermal discontinuity across the pipe cross section. Thermal stratification has caused leaks in normally stagnant lines connected to the reactor coolant loop: safety injection lines, reactor coolant system drain lines, residual heat removal suction lines, and excess letdown lines. In addition, stratification has caused cracking of steam generator feedwater nozzles and high piping displacements in pressurizer surge lines. A common criterion for evaluating thermal stratification in piping involves assessing the temperature difference between the top and bottom of the pipe. A significant temperature difference, exceeding a threshold like 28 °C, may indicate a need for detailed analysis or monitoring due to potential stress and fatigue risks (Kim et al. 2009).

4.6.4.2 Thermal Cycling

Thermal stratification by itself is generally not a concern for fatigue failure. However, if a mechanism for cycling exists, fatigue cracking can occur. Sources of thermal cycling include the following:

- Changes in interface level: If, for example, stratification is caused by a leaking valve and if the rate of leakage varies, the hot to cold fluid interface level will move up and down. A point in the pipe cross section near the interface can alternately be exposed to tensile and compressive stresses, as well as through-wall thermal gradient stresses, as the point cycles between cold and hot.
- Changes in temperature: If either the hot fluid or the cold fluid undergoes temperature fluctuations, the global bending moments will cycle and the stress distribution in the pipe cross section will vary.
- Turbulence penetration: Turbulence penetration is secondary turbulence in branch lines caused by flow turbulence from high-velocity fluid in the run pipe. As reactor coolant flow passes branch lines of connecting systems, some of the hotter reactor coolant periodically enters the branch lines. The hotter fluid can interact with cold leakage fluid and cycle at a high frequency, depending on the distance from the run pipe. The length that the turbulence penetrates into the branch varies with flow conditions, temperature, and randomness of the fluid dynamics. The maximum extent of turbulence penetration cycling length in the branch is approximately 25 branch pipes inside diameters.

4.6.4.3 Thermal Striping

Under certain thermal-hydraulic conditions, the interface between the hot and cold layers can become turbulent, leading to high-frequency cycling over a narrow range in the vertical profile of the pipe (Hu et al. 2003). This phenomenon generally occurs when the relative velocities of the hot and cold fluids are high. Testing has demonstrated that the temperature amplitude vs. frequency of cycling can be represented by a spectrum. The higher-frequency cycling occurs with the lower-amplitude temperature changes, such that in most cases, the contribution of striping to the overall fatigue usage is not large. In addition, striping stresses occur only at the pipe surface; thus, although striping could cause initiation of a crack, it cannot cause the crack to grow through wall. In general, this phenomenon will not occur in nominally stagnant lines.

4.6.4.4 Valve In/Out-Leakage

Two conditions need to be present for valve in-leakage to occur: (1) the isolation valve must leak, and (2) there needs to be a pressure gradient from the upstream to the downstream side of the isolation valve. As an example, the high-pressure safety injection system in Westinghouse plants is driven by charging pump discharge pressure, up to about 17.6 MPa, which is higher than reactor coolant system pressure, about 15.5 MPa. The case of concern is when the leakage is into an un-isolable, normally stagnant portion of the piping system, and the leakage interacts with the hot water from the reactor coolant system. The colder leakage fluid can interact with turbulence penetration from the reactor coolant system (RCS) and cause cycling at a high frequency, depending on the distance from the leaking valve to the RCS pipe.

The USNRC Bulletin 88-08 "*Thermal Stresses in Piping Connected to Reactor Coolant Systems*" (22 June 1988) was issued in response to a leak that occurred at Farley Unit 2 in 1987. The Farley leak was in the heat-affected zone of the weld between the first elbow and the horizontal run upstream of the RCS cold leg nozzle. The line rises vertically from the cold leg and then turns horizontally to the check valve. The crack location was about 1 m from the RCS cold leg nozzle. The orientation of the crack was circumferential, in the bottom third of the pipe. Farley is a three-loop Westinghouse plant. The cause of the crack was the interaction of hot RCS fluid with the cold valve leakage fluid that had stratified at the bottom of the pipe. Temperature monitoring performed after cracking indicated that downstream of the check valve, the difference in temperature between top and bottom of the pipe was 102 °C. The bottom temperature was not constant; the amplitude of the cycling was 21 °C with a cycle period between 2 and 20 min.

With respect to the valve out-leakage phenomenon, the leak path is from the reactor coolant loop, through an un-isolable branch line, to a lower-pressure system. The valve leak could be either past the isolation valve seat or past the stem packing and out through the leak-offline. Out-leakage stratification is generally steady with a limited number of cycles.

4.6.5 Corrosion Fatigue

Corrosion fatigue or "environmentally assisted fatigue" refers to the behavior of carbon steel and low-alloy steel materials under cyclic loading conditions and in a corrosive, high-temperature/high-pressure operating environment. The nuclear power plant operating experience with corrosion fatigue of piping is limited (Fig. 4.11). To date, the most significant failures involved ASME III Class 2 feedwater welds in some PWR plants in the United States in the late 1970s.

Corrosion fatigue is defined as the sequential stages of metal damage that evolve with accumulated load cycling in an aggressive environment. In simple terms,

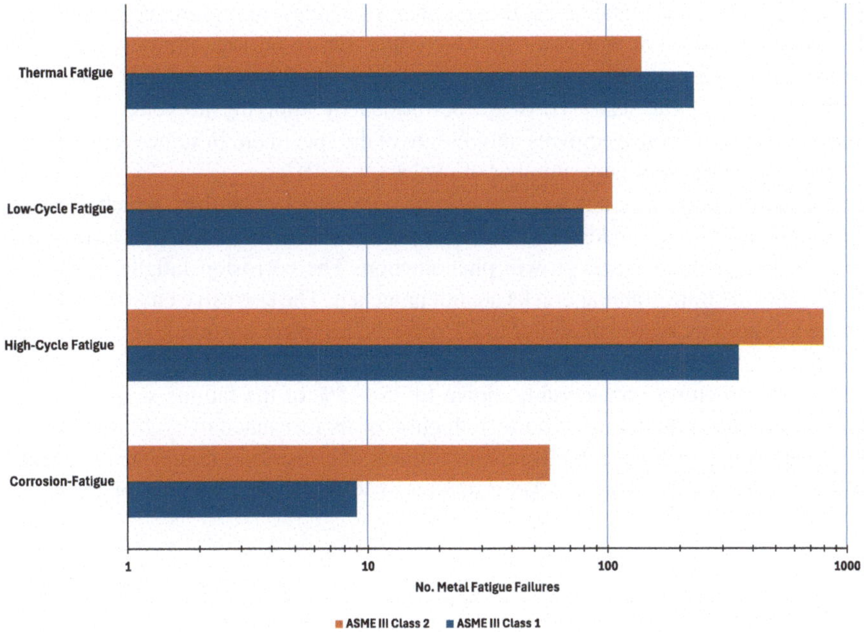

Fig. 4.11 Corrosion fatigue failures in perspective

corrosion fatigue refers to the interaction of environment and cyclic stress. Evolution of corrosion fatigue damage is divided into four different stages (Gangloff 2005):

1. Cyclic plastic deformation.
2. Microcrack initiation.
3. Small crack growth to linkup and coalescence.
4. Macro-crack propagation.

Small cracks are distinguished from short cracks. While the former are small in all dimensions (length, width, and opening), the latter are small only in two dimensions (length and opening). Laboratory tests have shown that light-water reactor coolant water can have a detrimental effect on the fatigue curve, S-N curve, and fatigue crack growth. Much lower failure stresses and much shorter failure times can occur in a corrosive environment compared to the situation where the alternating stress is in a noncorrosive environment. Codes and standards include procedures for estimating the remaining useful life of cracked reactor pressure vessel components. These procedures combine fatigue crack growth analysis with a maximum allowable flaw size. Fatigue properties of materials are often described using the S-N curve (or "Wöhler curve").[6] The S-N curve describes the relation between cyclic stress amplitude and number of cycles to failure. The figure below shows a typical S-N

[6] August Wöhler (1819–1914) was a German railway engineer and scientist who developed the S-N curve, also known as the Wöhler curve, in 1858.

curve. On the horizontal axis, the number of cycles to failure is given on a logarithmic scale. On the vertical axis (either linear or logarithmic), the stress amplitude (sometimes the maximum stress) of the cycle is given. S-N curves are derived from fatigue tests. Tests are performed by applying a cyclic stress with constant amplitude on specimens until failure of the specimen. In some cases, the test is stopped after a very large number of cycles ($N > 10^6$).

Corrosion fatigue is not to be confused with stress corrosion, which is crack initiation and growth under sustained load or residual stress. Corrosion fatigue is a mostly transgranular crack growth phenomenon. The corrosion fatigue fracture is brittle, and the transgranular cracks are not branched. The corrosive environment can cause a faster crack growth and/or crack growth at a lower tension level than in dry air. Even relatively mild corrosive atmospheres can reduce the fatigue strength of aluminum structures considerably, down to 75–25% of the fatigue strength in dry air. No metal is immune from some reduction of its resistance to cyclic stressing if the metal is in a corrosive environment. Control of corrosion fatigue can be accomplished by either lowering the cyclic stresses or various corrosion control measures.

4.6.6 *Operating Experience Insights*

This section summarizes selected fatigue failures of historical significance. The term "significance" reflects combinations of operational impact and the extent to which industry and regulatory agencies responded to an event in terms of finding acceptable solutions to long-term fatigue management. The summaries of selected significant corrosion fatigue, high-cycle fatigue, and thermal fatigue failures are listed in chronological order.

4.6.6.1 The VBWR Thermal Fatigue Failure (1962)

In September 1962, the Vallecitos Experimental Boiling Water Reactor (VBWR) experienced a mixing tee failure. The VBWR was located at General Electric's Vallecitos Atomic Laboratory near Pleasanton, California. The center included chemistry, metallurgy, and physics laboratories; a machine shop; and other needed support facilities. The reactor plant was the first privately funded and constructed nuclear power plant to supply power in megawatt amounts to an electric utility grid. The reactor was issued Power Reactor License No. 1 by the United States Atomic Energy Commission. The reactor achieved its first criticality on August 3, 1957. It was connected to the utility grid on October 19, 1957, and after 6 years of operation, it was permanently shut down on December 9, 1963. This incident is noteworthy due to similar thermal fatigue events in Sweden about two decades later.

On September 12, 1962, the reactor was shut down to look for a steam leak in the basement of the reactor containment building. On September 13, 1962, after removing some of the shielding blocks from around the reactor vessel, it was discovered that a circumferential crack had developed in one of the four suction pipes to the reactor recirculation pumps. The extent of this crack was about 64 mm in length on the outside diameter of the pipe and 152 mm in length on the inside diameter, and it was located between the reactor vessel nozzle and a DN25 feedwater injection line.

The VBWR recirculation piping system consisted of four pump suction lines and two discharge lines. The piping was DN250 (wall thickness of 16.5 mm), fabricated from ASTM A240, Type 304, stainless steel plate, which was rolled, welded, radiographed, solution heat treated, and liquid dye penetrant inspected. The pipe was bent hot, and the bends were subsequently solution heat treated and water quenched. The fabrication and erection were performed in accordance with the ASME Boiler and Pressure Vessel Code, Section I—Power Boilers. The four pump suction pipes which connect to the reactor vessel nozzles at approximately mid-vessel height are about 500 mm above the vessel support. A DN25 feedwater injection inlet was attached to each of the four pump suction lines about 200 mm from the reactor vessel nozzle at the 12 o'clock position. The feedwater inlets introduced feedwater at 21–54 °C into the piping containing water at 288 °F. The feedwater was introduced in such a manner that it may have cycled on and off more than 1000 cycles during operation.

The through-wall crack was in the DN250 stainless steel piping and was located well outside of the pipe-to-vessel-nozzle weld. Upon completion of the visual and nondestructive testing examination, a section of the pipe about 530 mm in length, which contained the observed crack and the crack indications, was removed for destructive testing and metallurgical evaluation. The pieces of pipe were decontaminated to reduce the activity level and pickled to remove the corrosion and oxide film. Visual examination revealed a "crazed" pattern of cracking on the inside of the pipe in an area which appears to be related to the flow of cold water from the DN25 feedwater inlet pipe through which feedwater was introduced into the pipe, which contains water at 288 °C. In addition to the random "crazed" cracks, a series of circumferential cracks exist in the upper half of the pipe, about 75 mm from the vessel nozzle. Even though the crack was apparent for 64 mm on the outside of the pipe, the crack was not continuous completely through the wall. The stress analysis and the metallurgical investigation very conclusively indicate that the cause of cracking was related to the very high thermal and mechanical strains imposed upon the pipe in the area of the failure. The mode or mechanism of failure is analyzed to be low-cycle fatigue caused by the thermal and mechanical strain cycling. It was theorized that corrosion aggravated the condition which caused cracking, but there was no evidence that corrosion was a major contributor (Smith and Viloette 1962).

As a result of the failure investigation, the feedwater system was redesigned in such a way that feedwater was introduced in a DN75 pipe through a thermal sleeve into a DN150 nozzle near the top of the reactor vessel and distributed inside the reactor vessel to a ring header with dip tubes. The arrangement and flow characteristics were designed to permit thorough mixing of the feedwater inside the reactor vessel without developing significant thermal gradients at any point in the pressure piping, the nozzle, the reactor vessel shell, or the structural internal supports.

4.6.6.2 Corrosion Fatigue Failure (1979)

In May 1979, the USNRC was informed of cracking in two feedwater lines in D.C. Cook Unit 2. Leaking circumferential cracks were identified in the DN400 lines in the immediate vicinity of the steam generator nozzles. Subsequent volumetric examination (radiography) revealed crack indications at similar locations in all feedwater lines of both Units 1 and 2. The most severe crack identified was a through-wall crack 90 mm long at the outer surface in steam generator No. 1 at D.C. Cook Unit 2. This crack was located at a stress riser caused by a discontinuity introduced in the piping elbow from machining the weld end preparation counterbore. Metallographic examination of samples removed from the failed component revealed other less severe cracks in the tapered transition section of the elbow (Shah et al. 1997). Also, liquid penetrant examination of the nozzle bore from the thermal sleeve to the nozzle end revealed light pitting and intermittent linear indications along the circumference of the nozzle bore. Fractographic and metallographic examinations of the cracked piping removed from the two reactor units revealed the following:

- The primary type of cracking was transgranular with multiple fracture origin sites.
- The progression of the primary cracks was intermittent as indicated by the presence of beach marks (crack arrest lines) on the fracture face. Fatigue striations werc observed on the fracture face.
- There was little oxidation on the primary cracks indicating that stress played a primary role in propagating the cracks. Corrosion may have played a role.

From the metallographic and fractographic analyses, the mode of failure was identified as fatigue assisted by corrosion at the D.C. Cook units. Similar characteristics were observed, and similar conclusions drawn regarding feedwater piping cracks at other PWR plants with deep cracks and at the majority of facilities with shallow cracks. In June of 1979, Beaver Valley Unit 1 reported the discovery of cracking in steam generator nozzle-to-feedwater inlet piping welds. Radiographs showed cracking to be present in all three inlet pipes. The inlet piping at these locations is 90° elbows. The lines are DN400 Schedule 80 (21.44 mm wall thickness). In each instance, the cracks originated at the shoulder of the counter-bore in the piping or approximately 14 mm from the root of the weld. Sections of the cracked piping were subjected to metallurgical evaluations, which confirmed that the deepest crack penetration was 10.2 mm or about 50% through-wall. Based on microscopic tests, it was concluded that corrosion fatigue was the main contributor to the cracking.

4.6.6.3 Olkiluoto 1 Thermal Fatigue Failure (1979)

A major pressure boundary failure occurred on 29 August 1979 when a DN150, Schedule 120 pipe in the reactor water cleanup (RWCU) system fractured, resulting in a spill of 5000 m^3 primary water outside the containment (Holmberg and Pyy 1994). The through-wall crack in the pipe was 150 mm long and 2 mm wide. The unit was in the commissioning phase of operation, and the pipe fracture resulted in reactor trip and containment isolation. The rupture occurred in a mixing tee between the RWCU and residual heat removal (RHR) systems. The apparent cause of the failure was thermal stratification, but the underlying cause was an operator error due to a deficiency in the system operating procedure. The fatigue crack evolved because of plant operation with an RWCU valve in an intermediate state, leaving two flows having considerably different temperatures (130 and 280 °C) mixed incompletely.

A deficient operating procedure caused the RWCU control valve to be in the wrong position. This in turn led to a significant temperature difference between two mixing flows and subsequent thermal fatigue. At the time of the thermal fatigue failure, Olkiluoto 1 (an Asea-Atom fourth-generation-design BWR) was undergoing the final stages of commissioning prior to commercial operation. The reactor had achieved initial criticality on 29 July 1978, and the first electrical grid connection was on 2 September 1978.

4.6.6.4 Barsebäck 2 Thermal Fatigue Failure (1980)

During the annual refueling outage in 1980, ultrasonic inspection of the Safety Class 1 feedwater piping at Barsebäck 2 (an Asea-Atom third-generation-design BWR) revealed cracks on the inner surface of two branch connections, where feedwater at 20–170 °C was mixed with primary water at 270 °C returning from the shutdown cooling water (RHR) system (Burkhart 1981). These indications prompted inspections of similar pipe branch connections at Barsebäck 1 and Oskarshamn 2 (both units of the same design as Barsebäck 2) and at Ringhals 1 (a second-generation Asea-Atom reactor design). The affected branch connections were cut out and submitted for metallographic examinations. These examinations confirmed that the cracks were caused by thermal fatigue resulting from turbulent mixing of cold and hot water in the branch connection. A maximum depth of 10 mm was measured, but the cracks were generally 1–6 mm deep. The cracked sections of stainless steel piping had a wall thickness of 25–30 mm. The affected plants operated for periods of 2.2–4.6 equivalent full power operating years (EFPY).

The branch connections were manufactured from forged material. Branch connections in Ringhals 1 feedwater system consisted of seamless piping. Stainless steel material corresponding to ASTM A304 with a maximum carbon content of 0.05% was used in all branch connections. The feedwater system operated at 180 °C during normal operating conditions. The decay heat removal was connected to each feedwater line inside the containment and operated at 270 °C during normal operations. The auxiliary feedwater system was connected to one of the decay heat removal lines and operated intermittently (e.g., during startup) at 30 °C.

When the first cracked branch connection was removed, liquid penetrant examinations revealed that the cracked areas were larger than the areas predicted by the ultrasonic examination. To define the extent of cracking more accurately in the examinations of subsequent branch connections, a spark-eroded notch, 0.5 mm deep and 10 mm long, was usually used as the calibration reflector. The rectangular notch is about 10 dB more sensitive than the 3.5-mm-radius half-circle reflector. Defect indications equal to or larger than 50% of the notch signal were recorded. To determine the effectiveness of using the notch calibration reflector, cladding weld material was examined, and the ultrasonic amplitudes of indications were noted. Areas containing these indications were investigated and correlated with the ultrasonic amplitudes.

A small portion of cracked material was removed and metallurgically examined. The examination revealed that the material was not sensitized. Cracks were perpendicular to the surface of the pipe and transcrystalline with a weak tendency to branching. Deeper portions of cracks were completely covered with oxide, indicating that crack growth had stopped or at least was very slow after a certain crack depth (5–7 mm) had been reached. Cracking was caused by thermal fatigue resulting from temperature gradients when water of different temperatures was mixed in the pipe branch connection.

4.6.6.5 Farley 2 and Tihange 1 Thermal Fatigue Failures

In 1987, a primary coolant leak occurred inside the containment of Farley Unit 2 (EPRI 2000). At the time of this event, the reactor unit had been in operation for slightly less than six equivalent full power years (EFPY). The leak was found to be in the un-isolable location of the safety injection line. The crack was on the inside surface of the weld and extended approximately 120° circumferentially around the underside of the pipe. About 25 mm of this crack was through-wall. The crack was caused by thermal fatigue and had developed slowly. The leak rate was 2.7 L/min. The monitoring of circumferential temperature distribution at the failed weld at Farley 2 was carried out after the leak event showed spatial and temporal fluctuations in the temperature. The circumferential temperature difference at the weld varied from 3 °C to as high as 120 °C. Based on these measurements, it was assumed that the temporal variations resulted from intermittent action of the check valve. There were, however, no test results supporting this assumption.

Experiments performed to simulate the Farley event showed that the temperature fluctuation in the safety injection line was not caused by the intermittent action of the check valve (i.e., fluctuation in the flow rate) but by the mixing of low-temperature leakage flow with high-temperature turbulent flow in the pipe downstream of the check valves. The test results also concluded that the thermal cycling was severe enough to cause high-cycle fatigue failure of the piping material when the leak flow rate was equal to or larger than 100 kg/h, as measured at Farley 2. A similar occurrence caused cracking in both base metal and a weld of Tihange 1 on 18 June 1988 (IAEA 2003; Dahlberg et al. 2011) (Fig. 4.12). Both Farley 2 and Tihange 1 are of similar design.

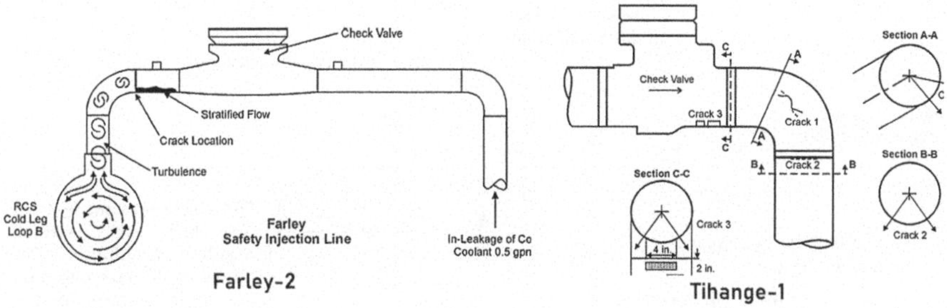

Fig. 4.12 Location of thermal fatigue failures

4.6.6.6 Loss-of-Coolant Incident at Pickering Nuclear Generating Station

While Pickering Unit 2 operated at full power, a fracture of a rubber diaphragm in a liquid relief valve resulted in a loss of primary coolant (Charlebois et al. 1995). As a result of the fractured rubber diaphragm, the valve failed to open, filling the bleed condenser followed by the actuation of the reactor-protective system. When pressure recovered, two spring-loaded safety relief valves opened and one of them chattered. The pressure pulsations cracked a 45-degree DN75 elbow in the inlet pipe to the chattering valve, and the subsequent loss of coolant triggered the emergency core cooling system.

The valve chatter, at about 60 Hz, damaged the valve and induced fluid and mechanical vibration in its inlet pipe. Emergency coolant injection was required after an instrumented reactor coolant pressure relief valve opened due to failure of its air supply line. The reactor began automatically to set back from full power. A boiler room entry was made to assess the extent of damage. Inspection revealed damage to the high-pressure instrument air supply to the heat transport relief valves. All supplies appeared to have been disconnected. In addition, damage had occurred to one of the bleed condenser relief valves' piping. This line was spraying water through a 270-degree circumferential crack into the area within the reactor containment building that houses the steam generator room, resulting in an accumulation of water in reactor building sumps.

4.6.6.7 Civaux 1 Thermal Fatigue Failure

On 12 May 1998, with the Civaux Unit 1 in intermediate shutdown, pressurizer level decreased, and a fire alarm occurred in the reactor building (OECD/NEA 1998, 2022). A large ca. 8.3 kg/s primary system leak had developed, and it was compensated for by the charging system. The time between the initiation of the crack and its development to a significant depth through the pipe wall was only about 1500 h. The system operated at a pressure of 3.6 MPa, with the hot leg containing water at 180 °C and the cold leg containing water at 20 °C. After rapid cooldown of the reactor, the

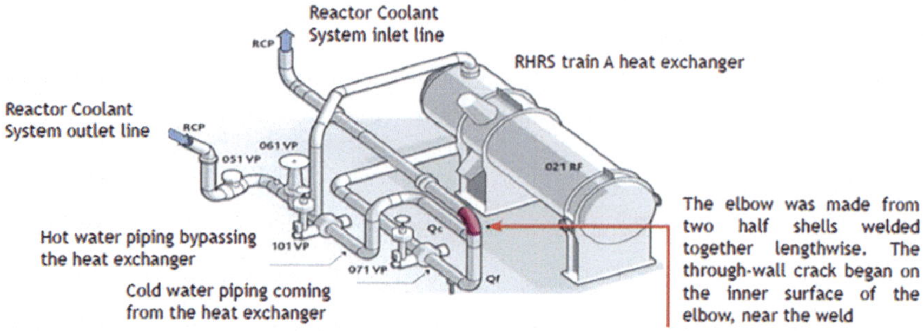

Fig. 4.13 Location of thermal fatigue failure. (Adapted from a presentation by Dr. D. Golfetto (ASNR) on September 5, 2018, at the 16th Working Group Meeting of the CODAP Project)

leak was identified to come from RHR train "A." About 10 h after the event occurred, the RHR train "A" was isolated, and the leak was stopped. The source of the leakage was a cracked weld on Safety Class 2 elbow downstream of a control valve (Fig. 4.13). The weld was a longitudinal weld; the elbow was fabricated from two half shells and welded together. The observed cracking was immediately downstream of the mixing tee in a DN250 Schedule 40 (9.3 mm wall thickness) 304L stainless steel seam-welded elbow. The worst cracking that led to the through-wall leak was at the extrados welded seam and had an outside surface length of 180 mm, with an inside surface length of 350 mm. The surrounding area of the elbow and locations between the elbow and the mixing tee exhibited widespread crazing-type cracking characteristic of thermal fatigue. Examination of this elbow and similar ones from other plants confirmed that high-cycle thermal fatigue was the mechanism. No fabrication defects were found. In general, deep cracks occurred along the longitudinal and circumferential welds, starting at the root of the weld on the inner surface. There were many small cracks, mainly in the weld counter-bore regions and where there was evidence of grinding.

4.7 Flow-Assisted Degradation

The term flow-assisted degradation (FAD) encompasses several material degradation processes, all of which result in loss of material, i.e., pipe wall thinning. These phenomena include erosion, erosion-cavitation, erosion-corrosion, flow-accelerated corrosion (FAC), and liquid droplet impingement erosion (LDIE). Figure 4.14 summarizes the distinguishing characteristics of different flow-assisted degradation mechanisms. Historically, the erosion-corrosion and flow-accelerated corrosion terms were used interchangeably to describe similar material degradation processes. Both types of damage involve destruction of a protective oxide film on the inside of the pipe wall. The removal of the oxide film is generally referred to as the erosion

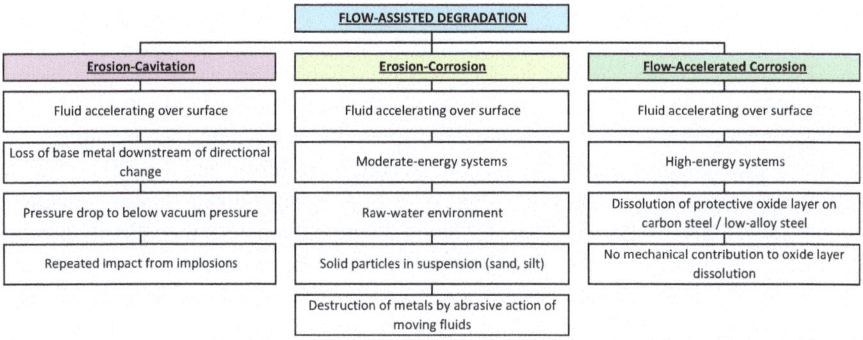

Fig. 4.14 Distinguishing characteristics of different FAD mechanisms

process. This is followed by electrochemical oxidation or corrosive attack of the underlying metal.

Differences between erosion-corrosion and flow-accelerated corrosion involve the mechanism by which the protective film is removed from the metal surface. In the erosion-corrosion process, the film is removed mechanically from the surface. In the flow-accelerated corrosion process, the oxide is dissolved or prevented from forming, allowing corrosion of the unprotected metal. Flow-accelerated corrosion occurs in two-phase flow conditions (e.g., water droplets in steam or steam bubbles in water) as well as in single-phase flow conditions.

4.7.1 *Erosion-Cavitation*

Erosion-cavitation is the process of surface deterioration and surface material loss due to the generation of vapor or gas pockets inside the flow of liquid. It is a form of localized material degradation which can occur after a relatively short time, less than 10^4 h. These vapor or gas pockets are formed due to low pressure well below the saturation vapor pressure of the liquid and erosion caused by the bombardment of vapor bubbles on the surface. Erosion-cavitation usually involves an attack on the surface by gas or vapor bubbles, creating a collapse due to a change in pressure near the surface. Low pressure (below the saturated vapor pressure) is generated hydrodynamically due to various flow parameters, such as liquid viscosity, temperature, pressure, and nature of flow. This deterioration is initiated by a sudden surge of bubbles hammering the surface, resulting in deformation, as well as pitting. Erosion-cavitation failures mostly occur downstream of flow control valves in piping of the chemical and volume control system, component cooling water system, residual heat removal system, and service water system. The cavitation coefficient σ (also referred to as Thoma's cavitation coefficient or '$Thoma - \sigma$') is used to measure the presence of cavitation (Fisher and Thoma 1932). It is defined by

$$\sigma = \frac{(P_2 - P_V)}{\Delta P}$$

In the above expression, P_2 is the absolute static pressure downstream of the section under consideration, P_V is the vapor pressure at the liquid temperature, and ΔP is the pressure drop across the section. According to Morel and Reynes (1992), a significant steam rate is observed when $\sigma < 5$. A steam bubble stream generated condenses rapidly. Within 5 to 10*D* (*D* being the pipe diameter), the velocity profile is sufficiently homogenized so that there are no fluid streams moving at a significant overspeed. The manifestations of erosion-cavitation damage are in the form of very localized loss of material and with pin-hole size pipe wall penetrations. Preventing reoccurrence of erosion-cavitation involves changes in system operating procedures and possibly coupled with system design changes to minimize hydrodynamic pressure differences in process flow streams.

4.7.2 Erosion-Corrosion

Erosion-corrosion is a mechanism of material loss by mechanical means due to impingement, abrasion, or impact, resulting from the movement of a liquid or gas over the surface of a metal coupled with corrosion. This type of degradation is characterized by small pits with bright surfaces free from corrosion products. These pits often have the form of a horseshoe with the nib pointing in the current direction. Erosion-corrosion may occur where the velocity of liquid is too high (Horowitz 2004). Most exposed are places where there are effects of turbulence, e.g., joints and bends. The corrosion rate will accelerate if the liquid contains gas bubbles and/or solid particles. Systems susceptible to erosion-corrosion include raw water-cooling systems, e.g., circulating water and service water systems.

Liquid droplet impingement erosion is a subset of erosion-corrosion. Liquid droplets are often generated in piping that operates in a two-phase flow condition and are due to the entrainment of liquid water from the upstream and by the heat transfer through the pipe wall. In the region behind the orifice and the valve in the pipeline, the velocity of the droplets is highly accelerated due to the contraction effect. This results in the occurrence of high impact pressure on the inner surface of the pipe due to the liquid droplet impingement. The impact pressure of the droplets increases as high as several hundred MPa, which is beyond the elastic limit of the pipe-wall material, so that the pipeline is often damaged by the impact pressure of droplets. In general, the liquid droplet impingement occurs on the dorsal side (extrados) of a bend or elbow, where the droplets cannot follow the steam flow due to the inertia of the droplets.

4.7.3 *Flow-Accelerated Corrosion*

Major pipe failures have occurred in piping due to single-phase or wet-steam flow-accelerated corrosion (FAC), resulting in injuries or loss of life as well as extensive plant downtime. Both nuclear and fossil power plant piping as well as petrochemical plant piping systems are susceptible to this form of flow-assisted degradation. FAC leads to wall thinning (metal loss) of steel piping exposed to flowing water or wet steam. The wall thinning is the result of the dissolution of the normally protective oxide layer formed on the surfaces of carbon and low-alloy steel piping (Malik Al-Abed Allah et al. 2026). The rate of metal loss depends on a complex interplay of several parameters including water chemistry, material composition, and hydrodynamics, but based on operating experience, the metal loss can be as high as 3 mm per year. FAC mainly affects feedwater piping and main steam line subsystems, including the high-pressure turbine exhaust piping, the turbine crossover piping, and the extraction steam lines. The moisture content in the main steam leaving the reactor pressure vessel is about 0.1% and increases as the steam reaches the main turbines. The high moisture content in the steam extraction and exhaust lines and turbine crossover lines makes these lines particularly susceptible to FAC. The main steam line pipes are not susceptible to FAC unless moisture is present.

Historically, the pipe wall thinning terminology was ambiguous since erosion-corrosion was used for both the chemical mechanism now known as FAC and the mechanisms in which the oxide is broken down mechanically by the impingement of particles, solids, or gaseous bubbles. There are also differences in the surface morphology of FAC and erosion-corrosion. Single-phase FAC has a scalloped or orange-peel appearance, and two-phase damage often has a characteristic pattern known as tiger striping. These surface features are absent in surfaces damaged by erosion mechanisms. Another difference is that FAC is often more widespread than the localized erosion damage. FAC was seen mainly as a problem in two-phase flow (e.g., wet steam) systems. The first case of single-phase FAC-induced pipe failure was reported in 1985. On March 9, 1985, Trojan Nuclear Power Plant was operating at 100% power when a DN350 heater drain pump discharge pipe made of SA-106 Grade B carbon steel failed catastrophically.[7] The failure caused the release of a steam-water mixture of approximately 180 °C into the turbine building, which in turn caused fire suppression system actuation by heat sensors in the turbine building and damaged secondary plant equipment.

A second case of single-phase FAC-induced pipe failure was reported in 1986. On December 9, 1986, a DN450 suction line to the main feedwater pump of Surry Unit 2 failed catastrophically. The line temperature at this location is approximately 185 °C, with a pressure of approximately 2.55 MPa. The ruptured elbow was made of ASTM A-234 Grade WPB carbon steel. Water flashing from the severed pipe

[7] Trojan Nuclear Power Plant, an early four-loop PWR designed by Westinghouse, was permanently shut down on 9 November 1992. At the time of its closure, the plant had been operating for about 16 years.

Table 4.5 Major FAC failures in nuclear power plants

Event descriptors	Trojan (PWR)	Surry 2	Mihama-3 (PWR)
Date of occurrence	3/9/1985	12/9/1986	8/9/2004
Location	356 mm heater drainpipe	450 mm 90° elbow following a tee off main feedwater suction header	560 mm condensate pipe between low-pressure heaters and deaerator
Temperature [°C]	173	190	142
Pressure [MPa]	3.11	2.55	0.93
Max fluid velocity [$m\ s^{-1}$]	7.3	5.2	2.2
Range of pH (typical)	(9.0)	8.9–9.0	8.6–9.3
Oxygen [$\mu g\ kg^{-1}$]	4	4	<5
Reducing agent	Hydrazine	Hydrazine	Hydrazine
Carbon steel grade	A106 Gr B	A106 Gr B	JIS G301SB42

engulfed equipment and personnel in the area. Within minutes of the pipe rupture, portions of the automatic fire protection system activated, opening 62 sprinklers to cool the atmosphere around the rupture. The water from the sprinklers seeped into electrical panels, shorted out several electrical circuits that controlled other fire suppression equipment, and activated some systems containing carbon dioxide. The carbon dioxide combined with other fire retardants and seeped into the control room. A third case of a major single-phase FAC failure occurred in Japan in 2004 (Table 4.5).

The events at Trojan and Surry are of historical significance. They demonstrated that significant FAC-induced pipe wall thinning can occur not only in wet steam lines (two-phase flow conditions) but also under single-phase flow conditions. The two events raised questions about the effectiveness of the then existing nondestructive examination (NDE) programs to monitor piping integrity for wall thinning and prevention of pipe failure. In July 1987, the United States Nuclear Regulatory Commission issued Bulletin No. 87-01, "*Thinning of Pipe Walls in Nuclear Power Plants*" (USNRC 1987). The Bulletin requested all licensees for nuclear power plants holding an operating license or construction permit to submit information concerning their programs for monitoring carbon steel pipe wall thickness.

The NRC staff completed its review of the responses in December 1987. Furthermore, at the end of September 1988, the NRC staff completed inspection of ten plants to assess the licensees' efforts toward implementing their FAC monitoring programs. NUREG-1344 (USNRC 1989a) and EPRI NP-6066 (EPRI 1988a, b) summarize the responses to Bulletin 87-01. In May 1989, the USNRC issued Generic Letter 89-08, which summarized the results of the regulatory review process as follows and noted that at the time of this communication, the term "erosion-corrosion" rather than the correct term FAC was still in use (USNRC 1989b):

> *The staff review of licensees' responses to the bulletin indicates that the pipe wall thinning problem is widespread for single-phase and two-phase high-energy carbon steel systems. The systems and components reported as having experienced pipe wall thinning are listed in Section 6 of the attachment to this letter. The staff review also showed that wall thinning in single-phase feedwater-condensate systems is more prevalent among pressurized water reactors (PWRs) but also occurs in boiling water reactors (BWRs).*
>
> *The staff audited ten operating plants (seven PWRs and three BWRs) in late 1988 to assess the implementation of erosion-corrosion monitoring programs by licensees and to ensure that adequate guidance was provided for corrective actions and other activities regarding repair and replacement of degraded piping and components. Detailed audit findings are described in Section 7 of NUREG-1344, which is enclosed with this letter. In general, all licensees have developed and put in place an erosion-corrosion monitoring program that meets the intent of NUMARC guidelines (Appendix A of NUREG-1344). In addition, all licensees have completed their initial examination as recommended by NUMARC. However, the staff found that none of these licensees has implemented formalized procedures or administrative controls to ensure continued long-term implementation of its erosion-corrosion monitoring program for piping and components within the licensing basis. Therefore, you should provide assurances that a program, consisting of systematic measures to ensure that erosion-corrosion does not lead to degradation of single-phase and two-phase high-energy carbon steel systems, has been implemented.*

Following the reported cases of flow-accelerated corrosion (FAC) failures in single-phase systems, EPRI developed the CHEC[8] family of computer codes to serve as predictive tools for plant operators. These tools support inspection planning and data evaluation to help prevent failures attributable to FAC. In March 1987, the Nuclear Management and Resources Council (NUMARC, now the Nuclear Energy Institute, NEI) established a working group on FAC. This group established a recommended industry program aimed at addressing both single-phase and two-phase FAC. Collaborative efforts between NUMARC and EPRI resulted in an inspection plan designed to monitor pipe wall thinning problems. That program is documented in "*Recommendations for an Effective Flow-Accelerated Corrosion Program*," NSAC-202L-R4 (EPRI 2013). In May 1989, the NRC issued Generic Letter 89-08, directing all licensees to confirm implementation of either the NUMARC program or an equivalent FAC program and to ensure the structural integrity of all high-energy carbon steel systems, encompassing both two-phase and single-phase flow. Licensees who had not yet adopted an FAC program were required to provide a scheduled implementation date.

---

[8] V.K. (Bindi) Chexal (1949–2001), the "C" in the CHEC® computer code, which was developed to assist plant owners in planning inspections and evaluating inspection data to prevent pipe failures caused by flow-accelerated corrosion (FAC). Before joining EPRI in 1982, he worked for the Atomic Energy of Canada, Ltd., and Quadrex Corporation. Chexal was the co-developer of the "Chexal-Horowitz erosion/corrosion" model for predicting piping system locations susceptible to flow-assisted material loss.

Basic research towards development of FAC monitoring and prediction strategies includes the theoretical and experimental work by Heitmann and Kastner (1982). This work resulted in the WATHEC computer code for the prediction of wall thinning in single- and two-phase water/steam systems. A literature survey of FAC predictive models noted that there remain uncertainties in the engineering analyses that are performed to assess structural integrity (Betova et al. 2010):

- Flaw modeling: In-service inspection techniques may not be sufficiently sensitive to characterize detected flaws, introducing the possibility of unconservative modeling results.
- Residual stress: Magnitude of residual stress remaining in components and its effect on the component integrity are difficult to quantify. In turn, the effect of these stresses on the kinetics of film growth and restructuring, as well as the kinetics of dissolution, could be an important parameter that has not been considered by the existing modeling approaches.
- Degradation of material properties: It is generally known and also observed in many actual cases that material properties may deteriorate due to operating conditions, leading also to a transformation of the surface oxides and hence their kinetics of dissolution which is one of the governing factors of flow-accelerated corrosion.

The variables that control the rate of FAC are temperature, fluid velocity, fluid pH, water amine, oxygen content, steam quality, void fraction of the fluid, piping geometry, and material composition (Uchida et al. 2008; Moon et al. 2024). The following subsections describe the different variable effects.

4.7.3.1 Effect of Temperature

An important variable affecting the FAC resistance of carbon and low-alloy steels is temperature. FAC damage is mostly reported between 80 and 230 °C in single-phase conditions and shifts to 140–260 °C under two-phase flow conditions. The exact location of the maximum wear rate changes with pH, oxygen content, and other environmental variables. Experience has shown that the wear rate is highest at around 150 °C and increases with fluid velocity. FAC can occur in low-temperature single-phase systems under unusual and severe operating conditions.

4.7.3.2 Effect of Flow Velocity

Flow rate of the liquid has been found to have a linear effect on the FAC wear rate. As higher velocities are experienced, higher wear rates are expected. Since the enhanced mass transfer associated with turbulent flows is the fundamental process in the accelerated dissolution of the pipe wall-protective oxide layer, the effect of flow is best described in terms of the mass transfer coefficient, which is a function of flow velocity and geometry. Local flow velocities can differ by a factor of 2–3 from the bulk flow velocity.

4.7.3.3 Effect of Fluid pH

FAC wear rates are strongly dependent on pH, potential of hydrogen. The pH relates to the acidity or alkalinity of a process medium. In general, increasing the pH value reduces the wear. The FAC wear rate of carbon steels increases rapidly in the pH range of 7–9 and drops sharply above pH = 9.2. As the fluid becomes more acidic, more pipe wall losses are expected. The pH value can be affected by the choice of control agents (e.g., morpholine or ammonia) and by impurities in the water. In two-phase flows, the critical parameter is the pH of the liquid phase. This can be significantly affected by the partitioning of the control agent between the steam and liquid phases. There is no adjustment of pH performed in BWR plants.

4.7.3.4 Effect of Oxygen

FAC rates are inversely affected by the amount of dissolved oxygen (DO) in the feedwater, and too low an oxygen level is harmful to carbon steel piping. The FAC rate decreases rapidly when the water contains more than 20 ppb oxygen, but the precise oxygen level required to prevent FAC depends on other factors such as pH and presence of contaminants.

In BWRs, hydrogen water chemistry (HWC) is applied with the main intention to suppress intergranular stress corrosion cracking (IGSCC) susceptibility and crack growth rate.[9] The FAC rate has been measured in a laboratory test to be higher for a period of 8 months after starting HWC. After this time, the FAC rate appears to be like that in a reference normal water chemistry (NWC) environment. General Electric guidelines consider an oxygen level of 20–50 ppb desirable for hydrogen in hydrogen water chemistry. Some plants must add oxygen in their feedwater when using HWC, while others do not. The effects of higher hydrogen levels under NWC conditions are plant specific and must be considered as for HWC conditions. The use of noble metals to reduce the quantities of hydrogen required to establish HWC conditions has to date not had a more pronounced effect on FAC than the application of HWC itself.

Main steam lines made of carbon steel are susceptible to FAC in the steam phase because most of the oxygen, being a gas, remains in the steam phase and does not partition to the liquid. For the same reason, injection of oxygen into the wet steam will not prevent FAC. Injection of hydrogen peroxide has been explored as a possible mitigation for FAC because most of the hydrogen peroxide partitions to

[9] The Swedish boiling water reactor Oskarshamn 2 was the first to use hydrogen water chemistry (HWC) in 1979.

the liquid phase and spontaneously decomposes into oxygen and water and thus enriches the liquid phase with oxygen. However, although the FAC rate is decreased, hydrogen peroxide injection is not as effective a remedy towards FAC as replacement of materials with low-alloy steel (alloyed with chromium) or presence of a stainless steel coating.

4.7.3.5 Effect of Alloy Additions

The FAC rate is highest in carbon steel piping with very low levels of alloying elements. The presence of chromium, copper, and molybdenum, even at low percentage levels, reduces the FAC rate considerably. The relative corrosion rate of steels is reduced by 80% at a chromium content as low as 0.2%. The FAC rate is decreased by a factor of 4 with the steel type 2–1/4% Cr and 1% Mo (2–1/4 Cr-1 Mo steel). Austenitic stainless steels are virtually immune to FAC.

4.7.3.6 The Entrance Effect

In the 1990s, a new FAC wear effect was documented. This effect has been called the "leading edge effect" or the "entrance effect" (EPRI 2007c). This effect occurs when flow passes from an FAC-resistant material to a nonresistant (susceptible) material, which causes a local increase in the corrosion rate. This effect is normally manifested by a groove up- or downstream of the attachment weld between the corroding and the resistant stainless steel material. In one, relatively recent example, significant wear was detected in an expander. The area in question consisted of a valve followed by a 150 mm by 200 mm expander attached to another 200 mm by 400 mm expander. Several FAC inspections had been performed before and after replacement of the upstream expander with resistant material due to high FAC wall thinning. An almost linear thinning rate of 1.77 mm per cycle was noted over the subsequent refuelling (RF) cycles (RF11–RF15); the downstream expander was replaced during the RF15 outage. A recent example of FAC wear near a dissimilar metal weld in DN300 extraction steam piping is documented in the Journal of Nuclear Materials (You et al. 2025).

4.7.4 Liquid Droplet Impingement Erosion

Liquid droplet impingement erosion (LDIE) or jet impingement erosion (JIE) is caused by the impact of high-velocity droplets or liquid jets on a pipe wall. LIE occurs when a two-phase stream experiences a high-pressure drop (e.g., across an orifice on a line to the condenser) (EPRI 2004a, b, c, d; Li et al 2011; Seo et al 2014). When this occurs, there is an acceleration of both phases with the liquid velocity increasing to the point that, if the liquid strikes a metallic surface, damage to the surface will result. The fluid is mostly "high-quality" steam with some liquid.

LDIE failures have been observed in the high-pressure feedwater heater system and lead to power reduction as well as possible spatial impacts on adjacent equipment in the turbine building. An investigation of an event at a CANDU-600 plant in 2008 used computational fluid dynamics analysis to simulate this wall thinning phenomenon (Hwang et al. 2014). A header pipe in the high-pressure feedwater heater vent system leaked at a Korean nuclear power plant. From the results of the analysis, the effect of the generated droplets from the vent pipe and the cause of wall thinning were identified.

4.7.5 Operating Experience Insights

This section presents summaries of historically important pipe failures. Here, "significance" refers to incidents that had major operational consequences and prompted notable responses by industry and regulators seeking ways to manage long-term pipe material degradation. The selected failures, attributed to flow-assisted degradation of piping pressure boundaries, are outlined below in chronological order.

4.7.5.1 Pipe Rupture at Surry 2

On December 9, 1986, a main steam isolation valve failed to close at Surry Unit 2 (PWR), and the resulting increased pressure in the steam generator collapsed the voids in the water. This caused the system pressure to surge beyond the normal operating pressure and led to a catastrophic failure of a 90-degree carbon steel (SA-234 Grade WPB) DN450 elbow in the suction line to the main feed pump. At the time of the event, the reactor was at full power, and the feedwater was single phase, with a flow velocity of about 4.3 m/s, a pH level in the range of 8.8–9.2, an oxygen content of about 4 ppb, and a coolant temperature and pressure of approximately 188 °C and 3.1 MPa, respectively. Ammonia was used for the feedwater treatment. The examination of the ruptured elbow showed that the wall thinning was relatively uniform except in some local areas. The wall thickness of the elbow was reduced from a nominal 13 to 0.38–1.22 mm in small local areas and to 2.3 mm in larger areas. Eight workers were burned by flashing feedwater, four of whom subsequently died. The flashing feedwater interacted with and disrupted the fire protection, security, and electrical distribution systems.

Observation of the damaged A MFW pump suction piping indicated significant movement of the piping following the rupture. The piping, attached to the pump suction, dropped and rotated away from the break point, pivoting on the elbow near the pump suction. Although the piping came to rest against a portion of the B MFW pump discharge piping, it did not appear to have damaged it significantly. Inspection of the area following the event also revealed that one piece of suction piping had ripped off and was blown some distance from the break point. The piece was about 2 ft by 3 ft in size. It appears that the joint between the suction pipe and condensate supply header provided lateral support of the suction piping assembly, including the suction isolation valve. The loss of this support, along with the weight distribution of the suction pipe assembly, probably contributed to the pivot and rotation of the assembly. It is also likely that the backflow by the B MFW pump through the damaged A MFW pump discharge check valve and out the broken suction pipe contributed to pipe whip motion of the feed pump suction pipe.

The suction line was completely separated and dislocated from the header. Immediately after the failure, the licensee initiated a comprehensive analysis to determine the reasons for the failure and the necessary countermeasures. The failed elbow was fabricated from ASTM A-106, Grade B, Extra Strong, carbon steel seamless pipe and ASTM A-234, Grade B, Extra Strong, WP8 carbon steel wrought fittings with a nominal wall thickness of 12.7 mm.

When the DN450 elbow failed, a fragment approximately 2 ft by 3 ft was ejected from the outside of the elbow. The free end of the severed line was displaced in a horizontal direction approximately 2 m and was wedged against the bottom of the B main feed pump discharge line. During the displacement, the A suction line rotated around the point of connection at the inlet to the pump, severely deforming that portion of the line. Field metallurgical replicas taken on the surface of the elbow revealed a microstructure typical of ASTM A-106 Grade 8 material, with no signs of strain. The fracture surface was typical of a ductile tearing mode failure. Tears, which appeared to be fracture origination points, were noted at two localized thin cavity areas of the fracture. Small defects, indicative of laps, laminations, and inclusions typical for A-106, Grade B, materials, were noted at the fracture surface. One of these small defects was noted at one of the thin overload tear areas and could have been the start of the fracture (EPRI 1987).

4.7.5.2 Fukushima-Daiichi Unit 1 Turbine Building Flooding

The emergency diesel generators (EDGs) and battery rooms in the basement floor of the turbine building and the emergency electrical switchgear on the first floor of the turbine building were vulnerable to flooding, as was demonstrated by a plant event involving flooding at Unit 1 on October 30, 1991. Seawater leaked from a corroded pipe inside the turbine building, flooding an EDG room through its door and cable penetrations. Given the vulnerability to flooding from internal and external hazards of the EDGs, battery rooms, and rooms containing emergency electrical switchgear, an assessment of flooding should have been performed and documented in the plant safety analysis (IAEA 2015).[10]

[10] Each reactor unit at Fukushima Daiichi had two emergency diesel generators (DGs). The EDGs of reactor units 1, 2, and 5 were located in the basement of respective turbine building. The reason for installing EDGs on the basement floor level had been the need to deal with the vibration and seismic

A seawater leak was detected during routine patrol at 5:55 pm on October 30, 1991, near the reactor feedwater pumps in the basement of the turbine building. The plant began manual shutdown at 6:30 pm and fully shut down by 11:45 pm to identify the cause. Afterward, one of the two EDGs was flooded at a rate of about 10 metric tons of seawater per hour, traced to a failed DN250 branch line off the main service water system.

As a result of the on-site investigation, it was confirmed that a through-hole of approximately 900 mm^2 had developed near the point where the failed pipe branches off from the main service water pipe buried under the floor near the electric motor-driven reactor feedwater pumps and supplies the air conditioner for the reactor feedwater pump. It was also confirmed that there was no leakage from any other part except for this part. In the area near the through-hole, the rust inhibitor on the outer surface of the pipe had almost peeled off, and the wall thickness was 2.0 mm compared to the nominal wall thickness of 5.5 mm and the required minimum wall thickness of 3.0 mm. The wall thickness was relatively thin, ranging from 3 to 3.7 mm. The wall thickness in other areas was roughly equivalent to the nominal thickness. Observation of the cross section near the through-hole confirmed that the wall thickness of the pipe gradually decreased from the inner surface towards the end of the through-hole.

An analysis of the pipe's inner surface showed that it was coated with a standard tar epoxy lining. Oxides, likely caused by metal thinning, were discovered between the lining and the metal near the through-hole. The root cause analysis found that the epoxy liner had been damaged, allowing seawater to seep through the affected area. This led to gradual internal corrosion and material erosion. Investigating the nearby equipment, it was confirmed that the Units 1–2 common emergency diesel generator had experienced partial flooding. The diesel generator was sent to the factory for inspection and repairs, followed by comprehensive functional tests on-site to ensure its integrity. The damaged seawater piping was replaced; since the leak occurred in a buried section, the new piping was installed overhead to allow easier inspection and maintenance. The inner surface of the replacement piping received a polyethylene coating.

4.7.5.3 Pipe Rupture at Mihama 3

While Mihama Unit 3 was in operation at full power, a fire alarm sounded in the central control room on August 9, 2004. The control room operators determined that the alarm-generated area was on the second floor of the turbine building and checked the area to find that the building was filled with steam. Thus, it was judged that there was a high possibility of steam or high-temperature water leakage from the secondary piping. The operator started emergency load reduction. While those operations

safety of large-size equipment. The service water pipe failure in 1991 was an important precursor event.

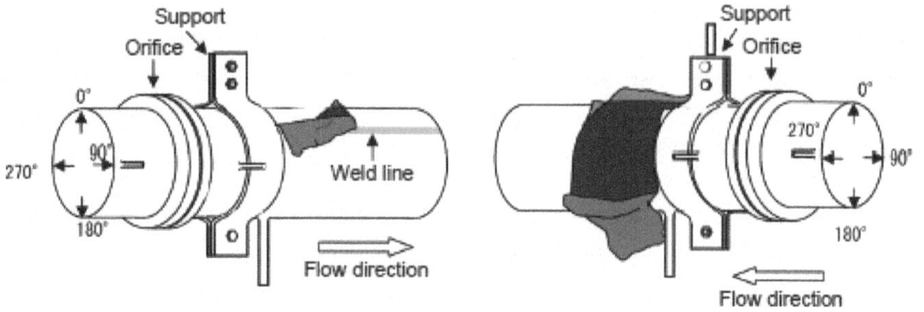

Fig. 4.15 Ruptured section of condensate pipe. (Reproduced from page 12 of the final incident investigation report, NISA (2005))

took place, a "3A Steam Generator Feedwater < *Steam Flow Inconsistency Trip1* >" alarm was generated, and the reactor and then the turbine shut down automatically. The operator inspected the turbine building and confirmed a ruptured DN500 condensate system pipe from the fourth low-pressure feedwater heater to the deaerator running near the ceiling on the deaerator side on the second floor of the turbine building (Fig. 4.15).

For the unit in question, the 21st periodical inspection was planned to start on August 14, 2004. At the time of occurrence of the accident, a total of 105 workers of Kansai Electric Power Company (the plant owner/operator) and affiliated companies were doing preparatory work in the turbine building for the periodical inspection. Of them, the affiliated company staffs working near the ruptured A-loop pipe were exposed to the steam and high-temperature water flowing out from the ruptured opening.

The affected section of piping had never been subjected to ultrasonic testing for wall thinning. Prior to this event, no regulatory requirements existed for nondestructive examination of piping systems considered to be susceptible to FAD mechanisms. In the immediate aftermath of the pipe rupture, the Japanese Agency for Natural Resources and Energy of the Ministry of Economy, Trade, and Industry (METI) requested that all nuclear plant license holders report on the implementation status of nondestructive examination of water and steam pipe wall thickness at the portions where thinning can occur and inspection implementation plans for the portions not yet subjected to inspections. The responses to this request were included as Appendix 10 ("*Pipe Wall Thickness Control at Thermal Power Stations*") of the final Mihama 3 incident investigation report (NISA 2005; Fujisawa 2025).

In association with the amendment of the Rules for the Enforcement of the Electricity Utilities Industry Law, the Nuclear and Industrial Safety Agency (NISA)[11] laid down the items concerning the selection of locations subject to

[11] NISA was the nuclear regulatory and oversight branch of the Agency for Natural Resources and Energy under the Ministry of Economy, Trade, and Industry (METI). It was created in 2001 during the 2001 Central Government Reform. After the Fukushima Daiichi nuclear accident, it was decided on June 20, 2012, that NISA would be abolished and that it would be replaced by a new

inspections, the selection of measuring points, the determination of the timing for the implementation of inspections, the measures that should be conducted depending on the calculated remaining life, etc. on the occasion of the implementation of the pipe wall thickness control, which was entrusted to licensees in the past, with which each licensee should comply on the occasion of the implementation of the inspections. This request was positioned as a provisional measure to be applied until the Japan Society of Mechanical Engineers (JSME) establishes technical standards concerning the pipe wall thickness control of nuclear power generation equipment and then the technical standards are positioned as criteria after NISA technical evaluations. In 2017, JSME published the "*Codes for Thermal Power Generation Facilities: Rules on Pipe Wall Thinning Management for Thermal Power Plants*," JSME S TB1-2016 (JSME 2017).

4.8 Stress Corrosion Cracking Mechanisms

Stress corrosion cracking (SCC) is the cracking induced by the combined influence of tensile stress and a corrosive environment. The required tensile stresses may be in the form of directly applied stresses or in the form of residual stresses. Cold deformation and forming, welding, heat treatment, machining, and grinding can introduce residual stresses. The magnitude and importance of such stresses have been underestimated. The time from crack nucleation to crack growth in the through-wall direction can vary significantly, from hundreds of hours to many years.

Residual stresses set up because welding operations tend to approach the yield strength. The buildup of corrosion products in confined spaces can also generate significant stresses and should not be overlooked. SCC usually occurs in certain specific alloy-environment-stress combinations. Ford (2010) defines the following sequence of events:

1. ***Precursor period*** during which specific metallurgical or environmental conditions may develop at the metal/solution interface that are conducive to subsequent crack initiation. For example, the condition at the end of the precursor period may be pitting corrosion that causes breakdown of the normally protective surface oxide film and which can lead to crack initiation. Other precursor phenomena might be associated with the creation of localized environments within crevices or on heat transfer surfaces. These "precursor" periods may be very short if stresses are high or if severe water impurity transients occur during initial reactor operations. An example would be the cracking of weld-sensitized Type 304 stainless steel BWR piping whose surface had been ground following welding to facilitate ultrasonic inspection. On the other hand, the precursor events may take years if they are associated with a change in metallurgical microstructure involving

agency, under the Ministry of the Environment. The Nuclear Regulation Authority (NRA) was founded on September 19, 2012.

thermal aging, irradiation embrittlement, grain boundary diffusion or oxidation, or creation of the necessary stress due to corrosion product formation and expansion. While the mechanism of degradation during the precursor period may be closely related to the processes that control cracking during subsequent stages, it can also be entirely different, depending on the details of the breakdown of the surface film. If this latter situation is the case, then analyses of times to failure can be complicated, especially if, for example, the system conditions are changed during the component lifetime.

2. ***Initiation of cracks*** when the local environment, microstructure, stress, and crack geometry conditions have reached a critical state. The morphology of the microscopic cracks that form during both this initiation period and the subsequent short crack growth stage depends on the precursor event and the initiating site. For instance, for cold-worked surfaces, or for cracks that initiated at randomly spaced pits, the cracking morphology is usually transgranular, while those cracks that are initiated at sites of intergranular attack will often propagate intergranularly. However, it is noted that, as the crack progresses, the cracking morphology may change depending on the specifics of the loading mode, the material microstructure, and the environmental conditions. Thus, it is not always correct to associate stress corrosion cracking of sensitized stainless steels in high-temperature water with exclusively an intergranular morphology, and this is especially true of more resistant materials like Alloy 690 and stabilized or L-grade stainless steels.[12]
3. ***Growth*** of the shallow individual cracks, followed by their coalescence to form a "major" crack. It is usually during this coalescence stage that the crack has reached dimensions that are resolvable by sensitive nondestructive examination techniques or there has been an observable drop in stress in a strain-controlled test or component. It is at this point that the crack is deemed to have "initiated." It is important to recognize, however, that the extent of damage at this stage is relatively large metallurgically (i.e., cracks may be much greater than grain size dimensions), and a lot has happened in terms of degradation phenomena during this "engineering initiation" time.
4. ***Propagation*** of a single dominant crack at a rate that, as will be discussed below, is very dependent on the material, stress, and environmental conditions. The growth rate is rarely constant; it may accelerate or decelerate over time depending on the stress and strain profiles (which can be complicated adjacent to welds) and material and irradiation conditions.

The historical development of SCC in nuclear power plants is told by Gorman (2015). Most failures, or about 80%, have been non-through-wall. Invariably, these failures have resulted in relatively lengthy plant outages. The through-wall failures have resulted in minor leak rates. Over 90% of the through-wall failures were discovered while preparing for nondestructive examination during scheduled

[12] "L"-grade stainless steel refers to low-carbon versions of standard stainless steel grades, like 304L or 316L, which have a carbon content limited to 0.03% or less.

maintenance and refueling outages, so-called perceptible leaks. SCC is mainly observed in the weld deposit and heat-affected zone, and it is considered that it occurs due to a synergistic effect of three factors of material, stress, and environments. SCC may occur when a susceptible material is subjected to stress in a corrosive environment. One example of a degradation scenario that might lead to SCC is one in which a weldment is (1) sensitized due to high heat input, (2) subjected to high local stresses such as welding residual stresses, and (3) subjected to a corrosive environment. There are five types of SCC mechanisms, and they are differentiated by crack morphology:

1. Interdendritic SCC in castings or welds in BWR and PWR operating environments.
2. Intergranular SCC in austenitic stainless steel heat-affected zones in BWR and PWR operating environments.
3. Primary water SCC, intergranular SCC that occurs during exposure of nickel-base alloys to high-temperature PWR primary water.
4. Strain-induced corrosion cracking of high-strength carbon steel and low-alloy steel.
5. Transgranular SCC (TGSCC), initiated in base metal from pipe inside or outside diameter in BWR and PWR operating environments.

Crack morphology refers to the study of the shape, form, and characteristics of cracks within metallic materials. Understanding crack morphology helps engineers design and analyze structures, predict potential failures, and develop strategies for improving material durability (Kitagawa et al. 1975). Factors affecting crack morphology are the following:

- Speed at which a crack propagates through a material is directly related to the morphology of the fracture surface. Smooth surfaces often indicate slow crack growth, while sharp, distinct features suggest rapid crack extension.
- Microstructure of a material significantly impacts crack morphology.
- Crack initiation and propagation: Cracks can initiate at various locations, such as grain boundaries or stress concentrators, and then propagate along specific pathways within the material.
- Cracks can propagate through different fracture modes, including ductile (deforming before fracturing) and brittle (fracturing without significant deformation).
- The shape and features of the crack tip (e.g., sharpness, branching) provide insights into the stress field, crack propagation mechanisms, and material toughness.
- Environmental factors, such as temperature, pressure, and corrosive environments, can also influence crack morphology and propagation.

4.8.1 Interdendritic Stress Corrosion Cracking

Interdendritic SCC is a form of corrosion that preferentially occurs along the dendritic boundaries within a metal structure, particularly in welds and cast alloys. It is a type of stress corrosion, where the combination of tensile stress and a corrosive environment causes crack formation and propagation. This type of cracking has been seen in austenitic stainless steels like AISI Type 316 and nickel-base alloys like Inconel 182.

Following the removal of reactor recirculation piping at BWR unit in the United States in late 1982, metallurgical analyses were performed to characterize the depth and mode of cracking in Type 316 stainless steel base metal. Surprisingly, at the time, these analyses revealed that in 2 of the 28-inch-diameter girth weld samples, the cracking had penetrated into the weld metal and propagated through the approximate midplane of a repair weld volume, thus posing questions about the possible contributory role of hot cracking in this weld defect. Since weld metal microfissuring or hot tearing tendencies are usually increased in such repair geometries, the extent and location of the cracking suggested a preferential crack path. The crack appears to have propagated in an "interdendritic" manner from ID-initiated IGSCC through a substantial amount of weld metal.

4.8.2 Intergranular Stress Corrosion Cracking

Intergranular SCC of stainless steels is a time-dependent type of material degradation phenomenon. The intergranular SCC microstructure is associated with the temperature/time fabrication conditions that gave rise to thermal sensitization and formation of chromium carbide precipitation and chromium depletion at the grain boundary. Reduction in chromium concentration adjacent to the grain boundary gives rise to a reduction in passivity and makes the material susceptible to intergranular SCC. The cracks originate from the water side of the pressure boundary; that is, it is the aqueous environment which is responsible for the corrosion factor in the cracking.

Research on intergranular stress corrosion cracking (IGSCC) has concentrated on crack propagation mechanisms, providing a framework for analyzing the influences of material properties, stress, and environmental conditions. Nevertheless, there remains a notable gap in data regarding crack initiation processes. Despite the introduction of low-carbon and stabilized grades of stainless steel, instances of IGSCC have been observed in materials that were demonstrably not sensitized. Studies indicate that susceptibility in these cases arises from cold work during fabrication. Frequently, initial cracking manifests as transgranular before transitioning to an intergranular mode, with early cracks often linked to surface layers affected by cold work from grinding.

Table 4.6 Chronology of significant IGSCC failures

IGSCC failures	Year of discovery
USA: Dresden 1: Weld in DN150 reactor recirculation loop piping	1965
Japan: Through-wall leak reported at the Japan Power Demonstration Reactor (JPDR) (Torikai et al. 1978; Fatamura et al. 1979)[a]	1972
Germany: The first IGSCC failure at KWW Würgassen	1973
Germany: The first IGSCC failure of large-diameter (DN600) reactor recirculation piping. Cracks were found in furnace-sensitized stainless steel piping	1978
USA: Extensive IGSCC in nickel-base alloy at Duane Arnold—Through-wall leak	1978
Spain: Extensive IGSCC in nickel-base alloy at Santa Maria de Garona—Through-wall leak	1980
USA: Nine Mile Point 1: Extensive IGSCC in DN700 piping in 1982 and again in 2023 in a single weld	1982
Germany: Extensive IGSCC found in stabilized austenitic stainless steel piping	1992
Japan: Extensive IGSCC found in non-sensitized low-carbon-stabilized austenitic stainless steel piping. Attributed to material surface hardening by machining and cold work or surface residual stress resulting from cutting and grinding	2002
China: Extensive cracking in various diameter elbows at the Daya Bay Units 1 and 2. The first reports of cracking occurred after about 17 years of operation in October 2010	2010
Japan: During ultrasonic examination performed at Ohi Unit 1, an indication was reported for a pipe weld between the primary coolant pipe nozzle and an elbow of the DN100 pressurizer spray valve	2020
France: IGSCC found in multiple locations in multiple plants in low-pressure safety injection (LPSI) and residual heat removal (RHR) pipe welds	2021

[a] The JPDR was a demonstration of BWR. It was first connected to the grid in 1963 and was permanently shut down in 1976. The JPDR played a crucial role in advancing the understanding of stress corrosion cracking

4.8.2.1 IGSCC in BWR Operating Environments

IGSCC has been observed in various BWR structural components since the 1960s due to the interactions of specific material, environmental, and stress conditions (Table 4.6). Bush and Dillon document observed preoperational failures, failures during startup tests, and operational failures in the period 1960–1973, a total of 88 cases (Bush and Dillon 1973).

The early failures were dominated by stress corrosion cracking on the inside surface of piping. In some cases, these cracks were detected after relatively short operating times. A major concern arose when IGSCC was observed in late 1965 at Dresden 1 due to IGSCC in the weld heat-affected zone (HAZ) of a DN150 Type 304 stainless steel recirculation bypass piping after about 6 years of operation. Initially, the Dresden 1 incident in 1965 was regarded as an "unusual" event, but its generic nature was revealed when numerous indications of IGSCC of welded Type 304 stainless steel piping occurred in the United States late in the 1974–1975 timeframe; 64 incidents of cracking were identified during this short period, and all of them occurred in piping < DN250. There were also numerous examples of stress corrosion cracking in sensitized Type 304 stainless steel components in the oxidizing BWR "normal water chemistry" environment being accelerated by the presence of surface cold work. In these cases, the cracking morphology started off as transgranular and transitioned to the intergranular mode at greater crack depths.

During 1978, IGSCC was observed at Duane Arnold in a large-diameter (610 mm) recirculating system pipe, and in 1982, extensive IGSCC of 710 mm recirculation piping was found at Nine Mile Point Unit 1 adjacent to the furnace-sensitized safe-end-to-pipe weld (USNRC 1982a, b; Delwiche et al. 1983). USNRC Bulletins in 1982 (USNRC 1982b) and 1983 (USNRC 1983) at Monticello, Browns Ferry 2, Quad Cities 1, Millstone 1, Hatch 1, Brunswick 1, Oyster Creek, and Duane Arnold nuclear power plants revealed similar cracking in large-diameter piping (USNRC 1983). These events posed considerable concern since replacement of such large-diameter lines would be difficult and costly and would require additional leak-before-break and degraded pipe behavior analysis.

The USNRC issued two Guidance Letters (GLs)[13] regarding pipe replacement and weld inspections in 1984. The first, GL 84-07 (USNRC 1984a), addressed the overall procedure for pipe replacements, and the second, GL 84-11 (USNRC 1984b), addressed the inspection procedures. At the same time, a long-term strategy for regulating these issues was defined in SECY 84-301 (USNRC 1984c).

Utilities in the United States formed the BWR Owners Group (BWROG) in 1979 to focus research into repair and mitigation techniques, and the USNRC issued further guidance in Generic Letter 88-01 (USNRC 1988) together with a detailed technical argument, outlining the staff position on materials, inspections, repairs, and mitigation actions for austenitic stainless steel piping with diameters DN100 or greater. Subsequently, in 1999, the industry, via the BWR Vessel and Internals Project (BWRVIP), submitted revisions to the Generic Letter and referred it to as the BWRVIP-75 report (EPRI 1999). The proposed revisions were based on the consideration of inspection results and service experience gained by the industry since the issuance of GL 88-01 and included additional knowledge regarding the benefits of improved BWR water chemistry. The BWRVIP-75 report also provided justification for the proposed inspection criteria. These revisions were accepted by the USNRC in the following year.

Since the late 1970s, the importance of water purity control became increasingly apparent, especially regarding creviced components (where the geometry and oxidizing conditions in the bulk environment could give rise to increased anionic activity in the creviced region), even though the bulk water purity was acceptable at that time. These observations led to the EPRI "*BWR Water Chemistry*

[13] A Generic Letter in reactor regulation is a formal communication from the USNRC that requests all licensees to act, provides new policy or technical guidance, or asks for information on a specific issue. These letters address safety issues that are common to multiple facilities and require a written response from the addressees.

Guidelines"[14] that defined the specifications for water purity and corrosion potential (Wada et al. 1999). Various mitigation actions were developed because of these evolving problems and centered on the following aspects:

- Alloy compositions and fabrication processes that minimize grain boundary sensitization.
- Fabrication processes that minimize the tensile residual stress adjacent to welds.
- Water chemistry control procedures to lower the corrosion potential of the stainless steel and to lower the non-OH^- impurity content, especially chlorides and sulfates. These water chemistry control approaches were particularly effective since they impacted on all wetted surfaces in the reactor circuit.

There have been situations where mitigation effectiveness has been shown to be compromised. Although the substitution of Type 304 stainless steels with Type 316L-grade and NG-grade alloys in both unirradiated and irradiated components has increased the operating time before cracking is observed, incidents can still occur (Horn et al. 1997; Ehrnsten et al. 2001). This has been observed, for example, in the 316NG piping at Forsmark in Sweden and the 316L core shroud at Fukushima 3 in Japan, especially if (1) the component has been severely surface cold worked by grinding, straightening, etc. during manufacture; (2) there is evidence of high weld shrinkage strains; or (3) heat treatment and/or compositional control to avoid sensitization has been inadequate. This observation has also been made for the stabilized stainless steels, following the observation in 1992 of stress corrosion cracking of Type 321 piping at the Würgassen and Brunsbüttel plants in Germany (Kilian et al. 1995; Erve et al. 1997).

During the 1970s, Japanese and German BWR operators implemented programs to replace conventional Type 304 stainless steel in their plants. The Japanese operators selected nuclear-grade Type 316 stainless steel, while the Germans used a stabilized Type 347 stainless steel. In Japan, the replacement effort included applications of induction heating stress improvement (IHSI) for selected larger-diameter pipes. Laboratory testing on these materials studied sensitization, effects of cold work, effects of off-normal water chemistries, and crack growth. Based on the study of several product forms of Type 304 and Type 316 stainless steel, the nuclear grades of Type 304 (304NG) and Type 316 (316NG) stainless steel were seen to offer the optimum combination of IGSCC resistance, mechanical properties, and availability. Nuclear grades are low-carbon (below 0.02%) versions of the conventional Type 304 and Type 316 stainless steel with the addition of nitrogen (0.06–0.10%) to maintain the same mechanical properties.

The incidents of stress corrosion cracking of stainless steels in BWRs have occurred in eight "stages," including the incidents starting in the mid-1960s associated with smaller-diameter sensitized Type 304 stainless steel piping which peaked in the early 1980s with the detection of similar cracking in larger-diameter piping. This was followed in the early 1990s by cracking of stabilized stainless steel (Type

[14]The first edition of these guidelines was published in April 1985, EPRI NO-3589-SR-LD.

347 and Type 321). In summary, from the point of industry and regulatory responses to the IGSCC problem, the eight stages of IGSCC are as follows:

1. Early indications of IGSCC in small-diameter primary system piping.
2. First IGSCC incidents in large-diameter primary system piping.
3. Mandatory inspections of all primary system pipe weld in response to Bulletins 82-03 (USNRC 1982b) and 83-02 (USNRC 1983), and Generic Letters 84-07 (USNRC 1984a) and 84-11 (USNRC 1984b).
4. Implementation of IGSCC mitigation measures. Improved water chemistry and chemistry control programs, lowering the electrochemical corrosion potential by the addition of hydrogen. Application of piping stress improvement processes and use of nuclear-grade stainless steels.
5. German IGSCC experience with respect to stabilized austenitic stainless steel applications.
6. Mostly non-US operating experience, which reflects the implementation of enhanced inspection programs and new NDE qualification procedures.
7. Japanese IGSCC experience with respect to low-carbon-content ($\leq$0.03%) austenitic stainless steel applications (IAEA 2011).
8. The post-2005 to present-day IGSCC experience: Recent operational data underscores the necessity of validating the effectiveness of IGSCC mitigation strategies for extended service life, including periods exceeding 40 and 60 years.

4.8.2.2 IGSCC in PWR Operating Environments

IGSCC failures in the PWR primary system are rare occurrences. In principle, these pipe runs are not affected by this degradation mechanism due to the high quality of the chemistry-controlled primary water. In 1978, the USNRC formed a second Pipe Crack Study Group (PCSG); the first PCSG, established in 1975, addressed cracking in BWRs only. One of the objectives of the second PCSG was to include a review of the potential for stress corrosion cracking in PWRs. The results of the study were presented in NUREG-0691 (1980) entitled "*Investigation and Evaluation of Cracking Incidents in Piping in Pressurized Water Reactors*." The report (USNRC 1980) stated:

> *The PCSG has determined that the potential for stress corrosion cracking in PWR primary system piping is extremely low because the ingredients that produce IGSCC are not all present. The use of hydrazine additives and hydrogen overpressure limit the oxygen in the coolant to very low levels. Other impurities that might cause stress-corrosion cracking, such as halides or caustic, are also rigidly controlled. Only for brief periods during reactor shutdown when the coolant is exposed to the air and during the subsequent startup conditions are even marginally capable of producing stress-corrosion cracking in the primary systems of PWRs. Operating experience in PWRs supports this determination. To date, no stress corrosion cracking has been reported in the primary piping or safe ends of any PWR.*

Except for safe ends, this expert opinion has largely proven to be accurate. However, IGSCC has affected Safety Class 1 piping connected to the primary piping. During an in-service inspection in October 2021 at Civaux Unit 1[15] PWR in France, the licensee found circumferential cracking at several locations near an elbow in the low-pressure safety injection system, near the reactor coolant system cold leg. The maximum depth of the cracking was 5.6 mm and, in one case, extended all the way around the inside circumference of the pipe. A few weeks later, similar cracking was found at other plants of the same design as Civaux Unit 1. The affected piping was about 300 mm in diameter with a thick 30 mm pipe wall. These lines are typically stagnant, but some circulation occurs in these sections of the SIS due to the proximity of the piping to the cold leg reactor coolant piping. The locations of the flaws were in a non-isolable section of the piping system that is susceptible to thermal fatigue. In most cases, the cracks were long, but very shallow. Through destructive examination, it was determined that the root cause of the cracking was IGSCC caused by stresses associated with thermal stratification. Weld residual stress (WRS) may have also played a role in the cracking (Kaijua et al. 2018; ACRS 2022; Hahn et al. 2023; Ehrnstén et al. 2024).

Subsequent analyses have attributed the IGSCC incidents to a combination of the presence of sensitized or cold-worked stainless steels with operational practices that involved injecting inadequately deaerated boric acid and/or activity reduction practices involving oxygenation using hydrogen peroxide of the primary circuit during cooldown. Welding processes have a strong influence on IGSCC susceptibility. As an example, narrow-gap welding generally results in lower residual stress and strain in the heat-affected zone than wide-gap welding.[16]

Factors promoting IGSCC include high temperature and an oxygenated environment. The presence of oxygen in the reactor cooling system (RCS) of PWRs is largely eliminated to enhance the corrosion resistance of stainless steels. To suppress oxygen in the RCS, scavengers such as hydrogen are added to the primary water. Consequently, because of the hydrogen injection, oxygen is not monitored in some plants when the reactor is in operation. In plants while monitoring oxygen in the RCS, the oxygen content is generally in the order of one to a few parts per billion (ppb). Cattant (2022) lists the following possible sources of oxygen ingress into the RCS:

- Addition of reactor make-up water-concentrated boric acid and/or blended (recycled and fresh) boric acid via the chemical and volume control system charging line induces an oxygen ingress.

[15] A PWR of type N4 entered commercial operation on 29 January 2002.

[16] Narrow-gap welding uses a smaller gap, typically with a V-groove angle between 0° and 8°, while wide-gap welding uses a larger gap, often with a bevel angle of 30–37.5°. The choice between them depends on factors like material thickness, desired weld properties, and cost considerations.

[17] Load-following nuclear power plants are those designed to adjust their power output in response to fluctuations in electricity demand.

- Load-following also results in oxygen ingress.[17] Boric acid is added to reduce the power load, and the RCS is diluted with make-up water to return to 100% power.
- Oxygen ingress can occur at power. Some plants have fully oxygenated reactor make-up water tanks.
- Oxygen increases can increase at power towards the end of cycle. For PWRs without a boron thermal regeneration system (BTRS), increasingly larger volumes of reactor make-up water are added at the end of cycle to dilute the boron concentration, carrying with it the risk of significant oxygen ingress into the RCS.
- Oxygen can also enter the RCS during boration under hot shutdown conditions. Under these conditions, part or all of the CVCS charging and seal injection flows can be oxygenated.
- Oxygen ingress can occur when aligning the residual heat removal system (RHRS) trains at less than 177 °C. The water contained in the RHRS can sometimes be fully aerated.
- Oxygen ingress can also happen when adding blended boric acid during cool-down to accommodate changes in density and on collapsing the pressurizer bubble.
- Oxygen can enter the pressurizer during shutdown. Typically, boric acid and blended boric acid can be added to the RCS via the auxiliary spray line; thus, local oxygenated conditions can be formed in the pressurizer.

4.8.2.3 IGSCC in Carbon Steel Piping

Carbon steel piping is susceptible to stress corrosion cracking. Evidence of IGSCC was first observed in component cooling water (CCW) piping in the mid-1990s (Fig. 4.16). The piping of the closed-loop CCW system is moderate energy, with a design temperature of about 65 °C and operating at around 30 °C. In subsequent years, from around year 2000 to present day, through-wall leaks have been observed at several plants, mainly in weld heat-affected zones in DN75, DN100, and DN150 Schedule 80, 120, or 160 seamless carbon steel piping, e.g., ASTM-A-106 GR B (Staehle 2002; EPRI 2004b).

The CCW system is designed to remove heat from heat exchangers, pumps, and other systems. Cooling water is pumped through these units in parallel flow circuits, removes heat from the various components, and flows to the cooling heat exchanger, which is cooled by the service water. In PWR plants, some CCW loops serve as an intermediate system between the reactor coolant system and the service water system. The water in CCW systems can be either fresh water that is chemically treated to inhibit corrosion and biological growth or high-purity demineralized water, to which an oxygen scavenger may be added. Water chemistry guidelines have been developed by EPRI (2004c) and VGB PowerTech® in Germany (VGB PowerTech 2000).

Failure analyses have determined that the stress corrosion cracking resulted from a combination of high residual stresses at welds and dissolved oxygen in the coolant, with possible contributions from slightly low concentrations of molybdate, presence

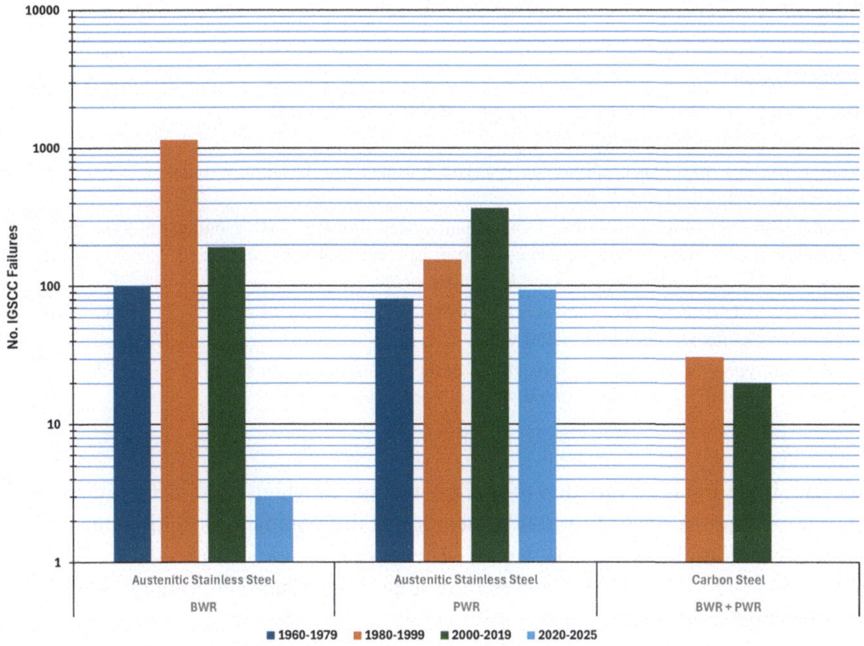

Fig. 4.16 Carbon steel stress corrosion cracking experience in perspective

of tolyl-triazole, and slightly low pH value. At high tensile stresses, such as residual weld stresses, SCC can occur even in pure, low-oxygen water. It is believed that SCC can initiate at stresses higher than about 0.5 times the yield strength. Cracks may initiate by stress-induced pitting, where there is a competition between a fracture of the protective oxide film due to local plasticity and re-passivation. Stress corrosion crack propagation rate is constant over a range of stress intensity factors for carbon steels with similar yield strength and specific environments and temperatures. The propagation rate increases exponentially with temperature: for pure, low-oxygen water: 2×10^{-5} mm/h at 100 °C and 1×10^{-4} mm/h at 150 °C.

4.8.3 Primary Water Stress Corrosion Cracking

The intergranular SCC of Ni-base alloys is a time-dependent material degradation phenomenon. Ni-base alloys, particularly Alloy 600, and weld metals Alloys 82, 132, and 182 proved to be generically susceptible to interdendritic and intergranular SCC in normal specification PWR primary water. This is commonly referred to as primary water SCC (PWSCC). A chronology of significant PWSCC failures is given in Table 4.7 (Gorman and Hunt 1993). Post-welding heat treatment (PWHT) is avoided by using weld buttering. According to the American Welding

Table 4.7 Chronology of significant PWSCC failures

PWSCC event	Year of discovery
USA: San Onofre Unit 3: First case of piping PWSCC of dissimilar metal pressurizer instrument nozzle weld	1986
France: through-wall leaks reported at two type N4 reactors—Dissimilar metal pressurizer instrument nozzle welds	1989
Sweden and USA: First cases of reactor coolant system hot leg to RPV outlet nozzle dissimilar metal welds. Leakage-monitoring records at V.C. Summer showed 0.02 kg/s unidentified leakage from all sources	2000
Japan: PWSCC detected at several reactor coolant system hot leg to steam generator inlet nozzle dissimilar metal welds	2006–2008
USA: Wall-penetrating PWSCC at reactor coolant system hot leg to steam generator inlet nozzle dissimilar metal weld at North Anna Unit 1	2012

Society, buttering is "a variation of surface welding in which weld metal is deposited on one or more surfaces to provide metallurgically compatible weld metal for the subsequent completion of the weld." The operating experience shows that the fabrication-induced residual stresses have a large influence on primary water SCC in Alloy 600 weld metal.

Primary water SCC in the weld metal grows along the grain boundaries of columnar crystal dendrite packets. Initiation in the weld metal is often thought to be the result of typical and nontypical fabrication processes leading to locally high residual stresses or surface stresses from, for example, grinding. To date, it has been found that the susceptibility to SCC of nickel-base alloy weld metal is higher than that of the base metal. Alloy 600 was developed in the 1950s for nuclear power plant applications. It was qualified for use because of its perceived resistance to SCC; it was viewed as an alternative to Type 304 or Type 316 austenitic stainless steels. An early, possibly the earliest recorded instance of SCC of Alloy 600 material is that of the failed inspection tubes in the Swedish Ågesta heavy water reactor in September 1964 (Grönwall et al. 1966).

Alloy 690, which has higher chromium content than Alloy 600, was developed in the 1960s and has been widely used since the 1980s to replace Alloy 600 components, beginning with thin-walled steam generator tubes and eventually including thick-section nozzle penetrations in reactor pressure vessel heads. This choice was made following numerous laboratory studies which confirmed the resistance to SCC in reactor coolant primary water for this alloy. Alloys 52 and 152 are the higher-Cr-content nickel-base weld metals that are used for joining Alloy 690 components and which are also used for weld overlays to mitigate the SCC susceptibility of Alloy 82 and 182 welds.

To date, there are no known cases of in-service primary water SCC of Alloy 690 and its weld metal Alloys 52 and 152. Scott and Combrade (2019) made the following statement regarding the use of Alloy 600 materials:

This story is something of a valedictory review as most Alloy 600 components in LWR service have now either been replaced during the last two to three decades with more SCC-resistant materials like Alloy 690 or been given a remedial treatment to eliminate one or more of the precursors of SCC. Today, due to the vulnerability of the Alloy 600 family of materials to SCC, especially in PWR primary water service, they would be most unlikely to be specified for any new plants or for replacement components in existing plants. Nevertheless, the scientific fascination with Alloy 600 and its kindred materials continues, especially with trying to elucidate the mechanism(s) of SCC. Indeed, it is justified in our opinion as an essential steppingstone to qualifying the required longevity of the replacement materials.

4.8.4 Transgranular Stress Corrosion Cracking

The earliest indications of cracking in unirradiated austenitic stainless steels occurred in the late 1960s in components where the temperature was <100 °C, and this was observed during storage, fabrication, and operation. The degradation mode was TGSCC on the outside surface of the pipe. TGSCC of austenitic stainless steels occurs primarily owing to chloride contamination. Other halide anions, such as fluorides, can also cause TGSCC. The problem initiates on the outside surfaces of austenitic stainless steel components, mainly owing to a lack of adequate cleanliness. Wetting due to condensation or nearby water leaks can allow an aqueous environment to form and lead to TGSCC accompanied by pitting or crevice corrosion. The implementation of known procedures that ensure adequate surface cleanliness is a continuing necessity that requires careful management attention at all stages of construction and operation.

Chloride-induced TGSCC can also occur in internal surfaces, generally in dead legs and stagnant regions due to the high probability of the simultaneous presence of chloride contamination and oxygen. One location that has been rather frequently affected by TGSCC in PWRs is at the canopy seals that ensure the pressure boundary of threaded connections in the control rod drive housings that are in the reactor pressure vessel head. Leaks from the canopy seals have caused serious boric acid corrosion of the upper-head low-alloy steel.

TGSCC has also occurred from inner surfaces, mainly in pipe sections containing stagnant two-phase coolant, where evaporation and concentration of chlorides can occur. Wetting due to condensation or nearby water leaks allows an aqueous environment to form that leads to TGSCC, usually accompanied by pitting or crevice corrosion. The stress required for chloride-induced TGSCC is relatively modest, with the threshold being close to the proportional yield strength of solution-annealed austenitic stainless steels. Implementation of the known adequate procedures to ensure appropriate surface cleanliness is a continuing necessity that requires careful management attention at all stages of construction and operation.

The external chloride-induced SCC (ECSCC) is TGSCC initiated on the outside surface of a component due to the presence of chloride in sea salt, coatings, etc. attached to the material surfaces and by perspiration. For ECSCC to take place, several factors are usually present: high carbon content, tensile stress, moisture, heat, aggressive chemical environment (e.g., chlorides or fluorides), and sensitized metal. Heat treatment or welding sensitizes the metal by depleting the chromium in grains and forming chromium compounds at grain boundaries.

4.8.5 Strain-Induced Corrosion Cracking

The term "strain-induced corrosion cracking" (SICC) refers to those corrosion situations in which the presence of localized dynamic straining is essential for crack formation to occur, but in which cyclic loading is either absent or restricted to a very low number of infrequent events. SICC has sometimes been referred to as "strain rate-sensitive SCC." Strain-induced corrosion cracking has been observed in pressurized components made of higher-strength ferritic carbon steel. This kind of degradation has caused circumferential cracking in feedwater nozzle regions and at welds and axial cracking in pipe bends but also in straight sections of thin-walled piping in German nuclear power plants.

The use of fine-grained, low-alloy higher-strength steels such as WB35 and WB36 allowed the use of thinner-walled piping without stress relief treatment of welds. The environmental conditions that aggravated the cracking susceptibility in these incidents were as follows (Kussmaul et al. 1984; Hickling 1990):

- Dynamic straining associated with, for instance, reactor startup or thermal stratification during low feedwater flow or hot standby conditions. Such operations lead to a wide range of applied strain rates and would be expected to increase the crack propagation rate.
- High local stress at or above the high temperature yields stress, thereby giving a lack of plastic constraint at the incipient crack tip and consequently an anomalous increase in crack propagation rate due to the effective increase in crack tip strain rate. Such high local stresses were attributed in failure analyses to welding defects (e.g., misalignment of weld edges), piping fit-up stresses, and, in some cases, inadequate pipe support at elbows. The combination of this high stress adjacent to the weld and the high applied strain rate led to a distribution of multiple cracks around the circumference of the pipe that was no longer confined by the asymmetric azimuthal distribution of weld residual stresses.
- Oxidizing conditions, in conjunction with intermediate temperatures and potential anionic impurities: The anions may influence the conductivity and pH of the coolant within a crack. This, in conjunction with environmental factors, was further aggravated by the fact that during reactor shutdown, stagnant water was sometimes left exposed to air in horizontal portions of piping; pitting and general corrosion occurred under these low-temperature conditions, and these pits were observed to act as crack initiators during subsequent operating cycles.

4.8.6 *Operating Experience Insights*

This section provides abstracts of notable SCC failures with historical relevance. The designation "significance" encompasses both the operational consequences and the degree of response from industry and regulatory bodies in developing effective strategies for the long-term management of pipe material degradation. Summaries of key SCC failures, presented in chronological order, are included below.

4.8.6.1 Duane Arnold Energy Center IGSCC Failure

Arguably the most significant IGSCC failure to date, the Duane Arnold Energy Center (DAEC)[18] experienced a primary system leakage in 1978. The control room operators had been monitoring a slowly increasing drywell unidentified leakage since May 1, 1978. On June 14, 1978, an increase in unidentified leakage was observed. The leakage increased from approximately 6.3×10^2 kg/s to approximately 0.2 kg/s during the day of June 14, 1978. The increased leakage was immediately observed by six independent methods, including the following: (1) recorders in the control room indicated increased drywell sump pumping activity; (2) radwaste control room operators noticed increased tank levels; (3) iodine levels in the drywell increased; (4) drywell temperature in local areas showed slight increases; (5) particulate radiation levels increased slightly; and (6) containment atmosphere weight increased.

An investigation was initiated to determine the possible cause of the increase in leakage. A decision was made late on June 13 to wait for a few days to see if the leakage would remain steady. The leakage at that time was about 0.2 kg/s in comparison to the technical specification limit of 0.3 kg/s. At 00:55 on June 17, 1978, during weekly control valve testing, an automatic reactor scram occurred due to problems in reactor protection system relays associated with the testing. The Chief Engineer was notified approximately 30 min after the shutdown, and he made the decision to reduce reactor pressure, de-inert the containment, enter the containment, and investigate the leakage. Based on the week's observation of drywell leakage and the fact that the plant was shut down, it seemed prudent to look for the leak rather than wait until the leakage approached the technical specification limit. The survey of the drywell revealed a leak in the N2A nozzle (1-of-8 recirculation inlet nozzle-to-safe-ends).

All safe ends were removed, and destructive examinations of the affected safe ends revealed that all eight safe ends had surface cracks which extended essentially completely around the circumference of the design; these cracks were in the crevice region of the safe-end design. The depth of the cracks typically ranged from 50% to

[18] Duane Arnold, a General Electric BWR/4 reactor, was commissioned in the early 1970s and entered commercial operation on February 1, 1975. The reactor was permanently shut down on October 12, 2020. However, as of this writing, plans are in place to recommission the plant.

Table 4.8 The results of the initial UT/RT examinations

Safe-end designation	Ultrasonic testing results	Radiographic testing results
A	Approximately 90° through-wall crack with additional indications for an additional 180° of the ID	Crack indication for approximately 270° of ID including through-wall crack for approximately 90°
B	Linear 360° indication on the ID in both axial directions	Crack indication on ID for 360°
C	Same as B	Crack indication for approximately 270° of ID with an added approximately 5 indication in the middle of remaining 90°
D	Same as B	Crack indication for approximately 320° of ID with no indications occurring at two locations for approximately 20° in each location
E	Intermittent traveling indication for 360° of the ID	Crack indication for approximately 160° of ID with intermittent slag indications throughout the thermal sleeve weld
F	Same as B	Crack indication on ID for 360°
G	Same as B	Crack indication on ID for approximately 260°
H	Same as B	Clear crack indications at two locations on the ID for a total of approximately 120°. Intermittent indications on ID for an additional approximate 30°

75% of the wall thickness, except for the leaking safe end where a through-wall crack was present at an 80-degree segment of the circumference. Based on radiography, crack indication was observed for approximately 360° of the inside diameter, including a through-wall crack for approximately 90° on the outside diameter. The crack was believed to have started at a crevice location with high residual stresses caused by an unusual 360-degree repair weld. Following the preliminary UT and RT examinations, safe ends A and E were sent to the Battelle Memorial Institute (Pasupathi et al. 1979) and the Southwest Research Institute (Burghard and Bursle 1979) for metallurgical examinations (Table 4.8).

These examinations confirmed that (1) the cracking was entirely on the safe-end side of the crevice formed by the welding of the thermal sleeve to the safe end, originating near the crevice tip; (2) the cracks began in the weld HAZ; (3) the cracks propagated entirely by intergranular stress corrosion with an absence of evidence of mechanical fatigue; (4) the weld repair on the outside surface of the safe end was not involved in crack initiation (and had little to do with the later states of cracking); (5) the cracks did encircle the safe-end inside diameter (at the crevice) around the full 360°; (6) there were a significantly large number of particles on the fracture surface which were compounds high in sulfur from an unidentified source.

4.8.6.2 RCS Hot Leg PWSCC Failures

PWSCC in PWR RCS hot leg dissimilar metal welds: Cracking in Alloy 82/182 butt welds was not observed in operating PWR plants until the year 2000, when several incidents occurred. The first cracking event was found in June of that year in Ringhals Unit 3, but the flaw indications were thought to be shallow artifacts, and the plant was allowed to remain in service without repair (Skånberg 2001). The second was in the reactor vessel outlet nozzle to safe-end weld of Ringhals Unit 4 in July of 2000. Several small axial cracks were found and removed with a boat sample by electrical discharge machining (EDM) (Table 4.9). In both cases, the welds were Alloy 182, and the cracks were axial.

The next major incident occurred in October 2000, when the Virgil C. Summer plant was found to have a through-wall flaw in the same region as the PWSCC indications at the Ringhals plant, the reactor vessel outlet nozzle to safe-end weld. In this case, the weld was a field weld that had experienced multiple repairs during the construction process.

On October 7, 2000, the Virgil C. Summer plant was shut down for a normal refueling outage. On the same day, plant personnel identified an accumulation of boric acid near the "A" loop of the reactor vessel. Subsequent inspections revealed small amounts of boron buildup on the weld between the vessel nozzle and the hot leg pipe. Within hours, the suspect area was cleaned, and a dye penetrant (PT) examination of the pipe identified a 100-mm-long indication at the weld approximately 1 m from the vessel, between the hot leg piping and the reactor vessel nozzle (Brust et al. 2002; EPRI 2004d). The flaw indication was located about 430 mm from the top of the pipe. This pipe has an inside diameter of 736.6 mm and wall thickness of 63.5 mm. After the insulation was removed, leakage-monitoring records were searched. Leakage records showed a nearly constant value of 0.02 kg/s of unidentified leakage from all sources, well below the plant technical specification limit of 0.06 kg/s. The outside-diameter through-wall flaw was characterized as a "weep hole" or a "pinhole" which indicated that the leak rate produced by the flaw was less than or much less than 1×10^{-3} kg/s. Ultrasonic inspection (UT) performed on the pipe from the inside surface initially revealed a single axial flaw near the top of the pipe. Follow-up exams conducted in the spring of 2002 revealed that there were several flaws, all but one of which was axial, and that the largest axial flaw was a through-wall flaw. The flawed region was removed, and a new spool piece welded in place, restoring this region to its original condition. The

Table 4.9 Ringhals 4 crack sizing results

Crack position relative to TDC	Depth per ultrasonic measurements (mm)	Depth per destructive ultrasonic measurements (mm)
91°	10 ± 3	9 ± 1
124°	9 ± 3	6 ± 1
305°	16 ± 3	22 ± 3
330°	13 ± 3	22 ± 3

Virgil C. Summer outlet nozzle to pipe weld was repaired with Alloy 52, for a portion of the thickness exposed to the reactor coolant, and the remainder of the weld was filled with Alloy 82.

4.9 Hydraulic Pressure Transients

Steam or water hammer is a phenomenon occurring in a piping system with valves used to control the fluid (or steam) flow; it occurs from the starting and stopping of pumps, from the opening and closing of valves, from changes in flow direction, or from water column separation and collapse. It is a result of a pressure surge (high shock wave) that propagates through a piping system because fluid in motion is forced to change its direction or is abruptly forced to stop moving. This abrupt change in fluid momentum makes a shock wave to travel back and forth between the cause that created it and a resistance point in the system, such as a restriction orifice, a valve downstream, or several elbows in between. If the intensity of a shock wave is high, physical damage to the system can be expected. Although the water hammer is extensive, relatively few water hammer events have resulted in significant pipe failures.

4.9.1 The Causes of Water Hammer Transients

Fluid transients result from disturbances, either from within a system or from without, which create pressure waves. These pressure waves may be either pressure increases or pressure decreases, which travel through a system. The result of such pressure waves depends on the magnitude of the pressure waves and the system within which these waves travel. As a minimum, the pressure waves cause localized stressing of the system and in severe cases can result in failure of the system pressure boundary or damage to and/or failure of adjacent components resulting from gross system movement. The pressure waves can be caused by the collapse of a void, a water slug against an area restriction or pipe bend, a pump start, control or relief valve actuation, or water suspended in a steam line. It is the physical interaction of fluid against fluid or fluid against a structure which causes pressure waves, and the magnitude of these pressure waves can be related directly to the amount of energy the fluid must gain or lose. There are four basic forms of hydraulic pressure transient phenomena (USNRC 1984a, b, c, d; EPRI 1996; Bergant et al. 2004; Leishear 2013).

1. **Colum Separation**: This type of transient results from the creation of voids due to either a local depressurization below the saturation pressure of the fluid or a local depressurization below the saturation pressure of dissolved air or other gases in the system. Once a void has been formed, an increase in system pressure, a change in fluid temperature, a change in the position of the void, or a change in

the flow pattern of the system could cause that void to collapse. Such a collapse is characterized by an acceleration of the void boundaries towards each other with a subsequent impact when the void has completely collapsed. If the difference in velocity of the fluid surfaces at the instant the void is collapsed is ΔV, a maximum pressure wave of $\alpha \times \Delta V/g$ can be expected where α is the velocity of the fluid, ρ is the density of the fluid, and g is the gravitational acceleration; this is the Joukowski equation which approximates the fluid transient pressure rise from an instantaneous change in fluid velocity.[19]

2. **Discharge into a Voided Line**: Plant operating conditions sometimes necessitate establishing flow in lines which are completely or partially voided. Oftentimes, the establishment of flow in these lines cannot be done to avoid transient interactions of the fluid and structures. As a result, pressure waves sometimes develop when the moving fluid front encounters a point of structural resistance, which could be in the form of an area change, pipe bends, or other such discontinuity in the system. The result of such an interaction is a change in velocity and energy of the moving fluid front, with a resultant pressure wave which is transmitted throughout the system.
3. **Vapor Bubble Collapse**: This type of transient is characterized by the entrapment of a steam bubble in a system which normally contains water. This is different from column separation in that a void is not created due to a local depressurization; rather, it is due to entrapment. The transient is created when the entrapped bubble condenses and essentially creates a very-low-pressure void. The fluid surrounding this void is accelerated to fill the void, and upon collapse of the void, very large pressure waves are generated. Plant conditions which could cause such a transient occur mostly in areas where vapor and liquid phases are in close proximity, such as in the feedwater inlet piping to a steam generator. If flow is interrupted in this line, it could drain and allow steam to enter the line. Upon recommencement of flow, the steam in the line could be trapped due to the piping configuration. Once trapped, the cooler feedwater inlet water would cause the steam bubble to collapse, and the resulting impact would cause a pressure wave to be generated and transmitted through the system.
4. **Steam/Water Flow Changes**: This type of transient is often identified as the classic definition of water hammer, in which the flow of fluid is stopped suddenly. This action results in a pressure wave which is transmitted through the system. This type of transient is the result of flow changes in a pipe being disturbed by some event, such as a valve closure or valve opening, causing the velocity and consequently the energy of the fluid to change. This change in energy results in a pressure wave which is transmitted throughout the system. As this pressure wave can be either positive or negative, the effect on a system for either condition must be evaluated. Positive pressure waves can be generated from such plant

[19]Nikolai Egorovich Zhukovskii (or Zhukovsky or Joukowski), 1847–1921. After receiving his master's degree, a special chair of mechanics was created for him at Moscow Technical School. He obtained a doctorate from Moscow University in 1882 for a dissertation on the stability of motion.

conditions as a rapid valve closure, whereas negative pressure waves can be generated downstream of a stopped pump or upstream of an opening discharge valve. In either case, the pressure wave will be transmitted throughout the system, and the effects of such a pressure wave must be considered.

A water hammer is a pressure surge that can arise in any pressurized system that undergoes an abrupt change in its flow rate from starting and stopping of pumps, from the opening and closing of valves, from changes in flow direction, or from water column separation and collapse. These types of water hammer events mostly occur under single-phase flow conditions. Using a service water (SW) piping system as an example, the latter phenomenon occurs as a result of draining of water from high points of a piping system to lower elevation branch lines, as would occur when the SW pumps shut down following a loss of electrical power. As the water is drained from the high points in the system, a partial vacuum condition inside the piping would result in separation of the water column into vapor and liquid phases (column separation). The separation occurs at high-point elevations because the water pressure at the top of the column falls below the vapor pressure of the liquid while the bottom of the column is open to the atmosphere. Pipe damage due to water hammer is caused by either a pressure wave traveling at sonic velocity in water or a fast-moving water slug. In severe cases, the pressure can exceed the piping design pressure resulting in plastic deformation and rupture.

Extensive theoretical and experimental work has been performed to study the characteristics of water hammer phenomena. In a six-volume set, EPRI compiled an extensive information resource on water hammer as it applies to nuclear power plant operation. At the time of publication, EPRI report NP-6766 represented state of knowledge as of 1996 and remains applicable:

- NP-6766 Volume 1 documents the US operating experience (EPRI 1992a).
- NP-6766 Volume 2 describes the root causes of reported events (EPRI 1992b).
- NP-6766 Volume 3 includes a discussion of experimental data on water hammer (EPRI 1992c).
- NP-6766 Volume 4 describes and evaluates analytical models and computer codes applicable to water hammer assessment (EPRI 1992d).
- NP-6766 Volume 5 presents guidelines for water hammer prevention, diagnosis, and assessment (EPRI 1992e, f, g).
- NP-6766 Volume 6 describes plant configurations and operating procedures for system susceptible to water hammer (EPRI 1992h).

4.9.2 Operating Experience Insights

This section presents abstracts of historically notable pipe failures. The term "significance" pertains to both operational impact and degree of response from industry and regulatory bodies in formulating solutions for long-term pipe material degradation management. The following examples, arranged chronologically, illustrate major pipe failures attributed to water hammer transients.

4.9.2.1 Severed Fire Water System Pipe Header

On June 17, 1998, multiple fire alarms were received in the Columbia Generating Station control room coincident with all four fire pumps starting and severe water hammer noises being heard from the control room. The water hammer caused a valve rupture in the fire protection system main header in the northeast stairwell of the building. The water in the stairwell quickly rose to approximately a depth of 5.7 m. Water from the stairwell entered residual heat removal (RHR) pump room "C" through an open watertight door and eventually submerged the pump motor. A ground developed on a 125 DV bus, which supplies control power to RHR pump "C." The ground was subsequently cleared. A valve located in a line connecting the sumps of RHR pump "C" and low-pressure core spray pump rooms failed to close, allowing water to enter the low-pressure core spray pump room. Water eventually crested just below the base of the low-pressure core spray pump motor.

High-water-level alarms for the RHR C room also alerted the control room about the water leakage in the reactor building. Due to the rising water level in the RHR C room, the keep-fill pump for RHR B and RHR C tripped on an electrical fault. Control room personnel immediately started RHR B (located in an unaffected room) in suppression pool cooling mode to maintain the system piping full and system operational. The RHR C pump room subsequently flooded to approximately 5 m above the floor, submerging the RHR C pump motor and other equipment in the room. As a result of the loud noise made by the water hammer, a shift supervisor in the area conducted a visual inspection and noted the flooding in the stairwell. It was estimated that during this event, the flooding rate from the failed valve into the stairwell was approximately 500–600 kg/s. The root cause of the water hammer event was determined to be due to water column separation and collapse. The voiding occurred during the transient as a heavy flow demand was placed on the system over a short duration that could not be satisfied by the jockey pump in operation. Startup of the diesel fire pump collapsed the voids causing a severe water hammer in the plant's tallest and largest riser that resulted in a catastrophic failure of its isolation valve (Arastu et al. 1999).

4.9.2.2 Pipe Rupture at Oconee Unit 2

On September 24, 1996, while operating at 60% power, Oconee 2 experienced a rupture of a DN450 drain line on the secondary side of the plant (Fig. 4.17). The pipe rupture was due to a water hammer which was a result of the following three root causes: (1) a lack of a general understanding of the potential consequences of the water hammers experienced in the past resulted in untimely development and implementation of modifications necessary to prevent water hammers; (2) there

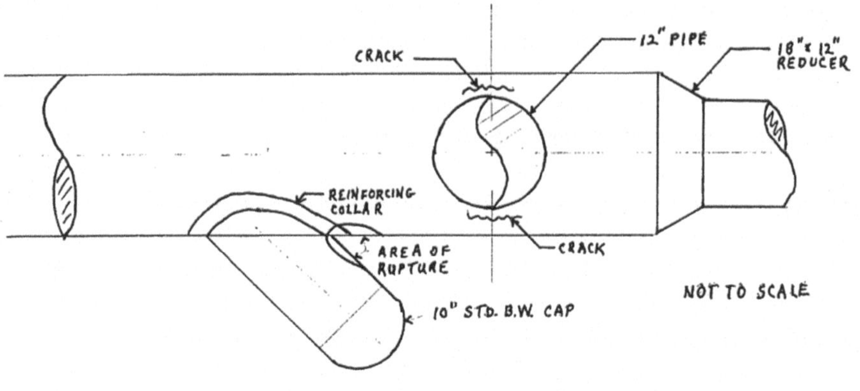

Fig. 4.17 Oconee Unit 2 piping in the area of pipe rupture. (Reproduced from USNRC Augmented Inspection Team Report 50-270/1996-15, November 4, 1996, NRC-ADAMS Accession No. ML15118A154)

was a failure to revise the moisture separator reheat procedure to provide the information and instructions necessary to successfully perform the required task; and (3) a human error of inadequate verbal communication resulted in incomplete and inaccurate instructions being provided to perform the required task.

The pipe rupture consisted of a through-wall, longitudinal split originating at the inside corner of the 45-degree 250 mm lateral. The rupture ran approximately 300 mm along each line (DN250 and DN450 lines). It ran through both the reinforcing saddle, connecting welds, and the underlying pipe of the DN450 line. It continued some distance past the edge of the reinforcing saddle before arresting in the pipe base material. In the DN250 pipe, the fracture ran from the same origin, along the entire length of the branch, arresting in the pipe cap. A second fracture ran about 90° around the edge of the reinforcing saddle of the DN450 line. It was a 100% ductile rupture. There was no evidence of wall thinning due to flow-accelerated corrosion.

4.9.2.3 Feedwater Pipe Failure in Maine Yankee (1983)

Following a reactor trip from full power on January 25, 1983, the Maine Yankee Nuclear Power Plant experienced a steam condensation-induced main feedline water hammer that resulted in a through-wall crack near the inlet nozzle to the no. 2 steam generator (SG). The crack in the main feedwater line to the no. 2 SG occurred at the bottom of the 20-degree elbow adjacent to the weld between the elbow and the SG nozzle safe end. The crack was initiated by thermal cycling of the feedwater line and the existence of a stress riser in the elbow that caused an internal crack to develop prior to the break. The extreme pressures generated by the steam condensation-induced water hammer caused the preexisting crack to become a through-wall crack approximately 280 mm in length and, on average, 1.6 mm wide. A 50 mm length in the center of the crack was about 2.4 mm wide (Heitner 1983).

The water hammer appeared to be related to the addition of a 100%-capacity turbine-driven main feedwater pump because, for the first time in its operating history, the plant experienced a reactor trip from full power without the availability of main feedwater. At the time of the reactor trip, both electric motor-driven feedwater pumps were unavailable, and the plant was operating with the 100%-capacity turbine-driven feedwater pump. Since this pump is driven by high-pressure extraction steam, it became unavailable when the main turbine tripped, causing the feedring to drain. As the feedring became resubmerged and began to fill with relatively cool auxiliary feedwater, a water hammer was induced by the condensation of the steam in the partially filled feedring.

An unsuccessful attempt to parallel the two motor generator sets that power the control element drive mechanisms (CEDMs) resulted in the power supply breakers to the CEDMs opening and the control element assemblies (CEAs) dropping into the core. The CEA insertion caused the reactor protection system to trip due to the axial flux offset. The main turbine and the turbine-driven main feedwater pump also tripped, as they were designed to do following a reactor trip. Auxiliary feedwater was automatically initiated due to the low steam generator level that normally follows a turbine trip, due to void collapse in steam generators. A loud noise was heard in the plant machine shop, located directly beneath the main feedwater lines, about 15 min after the trip, and operators noticed a rise in containment humidity, increasing containment sump level, and the actuation of a containment fire detection alarm. A containment entry was made, and the visual inspection revealed a substantial leakage from the main feedwater line feeding the no. 2 steam generator (SG), in the vicinity of the SG inlet nozzle.

The crack in the main feedwater line to the no. 2 SG occurred at the bottom of the 20° elbow adjacent to the weld between the elbow and the SG nozzle safe end. The crack was initiated by thermal cycling of the feedwater line and the existence of a stress riser in the elbow due to the sharp transition between the elbow counter-bore, machined to match the internal diameter of the "safe end" and the nominal internal diameter of the failed elbow. Visual, magnetic particle, and radiographic examination of the failed elbow showed that an internal crack existed prior to the break. The extreme pressures generated by the steam condensation-induced water hammer caused the preexisting crack to become a through-wall crack approximately 279 mm in length and, on average, 1.6 mm wide. Similar examination of the 20″ elbows in the feedwater lines serving SG nos. 1 and 3 indicated the existence of similar cracks. An examination of steam generator internals revealed minor deformation of a feedring support in SG no. 3 and an expansion of the thermal sleeve against the feedwater inlet nozzle in SG nos. 2 and 3. Pipe supports for the feedwater lines supplying all SGs also showed evidence of minor damage. Damage to the line feeding SG no. 3 was more extensive than the others, showing spalled and cracked concrete around some of the pipe supports, loose anchor bolts, and hanger deformation. The licensee thinks that the no. 3 SG feedwater line may have experienced more damage than the others because it is the longest run, in addition to other geometric and support differences. For example, the horizontal run of feedwater piping in the area between the crane wall and the containment was 65 m for SG no. 3, 20 m for SG no. 2, and 8 m for SG no. 1..

Analysis of the main feedline through-wall crack, although a direct result of the water hammer, revealed preexisting internal cracks in the main feedlines caused by fatigue induced by thermal cycling. Thermal cycling is due to the fact that at hot standby or hot shutdown, cold auxiliary feedwater from the demineralized water storage tank may be used to maintain steam generator level. The auxiliary feedwater flow velocity is generally low enough not to cause mixing with the hot fluid in the main feedwater lines. This caused a thermal stratification in the main feedwater lines. The cold auxiliary feedwater, running along the bottom of the feedlines due to its higher density, results in a thermal gradient in the pipe walls. It is thought that this thermal gradient is responsible for the preexisting internal cracks in the bottom of the main feedwater piping. Another reason for the internal crack was due to stress risers in the piping, caused by the sharp corners created by abrupt changes in internal diameter.

4.9.2.4 Pipe Rupture at Robinson 2 (1970)

On April 28, 1970, during hot hydraulic testing of the unfueled H.B. Robinson 2 (PWR), a pipe nozzle, counter-bored and tapered from Schedule 80 to Schedule 40, between a pressurized main steam line and a safety valve failed completely, a 360-degree circumferential break (Fig. 4.18) (Westinghouse Electric Co 1970). The rupture was a non-isolable break, and an uncontrolled cooldown of the plant occurred. Seven men were injured by the escaping steam. The fracture path coincided with the end of the machined taper adjacent to the weld inside the reduced section of the DN150 pipe nozzle which connected the safety valve to the DN650 main steam line.

The section of pipe which failed was insulated to a point approximately 75 mm above the failure, which minimized the temperature gradient across the failed area. A pneumatic test device was connected to the valve which loaded the spring to balance the seating pressure on the valve disc. The operator was opening the valve regulator to balance the spring force when the failure occurred. The failure pattern, surrounding damage, and trajectories of the valve and exhaust chute indicate that the pipe first opened on the west side, directly opposite the exhaust, emitting steam in a fan jet which sharply cut the insulation on an adjacent pipe. The pattern of the insulation cut and fracture face suggested that the jet had a fan angle of about 80° wide and was directed about 45° up from the horizontal. The safety valve apparently rotated as the pipe tore across its section. The exhaust chute broke off and was found under the loop #3 steam pipe. An inspection of the exhaust elbow revealed longitudinal markings on the inner surface of the extrados. Analyses of the failed piping showed extreme plastic strain on the fracture surface, which indicated an overload failure, but no plausible mechanism for the overloading was found. Stress analyses indicated

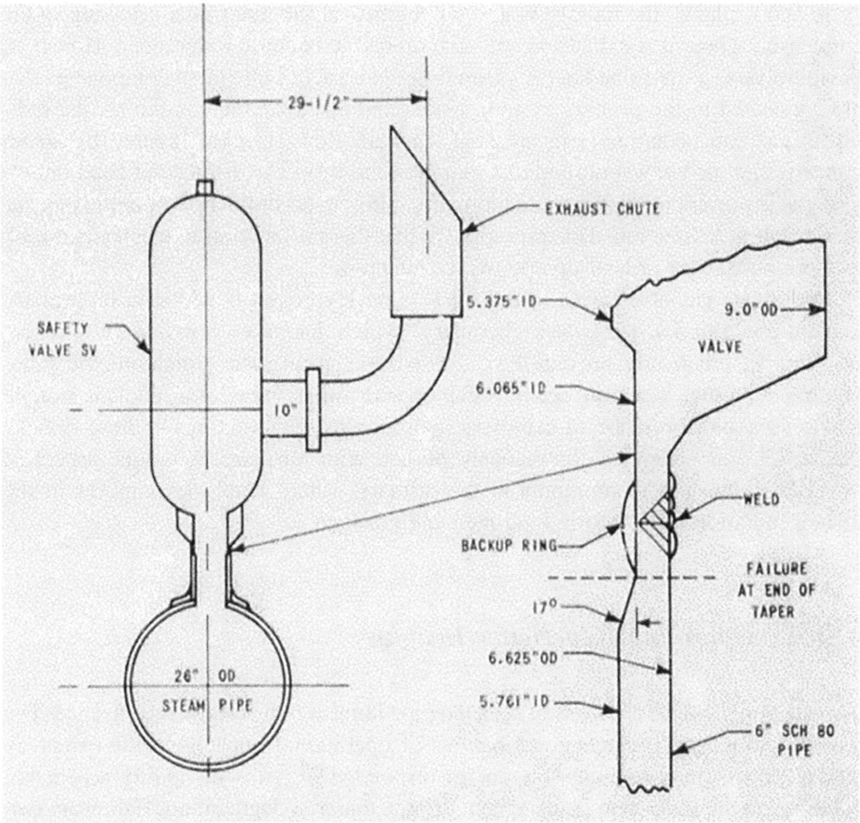

Fig. 4.18 Failed safety valve connection

that the branch line was undersized and that the stress calculated for a full-capacity discharge through the valve could exceed the ultimate strength of the material.

4.10 Hydrogen Deflagration

Hydrogen is used for primary water chemistry control and as a coolant for electric generators. It is stored as high-pressure gas in vessels and is supplied to various systems in the auxiliary building, reactor building, and turbine building through small-diameter piping. Leaks or breaks in the piping can result in the accumulation of a combustible or explosive mixture of air and hydrogen within a building structure. A hydrogen-oxygen gas mixture is called radiolysis gas if it is generated by dissociation of water under the influence of gamma and neutron radiation. Radiolysis gas can appear, for example, in safety-relevant piping (Astbury and Hawksworth 2007; Leishear 2010; Beaucourt and Georgescu 2021).

In BWR plants, the radiolysis gas is entrained in the main steam flowing to the condenser, where it is exhausted and recombined to become water again. However, the radiolysis gas entrained in the steam flow also reaches into plant components that are connected to the primary system. Under certain favorable conditions, the radiolysis gas can accumulate in areas of stagnant flow strongly favored by steam condensation or by evaporation due to a pressure drop. As the accumulated radiolysis gas is present at an ideal stoichiometric ratio, an accumulation of radiolysis gas may result in detonation. This presupposes that there is an ignition mechanism such as a pressure surge or heat-up upon valve operation.

Radiolysis gas also forms in the PWR core. Hydrogen is added to its primary coolant to optimize the water chemistry, which increases bonding of oxygen, resulting in practically no radiolysis gas present at an ideal stoichiometric ratio. Hydrogen is then released again, together with other gases, into coolant storage tanks. To avoid formation of explosive hydrogen/oxygen mixtures in these vessels, the vessel atmosphere is continuously purged with nitrogen. A unique aspect of WWERs is the use of ammonia in the primary water, which decomposes in the reactor radiation field to form hydrogen and nitrogen.

4.10.1 Operating Experience Insights

This section includes abstracts of selected pipe failures of historical significance. The term "significance" reflects combinations of operational impact and the extent by which industry and regulatory agencies responded in terms of finding acceptable solutions to the long-term management of pipe material degradation. Below are two examples of pipe failures caused by hydrogen deflagration.

4.10.1.1 Steam Condensation Line Rupture at Hamaoka 1

In November 2001, a high-pressure coolant injection (HPCI) pipe rupture occurred at Hamaoka 1, a Japanese BWR plant (NISA 2002). At the time of the event, the reactor had been in commercial operation for 26 years. A periodic, manual startup test of the HPCI system was performed. During the test, the HPCI automatically isolated, and fire detectors in ten locations of the reactor building were actuated. As a result, the reactor was shut down to investigate the cause of the event. Steam leakage stopped immediately with the automatic isolation of the HPCI. As a result of the field inspection, it was recognized that the elbow portion of the steam pipe, which branches from the HPCI to the residual heat removal system heat exchanger, ruptured, which led to the steam leakage (Fig. 4.19).

Metallurgical analysis of the fractured surface found that an excessive force on the affected portion of the pipe caused ductile fracture with plastic deformation. Furthermore, it was determined that high concentrations of hydrogen and oxygen stayed in the ruptured portion due to steam condensate. Prompt hydrogen burning

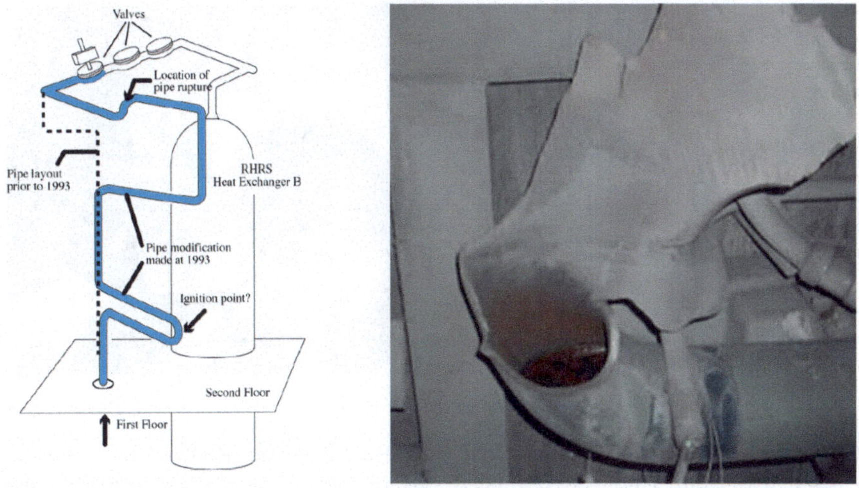

Fig. 4.19 Failed high-pressure coolant injection DN100 pipe elbow. (The piping layout drawing is reproduced from https://cnic.jp/english/?p=392)

caused the pipe rupture (Naitoh et al. 2003; Kishi et al. 2004). Experiments concluded that the occurrence of hydrogen detonation caused the pipe rupture. A three-dimensional hydrogen combustion analysis of the combustion behavior of hydrogen accumulated in the steam-condensing line was performed, assuming that the pipe wall and accumulated water surface were rigid. As a result, it was found that the pressure peak at the straight part of the pipe in the detonation region was around 120 MPa, and the pressure wave propagation speed was approximately 3000 m/s.

The peak pressure in the elbows was 2.0–2.5 times higher than in the straight part, due to reflection and overlapping of the pressure waves. In the elbow above the surface of the accumulated water, the pressure peak was approximately 300 MPa, due to the overlapping of the reflected pressure wave. This was the highest pressure in all positions in the pipe. The combustion mode was deflagration for about 3 m in length from the ignition point. In the deflagration region, the propagation velocity of the flame increased at elbows because the turbulent fluctuation occurred when the gas passed through the elbows. However, the increase of the flame propagation velocity did not affect the pressure wave propagation speed and the pressure in the detonation region. It was further determined that the strain in the pipe elbow above the surface of the accumulated water exceeded the critical strain, and the pipe ruptured.

4.10.1.2 Pipe Rupture at Brunsbüttel (KKB)

On December 14, 2001, the control room staff received indications of drywell leakage. These indications were initially interpreted as a flange leakage. Further

Fig. 4.20 Photograph showing the extent of damage to RPV head cooling pipe

evaluation of the control room indications resulted in the determination that the leakage was from a non-safety-related system. It was later determined that the affected system was the RPV head cooling system (TC), a system only used during reactor shutdown. The apparent leakage was an isolable section of DN100 TC piping on the afternoon of February 18, 2002. On February 21, 2002, an investigation team was chartered to evaluate the event. During the initial drywell inspection on February 18, 2002, it was discovered that a large section (length of 2–3 m) of DN100 stainless steel piping had ruptured (Fig. 4.20). The root cause analysis pointed to deflagration (hydrogen explosion) due to a local, high concentration of hydrogen in the pipe (Schulz et al. 2002).

References

ACRS (Advisory Committee on Reactor Safeguards) (2022) French PWR safety injection system cracking, November 16 meeting transcript. Nuclear Regulatory Commission, Washington, DC. NRC-ADAMS Accession No. ML22335A495

Arastu AH, LaFramboise WL, Noble LD, Rhoads JE (1999) Diagnostic evaluation of a severe water hammer event in the fire protection system of a nuclear power plant. In: FEDSM99-6891, Proceedings of the 3rd ASME/JSME joint fluids engineering conference. American Society of Mechanical Engineers, New York

Astbury GR, Hawksworth SJ (2007) Spontaneous ignition of hydrogen leaks: a review of postulated mechanisms. Int J Hydrog Energy 32(13):2178–2185

Atkinson JTN, Van Droffelaar H (1982) Corrosion and its control: an introduction to the subject. NACE International, Houston

Beaucourt J, Georgescu G (2021) Modeling hydrogen explosion in level 1 PSA. Nucl Technol 207(3):323–334

Bellamy LJ et al (1989) Evaluation of the human contribution to pipework and in-line equipment failure frequencies, contract report 15/1989. U.K. Health and Safety Executive, Liverpool

Bergant A, Simpson AR, Tusseling AS (2004) Water hammer with column separation: a review of research in the twentieth century. Department of Mathematics and Computer Science, Eindhoven University of Technology, Eindhoven

Betova I, Bojinow M, Saario T (2010) Predictive modelling of flow-accelerated corrosion—unresolved problems and issues, VTT-R-08125-10. Technical Research Centre of Finland (VTT), Espoo

Brust FW, Yang YP, Scott PM (2002) Evaluation of reactor pressure vessel (RPV) nozzle to hot-leg piping bimetallic weld joint integrity for the V. C. Summer Nuclear Power Plant, Part of Task 8 of the BINP Program (Contract Number-NRC-04-97-052, Job Code W6775), Battelle, Columbus

Burghard HC, Bursle AJ (1979) Metallurgical investigation of cracking in a reactor vessel nozzle safe-end, swill project 02 5389 001. Southwest Research Institute, San Antonio

Burkhart D (1981) Thermal fatigue cracking in BWR Feedwater Systems in Sweden, SWE810216-01T. AB Statens Anläggningsprovning, Stockholm

Bush SH (1985) Statistics of pressure vessel and piping failures. J Press Vessel Piping Technol. ASME pressure vessel and piping technology—a decade of progress, Chapter 8, pp 875–893

Bush SH, Dillon RL (1973) Stress corrosion in nuclear systems, BNWL-SA-4668. Pacific Northwest Laboratories, Richland

Cattant F (2022) The underestimated role of the oxygen on RCS component failures. In: Proceedings of the Fontevraud10 international symposium: contribution of materials investigations and operating experience to LWRs' safety, performance and reliability, September 19–22, Avignon, France

Charlebois PR, Clarke TR, Goodman RM, McEwan WF (1995) The 1994 loss of coolant incident at pickering NGS. In: Proceedings of the 16th annual conference of the Canadian Nuclear Society, Saskatoon, June 4-7

Dahlberg M, Nilsson K-F, Taylor N, Faidy C, Wilke U, Chapuliot S, Kalkhof D, Bretherton I, Church M, Solin J, Catalano J (2011) Development of a European procedure for assessment of high cycle thermal fatigue in light water reactors: final report of the NESC-thermal fatigue project, EUR-22763-EN, network for evaluating structural components. European Commission, Joint Research Centre, Petten

Delwiche DE et al (1983) Nine Mile Point 1 pipe crack task force report. Niagara Mohawk Power Corporation, Syracuse. (NRC-ADAMS Accession No. ML20072M138)

Ehrnsten U et al (2001) Intergranular cracking of AISI 316NG stainless steels in BWR environment. In: Proceedings of the 10th international conference on environmental degradation in nuclear power systems—water reactors. National Association of Corrosion Engineers, Houston. ISBN: 978-9992836873

Ehrnstén U, Andresen PL, Que Z (2024) A review of stress corrosion cracking of austenitic stainless steels in PWR primary water. J Nucl Mater 588:154815

EPRI (1977) Characteristics of pipe system failures in light water reactors, NP-438, Palo Alto

EPRI (1987) Fracture mechanics evaluation of feedwater line failure at Surry-2, NP-5462, Palo Alto

EPRI (1988a) Sourcebook for microbiologically influenced corrosion in nuclear power plants, NP-5580, Palo Alto

EPRI (1988b) Utility response to NRC bulletin 87-01, thinning of pipe walls in nuclear power plants, final report, NP-6066, Palo Alto

EPRI (1992a) Water hammer prevention, mitigation, and accommodation, volume 1: plant water hammer experience, NP-6766, Palo Alto

EPRI (1992b) Water hammer prevention, mitigation, and accommodation, volume 2: root cause analysis for plant water hammer experience, NP-6766, Palo Alto

EPRI (1992c) Water hammer prevention, mitigation, and accommodation, volume 3: experimental and engineering data, NP-6766, Palo Alto

EPRI (1992d) Water hammer prevention, mitigation, and accommodation, volume 4: review of analytical models and computer codes, NP-6766, Palo Alto

EPRI (1992e) Water hammer prevention, mitigation, and accommodation, volume 5—part 1: water hammer assessment guidelines, NP-6766, Palo Alto

EPRI (1992f) Water hammer prevention, mitigation, and accommodation, volume 5—part 2: water hammer prevention guidelines, NP-6766, Palo Alto

EPRI (1992g) Water hammer prevention, mitigation, and accommodation, volume 5—part 3: water hammer diagnostic guidelines, NP-6766, Palo Alto

EPRI (1992h) Water hammer prevention, mitigation, and accommodation, volume 6: plant water hammer experience, NP-6766, Palo Alto

EPRI (1996) Water hammer handbook for nuclear power plant engineers and operators, TR-106438, Palo Alto

EPRI (1999) BWR vessel and internals project technical basis for revisions to generic letter 88-01 inspection schedules (BWRVIP-75), TR 113932, Palo Alto

EPRI (2000) Operating experience regarding thermal fatigue of unisolable piping connected to PWR reactor coolant systems (MRP-25), 1001006, Palo Alto

EPRI (2004a) Recommendations for controlling cavitation, flashing, liquid droplet impingement, and solid particle erosion in nuclear power plant piping systems, 1011231. Electric Power Research Institute, Palo Alto

EPRI (2004b) Stress corrosion cracking in PWR and BWR closed cooling water systems, 1009721, Palo Alto

EPRI (2004c) Closed cooling water chemistry guideline, revision 1, 1007820, Palo Alto

EPRI (2004d) MRP Butt Weld Safety Assessment Working Group (2004), Materials Reliability Program: alloy 82/182 Butt Weld Safety Assessment for US PWR Plant Designs (MRP-113NP), 1007029-NP, Palo Alto

EPRI (2007a) Flow-accelerated corrosion—the entrance effect, 1015072, Palo Alto

EPRI (2007b) Investigation into flow-accelerated corrosion at low temperatures, 1015170, Palo Alto

EPRI (2007c) Flow-accelerated corrosion—the entrance effect, 1013013, Palo Alto

EPRI (2013) Recommendations for an effective flow-accelerated corrosion program (NSAC-202L-R4), 3002000563, Palo Alto

Erve M, Wesseling U, Kilian R, Hardt R, Brümmer G, Maier V, Ilg U (1997) Cracking in stabilized austenitic stainless steel piping of German boiling water reactors—characteristic features and root cause. Nucl Eng Des 171:113–123

Fatamura Y, Torikai K, Ogawa Y (1979) Piping cracks in JPDR, (IV): analysis of cause of cracks found near weld joints connecting reactor vessel nozzle safe-end to pipe. J Nucl Sci Technol 16(2):137–146

Féron D, Staehle RW (2016) Stress corrosion cracking of nickel-based alloys in water-cooled nuclear reactors. The Couriou effect. European Federation of Corrosion Publications Number 67. Woodhead Publishing Ltd., Kidlington. ISSN 1354-5116

Féron D, Guerr C, Martin F (2019) Historical review of alloy 600 stress corrosion cracking: from the "Coriou effect" to the quantitative micro-nano approach. Corros J 75(3):267–273

Fisher K, Thoma D (1932) Investigation of flow conditions in centrifugal pumps. Trans ASME Hydraulics 54:141–155

Ford FP (2010) Stress corrosion cracking of stainless steels in boiling water reactors, technical report prepared for the OECD Nuclear Energy Agency under Contract 09/00056277B, Rexford

Fujisawa N (2025) Review: pipeline layout effect on the wall thinning of Mihama Nuclear Power Plants. J Nucl Eng 6:19. https://doi.org/10.3390/jne6020019

Gangloff RP (2005) Chapter 26: Environmental cracking—corrosion fatigue. In: Baboian R (ed) Corrosion tests and standards: application and interpretation. ASTM International, West Conshohocken

Gorman JA (2015) 2015 frank Newman speller award: stress corrosion cracking and nuclear power. Corrosion 71(12):1414–1433

Gorman JA, Hunt ES (eds) (1993) Proceedings: 1992 EPRI workshop on PWSCC of alloy 600 in PWRs, TR-103345. EPRI, Palo Alto

Grönwall B, Ljungberg L, Hübner W, Stuart W (1966) Intercrystalline stress corrosion cracking of Inconel 600 inspection tubes in the Ågesta Reactor, AE-245. Aktiebolaget Atomenergi, Stockholm

Hahn E-H, Zhang Z, Wang J (2023) Root cause analysis for cracked pipe in auxiliary system in nuclear power plant. Eng Fail Anal 150:107386

Hattori S, Sasakawa H, Handa N (1988) Human error analysis related to structural reliability. Nucl Technol 85(1):22–32

Heitmann HG, Kastner W (1982) Erosion-corrosion in water-steam cycles—causes and countermeasures. VGB Kraftwerkstechnik 62(3):180–187. (in German)

Heitner KL (1983) Summary of 830209 meeting with utility regarding 830125 Feedwater Line Break Event. USNRC, Washington, DC. NRC-ADAMS Accession No. ML20083Q272

Henriksson S (1981) Operational experience of stainless steels in seawater-cooled systems, SVF-109 (in Swedish). Swedish Corrosion Institute, Stockholm

Hickling J (1990) Strain-induced corrosion cracking: relationship to stress corrosion cracking/ corrosion fatigue and importance for nuclear plant service life. In: Proceedings of the 3rd international Atomic Energy Agency specialists' meeting on subcritical crack growth, NUREG/CP-0112, vol 2. USNRC, Washington, DC

Holmberg J, Pyy P (1994) Example of a PSA-based analysis of an occurred external pipe break at TVO I, NKS/SIK-1(93)17. VTT Industrial Automation, Espoo

Horn RM, Gordon GM, Ford FP, Cowan RL (1997) Experience and assessment of stress corrosion cracking in L-grade stainless steel BWR internals. Nucl Eng Des 174(3):315–325

Horowitz J (2004) Recommendations for controlling cavitation, flashing, liquid droplet impingement, and solid particle erosion in nuclear power plant piping systems, 1011231. EPRI, Palo Alto

Hu L-W, Lee J, Saha P, Kazimi MS (2003) Nuclear systems enhanced performance program: thermal striping in LWR piping systems, MIT-NSP-TR-117, Cambridge

Hu X, Jiang H-Y, Luo Y, Jin Q, Peng W, Yi C-M (2020) A study on microstructure, residual stresses and stress corrosion cracking of repair welding on 304 stainless steel: part II—effects of reinforcement height. Materials 13:2434. https://doi.org/10.3390/ma13112434

Hurst JW, Bellamy LJ, Geyer TAW, Astley JA (1991) A classification scheme for pipework failures to include human and sociotechnical errors and their contribution to pipework failure frequencies. J Hazard Mater 26:159–186

Hwang, K.M et al (2014), "A Study on the Cause Analysis for the Wall Thinning and Leakage in Small Bore Piping Downstream of Orifice," World Journal of Nuclear Science and Technology, 4:1–6.

IAEA (2003) Assessment and management of ageing of major nuclear power plant components important to safety. Primary piping in PWRs, IAEA-TECDOC-1361, Vienna, Austria

IAEA (2011) Stress corrosion cracking in light water reactors: good practices and lessons learned. Nuclear energy series no. NP-T-3.13, Vienna, Austria. ISBN: 978-92-0-117210-5

IAEA (2014) Approaches to ageing management for nuclear power plants: international generic ageing lessons learned (IGALL) final report, IAEA-TECDOC-1736, Vienna, Austria

IAEA (2015) The Fukushima Daiichi accident, technical volume 2, safety assessment, STI/PUB/ 1710, Vienna, Austria

Janzen F (1981) A study of piping failures in U.S. Nuclear Power Stations, AECL-Misc-204. Atomic Energy of Canada, Chalk River

JSME (2017) Thermal power equipment piping thinning management technical standard, 2016 edition, Tokyo, Japan. ISBN: 978-4-88898-271-9

Kaijua Y, Wua T, Minc Z, Jun X, Pengcheng S, Qi X, Rui T, Shaoyu Q, Yong C (2018) Failure analysis of elbow pipe cracking in a residual heat-removal system of a nuclear power plant. Eng Fail Anal 84:101–108

Kaneko S et al (eds) (2014) Flow-induced vibrations: classifications and lessons learned from practical experience, 2nd edn. Elsevier, Amsterdam. ISBN 978-0-08-101318-2

Kilian R, Brümmer G, Ilg U, Maier V, Teichmann H, Wachter O (1995) Characterization of sensitization and stress corrosion behavior of stabilized stainless steels under BWR conditions. In: Proceedings of the 7th international symposium on environmental degradation of materials in nuclear power systems—water reactors. Minerals Metals and Materials Society, Pittsburgh

Kim K-C, Lim J-H, Yoon J-K (2009) Thermal fatigue estimation due to thermal stratification in the RCS branch line using one-way FSI scheme. J Mech Sci Technol 22:2218–2227

Kishi T, Kikuchi H, Miura S, Fukuda M, Hirano M, Watanabe N (2004) Application of probabilistic safety assessment to the pipe rupture incident at Hamaoka Unit-1. J Nucl Sci Technol 41(1): 77–85

Kitagawa H, Yuuki R, Ohira T (1975) Crack-morphological aspects in fracture mechanics. Eng Fract Mech 7(3):515–529

Krishnaswamy RK (2005) Analysis of ductile and brittle failures from creep rupture testing of high-density polyethylene (HDPE) pipes. Polymer 46(25):11664–11672

Krishnaswamy P (2007) Review of literature on the use of polyethylene (PE) piping in nuclear power plant safety-related class 3 service water systems, report prepared for the U.S. Nuclear Regulatory Commission by Engineering Mechanics Corp. of Columbus (Emc2), Columbus. NRC-ADAMS Accession No. ML14266A208

Krishnaswamy RK, Lamborn MJ, Sukhadia AM, Register DF, Maeger PL (2006) Rapid crack propagation failures in HDPE pipes: structure–property investigations. Polym Eng Sci 46(10): 1358–1362

Kussmaul K, Blind D, Jansky J, Rintamaa R (1984) Formation and growth of cracking in feed water pipes and RPV nozzles. Nucl Eng Des 81:105–119

Lee D-M, Kim S-J, Lee H-J, Kim Y-J (2021) Fatigue life evaluation of socket welded pipe within complete penetration defect: I-test and FE analysis. Nucl Eng Technol 53:3852–3859

Leishear RA (2010) Hydrogen ignition mechanisms for explosions in nuclear facility pipe systems. In: PVP2010-25261, Proceedings of the ASME 2010 pressure vessels & piping division conference. American Society of Mechanical Engineers, New York

Leishear RA (2013) Fluid mechanics, water hammer, dynamic stresses, and piping design. ASME Press, American Society of Mechanical Engineers, New York

Li R, Ninokata H, Mori M (2011) Parametric investigation on the effect factors for liquid droplet impingement erosion. In: AJK2011-03034, Proceedings of the ASME-JSME-KSME 2011 joint fluids engineering conference. ASME, New York

Luo Y, Gu W, Peng W, Jin Q, Qin Q, Yi C (2020) A study on microstructure, residual stresses and stress corrosion cracking of repair welding on 304 stainless steel: part I—effects of heat input. Materials 13:2416. https://doi.org/10.3390/ma13102416

Malik Al-Abed Allah M, Arroussi M, Siddiqui OK, Yusuf M, Hunter TN, Shams A, Toor I (2026) Flow accelerated corrosion in nuclear power plants: a detailed review on mechanisms, mitigation, and management. Eng Fail Anal 183:110203. https://doi.org/10.1016/j.engfailanal.2025.110203

Merikoski M (2017) Pipe vibrations in nuclear applications, report 2017-451. Swedish Energy Research Centre, Stockholm. ISBN 978-91-7673-451-3

Miller NA, Boulton LH (1982) Using stainless steel. DSIR industrial information series no. 2. Auckland Industrial Development Division, Department of Scientific and Industrial Research, Auckland

Moon S, Lee JY, Kim K-M, Han S-W, Lee G-G, Maeng W-Y, Oh S, Kim D-J (2024) Assessment of flow-accelerated corrosion-induced wall thinning in SA106 pipes with elbow sections. Nucl Eng Technol 56:1244–1249

Morel AR, Reynes LJ (1992) Short-term degradation mechanisms of piping. Nucl Eng Des 133:37–40

Naitoh M, Kasahara F, Mitsuhashi T, Ohshima I (2003) Analysis of pipe rupture of steam condensing line at Hamaoka-1, (I) accumulation of non-condensable gas in a pipe. J Nucl Sci Technol 40:1032–1040

Nedzinskas L, Khmagauskas A (2003) Intergranular stress corrosion cracking of Ignalina NPP austenitic piping of outside diameter 325 mm. In: Proceedings of the international conference on strength, durability and stability of materials, Klaipeda, Lithuania

NISA (Nuclear and Industrial Safety Agency) (2002) Investigation report on pipe rupture incident at Hamaoka Nuclear Power Station Unit-1. Ministry of Economy, Trade and Industry (METI), Tokyo

Nuclear and Industrial Safety Agency (NISA) (2005), Secondary Piping Rupture Accident at Mihama Power Station, Unit 3, of the Kansai Electric Power Company, Final Report, Tokyo, Japan

OECD/NEA (1998) Specialists meeting on experience with thermal fatigue in LWR piping caused by mixing and stratification, NEA/CSNI/R(1998)8, Boulogne-Billancourt, France

OECD/NEA (2022) A review of the post-1998 experience with thermal fatigue in heavy water and light water reactor piping components, NEA/CSNI/R(2019)13, Boulogne-Billancourt, France

OECD/NEA (2023) Evaluation of design and monitoring requirements for vibrations in safety-related fluid systems, NEA/CSNI/R(2020)2, Boulogne-Billancourt, France

Pasupathi V, Smith GP, Farmelo DR, Perrin JS (1979) Examination of Inconel safe end from Duane Arnold, BCL-585-9. Battelle Columbus Laboratories, Columbus

Schulz H, Voswinkel A, Reck H (2002) Insights and lessons learned from the Brunsbüttel pipe failure event. In: Proceedings of the EUROSAFE forum 2002, November 4–5, Berlin, Germany

Scott PM, Combrade P (2019) General corrosion and stress corrosion cracking of alloy 600 in light water reactor primary coolants. J Nucl Mater 524:340–375

Seo H, Hwang K, Cho C (2014) Numerical case study on the LDIE degradation occurred in the vent header of a high-pressure feedwater heater. World J Nucl Sci Technol 2014(4):26–32

Shah VN, Ware AG, Porter AM (1997) Review of industry efforts to manage pressurized water reactor feedwater nozzle, piping, and feedring cracking and wall thinning, NUREG/CR-6456. USNRC, Washington, DC

Siba M, Wanmahmoodi W, Zaki Nuaw M, Rasani R, Nassir M (2016) Flow-induced vibration in pipes: challenges and solutions—a review. J Eng Sci Technol 11(3):362–382

Skånberg L (2001) PWSCC observations in Ringhals 3 and 4 safe ends. In: Presentation WO-06, transactions of SMiRT-16, structural mechanics in reactor technology, August 12–17, Washington, DC

Slater G (1983) The effect of repair welds on service performance, LD-22827. The Welding Institute, Cambridge

Smith WB, Viloette JB (1962) The cause and failure and the repair of the VBWR recirculation piping, APED-4116. General Electric Co., Atomic Products Energy Division, Vallecitos

Staehle RW (2002) Assessment of stress corrosion cracking (SCC) in the component cooling water system (CCW) of the Wolf Creek Pressurized Water Reactor (WCNOC). Staehle Consulting, North Oaks

STP Nuclear Operating Company (2016) Overview and bases for license renewal aging management program outlined in license renewal application Appendix A1.37, B2.1.37, and PSALBZ "selective leaching of aluminum bronze", NOC-AE-16003378, Wadsworth

Taylor JR (1974) Design errors in nuclear power plants, Risø-M No. 1742. Risø National Laboratory, Roskilde

Torikai K, Kinoshita T, Futamu Y (1978) Piping cracks in JPDR, (I): relation between crack and stress. J Nucl Sci Technol 15(9):697–703

Uchida S, Naitoh M, Uehara Y, Okada H, Hiranuma N, Sugino W, Koshizuka S (2008) Evaluation methods for corrosion damage of components in Cooling Systems of Nuclear Power Plants by coupling analysis of corrosion and flow dynamics (II): evaluation of corrosive conditions in PWR Secondary Cooling System. J Nucl Sci Technol 45:1275–1286

USNRC (1980) Investigation and evaluation of cracking incidents in piping in pressurized water reactors, NUREG-0691, Washington, DC

USNRC (1982a) Summary of April 22, 1982, Meeting on replacement of safe-ends at the Nine Mile Point Nuclear Station, Unit 1, Washington, DC. NRC-ADAMS Accession No. ML20080N153

USNRC (1982b) Stress corrosion cracking in thick-wall, large-diameter, stainless steel, recirculation system piping at BWR plants, IE Bulletin 82-03 revision 1, Washington, DC

USNRC (1983) Stress corrosion cracking in large-diameter stainless steel recirculation system piping at BWR plants, IE Bulletin 83-02, Washington, DC

USNRC (1984a) Procedural guidance for pipe replacement at BWRs, Generic Letter No. 84-07, Washington, DC

USNRC (1984b) Inspections of BWR stainless steel piping, Generic Letter No. 84-11, Washington, DC

USNRC (1984c) Staff long range plan for dealing with stress-corrosion cracking in BWR piping, SECY-84-301, Washington, DC

USNRC (1984d) Evaluation of water hammer occurrence in Nuclear Power Plants, NUREG-0927, revision 1. USNRC, Washington, DC

USNRC (1987) Thinning of pipe walls in nuclear power plants, Bulletin No. 87-01, Washington, DC

USNRC (1988) NRC position on IGSCC in BWR austenitic stainless steel piping, Generic Letter 88-01, Washington, DC

USNRC (1989a) Erosion/corrosion-induced pipe wall thinning in U.S. Nuclear Power Plants, NUREG-1344, Washington, DC

USNRC (1989b) Erosion/corrosion induced pipe wall thinning, Generic Letter 89-09, Washington, DC

VGB PowerTech (2000) Water treatment and use of materials, VGB-R455Pe, First Edition, Essen, Germany

Villabon TAA, Valderrama JS, Garzon PJ, Rodríguez CE, Álvarez GEÁ (2024) Reliability analysis of slow crack growth in HDPE pipes: impact of buried pipeline design and soil characteristics. J Pipeline Sci Eng 5(16):100247. https://doi.org/10.1016/j.jpse.2024.100247

Wada Y, Uchida S, Nakamura M, Akamine K (1999) Empirical understanding of the dependency of hydrogen water chemistry effectiveness on BWR designs. Nucl Sci Technol 36(2):169–178

Westinghouse Electric Co (1970) Incident report: H.B. Robinson, Unit 2 Steam Pipe Break, Pittsburgh, PA. (NRC-ADAMS Accession No. ML20086C373)

You Y, Zhao L, Wang J, Zheng H, Chen K, Zhang L (2025) Flow-accelerated corrosion and wall thinning mechanism of a dissimilar metal welded joint in the secondary loop piping system of pressurized heavy water reactors. J Nucl Mater 616:156026. https://doi.org/10.1016/j.jnucmat.2025.156026

Chapter 5
Piping Pressure Boundary Integrity

5.1 Materials

Materials for nuclear power plant piping must simultaneously withstand the effects of pressure and temperature, local flow conditions, and potential corrosiveness of process media. The requirements for structural materials include good mechanical properties under all operating conditions, low rate of corrosion, and ease of fabrication. In response to operating experience, original piping materials have been replaced with new material types offering improved chemical and mechanical properties. Codes and standards, owner/operator technical requirements, regulations, and cost/risk-benefit considerations play a role in the selection of replacement materials.

Many of the currently operating commercial nuclear power plants were designed, constructed, and commissioned in the period 1970–1990. Most, if not all, plants have undergone degrees of piping design changes or modification, including the use of new material grades and improved piping layouts as part of ongoing plant aging management, plant modernization, and operability improvement programs. There are four basic types of metallic materials: carbon steels whose properties are mainly dependent on the percentage of carbon present, and three types of alloys that are comprised of one or two principal elements with their properties modified by small alloying additions. The principal metallic material types are as follows:

- Carbon steels are used extensively throughout a plant. The primary heat transport system of heavy water reactors consists of carbon steel pipes. In some PWR designs, the reactor coolant system consists of large-diameter stainless steel-clad carbon steel piping. Chemical properties of the most common types of carbon steel are given in Table 5.1.
- Low-alloy steels are used extensively for balance-of-plant piping systems such as feedwater, condensate, and steam systems to ensure resistance to flow-assisted material degradation. The term “low alloy” refers to a type of steel that contains a low percentage of alloying elements, typically less than 5% by weight. Total alloy

B. Lydell, *Knowledge Preservation in Nuclear Safety*,
https://doi.org/10.1007/978-3-032-25463-4_5

Table 5.1 Selected carbon steels

National standard		Chemical composition (weight %)				
ASTM	EN/DIN	C	Mn	P	Cr	Mo
106 Gr. A	P235GH-TR1/2/St 35.8	0.17			—	—
106 Gr. B	P265GH-TR1/2/St 45.8	0.30	1.06	0.048	0.30	0.12
106 Gr. C		0.35	1.20	0.050	0.40	0.15
53 Gr. B	P265TR1/2/St. 37.2	0.30	1.20	0.05	—	—
334 WP22		0.15	0.60	0.04	2.60	1.13
335 P12	13 CrMo 44	0.15	0.61	0.045	1.25	0.65
335 P22	10 CrM 9 10	0.15	0.60	0.030	2.60	1.13
B179 170.1	17 MnMoV 6 4/WB 35	0.21	1.80	0.035	—	0.55
—	15 NiCuMoNb5S1/WB 36	0.17		0.016	—	0.40

Table 5.2 Selected nickel-base alloys

Mechanical properties									
Temperature		Alloy 690				Alloy 600			
		Yield strength		Tensile strength		Yield strength		Tensile strength	
°F	°C	ksi	MPa	Ksi	MPa	ksi	MPa	Ksi	MPa
1000	565	45.5	314	105.5	727	28.5	196	84.0	579
1200	650	46.1	318	108.5	748	26.5	183	65.0	448
Chemical properties									
Nickel		≥ 58%				≥ 72%			
Chromium		27.0–31.0%				14–17%			
Iron		7.0–11.0%				6.0–10.0%			
Carbon		≤ 0.05%				≤ 0.15%			

content can range from 2% up to levels just below that of stainless steels, which contain a minimum of 10% Cr. The primary function of the alloying elements is to increase hardenability in order to optimize mechanical properties and toughness after heat treatment. In some cases, however, alloy additions are used to reduce environmental degradation under certain specified service conditions. Typical alloying elements are chromium and molybdenum (Mo). Two types of low-alloy steels are 1¼Cr½Mo (grade 11) and 1Cr½Mo (grade 12), "chrome-moly" materials (Odette and Zinkle 2019).

– Nickel-base alloys such as Alloy 600 and its weld metals Alloy 82/182 were used mainly in primary systems, i.e., piping connected to the reactor pressure vessel. Throughout this text, the term "nickel base" is used instead of "nickel-based." Chemical properties of nickel-base alloys are given in Table 5.2. Alloys 600/82/182 were found to be highly susceptible to stress corrosion in high-temperature and high-weld-residual-stress environments, however. Alloys 690/52/152 are the preferred replacement materials.
– Stainless steels, also known as inox, corrosion-resistant steels, or "rust-less" steels, are steels containing at least 10.5% chromium. Chemical properties of

Table 5.3 Selected stainless steels (The Unified Numbering System for Metals and Alloys (UNS) provides a means of correlating many internationally used metal and alloy numbering systems administered by societies, trade associations, and individual users and producers of metals and alloys.)

Common designation	UNS number	Chemical composition [weight %]						
		C	N	Cr	Ni	Mo	Cu	Ti
Type 304	S30400	0.08	0.10	18.0–20.0	8.0–10.5	—	—	—
Type 316	S31600	0.08	0.10	16.0–18.0	10.0–14.0	2.0–3.0	—	—
Type 316 L	S31603	0.03	0.10	16.0–18.0	10.0–14.0	2.0–3.0	—	—
Type 316NG	S31653	≤ 0.02	0.06–0.12	16.0–18.0	10.0–14.0	2.0–3.0	—	—
Type 321	S32100	≤ 0.08	See "Ti" column	17.0–19.0	9.0–12.0	—	—	≥ 5× (C + N)
Type 347	S34700	≤ 0.08		17.0–19.0	9.0–13.0	—	—	≥ 10× (C + N)
High-performance stainless steels								
254-SMO®	S31254	0.02	0.18–0.22	19.5–20.5	17.5–18.5	6.0–6.5	0.50–1.50	—
AL-6XN®	N08367	0.03	0.18–0.25	20.0–22.0	23.5–25.5	6.0–7.0	0.75	—
654-SMO®	S32654	0.02	0.45–0.55	24.0–26.0	21.0–23.0	7.0–8.0	0.30–0.60	Mn: 2.0–4.0

Table 5.4 Mechanical properties of selected carbon steels and stainless steels

Name	UNS number	Tensile strength (Min.) [MPa]	Yield strength (Min.) [MPa]	Charpy V-Notch impact @ 20 °C [Nm]
Carbon steels				
ASTM A-106 Gr. B	K03006	414	241	27
ASTM A-106 Gr. C	K03501	483	276	41
Standard and high-performance stainless steels				
Type 304	S30400	607	311	65–100
Type 316	S31600	593	290	65–100
Type 316 L	S31603	483	173	65–100
254-SMO®	S31254	656	311	> 71
AL-6XN®	N08367	690	359	> 140
654-SMO®	S32654	752	428	> 106

selected stainless steels are given in Table 5.3. Mechanical properties of stainless steel and carbon steel are given in Table 5.4.

– High-performance stainless steels are especially developed for highly corrosive raw water environments.

The main grades used for fabricating stainless steel piping are Type 304 and Type 316.[1] There are two types of austenitic stainless steel, "stabilized" and "unstabilized" stainless steel. Stabilization, also known as homogenization, is a heat treatment process for austenitic stainless steels that contain titanium or niobium. The process involves heating the steel to about 885 °C and then slowly cooling it to charge the titanium or niobium carbide that dissolves in the austenite (Rosenberg and Darr 1948). Stabilization improves the weldability of the material, making it resistant to sensitization of the weld heat-affected zone (HAZ). Sensitization occurs when the heat from welding causes chromium carbide precipitation at grain boundaries in stainless steel. This reduces the amount of chromium in solution, making the stainless steel susceptible to stress corrosion cracking.

With the accumulation of operational experience and the advancement of knowledge regarding material degradation mechanisms, research into alternative materials has accelerated. This investigation emphasizes a deeper understanding of variables influencing degradation susceptibilities, including the chemical and mechanical properties of various materials, as well as the impact of manufacturing, fabrication, and welding processes. Illustrative examples of somewhat recently developed material types include the following[2]:

- Alloy 690 (and the corresponding weld metals Alloy 52 and 152) was developed to resist stress corrosion cracking, and it has become the preferred replacement for Alloy 600/82/182. As of this writing, there are no known failures after several decades of field experience with Alloy 690.
- The corrosion resistance of austenitic stainless steel is enhanced by adding chromium, molybdenum, nitrogen, and titanium. Included in this class of steels are the high-performance "super-austenitic" stainless steels such as UNS S32654 grade (24Cr, 22Ni, 7.3Mo, 3Mn, 0.5 N) especially developed for raw water environments.[3] These steels were invented in the 1970s. Tables 5.3 and 5.4, respectively, include the chemical and mechanical properties of these materials.

The term "nuclear grade" (NG) implies that chemical compositions, for example carbon content and nitrogen content, are very carefully controlled during the manufacturing process. For austenitic stainless steels, the carbon content is kept as low as possible (<0.02%) and about a maximum of 0.1% nitrogen is added for improved strength. Copeland et al. (1980) defined NG stainless steels as having

[1] *The designations for 304 and 316 stainless steel originated from the three-digit numbering system established by the American Iron and Steel Institute (AISI) in the 1930s to standardize stainless steel grades.*

[2] *Alloy 690 was developed and patented by INCO Alloys International in the late 1960s. Super-austenitic stainless steels were developed in the mid-1970s. These steels did not find widespread nuclear industry use until decades later.*

[3] *The term "super-austenitic" comes from a graphical classification, specifically the Schaeffler diagram, where these alloys plot in a position "super" to standard austenitic steels due to their enhanced corrosion resistance. This performance is achieved through higher additions of elements like chromium, molybdenum, and nitrogen.*

Table 5.5 Chemical properties of selected materials for service water piping

Name	UNS number	Chemical composition (weight %) C	N	Cr	Ni	Mo	Cu	Other
Carbon steels								
ASTM A-106 grade B	K03006	≤ 0.30	—	≤ 0.40	≤ 0.40	≤ 0.15	≤ 0.40	Mn: 0.29–1.06
Standard and high-performance stainless steels								
Type 304	S30400	0.08	0.10	18.0–20.0	8.0–10.5	—	—	—
Type 316	S31600	0.08	0.10	16.0–18.0	10.0–14.0	2.0–3.0	—	—
Type 316 L	S31603	0.03	0.10	16.0–18.0	10.0–14.0	2.0–3.0	—	—
2205	S32205	0.03	0.14–0.20	22.0–23.0	4.5–6.5	3.0–3.5	—	—
254-SMO®	S31254	0.02	0.18–0.22	19.5–20.5	17.5–18.5	6.0–6.5	0.50–1.50	—
AL-6XN®	N08367	0.03	0.18–0.25	20.0–22.0	23.5–25.5	6.0–7.0	0.75	—
654-SMO®	S32654	0.02	0.45–0.55	24.0–26.0	21.0–23.0	7.0–8.0	0.30–0.60	Mn: 2.0–4.0

controlled low carbon (0.02% max.) and nitrogen (0.1% max) and meeting all requirements, including mechanical property requirements of the ASME specifications for regular grades of Type 304 or 316 stainless steel. Although factors of improvement (FOI) in service life of much greater than 20 have been observed in controlled laboratory conditions, it is important to note that the nuclear grade stainless steels are not totally immune to stress corrosion cracking.

5.1.1 High-Performance Stainless Steels

Standard-grade stainless steels are those having relatively low chromium and molybdenum contents, e.g., Types 304 and 316. The "high-performance" stainless steels, also referred to as "super-austenitic" stainless steels, are defined as those with substantially higher chromium, molybdenum, and nitrogen content, such as the 6% molybdenum austenitic stainless steels. The chemical properties of the three materials are summarized in Table 5.5. The high levels of molybdenum and nitrogen in high-performance stainless steels, which promote corrosion resistance, also provide enhanced mechanical properties (Table 5.6) relative to standard stainless steels. The effect of composition on corrosion resistance is mainly due to chromium, molybdenum, and nitrogen. The combined influence is represented by the pitting resistance equivalent number (PREN), which is expressed by

Table 5.6 Mechanical properties of selected service water piping materials

Name	UNS number	Tensile strength (Min.) (psi)	Yield strength (Min.) (psi)	Charpy V-Notch impact @ 20 °C (ft-lb)
Carbon steels				
ASTM A-106grade B	K03006	60,000	35,000	20
ASTM A-106grade C	K03501	70,000	40,000	30
Standard and high-performance stainless steels				
Type 304	S30400	88,000	45,000	65–100
Type 316	S31600	86,000	42,000	65–100
Type 316 L	S31603	70,000	25,000	65–100
254-SMO®	S31254	95,000	45,000	> 71
AL-6XN®	N08367	100,000	52,000	> 140
654-SMO®	S32654	109,000	62,000	> 106

$$PREN = \%Cr + 3.3 \times \%Mo + 30 \times \%N \tag{5.1}$$

$$PREN(106\ GrB) = 0.4 + 4.95 = 5.35 \tag{5.2}$$

$$PREN(654 - SMO) = 24 + 23.1 + 13.5 = 60.6 \tag{5.3}$$

Stainless steels with higher PREN have better resistance to crevice corrosion and pitting and, with PREN >40, are generally preferred for harsh environments.

5.1.2 HDPE Materials

While not a predominant replacement material, high-density polyethylene (HDPE) materials are being used in belowground applications, for example, for belowground firewater supply piping and service water supply and discharge piping (i.e., from/to the ultimate heat sink). Properties of HDPE are given in Table 5.7.

5.2 Pipe Dimensions

Primary pipe dimensions are the inside (or hydraulic) diameter, outside diameter, and wall thickness. There are national standards for dimensions, permissible stress values as functions of temperature, and formulas for determination of pipe wall thickness based on pressure and material at a given temperature. Pipe size is given by a nominal pipe size, which is a "round number" or approximation of the pipe diameter. Table 5.8 lists the nominal pipe size and the inside and outside diameters

Table 5.7 Comparison of selected carbon steel and HDPE properties

Parameter	Moderate-energy service: Carbon steel (isotropic, linear elastic) [a]	Moderate-energy service: HDPE (viscoelastic) [b]
Design pressure [MPa]	≈ 1.17	≈ 1.17
Operating temperature [°C]	≈ 80	≈ 80
Typical pipe diameters [DN]	≤ 750	≤ 900
Typical wall thickness [mm]	9.5 (schedule 40)	≤ 100
Mean diameter-to-thickness ratio (DR)	≈ 80 (thin wall >20)	≈ 8.5 (thick wall <20)
Density [kg/m^3]	7833	969
Modulus of elasticity I (MPa) [c]	1.92	193
Poisson's ratio (ν) [d]	0.3	0.4 to 0.45
Coefficient of thermal expansion	1.17×10^{-5}/°C	1.08×10^{-4}/°C
Thermal conductivity (K) (kcal/hr-m-°C)	38.7	0.37
Allowable stress (MPa)	103	2.4

[a] Stress is proportional to strain, and the material returns to its original shape when the loads are removed
[b] When subjected to constant stress that is maintained for a sufficiently long period, the material experiences viscoelastic creep
[c] Ratio of applied tensile stress to tensile strain
[d] Ratio of the transverse strain to axial strain

Table 5.8 Selected nominal pipe sizes

Nominal pipe size (inch)	Nominal pipe size (mm)	Outside diameter (inch)	Outside diameter (mm)	Inside diameter (inch)	Inside diameter (mm)	Schedule designations (ASME)	Wall thickness (inch)	Wall thickness (mm)
6"	150	6.625	168.28	6.065	154.1	STD/40/40S	0.28	7.11
6"	150	6.625	168.28	5.761	146.3	XS/80/80S	0.432	10.97
6"	150	6.625	168.28	4.897	136.7	160	0.719	18.26
8"	200	8.625	219.08	7.981	202.7	STD/40/40S	0.322	8.18
8"	200	8.625	219.08	7.625	193.7	XS/80/80S	0.5	12.7
8"	200	8.625	219.08	6.813	173.1	160	0.906	23.01
12"	300	12.75	323.85	12	304.8	STD/40S	0.375	9.53
12"	300	12.75	323.85	11.938	304.2	40	0.406	10.31
12"	300	12.75	323.85	11.75	298.5	XS/80S	0.5	12.7
12"	300	12.75	323.85	11.372	288.9	80	0.688	17.48
12"	300	12.75	323.85	11.062	281.1	100	0.844	21.44
12"	300	12.75	323.85	10.75	273.1	120/XX	1	25.4
12"	300	12.75	323.85	10.126	257.2	160	1.312	33.32

for a small selection of standard pipes. When the nominal pipe size is 14 inches (350 mm) or larger, the nominal pipe size (NPS) and the outside diameter are equal. When the nominal pipe size is less than 350 mm, the nominal pipe size approximates the pipe inside diameter, depending on the pipe wall thickness. Imperial units continue to be used throughout North America as well as in countries with nuclear power plants of US design.

Table 5.8 is an incomplete pipe dimensional data table. Other parameters may include cross-sectional area, weight, and volume per running length of pipe. The designations STD, XS, and XX refer to "standard," "extra strong," and "double extra strong pipe," respectively. Schedule numbers allow for specifying more exactly the wall thickness and, therefore, the pipe strength needed to meet actual use requirements. The schedule number (SN) refers to the pipe wall thickness, and it is determined by the following equation:

$$SN = 1{,}000 \times \frac{P}{SE} \tag{5.4}$$

P is the operating pressure in pounds per square inch (psi), and SE is the allowable stress range (in psi) of pipe material, which is obtained from tables prepared by standardization organizations, e.g., ASME B31.3 (ASME Code for Pressure B31.3), Appendix A (*Allowable Stresses and Quality Factors for Metallic Piping and Bolting Materials*). ASME standardizes dimensions, sets permissible stress values as functions of temperature, and establishes working formulas for determination of wall thickness based on pressure and material at a given temperature and, in general, sets up a code of minimum requirements for the safety and reliability of a system: ASME B36.10 M-2018 for carbon steel (ASME 2018) and ASME B36.19–2022 for stainless steel (ASME 2022). The NPS and DN numbers are not always the same as the hydraulic (or inside) pipe diameter.

5.3 High-Energy Versus Moderate-Energy Piping

Consequences of pipe failure can include pipe whip and spray impacts on adjacent piping, electrical and mechanical components, and structures. Piping systems are differentiated by the fluid operating pressure and temperature. The "*Standard Review Plan: Branch Technical Position 3-3 – Protection Against Postulated Piping Failures in Fluid Systems Outside Containment*" defines the following types of fluid systems[4]:

[4]*A standard review plan provides guidance for reviewing an application to obtain a license to construct or operate a nuclear facility. A branch technical position (BTP) means a final report prepared by USNRC staff setting forth recommendations or a method considered acceptable for performing analyses, calculations, or other technical evaluations that are used to satisfy or to evaluate or demonstrate compliance with NRC regulatory requirements. BTP 3–3 is available from NRC-ADAMS, Accession No. ML070800008.*

– High-energy fluid system: Fluid system that, during normal plant conditions, is either in operation or maintained pressurized under conditions where either or both of the following are met:

  Maximum operating temperature exceeds 95 °C.
  Maximum operating pressure exceeds 1.9 MPa.

– Moderate-energy fluid system: Fluid system that, during normal plant conditions, is either in operation or maintained pressurized (above atmospheric pressure) under conditions where both of the following are met:

  Maximum operating temperature is 95 °C or less.
  Maximum operating pressure is 1.9 MPa or less.

5.4 Safety Classification Schemes

With specific reference to pressure boundary components, the USNRC developed a quality classification system in the early 1970s to provide licensees with guidance for satisfying General Design Criterion 1 (*Quality Standards and Records*),[5] which states that structures, systems, and components important to safety shall be designed, fabricated, erected, and tested to quality standards commensurate with the importance of the safety functions to be performed. The quality classification system consists of four quality groups, A through D; methods for assigning components to those quality groups; and specific quality standards applied to each quality group. When the USNRC issued Revisions 2 and 3 of Regulatory Guide (RG) 1.26, 10 CFR 50.55a required that only components of the reactor coolant pressure boundary be designed, fabricated, erected, and tested to the highest available national standards; this corresponded to the quality standards for Quality Group A of the USNRC system. The regulator issued Revision 4 to align this RG with the regulations, published on March 15, 1984, amending 10 CFR 50.55a to incorporate by reference the criteria in Section III of the ASME BPV Code as they relate to the design and fabrication of Class 2 and Class 3 components (Quality Group B and Group C components, respectively) (USNRC 1984a, b).

Components of Quality Group A must meet the requirements for Class 1 components in Section III of the ASME Boiler and Pressure Vessel Code (ASME 2023a, b). This category is limited to components that are part of the reactor coolant pressure boundary, except for the portions excluded by 10 CFR 50.55a(c)(2) that are included in Quality Group B below. This exclusion applies to components whose failure

[5] *General Design Criterion 1 (GDC 1) specifically addresses the need for quality standards commensurate with the importance of safety functions. While the exact date of GDC 1's initial development is not explicitly stated in the provided context, it falls within the overall timeframe of Appendix A's development, which was finalized on June 27, 1970. The NRC has continued to refine and update its approach to quality assurance, including the endorsement of standards like NQA-1 and the development of regulatory guides like RG 1.28.*

would not prevent the reactor from being shut down and cooled down in an orderly fashion with normal make-up and components that are or can be isolated from the reactor coolant system by two valves in series, with automatic closure of open valves.

Quality Group B (ASME III Class 2) applies to water- and steam-containing pressure vessels, heat exchangers (other than turbines and condensers), storage tanks, piping, pumps, and valves that are either (1) part of the reactor coolant pressure boundary, as defined in 10 CFR 50.2, "Definitions," but excluded from the requirements of 10 CFR 50.55a(c)(1) for reactor coolant pressure boundary components pursuant to 10 CFR 50.55a(c)(2), or not part of the reactor coolant pressure boundary but part of the following:

- Systems or portions of systems important to safety that are designed for emergency core cooling, post-accident containment heat removal, or post-accident fission product removal.
- Systems or portions of systems important to safety that are designed for reactor shutdown or residual heat removal.
- Those portions of the steam systems of boiling water reactors extending from the outermost containment isolation valve up to but not including the turbine stop and bypass valves, and connected piping up to and including the first valve that is either normally closed or capable of automatic closure during all modes of normal reactor operation; alternatively, for boiling water reactors containing a shutoff valve (in addition to the two containment isolation valves) in the main steam line and the main feedwater line, those portions of the steam and feedwater systems extending from the outermost containment isolation valves up to and including the shutoff valve or the first valve that is either normally closed or capable of automatic closure during all modes of normal reactor operation.
- Those portions of the steam and feedwater systems of pressurized water reactors extending from and including the secondary side of steam generators up to and including the outermost containment isolation valves and connected piping up to and including the first valve (including a safety or relief valve) that is either normally closed or capable of automatic closure during all modes of normal reactor operation.
- Systems or portions of systems that are connected to the reactor coolant pressure boundary and are not capable of being isolated from the boundary during all modes of normal reactor operation by two valves, each of which is either normally closed or capable of automatic closure.

Quality Group C (ASME III Class 3) applies to water-, steam-, and radioactive waste-containing pressure vessels; heat exchangers (other than turbines and condensers); storage tanks; piping; pumps; and valves that are not part of the reactor coolant pressure boundary or included in Quality Group B but are part of the following:

- Cooling water and auxiliary feedwater systems or portions of those systems important to safety that are designed for (1) emergency core cooling,

(2) post-accident containment heat removal, (3) post-accident containment atmosphere cleanup, or (4) residual heat removal from the reactor and from the spent fuel storage pool (including primary and secondary cooling systems), although Quality Group B includes portions of those systems that are required for their safety functions and that (1) do not operate during any mode of normal reactor operation and (2) cannot be tested adequately.

- Cooling water and seal water systems or portions of those systems important to safety that are designed for the functioning of components and systems important to safety, such as reactor coolant pumps, diesels, and the control room.
- Systems or portions of systems that are connected to the reactor coolant pressure boundary and are capable of being isolated from that boundary during all modes of normal reactor operation by two valves, each of which is either normally closed or capable of automatic closure.
- Systems, other than radioactive waste management systems, that contain or may contain radioactive material and whose postulated failure would result in conservatively calculated potential offsite doses that exceed 0.1 rem total effective dose equivalent.

Quality Group D applies to water- and steam-containing components that are not part of the reactor coolant pressure boundary or included in Quality Groups B or C but that are part of systems or portions of systems that contain or may contain radioactive material. Group D corresponds to ASME III non-safety-related components. The American Nuclear Society (ANS) developed ANSI/ANS-58.14-2011, "*Safety and Pressure Integrity Classification Criteria for Light Water Reactors*" (ANSI 2022), which applicants or licensees can use as an alternative classification method to meet USNRC regulations such as General Design Criterion (GDC) 1[6] and 10 CFR 50.55a, "*Codes and Standards*."

5.5 Failure Definitions and Failure Modes

Pipe failure mode terminology can be ambiguous. Essentially, a pipe failure means that the integrity of a pressure boundary is compromised, necessitating repair or replacement. A pipe can fail when the generated stress exceeds the nominal material strength or when the stress intensity generated at a critical defect (for instance, as a result of stress corrosion) exceeds the material toughness or possibly as a combination of both. The degradation of pipe material and its consequences affect structural reliability modeling. Pipe failure refers to any degraded pressure boundary, from a flaw needing further evaluation to a structural failure requiring immediate repair. The relationships between pipe failure modes and material degradation mechanisms are

[6] https://www.nrc.gov/reading-rm/doc-collections/cfr/part050/part050-appa

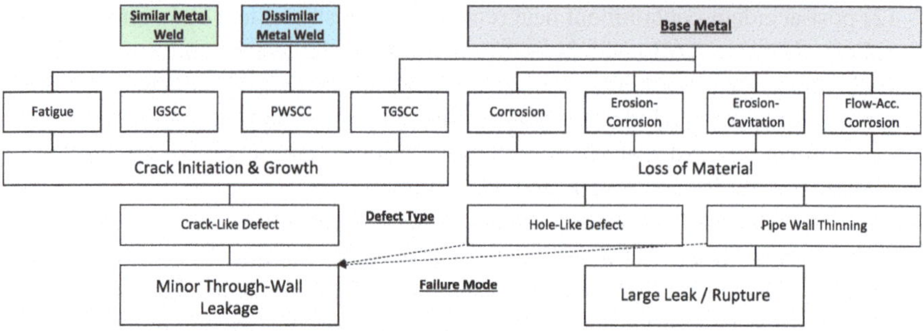

Fig. 5.1 Degradation mechanisms versus mode of failure

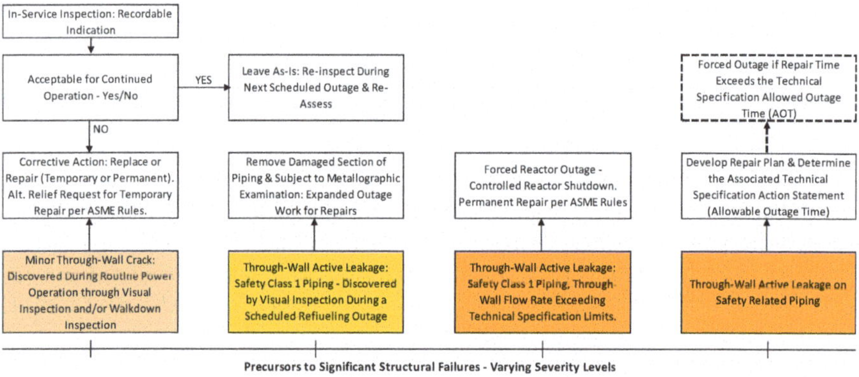

Fig. 5.2 Pipe failure mode definitions

illustrated in Fig. 5.1. Pipe failure modes and severity levels are illustrated in Fig. 5.2.

In probabilistic safety assessment, the definition of the consequence of a pipe failure is important. Questions like "what is the frequency of a pipe leak?" or "what is the frequency of pipe rupture?" are not very meaningful without elaborations. In probabilistic terms, the definition of failure versus its consequences affects the modeling approach as well as modeling of uncertainty. The consequence of a pipe failure can be characterized in terms of the through-wall mass or volumetric flow rate. EBS is determined based on engineering analyses that rely on fluid dynamics. Empirical data on the conditional failure probability versus observed through-wall flow rate is shown in Fig. 5.3.

Probabilistic fracture mechanics (PFM) defines a leak as a stable, wall-penetrating defect (IAEA 1976; Yoshimura and Kanto 2020; USNRC 2021). A wall-penetrating defect which fails due to the applied loads is considered a rupture with unstable crack growth rate. The crack opening area (COA, Fig. 5.4) is an important term in PFM (EPRI 1988; Rahman et al. 1998). Estimates of the crack opening area for a postulated through-wall cracking can vary widely depending on how the crack is idealized, which crack opening model is used, and what material properties are

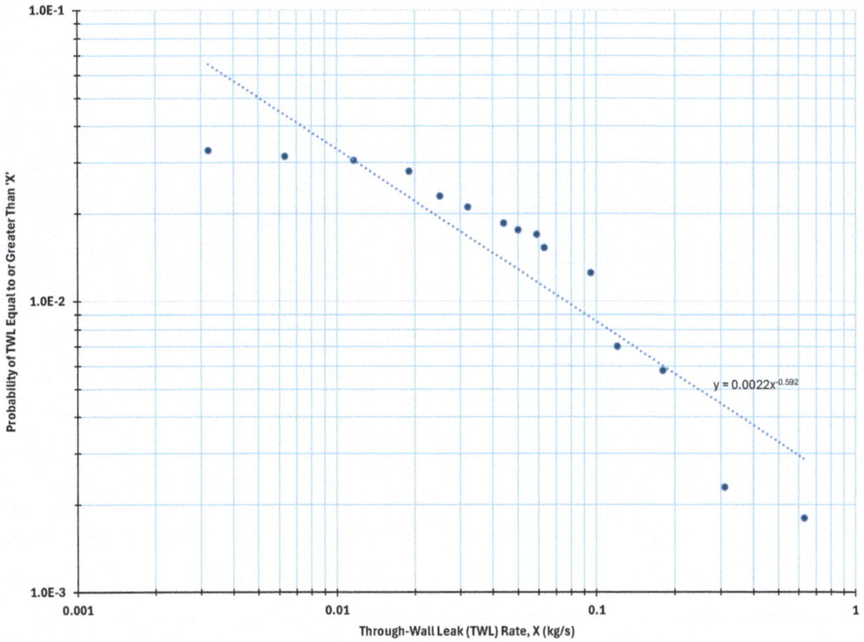

Fig. 5.3 Empirical conditional failure probability of through-wall leak in BWR piping

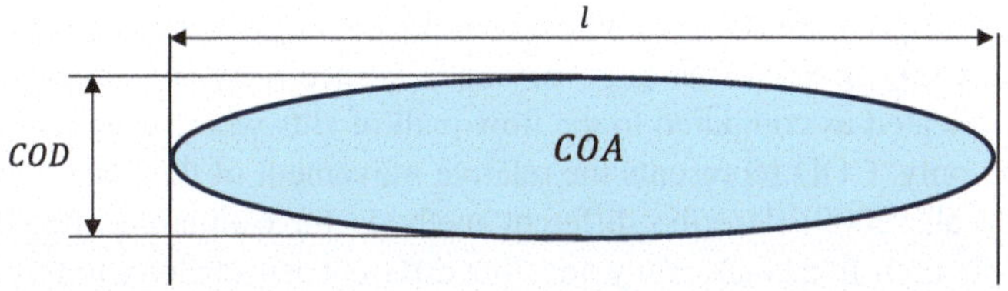

Fig. 5.4 Crack opening area (COA) versus crack opening displacement (COD)

assumed. A range of published solutions are available for idealized notch-like cracks in simple geometry subject to basic loading (pressure, membrane, and bending). Their accuracy varies with pipe diameter and wall thickness, crack size, and applied loads.

The nature of roughness of the crack surface depends on crack opening displacement (COD), which is defined as the distance between the opposite faces of a crack tip at the 90° intercept position. For a large COD, the mean flow path is straight, and global roughness acts as surface roughness (SR), as against global and local roughness accounted for together, for small COD with a "zigzag" flow path. The surface roughness parameter defines the peak-to-peak roughness of the crack face surface. This is used in the calculation of friction factor and pressure loss due to friction for fluid flow through a crack in a pipe. Earlier, surface roughness was assumed to be invariant with respect to crack opening displacement and crack width (CW).

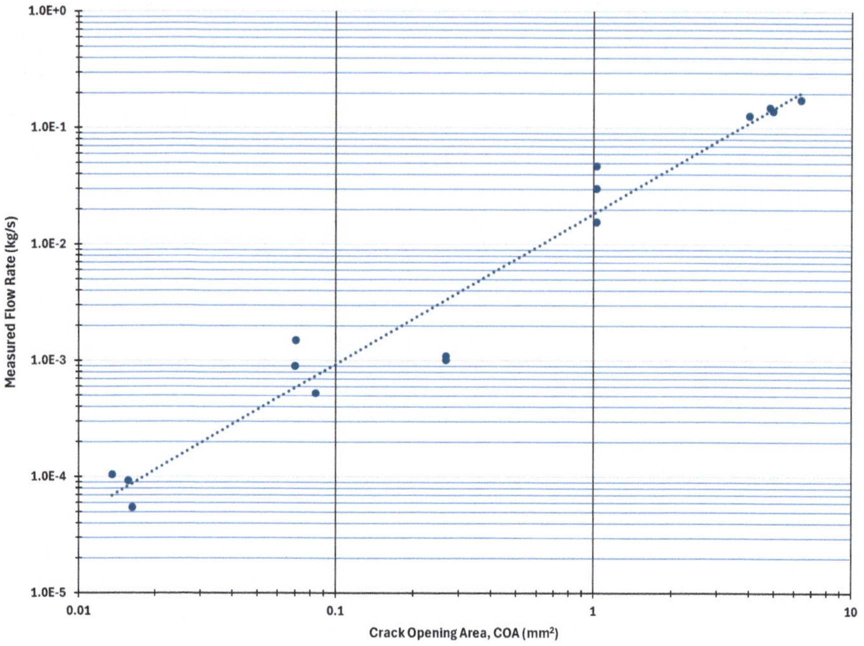

Fig. 5.5 COA vs. measured through-wall flow rate based on experiments

However, appropriate surface roughness could be large or small depending on whether the COD or CW was large or small, respectively. For cracks, the flow path is complicated as compared to the flow path of slits where roughness is due to wall friction only. COD represents the relative movement of the crack faces.

Ansari et al. (2000) describe different methods for estimating the COA for a through-wall notch-like crack. They note that different aspects have to be considered that influence the results, including the "cross section of the leak in terms of COA or COD, the morphology of the leak and the resulting fluid-dynamical resistance, and the fluid motion together with the evaporation." A longitudinal break is parallel to the pipe axis and oriented at any point around the pipe circumference. The break area is equal to the effective cross-sectional flow area upstream of the break location with the length of the break equivalent to twice the inside pipe diameter. Results from experimental work are illustrated in Fig. 5.5, which shows measured through-wall leak rates as a function of the COA.

5.5.1 Branch Technical Position 3–4

A branch technical position (BTP) is a document issued by the U.S. Nuclear Regulatory Commission that provides guidance on how to address specific technical issues related to nuclear safety. BTPs are used by licensees to demonstrate compliance with NRC regulations. BTP 3–4, initially issued in September 1975, focuses on postulating pipe rupture locations in nuclear power plants, particularly concerning the criteria for higher stress range. BTP 3–4 differentiates between three types of pipe failure modes (USNRC 2007):

1. **Circumferential breaks** in high-energy piping: Such breaks should be assumed to result in pipe severance and separation amounting to at least a one-diameter lateral displacement of the ruptured piping sections unless physically limited by piping restraints, structural members, or piping stiffness as may be demonstrated by inelastic limit analysis (e.g., a plastic hinge in the piping is not developed under loading).
2. **Longitudinal pipe breaks** in high-energy piping: Such breaks should be assumed to result in an axial split without pipe severance. Splits should be oriented (but not concurrently) at two diametrically opposed points on the piping circumference such that the jet reactions cause out-of-plant bending of the piping configuration. The dynamic force of the fluid jet discharge should be based on a circular or elliptical (2D x ½D) break area equal to the effective cross-sectional flow area of the pipe at the break location and on a calculated fluid pressure modified by an analytically or experimentally determined thrust coefficient as determined for a circumferential break at the same location.
3. "**Leakage cracks**" in high- or moderate-energy piping: Such pressure boundary failures should be postulated to be in those circumferential locations that result in the most severe environmental consequences. Fluid flow from a "leakage crack" should be based on a circular opening of area equal to that of a rectangle one-half pipe diameter in length and one-half pipe wall thickness in width $(\frac{1}{2}D \times t_{Wall\ thickness})$. The flow from the "leakage crack" should be assumed to result in an environment that wets all unprotected components within the compartment, with consequent flooding in the compartment and communicating compartments. Flooding effects should be determined on the basis of a conservatively estimated time period necessary to effect corrective actions, i.e., leak isolation.

5.5.2 *Fluid Dynamics*

The theory of flow through a crack or hole in a pipe is based on fluid dynamics (Benedict 1980; EPRI 1983; Paul et al. 1991; OECD/NEA 2022). When reviewing operating experience, it is often necessary to estimate pipe flow rates using basic engineering principles. Flow assessments depend on flaw size and pressure, whether the defect is a pinhole, cylindrical, or elliptical hole. Failure reports provide measurements for these defects, which are used to calculate an equivalent break size (EBS). The EBS is the diameter of a circular conduit that gives the same area as a noncircular conduit and is substituted for diameter in equations where flow rate and velocity are calculated. With knowledge of the operating pressure at the time of pipe failure, it becomes straightforward to obtain a reasonable estimate of the through-wall mass or volumetric flow rate. The following equations are from the Crane Handbook (Crane 2022). The dimensional analysis is included for clarity:

$$q = C \times A \times \sqrt{2gh_l} = C \times A\sqrt{\frac{2P}{\rho}} \tag{5.5}$$

$$q = C\pi\frac{d^2}{4}\sqrt{2}\sqrt{\frac{P}{\rho}} \tag{5.6}$$

$$\left[{}^{m^3}\!/_{s}\right] = C \times \pi\sqrt{2} \times \left[\frac{mm^2}{4 \times 10^6}\right]\sqrt{\left[\frac{kg\ m}{s^2}\right] \times \left[\frac{m^3}{kg}\right]} \tag{5.7}$$

$$q = 1.11 \times 10^{-6} d^2 C\sqrt{\frac{P}{\rho}} \tag{5.8}$$

$$Q = 6.66 \times 10^{-2} d^2 C\sqrt{\frac{P}{\rho}} \tag{5.9}$$

In the above equations, q is the volumetric flow rate in m^3/s and Q the volumetric flow rate in liter per minute. The diameter of the orifice is denoted as *d*, ρ is the mass density of fluid (kg/m^3), *P* is the pressure maintained in the pipe (Pascal), and *C* is the dimensionless flow coefficient. The flow coefficient is a relative measure of its efficiency at allowing fluid flow. It describes the relationship between the pressure drop across an orifice and the corresponding flow rate. Flow coefficient look-up tables and charts are available from engineering handbooks. The flow rate, *Q* (in L/min), from a pipe failure with an EBS of D (diameter of the break, in mm) and a system pressure of *P* (in kilopascals) can be estimated using Eqs. 5.5–5.10:

$$Q = 0.068 \times D^2\sqrt{P} \tag{5.10}$$

To convert to an EBS, the analyst must find an equivalent diameter that would yield a flow of 2*A*. Since the total area and break area of the pipe are different by a factor of 2, an analyst can solve for a coefficient that must be multiplied by the original pipe diameter to result in an equivalent break size for a double-ended guillotine break (DEGB). For a pipe with a diameter d, the hydraulic diameter, the coefficient C can be derived from

$$DEGB = 2A = 2 \times \pi \times (d/2)^2 = \frac{\pi}{2} \times d^2 \tag{5.11}$$

$$DEGB = \pi \times (C \times D/2)^2 \tag{5.12}$$

$$DEGB: \rightarrow \pi/2 \times d^2 = \pi/4 \times (C \times d)^2 \tag{5.13}$$

$$C = \sqrt{2} \tag{5.14}$$

$$DEGB = \sqrt{2} \times d \tag{5.15}$$

That is, the maximum flow rate from a DEGB from a pipe with hydraulic diameter d (i.e., the inside diameter) can be represented by a cross-sectional area with an equivalent break size (EBS) equal to $\sqrt{2} \cdot d$.

Reports on pipe failure sometimes characterize the observed leak rate in terms of drops per minute. As a rule of thumb, approximately 4000 drops equal 1 liter of water. Therefore, a water leak from a flawed pipe of 60 drops per minute is approximately equal to $2.5 \cdot 10^{-4}$ kg/s, a very minute leakage. Computational fluid dynamics (CFD) improves the modeling of consequences of a postulated piping pressure boundary failure (Nayak and Joshi 2019). CFD allows precise modeling of fluid behavior, enabling more detailed internal flooding PSA models, loss-of-coolant accident models, and high-energy line break models. CFD's capability to simulate complex fluid dynamics and visualize results provides a useful tool for risk assessment, leading to more accurate evaluations of the spatial impact of pressure boundary failures.

5.6 Leak-Before-Break and Break Preclusion

The leak-before-break (LBB) concept is based on the analysis which demonstrates by deterministic fracture mechanics analysis that a crack would grow through the pipe wall, resulting in a leak, and that this postulated small through-wall pipe defect would be detected by leakage monitoring systems long before the defect could grow to an unstable size (Wilkowski and Chao 1990). Leakage exceeding the technical specification limits requires identifying the leak location and isolating it if feasible by closing isolation valves or initiating a controlled plant shutdown. The margins for leak detection and crack length, respectively, are indicated in an LBB diagram. When the length of a through-wall crack approaches a certain value, corresponding to an "identified" or "unidentified" leakage, operator action is required. When the length of a through-wall crack approaches a certain value, e.g., 50% of the outside pipe diameter, an unstable crack growth may occur. In fracture mechanics, leakage occurs if the ligament zone becomes unstable.

Historically, DEGB was considered the most severe loss-of-coolant accident and led to heavy pipe restraints and jet shields due to lack of alternatives. DEGB was extended to safety-equipment environmental design. However, the nuclear industry found DEGB highly unlikely and too restrictive. Research indicated the near impossibility to have a full DEGB in reactor coolant pressure boundary piping under Service Level D conditions, even down to DN25 piping. Service Level D involves gross deformation requiring repair or replacement. The different service levels are as follows:

- Service Level A: This level represents normal operating conditions. It includes the loadings that a component is expected to experience during regular service.
- Service Level B: This level covers loadings that may exceed normal conditions but are still expected to be within the design basis. Components must withstand these loads without requiring repair.
- Service Level C: This level includes loadings that could cause significant deformations and potentially require inspection or repair of the component. However, the component should still be operable during normal events postulated by the design.
- Service Level D: This level represents the most severe and least frequent loadings, such as design basis accidents. Components may experience gross deformation and damage, potentially requiring removal from service for repair.

Leak-before-break has two basic connotations. First, it means that a piping system and its components are sufficiently robust that a through-wall leak will be detected early and that a major pressure boundary failure is incredible, the "LBB concept." Demonstration of LBB is accomplished through adherence to relevant codes, standards, and regulations. Secondly, it means an application of a methodology for enhancing nuclear safety through the removal of certain pipe whip restraint devices and jet impingement shields to increase access for and to improve the quality of in-service inspections. The LBB methodology is concerned with two structural integrity considerations:

1. Demonstration of an adequate margin towards experiencing a circumferential through-wall crack reaching a critical size.
2. Demonstration of an adequate cumulative fatigue usage factor (CUF) margin: The theory of cumulative damage assumes that a stress cycle with an alternating stress above the endurance limit inflicts measurable permanent damage to a structure. It also assumes that the total damage caused by several stress cycles is equal to the summation of damage caused by the individual stress cycles. A regulatory stipulation is that CUF must be kept below 1.0. A current US regulatory position is that the cumulative usage factor (CUF) should be less than 0.1 (Herrmann et al. 2011; Benson 2019). For new reactor design certification reviews, the regulator has considered a CUF limit of 0.4 to be acceptable when the effects of environmentally assisted fatigue (EAF) are considered in piping design.

LBB methodology involves deterministic or probabilistic fracture mechanics analyses to demonstrate that a predefined pipe flaw at most will result in stable through-wall crack growth and that the probability of a pipe rupture given the presence of a flaw is "extremely low." LBB is ensured by demonstrating three levels of safety against a DEGB:

- Level 1 is inherent in the design philosophy of ASME BPVC Section III, which is normally followed in piping design. Ductile and tough materials are widely used in nuclear power plant piping, because of their high resistance to catastrophic rupture.

- Level 2 consists of postulating a surface crack, which may go undetected during nondestructive testing. It is then shown through fatigue analysis that there is insignificant crack growth of this surface crack, during the entire reactor life.
- Level 3 postulates a through-wall leaking crack at a location of maximum stress with the worst material properties. The length of the through-wall crack should be such that the calculated leakage rate of fluid discharged from the crack under normal operating load is equal to an immediate shutdown action limit. For BWR plants, this implies a primary pressure boundary leak rate greater than 0.13 kg/s if the exact leak location cannot be identified. For PWR plants, this implies a primary pressure boundary leak rate greater than 0.063 kg/s if the exact leak location cannot be identified.[7] A linear elastic fracture mechanics (LEFM) or an elastic-plastic fracture mechanics (EPFM) assessment is performed to demonstrate that adequate margins exist against the onset of unstable crack growth. The minimum margins which must be met are twice on the size (length) of crack and a square root of two ($\sqrt{2}$) on the extreme load, acting on the pipe containing the through-wall crack. Codes and standards provide detailed guidance for the application of LEFM and EPFM to evaluations of indications and flaws detected during in-service inspection of piping.

Typical inputs for LBB evaluations include pipe geometry, material properties, crack morphology, cyclic loads, operating pressure, and temperature. The major steps in an LBB evaluation are described as follows (EPRI 1986; Bourga et al. 2015):

1. Characterize/postulate the initial flaw: Flaw dimensions have to be defined for surface or through-wall flaws. Depending on the procedure used, surface defects may be assessed in addition to through-wall defects, taking crack growth into account.
2. Determine the critical length of the through-wall flaw: This refers to the length at which the through-wall defect becomes unstable, based on fracture mechanics calculations, assuming that stresses are load-controlled.
3. Estimate the flaw length at breakthrough: This is carried out by calculating the surface flaw length at which ligament failure is predicted to occur and re-characterizing this flaw as a through-wall flaw.
4. Determine detectable leakage length of the through-wall flaw: This includes calculation of the crack opening area (COA) associated with the crack length and calculation of the resulting detectable leak rate appropriate to the leak detection system capabilities. Time to detect the leak should be considered. The leak rate may be estimated from relevant experimental data, if available, or computer codes which predict leakage rates for single- or two-phase flows for a

[7] *The BWR- and PWR-specific limits for reactor coolant system operational leakage are detailed in the technical specification bases documents;* https://www.nrc.gov/reactors/operating/licensing/techspecs/current-approved-sts.html

wide range of through-wall defects that appropriately account for the surface roughness, number of turns, etc. for the crack mechanisms of interest.
5. Assess the results: A case for LBB is established, provided that the calculations in previous steps show that (a) the flaw length at breakthrough is less than the critical length of the through-wall flaw, and (b) the time to detect the leak is less than the time for the flaw to grow to the critical length.

A DEGB is highly improbable, even during severe accidents, making the design-basis LOCA based on DEGB unnecessary. The LBB methodology could be accepted as a technically justifiable approach for eliminating postulated DEGB in high-energy piping systems. This conclusion resulted from extensive research, development, and rigorous evaluations by the USNRC, the German Commission on Reactor Safety (RSK),[8] and several other industry organizations since the early 1970s. Recommendations for the application of LBB in the USNRC licensing process were published in 1984 as NUREG-1016 Volume 3; in it, a "*Pipe Break Task Group*" gave the following recommendations regarding limitations and technical guidance:

Limitations (NUREG-1061 Volume 3, pp. 5–1 to 5–2) (USNRC 1984a, b)

(a) For specifying design criteria for emergency core coolant systems, containments, and other engineered safety features, loss of coolant shall be assumed in accordance with existing regulations, i.e., to be through an opening equivalent to twice the pipe flow area up to and including the largest diameter pipe in the system. The evaluation of environmental effects should be considered on a case-by-case basis.
(b) The LBB approach should not be considered applicable to high-energy fluid system piping, or portions thereof, that operating experience has indicated susceptibility to failure from the effects of corrosion (e.g., intergranular stress corrosion cracking), water hammer, or low- and high-cycle (i.e., thermal, mechanical) fatigue.
(c) For plants for which there is an operating license or construction permit, component (e.g., vessels, pumps, valves) and piping support structural integrity should be maintained with no reduction in margin for the Final Safety Analysis Report (FSAR) or Preliminary Safety Analysis Report (PSAR) loading combination that governs their design.
(d) The LBB approach should not be considered applicable if there is a high probability of degradation or failure of the piping from more indirect causes such as fires, missiles, and damage from equipment failures (e.g., cranes) and failures of systems or components in proximity.

[8] *The German Reactor Safety Commission (*https://rskonline.de/*) was created in 1958. It advises the Federal Ministry for the Environment, Nature Conservation, Nuclear Safety and Consumer Protection (BMUV) in safety-related matters.*

(e) The LBB approach should not be considered applicable to high-energy piping, or portions thereof, for which verification has not been provided that the requirements of IE Bulletin 79–14 have been met (USNRC 1979).
(f) The LBB approach described in this report is limited in application to piping systems where the material is not susceptible to cleavage-type fracture over the full range of system operating temperatures where pipe rupture could have significant adverse consequences.

General Technical Guidance (NUREG-1061 Volume 3, pp. 5–2 to 5–4)

(a) Provide a discussion to support the conclusion that this piping run or system does not fall within the limitations delineated under "Limitations."
(b) Specify the type and magnitude of the loads applied (forces, bending, and torsional moments), their source(s), and method of combination. Identify the location(s) at which the highest stresses coincide with the poorest material properties that occur for base materials, weldments, and safe ends.
(c) Identify the types of materials and material specifications used for base metal, weldments, and safe ends, and provide the material properties including appropriate toughness and tensile data, long-term effects such as thermal aging, and other limitations.
(d) Postulate a flaw at the location(s) specified in (b) above that would be permitted by the acceptance criteria of Section XI of the ASME Boiler and Pressure Vessel Code. Demonstrate by fatigue crack growth analysis for Code Class 1 piping that the crack will not grow significantly during service.
(e) Postulate a through-wall flaw at the location(s) specified in (b) above. The size of the flaw should be large enough so that the leakage is assured of detection with margin using the installed leak detection capability when the pipes are subjected to normal operating loads. If auxiliary leak detection systems are relied on, they should be described.
(f) For geometrically complex lines or systems, performance of a system evaluation should be considered.
(g) Assume that a safe shutdown earthquake (SSE) occurs prior to detection of the leak to demonstrate that the postulated leakage flaw is stable under normal operating plus SSE loads for a long period of time; that is, crack growth if any is minimal during an earthquake.
(h) Determine flaw size margin by comparing the selected leakage size flaw (Item e) to critical size crack. Using normal plus SSE loads, demonstrate that there is a margin of at least 2 between the leakage size flaw and the critical size crack to account for the uncertainties inherent in the analyses and leak detection capability.
(i) Determine margin in terms of applied loads by a crack stability analysis. Demonstrate that the leakage-size cracks will not experience unstable crack growth even if larger loads (at least $\sqrt{2}$ times the normal plus SSE loads) are applied. Demonstrate that crack growth is stable, and the final crack size is limited such that a DEGB will not occur.

(j) The piping materials' toughness (J-R curves) and tensile (stress-strain curves) properties should be determined at temperatures near the upper range of normal plant operation. The test data should demonstrate ductile behavior at these temperatures.

(k) Ideally, the J-R curves should be obtained using specimens whose thickness is equal to or greater than that of the pipe wall. The specimen should be large enough to provide crack extensions up to an amount consistent with J/T condition determined by analysis for the application. Because practical specimen size limitations exist, the ability to obtain the desired amount of experimental crack extension may be restricted. In this case, extrapolation techniques may be used if appropriate as described in Section A2.4.3 of Appendix A, NUREG-1061, Volume 3. The stress-strain curves should be obtained over the range from the proportional limit to maximum load.

(l) Ideally, the material tests should be conducted using archival material for the pipe being evaluated. If archival material is not available, tests should be conducted using specimens from three heats of material having the same material specification. Test material should include base and weld metals.

(m) At least two stress-strain curves and two J-resistance curves should be developed for each of a minimum of three heats of materials having the same material specifications and thermal and fabrication histories as the in-service piping material. If the data are being developed from an archival heat of material, a minimum of three stress-strain curves and three J-resistance curves from that one heat of material is sufficient. The tests should be conducted at temperatures near the upper range of normal plant operation (e.g., 288 °C). Tests should also be conducted at a lower temperature, which may represent a plant condition (e.g., hot standby) where pipe break would present safety concerns similar to normal operation. These tests are intended only to determine if there is any significant dependence of toughness on temperature over the temperature range of interest. One J-R curve and one stress-strain curve for one base metal and weld metal are considered adequate to determine temperature dependence.[9]

(n) As indicated in Section 5.9.1 [*of NUREG-1061, Vol. 3*], there are certain limitations that currently preclude the generic use of limit-load analyses to evaluate leak-before-break conditions for eliminating pipe restraints. However, the Task Group believes that limit-load analysis can be used to demonstrate acceptable leak-before-break margins for the application, provided that the limit moment is greater than the applied (normal operation plus safe shutdown earthquake (SSE)) moment at any location in the pipe run by a factor of at least three. Limit moment should be determined from Eq. (A-19) in Appendix A where the flow stress is determined from ASME code minimum properties. Data obtained from future tests (see Section 10.0) may provide information that

[9]*A J-R curve, also known as a crack growth resistance curve, is a graphical representation of a material's resistance to ductile crack propagation. It plots the J-integral value, a measure of the stress intensity at the crack tip, against the amount of crack growth.*

would allow less restrictive use of limit-load analyses for justifying elimination of pipe restraints.

The original application of LBB was for the elimination of protective structures such as pipe whip restraints, jet impingement shields, and other applications, which provide protection against high-energy line breaks. LBB has been used as part of overall safety assessments for leaving real or postulated flaws in service for some operating periods, usually not beyond the next planned maintenance and refueling outage. Released in November 1987, "*Modification of General Design Criterion 4 (GDC-4): Requirements for Protection Against Dynamic Effects of Postulated Pipe Ruptures*" uniquely defines the leak-before-break case and assigns legislative consequences to the LBB status.[10] For plant owners that can prove the LBB case according to GDC-4, extensive measures to handle the dynamic consequences of the DEGB accident are no longer obligatory. This had practical implications for the layout of the primary piping; for example, pipe whip restraints and jet shields are no longer needed. The LBB approach cannot be applied to piping that can fail in service from water hammer, corrosion, creep, flow-assisted degradation, or excessive fatigue. The rationale is that these failure mechanisms challenge the assumption in the LBB acceptance criteria. For example, water hammer may introduce excessive dynamic loads that are not accounted for in the LBB analyses, and corrosion and fatigue crack growth may introduce flaws whose geometry may not be bounded by the postulated "through-wall" flaw in the LBB analyses.

To demonstrate that the candidate piping is not susceptible to failure from these degradation mechanisms, the operating history and measures to prevent their operation or mitigate their effects must be reviewed. After the LBB candidate piping has been reviewed for degradation mechanisms and found to be acceptable, the piping is subjected to a fracture mechanics evaluation. The purpose of this evaluation is to show that there is flaw stability and the resulting leakage will be minor. National codes and standards and regulatory guidelines exist for LBB applications. In the United States, the regulator began accepting LBB for large-diameter high-quality piping systems in the mid-1980s. A draft Standard Review Plan (SRP) entitled "*Leak-Before-Break Evaluation Procedures*" was published in 1987. Because of ongoing research, a final SRP was not issued until 2007: SRP 3.6.3 Revision 1. The following SRP acceptance criteria apply[11]:

- Compliance with General Design Criterion (GDC) 4 requires that components important to safety be designed to accommodate the effects of, and be compatible with, environmental conditions associated with normal operation, maintenance, testing, and postulated accidents, including loss-of-coolant accidents. Safety-related components should be protected against dynamic effects, including the effects of missiles, pipe whipping, and discharging fluids that may result from equipment failure or events and conditions outside the nuclear power unit.

[10] https://www.nrc.gov/docs/ML2023/ML20236L715.pdf

[11] https://www.nrc.gov/docs/ML0636/ML063600396.pdf

- LBB analyses should demonstrate that the probability of pipe rupture is extremely low under conditions consistent with the design basis for piping. A deterministic evaluation of the piping system that demonstrates sufficient margins against failure can be assumed to satisfy the extremely low probability criterion.

The German break preclusion (BP) concept is based on three core principles: (1) basic safety, (2) leak-before-break, and (3) surveillance (Table 5.9). Basic safety is achieved through requirements placed on design, material selections, product forms, manufacturing, and quality assurance. Leak-before-break is demonstrated through fracture mechanics analyses that model potential crack growth from initial flaws (like weld defects) under operational stresses. The analysis shall prove that the crack will transition from an undetectable flaw to a detectable leak (through-wall crack) long before critical instability and rupture. Surveillance measures include fatigue monitoring, water chemistry control, leakage detection, and in-service inspection. The original break preclusion concept was developed in the 1970s, finalized with the publication of the RSK Guidelines for PWRs in 1981, and subjected to further refinements (Goetsch et al. 1995; Schulz 1995).[12] It was adopted for the design of the European pressurized water reactor (EPR), for which the full Basic Design Report was issued in 1997.

5.7 Transition Break Size

Regulatory requirements in the United States for the design, construction, and operation of emergency core cooling systems (ECCSs) for light-water nuclear reactors are contained in 10 CFR 50.46, "*Acceptance Criteria for Emergency Core Cooling Systems for Light-Water Nuclear Reactors*"; Appendix K to 10 CFR Part 50, "ECCS Evaluation Models"; and Appendix A to 10 CFR Part 50, "General Design Criteria (GDC) for Nuclear Power Plants" (e.g., GDC 35, "*Emergency Core Cooling*"). These requirements state, in part, that the ECCS shall be sized to provide adequate make-up water to compensate for a break of the largest diameter pipe in the primary system, i.e., DEGB. However, the DEGB was recognized as an extremely unlikely event. In order to risk-inform the break size, the USNRC proposed the concept of the "transition break size" (TBS), as described in SECY-05-0052, Proposed Rulemaking for "*Risk-Informed Changes to Loss-of-Coolant Accident Technical Requirements,*" dated March 29, 2005, which included the following statement[13]:

> *The proposed rule would divide the current spectrum of LOCA break sizes into two regions. The division between the two regions is determined by a "transition break size" (TBS). The first region includes small breaks up to and including the TBS. The second region includes breaks larger than the TBS up to and including the double-ended guillotine break (DEGB) of*

[12] https://www.rskonline.de/en/results

[13] *NRC-ADAMS Accession No. ML052130023*

Table 5.9 Leak-before-break vs. break preclusion concepts

Leak-before-break (LBB)	Break preclusion (BP)	
Design and operation principle according to which available ISI, maintenance and monitoring systems including leak detection, give early and reliable warning of a defect in piping component, so that necessary actions may be undertaken timely to avoid its failure as a consequence of any design basis condition	Safety concept according to which advanced construction and operation practices, assured by measures to limit the bounding conditions, enable the local protection against pipe failures to be designed based on LBB behavior	
Break preclusion defense-in-depth requirements		
Step	**Objective**	**Not-to-exceed annual target failure frequency**
1	**Basic safety concept: Prevention of material degradation**	$\leq 10^{-4}$
1.1	Design/manufacture/construction in such a way that pipe flaws are prevented	
1.2	Commissioning and operations without relevant material degradation	
2	**Fracture mechanics (FM): Effect of non-detected but possible flaws to be assessed using FM**	$\leq 10^{-1}$
2.1	During plant design life (40 years), a flaw will not penetrate the pipe wall	
2.2	After many "plant lives" (i.e., plant in extended period(s) of operation), a wall-penetrating crack will exhibit LBB characteristics	
3	**Reliability and integrity management (RIM): Ability to detect pipe flaws**	$\leq 10^{-1}$
3.1	Continuous monitoring and surveillance	
3.2	In-service inspection using qualified techniques	
4	**Leak detection: All operational leakages due to defects have to be detected**	$\leq 10^{-1}$
4.1	Diverse leak detection systems: Humidity, temperature, in-flow to sumps	
4.2	Acoustic leak detection system	
5	**Major structural failure: Large loss-of-coolant accident**	$\leq 10^{-7}$

the largest reactor coolant system pipe. The term, "break," in the TBS does not mean a double-ended guillotine break; rather it refers to an equivalent opening in the reactor coolant system boundary.

Similarly, the proposed rule defined the TBS as follows: Transition break size (TBS) is a break of area equal to the cross-sectional flow area of the inside diameter of specified piping for a specific reactor. The specified piping for a boiling water reactor is the larger of the feedwater line inside containment or the residual heat removal line inside containment. The specified piping for a pressurized water reactor is the largest piping attached to the reactor coolant system:

- *TBS Definitions (2005)*

  For BWRs, the size of the largest pipe in either of the RHR or feedwater systems inside primary containment ($\approx$ 500 mm).
  For PWRs, the size of the largest pipe attached to the cold or hot leg main loop piping ($\approx$ 300 mm).

- *TBS Supporting Rationale (2005)*

  Next larger pipes are significantly less likely to break.
  Piping sizes < TBS have experienced most significant degradation.
  Accommodates uncertainties and provides regulatory stability as variation in future LOCA frequency estimates not likely to require new TBS definition to maintain acceptable risk.

A proposed new regulatory guide was prepared to support implementation of proposed rulemaking, setting forth an alternate approach for evaluating the performance of an emergency core cooling system (ECCS). The proposed rule, 10 CFR 50.46a, "*Risk-Informed Changes to Loss-of-Coolant Accident Technical Requirements*," was published in the Federal Register on August 10, 2009. Subsequently, a draft guide, DG-1216,[14] described a method that the USNRC considered acceptable for demonstrating that the generic transition break size (TBS) specified in the proposed 10 CFR 50.46a is applicable to a specific plant. DG-1216 reflected the state of knowledge regarding material degradation, effectiveness and implementation of the mitigation process, probabilistic fracture mechanics, and operating experience at the completion of the expert elicitation supporting NUREG-1829 (USNRC 2008).

However, in response to internal and external feedback, the NRC announced the discontinuation of further NRC action on draft guide DG-1216, "*Plant-Specific Applicability of Transition Break Size Specified in 10 CFR 50.46a*," and its withdrawal. This draft guide described a method that the USNRC considered acceptable for demonstrating that the generic transition break size specified in the proposed 10 CFR 50.46a ECCS rule was applicable to a specific plant. The NRC provided an opportunity for public comment on DG-1216 in the Federal Register on June

[14] *NRC-ADAMS Accession No. ML100430445*

Table 5.10 Results of reexamination of NUREG-1829

		Plant-level LOCA frequency (1/year)			
		BWR—piping		PWR—piping	
LOCA category	Equivalent break size (mm)	NUREG-1829	2024 update	NUREG-1829	2024 update
Cat4	175	7.6E-07	2.4E-08	7.6E-07	6.0E-08
Cat5	350			1.3E-07	2.6E-08
	450	1.3E-07	4.3E-09		
Cat6	775	N/A	N/A	1.2E-08	4.0E-10

28, 2010 (75 FR 36698). The DG package (ADAMS Accession No. ML100430352) consists of DG-1216 (ADAMS Accession No. ML100430356), a Federal Register Notice (FRN) (ADAMS Accession No. ML100430445), and a regulatory analysis (ADAMS Accession No. ML101530472).

Draft Regulatory Guide DG-1216 was a proposed new regulatory guide written to provide implementing guidance for a proposed ECCS rulemaking which would have provided a voluntary, risk-informed alternative to the existing, deterministic requirements for evaluating ECCS performance. The proposed ECCS rule was published in the Federal Register on November 7, 2005 (70 FR 67597), with a supplemental proposed rule published on August 10, 2009 (74 FR 40006). In SECY-16-0009, "Recommendations Resulting from the Integrated Prioritization and Re-Baselining of Agency Activities," dated January 31, 2016 (ADAMS Accession No. ML16028A189), the staff recommended that the ECCS rulemaking be discontinued. In the Staff Requirements Memorandum for SECY-16-0009, dated April 13, 2016 (ADAMS Accession No. ML16104A158), the Commission approved discontinuation of the ECCS rulemaking.

Building on DG-1216, the USNRC performed a reexamination of NUREG-1829 in 2024 by accounting for piping operating experience from the beginning of 2005 until the end of 2024 and building the current probabilistic piping reliability state of knowledge (Table 5.10 and Fig. 5.6) (Sallaberry et al. 2025; NEI 2025). Results of this reexamination were incorporated in a new draft regulatory guide DG-1428, "*Plant-Specific Applicability of the Transition Break Size*" (USNRC 2024).

Nearly 20 years after NUREG-1829, USNRC staff reassessed the relevance of the TBS concept by analyzing new operating history, research, and advances in probabilistic fracture mechanics. These analyses confirmed that NUREG-1829 remains applicable. To ensure its continued validity, the staff carried out the following activities:

- Considered the impact of recent operational experience and its relevance to the TBS.
- Conducted a detailed probabilistic fracture mechanics (PFM) study to verify the NUREG-1829 base case frequencies, which were used to anchor the LOCA frequencies.

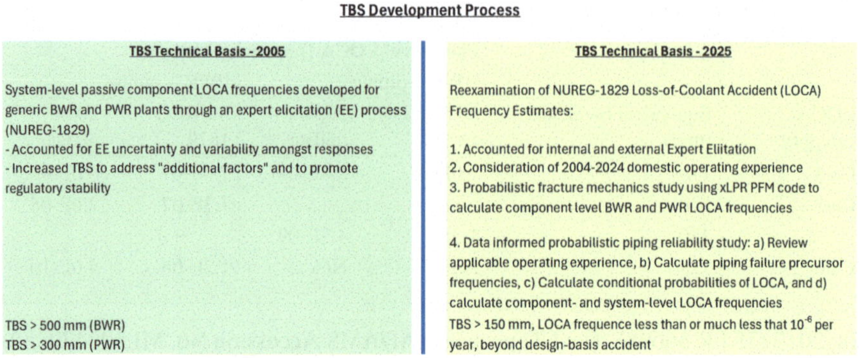

Fig. 5.6 The TBS development process 2005–2025

- Conducted detailed probabilistic piping reliability study to very precursor frequencies, conditional failure probabilities to verify NUREG-1829 base case frequencies, and BWR and PWR system-level LOCA frequencies.
- Performed an internal and external elicitation to identify possible scenarios not considered, or underestimated, in NUREG-1829 that could result in primary pressure boundary breaches that are larger than the TBS in both PWRs and BWRs.
- Completed a database study to estimate whether domestic and international operational experience would modify the LOCA frequencies.

5.8 Piping Manufacturing Processes

Steel pipes are produced by two distinct methods which result in a wrought pipe, either a seamless pipe or a seam-welded pipe. In both methods, raw steel is first cast into a more workable starting form. It is then made into a pipe by stretching the steel out into a seamless tube or forcing the edges together and sealing them with a weld. The two types of pipe used in commercial nuclear power plants are seamless pipe and seam-welded pipe (Fig. 5.7).

5.8.1 *Seamless Pipe*

Square billets are heated and molded into a cylinder and then heated white-hot in a furnace. High-pressure rolling stretches the billet, creating a hole in the center. A piercer point shapes this hole while rolling. Finally, the pipe is passed through rolling mills to correct its thickness and shape. An early method for producing metal tubes was patented by the English inventor James Russell in 1824. In his

Seam Welded Pipe		Seamless Pipe	
Single Longitudinal Weld	Double Longitudinal Weld	Standard	Cold Formed Using a Bending Machine
Single seam welded pipe is used for diameters less than 900 mm. This piping is made out of a flat steel plate and then formed and prepared for welding. Double seam welded pipe is used for diameters greater than 900 mm. This piping has two longitudinal welds 180° apart.		Seamless pipe comes in standard lengths of 6, 9 or 12 meters. The length of steel pipe is often related to its diameter. Generally, the thicker the pipe, the longe it is. Small-diameter pipes are mostly below 6 meters, pipes with diameters over DN500 are typically 10 to 12 meters in lenght.	
		Sections of seamless pipe are butt welded together to form a piping system	

Fig. 5.7 The different types of pipe

method, tubes were created by joining together opposite edges of a flat iron strip. The metal was first heated until it was malleable. Using a drop hammer, the edges were folded together and welded. The pipe was finished by passing it through a groove and rolling mill. Square billets were heated and molded to form a cylinder shape, also called a round. The round was then put in a furnace where it was heated white-hot. The heated round was then rolled with great pressure. This high-pressure rolling caused the billet to stretch out and a hole to form in the center. Since this hole was irregularly shaped, a bullet-shaped piercer point was pushed through the middle of the billet as it was being rolled. To correct any irregularities in wall thickness, the pipe was passed through another series of rolling mills.

Russell's method was not used for long. In 1825, Cornelius Whitehouse developed an improved method for making metal tubes.[15] This process, called the butt weld process, is the basis for current pipe-making procedures. In his method, thin sheets of iron were heated and drawn through a cone-shaped opening. As the metal went through the opening, its edges curled up and created a pipe shape. The two ends were welded together to finish the pipe.

In 1885, Reinhard and Max Mannesmann of Germany invented the "mandrel mill" process which uses a high-temperature rotary hearth furnace to heat a steel billet.[16] Next, a cylindrical hollow is produced with the help of a rotary piercer and a set of roller arrangements that keeps the piercer at the center of the billet. The outside diameter of the piercer is approximately that of the inside diameter of the finished pipe. A secondary roller arrangement achieves the desired outside diameter and thickness. In the 1880s, while they were rolling starting material for the family's file factory, the Mannesmann brothers had noticed that rolls arranged at an angle to each other could loosen the core of an ingot and cause it to break open. Based on the bold idea of putting this phenomenon to systematic use, they managed to produce a seamless hollow body from a solid ingot, initially by rolling alone. They optimized

[15] *Cornelius Whitehouse (22 July 1795–7 August 1883) was an English engineer, businessman, and inventor of a method of manufacturing tubes cheaply and accurately.*

[16] https://www.referenceforbusiness.com/history2/28/Mannesmann-AG.html#:~:text=The%20history%20of%20Mannesmann%20began,and%20at%20home%20in%20Remscheid

the rolling process by using a plug to ensure more uniform piercing and a smoother inside surface.

Patented in 1886, the German Mannesmann[17] pierce rolling process differs from the plug mill and the original mandrel mill processes in that the inside diameter is achieved through a multistage reduction. A solid steel billet is heated to between 1200 and 1300 °C and then positioned between two rolls arranged at an angle to each other and operating in the same direction, thus locking the billet into a kind of gear mechanism. As the billet progresses lengthwise through the rolls in a helical motion, it is pierced by a plug which is supported at the run-out end of the piercing mill by a freely rotating bar. The alternating tension and compression generated by the helical rotating motion loosen the material's microstructure in the core, and this facilitates the internal "drilling" by the plug to produce a smooth inside surface.

In 1890, Reinhard and Max Mannesmann developed the "pilger rolling process" that rolls the pierced hollows into thinner-walled tubes of significant length.[18] In the pilger rolling stand, a pair of conical-shaped rolls, one arranged above the other, operates in the opposite direction to the material flow. The thick-walled hollow body, with a cylindrical mandrel inside it, is guided towards the pilger rolls. As soon as it is gripped by the tapered portion of the work pass, a small material wave is sheared off the hollow. This wave is forged to the desired wall thickness on the pilger mandrel by the smoothing portion of the work pass, with the hollow body plus mandrel moving backwards in the same direction as the rolls are rotating, i.e., in the opposite direction to the rolling, until they reach the idler pass of the rolls and are released. As the hollow is rotated, it is once again pushed forward between the rolls, and a new rolling cycle begins.

5.8.2 Seam-Welded Pipe

Seam-welded piping is relatively commonplace, but the type and extent of applications vary across the different reactor design generations and plant systems. Two types of seam-welded piping are used. For diameters less than 914 mm, single-seam-welded piping is used. This piping is made of a flat steel plate and then formed and prepared for welding using the submerged arc welding (SAW) method. For diameters greater than 914 mm, double-seam-welded piping is used. This piping has two

[17] In 1886, the German brothers Reinhard (1856–1922) and Max Mannesmann (1857–1915) received the world's first patent for their invention of a process for rolling seamless steel pipes (the "mannesmann process"). Between 1887 and 1889, they founded tube mills with several different business partners in Germany, the Czech Republic (at the time, Bohemia), and Wales (UK). In 1890, due to technical and financial start-up problems, the tube and pipe mills folded into Deutsch-Österreichische Mannesmannröhren-Werke AG. The new company had its headquarters in Berlin. The company was renamed Mannesmannröhren-Werke AG in 1908.

[18] https://geschichte.salzgitter-ag.com/en/history-of-the-business-units-and-sites/steel-processing-business-unit/timeline-in-the-history-of-mannesmann.html

longitudinal welds 180° apart. The indicated below/above 914 mm division is not strict, since the fabrication practices also are a function of wall thickness.

For large-diameter seam-welded piping, the double SAW (DSAW) method is preferred. The edges of the rolled steel plate are prepared so that V-shaped grooves are formed on the interior and exterior surfaces at the location of the seam. The pipe seam is then welded by a single pass of an arc welder on the interior and exterior surfaces (hence double submerged). The welding arc is submerged under flux. An advantage of DSAW is that welds penetrate 100% of the pipe wall and produce a strong bond of the pipe material. Seam-welded elbows are formed by two half-shells joined together by two longitudinal welds. Depending on the diameter, the SAW or DSAW method is used to join the seam(s).

Seamless elbows are formed by two half-shells joined together by two longitudinal welds. Depending on the diameter, the SAW or DSAW method is used to join the seam(s). Welded elbows are made from steel plates, so there is a welded seam on it. Use a mold and press the steel plate into the shape of the elbow, and then weld the seam to be a finished steel elbow. It is the old production method for elbows. In recent years, small-size elbows are almost exclusively manufactured from steel pipes. For large-size elbows, for example, it is very difficult to produce elbows over 914 mm outside diameter from steel pipes. So, it is commonly made from steel plates, pressing the plate into the shape of a half elbow and welding the two halves together.

ASME Section XI requires one pipe diameter in length, but no more than 300 mm, to be examined for Class 1 longitudinal welds (examination category B-J, Item No. B9.12). Class 2 longitudinal welds are required to be examined for a length of 2.5·t, where "t" is the thickness of the weld. These weld lengths are measured from the intersection of the circumferential weld and longitudinal weld. In some cases, the owner may perform 100% elective welding examination. This may be the case for specific Class 1 fittings such as elbows or tees.

ASME Code Case N-524, *Alternative Examination Requirements for Longitudinal Welds in Class 1 and 2 Piping* (ASME, 2001), significantly reduces the required examination volume or surface area. It does this by limiting the examination requirements of the intersecting circumferential weld. This includes the weld and 0.5 inch on both sides of the weld crown for surface examinations and one-quarter inch on both sides of the weld crown for volumetric examinations in the lower 0.33-inch weld volume. Code Case N-524 directs the examination effort at the high-risk area with the weld intersections. It eliminates low-risk areas on the longitudinal weld from examination, significantly reducing examination time and radiation exposure to examination personnel.

Based on ASME Code Case N-524, there is no need to separately address these welds in risk-informed ISI (RI-ISI) applications. These welds are not required to be examined beyond the lengths that are normally examined during the inspection of intersecting circumferential welds. Seamless piping is used across all code classes. Some examples are given below:

- Seam-welded piping and fittings (elbows, tees) are found in Code Class 1 and 2 systems with emphasis on those systems with relatively mild operating conditions (e.g., low temperature and/or absence of severe operating transients). In BWR plants, systems with seam-welded piping include core spray (CS), reactor recirculation (RR), residual heat removal (RHR), and reactor water cleanup (RWCU).
- The review identified applications of seamless piping in PWR plants for Code Class 1 and 2 systems. Examples of Class 1 applications include elbows and tees in the reactor coolant system. Other systems with seam-welded piping include containment spray (CS), main steam (MS), RHR, and low head safety injection (LHSI) suction piping from the refueling water storage tank (RWST) and LHSI pump discharge to the high head safety injection (HHSI) pump, all Code Class 2 systems.

5.8.3 Pipe Elbow Manufacturing Processes

According to the fluid flow direction of the pipes, elbows can be divided into different degrees, such as 45 degrees and 90 degrees, which are the most common elbows. Elbows are split into two groups, which define the distance over which the flowing fluid changes direction: the center line of one end to the opposite face. This is known as the "center-to-face" distance and is equivalent to the radius through which the elbow is bent. As per the connection with the pipe, elbows could be classified as butt-welded elbow, socket-welded elbow, and threaded elbow.

- Butt-welded elbow: It is formed by hot pressing or forging. Its connection form is to directly weld the elbow and steel pipe. Butt-welded elbows have beveled ends to allow for ease of welding. This bevel allows for full penetration weld in most cases. Butt-welded elbows are mainly used for elbow with higher pressure and temperature.
- Socket-welded (SW) elbow: It is welded to the pipe and fittings end. Unlike the butt weld elbow, the socket weld elbow has a trapezoidal area at the end. We can insert pipe end into this area (the diameter of SW elbow is matched with the outer diameter of pipe) and then do welding work to connect them together. Therefore, the actual diameter of SW elbow end is greater than the elbow diameter.
- Threaded elbow: The shape of threaded elbow is like the SW elbow, except that the inner surface of the trapezoidal area has been machined into thread.

If the radius is the same as the pipe diameter (center-to-face dimension of 1.0·diameter), it is called "short radius elbow" (SR elbow), used normally for low-pressure and low-flow-rate pipelines or in tight areas where clearance is the main issue. If the radius is larger than the pipe diameter (center-to-face dimension of 1.5· diameter), it is called a long-radius elbow (LR elbow), used for high-pressure and high-flow-rate pipelines. Pipe elbows are mainly manufactured by hot mandrel bending from the raw material of straight steel pipe. After heating the raw material, it is pushed over a die called a "mandrel" which allows the pipe to expand and bend simultaneously. Characteristics of mandrel bending strongly depend on the integrated shape and dimensions of the mandrel. Elbows manufactured by using hot mandrel bending have the advantages of small thickness deviation and a shorter bending radius than those of any other bending method.

5.8.4 Socket Fittings

A socket fitting is a cast or forged carbon steel or stainless steel fitting for connecting pipes to form a piping system. A pipe is inserted into a recessed area of an elbow, flange, valve, or tee, allowing a gap of at least the code-required 1.5 mm to allow for thermal expansion of the pipe under welding pressure and to prevent cracking. Socket welds are mainly used for small-diameter piping whose nominal diameter is DN50 or smaller. Once the pipe is in position, it can be welded in place with a series of fillet-type welds.

Socket weld connections were typically welded with a 1:1 leg length fillet weld with a leg length meeting the ASME code minimum requirement, corresponding to $l_1 = l_2$ in Fig. 5.8. However, such a profile has proven susceptible to high-cycle fatigue damage. Test data and field experience have demonstrated that an uneven 2:1 weld leg has improved fatigue life (EPRI 1999a, c). ASME Boiler and Pressure Vessel Code (BPVC) Section III (*Rules for Construction of Nuclear Facility Components*) stipulates that socket-welded piping joints shall be limited to pipe sizes of DN50 and less. The use of socket welds in Safety Class 1 systems can range from approximately fifty to many hundreds of socket welds.

For a nuclear power plant to obtain a license renewal or subsequent license renewal in the United States, ASME III Safety Class 1 socket welds must be subjected to volumetric or destructive examination as indicated in Table 5.11. National differences exist in the use of socket welds in safety-related piping. As an

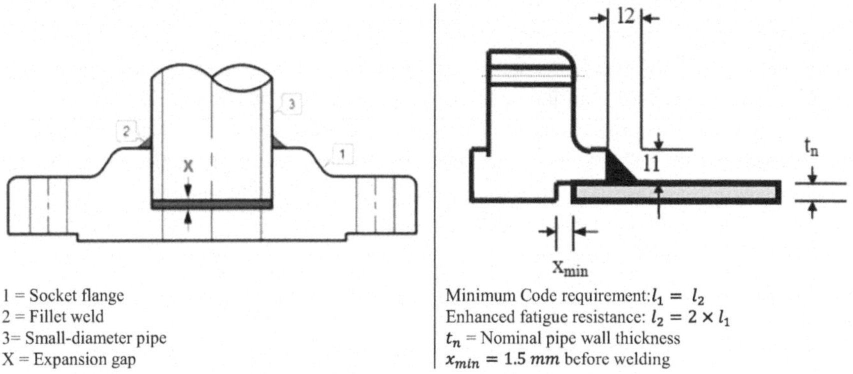

Fig. 5.8 Sectional view of a socket weld

Table 5.11 Socket weld examination requirements per NUREG-2191 (GALL-SLR)

Examination category	Operating experience	Mitigation	Examination schedule	Sample size	Examination method
A	No age-related cracking	N/A	One time: Completed within 6 years prior to the start of the subsequent period of extended operation	3% of total population per unit, up to 10 [3]	Volumetric or destructive
B	Age-related cracking	Yes [a]	Same as for category A	10% of total population per unit, up to 25 [3]	Volumetric or destructive
C	Age-related cracking	No	Periodic: First examination completed within the 6 years prior to the start of the subsequent period of extended operation with subsequent examinations every 10 years thereafter	10% of total population per unit, up to 25 [b]	Volumetric or destructive

[a] Actions must have been taken to mitigate the cause of the cracking. These actions, such as design changes, would generally go beyond typical repair or replacement activities. If welds that have been redesigned or repaired and the applicant could demonstrate through OE that no additional failures have been reported for the last 30 years, then the inspection sample size could follow the guidance in category A

[b] The welds to be examined are selected from locations that are determined to be the most risk significant and most susceptible to cracking. Other factors, such as plant-specific and industry OE, accessibility, and personnel exposure, can also be considered in selecting the most appropriate locations for the examinations

example, the French nuclear regulatory authority, Autorité de Sûreté Nucléaire (ASN), issued a directive in 2002 concerning socket weld integrity (DGSNR-BCCN/OT/VF No. 020406) (ASN 2002). According to this directive, Safety Class 1 socket welds not meeting the requirements for weld dimensions and/or weld integrity as specified in the French "Design and Construction Rules for Mechanical Components of PWR Nuclear Islands" (RCC-M) shall be replaced with butt welds. Per German codes and standards, socket welds were not used in reactor coolant pressure boundary piping. Hence, few high-cycle fatigue failures were reported in the period 1967–2024 when the last three reactor units entered into the decommissioning phase. The Korean approach to socket weld integrity management is described in a technical paper by Yoon et al. (2014).

5.9 Piping Design Process

The design of a piping system includes an evaluation of layout and dimensions and thermal-hydraulic and structural integrity considerations. Codes and standards apply to the design process. They standardize dimensions; set permissible stress values as functions of temperature; establish working formulas for determination of wall thickness as determined by pressure and material at a given temperature; specify general character of construction of valves and fittings (e.g., elbows, reducers); deal with the support, anchoring, and flexibility of a piping system; and, in general, set up a code of minimum requirements for the safety and reliability of a system. The design process utilizes computational tools for performing thermal-hydraulic analyses and stress analyses in determining pipe diameters, arrangements, layouts, and functionalities (Scheaffer 1955; Nayyar 2000; Pullarcot 2023).

Sizing of a piping system refers to the completion of two independent design functions: (1) the fluid flow design and (2) the pressure-integrity design. The purpose of the fluid flow design is to determine the minimum acceptable inside diameter of the various segments of a piping system. The piping inside diameter must accommodate the design flow rate while maintaining the pressure drop and flow velocity within reasonable limits. Piping system pressure drops must be maintained within reasonable values to limit the installed size of the system pumps. System flow velocities are limited by design to avoid potential operating problems such as cavitation or flow-induced vibrations. For piping that forms part of the steam cycle, a heat balance diagram (HBD) provides detailed information on pressure, temperature, enthalpy, and mass flow rate at every junction of the power generation cycle.

The purpose of the pressure-integrity design is to determine the minimum acceptable pipe wall thickness and the pressure ratings of the in-line components. Major steps in the process to consider are as follows:

1. Using the minimum inside diameter determined from the fluid flow evaluation, select the next larger standard nominal or outside diameter (OD) size pipe.
2. Based upon the fluid and service, select a suitable piping material, and if necessary, determine the required allowance for mechanical strength.
3. Using equations provided in a design code, calculate the required minimum wall thickness to provide for pressure integrity and allowances.
4. Refer to an applicable standard to select an appropriate nominal wall thickness or schedule.
5. Confirm that the standard manufacturing tolerance will not reduce the nominal wall thickness below the minimum required, as calculated in step 3.
6. Confirm that the inside diameter of the pipe selected is compatible with the minimum inside diameter requirements obtained from the fluid flow evaluation.

A structural evaluation of piping involves consideration of loading conditions, ranging from thermal expansion and weight of the piping and its contents to complex dynamic loads caused by seismic events or fluid transient loads. The stress and loads developed in the piping system are dependent upon several different variables. Piping characteristics such as inherent flexibility, location and types of supports, damping characteristics of the piping system and its supports, pipe support gaps, and linkage free play, along with material nonlinearities, are some of the factors that affect stress levels developed in piping systems. This complex material and structural behavior is typically analyzed using linear elastic analysis methods such as finite element analysis (FEA).

ASME Section III provides design rules for three classes of nuclear piping components that are matched to their safety significance. ASME III Class 1 is provided for components of the highest nuclear safety significance, and Class 3 for those of the lowest nuclear safety significance. There is then a hierarchy of design code requirements for pressurized components, starting with nonnuclear codes as the minimum and progressing through the ASME III nuclear Classes 3, 2, and 1. In establishing and assessing the safety justifications of nuclear plants, it is important to have an appreciation of the gradation of requirements in the ASME III design rules and how these go beyond nonnuclear component design rules. There are two broad aspects to the structural integrity of pressurized components, namely the achievement of integrity and the demonstration of integrity. The technical requirements of design codes are associated with achieving integrity, while the documentary aspects are usually associated with demonstrating integrity through analysis using advanced stress analysis methodologies. In practice, documents also have a part in achieving integrity in the communication of information between different organizations and personnel involved in the design process.

Computer-aided design (CAD) and computational fluid dynamics (CFD) tools are used in the design of piping. The former tool provides for an interactive layout of piping in 3D computer models. Computer models allow for determining possible pipe-to-pipe interference, pipe-to-grating interference, personnel access for routine visual inspection and nondestructive examination, equipment removal spaces, and construction access. CAD allows for automated generation of isometric drawings and bills of material for piping fabrication and installation. Many computer-aided design systems include interfaces with pipe stress analysis software systems and piping fabrication equipment such as numerically controlled pipe bending systems. CFD tools are used extensively in the visualization and analysis of dynamic effects of high-energy pipe breaks and simulation of water hammer phenomena and resulting pipe stresses.

5.9.1 Heat Balance Diagram

A heat balance diagram (HBD) is a detailed schematic representation of the whole steam cycle from the reactor (BWR) or steam generators (PWR) to high-pressure turbines, intermediate-pressure turbines, and low-pressure turbines to condenser, to condensate pump and condensate booster pump, to feedwater heaters and feedwater pumps, and again to the reactor (BWR) or steam generators (PWR) (Bowman and

Bowman 2021). The HBD is a representation of the steam cycle performance as a function of the temperature of the ultimate heat sink and the pressure and temperature of the steam at the admission valve. It shows the steam/condensate flow streams associated with the power generation cycle. In addition, pressures, temperatures, and enthalpies of the various flow streams are also represented. A complete heat balance provides enough information to balance the energy distribution within the power station steam cycle. Also indicated on an HBD are the extraction steam line end points and used energy end point of the steam as it exhausts to the condenser. Based on this information, a process engineer performs an energy balance for the major equipment associated with the turbine, feedwater, condensate, and heat rejection systems.

Development of a flow-accelerated corrosion (FAC) monitoring program starts with the development of an HBD that provides details on pressure, temperature, enthalpy, and mass flow rate at every junction of the power generation cycle. Next, hydrodynamics, water chemistry, and materials data are used to predict pipe wall wear rates. Balance-of-plant piping systems are designed to meet ASME B31.1 pressure and temperature requirements. The sizing of piping is done to meet the pressures and temperatures that are indicated on a plant-specific HBD. The diagram allows for mass flow calculations.

5.9.2 Piping and Instrumentation Diagram

A piping and instrumentation diagram (P&ID) is a detailed diagram which shows the piping and process equipment together with the instrumentation and control devices.[19] P&IDs are originally drawn up at the conceptual design stage from a combination of process flow sheet data, the mechanical process equipment design, and the instrumentation engineering design. During the design stage, the diagram also provides the basis for the development of system control schemes, allowing for further safety and operational investigations. Some piping reliability analysis schemes use P&IDs as a basis for defining the pipe segments for which leak and rupture frequencies are to be calculated. P&IDs indicate piping (including diameter) and associated components (e.g., valves, fittings), instrumentation (e.g., sensors, transmitters, controllers), process equipment (e.g., pumps, tanks), and flow directions.

[19] *P&IDs are crucial in hazard and operability (HAZOP) studies because they provide a detailed visual representation of the process system, including piping, equipment, instrumentation, and control systems. This visual aid helps HAZOP teams understand the process flow and identify potential hazards and operability issues. A comprehensive overview of HAZOP is found in the three-volume "Loss Prevention in the Process Industries: Hazard Identification, Assessment and Control" by Frank Lees.*

5.9.3 Layout Drawings and Isometric Drawings

A piping layout drawing, or general arrangement drawing, shows all major equipment, and all piping leading to and from equipment is developed by piping designers. All main piping items (pumps, valves, fittings, etc.), instrumentation, access ladders, and platforms are shown. The layout drawing usually shows a plan (top) view with elevations (side) and sectional drawings with piping dimensions and details, including line numbers, size, specification, direction of flow, etc., to help the piping designers to extract all necessary information for the fabrication preparation. The piping designer prepares isometric drawings that give the necessary specification and outline all the requirements needed for the fabrication and installation. There are three types of isometric drawings: (1) fabrication isometrics, (2) in-service inspection (ISI) isometrics, and (3) stress analysis isometrics. A fabrication isometric drawing consists of three sections: (1) three-dimensional representation of a pipeline route within a building structure, (2) design parameters and fabrication specifications with reference to welding process specifications, and (3) an itemized "piece-part" list including all the specific fittings and associated dimensions and material specifications.

Computer-aided design (CAD) is an essential tool in the piping industry. Using drafting software, designers create 2D and 3D models of piping systems. Models easily can be manipulated, modified, and shared with others, making it easier to collaborate with engineers. A benefit of CAD is its ability to create precise drawings that include all the necessary details, such as pipe sizes, angles, and connections. This level of detail helps to minimize errors during construction. It has become an essential tool for the development of in-service inspection programs, as well as a planning tool in support of operability determinations and repair and replacement processes. The 3D models of piping systems are important to analysts performing internal flooding probabilistic safety assessments.

An in-service inspection isometric drawing is a simplified version of a fabrication drawing. It includes line numbers and volumetric examination locations by weld identification numbers, weld details, and pipe restraint locations. A stress analysis isometric drawing is like a fabrication isometric, but it includes additional information relating to pipe support locations and how the supports respond to excessive or unusual piping displacements and loads. The stress isometric often summarizes the piping design data as gathered from other documents, such as the line list, piping specification, and piping drawing. Design data typically required in order to do pipe stress analysis consists of pipe materials and sizes; operating parameters, such as temperature, pressure, and fluid contents; code stress allowable; and loading parameters, such as insulation weight, external equipment movements, and seismic criteria. Points of interest on the stress isometric are identified by node points. Included are the node numbers for which stress analyses have been performed.

5.10 Fabrication of Piping

Pipe fabrication is the manufacture of straight lengths of metallic pipe and pipe fittings such as wrought or forged elbows, tee and reducer fittings, and forged flanges to provide a piping system for the transport of reactor coolant, treated and untreated water, steam, instrument air, etc. Fabrication of a piping system includes cutting, beveling, welding, and bending (or other pipe-forming operations). Some piping is fabricated on a job site close to where it will be installed. Whether the piping is fabricated in a shop or on-site depends on the size and scope of the project.

All engineering information, such as isometric drawings, inspection and testing requirements, and applicable codes and standards related to the project, is contained in a pipe fabrication specification. All pipes and their fittings must be manufactured and assembled according to the spool drawings in the pipe fabrication specification. Undetected fabrication defects and flaws can have a very fundamental impact on piping pressure boundary integrity if left undetected during prefabrication inspection of materials and pre-service inspection (PSI) of the installed piping system. Once the pipe fabrication specification is finalized, the pipe fabrication process involves the following steps:

- Pipe materials are marked and verified according to the pipe fabrication specification.
- Pipe materials are tagged using a stamp or paint to ensure that pipe heat numbers and schedule numbers are displayed on all sections before cutting.
- Pipe-cutting methods, such as a saw, gas torch, grinding, or plasma cutting, are chosen according to the type of steel or other metals in use.
- **End preparation:** Beveling and fit-ups are completed following an approved specification and welding process specification (WPS).
- **Welding:** Pipe welding planning ensures that all seams will be positioned and aligned correctly in the final assembly.
- **Pipe and fitting welding:** Pipes and fittings are supported and secured for a fit-up inspection. If passed, the joints are welded.
- **Details and marking:** Various identifying details of the piping system, inspection details and signature, and welding date are marked near joints. The pipe spool number is marked with a paint marker and aluminum tag.
- **Heat treatment:** Preheating and post-weld heat treatment (PWHT) may be completed depending on project requirements. Also known as stress relieving, PWHT is a heat treatment process performed on welded structures after welding to reduce residual stresses and improve the mechanical properties of the weld. This process involves heating the weldment to a specific temperature, holding it there for a period, and then cooling it down slowly.
- **Inspection:** Nondestructive testing ensures that all steps of the fabrication process have been completed and meet or exceed the project requirement and regulations. Piping found to have no defects is released for construction. Piping found to have defects must be repaired and reinspected.

5.10.1 Welding Technology

Section IX of the ASME Boiler and Pressure Vessel Code (BPVC) specifies the requirements for the qualification of welders, welding operators, and the procedure employed in welding in accordance with ASME B31.1, ASME Sections III and XI. Currently, the most used welding processes for fabrication of piping are shielded metal arc welding (SMAW), submerged arc welding (SAW), gas tungsten arc welding (GTAW), gas metal arc welding (GMAW), and flux core arc welding (FCAW). Some special applications may involve plasma arc welding (PAW) or electron beam welding (EBW), but their application to piping is less common (Sinnes 2018).

SMAW is a process where an arc is manually struck between the work area and a flux-coated electrode, which is consumed in the weld. The core wire serves as the filler material, and the flux coating disintegrates to provide shielding gases for the molten metal, scavengers, and deoxidizers for the weld puddle and a slag blanket to protect the molten metal until it is sufficiently cool to prevent oxidation. It can be used in all positions, for upward or downward progression and for root pass welding depending on the flux composition. Each weld pass is about 3.2 mm, and before subsequent passes are made, the slag must be removed and the surface prepared by removing irregularities which could entrap slag during subsequent passes.

Unlike SMAW, SAW is an automatic or semiautomatic process. For circumferential welds in pipe, the welding head is fixed for flat welding, and the work is rotated under the head (1G position). It is used most efficiently in groove butt welds in heavy wall materials with pipe sizes DN150 and larger. The arc is created between the work and a bare solid wire or composite electrode, which is consumed during the operation. The electrode comes in coils. Shielding is accomplished by a blanket of granular, fusible material called a flux which covers the arc and molten metal by forming a slag blanket to prevent oxidation of the molten metal until it has sufficiently cooled. Specific wire-flux combinations are required to assure that the deposited weld has the needed chemical and physical properties. This process has the greatest deposition rate and accordingly is the preferred process wherever possible. Because of the high heat input, care must be taken to assure that the interpass temperature is controlled to minimize sensitization in austenitic stainless steels or loss of notch toughness in ferritic steels.

Gas tungsten arc welding (GTAW), also known as tungsten inert gas (TIG) welding, is a form of gas-shielded arc welding where the arc is generated between the work and a tungsten electrode, which is not consumed. The filler metal must be added from an external source, usually as a bare filler rod or a preplaced consumable insert. The filler metal is melted by the heat of the arc, and shielding gases are usually argon or helium. Alloying elements are always in the filler material. GTAW is considered the most desirable process for making root welds of the highest quality. Techniques using added filler metal or preplaced filler metal as inserts are equally effective in manual and automatic applications.

Table 5.12 Examples of welding position symbols

1F	Fillet weld—flat position
2F	Fillet weld—Horizontal position
3F	Fillet weld—Vertical position
4F	Fillet weld—Overhead position
1G	Groove weld—Flat position
2G	Groove weld—Horizontal position
4G	Groove weld—Overhead position

Gas metal arc welding: GMAW is a type of gas-shielded welding generally used in manual mode but adaptable to automation. The filler wire is the electrode and is furnished in coils or spools of solid wire. It is fed automatically into the joint, melted in the arc, and deposited in the weld groove. Alloying elements are in the wire, and shielding gas may be argon, helium, nitrogen, carbon dioxide, or combinations thereof, depending on the application. Depending on the equipment and the heat input settings, filler metal can be transferred across the arc in several modes. In short-circuiting transfer, the electrode touches the work where it short-circuits, melts, and restarts the arc. This process has low heat input and accordingly low penetrating power. It can often result in lack of fusion. Because of the low heat input, however, it can be effectively used for open-butt root pass welding.

In spray transfer, the heat input parameters are sufficiently high to transfer the molten electrode across the arc as small droplets. Argon or argon-rich gases are used for shielding, resulting in a very stable, spatter-free arc. Because of the high arc energy, it is normally used in the flat (1G) position (Table 5.12). For all-position welding, a procedure which superimposes high-amplitude pulses of current on a low-level steady-state current at regular intervals is often used. This results in a discrete transfer of metal with lower heat input needed for all-position welding.[20]

Flux core arc welding: FCAW is a variation of GMAW where a composite electrode is substituted for the solid wire. The electrode is a tubular wire containing a flux material. Depending on the application, the arc may be self-shielding or shielding gases may be used. Because of its high deposition rate, this process is rapidly being developed for shop and field welding of piping.

High heat input can result in excessive penetration, so this process cannot be used effectively for root pass welding unless the root is deposited against a backing ring or sufficient backing is provided by two or more weld passes made by the shielded metal arc or a gas-shielded arc process. The term gas-shielded arc welding applies to those welding processes where the arc and molten metal are shielded from oxidation by some type of inert gas rather than by a flux. In welding, a root pass is the initial weld bead deposited at the base of a joint, forming the foundation for subsequent

[20] *Welding position symbols and weld symbols are standardized to ensure clear communication in fabrication of piping,* e.g., *American Welding Society (AWS A2.4:2020, Standard Symbols for Welding, Brazing, and Non-Destructive Examination) and ISO 2553 (Welding and Allied Processes—Symbolic Representation on Drawings—Welded Joints).*

passes. It is typically used to fill the gap between the two pieces being joined and ensures proper penetration.

For shop welding, efficiency in all welding processes is attained when the pipe axis is horizontal and the piece is rotated so that welding is always done in the flat position. This is referred to as the 1G position or flat position, where the letter "G" stands for groove. Other positions are 2G (pipe vertical and fixed, weld horizontal), 5G (pipe horizontal and fixed, weld a combination of flat, vertical, and overhead), and 6G (pipe inclined at 45° and fixed).

Micro-fissuring is a problem in welding fully austenitic stainless steel. To prevent this, filler material composition is adjusted to produce welds with small amounts of ferrite. ASME III requires that filler materials used in welding austenitic stainless steels contain a minimum of 5% ferrite. Ferrite, however, can be a problem at high temperatures. For very high temperatures, ferrite in the weld may convert to a brittle phase called sigma. For this reason, applications over about 425°C usually require a minimum of 3% ferrite for weldability, but not exceeding 7% to minimize sigma-phase formation. Sigma phase is a detrimental intermetallic compound that can form in stainless steel when cooled slowly through a specific temperature range, typically around 550 °C–1000 °C. This phase is characterized by its brittle, chromium-rich structure, which reduces the material's toughness and corrosion resistance.

Ferritic materials undergo metallurgical phase changes when cooling from welding to ambient temperature. Mild steels which contain no more than 0.20% carbon and 1% manganese can be welded without preheat when the thickness is 25 mm or less. As the chemical composition changes by increases of carbon, manganese, and silicon or addition of chromium and certain other alloying elements, preheating becomes increasingly important since the higher carbon and chrome molybdenum steels can develop more crack-sensitive martensitic, martensitic-bainitic, and other mixed-phase structures when cooled rapidly from welding temperatures. There is also potential for hydrogen from SMAW electrode coatings or from moisture on the base metal surface to be dissolved in the weld. Also, as the weld cools, stresses caused by shrinkage are imposed on the parts and distortion can result, and as thickness increases, thermal shock from the heat of welding can induce cracking more readily.

Preheating slows the cooling rate of the weld joint and results in a more ductile metallurgical structure in the weld metal and HAZ. It permits dissolved hydrogen to diffuse more readily and helps to reduce shrinkage, distortion, and possible cracking caused by the resultant weld residual stresses (WRSs). It raises the temperature of the material sufficiently high to be above the brittle fracture transition zone for most materials. The codes vary regarding preheat requirements. Some have mandatory requirements, while others give suggested levels. For example, for carbon steel welding, the B31.1 Code requires preheating to a temperature of 80 °C when the carbon content exceeds 0.30% and the thickness of the joint exceeds 25 mm. B31.3 recommends preheating to 80 °C when the base metal specified strength exceeds 489 MPa or the wall thickness is equal to or greater than 25 mm. ASME Section III suggests a preheat of 95 °C when the maximum carbon content is 0.30% or less and the wall thickness exceeds 37.5 mm for P No. 1 Gr. No. 1 (e.g., ASTM A-106 Grade

B carbon steel) or 25 mm for P No. 1 Gr. No. 2 (e.g., ASTM A-105 carbon steel). It also suggests a 120 °C preheat for materials with more than 0.30% carbon and wall thicknesses exceeding 25 mm. The ASME B31.4 and B31.8 Codes require preheat based on carbon equivalents. When the carbon content (by ladle analysis) exceeds 0.32%, or the carbon equivalent (C $^{1}/_{4}$ Mn) exceeds 0.65%, preheating is required.

5.10.2 Butt Welding

The butt weld, or full-penetration weld, is the most common type of joint used in fabricating piping systems. It is defined as one in which two pipe sections are joined and in the same plane. The circumferential butt joint is the most universally used method of joining pipe to itself, fittings, flanges, valves, and other equipment. The type of end preparation may vary depending on the preferences of the individual, but in general, the bevel shape is governed by a compromise between a root sufficiently wide to assure a full-penetration weld, but not so wide as to require a great deal of filler metal.

In the shop, the inside surface of large-diameter pipe joints is often accessible. In this case, the joint is most often double-welded (welded from both sides), and a double-V bevel is used. For heavier walls, machined double-U bevels can be used. However, most piping butt welds must be made from one side only. For this situation, the most frequently specified shapes are the V bevel, compound bevel, and U bevel, all of which can have varying angles, lands, and tolerances. Multi-pass welding is typically required when thick-walled pipe sections are required to be welded. This type of welding consists of multiple passes of welding in a joint. One pass of the weld is generally allowed to cool before depositing another pass. Due to more welding metal being involved, the strength of the joint is higher as compared to a single-pass weld.

For single-welded joints, achieving proper alignment can be more challenging due to the inaccessibility of the inside surface. The extent of misalignment is affected by numerous factors, and its significance varies based on the specific service application. Industry codes impose limits on inside-diameter misalignment to ensure that the resulting stress intensification remains within acceptable levels. The B31.1 Code requires that the misalignment between ends to be joined not exceed 2.0 mm unless the design specifically permits greater amounts. The B31.4 and B31.8 Codes do not require special treatment unless the difference in the nominal walls of the adjoining ends exceeds $^{3}/_{32}$ inch (2.5 mm) arc cutting. Other bevel shapes, such as the compound V, U, and J bevels or combinations thereof, require machining in lathes or boring mills. The term "transition weld joint" refers to the weldment made from two different metals. Another term is dissimilar metal weld (DMW).

5.10.3 Girth Welding

A girth weld is a circumferential weld joining two pipes end to end, commonly used in oil and gas pipeline construction to create a continuous flow path. ASME Section III, *Rules for Construction of Nuclear Facility Components*, does not prohibit girth welding but instead establishes a framework for its proper execution and verification, including qualifying welding procedures and performing necessary examinations to ensure weld integrity and safety. The following principles apply:

- ASME Section III focuses on the integrity of welds, regardless of their orientation. It provides rules for design, fabrication, and examination to ensure the safety and reliability of components in nuclear applications.
- Welding procedures for girth welds must be qualified in accordance with ASME Section IX, *Welding, Brazing, and Fusing Qualifications*.
- ASME codes outline specific requirements for examining girth welds. For example, ASME B31.1, *Code for Pressure Piping*, requires a percentage of circumferential butt and miter groove welds in normal fluid service piping systems to be examined by radiography or ultrasound.

In girth welding, the pipe ends are placed in a stationary clamp and an electric current is passed through them. This causes the heated metal to melt, which flows over the edges of both pipes and solidifies into one continuous piece. A solid metal rotating ring is placed between the pipe ends to be joined. The ring rotates at high speed with applied axial pressure, generating friction and heat to reach the forge temperature at the contact areas. The ring's rotation is then stopped, and increased axial force forges the pipes together. After welding, excess ring material and flashes are removed (Faes et al. 2009; Pissanti et al. 2019).

5.10.4 Saddle Weld and Stub-in/Stub-on Welds

A saddle weld is a type of weld used to connect a pipe to another pipe or surface at an angle, creating a branch or connection point. It involves cutting a saddle-shaped piece of pipe to fit snugly against the main pipe and welding it in place. This method is commonly used for creating pipe branches in pipelines. Stub-in and stub-on refer to two methods of connecting a branch pipe to a main pipe. In a stub-in connection, the branch pipe extends into the inside of the main pipe, often requiring a beveled hole in the main pipe for welding. In a stub-on connection, the branch pipe sits on the outside of the main pipe, typically joined with a fillet weld. Stub-in connections are generally stronger and more durable due to the full-penetration weld, while stub-on connections are simpler and faster to install.

5.10.5 Repair Welds

Welding processes introduce residual stresses, which can be tensile and contribute to the initiation or propagation of stress corrosion cracking. Repair welds can introduce new flaws or modify existing ones. There are many examples where a weld failure attributed to stress corrosion cracking has been initiated from a poorly executed repair weld during plant construction. When a nondestructive examination detects a weld discontinuity, the ensuing repair is performed by removing the weld through a grinding process. Unless performed using qualified methods, this process induces higher levels of residual stresses and hardening, thus making the weld susceptible to intergranular stress corrosion cracking. Numerous studies have been conducted to investigate the effect of repair welding heat input on microstructure, residual stresses, and stress corrosion cracking (SCC) susceptibility (Luo et al. 2020; Schröpfer et al. 2024).

5.10.6 Heat Input and Weld Residual Stresses

Too much heat during welding can be a problem (Dadkhah et al. 2025; Shridhar et al., 2025). Where notch toughness is a requirement, prolonged exposure to temperatures exceeding 316 °C can temper the base metal. Notch toughness in welding refers to a material's ability to absorb energy and resist fracturing when a stress concentrator is present. Controlling the interpass temperature is required to minimize this problem. Interpass temperature control means allowing the temperature of the joint to cool below some specified level before the next pass is deposited. Because of its martensitic structure, a maximum interpass temperature of 316 °C should be observed when welding 9Cr-1Mo-V material, which is a high-temperature steel alloy primarily used in high-temperature applications. In welding of austenitic stainless steels, sensitization of the base metal HAZ will result from the heat of welding. Here, the solution is to weld with as low a heat input as possible at the highest possible speed to minimize the precipitation of carbides (sensitization). A maximum interpass temperature of 149°C to 177°C is used.

Heat input in different welding processes varies based on factors like the welding process used, material, and welding parameters such as voltage, current, and travel speed. Heat input influences residual stress in welds. Increasing heat input generally leads to higher thermal gradients and greater expansion/contraction, resulting in higher stresses, but can also increase plastic deformation, which may release stress in some cases. Optimizing heat input is crucial to mitigate residual stresses, alongside techniques like preheating and multi-pass welding, which can reduce stress concentration and improve joint reliability.

Root cause analyses of pipe failures invariably point to the contributing factors that are associated with welding processes. As an example, too much heat input in multi-pass welds is such a factor. As an example, TIG (GTAW) welding typically results in lower heat input and potentially lower residual stresses compared to SMAW due to its precise, nonconsumable electrode, leading to narrower heat-affected zones (HAZs) and finer microstructures, especially on thin materials. In contrast, SMAW, with its consumable electrode, generally produces higher heat input, wider HAZs, and often higher residual stresses further from the weld center.

Heat input and resulting residual stresses depend on the specific welding process parameters, material, joint design, and welder skill, with some studies finding higher stresses with TIG on certain materials or when post-weld heat treatment is not applied. In other words, welds in a specific pipe spool can exhibit different susceptibilities to environmental degradation, and it is a source uncertainty to be considered in probabilistic piping reliability. A recent operating experience[21] from August 2020 involving IGSCC in austenitic stainless steel piping points to the combination of high heat input and an inexperienced welder as a primary cause of crack initiation. The affected location was a multi-pass weld with the first layer applied using TIG welding and subsequent layers applied using SMAW.

5.11 Piping Integrity Management and Defense-in-Depth

"Reliability and integrity management" (RIM) is a term which refers to programmatic and technical processes that are implemented to ensure piping integrity throughout a nuclear plant's life cycle. RIM includes prefabrication and pre-service inspections (PSI),[22] pipe stress improvements, and risk-informed in-service inspection programs. RIM has two meanings. First, it involves all regulatory requirements and nuclear industry processes implemented to ensure a very high level of piping pressure boundary structural integrity, i.e., no-in-service through-wall leaks. Included in this concept are piping codes and standards, rules for prefabrication inspection, pre-service inspection, and in-service inspection.

A second meaning relates to relatively recent activities to develop processes and an analytical framework for RIM, which complies with ASME requirements for advanced reactor implementations. These ASME/RIM requirements contain provisions beyond a traditional in-service inspection program, such as significant use of probabilistic safety assessment (PSA) to develop reliability targets for systems, structures, and components (SSCs) within the scope of a program. ASME/RIM also relies on establishing nondestructive examination, repair, and replacement to

[21] *In August 2020, intergranular cracking was discovered in the pressurizer spray line piping of Ohi Unit 3, which started commercial operation in December 1991. The piping is DN100, Schedule 160 Type 316 stainless steel exposed to 290 °C PWR primary water.*

[22] *PSI is a preoperational examination performed to provide a baseline for subsequent in-service inspections. It must be performed prior to placing the system in operation. ASME XI IWB-3131(c) only requires the results of ISI be compared with PSI and previous ISI examinations and differences recorded.*

maintain the reliability of components based on the degradation mechanisms that may exist throughout the life cycle of a plant.

A technical letter report prepared for the USNRC offers insights gained from the development of ASME/RIM programs for high-temperature gas-cooled reactors and liquid metal-cooled reactors (Anselmi et al. 2023). The report concluded that the use of such a framework would be "*extremely beneficial because it allows for optimized selection of strategies for plant performance monitoring that ensure safety and reliability*." RIM builds on detailed knowledge of material degradation mechanisms and corresponding pipe failure propensities. Probability of a pressure boundary failure is compared against a reliability target in defining an appropriate strategy for inspection. Consideration is given to probabilistic rate of flaw occurrence, flaw temporal progression, frequency of periodic inspection, probability of detecting a flaw, and time to make corrective action.

Guidance for implementation of an ASME/RIM methodology for advanced reactors is found in Nuclear Energy Institute (NEI) Technical Report NEI-18-04 Revision 0 (NEI 2019, 2021). It is referred to as "*Technology-Inclusive, Risk-Informed, and Performance-Based (TI-RIPB) Process*" for selection of licensing basis events (LBEs) (Table 5.13); safety classification of structures, systems, and components (SSCs) and associated risk-informed special treatments; and determination of defense-in-depth (DID) adequacy for non-LWRs. A 2023 EPRI technical report expands on NEI-18-04 by providing an "*Enhanced Risk-Informed Categorization Methodology for Pressure Boundary Components*" (EPRI 2025).

5.11.1 Piping Design Codes and Standards

A code is a legally enforceable standard. A standard is a set of rules that knowledgeable people recommend for others to follow. It consists of detailed guidance, including justifications for how to meet acceptance criteria. When a regulatory body adopts a standard, it becomes a code-in-the-code. As an example, in the United States, the Code of Federal Regulations (CFR) Title 10, Part 50.55(a), lists the ASME standards that have been approved for incorporation by reference (IBR).[23]

Codes and standards set the criteria to maintain pressure boundary integrity, along with design guidelines to meet these requirements. According to the *Piping Handbook* (Nayyar 2000), codes and standards are not synonymous. Piping codes provide specific design criteria such as permissible materials of construction, allowable working stresses, and load sets that must be considered in design. In addition, rules are provided to determine the minimum wall thickness and structural behavior due to the effects of internal pressure, deadweight, seismic loads, live loads, thermal expansion, and other imposed internal or external loads.

[23] https://www.nrc.gov/reading-rm/doc-collections/cfr/part050/part050-0055a.html

Table 5.13 Definition of licensing basis events

Event type	Current USNRC definition or common use	NEI guidance document definition
Anticipated operational occurrences (AOOs)	"Conditions of normal operation that are expected to occur one or more times during the life of the nuclear power unit and include but are not limited to loss of power to all recirculation pumps, tripping of the turbine generator set, isolation of the main condenser, and loss of all offsite power" [standard review plan, SRP 15.0 and 10 CFR 50 appendix A]. SRP 15.0 further breaks down AOOs into events with "moderate" frequency (i.e., events expected to occur several times during the plant life) and "infrequent"	Anticipated event sequences expected to occur one or more times during the life of a nuclear power plant, which may include one or more reactor modules. **Event sequences with mean frequencies of 1 × 10^{-2}/plant-year and greater are classified as AOOs**. AOOs consider the expected response of all SSCs within the plant, regardless of safety classification
Design-basis events (DBEs)	"Conditions of normal operation, including AOOs, design-basis accidents, external events, and natural phenomena, for which the plant must be designed to ensure functions of safety-related electric equipment that ensures the integrity of the reactor coolant pressure boundary; the capability to shut down the reactor and maintain it in a safe shutdown condition; or the capability to prevent or mitigate the consequences of accidents that could result in potential offsite exposures"	Infrequent event sequences that are not expected to occur in the life of a nuclear power plant, which may include one or more reactor modules, but are less likely than AOOs. **Event sequences with mean frequencies of 1 × 10^{-4}/plant-year to 1 × 10^{-2}/ plant-year are classified as DBEs**. DBEs consider the expected response of all SSCs within the plant regardless of safety class
Design-basis accidents (DBAs)	"Postulated accidents that are used to set design criteria and limits for the design and sizing of safety-related systems and components" "A postulated accident that a nuclear facility must be designed and built to withstand without loss to the systems, structures, and components necessary to ensure public health and safety" (NUREG-2122, Glossary of Risk-Related Terms in Support of Risk Informed Decisionmaking, 2013), https://www.nrc.gov/reading-rm/doc-collections/nuregs/staff/sr2122/index	Postulated event sequences that are used to set design criteria and performance objectives for the design of safety-related SSCs. DBAs are derived from DBEs based on the capabilities and reliabilities of safety-related SSCs needed to mitigate and prevent event sequences, respectively. DBAs are derived from the DBEs by prescriptively assuming that only safety-related SSCs are available to mitigate postulated event sequence consequences to within the 10 CFR 50.34 dose limits
Beyond design-basis events (BDBEs)	"This term is used as a technical way to discuss accident sequences that are possible but were not fully considered in the design process because they were judged to be too unlikely. (in that sense, they are considered	Rare event sequences that are not expected to occur in the life of a nuclear power plant, which may include one or more reactor modules, but are less likely than DBE. **Event sequences with mean frequencies**

Table 5.13 (continued)

Event type	Current USNRC definition or common use	NEI guidance document definition
	beyond the scope of design-basis accidents that a nuclear facility must be designed and built to withstand.) as the regulatory process strives to be as thorough as possible, 'beyond design-basis' accident sequences are analyzed to fully understand the capability of a design"	**of 5×10^{-7}/plant-year to 1×10^{-4}/plant-year are classified as BDBEs.** BDBEs consider the expected response of all SSCs within the plant regardless of safety classification

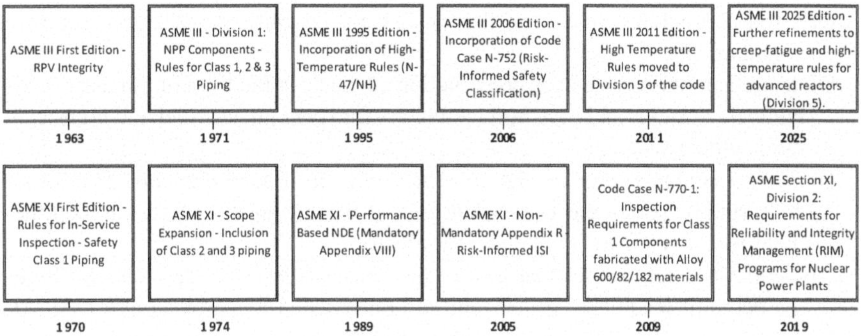

Fig. 5.9 Abridged version of the evolution of ASME codes for piping pressure boundary integrity

There are two general types of standards: (1) for dimensions and (2) for pressure integrity. Dimensional standards provide configuration control parameters for components. The main purpose of dimensional standards is to ensure that similar components manufactured by different suppliers will be physically interchangeable. Pressure integrity standards provide uniform minimum-performance criteria. Components designed and manufactured to the same standards will function in an equivalent manner.

5.11.1.1 ASME Boiler and Pressure Vessel Code (BPVC)

The first edition of the BPVC was published in 1915. It developed a systematic approach to the application of requirements that would provide assurance of pressure boundary integrity of piping. A premise of all ASME pressure boundary codes is to ensure "*no pressure boundary failure*." The code establishes rules of safety governing the design, fabrication, and inspection of boilers, pressure vessels, and piping (Fig. 5.9). The BPVC consists of 11 sections; of relevance to this book are Sections III, V, and XI:

- Section III, Rules for Construction of Nuclear Facility Components: The first edition was published in 1963 for nuclear pressure vessels. It was completely revised in 1971 and entitled "*Nuclear Power Plant Components*." The 1971 edition of ASME BPVC Section III marked the transition of the code from one covering only the reactor pressure vessel to covering the entire primary pressure boundary in a nuclear power plant.

  Division 1, *Nuclear Power Plant Components*, contains the rules of construction of ASME Class 1, 2, and 3 piping components and their supports and reactor core support structures. These rules apply to time-independent material strength and deformation, with a maximum allowable temperature of 370 °C for some materials and 425 °C for others.

  Division 2 contains rules for the design of pressure vessels and concrete containments.

  Division 3 contains rules for construction of nuclear facility components: containment systems and transport containers for spent nuclear fuel and high-level radioactive waste.

  Division 4 contains rules for the construction of fusion components. The fusion code rule development is focused on two basic concepts: magnetic confinement fusion and inertial confinement fusion.

  Division 5 is a set of rules that govern the construction of components for high-temperature reactors. The current edition extends the provisions of ASME code, Section III, Division 1, to allow the construction of metallic components that would operate within the material strength and deformation time-dependent regime and address elevated temperature conditions.

- Section V, Nondestructive Examination (NDE): Updated every 2 years, it deals with NDE requirements, including personnel qualifications, procedures, equipment, calibrations, and demonstration of NDE procedures.
- Section XI, Rules for In-Service Inspection of Nuclear Power Plant Components: The first draft was issued in 1968 (ASME 1968), and the final version was published in 1970. These inspection rules represent a defense-in-depth with respect to pressure boundary integrity. Section XI is updated every 2 years. The current edition consists of:

  Division 1, Rules for Inspection and Testing of Components of Light-Water-Cooled Plants.

  Division 2, Requirements for Reliability and Integrity Management (RIM) Programs for Nuclear Power Plants. It applies to all types of nuclear reactors.

The BPVC contains 29 books and several code case books. Additionally, three companion guides provide technical context for the code requirements and function as user manuals (Rao 2009a, b, 2023). Volume 1 includes chapters that mainly address pressure vessel integrity. Volume 2 includes chapters on repair/replacement activities, flaw acceptance criteria, fatigue crack growth, and stress corrosion crack growth. Volume 3 includes chapters on material degradation mechanisms, international codes and standards related to ASME BPVC, PSA, and risk-informed analyses.

5.11.1.2 ASME B31.1 Process Piping Code

The ASME B31 Code for Pressure Piping (ASME 2024) consists of several individually published sections, each an American National Standard, under the direction of ASME Committee B31, Code for Pressure Piping. Rules for each section reflect the kinds of piping installations considered during its development. When the first commercial nuclear power plants were being developed, specific codes for nuclear applications did not exist. Engineers initially used the most relevant codes available, which included ASME B31.1 for balance-of-plant (nonnuclear safety-related) systems and ASME Boiler and Pressure Vessel Code, Section VIII, for pressure vessels.

In 1958, the ASME Boiler and Pressure Vessel Committee formed a special committee to develop rules for high-quality pressure vessels, a move that eventually led to the development of a standard specifically for nuclear vessels. Eventually, the responsibility for nuclear piping systems was officially moved from B31.1 to Section III of the ASME B&PV Code. In the period 1960–1962, several B31.1 code cases were developed to provide more detailed provisions for stress analysis, material selection, welding practices, and testing methods.

5.11.1.3 ASME BPVC Section III, Division 1

Relative to ASME B31.1, Section III (ASME 2025) of the BPVC tightened the control of the design and manufacture of the pressure boundary items by defining specific responsibilities for the plant owner, the designer, and the manufacturer of piping pressure boundary components. The safety classification of ASME III is based on the concept of providing greater assurance of pressure boundary integrity for components which have the potential for a greater impact on the safety of the public. This classification approach is divided into three classes, Class 1, Class 2, and Class 3. There is a hierarchy of specific requirements for each class, which are defined in the ASME BPVC by Subsection NB (Class 1), Subsection NC (Class 2), and Subsection ND (Class 3) (Table 5.14). Generally, Class 1 is applied to the systems associated with cooling the fuel, and these systems can experience high levels of radioactivity. Class 2 is mainly used for piping providing for containment integrity, and Class 3 is applied to safety-related support systems such as closed-loop component cooling systems and essential service water systems. Piping of DN20 within the primary pressure boundary may be classified as Safety Class 2.

Within each subsection (Code Class) in Section III, the construction of a component is divided into seven areas, material, design, fabrication, welding, examination, testing, and overpressure protection, and each subsection contains requirements to be met in each of these areas. A similar approach is used for the ASME B31.1 and

Table 5.14 Hierarchy of ASME III piping design rules

Mandatory requirements	Safety Class 1	Safety Class 2	Safety Class 3
Design and analysis	Design by analysis: Requires detailed stress analysis to identify all critical areas	Design-by-rule analysis: Use established rules (part 4) or, if needed, design by analysis (part 5), often with finite element analysis (FEA)	Design: Components must be designed to withstand specified service loadings, including normal, upset, emergency, and faulted conditions. The design rules in subsection NCD provide specific requirements for class 3 items like vessels, pumps, valves, piping, and supports
	Thermal stresses and fatigue: Thermal stresses must be explicitly considered, and the design must address potential failure modes such as fatigue and brittle fracture. This involves considering cyclic life and meeting specific fatigue life criteria	Conservative stress limits: Design must account for normal, upset, and faulted conditions using allowable stresses lower than class 1	
	Service levels: The design must account for various service levels (e.g., normal, upset, emergency, and faulted events) and their associated stress limits	Pressure boundary definition: Clear definition of parts resisting design/test	
	Pressure relief: Specific design criteria for pressure relief devices and associated piping are required to ensure that the system can manage pressure surges safely. Key requirements are detailed in article NB-7000, "overpressure protection." acceptable types of pressure relief devices (PRDs) include safety valves, relief valves, safety relief valves, and pilot-operated pressure relief valves, provided that they meet specific design and performance criteria	Pressure relief: The aggregate capacity of all pressure relief devices must be sufficient to prevent the pressure within the system from rising above specified limits for different operating conditions (e.g., normal, anticipated operational occurrences (AOOs), infrequent events)	Overpressure protection: Systems must include provisions for overpressure protection devices, such as pressure relief valves, as detailed in the code (e.g., appendix O for discharge systems)
Materials	Approved materials: Materials used must meet specific chemical and mechanical property	Certified materials: Pressure-retaining materials must come from qualified suppliers with	Materials: Materials must conform to specific chemical and mechanical property

Table 5.14 (continued)

Mandatory requirements	Safety Class 1	Safety Class 2	Safety Class 3
	requirements outlined in the code and are generally listed in section II parts A and B of the BPVC	quality system certificates or approved QA programs	requirements outlined in the code, including various steels and alloys suitable for nuclear service. Material traceability from raw stock to the finished product is required
	Material traceability: Strict documentation, including material certificates, is required to ensure full traceability of all materials used in construction	Material traceability: Materials need heat/lot traceability, often via certified material test reports (CMTRs)	
	Material compatibility: Welded materials and attachments must be compatible	Material compatibility: Welded materials and attachments must be compatible	
Fabrication, welding, and installation	Quality assurance (QA): A comprehensive quality assurance program, consistent with the requirements of ASME NQA-1, is mandatory for all stages of work	QA: A comprehensive program covering all aspects of construction	QA: Rigorous quality assurance program is mandatory. This program must meet the requirements of NCA-4110(b) and is typically aligned with ASME NQA-1 standards
	Qualified procedures and personnel: All welding procedures must be qualified, and all welders and welding operators must be certified to ensure the integrity of joints. Welder performance qualifications (WPQs) must be documented and kept on file	Qualified welders: Welds must be performed by certified personnel	Controls: Strict manufacturing controls and installation procedures must be followed. This includes qualified welding procedures and personnel, ensuring proper joint types, alignment, and environmental controls during processes like HDPE piping fusion (if applicable)
	Nondestructive examination (NDE): Mandates extensive NDE methods, such as radiography, ultrasonic testing, and dye penetrant inspection, to detect flaws before a component enters service	NDE: Radiography (RT), ultrasonic (UT), dye penetrant (PT), etc., as required by code	Examination: Nondestructive testing (NDE) and other examinations are required during fabrication and installation to ensure integrity
	Post-weld heat treatment (PWHT): PWHT may be	PWHT: Required where specified	

Table 5.14 (continued)

Mandatory requirements	Safety Class 1	Safety Class 2	Safety Class 3
	required to relieve residual stresses from welding		
	Third-party oversight: An authorized inspection agency and an authorized nuclear inspector provide third-party oversight to verify compliance with all code requirements	Authorized inspection: Performed by an independent authorized inspector (AI)	
Testing and certification	Pressure testing: The completed component or system must undergo rigorous pressure testing, typically hydrostatic or pneumatic tests, with detailed records maintained	Hydrostatic testing: Mandatory pressure test (e.g., 1.1x design pressure) to verify integrity	Testing: Components and systems must undergo testing, such as hydrostatic or pneumatic pressure testing, to verify their integrity and performance. Test records, including date, fluid, pressure, and certification by the examiner, are required
	Documentation and certification: Meticulous recordkeeping is required for all aspects of design, fabrication, and testing. Manufacturers must obtain an appropriate ASME nuclear certificate of authorization (e.g., "N," "NA," "NPT," "NV" stamps) and apply the ASME certification mark to the item, signifying full compliance with all applicable code requirements	Certifications: Manufacturer must hold an ASME certificate of authorization	Data reports: Documentary evidence, in the form of specific data report forms (specified in table NCA-8100-1), must be available at the site before installation. These reports are prepared by the certificate holder or owner and signed by them and an authorized inspector. Completed items must be stamped with the ASME certification mark (typically an "N"-type stamp) to signify conformance to all code requirements
	Regulatory compliance: In jurisdictions such as the United States, these standards become legally mandatory when incorporated into regulations by the USNRC	Compliance: In essence, class 2 requires rigorous documentation and verification to ensure that components perform their safety functions, even if the design rules are less stringent than class 1.	Third-party oversight: The requirements are enforced through third-party oversight by an authorized inspection agency and the duties of an authorized nuclear inspector

B31.3 piping codes. Section III also provides a general guidance list of potential loads/root causes of failure. These are not all-inclusive but are ones "*the designer should consider*." Class 1 uses a design by analysis approach, differing from Classes 2 and 3. Its requirements are stricter due to reduced basic margins, but this is offset by using highly ductile, proven materials, defining specific design conditions, more thorough examinations (particularly volumetric), and tighter acceptance criteria.

5.11.1.4 ASME BPVC Section III, Division 5

The 2025 edition Section III, Division 5, of the BPVC has the title: ASME Boiler and Pressure Vessel Code Section III Division 5: High Temperature Reactors. It governs the construction of vessels, piping, pumps, valves, supports, core support structures, and nonmetallic core components for use in high-temperature reactor systems and their supporting systems. Division 5 is organized by code classes:

- Class A, Class B, Class SM for metallic components.
- Class A is analogous to Class 1 in Section III, Division 1.
- Class B is analogous to Class 2 in Section III, Division 1.
- Class SM is for metallic core supports.
- Class SN for nonmetallic components, e.g., graphite core supports.

Furthermore, Section III Division 5 is the only design code that provides construction rules for graphite materials. It is a design-by-analysis code, where the designer performs a stress analysis of the component, e.g., with finite element analysis methods, and determines the acceptability of the design by applying criteria based on those stresses. It addresses the anticipated high-temperature failure modes and acceptable metallic material selections. A limited set of materials is as follows:

- Type 304H and Type 316H stainless steel. Minimum carbon content of 0.04 weight % required for better high-temperature properties.
- Alloy 800H for steam generator tubes.
- 2.25Cr-1Mo
- 9Cr-1Mo-V (Grade 91)
- Alloy 617 (Code Cases N-872 and N-898).

Design parameters are mostly self-contained in Division 5, except the following contained in Section II:

- Elastic constants.
- Thermal properties.
- Part of yield strength (S_y) table.
- Part of ultimate tensile strength (S_u) table.

Code Case N-872 (ASME, 2016) provides material properties and allowable stresses for using Alloy 617 (UNS N06617) in components constructed to the requirements of Section III, Division 5, Subsection HB, Subpart A, Class A, and Subsection HC, Subpart A, Class B. N-872 allows the use of Alloy 617 up to a maximum temperature of 425 °C. This temperature range is appropriate for the lower-temperature components of high-temperature reactors constructed to Section III-5. ASME Code Case N-872 provides rules for using nickel-base Alloy 617 (UNS N06617) for low-temperature nuclear service components under ASME Section III, Division 5, offering specific allowable stresses and properties, complementing Code Case N-898 for high-temperature applications, and is part of the framework for advanced nuclear reactors like high-temperature reactors.

ASME Code Case N-898 (ASME 2019a, b) provides rules for using nickel-based Alloy 617 for high-temperature nuclear reactor components (Class A) under ASME Section III, Division 5. It introduces specific design rules, material properties, constitutive models (for creep/plasticity), and analysis methods (like elastic-perfectly plastic) for this alloy, allowing its use in advanced reactor designs by extending material data and addressing high-temperature concerns like stress relaxation cracking.

When using ASME Code, Section III, Division 5, where Division 5 references ASME Code, Section III, Division 1, applicants and licensees should follow any applicable conditions for Division 1 that are identified in 10 CFR 50.55a (USNRC 2025):

- The phrase "items commensurate with their contribution to safety or risk" appears in Code Case N-940, HBA-2610, and HBB-2610 and within ASME Code, Section III, Division 1. In the application of ASME Code, Section III, Division 5, design rules, clarification is needed for which SSCs these rules are appropriate. The alternate requirements in these sections should not be applied to safety-related SSCs but may be appropriate for use for SSCs categorized as NSRST under RG 1.233.[24] The justification for use of alternate requirements in these sections as a special treatment to achieve the reliability and capability targets specified for the NSRST SSC is subject to NRC review and approval. The NRC may review classification of NSRST SSCs in accordance with the approved methodology in RG 1.233.

5.11.1.5 KTA Rules

Established in 1972, the Nuclear Safety Standards Commission (Kerntechnische Ausschuss, KTA) of Germany has the task to issue nuclear safety standards for topics in nuclear technology; the KTA rules are available at https://www.kta-gs.de/common/regel_prog1.htm. The KTA standards are issued where a consensus between experts of the manufacturers, operators of nuclear power plants, authorized experts, and state officials is apparent and supports their application. The KTA design rules were originally derived from ASME rules. The basic ideas about design

[24] USNRC (2020), *Guidance for a Technology-Inclusive Risk-Informed, and Performance-Based Methodology to Inform the Licensing Basis and Content of Applications for Licensees, Certifications, and Approvals for Non-Light Water Reactors, Regulatory Guide 1.233, Washington, DC.*

Table 5.15 KTA rules for piping

KTA 3201—primary system piping		KTA 3211—Safety Class 2 Piping	
Section	Title	Section	Title
3201.1	Materials and product forms	3211.1	Materials and product forms
3201.2	Design and analysis	3211.2	Design and analysis
3201.3	Manufacture	3211.3	Manufacture
3201.4	ISI and surveillance	3211.4	ISI and surveillance

analysis have been kept while adding some design rules from German industrial experience with pressure equipment. KTA design standards are generally compatible with the ASME code but are restricted to specific systems of light water reactors. KTA 3201 (Table 5.15) is restricted to design rules for the primary loop piping, whereas KTA 3211 is restricted to design rules for pressure-retaining components and piping outside the primary loop for light water reactors. In contrast, the ASME Boiler and Pressure Vessel Code has a broader scope and is not restricted to specific systems or types of power plants.

The German BPVC Section XI International Working Group serves as a forum to provide for the participation of German-speaking BPV Section XI technical experts and to create a means of communication and collaboration between ASME Section XI, the German Nuclear Safety Standards Commission (KTA), and all stakeholders and technical experts in German-speaking countries (Germany, Switzerland, Austria) as well as from other European countries. The German BPV XI International Working Group reviews and comments on proposed changes or additions to the BPV XI Code, initiates and processes proposed standards actions for the BPV XI Code, conducts activities in German-speaking countries related to codes and standards, facilitates discussion of ASME standards issues and experiences among local stakeholders, contributes to development and local stakeholder comprehension of the standard to improve the usability and acceptance of the standard, and coordinates German BPV XI International Working Group activities with the parent BPV XI Executive Committee and ASME Staff.

5.11.1.6 RCC-M and RSE-M Rules

The RCC-M and RSE-M codes are developed by AFCEN (French Association for Design, Construction, and Surveillance Rules of Nuclear Power Plant Components). This association was mainly formed by EDF-AREVA-CEA and was opened to many national and international companies (around 50 companies and 350 active participants in charge of seven nuclear codes, including RSE-M). The first edition was published in 1990; at the time of this writing, the most recent edition was published in 2022 in French and in English. Table 5.16 compares the ASME and AFCEN codes. RSE-M specifies the methods to be used. In summary:

Table 5.16 ASME vs. RCC-M rules

ASME—American Society of Mechanical Engineers, Boiler and Pressure Vessel Code (BPVC)		AFCEN—French Association for Design, Construction, and In-Service Inspection Rules for Nuclear Components	
Section III, Division 1: Rules for Construction of Nuclear Facilities Components	Section XI: Rules for In-Service Inspection of Nuclear Power Plant Components	RCC-M: Design and Construction Rules for Mechanical Components of PWR Nuclear Islands	RSE-M: In-Service Inspection Rules
Division 1 has six subsections that deal with class 1, class 2, and class 3 components, containment vessels, supports, and core support structures, respectively. The first edition of section III was published in 1963	Section XI provides requirements for examination, testing, and inspection of components and systems and repair/replacement activities in a nuclear power plant. The first edition of section XI was published in 1970	The RCC-M: Design and construction rules for pressure vessels, reactor internals, piping, and equipment support structures. A first edition was issued in 1980	Section A: In-service inspection. • Section B: Rules for class 1 components • Section C: Rules for class 2 and 3 components

- The RCC-M is the French design and construction code for mechanical components of PWR nuclear islands. Its full title is *Règles de Conception et de Construction des Matériels Mécaniques des Ilôts Nucleaires PWR*. The code provides rules for design, fabrication, and quality control of pressure equipment and some nonpressure components within nuclear safety classes 1, 2, and 3, ensuring safety through experience gained from operating PWRs worldwide.
- RSE-M stands for "*Rules for In-service Inspection of Mechanical Components*" and is a French nuclear code that defines the mandatory in-service inspection and surveillance requirements for metal components in pressurized water reactor nuclear power plants. It covers everything from inspection classification and methods for flaw evaluation and recordkeeping.

The French rules state that all in-service inspection examinations shall be performed in accordance with the instructions given in a set of duly identified documents, which must comply with the requirements of the present code. The extent of the areas subject to examination is defined for each examination method. If there is any obstacle to straightforward, direct application of the general conditions defined in this code, an analysis must be provided to justify that the aims of this code will be met, subject to compensatory measures if necessary. Nondestructive examination methods are based on the use of specific physical principles. The nondestructive examination technique concerns the specific use of a nondestructive examination method. The NDE process covers both the equipment and the supporting processes and procedures required to implement nondestructive examinations. Metallography, physical/chemical analyses, mechanical tests, etc. may also be used to determine the type and cause of the defects observed and defect growth.

5.11.1.7 HDPE Pipe Codes and Standards

The Nuclear Energy Standards Coordination Collaborative (NESCC) in the United States was established in 2009 to address the requirements of the nuclear industry. Focus areas included the utilization of polyethylene material in ASME Code Class 3 piping applications for nuclear power plants, identifying existing gaps in HDPE piping standards for nuclear applications, determining a reasonable mechanism and timeline to address these gaps, and developing a road map for the application of HDPE piping in safety-related nuclear applications and the anticipated needs for codes and standards. The following are examples of industry organizations and standards organizations that have been developing standards and specifications that are relevant to HDPE piping:

1. ASTM International (formerly American Society for Testing and Materials): ASTM F2620, "*Standard Practice for Heat Fusion Joining of Polyethylene Piping and Fittings*," is applicable for nonnuclear HDPE installations (ASTM 2013, 2024).
2. The International Organization for Standardization has published ISO 21307, "*Plastics Pipes and Fittings — Butt Fusion Jointing Procedures for Polyethylene (PE) Piping Systems*" (ISO 2017).
3. American Society of Mechanical Engineers (ASME): The ASME Boiler and Pressure Vessel (BPV) Code Committee has issued rules for the construction of buried Section III, Division 1, Class 3 HDPE piping systems in Code Case N-755 (ASME 2010; Haddad 2013). Code Case N-755 was incorporated into ASME Code Section III, 2015 edition, as Mandatory Appendix XXVI, which addresses the use of HDPE piping in buried safety-related ASME Code Class 3 piping systems.
4. ASME Section III, Appendix XXVI, specifically provides rules for the construction of Class 3 buried polyethylene (HDPE) pressure piping systems, primarily for service water applications. These rules are applicable to the buried portions of these systems and focus on the use of PE4710 high-density polyethylene materials.
5. The Nuclear Energy Standards Coordination Collaborative (NESCC) is a joint initiative of the American National Standards Institute (ANSI) and the National Institute for Standards and Technology (NIST) to identify and respond to the current needs of the nuclear industry. In July 2010, NESCC formed a task group "Polymeric Piping for Nuclear Power Plants Task Group," referred to as the "Polymer Piping Task Group" (PPTG) (NIST 2013).
6. Standardization Administration of China (SAC): Two national standards on nondestructive examination (NDE) and safety assessment of electrofusion (EF) of HDPE pipes for natural gas distribution (AQSIC 2012a, b).

5.11.1.8 Impact of Different Piping Design Codes on Piping Integrity

The piping design code of reference for operating nuclear power plants commissioned prior to 1975 is ASME B31.1 (or an equivalent code). An occasionally raised question refers to the relationship(s) between piping design code and piping pressure boundary integrity achieved in service, i.e., going from B31.1 to ASME Section III, Subsection NB (ASME III Class 1 components), or NC (ASME III Class 2 components). A difference between B31.1 and ASME Section III lies in the approach to fatigue damage evaluations. ASME Code Section III fatigue analysis introduced the cumulative usage factor (*CUF*) as the sum of the individual fatigue usage factors (U_i) for all load set pairs and with a requirement that the *CUF* must not exceed unity for acceptable fatigue design at each location of the pressure boundary components. Table 5.16 identifies key differences between B31.1 and ASME III (Table 5.17).

Studies have been performed to determine whether piping designed according to B31.1 as opposed to ASME III exhibits less structural integrity over time. Electric Power Research Institute (EPRI) published a "*Fatigue Comparison of Piping Designed to ANSI B31.1 and ASME Section III, Class 1 Rules*" (EPRI 1993). A technical position at the time the report was published asserted, in part, that the licensing basis for fatigue for Class 1 reactor coolant pressure boundary piping in early plants was adequate for license renewal, with clearly defined exceptions. These exceptions were associated with component locations having "*severe geometric and loading discontinuities*," such as socket welds and slip-on flanges, and regions affected by step-change thermal transients. Technical evaluations showed that the B31.1-designed systems had only very limited areas with high fatigue usage. In both systems, the locations of indicated high fatigue usage were those with geometric discontinuities that were also affected by severe step-change thermal transients. The evaluations also showed that the evaluation per the requirements of the 1993 version of ASME III will produce significantly less indicated fatigue usage than earlier code versions used for the design of most nuclear plants in service in the United States, with the reduced fatigue usage due to more realistic reclassification of through-wall thermal gradients.

The Atomic Energy Control Board (AECB) of Canada published a report in 1995 on "*Failure Rates in Piping Manufactured to Different Standards*" (Barnes and Cooper 1995.[25] The investigators performed evaluations of 65 pipe failures. The root causes for this dataset were identified to determine whether the use of ASME Section III as opposed to B31.1 would have prevented these failures from occurring. Adapted from this Canadian study, the conclusions were as follows:

- From a sample of 65 recorded failures, over 75% of the root causes of failure were addressed in all codes either through specific rules or by general reference. The design requirements in one code are usually quite like the others. Therefore, the

[25] *The Atomic Energy Control Board (AECB) of Canada was established in 1946 under the Atomic Energy Control Act. In March 1997, the name of the AECB was replaced by the Canadian Nuclear Safety Commission (CNSC).*

Table 5.17 Key differences between piping design codes

Piping design criteria	B31.1	ASME III Class 1
General design:		
Basic allowable stress	Lower	Higher
Allowable stress basis	Maximum stress	Stress intensity and stress intensity range
Local effects	Allowable stress basis	Stress intensification factors
Fatigue basis:		
Consideration of geometric differences	Limited evaluation	Complete range
General thermal expansion and secondary stresses	Stress range reduction factors (which consider number of thermal expansion cycles)	$3 \times S_m$ (but may be exceeded with additional evaluation of cyclic operations)
Bimetallic and adjoining thickness difference effects	Not considered	Secondary stress intensity
Through-wall transient stresses	Not considered	Peak stress intensity Secondary stress intensity
Maximum thermal expansion stress	$1.25 \times S_c + 1.25 \times S_h$ ($\approx 1.6 \times S_y$, *Ferritic*) ($\approx 2.2 \times S_y$, *Austenitic*)	$3 \times S_m$ ($\approx 2 \times S_y$, *Ferritic*) ($\approx 2.2 \times S_y$, *Austenitic*)
Anchor movements:	No consideration (prior to 1973)	Included
Weld gap for socket weld	Left to engineer of record along with concurrence from owner	Gap of approximately 1.5 mm to be provided between the inserted pipe and the bottom of the socket before welding to prevent crack-inducing stress from thermal expansion
Root weld gap	Specified in welding process specifications and on fabrication isometric drawings	
Notation:		
S_c	Allowable stress at cold condition	
S_h	Basic material allowable stress at maximum temperature from allowable stress tables	
S_m	Basic allowable design stress intensity value	
S_y	Material yield strength value taken at average fluid temperature of the transient under consideration	

application of Section III design requirements rather than the requirements of B31.1 would not have been a significant factor in preventing failure.

– Failures in piping do occur and will continue to occur because of the occurrence of situations that were not covered by the design specification, aging processes, and human errors which cause the system to be operated outside the design envelope.

- Section III requirements have merit because of the control, examination, and repair of materials and the requirements for welding, manufacturing, and installation. Section III also limits the material use to piping specifications that specify high ductility. This will tend to mitigate the type of failures that occur towards leaks rather than explosive-type ruptures.
- The review of the full database collected for the study indicated that there is a significant correlation between the root cause of the failure and the service for which the piping is used. For this sample, the correlation was also translated to pipe sizes, probably because for the industries from which this database was collected, certain pipe sizes are associated with the processes and services that result in these failure types. This finding was also confirmed by the other databases referenced in this study, which would seem to indicate that certain processes and services across a broad range of industries use similar pipe sizes for these services.
- In summary, ASME Section III does have an advantage in the detection of material and manufacturing defects over the other piping codes and therefore will make a difference. It must once again be kept in mind that these findings are the result of a very small sample of pipe failures, and the conclusions reached might change if another set of failures were studied.

5.11.2 In-Service Inspection

This section summarizes ISI requirements as formulated in national codes and standards. Also included is an overview of selected owner-defined in-service inspection programs that go beyond inspection requirements as formulated in codes and standards.

5.11.2.1 ASME Section XI

The first edition of ASME Section XI, "*Rules for In-Service Inspection of Nuclear Reactor Coolant Systems*," was issued in 1971 and consisted of about 50 pages of rules text. In contrast, the 2023 edition consisted of over 500 pages of rules and recommendations. The plant systems to be subjected to examination were the following: (1) reactor coolant system, (2) portions of the associated auxiliary systems connected to the reactor coolant system, and (3) portions of the emergency core cooling systems connected to the reactor coolant system. A 10-year inspection cycle, "the interval," was chosen based on historical failure rate data for nonnuclear steam power and petrochemical plant systems (Dau 1977). In a letter from the Advisory Committee on Reactor Safeguards (ACRS) dated 17 November 1970, the following statement was made (Bush 1970):

> *In a previous report to you* [Dr. G. Seaborg, Chairman of the United States Atomic Energy Commission] *the ACRS recommended the institution of periodic, comprehensive reviews of operating power reactors at intervals of approximately ten years. Such reviews were recommended as a mechanism for providing additional assurance that each reactor can continue to be operated without undue hazard to the health and safety of the public.*

In 1974, the title of Section XI was changed to "*Rules for In-Service Inspection of Nuclear Power Plant Components.*" Section XI examination categories are listed in Table 5.18.

Table 5.18 ASME Section XI examination categories

Examination Category B-F, Dissimilar Metal Pressure Retaining Welds in Piping (Safety Class 1)	
Parts examined	Examination method
Reactor vessel	
DN 100 or larger nozzle-to-safe-end butt welds	Surface and volumetric
Less than DN 100 nozzle-to-safe-end butt welds	Surface
Nozzle-to-safe-end socket welds	Surface
Pressurizer	
DN 100 or larger nozzle-to-safe-end butt welds	Surface and volumetric
Less than DN 100 nozzle-to-safe-end butt welds	Surface
Nozzle-to-safe-end socket welds	Surface
Steam generator	
DN 100 or larger nozzle-to-safe-end butt welds	Surface and volumetric
Less than DN 100 nozzle-to-safe-end butt welds	Surface
Nozzle-to-safe-end socket welds	Surface
Examination category B-J, pressure retaining welds in piping (safety class 1)	
Parts examined	Examination method
DN100 or larger circumferential welds	Surface and volumetric
Less than DN100	
Circumferential welds other than PWR high-pressure safety injection systems	Surface
Circumferential welds of PWR high-pressure safety injection systems	Volumetric
Branch pipe connection welds	
DN100 or larger	Surface and volumetric
Less than DN100	Surface
Socket welds	Surface
Examination categories C-F-1 (pressure retaining welds in austenitic stainless steel or high alloy steel piping) and C-F-2 (pressure retaining welds in carbon steel or low alloy steel piping, safety class 2)	

Table 5.18 (continued)

Examination Category B-F, Dissimilar Metal Pressure Retaining Welds in Piping (Safety Class 1)	
Parts examined	Examination method
Parts examined	Examination method
Piping welds $\geq$10 mm nominal wall thickness for piping > DN100	
Circumferential weld	Surface and volumetric
Piping welds $\geq$5 mm nominal wall thickness for piping > DN50 and less that DN100	
Circumferential weld	Surface and volumetric
Socket welds	Surface
Pipe branch connections of branch piping $\geq$ DN 50	
Circumferential weld	Surface

Subsequent editions have provided more detailed guidance and rules for the selection of areas to be inspected. The history of how areas are selected for inspection has evolved from a prescriptive approach (1970s) to a focused, risk-informed approach that targets components with the highest probability of failure:

Initial Conception (1970–1974): Deterministic and representative sampling with a focus on areas with the highest service factors, specifically, locations with high operating stresses. The code emphasized volumetric examination of welds and adjacent base metal to detect service-induced flaws (e.g., fatigue). Initial efforts primarily targeted Class 1 systems.

Evolution and Refinement (1974–1980s): Scope was expanded to include Class 2 and 3 systems. Realization that many plants designed before Section XI were not accessible for 100% volumetric inspection, leading to the "as far as practical" philosophy and the frequent use of relief requests. Development of better volumetric techniques, culminating in the 1989 edition's Appendix VIII (Performance Demonstration Initiative).

Shift to Risk-Informed In-Service Inspection (RI-ISI) (1990s–2000s): By the 1990s, experience showed that traditional sampling (like 25% of Class 1 welds) was not optimized. The industry moved towards using probabilistic safety assessments (PSAs) to identify critical piping locations based on specific material degradation mechanisms. Code Case N-560/N-716 allowed a reduction of the inspection sample for Class 1 (Category B-J) piping welds from 25% to 10%, focusing only on those with high risk of failure, without increasing overall risk.

Modern Era (2010s–Present): The 2019 edition introduced a new Division 2, "Requirements for Reliability and Integrity Management (RIM) Programs," which moves away from strictly prescriptive scheduling towards managing the reliability of components and structures.

5.11.2.2 ASME Section XI Code Cases

ASME Section XI specifies code-acceptable repair methods for flaws that exceed code acceptance limits in piping that is in service. A code repair is required to restore the structural integrity of flawed ASME code piping, independent of the operational mode of the plant when the flaw is detected. Those repairs not in compliance with Sect. XI of the ASME code are "non-code repairs." However, the required code repair may be impractical for a flaw detected during plant operation unless the facility is shut down. In 1990, the USNRC issued Generic Letter 90–05 (Guidance for Performing Temporary Non-Code Repair of ASME Code Class 1, 2 and Piping) (USNRC 1990). This generic letter provided guidance that would be considered by the regulator in evaluating relief requests submitted by licensees for temporary non-code repairs of Code Class 3 piping. Temporary non-code repairs are applicable until the next scheduled outage exceeding 30 days, but no later than the next scheduled refueling outage.

A code repair requires the removal of the flaw and a subsequent weld repair. The repair weld is subject to a post-repair nondestructive examination, and a post-repair pressure test may also be required. A code repair is practical during a scheduled shutdown. If a flaw is detected during plant operation, the plant may have to be shut down to perform a code repair. To avoid a plant shutdown and to limit the leakage from a through-wall flaw, some licensees have used temporary non-code repairs such as clamps with rubber gasketing, encapsulation of leaking pipes in cans using liquid sealants, or weld overlays. Temporary non-code repairs are not permitted on ASME code piping without regulatory approval.

For Code Class 1 and 2 piping, a licensee is required to perform code repairs or request regulatory relief for temporary non-code repairs on a case-by-case basis. Temporary non-code repairs of Code Class 1 and 2 piping must have load-bearing capability like that provided by engineered weld overlays or engineered mechanical clamps. Engineered weld overlays or engineered mechanical clamps are designed to meet the load-bearing requirements of the piping, assuming that the flaw is completely through the wall for 360 degrees, all around the pipe circumference, at the location of the flaw. For Code Class 3 piping, a licensee is also required to perform code repairs or request relief for temporary non-code repairs.

Several alternatives are available for evaluating and repairing piping wall thinning and pitting, including through-wall leaks (Table 5.18). Where Code Class 3 SW piping is included in a plant's ISI program for flow-accelerated corrosion (FAC), ASME Code Case N-513, N-523-1, N-561, N-562, N-597, or N-661 may be applied in lieu of Generic Letter 90–05. These ASME code cases are not available for application at nuclear power plants without specific USNRC approval. Upon approval, a formal fitness for service evaluation is required and must be available for on-site inspection.

5.11.2.3 USNRC Generic Letters 90–05 and 91–18

The Code of Federal Regulations (CFR) 10 CFR 50.55a(g) requires nuclear power plant piping to meet the applicable requirements of Section XI of the ASME Boiler and Pressure Vessel Code. This section of the code specifies acceptable repair methods for flaws that exceed Section XI acceptance limits for piping that is in service. A code repair is required to restore the structural integrity of the flawed code piping, independent of the operational mode of the plant when the flaw is detected. Article IWA-4000 of Section XI of the ASME code describes the code repair procedures. A code repair requires the removal of the flaw and a subsequent weld repair. The repair weld is subject to post-repair nondestructive examination, and a post-repair pressure test may also be required. A code repair is practical during a scheduled shutdown. If a flaw is detected during plant operation, the plant may have to be shut down to perform a code repair. Those repairs not in compliance with Section XI are non-code repairs, also known as "housekeeping" or "temporary" repairs.

Effective June 15, 1990, Generic Letter 90–05 provides guidance for preparing a relief request for temporary non-code repair of ASME Code Class 1, 2, and 3 piping. On November 7, 1991, the U.S. NRC issued Generic Letter 91–18 with information regarding two NRC Inspection Manual Sections on "*Resolution of Degraded and Nonconforming Conditions and on Operability*" (USNRC 1991; 2003). This Generic Letter, with the attached NRC Inspection Manual Part 9900, provides detailed discussions of specific operability determinations, one of which is operational leakage. In this regard, Section 6.15 of the Inspection Manual states the following:

> *Upon discovery of leakage from a Class 1, 2, or 3 component pressure wall (i.e., pipe wall, valve body, pump casing, etc.) the licensee should declare the component inoperable. The only exception is Class 3 moderate energy piping as discussed in Generic Letter 90-05. For Class 3 moderate energy piping, the licensee may treat the system containing the through-wall flaw(s), evaluated, and found to meet the acceptance criteria in Generic Letter 90-05, as operable until relief is obtained from the NRC.*

Generic Letter 90–05 develops the regulatory position on temporary non-code repairs. Such repairs are applicable until the next scheduled outage exceeding 30 days, but no later than the next scheduled refueling outage. The guidance in the Generic Letter applies when a flaw is detected during plant operation. If a flaw is detected during a scheduled shutdown, such as a planned maintenance outage or a refueling outage, a code repair is required before the plant restart. The letter states that temporary non-code repairs of Class 3 piping must be assessed every 3 months using NDE methods like ultrasonic or radiographic testing. Weekly inspections should check for leaks to determine structural integrity. An engineering evaluation is needed to assess degradation and identify necessary remedial actions. If structural integrity cannot be assured, the temporary repair is invalid. A four-part guideline on temporary repairs is as follows:

1. Flaw detection during plant operation and impracticality determination: “Impractical” means that the affected section of piping cannot be isolated for completing a code repair within the Technical Specification Limiting Condition for Operation (LCO) without a plant shutdown. For example, an LCO may prescribe that an affected SW train must be restored to operable status within 72 h or be at least in hot standby within the next 6 h.
2. Root cause determination and flaw characterization: This determination should include a positive identification of root cause and an identification of the most susceptible locations in the piping system.
3. Flaw evaluation: The flawed piping should satisfy the criteria of one of the two approaches, the “through-wall flaw” approach or the “wall-thinning” approach for non-through-wall flaws.
4. Augmented inspection: If the flaw is evaluated and found acceptable, the plant operator should perform an augmented inspection via UT or RT to assess the overall degradation of the system. The augmented inspection, performed within 15 days of detection of the flaw which results in temporary non-code repair, is part of the relief acceptance criteria of the repair. The inspection should be performed for at least the five most susceptible (and accessible) locations.

5.11.2.4 ASME Code Case N-752

Approved in 2019, ASME Code Case N-752 provides for risk-informed categorization requirements for performing repair/replacement activities on Class 2 and 3 pressure-retaining items (ASME 2019a, b). It uses the consequence evaluation portion of EPRI’s risk-informed ISI methodology supplemented with “additional considerations” to define high and low safety significance (Table 5.19). Code Case N-752 is not applicable to piping within the break exclusion region (> DN 100) for high-energy piping systems as defined by the owner and the portion of Class 2 feedwater piping (> DN 100) of PWRs from the steam generator to the outer containment isolation valve (Table 5.20).

The methodology relies on the conditional core damage and large early release probabilities (CCDP and CLERP, respectively) associated with postulated ruptures. The risk-informed process categorizes components solely based on consequence, which measures the safety significance of the component given that it ruptures. It additionally applies deterministic considerations (e.g., defense in depth, safety margins) in determining safety significance. The risk-informed process categorizes components as either high-safety-significant (HSS) or low-safety-significant (LSS). HSS components must continue to meet ASME Section XI rules for repair/replacement activities.

LSS components are exempt from ASME Section XI repair/replacement requirements and can be repaired/replaced in accordance with treatment requirements established by the owner. The treatment requirements must provide reasonable confidence that each LSS item remains capable of performing its safety-related functions under design-basis conditions. Component supports, if categorized, are

Table 5.19 Selected ASME code cases

Code case	Title	Applicability
N-513 (1997)	Evaluation criteria for temporary acceptance of flaws in moderate energy class 3 piping	Permits through-wall flaws in moderate-energy class 3 straight piping for a limited time not exceeding the time to the next scheduled outage if it can be demonstrated that adequate pipe integrity and leakage containment are maintained. N-513 allows the licensee to perform a flaw growth analysis to establish the allowable time for temporary operation. Periodic examinations of no more than 90-day intervals shall be conducted to verify the analysis
N-513-1 (2001)		N-513-1 expands the temporary acceptance methodology from class 3 moderate-energy piping to class 2 and 3 moderate-energy piping. Provides requirements which may be followed for temporary acceptance of flaws in ASME III, ANSI B31.1, and ANSI B31.7 class 2 or 3 piping. The allowable time for the temporary repair may not exceed the time to the next scheduled refueling outage
N-513–4 (2014)		Code case N-513–4 provides guidance for evaluation of flaws in locations not covered by code case N-513–3. The scope expanded to include components other than straight pipe and tanks. N-513–4 also includes time limitations until repair/replacement is required following detection of component leakage. Daily walkdown requirement for through-wall leaks
N-523-1 (1999)	Mechanical clamping devices for class 2 and 3 piping	The clamping device limits leakage by preventing separation of piping components. This code case was developed to address the criteria for temporary acceptance of flaws (including through-wall leaking) of moderate-energy class 3 piping where a section XI code repair may be impractical for a flaw detected during plant operation (i.e., a plant shutdown would be required to perform the code repair)
N-561 (1996)	Alternative requirements for wall thickness restoration of class 2 and high energy class 3 carbon steel piping	N-561 allows carbon steel piping experiencing internal thinning or pitting to be restored by means of a weld-deposited carbon or low-alloy steel reinforcement (i.e., weld overlay) on the outside surface of the piping. According to a USNRC

Table 5.19 (continued)

Code case	Title	Applicability
		evaluation in 2000, it was concluded that section XI and N-561 do not have sufficient rules for determining the rate or extent of the degradation of the repair or the surrounding-based metal. In addition, reinspection requirements are not specifically defined to verify the structural integrity of the component since the root cause may not be mitigated
N-562 (1996)	Alternative requirements for wall thickness restoration of class 3 moderate energy carbon steel piping	Relates to internal wall thinning of class 3 piping systems which have experienced degradation mechanisms such as MIC that would provide an acceptable repair configuration. The primary purpose for implementing weld overlay repair is to allow for adequate time for additional examination of adjacent piping so that pipe replacement can be planned to reduce the impact on system availability including maintenance rule applicability, availability of replacement materials, and cost. This alternative repair technique involves the application of additional weld metal on the exterior of the piping system, which restores the wall thickness requirement
N-660 (2002)	Risk-informed safety classification for use in risk-informed repair/replacement activities	Because of limitations in the methods for estimating failure probabilities, ASME section XI decided to base the process on conditional consequence given a failure, which removed the concern with large uncertainties regarding failure estimates. The proposed process initially sets the failure probability of any component to unity (1.0). The conclusion does not contradict the previous risk-informed ISI work, as it is acknowledged for inspection purposes that failure assessment is necessary to allow examination resources to be allocated to locations within a piping system most likely to degrade and fail. The process has been limited specifically to considerations of code repairs and replacements and has been refined to eliminate consideration of failure probability
	Additional examinations for PWR pressure retaining welds in class 1 components	Code case N-722 requires bare metal visual examinations of class 1 welds in

Table 5.19 (continued)

Code case	Title	Applicability
N-722-1 (2009)	fabricated with alloy 600/82/182 materials, section XI, division 1	alloy 600/82/182 materials. The requirements have criteria for when these examinations need to be performed. Unmitigated reactor pressure vessel bottom head (BMI) nozzle penetrations must be subjected to visual examination every other outage. The inspection requirements of ASME code case N-722-1 do not apply to components with pressure-retaining welds fabricated with alloy 600/82/182 materials that have been mitigated by weld overlay or stress improvement
N-752-1 (2019)	Risk-informed categorization and treatment for repair/replacement activities in class 2 and 3 systems section XI, division	This code case provides alternative rules for repair/replacement of certain class 2 and 3 components and their supports. Code case N-752 provides for risk-informed categorization as high safety significance (HSS) or low safety significance (LSS). LSS components are exempted from the code requirements listed above; in lieu of these requirements, code case N-752, paragraph –1420, requires the owner to define alternative treatment requirements, which will confirm with reasonable confidence that each LSS item remains capable of performing its safety-related function. These treatment requirements must cover items such as design control, procurement, installation, configuration control, and corrective actions. The risk categorization process is based on founding principles in EPRI report TR-112657, revision B-A. N-752 characterizes risk using CCDP and incorporates a multidisciplinary review to evaluate defense in depth
N-921 (2022)	Alternative 12-year inspection interval duration, section XI, division 1	This case provides an alternative to the inspection program requirements of IWA-2431 and provides requirements for an alternative 12-year ISI inspection interval. The code case establishes 4-year periods (4–4-4 year breakdown) over the course of a 12-year inspection interval. This change supports LWR plants that have transitioned from 18-month refueling cycles to 24-month refueling cycles and ensures that each inspection period during a 12-year interval will have at least two refueling outages (2–2-2

Table 5.19 (continued)

Code case	Title	Applicability
		format). A proposed regulatory condition in the United States was to require licensee update to the 2019 edition of ASME section XI or later before applying the code case

assigned the same safety significance, HSS or LSS, as the highest passively ranked segment within the bounds of the associated analytical pipe stress model.

The risk-informed methodology of Code Case N-752 may be applied on a system basis or on individual items within selected systems. Paragraph "-1100" of Code Case N-752 states: "*This case may be applied on a system basis, including all pressure-retaining items and their associated supports, or on individual items categorized LSS within the selected systems.*" While this is the case, the risk-informed methodology is applied to the pressure boundary function of the individual components within the system. The risk-informed methodology contained in Code Case N-752 requires that the component's pressure boundary function be assumed to fail with a probability of 1.0, and all impacts caused by the loss of the pressure boundary function be identified. This would include identifying impacts of the pressure boundary failure on the component under evaluation, identifying impacts of the pressure boundary failure of the component on the system in which the component resides, as well as identifying impacts of the pressure boundary failure of the component on any other plant SSC. This includes direct effects (e.g., loss of the flow path) of component failure and indirect effects of the component failure (e.g., flooding, spray, pipe whip, loss of inventory).

The categorization process of Code Case N-752 is delineated in Appendix I of the code case. This categorization process is technically based on the founding principles in EPRI Report TR-112657, Revision B-A, "*Revised Risk-Informed Inservice Inspection Evaluation Procedure*," and the categorization process of Code Case N-660, but with improvements and lessons learned from trial applications. Section 5.13.3 offers details on the "founding principles." Additional details of the risk categorization process are found in two documents, Nuclear Energy Institute Report NEI 00–04 (NEI 2005) and EPRI Technical Report No. 3002033536 (EPRI 2025).

Risk-informed categorization is based on the conditional consequence of a postulated pipe failure. A consequence category for each piping segment or component is determined through failure modes and effects analysis (FMEA), which considers pressure boundary equivalent break sizes, ability to isolate a pressure boundary breach, indirect effects, initiating events, system impact or recovery, and system redundancy. Results are partitioned into core damage impact groups based on postulated piping failures that (1) cause an initiating event, (2) disable a system/train/loop without causing an initiating event, or (3) cause an initiating event and disable a system/train/loop. Failures are also evaluated for their importance relative to

Table 5.20 Scope of systems classified as high-safety-significant

No.	Scope of piping systems considered as HSS
1	Reactor coolant pressure boundary piping. Includes interface systems besides the RCS (PWR) and RR (BWR) such as the emergency core cooling and decay heat removal
2	Applicable portions of the shutdown cooling pressure boundary function. Class 1 and 2 components of systems or portions of systems needed to use the normal shutdown cooling flow path: (a) as part of the RCPB from the reactor pressure vessel (RPV) to the second isolation valve (that is, farthest from the RPV capable of remote closure) or to the containment penetration, whichever encompasses the larger number of welds, or (b) other systems or portions of systems from the RPV to the second isolation valve (that is, farthest from the RPV) capable of remote closure or to the containment penetration, whichever encompasses the larger number of components
3	Class 2 portions of steam generators and class 2 feedwater system components greater than DN 100 of PWRs from the steam generator to the outer containment isolation valve
4	Components larger than DN 100 within the break exclusion region (BER) for high-energy piping systems
5	Portions of the ultimate heat sink (UHS) flow path (for example, service water) whose failures will fail both trains (that is, fail the UHS function). Even if piping is isolated/independent, the service water pump house [for example, reservoir, bay] would be expected to be HSS
6	Tanks/vessels and connected piping and components up to the first isolation valve that support/provide inventory to multiple systems/functions (for example, RWST for PWRs, containment sumps, pressure suppression pools for BWR)
7	CST for AFW/EFW in a PWR unless there is a redundant independent reliable source (for example, auto switchover to service water supply to each train of AFW/EFW suction)
8	For PWR plants, low-volume, intermediate safety systems that typically consist of two physically independent (for example, component cooling water) trains that are, on a plant-specific basis, physically connected. For example, loss of pressure boundary integrity on train A will drain train B as well
9	Any piping or component (including piping segments or components grouped or subsumed within existing plant initiating event groups) whose contribution to CDF is greater than 1E-06/year or whose contribution to LERF is greater than 1E-07/year, based upon a plant-specific PSA of pressure boundary failures (for example, pipe whip, jet impingement, spray, and inventory losses). This may include class 1 and 2 and class 3 or non-class components
10	Any piping or component (including piping segments or components grouped or subsumed within existing plant initiating event groups) whose contribution to CDF is greater than 1E-08/year and the product of its CDF contribution times its associated CCDP is greater than 1E-08/year, based upon a plant-specific PSA of pressure boundary failures
11	Any piping or component (including piping segments or components grouped or subsumed within existing plant initiating event groups) whose contribution to LERF is greater than 1E-09 and the product of its LERF contribution times its associated CLERP is greater than 1E-09, based upon a plant-specific PSA of pressure boundary failures

A break exclusion region (BER) is a section of piping in a nuclear power plant where breaks are not postulated for safety analysis purposes, often due to its low stress levels or the application of leak-before-break (LBB) or break preclusion concepts. These areas are considered to have a very low probability of failure.

containment performance. In addition, the consequence rank is reviewed and adjusted to reflect the pressure boundary failure's impact on plant operation during shutdown and on the mitigation of external events.

Piping segments/components ranked as "high-safety-significant" (HSS) require no further review. Piping segments/components ranked as medium, "low-safety-significant" (LSS), or "none" shall be determined to be HSS or LSS by evaluating the additional categorization considerations or conditions outlined in paragraph I-3.4.2(b) of Code Case N-752. If any of these conditions are not met, then HSS shall be assigned. If all conditions are met, then LSS may be assigned. Finally, if LSS is assigned, the categorization process shall verify that there are sufficient margins to account for uncertainty in the engineering analysis and supporting data. If sufficient margin exists, then LSS should be assigned. If sufficient margin does not exist, then HSS shall be assigned. HSS components will include the following:

1. Class 1 portions of the RCPB: Piping as described below in "a" and "b" might have been optionally classified as Class 1 or "Class 1–exempt." Reclassifying this piping other than Class 1 to gain the full benefit of a 10 CFR 50.69 application should be considered. This change would obviously need to follow the applicable commitment change control process (such as 10 CFR 50.59). LSS shall not be assigned to the piping described in "a" and "b" below until this piping has been classified as non-Class 1.

   (a) In the event of postulated failure of the component during normal reactor operation, the reactor can be shut down and cooled down in an orderly manner.
   (b) The component is or can be isolated from the reactor coolant system (RCS) by two valves in series.

2. Any piping or component, including piping segments or components grouped or subsumed within existing plant initiating event groups (main feedwater breaks inside containment; main steam line breaks outside containment; service water flooding events; interfacing system LOCAs; failures of non-Class 1 RCPB connections, such as instrumentation lines; and so on), whose contribution to CDF is greater than 1.0×10^{-6} per year or whose contribution to LERF is greater than 1.0×10^{-7}/year, based on a plant-specific PSA that includes considerations of pressure boundary failures.
3. Any piping grouped or subsumed within existing plant initiating event groups (main feedwater breaks inside containment, main steam line breaks outside containment, service water flooding events, interfacing system LOCA, non-Class 1 RCPB connections, such as instrumentation line failures) whose contribution to CDF is greater than 1.0×10^{-8} per year, and the product of its CDF contribution times its associated CCDP is greater than 1.0×10^{-8} per year:

   (a) CDF > 1.0E-6 = HSS, regardless of CCDP.
   (b) CDF < 1.0E-8 = LSS, regardless of CCDP.
   (c) CDF × CCDP ≤1.0E-8 = LSS.
   (d) CDF × CCDP >1.0E-8 = HSS.

5.11.2.5 Code Case N-770-1

At the request of the USNRC, ASME developed Code Case N-770-1 (ASME 2009), which provides inspection requirements to address primary water stress corrosion cracking (PWSCC) in Class 1 butt welds containing Alloy 82/182. Code Case N-770-1 has requirements for inspection of unmitigated as well as mitigated Alloy 82/182 RCS butt welds. As such, specific inspection requirements for welds mitigated by weld overlays (WOLs) are contained in the code case. The NRC incorporated ASME Code Case N-770-1 by reference into §50.55a (Federal Register, Volume 76, Number 119, p. 36278) in June 2011.[26] Operators of PWRs with dissimilar metal nozzle safe-end welds, which are susceptible to PWSCC, are required to examine those welds with increased frequency unless mitigating actions are taken. ASME published Code Case N-770 on January 26, 2009, "*Alternative Examination Requirements and Acceptance Standards for Class 1 PWR Piping and Vessel Nozzle Butt Welds Fabricated with UNS N06082 or UNS W86182 Weld Filler Material with or Without Application of Listed Mitigation Activities Section XI, Division 1*." On December 25, 2009, ASME modified ASME Code Case N-770 and issued version 1 of the code case as N-770-1. Requirements for these examinations are specified in ASME Section XI, Code Case N-770-1, as modified by 10CFR50.55a.

USNRC issued rulemaking (76 FR 36278) in June 2012 that implemented ASME Code Case N-770, "*Alternative Examination Requirements and Acceptance Standards for Class 1 PWR Piping and Vessel Nozzle Butt Welds Fabricated with UNS N06082 or UNS W86182 Weld Filler Material with or Without the Application of Listed Mitigation Activities, Section XI, Division 1*," United States Code of Federal Regulations, 10CFR50.55a(g)(6)(ii)(F)(1), effective date August 22, 2011, states "*licensees of existing, operating PWRs as of July 21, 2011, shall implement the requirements of ASME Code Case N-770-1, subject to the conditions specified in paragraphs (g)(6)(ii)(F)(2) through (g)(6)(ii)(F)(10) of this section, by the first refueling outage after August 22, 2011*." The Code Case 770–1 inspection requirements are summarized in Table 5.21.

Welds in Inspection Items A-1, A-2, and B are required to be examined by visual examination and UT. Inspection Item A-1 welds, hot leg welds with an operating temperature greater than 325 °C (i.e., pressurizer nozzle dissimilar metal welds, DMWs), are required to be examined visually each refueling outage and volumetrically every second refueling outage. Inspection Item A-2 welds, hot leg welds with an operating temperature less than or equal to 325 °C, are required to be examined visually each refueling outage and volumetrically every 5 years. Inspection Item B welds, cold leg welds with an operating temperature greater than or equal to 274 °C and less than or equal to 304 °C, are required to be examined visually once per

[26] *USNRC Regulations Title 10, Code of Federal Regulations (CFR), Part 50.55a, lists the documents that have been approved for incorporation by reference,* https://www.nrc.gov/reading-rm/doc-collections/cfr/part050/part050-0055a.html

Table 5.21 Code Case N-770-1 inspection requirement summary

Inspection item		Inspection location	Extent and frequency of examination
Unmitigated alloy 82/182 butt welds	A-1	Butt welds at hot leg operating temperature > 329 °C	Bare metal visual examination each refueling outage. Volumetric examination every second refueling outage
	A-2	Butt welds at hot leg operating temperature ≤ 329 °C	Bare metal visual examination each refueling outage. Volumetric examination every 5 years
	B	Butt welds at cold leg operating temperature ≥ 274 °C and < 304 °C	Bare metal visual examination once per interval. Essentially 100% volumetric examination for axial and circumferential flaws in accordance with the applicable requirements of ASME section XI, appendix VIII, every second inspection period not to exceed 7 years. Baseline examinations shall be completed by the end of the next refueling outage after January 20, 2012
Mitigated alloy 82/182 butt welds	C	Uncracked butt welds reinforced by a full structural weld overlay (FSWOL) of alloy 52/152 material	If volumetric examination is not performed on the weld prior to FSWOL, the weld shall be assumed cracked and shall be classified inspection item F. If cracking is not observed during post-FSWOL pre-service volumetric examination performed from the outside surface of the overlay, axial and circumferential cracks at least 75% through the original wall thickness are required to be assumed. Inspection item C welds are required to be placed in a population to be examined on a 25% sample basis. A once-per-interval inspection frequency applies to this sample
	D	Uncracked butt welds mitigated by mechanical stress improvement process (MSIP®)	No visual inspection requirements. Volumetric examination shall be performed within 10 years following stress improvement. If the examination volumes of the welds show no indication of cracking, the welds shall be placed into a population to be examined on a sample basis. Twenty-five percent of this population shall be examined once each interval

Table 5.21 (continued)

Inspection item		Inspection location	Extent and frequency of examination
	E	Cracked butt welds mitigated by MSIP®	No visual inspection requirements. Volumetric examination shall be performed once during the first or second refueling outage following stress improvement. If the examination volumes of the welds show no indication of crack growth or new cracking, the welds shall be placed into a population to be examined once each interval
	F	Cracked butt welds reinforced mitigated by an FSWOL of alloy 52/152	Inspection item C welds are required to be placed in a population to be examined on a 25% sample basis. A once-per-interval inspection frequency applies to this sample. The FSWOL technique serves a defense-in-depth monitoring function
	G	Uncracked butt weld mitigated with an inlay of alloy 52 or alloy 152	A volumetric examination shall be performed immediately before application of inlays or onlays and after application as a pre-service baseline inspection. Volumetric and surface examinations shall be performed no sooner than the shorter of 10 years following the application of the inlay or onlay and the design life of the inlay or onlay. Examination volumes that show no indication of cracking are to be placed into a population to be examined on a sample basis
	H	Uncracked butt welds mitigated by a weld onlay constructed of alloy 52 l	
	J	Cracked butt welds mitigated with an inlay of alloy 52 or alloy 152	Volumetric and surface examinations shall be performed once during the first or second refueling outage following application of the inlay or onlay. Examination volumes that show no indication of cracking are to be placed into a population to be examined on a sample basis
	K	Cracked butt welds mitigated by a weld onlay constructed of alloy 52 material	

interval and volumetrically every second inspection period, not to exceed 7 years. For certain components, 100% volumetric coverage is required. Examples of such components include the pressurizer surge line nozzle and the reactor coolant system cold leg and hot leg nozzles.

5.11.3 Pressure Testing

Pressure tests are performed to ensure the leak tightness of a piping system. A pressure test is required for a new piping system before use or an existing system after repair or alteration. There are two methods for pressure tests: hydrostatic and pneumatic. A hydrostatic test is performed by using water as the test medium, whereas a pneumatic test uses air or nitrogen. ASME hydrostatic test requirements for process piping are found in ASME XI Article IWA-5000 (System Pressure Tests ASME B31.1, and ASME B31.3). For ASME III piping test, pressure is often $1.25 \times P_{Design}$ to which are typically added a 6% code margin and a 0.03 MPa test margin. For B31.1 piping systems, the hydrotest pressure is often $1.5 \times P_{Design}$ to which is added a 0.03 MPa test margin. In either case, ASME III or ASME B31.1, the hoop stress caused by the test pressure should not exceed yield. For pneumatic testing, the test pressures for ASME III are typically $1.2 \times P_{Design}$, and for B31.1 piping, it is $1.25 \times P_{Design}$.

Pressure testing has proven effective in detecting pipe defects and potential failure points prior to the commissioning of a system. However, there have been examples of significant structural failures occurring as a result of improperly applied test procedures or as a result of ISI failures (Fig. 5.10).

ASME codes state that the pressure during the hydrostatic test should never fall below one and a half times the pressure that the system is designed to hold. During the hydrostatic test, pressure must be maintained for at least 10 min. After this period, the test pressure is reduced to match the design pressure, followed by a visual examination for leaks. Performing a hydrostatic pressure test is a complex undertaking. Relatively soon after ASME Section XI was first published, ASME Code Committees performed studies regarding the benefits of performing system hydrostatic pressure tests versus system operating leakage tests at normal operating pressure for both repairs/replacements and 10-year inspection interval requirements.[27] These studies resulted in the issuance of two code cases:

[27] *S. H. Bush and R. R. Haccary (1971), "Development of In-Service Inspection Safety Philosophy for U.S.A. Nuclear Power Plants," ASME: "The system hydrostatic test, as stipulated in Section XI, is not a test of the structural integrity of the system but rather an enhanced leakage test. Hydrostatic testing only subjects the piping components to a small increase in pressure over the design pressure: therefore, piping dead weight, thermal expansion, and seismic loads present far greater challenges to the structural integrity of a system. Consequently, the Section XI hydrostatic pressure test is primarily regarded as a means to enhance leak detection during the examination of components under pressure, rather than as a method to determine the structural integrity of the*

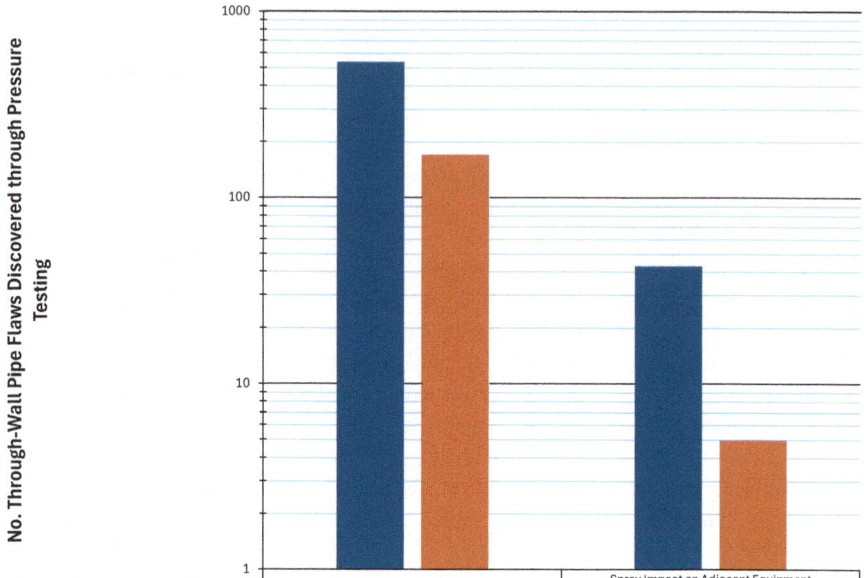

	No Spatial Impact	Spray Impact on Adjacent Equipment /Flooding of Equipment Area
Hydrotesting / Surveillance Testing @ Power	537	43
Hydrotesting (ASME XI)	171	5

Fig. 5.10 Piping operating experience pertaining to pressure testing

- Code Case N-498, *Alternative Rules for 10 year Hydrostatic Pressure Testing for Class 1 and 2 Systems*, permits use of a system leakage test in lieu of the 10-year hydrostatic test.
- Code Case N-416, *Alternative Rules for Hydrostatic Testing of Repairs or Replacements*, recognizes that a piping/component hydrostatic pressure may be deferred without significantly impacting the safety of the system.

5.11.4 Pipe Repair

The repair of a flawed or leaking pipe is governed by national codes and standards, which differentiate permanent repairs from temporary repairs. The latter type refers to repairs performed to provide a leak barrier and in some cases performed online with the reactor operating at normal power. Examples of different repair technologies are given in Table 5.22.

---

components. In addition, industry experience indicates that leaks are not being discovered as a result of hydrostatic test pressures causing a preexisting flaw to propagate through the wall; in most cases leaks are being found when the system is at normal operating pressure."

Table 5.22 Examples of pipe repair technologies

Pipe repair technology	Flaw type				
	General wall thinning	Local wall thinning	Pitting	Circumferential cracks	Longitudinal cracks
Butt-welded insert plates	Yes (Y)	Y	Y	Y	Y
External weld overlay repair methods for internal thinning	No (N)	Y	Y	N	N
Welded leak box repair (A "welded leak box repair" refers to a method of temporarily fixing a leak in a pipe or vessel by welding a custom-made, boxlike enclosure around the leak area, essentially creating a sealed "box" to contain the fluid leakage while the system remains operational. The repair is typically considered temporary until a permanent solution can be implemented.)	N	Y	Y	(Y)	(Y)
Full encirclement steel reinforcing sleeves for piping	Y	Y	Y	Y	Y
Weld buildup, weld overlay, and clad restoration	Not applicable (N/A)	Y	(Y)	N	N
Full structural weld overlay (FSWOL)	N/A	N/A	N/A	Y	Y
Fillet weld patches	N/A	Y	Y	(Y)	(Y)
Metallization arc spray process	Y	Y	Y	N	N
Flaw excavation and weld repair	N/A	N/A	Y	Y	Y
Mechanical clamp repair	N/A	Y	Y	(Y)	(Y)
Nonmetallic composite repair system	Y	Y	Y	(Y)	(Y)
Nonmetallic internal lining for pipe, sprayed form for buried pipe	Y	Y	Y	Y	Y
Nondestructive examination in lieu of pressure testing for repairs and alterations	N/A	N/A	N/A	N/A	N/A
Optimized weld overlay (OWOL)	N/A	N/A	N/A	Yes	Yes

Regulatory requirements determine the acceptability of these technologies. As a rule of thumb, without regulatory approval, temporary repair of safety-related piping is inadmissible. Many of the listed pipe repair technologies have a long history with respect to design and engineering validation processes and widespread nuclear industry applications. Weld overlay repairs (WORs) of Safety Class 1 BWR piping were performed in the early 1970s. Hundreds of such repairs have been performed. WOR technology improvements have been made. In response to primary water stress corrosion failures, the so-called full structural weld overlay (FSWOL) process was introduced. The FSWOL process may be applied where flaws do not exceed 75% through-wall.

Optimized weld overlays (OWOLs), a subset of the FSWOL, have been proposed for larger geometries such as PWR reactor coolant system hot leg and cold leg nozzles, where FSWOL application becomes too time consuming for a typical refueling outage. The optimized weld overlay thickness is less than that of a full structural weld overlay in order to allow completion in the time available in a typical refueling outage for larger geometries. The required minimum axial length of the OWOL is defined in ASME Code Case N-754 and is the same as that required for an FSWOL. The code case also says that the thickness must be sufficient to provide compressive residual stresses to preclude stress corrosion crack growth. The nuclear industry has proposed that an OWOL design is sufficient if it reduces tensile stresses in the dissimilar metal weld area to less than 69 MPa with operating temperature and loads applied. Each OWOL must be designed for a specific geometry and application and must be evaluated on a case-by-case basis to prove that the overlay is sufficient to mitigate the possibility of PWSCC. The code case provides requirements for installing optimized structural weld overlays on the outside surface of ASME Class 1 heavy-wall, large-diameter piping composed of ferritic and austenitic stainless steel and nickel-base alloy materials in PWRs as a mitigation measure where no known defect exists or the defect depth is limited to 50% through wall. The upper 25% of the original pipe wall thickness is credited as a part of the OWOL design in the analyses performed in support of these repairs.

Raw water piping often suffers from corrosion, necessitating ongoing mitigation or repair for reliable operation throughout the license period. Various nonmetallic options offer cost-effective and efficient solutions for repairing degraded piping. Conventional welded and metallic repairs are effective, but in some instances, their installation can be more complex, costly, and time-consuming than certain nonmetallic alternatives. Additionally, when repairing damage caused by corrosion, unless a material change is implemented, a metallic repair will be susceptible to the same deterioration that caused the defect it was installed to repair.

Nonmetallic repair options offer corrosion-resistance advantages in corrosive environments like raw water service in coastal environments. Primary technologies are high-density polyethylene (HDPE) replacement, carbon fiber-reinforced polymer (CFRP) repair, and cured-in-place pipe (CIPP). For moderate-energy safety-related piping, different types of nonmetallic pipe repair processes have been developed, some of which are applied to the outside pipe wall and others to the inside pipe wall. These repairs can function as a liner, a corrosion barrier, as well as provide structural reinforcement and even full structural replacement. Gabriel et al. (2015) offer details on the CFRP and CIPP technologies.

5.12 Pipe Stress Improvement Processes

The purpose of stress improvement is to reduce the residual stress on the inner surface of a pipe in the heat-affected zone near a weld. This is the region susceptible to stress corrosion cracking. There are several techniques in use, most of which were developed in the early 1980s in response to the IGSCC and metal fatigue issues. Types of stress improvement include (1) weld overlay, (2) induction heat stress improvement, (3) surface stress improvement, and (4) mechanical stress improvement.

5.12.1 Weld Overlay

The weld overlay repair (WOR) technique was developed in 1982 as an interim repair for IGSCC-flawed welds in BWR stainless steel piping. Weld overlay repairs serve several design functions. These include structurally reinforcing the flawed location to restore code margins to failure, providing an IGSCC-resistant barrier to limit further crack growth, and imposing a favorable residual stress distribution in the inner portion of the flawed component, which will inhibit flaw initiation and propagation.

There are two types of weld overlays: full structural weld overlay (FWSOL), which has a minimum thickness of one-third the pipe wall thickness, and optimized weld overlay, which has different design requirements and results in less applied weld overlay material (EPRI 2008; Fredette and Scott 2010). Weld overlays extend in both axial directions some distance beyond the weld. Weld overlays extend 360° circumferentially around the pipe. For FWSOLs, no structural credit is taken for the original pipe thickness. ASME has developed Code Cases N-740-2 (ASME 2015) and N-754-1 (ASME 2013) for application of full structural weld overlays and optimized weld overlays, respectively. NUREG-0313 defines the following weld overlay designations (USNRC 1988a, b):

- **Standard overlay** design based upon the assumption that the flaw is circumferential in orientation and extends entirely through the original pipe wall, 360 degrees around the pipe. This design takes no credit for the original pipe wall in meeting the allowable flaw size criteria of ASME Section XI.
- **Designed overlay**: In cases where cracks are perpendicular to the weld (axial) or short in the circumferential direction, even a small amount of overlay will prevent further growth in the length direction, because high compressive stresses are induced at the inner surface of the pipe. In such cases, the overlay will also act to

prevent leakage. Weldments with a total length of circumferential cracking less than approximately 10% of the circumference, with no more than four axial cracks, are considered appropriate for repair by a designed overlay. A standard overlay should be used for more severe cracking. The thickness of the designed overlay should be at least two layers of weld metal after the surface has passed surface examination by penetrant inspection (PT). If credit is taken for the thickness of the first layer, it should be shown by actual test to contain a minimum of 7.5% ferrite, and the original surface must have passed penetrant testing (PT).
- **Limited service overlay**, which was considered suitable for only a single fuel cycle of operation: If an overlay falling into this category is identified in the field, it may be necessary to upgrade the overlay to one of the other design categories by adding weld material and/or surface conditioning for inspection, in order to justify longer-term operation.

The weld overlay repair technique for pipe welds containing intergranular stress corrosion cracking (IGSCC) flaws is based upon the application of weld metal to the outside pipe surface over and to either side of the flawed location, extending circumferentially 360°. The weld overlay repair performs a number of design functions (Riccardella et al. 1988):

- Provide structural reinforcement of the flawed location, such that adequate load-carrying capability is present, either in the overlay by itself or in some combination of the overlay and the original pipe wall thickness.
- Provide a barrier of resistant material to prevent crack propagation into the overlay weld metal.
- Produce a compressive residual stress distribution in, at least, the inner portion of the pipe wall, which will inhibit IGSCC initiation and propagation in the original pipe joint.
- Prevent leakage from small axial flaw.

5.12.2 Induction Heat Stress Improvement

Originally developed in Japan, the induction heat stress improvement (IHSI) process changes the normally high tensile stress present on the pipe inside surface of weld heat-affected zones to a beneficial compressive stress (Bertossa, 1980) (Shimizu et al. 1980; Ando et al. 1982; Shadley et al. 1982; Offer 1984). The process involves induction heating the outer pipe surface of completed girth welds to approximately 400 °C while simultaneously cooling the inside surface, preferably with flowing water. Thermal expansion caused by the induction heating plastically yields the outside surface in compression, while the cool inside surface plastically yields in tension. After cooldown, contraction of the pipe outside surface causes the stress state to reverse, leaving the inner surface in compression and the outside surface in tension.

This process is used for Type 304 stainless steel pipes, with the outer surface temperature limited to 400 °C to reduce sensitization inside. As a result, slight tensile stress may remain on the inner surface. In Japan, the method was adapted for nuclear grade Type 316 stainless steel, allowing full compressive stresses by raising the outer surface temperature to 650 °C without causing sensitization inside. The effectiveness of induction heating stress improvement is qualified by the following criteria:

– Induction heating stress improvement treatment reduces normally high tensile inside surface residual stresses to a zero compressive state.
– Full-size environmental pipe testing and residual stress tests have demonstrated that these beneficial residual stresses result in a large improvement in intergranular SCC resistance.
– Metallurgical investigations have shown that the IHSI produces no increase in sensitization, and no significant variation in mechanical properties occurs.

5.12.3 Surface Stress Improvement

Surface deformation to produce compressive stress on the surface has long been applied to piping that is subject to degradation due to the presence of preexisting tensile residual stress. This has been effective, assuming of course that the surface is accessible, by shot peening, laser peening, or cavitating jet (Soyama 2004, 2017, 2020; Soyama and Korsunsky 2022). Some concern has been raised regarding this approach, associated with the effect of the associated tensile stress component deeper beneath the surface and the relaxation of the compressive stress due to a subsequent applied tensile or compressive stress cycle.

Peening is a cold working process in which the surface of the component is deliberately deformed, in the basic method, by hammering. During peening, the surface layer attempts to expand laterally but is prevented from doing so by the elastic nature of the subsurface bulk material. This results in the development of beneficial, compressive residual stresses in the surface layer, which are balanced by tensile residual stresses elsewhere. The layer of compressed surface material produced by peening resists the development and propagation of cracks and increases resistance to fatigue failure, corrosion fatigue, stress corrosion, and cavitation erosion. The application of peening to surfaces of metals to develop a surface layer with compressive stresses can significantly increase resistance to SCC. Shot peening was applied to steam generator Alloy 600 tubes starting in 1985 as a mitigation method against the occurrence of PWSCC. Based on this experience and recognizing that both BWRs and PWRs had other reactor coolant system and reactor internal components that were susceptible to SCC, several Japanese firms began development in the 1990–1995 time period of methods to peen surfaces of susceptible core internal and pressure boundary components in BWRs and PWRs. These methods began to be applied in BWRs in 1999 and in PWRs in 2001.

The peening methods used in Japan in BWRs and PWRs have included the water jet peening (WJP) and underwater laser peening (ULP) methods. In addition, a third peening process called ultrasonic shot peening (USP) has been developed and used in PWRs in Japan. USP involves peening using 3 to 4 mm diameter nickel-base alloy shot that is locally accelerated in a small chamber held to the surface of the part using a piezoelectric element. This method was reported in 2009 as having been used at SG inlet or outlet nozzles at PWRs in Japan, at the inlet and outlet nozzle-to-safe-end welds of the reactor vessel at a PWR under construction, and at the control rod drive mechanism (CRDM) to reactor vessel head (RVH) J-groove welds[28] of several PWRs. It results in a compressive stress layer about 1 mm deep.

5.12.4 Mechanical Stress Improvement

Like the IHSI process, the mechanical stress improvement process (MSIP) is identified as a residual stress improvement technique, but along with application differences, there are some fundamental differences between the IHSI and MSIP processes (Beckjord 1984; Anderson et al. 2008). With IHSI, large thermal stresses are produced in the weld region which cause the material near the inside surface of the weld to yield in tension. Removal of the thermal load results in reversal of the stress state which puts the inside surface in compression. During MSIP, on the other hand, the material near the weld inside surface is loaded directly in compression, up to and beyond yield. There is some relaxation upon removal of the MSIP load. However, MSIP has been shown in both tests and analyses to leave substantial compressive residual stress at the weld inside surface. MSIP removes residual tensile stresses from weldments, thus preventing the initiation of cracks and retarding the growth of preexisting flaws in piping systems. It uses a hydraulically operated clamp system to uniformly compress the entire pipe at a location near the weld joint.

MSIP also causes slight plastic strain, and the residual stresses remaining after the treatment are compressive in the location susceptible to IGSCC because of weld sensitization. The process involves a slight permanent contraction of the pipe on one side of the weldment. The resulting plastic flow redistributes the residual as-welded stresses and generates beneficial compressive stresses at the inner pipe surface in the region of the weldment, including the weld metal and heat-affected zones. MSIP imposes only monotonic compressive plastic strains and does not require severe temperature gradients. The process can be applied to the piping system either filled with water or empty.

The mechanical stress improvement process (MSIP) has been used to prevent stress corrosion crack growth in BWR, PWR, and RBMK plants. It was first applied

[28] *A "J-groove weld" is a type of groove weld where the edge of one workpiece is prepared with a J-shaped bevel while the other workpiece is left flat. This configuration allows for deeper penetration of the weld and potentially uses less filler metal compared to a square groove weld.*

to weld heat-affected zones in a US BWR plant in the 1985–1986 timeframe. According to a letter from Commonwealth Edison (CECO) to the USNRC dated July 19, 1985, MSIP evolved from the "Pipelock" technology developed by O'Donnell Consulting Engineers Inc. in a cooperative program with CECO. Subsequently developed further, MSIP® is a patented process and has regulatory approval for use in the United States (USNRC 1988a, b). Pipelock technology refers to a type of mechanical joint restraint coupling used to connect and restrain plain-end pipes, especially those made of ductile iron or steel. It is a way to ensure a strong and reliable connection without welding or threading.

5.13 Risk-Informed in-Service Inspection

The role of periodic in-service inspection in reducing the probability of significant structural failure of piping components was the subject of studies in the early 1970s. Development of risk-informed in-service inspection (RI-ISI) methodologies began in earnest in the 1980s, and aspects of these methodologies have found use in reliability and integrity strategies for advanced reactors. With reference to piping pressure boundary integrity, the evolution of risk-informed initiatives is illustrated in Fig. 5.11.

RI-ISI method development and practical applications thereof evolved in multiple phases:

- **1970–1979**: Initial, conceptual quantitative studies on the role of in-service inspection in reducing pipe failure propensities.[29]
- **1980–1999**: Initial methodology development, pilot studies, codes and standards development, and regulatory review and approval processes (Simonen 1995).
- **2000–2005, Part 1**: Regulatory acceptance in Spain and the United States (OECD/NEA 2004). Widespread applications to Safety Class 1 and 2 systems, and in a few cases Class 3 systems. Continued international pilot applications. A limited benchmark of four different methodologies, referred to as SKIFS, EPRI RI-ISI, and Westinghouse Owners Group (WOG), the RIBA Project.[30] ASME Boiler and Pressure Vessel Code, Section XI, Nonmandatory Appendix R, *Risk-Informed Inspection Requirements for Piping*, was published in 2005.

---

[29] *In 1979, H.M. Thomas published a paper titled "Inservice Inspection Is Overrated" in the Journal of the Institution of Nuclear Engineers. This work likely presented a critical perspective or analysis regarding the efficacy and necessity of certain aspects of in-service inspection practices, potentially sparking debate or contributing to evolving approaches within the field (Journal of the Institution of Nuclear Engineers 20(2):43–56). WASH-1400 (1975) highlighted the need to focus on safety-significant components rather than relying solely on time-based or deterministic rules to ensure a high degree of reliability and safety.*

[30] *The RIBA project was established in November 1999 as a 24-month study contract funded by the European Commission. The participants were EDF (France), Ringhals AB (Sweden), Tecnatom SA (Spain), and Westinghouse Electric Europe.*

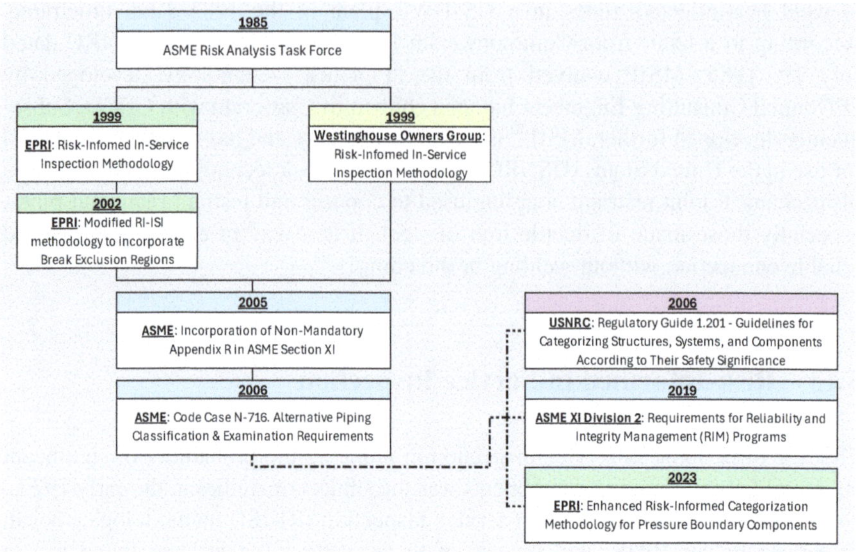

Fig. 5.11 Evolution of key risk-informed initiatives to ensure pressure boundary integrity

- **2000–2005, Part 2**: In response to user feedback, and insights obtained from the review of licensee submittals to regulatory authorities, EPRI developed a streamlined RI-ISI methodology to make future RI-ISI program updates easier to perform and implement. EPRI expanded the methodology to also include the break exclusion region (BER).[31]
- **2006–2015**: Approval of ASME Code Case N-716, "Alternative Piping Classification and Examination Requirements," also referred to as the "Streamlined RI-ISI Methodology." With reference to ASME Section XI, Nonmandatory Appendix R, a regulatory review and approval are no longer required in the United States prior to implementation of an RI-ISI program. Introduction of a "streamlined" approach to RI-ISI program development and implementation.
- **Beyond 2015**: Applications to advanced evolutionary reactor designs (Wang et al. 2025) and non-light water reactors, risk-informed high-energy line break analysis, and risk-informed repair and replacement requirements. These applications are evolutions of the EPRI "streamlined RI-ISI methodology."

[31] *The "break exclusion region" (BER) refers to a specific area around high-energy piping where a pipe rupture, which could cause damage to safety-related components, is not postulated to occur because of specific design features or application of the leak-before-break (LBB) concept. This allows for a reduction in the design and analysis of protective structures and inspections, by demonstrating that the dynamic effects of a postulated break are not a design-basis concern in that location. Practically, this includes Safety Class 2 components and piping (e.g., feedwater, main steam, auxiliary feedwater).*

5.13.1 Early Developments

While ASME should be credited as being the organization pioneering the work to implement risk-informed ISI procedures, discussion to develop a more systematic approach to defining locations to be inspected began during the early stages in the commercialization of nuclear power plants, The reader is encouraged to revisit NUREG/CP-0017, "*Executive Seminar on the Future Role Risk Assessment and Reliability Engineering in Nuclear Regulation.*" The seminar was held at Bethesda (MD) on January 28, 1980, with discussions during the seminar tape recorded and transcribed for eventual inclusion in the NUREG/CP-0017 report. Pages 8–19 to 8–20 of the report give insight into early thoughts that eventually led to implementation of RI-ISI; two quotations follow (van Erp 1981)[32]:

> *[W]e require through several standards, primarily the ASME code, a great deal of inspections and testing of valves, pumps, people going in doing hands-on inspection of welds and pipes, and these introduce another factor which means that every time somebody does this, there is certainty of a certain amount of exposure. We're wondering where the tradeoff starts. How many people are we burning up - absolutely - to avoid the potential for some minor exposures in the future and we have requested in the past, and we intend to offer further requests more formally and more extensively, for the application of probabilistic analysis to help us determine where to quit, where to start, and where to emphasize, with this type of activity in an operating plant. I'd like to hear if anybody can speak to that.*

> *.... [Y]es, I think I can help you on that. It's a tough job, especially if you are talking about in-service inspection because you have to first look at a good model for weld failure and that is principally our problem right now. If, let's say, you eliminate a weld inspection, the question becomes "Well, what does this do in terms of your perception of the safety of the plant?" And we don't have a good kind of information in terms of those models yet, but once we do, and I think there are many analyses afoot that will help us in this area, once we get good weld failure models, then I think we can progress along this line.*

ASME formed a risk analysis task force in 1985 to initiate the use of "*risk-based*" methods in the formulation of policies, codes, and standards. At the suggestion of the task force, the ASME Codes and Standards Research Planning Committee of the ASME Center for Research and Technology Development (CRTD) recommended that a research program be initiated to determine how risk-based methods could be used to establish inspection requirements and guidelines (Balkey et al. 1992). The committee initiated the project out of recognition that recent significant structural failures occurring across many industries highlighted the need to relate risk more explicitly with inspection programs. The task force provided a means for a cooperative multidisciplinary research effort to be performed that would have regulatory, academic, and industry participation. The research task force met four times during

[32]*This executive seminar was jointly sponsored by the US-NRC Office of Nuclear Regulatory Research and Argonne National Laboratory (Division of Educational Programs) for the purpose of informing NRC's staff on current thinking in the area of risk assessment and reliability engineering. Six presentations are included, along with their associated question-and-answer sessions:* https://www.osti.gov/biblio/6402322

the period 1988–1995. The group sought knowledge from several sources: ongoing research within the respective organizations of each task force member. Examples included the risk-based ISI reliability program at the Pacific Northwest National Laboratory (PNNL) (Vo et al. 1989, 1990, 1994), information from other domestic and foreign research programs (e.g., an advanced risk-based inspection strategy, championed by Dr. Vic Chapman (1983, 1989) of Rolls-Royce and Associates Limited, that had been implemented in the UK nuclear submarine program), experience from working with code committees that were developing new inspection requirements (e.g., ASME Section XI Subgroup on Core Support and Internal Structures), and related technical literature.

As part of the "*Evaluation and Improvement of Nondestructive Evaluation (NDE) Reliability for Inservice Inspection of Light Water Reactors Program*" sponsored by the USNRC, PNL developed and applied a method using risk-based techniques to establish ISI plans for nuclear power plant components. This approach determined ISI priorities for systems and components by estimating the probabilities of component failures through expert judgment elicitation and plant-specific PSA results, combined with a failure mode and effect analysis (FMEA). An approach for determining target risk and target rupture probability values was based on the philosophy that ISI should ensure that the risk of core damage resulting from pressure boundary and structural failures be a small fraction of the total core damage risk estimated in a plant-specific PSA. The ASME CRDT had recommended that the risk due to pressure boundary and structural failures, the "target risk," should be no more than 5% of the total PSA estimated risk from internal events. It was further recommended that this target risk be apportioned unequally among the risk-important components by considering the estimated risk associated with rupture of each component.

The ASME Research Foundation continued this work to develop procedures necessary to perform risk-informed in-service inspection evaluations (Gore and Balkey 1992). The early work led to R&D in two technical focus areas: (1) piping operating experience reviews and (2) probabilistic piping reliability analysis to explicitly assess the influence of NDE and leak detection. In summary, the motivations behind efforts to transition from a deterministic framework for ISI program implementation to a risk-informed approach were twofold:

1. Introduce a more structured approach to ISI program development and to optimizing the inspection location selection process.
2. Improve the quality of ISI by focusing the inspections on locations most susceptible to environmental degradation. In most cases, the application of RI-ISI demonstrated possibilities for reducing the scope of ISI without any negative impact on safety.

The Nuclear Regulators' Working Group (NRWG) in Europe established a Task Force in November 1996 to develop principles for risk-informed in-service inspection (RI-ISI) of nuclear power plant components, ensuring safety and minimal public radiation exposure. The Task Force reviewed existing RI-ISI methods and published the "*Report on Risk-Informed In-Service Inspection and In-Service Testing*" (EUR 19153 EN) in 1999 (NRWG 1999). In 2002, the Task Force reconvened and released the "*Report on the Regulatory Experience of Risk-Informed In-Service Inspection of Nuclear Power Plant Components and Common View*" (EUR 21320 EN) in 2004, outlining consensus guidelines for regulatory bodies in the EU when evaluating RI-ISI applications (NRWG 2004; NUGENIA 2024).

5.13.2 Formalizing the RI-ISI Methodology

Two methods evolved based on the early work and recommended processes. The method adopted by, as it were, the Westinghouse Owners Group (WOG; now the PWR Owners Group, PWROG) was based on the use of probabilistic fracture mechanics (PFM) for evaluating pipe failure potentials in terms of probabilities of failure given a location-specific material degradation susceptibility. Traditional quantitative PSA results were used to establish levels of consequence and the incremental contribution of these consequences to plant risk. The procedure relied on the use of an expert panel to assess the numerical results and to perform the final ranking and determine inspection locations.

A procedure developed by EPRI was based on a combination of probabilistic and deterministic methods and built on the Swedish "*SKIFS Methodology*."[33] In the EPRI approach, the consequence evaluation relied on a series of tables, based on plant-specific PSA results, in which the consequence of a pipe segment failure is categorized as resulting in a high, medium, or low consequence. The failure evaluation was based on operating experience and industry data on pipe and weld failure rates associated with different material degradation mechanisms. The possible presence of these degradation mechanisms at locations within each segment was evaluated based on material and thermal-hydraulic conditions.

These two methods formed the basis of three ASME code cases that were reviewed and approved for inclusion in ASME BPVC Section XI. Pilot plant studies were undertaken to demonstrate the applicability of each code case as the basis for revised in-service inspection programs.

- Code Case N-560, "*Alternate Exam Requirements for Class 1, Category B-J Piping Welds (Section XI, Division 1, Appendix I),*" used the EPRI approach and was meant to serve as an alternative to the current B-J inspection location selection criteria. The pilot plants for this case were Vermont Yankee (BWR) and Arkansas Nuclear One Unit 1 (PWR).
- Code Case N-577, "*Risk-Informed Requirements for Class 1, 2 and 3 Piping, Method A, Section XI, Division I*" Appendix I-Risk-Informed Process Based on

[33] *The Swedish SKIFS methodology is a qualitative, risk-informed in-service inspection (RI-ISI) approach for nuclear power plants that uses a risk matrix to rank components into three inspection groups (A, B, and C). It assesses the qualitative probability of degradation (damage index) and the potential consequences of failure (consequence index) to determine inspection priorities.*

the PWROG Methodology, is based on results of the initial ASME study. The pilot plants were both Westinghouse design PWRs, Surry-1 and Millstone-2.
- Code Case N-578, "*Risk-Informed Requirements for Class 1, 2, and 3 Piping, Method B, Section XI, Division I*" Appendix I-Risk-Informed Process Based on EPRI Methodology, followed the EPRI procedure. Pilot plants included the CE-designed PWR ANO-2 and the General Electric Mark I BWR Fitzpatrick.

The USNRC approved the EPRI and PWROG methodologies in 1999 as valid alternatives to ASME Section XI. RI-ISI was subsequently implemented at the majority of US nuclear power plants. Outside the United States, varying regulatory positions existed, ranging from country-specific methodology development to direct adaptation of the EPRI or PWROG methodology.

5.13.3 EPRI RI-ISI Methodologies

EPRI began developing RI-ISI methodologies in the mid-1990s. In a technical collaboration with the Swedish regulatory authority (SKI 1994), EPRI's approach was built on SKIFS 1994:1. One of the results of this collaboration was a database on the US piping operating experience. SKI began an effort in 1995 to develop a piping failure and degradation database. The objective was to use this information in conjunction with data on Swedish plant piping systems to assist in improving in-service inspection activities at the Swedish nuclear power plants, including the requirement for risk-informed inspections. As part of this program, over 1500 pipe failures including leaks and ruptures were identified from nonproprietary sources. The resulting database and the analysis findings were discussed in the two technical reports, one published by SKI (Bush and Chockie 1996), and another published jointly by EPRI and SKI (Bush et al. 1998).[34]

In April 1999, EPRI submitted its topical report TR-112657, "*Revised Risk-Informed In-Service Inspection Procedure*" (EPRI 1999a, c), for review and approval by the staff of the USNRC. This version of EPRI's RI-ISI methodology was referred to as the base approach, which subsequently evolved into two variations that provided a user with some flexibility in program development and implementation. These variations are referred to as the "base RI-ISI methodology" and the "expanded or optimized approach," respectively.

[34]*During this period, the Swedish Nuclear Power Inspectorate ran two parallel, independent activities that dealt with piping operating experience and use of such data for PSA and RI-ISI. One of these activities resulted in SKI Report 96:20 by S.H. Bush and A. Chockie. The other activity, a 5-year R&D project, ultimately resulted in an international collaborative effort known as the OECD Pipe Failure Data Exchange Project (2002–2011).*

5.13.3.1 EPRI Base RI-ISI Methodology

The "EPRI base" approach, or "EPRI traditional" approach, does not require probabilistic piping reliability analyses to be performed. It is a qualitative approach to RI-ISI program development, which builds on PSA insights and results. The pipe failure potential evaluation step uses insights from operating experience reviews together with a detailed material degradation mechanism analysis. Table 5.23 shows

Table 5.23 Comparison of EPRI and SKIFS pipe damage indices

Damage/degradation mechanism criteria	
EPRI pipe failure potential	SKIFS 1994:1 Damage index (DI)
3—High FAC, vibration-fatigue, water hammer	*DI-1* • Thermal sleeves susceptible to crevice corrosion • FAC • Vibration-fatigue • Thermal fatigue of mixing tees with $\Delta T \cong 100$ °C • Welds including weld-HAZ susceptible to IGSCC or PWSCC • For a specified pipe segment, data on observed pipe failures is available (plant specific or fleet specific)
2—Medium Corrosion (MIC, pitting), erosion-cavitation, SCC, thermal fatigue	*DI-2* Service conditions (pressure, temperature, flow) that are such that a structural failure (e.g., large through-wall flow rate given a pipe flaw) is considered unlikely (e.g., moderate-energy piping systems)
1—Low No damage/degradation mechanism	*DI-3* No damage/degradation mechanism
3—High FAC, vibration-fatigue, water hammer	*DI-1* • Thermal sleeves susceptible to crevice corrosion • FAC • Vibration-fatigue • Thermal fatigue of mixing tees with $\Delta T \cong 100$ °C • Welds including weld-HAZ susceptible to IGSCC, PWSCC • For a specified pipe segment, data on observed pipe failures is available (plant specific or fleet specific)
2—Medium Corrosion (MIC, pitting), erosion-cavitation, SCC, thermal fatigue	*DI-2* Service conditions (pressure, temperature, flow) that are such that a structural failure (e.g., large through-wall flow rate given a pipe flaw) is considered unlikely (e.g., moderate-energy piping systems)
1—Low No damage/degradation mechanism	*DI-3* No damage/degradation mechanism

a comparison of the "Damage Index" of SKIFS 1994:1[35] with the "pipe failure potential" of the EPRI traditional approach.

A pipe failure potential evaluation was developed as an alternative to formal structural reliability analysis using probabilistic fracture mechanics or other methods. The degradation mechanisms in a pipe segment are identified by comparing actual piping design and operating conditions to a well-defined set of material and environmental attributes. Bounding pipe failure frequencies (1/reactor operating year) are used in lieu of structural reliability analysis, and the underlying technical basis is documented in EPRI TR-110157 (EPRI 1998a). It was noted that these failure frequencies were not intended to be applied in a strict manner.

The methodology was developed to be implemented on a system-by-system basis. In order to conduct and document the analysis, the piping systems are divided into segments based both on the pipe rupture potential and its consequences. While the analysis is conducted on a segment basis, it is for ease of use rather than being a technical component of the analyses. As such, differences in segment definition or segment boundary definition will have no impact on the final results for applications using the EPRI RI-ISI methodology. Each segment, which includes all the elements within the segment, is placed in the appropriate place on the risk characterization matrix. The failure potential category is determined on the basis of identified degradation mechanism. The consequence category is determined from the plant-specific PSA by calculating the conditional core damage probability (CCDP) and the conditional large early release probability (CLERP): high = CCDP $>10^{-4}$; medium = 10^{-6} < CCDP $<10^{-4}$; and low = CCDP $<10^{-6}$. For CLERP, the boundary values are one order of magnitude smaller.

The risk categories were combined into three risk regions for more robust and more efficient utilization. For risk category 1, 2, or 3, the minimum number of inspection elements in each category should be 25% of the total number of elements in each risk category (rounded up to the next higher whole number). For risk category 4 or 5, the number of inspection elements in each category should be 10% of the total number of elements in each risk category (rounded up to the higher whole number). Pressure testing and leak inspection requirements remain in effect regardless of risk category, i.e., risk category 1 through 7. The EPRI methodology is implemented in six steps:

1. Selection of RI-ISI program scope, that is, ASME III safety classes and systems to be included.
2. Failure mode and effect analysis (FMEA) of pipe segments to determine the consequences of a pipe failure (Table 5.24).
3. Risk characterization of pipe elements through a qualitative or quantitative assessment of pipe failure potential.
4. Selection of inspection locations and examination methods.

[35] *Swedish Nuclear Power Inspectorate (1994), Regulations and General Advice on Mechanical Devices in Certain Nuclear Installations, SKIFS 1994:1, Stockholm, Sweden.*

Table 5.24 Consequences of pipe failure

Impact group	Configuration	Consequence
Initiating event	Operating	A pressure boundary failure (PBF) occurs in an operating (pressurized) system resulting in an initiating event, e.g., loss-of-coolant accident or high-energy line break
Loss of mitigating ability	Standby	A PBF occurs in a standby system and does not result in an initiating event but degrades the mitigating capabilities of a system or train. After failure is discovered, the plant enters the allowed outage time (AOT) defined in the technical specifications
	Demand	A PBF occurs when system/train operation is required by an independent demand
Combination event	Operating	A PBF causes an initiating event with an additional loss of mitigating ability (in addition to the expected mitigating degradation due to the initiator)
Containment	Any	A PBF, in addition to above impacts, also affects containment performance

5. Evaluation of risk impacts of inspection program changes.
6. Final RI-ISI program definition.

Once an approved RI-ISI program has been implemented, the results of subsequent inspections, PSA updates, and continued monitoring of plant-specific and generic industry operating experience with piping systems will be employed to make any necessary adjustments to the inspection program. These adjustments are made as appropriate to ensure the risk management effectiveness of the inspection program, as with any effective and risk-informed process. There are several scope options available in the RI-ISI program. These options range from full scope assessments that include all ASME Class 1, 2, and 3 systems and other piping systems important to plant safety (ASME Code Case N-578-1), limited-scope applications for ASME Section XI Category B-J welds (similar metal butt welds) in Class 1 systems (ASME Code Case N-560), and application to one or more selected systems (ASME Code Case N-578-1). The N-578-1 expanded scope applications include all Class 1 and 2 systems.

5.13.3.2 Expanded EPRI RI-ISI Methodology

Lessons learned through application of the EPRI base methodology in pilot studies and through USNRC and third-party independent reviews provided opportunities to refine the methodology. These refinements were as follows:

- The role of the existing PSAs and the number and scope of the PSA calculations that are needed to support the consequence assessment were clarified as a result of insights gained through the pilot studies and their reviews.
- The technical basis for the EPRI RI-ISI approach to assessing failure potential was strengthened by correlating numerical estimates of failure rates and rupture frequencies to the presence of active degradation mechanisms.

- Through confirmatory research in the development and application of piping system reliability models and supporting efforts to estimate pipe failure rates and rupture frequencies, the basis for establishing that the RI-ISI process will yield reductions in core damage frequency (CDF) and large early release frequency (LERF) was strengthened. Databases and analytical tools were developed to perform quantitative risk impact assessments.
- A step was added to perform a risk impact assessment to ensure that the cumulative changes to the inspection program will bring about an acceptable impact on CDF and LERF.

The process of assigning pipe segments to the risk matrix and the process of selecting the inspection locations would be expected to result in a net reduction in risk. Any changes in risk would be indicated by potential changes in CDF or LERF. For segments in the high-risk categories that have their inspection priority increased due to RI-ISI, there should be a reduction in both leak and rupture frequencies associated with degradation mechanisms, since the flaws that could be identified during inspection could lead to the preemption of either a leak or a rupture associated with that flaw. While it is conceivable that some segments that are no longer inspected due to rearranging the priorities in a risk-informed process could have a finite probability of increased leakage, the risk impact of such leakage would be negligible. However, to ensure that this is the case, a final risk comparison is performed to ensure that the changes to the inspection program result in either a reduction in risk or at most negligible increases in risk. In most cases, this assessment can be accomplished via a simple qualitative assessment. However, there may be cases in which a quantitative assessment should be performed to verify that the impact of changes to the inspection program is either a negligible increase in CDF and LERF or an actual decrease in these risk measures.

Qualitative evaluation of risk impacts of changes to the inspection program first requires that the changes to the inspection program be delineated. This can be accomplished by listing, for each system, all segments and locations in each risk category. Then, for each risk category, it can be established whether the number of locations in the inspection program increased or decreased. The qualitative evaluation is guided by the following rules:

- For segments classified as low risk, any changes in inspection locations will have negligible risk impacts. This is based on very conservative bounding estimates of pipe rupture frequencies and conditional core damage probabilities for low-risk segments. These bounding estimates envelop the quantitative estimates of pipe rupture frequencies that cover a full range of systems, failure mechanisms, and pipe sizes.
- Some of the existing weld inspection locations are satisfied with external surface examinations. A comprehensive review of operating experience indicates that surface exams are not effective for most degradation mechanisms. Therefore, elimination of surface exams would have negligible risk impact even if located on an element classified as high or medium risk.

- The qualitative risk evaluation should carefully review changes in weld locations selected for any high-or medium-risk category. If each high-risk or medium-risk category has an increased number of locations selected for inspection, or if it has a comparable number of locations that are redirected to locations that are more likely to identify failure precursors based on the characteristics of the identified damage mechanisms, the risk impacts associated with an implementation of RI-ISI are a net risk reduction or, at worst, risk neutral. Keeping the locations fixed, in number and location, increasing the locations, or changing the locations to areas more likely to find precursors would have at most positive safety impacts.
- If there are significant reductions in the number of locations selected for any high- or medium-risk category that are not offset by quantitative or qualitative enhancements to the proposed RI-ISI program that would clearly offset the risk impacts of these reductions, then further quantitative evaluation is required to establish risk impact acceptability as described in the next section.

The effect of implementing an RI-ISI program is to alter the number and effectiveness of locations that are inspected in relation to the existing ASME Section XI-based program. Hence, there are two strategies: the Section XI-based program and the RI-ISI program. The impact of the strategies on a specific location could include adding locations to the inspection program, removing locations, or changing the effectiveness of the program. When an element is selected for inspection, the effectiveness of the program is expected to increase because of the knowledge gained from the RI-ISI evaluation to determine which damage mechanisms are most likely to be present at that location. Hence, when the strategy is switched from Section XI to RI-ISI, there are three possibilities for each location to change its rupture frequency:

- If the location was inspected in Section XI and retained in RI-ISI, the change in the pipe rupture frequency, if any, would be to reduce the pipe rupture frequency due to the capability to inspect for the most probable degradation mechanisms.
- If the location was not inspected in Section XI and added to the RI-ISI program, as frequently occurs in medium- or high-risk segments, the pipe rupture frequency will decrease.
- If the location was inspected in Section XI and not retained in the RI-ISI program, there may be increases in pipe rupture frequency.

An evaluation starts by considering qualitative criteria: the risk impacts are acceptable, unless there is a reduction in locations to be inspected in high-risk and medium-risk categories. A realistic quantitative analysis is performed for all elements within the RI-ISI scope, including the low-risk category elements. The numerical risk impact criteria assume that a full-plant-level RI-ISI program should strive to ensure that the cumulative risk impacts for the full plant are maintained at levels less than 1.0×10^{-6} per year for core damage frequency (CDF) and 1.0×10^{-7} per year for large early release frequency (LERF).

Two approaches have been successfully applied, one based on a simple approach with reasonable assumptions, and another, more detailed approach that explicitly models the interactions between degradation mechanisms and inspection and repair process. This model is expressed in terms of the following equation:

$$F_{Aj} = F_{0j} I_{Aj} \tag{5.16}$$

F_{Aj}	Frequency of SSF at location *j*, subjected to inspection strategy *A*
F_{0j}	Frequency of SSF at location *j*, subjected to no inspection
I_{Aj}	Inspection effectiveness factor of reduction in pipe rupture frequency at location j due to implementation of inspection strategy A. The effectiveness factor can range from 1.0 to 0.0. An effective inspection would have an effectiveness factor close to 0, while an ineffective inspection or omitting the location from the inspection program will have an effectiveness factor close to 1

The change in the risk of core damage at location *j* that is impacted by the changes in the RI-ISI program can be estimated as

$$\Delta CDF_j A = \left(F_{rj} - F_{ej}\right) \times CDF_j = \left(I_{rj} - I_{ej}\right) \times F_{0j} \times CCDP_j \tag{5.17}$$

where the subscript "*rj*" refers to the risk-informed inspection at location *j*, and the subscript "*ej*" refers to the existing inspection program at location *j*. The term $CCDP_j$ is the conditional core damage probability from a pipe rupture at location j, which is independent of the inspection strategy. CCDP values may be set using the upper limit of the consequence range (e.g., assign medium-consequence locations 1.0×10^{-4}) or may be estimated from a plant-specific PSA.

Quantifying the above equation requires the capability to predict the impact that inspections make on pipe rupture frequency, as well as the baseline rupture frequency without crediting inspections. A Markov model was used to model the interactions between degradation mechanisms, pipe inspection, and leak detection processes that cause and mitigate pipe cracks, leaks, and ruptures. Markov models generate a time-dependent hazard rate which measures the time-dependent growth in pipe rupture frequencies due to degradation mechanisms. The Markov process models the role of the inspection and leak detection programs in reducing the likelihood of pipe ruptures.

A supporting pipe failure metrics database was developed for use with the Markov model. The database consisted of pipe failure precursor rate and rupture frequency look-up tables that were organized by plant type (General Electric BWR, Babcock & Wilcox PWR, Combustion Engineering PWR, and Westinghouse PWR), plant system, and degradation mechanism (EPRI 1998b; 1999b). A Bayesian analysis was performed to support a quantification of the uncertainties in pipe failure rates and rupture frequencies.

5.13.3.3 EPRI "Streamlined" RI-ISI Methodology

ASME Code Case N-716 was approved on April 19, 2006, as an alternative to ASME Section XI Nonmandatory Appendix R, which was first published in

Section XI 2005 addenda (EPRI 2009; ASME 2019a, b). The current edition of this code case was approved on November 7, 2016; it is referred to as the "2017 Edition" or N-716-2. A key motivation behind the code case was the perceived and real complexities and limitations in the methods for estimating pipe failure probabilities. ASME Section XI instead decided to base the RI-ISI process on conditional consequence given a failure, which removed the concern with large uncertainties regarding pipe failure frequency estimates.

Referred to as a "streamlined approach," it differs from the traditional approaches in two respects. First, the consequence assessment is not required, since it is replaced with a predetermined set of high-safety-significant (HSS) locations (e.g., reactor coolant system, break exclusion area) and a plant-specific assessment of the impact of pressure boundary failure using the plant-specific PSA directly. The second departure is that partial scope application, which is allowed by previous RI-ISI approaches, is not allowed by N-716. According to the process, the inspection selection should be equal to 10% of the high-safety-significant (HSS) welds, plus augmented programs for flow-accelerated corrosion, localized corrosion (e.g., MIC), and IGSCC in BWRs.

HSS welds are selected as follows: (1) a minimum of 25% of the population identified as susceptible to each degradation mechanism and degradation mechanism combination; (2) for the RCPB, at least two-thirds of the examinations shall be located between the first isolation valve (i.e., isolation valve closest to the RPV) and the reactor pressure vessel; (3) a minimum of 10% of the welds in that portion of the RCPB that lies outside containment (e.g., portions of the main feedwater system in BWRs) shall be selected; and (4) a minimum of 10% of the welds within the break exclusion region (e.g., high-energy piping penetrating containment) shall be selected.

The change in risk analysis uses a bounding type of pipe failure frequencies that are shown in Table 5.25. The equations below are used to estimate the change in risk that results from a new inspection volume selection. One estimate is to be based on core damage frequency (CDF) and one on large early release frequency (LERF). The parameters of the three equations are defined in Table 5.26.

$$\Delta R_{CDF} = \sum_j \left(I_{rj} - I_{ej}\right) \times PF_j \times CCDP_j \tag{5.18}$$

$$\Delta R_{CDF} = \sum_j \left(POD_{ej} - POD_{rj}\right) \times PF_j \times CCDP_j \tag{5.19}$$

$$\Delta R_{CDF} = \left[\left(POD_e \times N_{efc} - POD_{rfc} \times N_{rfc}\right)\right] \times PF_f \times CCDP_c \tag{5.20}$$

5.13.4 PWRWOG RI-ISI Methodology

In the PWROG methodology, the piping in the scope of application is divided into segments. Section 3.3 (*"Probabilistic Risk Assessment,"* pp. 14–16) of WCAP-

Table 5.25 Degradation mechanism categories and bounding pipe failure frequencies

Failure potential	Conditions	Degradation category	Degradation mechanism	Pipe failure frequency (1/weld and ROY)
High [a]	Degradation mechanism likely to cause a large pipe break	Large break	Flow-accelerated corrosion	2.0×10^{-6}
Medium	Degradation mechanisms likely to cause a small leak	Small leak	Thermal fatigue, erosion-cavitation, corrosion, stress corrosion cracking	2.0×10^{-7}
Low	None	None	None	1.0×10^{-8}

[a] Segments having degradation mechanism listed in the small leak category shall be upgraded to the high failure potential large/break category if the pipe segments also have the potential for water hammer loads

Table 5.26 The parameters of Code Case N-716-1 change-in-risk estimation

Parameter	Definition
ΔR_{CDF}	Change in CDF due to replacing the prior deterministic ISI program with the ISI program developed in accordance with N-716
$\sum_j$	Summation of locations selected for examination
I_{rj}	Factor of reduction in pipe rupture frequency at location j associated with the ISI program developed by N-716
I_{ej}	Factor of reduction in pipe rupture frequency at location j associated with the prior deterministic ISI program
PF_j	Piping failure frequency at location j without examination
$CCDP_j$	Conditional core damage probability at location j
POD_j	$1 - I_j$; probability of detection
POD_{ej}	Probability of detection at location j associated with the prior deterministic ISI program
POD_{rj}	Probability of detection at location j associated with the ISI program developed in accordance with code case N-716
N_{efc}	Number of examination locations in the consequence f and failure frequency c categories associated with the prior deterministic ISI program
PF_f	Probability of detection in the ISI program developed by this case (could be specific for certain degradation mechanism)
N_{rfc}	Number of examination locations in the consequence f and failure frequency c categories associated with ISI program developed in accordance with code case N-716

14572 Revision 1 provides a definition for piping segments based on the following considerations (Balkey et al. 1999; Bishop 1999; Perdue 1998):

– Piping failures that lead to the same consequence determined from the plant-specific PSA and other considerations (e.g., loss of a residual heat removal (RHR)

train, loss of a refueling water storage tank (RWST), inside or outside containment consequences) where flow splits or joins.
- Piping to a point where a pipe break could be isolated. (This includes check valves and motor-operated or air-operated valves. No credit is generally given for manual valves; however, situations may occur where manual valves can be used to isolate a failure by plant operators, and, in these cases, the decision for crediting manual valves is made by the plant expert panel and documented as such.)
- Pipe size changes.

In defining pipe segments, the possibility of check valves and other isolation valves failing to close is not considered; that is, proper operation of the valves is assumed when defining segment boundaries. The treatment of automatic isolation valves is clarified as follows:

> *Automatic isolation valves are assumed to close if the pipe failure in question would create a signal for the valves to close. Containment isolation valves should be carefully considered for segments which contain the containment penetrations. If the segment consequences are significantly different, assuming an automatic and/or containment isolation valve failure, then the piping segment definition should be reviewed and, if necessary, the piping segment should be further combined or subdivided such that the failure of the valve, under pipe failure conditions, would be considered in conjunction with the change in consequences.*

Failure probabilities are developed using the probabilistic fracture mechanics code (SRRA) for each of the consequences on the segments. The results of the PSA model and estimated failure probabilities are combined in the risk evaluation to develop risk metrics both with and without uncertainty. The risk reduction worth (RRW) is the main quantitative importance measure.[36] The probabilistic and additional deterministic data are presented to an expert panel to determine the final safety significance of the segment. Based on the safety significance and failure importance, the segments are placed in the structural element selection matrix. All high-safety-significant (HSS) elements affected by an active degradation mechanism are inspected. For the remaining HSS elements, a statistical model (Purdue model) is used to determine the minimum number of inspection locations. Pressure leakage testing requirements remain in effect regardless of the region in the structural element selection matrix. The acceptability of implementing the RI-ISI program is demonstrated by calculating the change in risk for both CDF and LERF and by conducting a review for defense-in-depth philosophy.

Piping failure potential is determined based on failure probability estimates from the SRRA software program (WCAP-14572, Supplement 1). An expert panel finalizes the appropriate input data required by the SRRA code. This process is described in Sections 3.3 and 3.4 and Attachment 3 of the program submittal. Details

[36] *Measures the amount that the total risk would decrease if a basic event's failure probability were 0* (i.e., *never fails). Calculated as either ratio or difference between the value of all cut sets representing the total risk (baseline risk) and the value of the total risk with the failure probability for the basic event of interest set to 0.*

regarding the methodology, process, and rationale were provided by the licensee in an enclosure (Engineering Transmittal ET No. MAT-97-0014, *Estimated Failure Probabilities for Risk-Based ISI Surry Unit 1, Rev 0*) of a submittal in 1998, which contains the guidance used by the engineering subpanel in determining the various inputs to the SRRA program, and Enclosure 2, which contains the SRRA data sheets for the failure estimates. Further details regarding the failure probability assessment were also provided by the licensee in its August 13, 1998, submittal.[37] An example of the regulatory requests for additional information (RAI) is as follows:

- **RAI:** *Pipe failure probabilities are calculated for various pipe segments such as small leaks and large leaks. Clarify if these two leak categories correspond, respectively, to a through-wall flaw and system-disabling leak. If so, were full break probabilities calculated for segments subject to pipe whip?*
- **Response to RAI:** *Small leak does correspond to a through-wall flaw, and large leak does correspond to system-disabling leak. Full break probability was calculated if required for pipe whip consideration. In most cases, pipe whip was determined to be an insignificant addition to the total GDF of a segment due to the comparatively small failure probability when combined with the small leak and large leak terms of a segment (consequence remaining the same). The RC system considered the effects of each possible LOCA event for a given pipe size. The large loop piping could probabilistically have a small break LOCA, or a medium break LOCA, or a large break LOCA during plant life. Failure probabilities were calculated based upon the LOCA size (5001 gpm for large LOCA, 1501 gpm for medium LOCA, and-100 gpm for small LOCA).*

The SRRA program was benchmarked on the basis of comparisons of calculated piping failure probabilities with those results using the pc-PRAISE program. A range of input parameters were used in benchmarking the small leak, large leak, and full break probabilities. Westinghouse conducted sensitivity studies on parameter uncertainties and found that the range of uncertainties of the SRRA-calculated piping failure probabilities was about two to five orders of magnitude around the best estimate value. In the application to the Surry RI-ISI pilot program, plant-specific Surry data were used as input to the SRRA. Although the use of the SRRA code was found acceptable for this plant-specific application, the use of the SRRA code requires analyst's judgment, and its methodological application is addressed in the NRC review of WCAP-14572. In the application to the Ringhals-2/3/4 RI-ISI program, the SRRA code underwent a detailed and protracted review by the Swedish Nuclear Power Inspectorate (Brickstad 2006a). This review raised concerns regarding cliff-edge effects when applying the code to model flow-accelerated corrosion (Brickstad 2006b).

---

[37] *Forwards revised responses to questions provided in RAI on risk-informed (RI) ISI pilot program. Updated RI ISI plan, encl. w/comparison of differences between current ASME Sect. XI inspections and proposed RI ISIs. NRC-ADAMS Accession No. ML18151A649.*

5.13.5 ASME Section XI Nonmandatory Appendix R

Nonmandatory Appendix R, Risk-Informed Inspection Requirements for Piping, was first published in Section XI 2005 Edition. Appendix R incorporates the risk-informed selection and examination provisions of ASME Code Cases N-560,[38] N-577,[39] and N-578[40] into Section XI, Division 1. Regulatory approval was needed for implementing Appendix R. Article R-2000, "*Examination and Inspection*," explains the duties of the inspector as follows:

> *Duties of the Inspector shall be in accordance with IWA-2110 except that the Owner's augmented programs, such as the Intergranular Stress Corrosion Cracking (IGSCC) program, the Microbiologically Influenced Corrosion (MIC) program, or the Flow Accelerated Corrosion (FAC) program, referenced in Table R-2500-1, Item Nos. R1.16, R1.17, or R1.18, are exempt from these requirements, unless an Owner chooses to select and credit structural elements for examination by one of the risk-informed selection processes defined in Supplement 1 or 2. If selected and credited structural elements are examined for one or more of the degradation mechanisms covered under these Owner's augmented programs, the requirements of IWA-2110 apply to the selection and examination of the credited structural elements*

Nonmandatory Appendix R provided two separate methodologies, "Method A" and "Method B," for performing risk-informed applications in Supplements 1 and 2. Supplement 1 incorporates the criteria of Code Case N577, while Supplement 2 addresses the methodologies that formed the bases for Code Cases N-560 and N-578. Both supplements have been updated by ASME Section XI on an ongoing basis to incorporate lessons learned from risk-informed applications.

5.13.6 European RI-ISI Guidelines and Recommended Practices

The European Network for Inspection and Qualification (ENIQ) was set up in 1992 in recognition of the importance of qualification of NDE inspection procedures. In

---

[38] *Code Case N-560, "Alternative Exam Requirements for Class 1, Category B-J Piping Welds (Section XI, Division 1, Appendix I)," used the EPRI approach and was meant to serve as an alternative to the current B-J inspection location selection criteria. The pilot plants for this case were Vermont Yankee (BWR) and Arkansas Nuclear One Unit 1 (PWR). The code case was approved on August 9, 1996.*

[39] *Code Case N-577, "Risk-Informed Requirements for Class 1, 2 and 3 Piping, Method A, Section XI, Division l" Appendix I-Risk-Informed Process Based on the PWROG Methodology, is based on results of the initial ASME study. The pilot plants were both Westinghouse design PWRs, Surry-1 and Millstone-2. The code case was approved on March 28, 2000.*

[40] *Code Case N-578, "Risk-Informed Requirements for Class 1, 2, and 3 Piping, Method B, Section XI, Division l" Appendix I-Risk-Informed Process Based on EPRI Methodology, followed the EPRI procedure. Pilot plants included the CE-designed PWR ANO-2 and the General Electric Mark I BWR Fitzpatrick. The code case was approved on March 28, 2000.*

2012, ENIQ joined the European R&D association on Gen II & III reactors, NUGENIA, the Nuclear Generation II and III Association, making ENIQ its eighth technical area. ENIQ Task Group for Risk (TGR) published the "*European Framework Document on RI-ISI*," EUR 21581 EN, in 2005 and has also developed a series of recommended practices (RPs). The RPs are consensus documents containing harmonized principles, methodologies, approaches, and procedures (ENIQ 2005; NRWG 2004; OECD/NEA 2004):

- Recommended Practice 9 (RP-9): Verification and Validation of Structural Reliability Models and Associated Software to be Used in Risk-Informed In-Service Inspection Programs.
- Recommended Practice 11 (RP-11): Guidance on Expert Panels in RI-ISI.

Like the EPRI and PWROG RI-ISI methodologies, the European RI-ISI framework recognizes that an assessment of pipe failure potential consists of a qualitative element, the degradation mechanism assessment, and a quantitative element that calculates an appropriate set of probabilistic failure metrics. According to the framework, ideally, the probability of failure should be calculated by using probabilistic fracture mechanics (PFM). It furthermore recognizes that PFM models do not exist for all the potential degradation mechanisms that affect nuclear power plants. For degradation mechanisms that do have a viable PFM model, there is only a limited acceptance that these estimates can be seen as representing some form of true or absolute value. RP-9 summarizes the verification and validation requirements that a PFM model and associated software should satisfy to be suitable for RI-ISI applications.

The European RI-ISI framework document recommends that an expert panel be used to review the proposed sites. This panel should review the information, analysis, and insights that have been used to identify the safety-significant sites. It is also important to investigate alternative possibilities for mitigation against the risk. It is, therefore, necessary to look at the nature of the risk associated with each site. It may be possible to identify ways other than inspection to address the risk. RP-11 gives guidance on the composition and responsibilities of an expert panel. It also gives guidance on planning and preparation, conduct, and documentation of the expert panel.

The Swedish regulatory authority implemented a qualitative RI-ISI requirement through regulation SKIFS 1994:1, which was subsequently incorporated into regulatory guide SSMFS 2008:13. The methodology for establishing an ISI program must be submitted for regulatory review and approval. This approach utilizes a deterministic, risk-informed process that employs both damage and consequence indices to prioritize various systems. Piping segments are categorized into three control groups based on relative risks associated with nuclear fuel damage, release of radioactive materials, accidental chain reactions, and safety-level deficiencies. Typically, risks are evaluated as potential outcomes stemming from damage to mechanical components or systems.

The risk-informed methodology to establish the ISI program became a regulatory requirement in Finland in 2004. Loviisa nuclear power plant has an RI-ISI program for piping systems based on a PSA. Each piping system is divided into segments for degradation classification, if the conditional core damage probability (CCDP) is greater than 10^{-6}. Based on the RI-ISI analysis, many segments other than nuclear safety Class 1 or 2 systems have been included in the ISI program.

5.13.7 Periodic RI-ISI Program Reviews and Major Updating

An RI-ISI program should be updated minimally for the periods that coincide with the plant ISI inspection cycle per ASME Section XI requirements. These updates should be performed more frequently if dictated by any changes to the technical basis for the RI-ISI program. As changes to plant design are implemented, changes to the inputs associated with the RI-ISI program segment definition and element selections may occur. It is important to address these changes to the inputs used in any assessment that may affect resultant pipe failure potentials used to support the RI-ISI segment definition and element selection. In addition, plant design changes could result in significant changes to a plant's PSA results for CDF or LERF, which, in turn, could result in a change as a consequence of failure for system piping segments. This could impact the risk ranking if the change in CDF or LERF were large and would impact risk evaluation.

The major RI-ISI program updates should be conducted once per ASME Section XI inspection interval. The periodic reviews are qualitative evaluations to determine whether any changes to the RI-ISI evaluation need to be made and to justify the decision to defer any minor changes that are identified to the next major update. A flowchart for performing periodic reviews and major RI-ISI updates is shown in Fig. 5.12.

Major RI-ISI program updates are accomplished by essentially updating each step of the RI-ISI methodology from beginning to end to account for any changes to the inputs of the RI-ISI evaluation. These updates potentially impact the scope of the piping systems that are included, definition of degradation mechanism, consequence and risk segments, assignment of pipe elements to each part of the RI-ISI risk matrix, element selection, risk impact evaluation, and associated RI-ISI program documentation. RI-ISI program updates are performed when the need to do so is indicated in the results of the periodic reviews of the RI-ISI programs, and no less frequently than the 10-year ISI program cycle. Table 5.27 lists the steps of the RI-ISI evaluation and the factors that influence each step that are subject to change and need for reevaluation.

5.14 RIM Program Development and Qualification

All piping in a nuclear power plant is subjected to inspection and surveillance techniques and processes to ensure a high level of pressure boundary integrity and to prevent significant structural failures as well as to prevent through-wall leaks

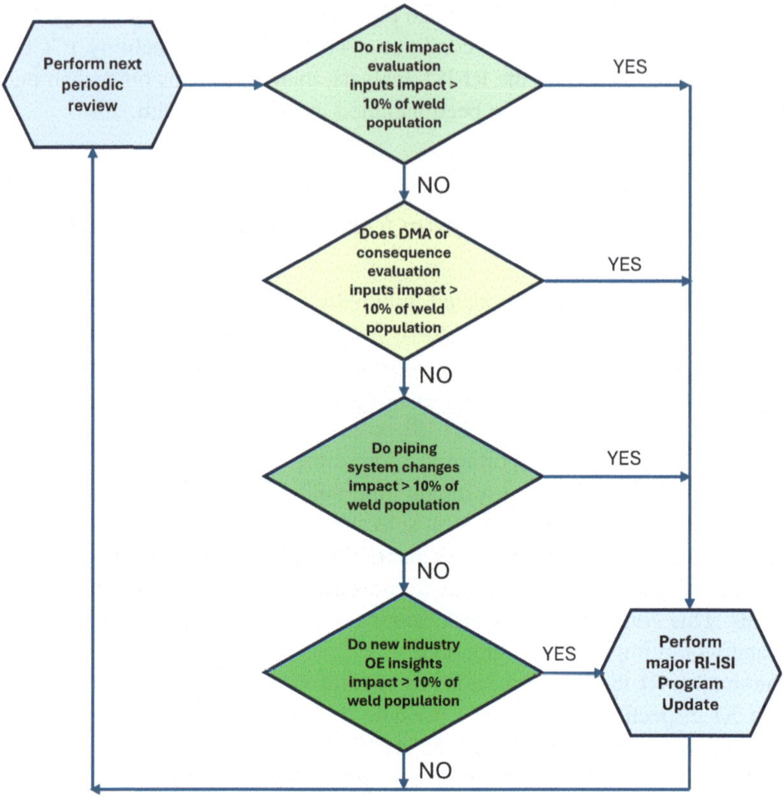

Fig. 5.12 RI-ISI update decision-making process

(minor to major). Reliability and integrity management (RIM) encompasses the following processes:

- Prefabrication inspection of pipes and fittings.
- Pre-service inspection (PSI) of installed piping systems: PSI includes visual and nondestructive examinations and pressure testing.
- In-service inspection, including visual examinations and nondestructive examinations using ultrasonic testing technologies, radiographic testing, liquid penetrant testing, and magnetic particle testing.
- Daily and weekly walk-down inspections of piping systems.
- Leak detection.
- Implementation of monitoring and diagnostic systems for thermal fatigue and high-cycle vibratory fatigue of primary pressure boundary piping.
- Application of pipe stress improvement process such as peening, induction heat stress improvement, and full structural weld overlays.

Division 2 of ASME XI provides the requirements for the development of a reliability and integrity management (RIM) program for advanced nuclear reactor

Table 5.27 The RI-ISI program update steps

Step	Activity	Element subject to change	Expected magnitude and frequency of change
1	RI-ISI scope	Enhancement of scope will require a separate evaluation	Could occur especially if RI-ISI expanded to include additional inspections
2A	Assess consequences	A PSA update impacts the consequence evaluation; these in turn are driven by changes to design, operation, and equipment performance	PSAs are routinely updated; most changes are not expected to impact RI-ISI program; only significant changes in CDF and LERF are likely to have an impact
2B	Consequence segments	Some design changes could lead to need to redefine pipe segments	This is expected to be very rare
2C	Assess degradation mechanisms (DMs)	Error may be identified in the inputs to the DM assessment or ISI may reveal presence of a new or different DM	This is expected to be rather infrequent to rare
2D	DM segments	Some design changes could lead to need to redefine pipe segments	This is expected to be very rare
3	Risk characterization	Any change in items 2A through 2D could impact risk characterization	Determined by frequency of changes driving consequence or DM assessment
4	Element selection	Any change in item 3 could impact element selection or ISI program changes to optimize cost of ISI or to reflect information obtained in inspections	Minor changes in element selection can be made routinely as long as RI-ISI criteria are met. Major changes require program updates
5	Risk impact evaluation	Any changes in the above items, ISI program, service experience insights, updates to failure rates, and rupture frequencies	Determined by frequency of updates of inputs to risk impact evaluation or at least every RI-ISI program update every 10 years
6	Long-term monitoring	Operating experience is monitored for trends in failure occurrences, new degradation mechanisms	Plant-specific service experience should be tracked continuously

designs (ASME 2023b). The RIM program addresses the entire life cycle, and it requires a combination of monitoring, nondestructive examination, tests, operation, and maintenance requirements that ensure that each structure, system, and component (SSC) meets plant risk and reliability goals that are selected for the RIM program. The RIM process consists of the following steps:

1. Program scope definition.
2. Degradation mechanism assessment (DMA) in the absence of operating experience applicable to a specific advanced reactor design.
3. Plant and SSC reliability target allocations originating from a plant-specific advanced reactor PSA. A ranking for risk contribution, known as a reliability target value, is assigned to each SSC (Kurisaka et al. 2011).
4. Identification and evaluation of RIM strategies.

5. Evaluation of uncertainties.
6. RIM program implementation.
7. Performance monitoring and RIM program updates to account for new operating experience and piping design modifications.

5.14.1 Historical Perspectives

The work behind Division 2 began in 2002. In the period 2002–2008, a methodology for RIM program development and implementation effort originated from South Africa's Pebble Bed Modular Reactor high-temperature project undertaken between 1994 and 2010 (Fleming et al. 2008; Fleming and Smit 2008). The insights from the PBMR work became the basis for a rewrite of ASME XI, Division 2, to apply to advanced non-light water reactor designs. By 2010, the Division 2 rewrite was near completion when the Republic of South Africa withdrew the funding for the PBMR project.[41]

The ASME and the American Nuclear Society (ANS) formed a Joint Committee on Nuclear Risk Management (JCNRM) in 2010. The JCNRM is responsible for several standard projects including ASME/ANS RA-S-1.4-2021, "Probabilistic Risk Assessment Standard for Advanced Non-LWR Nuclear Power Plants" and ASME/ANS RA-S 1.5, "Advanced Light Water Reactor PRA Standard." In 2012, ASME Section XI redirected the Special Working Group on Reliability and Integrity Management (RIM) Program (BPV XI) to rewrite Division 2 to address the (SMR/IPWR) designs. Section XI also requested that the JCNRM reclassify the non-light water reactor (LWR) standard as "technology neutral," for application to new SMR/IPWR designs including water-cooled designs. In August 2018, the Special Working Group RIM was elevated to become a standing ASME Subgroup concurrent with the anticipated publication of RIM. Division 2 (RIM) includes technology-specific appendices intended to account for different reactor designs for ISI parameters. The 2025 edition of Division 2 has the title "*BPVC.XI.2 - BPVC Section XI-Rules for Inservice Inspection of Nuclear Reactor Facility Components, Division 2, Requirements for Reliability and Integrity Management (RIM) Programs for Nuclear Reactor Facilities*"; it is reactor technology neutral.[42]

The Subgroup on "Reliability and Integrity Management Program" was formed to develop, review, and maintain, for publication in Section XI of the Boiler and Pressure Vessel Code, requirements for protecting the integrity of structures, systems, and components (SSCs) that affect reliability. It is applicable to all advanced

[41] *The Pebble Bed Modular Reactor (PBMR) is a high-temperature gas-cooled reactor based on the evolutionary design of the German AVR, THTR, and HTR-Modul designs. Development of the PBMR stopped in 2010. More details are available from* https://aris.iaea.org/sites/GCR.html

[42] https://www.asme.org/codes-standards/find-codes-standards/bpvc-section-xi-div-2-requirements-for-reliability-and-integrity-management-programs/2025/print-book

design reactor technologies. RIM establishes requirements to determine credible degradation mechanisms that could lead to the failure of SSC and imposes monitoring and NDE criteria to detect the onset of those degradation mechanisms. It is a "technology-neutral standard" that may be applied to any reactor technology, including light water-cooled, gas-cooled, liquid metal-cooled, as well as other advanced nuclear power plant designs. The USNRC published the Regulatory Guide 1.246 in October 2022 to provide an approach to RIM program development and implementation that is acceptable to the regulator (USNRC 2022). Similarly, in 2025, EPRI developed the document titled "*Initial Probabilistic Risk Assessment (PRA) Technical Guidance for ASME Section XI Reliability and Integrity Management Programs.*"[43]

5.14.2 Methodology for Establishing Reliability Targets

A reliability target is defined as a performance goal established for the probability that a passive component will complete its specified function in order to achieve plant-level risk and reliability goals. To enable such an approach, specification of integrity target reliability levels is required in order to benchmark the safety level of an existing piping system. The probability of failure per pipe segment or unit length is quantified and then checked against an integrity permissible probability of failure or integrity target reliability. This check against a specified reliability target allows for determining whether a segment of pipe is safe at current operating conditions while considering identified uncertainties. However, the main challenge around reliability targets is how to select such targets.

Regulatory Guide 1.246, "*Acceptability of ASME Code, Section XI, Division 2, Requirements for Reliability and Integrity Management (RIM) Programs for Non-Light Water Reactor*," describes an approach that is acceptable to the staff of the regulator for the development and implementation of a pre-service inspection (PSI) and in-service inspection (ISI) program for non-light water reactors (non-LWRs). It endorses, with conditions, the 2019 edition of the ASME Boiler and Pressure Vessel Code (ASME Code), Section XI, "Rules for In-service Inspection of Nuclear Power Plant Components," Division 2, "Requirements for Reliability and Integrity Management (RIM) Programs for Nuclear Power Plants." Among the conditions are:

- Documenting the methodology for establishing the reliability targets.
- Documenting the methodology for determining that the reliability targets will be satisfied by the RIM strategies.
- Documenting the reliability targets for each SSC that is in the program.

[43] *The EPRI report is available for funding members only.*

– Documenting the results of the degradation assessments that support the RIM program development.

An application of an RIM process to the Pebble Bed Modular Reactor (a high-temperature reactor design) defined the following reliability targets:

– Plant-level reliability target no. 1: The frequency of events and event sequences involving loss of reactor pressure vessel (RPV) structural integrity in the control of core heat removal and core heat generation shall be less than 10^{-8} per reactor year.
– Plant-level reliability target no. 2: The frequency of events and event sequences involving main power system depressurization due to failures of helium pressure boundary (HPB) SSCs with equivalent break size greater than 10 mm shall be less than 10^{-2} per reactor year.
– Plant-level reliability target no. 3: The frequency of events and event sequences involving main power system (MPS) depressurization due to failures of HPB components in the power conversion unit (PCU) with equivalent break size greater than 230 mm shall be less than 10^{-6} per reactor year.
– Plant-level reliability target no. 4: The frequency of events and event sequences involving MPS depressurization due to failures of helium pressure boundary (HPB) components connected to the top of the RPV and communicating with the core cavity with equivalent break size greater than 50 mm shall be less than 10^{-6} per reactor year.
– Plant-level reliability target no. 5: The frequency of events and event sequences involving significant flooding in the reactor cavity from the reactor cavity cooling system (RCCS) involving either long outages or loss of plant investment shall be less than 10^{-5} per reactor year.

References

Anderson MT, Cumblidge SE, Crawford SL (2008) Analysis of ultrasonic data on piping cracks at Ignalina nuclear power plant before and after applying a mechanical stress improvement process, PNNL-17367. Pacific Northwest National Laboratory, Richland, WA

Ando Y, Yagawa G, Hayase Y (1982) Evaluation of induction heating stress improvement (IHSI) treatment applied to nuclear primary piping. Int J Pressure Vessel Piping 10(5):399–406

Ansari NA, Patil S, Ghosh B, Chattopadhyay J, Bandyopadhyay SK, Gupta SK, Kushwaha HS (2000) Evaluation of crack opening area and leak rate in various PHT piping for LBB analysis of Indian PHWRs, BARC/2000/E/034. Bhabha Atomic Research Centre, Mumbai

Anselmi T, Lawrence S, Vetter D (2023) Reliability and integrity management scoping study, TLR-RES/DE/REB-2023-07. USNRC, Washington, DC

ANSI/ANS (2022) Safety and pressure integrity classification criteria for light water reactors, ANSI/ANS-58.14-2011 (R2022). American National Standards Institute, Washington, DC

AQSIC (2012a) Inspection and quarantine of the PRC (AQSIC), safety assessment for electrofusion joint of polyethylene pipes containing defects. GB/T 29460-2012. General Administration of Quality Supervision, Beijing

AQSIC (2012b) Ultrasonic testing for electrofusion joint of polyethylene pipe. In: GB/T 29461–2012. General Administration of Quality Supervision, Beijing
ASME (1968) Draft USA standard: code for Inservice inspection of nuclear reactor coolant systems. ASME, New York, NY
ASME (2009) Code case N-770-1, alternative examination requirements and acceptance standards for class 1 PWR piping and vessel nozzle butt welds fabricated with UNS N06082 or UNS W86182 weld filler material with or without the application of listed mitigation activities, section XI, division 1. ASME, New York, NY
ASME (2010) Use of polyethylene (PE) class 3 plastic pipe section III, division I and section XI supplement 6, code case N-755–1 (inactive). ASME, New York, NY
ASME (2013) Optimized structural dissimilar metal weld overlay for mitigation of PWR class 1 items, section XI, division 1, code case N-754-1. ASME, New York, NY
ASME (2015) Full structural dissimilar metal weld overlay for repair or mitigation of class 1, 2, and 3 items section XI, division 1, code case N-740-2. ASME, New York, NY
ASME (2018) Welded and seamless wrought steel pipe, ASME B.36.10M-2018. ASME, New York, NY
ASME (2019a) Risk-informed categorization and treatment for repair/replacement activities in class 2 and 3 systems, code case N-752. ASME, New York, NY
ASME (2019b) Alternative piping classification and examination requirements, section XI, division 1, code case N-716. ASME, New York
ASME (2022) Welded and seamless wrought stainless steel pipe, ASME B36.19–2022. ASME, New York, NY
ASME (2023a) BPVC section III-rules for constructions of nuclear facility components-subsection NCA-general requirements for division 1 and division 2, ISBN: 978–0-791875674. ASME, New York, NY
ASME (2023b) BPVC section XI-rules for Inservice inspection of nuclear reactor facility components, division 2, requirements for reliability and integrity management (RIM) programs for nuclear reactor facilities, BPVC.XI.2–2023. ASME, New York, NY. ISBN: 97800-791875889
ASME (2024) Process piping code. ASME code for pressure piping, B31.3. ASME, New York, NY. ISBN: 978-0-791876725
ASME (2025) BPVC.III.NCA—BPVC section III-rules for constructions of nuclear facility components—subsection NCA-general requirements for division 1 and division 2. In: BPVC. III.NCA-2025. ISBN: 978-0-791877074. ASME, New York, NY
ASN (2002) Assemblages "socket welding" du circuit Primaire principal des REP, Décision DGSNR-BCCN/OT/VF No 020406. Autorité de Sûreté Nucléaire, Paris. October 10, 2002
ASTM International (2013) Standard specification for high-density polyethylene (PE) line pipe, ASTM F2619/F2619M-13. ASTM International, West Conshohocken, PA
ASTM International (2024) Standard practice for heat fusion joining of polyethylene pipe and fittings, ASTM F2620–20ae2. ASTM International, West Conshohocken, PA
Balkey KR et al (1992) Risk-based inspection—development of guidelines, volume 2—part 1: light water reactor (LWR) nuclear power plant components, CRTD-Vol. 20–2. American Society of Mechanical Engineers, New York, NY
Balkey KR et al (1999) Westinghouse owners group application of risk-informed methods to piping Inservice inspection topical report, WCAP-14572, revision 1-NP-A. Westinghouse Energy Systems, Pittsburgh, PA
Barnes RW, Cooper GD (1995) Failure rates in piping manufactured to different standards, INFO-0607. Atomic Energy Control Board, Ottawa, ON
Beckjord ES (1984) Evaluation of the mechanical stress improvement process, research information letter no. 149. Washington, DC, USNRC
Benedict RP (1980) Fundamentals of pipe flow, ISBN 0-471-03375-8. John Wiley & Sons, New York, NY
Benson M (2019) Criteria for postulating pipe rupture locations: background and history technical, letter report. USNRC, Washington, DC

Bertossa DC (1980) Induction heating stress improvement qualification. In: Proceedings of seminar on countermeasures for pipe cracking in BWRs, EPRI-WS--79-174-Vol. 1. EPRI, Palo Alto, CA
Bishop BA (1999) Westinghouse structural reliability and risk assessment (SRRA) model for piping risk-informed Inservice inspection, WCAP-14572, revision 1-NP-A supplement 1. Westinghouse Energy Systems, Pittsburgh, PA
Bourga R, Moore P, Janin Y-J, Wang B, Sharples J (2015) Leak-before-break: global perspectives and procedures. Int J Pressure Vessel Piping 129-130:43–49
Bowman CF, Bowman SN (2021) Thermal engineering of nuclear power stations balance-of-plant systems, ISBN: 9780367502362. CRC Press, Boca Raton, FL
Brickstad B (2006a) Review of renewed application regarding transition to use of risk-informed inspection of piping systems at Ringhals 2, project RIVAL phase 2, SKI-2005/1401 (restricted). Swedish Nuclear Power Inspectorate, Stockholm
Brickstad B (2006b) Review of the revised plan to implement risk-informed Inservice inspection at Ringhals unit 2, SKI-2005/1401 (in Swedish, proprietary)
Bush SH (1970) Letter on the subject of "safety of operating reactors". Advisory Committee on Reactor Safeguards, Atomic Energy Commission, Washington, DC. NRC-ADAMS Accession No. ML20234C580
Bush SH, Chockie AD (1996) Failure frequencies and probabilities applicable to BWR and PWR piping, SKI report 96:39. Swedish Nuclear Power Inspectorate, Stockholm
Bush SH, Chockie AD, Nicholson W, Slavich A (1998) Nuclear reactor piping failures at U.-S. Commercial LWRs: 1967–1997, TR-110102. EPRI, Palo Alto, CA
Chapman OJV (1983) A statistical approach to the analysis of ISI data using the Bayes method. In: Paper D-1/7, Transactions of SMiRT 7, Chicago, IL. August 22–26
Chapman OJV (1989) Probabilistic-based ISI and life extension. In: Transactions of SMiRT 10, Anaheim, CA, August 22–27. https://repository.lib.ncsu.edu/items/59fc4bc6-d56b-4ca2-9475-ea777ebd654e
Copeland JF, Lin J, Alexander JE, Sayre ED (1980) Production and Implementation of Nuclear Grade Stainless Steel Piping. In: Proceedings: Seminar on Countermeasures for Pipe Cracking in BWR, W5–79-174, vol 4. EPRI, Palo Alto, CA
Crane Valves North America (2022) Flow of fluids through valves, fittings, and pipe. In: Technical paper no. 410M (metric version). Crane Valves North America, The Woodlands, TX
Dadkhah A, Khorrami MS, Kashani-Bozorg SF, Miresmaeili R (2025) The effect of heat input on the residual stress distribution in gas-metal arc welding of carbon steel: simulation and experimental methods. J Adv Joining Processes 12:100333. https://doi.org/10.1016/j.jajp.2025.100333
Dau GJ (1977) Rationale for incorporating inservice inspection requirements during nuclear power plant design/fabrication. Nucl Technol 40(2):122–128
ENIQ (2005) European framework document for risk-informed in-service inspection, ENIQ report 23, EUR 21581 EN. European Commission, Luxembourg
EPRI (1983) Calculation of leak rates through cracks in pipes and tubes, NP-3395. EPRI, Palo Alto, (CA)
EPRI (1986) Application of the leak-before-break approach to BWR piping, NP-4991. EPRI, Palo Alto, CA
EPRI (1988) Crack-opening area calculations for circumferential Through-Wall pipe cracks, NP-5959SR. EPRI, Palo Alto, CA
EPRI (1993) Fatigue comparison of piping designed to ANSI B31.1 and ASME section III, class 1 rules, TR-102901. EPRI, Palo Alto, CA
EPRI (1998a) Evaluation of pipe failure potential via degradation mechanism assessment, TR-110157. EPRI, Palo Alto, CA
EPRI (1998b) Piping system reliability and failure rate estimation models for use in risk informed in-service inspection applications, TR-110161. EPRI, Palo Alto, CA

EPRI (1999a) Vibration fatigue testing of socket welds (PWRMRP-07): PWR materials reliability project (PWRMRP), TR-113890. EPRI, Palo Alto, CA

EPRI (1999b) Revised risk-informed inspection evaluation procedure (PWRMAP-05-a), TR-112657 rev. B-A. EPRI, Palo Alto, CA

EPRI (1999c) Piping system failure rates and rupture frequencies for use in risk informed in-service inspection applications, TR-111880. EPRI, Palo Alto, CA

EPRI (2008) Materials reliability program: technical basis for preemptive weld overlays for alloy 82/182 butt welds in PWRs (MRP-169) revision 1, 1016602. EPRI, Palo Alto, CA

EPRI (2009) Nondestructive evaluation: conversion of RI-ISI programs to code case N-716, 1019129. EPRI, Palo Alto, CA

EPRI (2025) Enhanced risk-informed categorization methodology for pressure boundary components (revision 1-A), 3002033536. EPRI, Palo Alto, CA

Faes K, Dhooge A, De Baets P, Afschrift P (2009) New friction welding process for pipeline girth welds—welding time optimisation. Int J Adv Manuf Technol 43:982–992

Fleming KN, Smit K (2008) Evaluation of design, leak monitoring, and NDE strategies to assure PBMR helium pressure boundary reliability. In: HTR2008–58037, proc. 4th international topical meeting on high temperature reactor technology HTR-2008 ISBN: 978–0–7918-4854-8. ASME, New York, NY

Fleming KN, Fletcher J, Broom N, Gamble R, Gosselin SR (2008) Reliability and integrity management program for PBMR helium pressure boundary components. In: HTR2008–58036, proc. 4th international topical meeting on high temperature reactor technology HTR-2008., ISBN: 978-0-7918-4854-8. ASME, New York, NY

Fredette L, Scott P (2010) Evaluation of full structural and optimized weld overlays as mitigation strategies for primary water stress corrosion cracking in pressurized water reactors, technical report prepared for the USNRC, Battelle, Columbus (OH). NRC-ADAMS accession no. ML101260540

Gabriel J, Marchesi J, Pace R, Rengepis J, Roy V, Charest J (2015) Guidelines for Relief Request for Use of Nonmetallic Repairs of ASME Class 2 and 3 Piping, 3002005334. EPRI, Palo Alto, CA

Goetsch D, Bieniussa K, Schulz H, Jalouneix J (1995) Integrity of the reactor coolant boundary of the European pressurized water reactor: requirements for the application of the break precluding concept for the Main coolant lines. In: Proceedings seminar on leak before break in reactor piping and vessels, NUREG/CP-0155. USNRC, Washington, DC, pp 16–29

Gore BF, Balkey KR (1992) ASME development of risk-based inspection guidelines for nuclear power plants. In: CONF-9208103-2, proceedings INTER-RAMQ conference for electric power industry. ASME, Philadelphia, PA

Haddad AN (2013) ASME code development roadmap for HDPE pipe in nuclear service, STP-NU-057. ASME Standards Technology, LLC, New York, NY

Herrmann TJ, Harris DO, Hirschberg P, Deardorff AF, Lange CH (2011) Improved basis and requirements for break location postulation, 1022873. EPRI, Palo Alto, CA

IAEA (1976) Fracture mechanics applications: implications of detected flaws. IAEA-189, Vienna, Austria

ISO (2017) Plastics pipes and fittings—butt fusion jointing procedures for polyethylene (PE) piping systems, ISO 21307:2017(E), 3rd edn. International Organization for Standardization, Geneva

Kurisaka K, Nakai R, Asayama T, Takaya S (2011) Development of system based code (1): reliability target derivation of structures and components. J Power Energy Syst 5(1):19–32

Luo Y, Gu W, Peng W, Jin Q, Qin Q, Yi C (2020) A study on microstructure, residual stresses and stress corrosion cracking of repair welding on 304 stainless steel: part I—effects of heat input. Materials 13:2416. https://doi.org/10.3390/ma13102416

Nayak AK, Joshi J (eds) (2019) Advances of computational fluid dynamics in nuclear reactor design and safety assessment, ISBN: 978-08-10237-2. Elsevier Science, Amsterdam

Nayyar ML (2000) Piping handbook, 7th edition, ISBN 0-07-047106-1. McGraw-Hill, New York, NY

NEI (2005) SSC Categorization Guideline. NEI 00-04, Rev. 0,. Nuclear Energy Institute, Washington, DC
NEI (2019) Modernization of technical requirements for licensing of advanced non-light water reactors: risk-informed performance-based technology inclusive guidance for non-light water reactor licensing basis development, NEI-18-04 revision 1. Nuclear Energy Institute, Washington, DC
NEI (2021) Technology inclusive guidance for non-light water reactors: safety analysis report content for applicants using the NEI 18–04 methodology, NEI 21–07 revision 0. Nuclear Energy Institute, Washington, DC
NEI (2025) NEI white paper: risk significance of loss of coolant accidents (LOCAs) and justification for a risk-informed transition break size (TBS) methodology. Nuclear Energy Institute White Paper, Washington, DC
NIST (2013) Polymer piping codes and standards for nuclear power plants: current status and recommendations for future codes and standards development. National Institute of Standards and Technology, Gaithersburg, MD
NRWG (1999) Report on risk-informed in-service inspection and in-service testing, final report, EUR 19153 EN. The Nuclear Regulators' Working Group, Luxembourg
NRWG (2004) Report on the regulatory experience of risk-informed Inservice inspection of nuclear power plant components and common views, EUR 21320 EN. The Nuclear Regulators' Working Group, Luxembourg
NUGENIA (2024) ENIQ position on safety benefits of risk-informed in-service inspection, ENIQ report no. 70. NUGENIA, Brussels
Odette GR, Zinkle SJ (eds) (2019) Structural alloys for nuclear energy applications, ISBN: 978–0–12-397046. Elsevier Inc, Cambridge, MA
OECD/NEA (2004) International Developments and Cooperation on Risk-Informed In-Service Inspection (RI-ISI) and Non-Destructive Testing (NDT) Qualification, NEA/CSNI/R(2004)9. OECD/NEA, Boulogne-Billancourt
OECD/NEA (2022) NEA Leak Rate Benchmark Final Report, NEA/CSNI/R(2022)9. OECD/NEA, Boulogne-Billancourt
Offer HP (1984) Qualification of induction heating stress improvement as a pipe remedy. In: Proceedings second seminar on countermeasures for pipe cracking in BWRs, vol 2. Remedy development, EPRI-NP-3684-SR-VOL.2,. EPRI, Palo Alto, CA
Paul DD, Ahmad J, Scott PM, Flanigan LF, Wilkowski GM (1991) Evaluation and refinement of leak rate estimation models, NUREG/CR-5128. U.S. Nuclear Regulatory Commission, Washington, DC
Perdue RK (1998) A spreadsheet model for the evaluation of statistical confidence in nuclear inservice inspection plans. Westinghouse Science & Technology Center, Pittsburgh, PA
Pissanti DR, Scheid A, Kanan LF, Dalpiaz G, Kwietniewski CEF (2019) Pipeline girth friction welding of the UNS S32205 duplex stainless steel. Mater Des 162:198–209
Pullarcot S (2023) Process plant piping: a practical guide to fabrication, Installation, Inspection, Testing and Commissioning., ISBN: 978–1032357072. CRC Press, Boca Raton, FL
Rahman S, Brust FW, Ghadiali N, Wilkowski GW (1998) Crack-opening analyses for circumferential Through-Wall cracks in pipes—part 1: analytical models. Int J Pressure Vessel Pip 75: 357–373
Rao KR (ed) (2009a) Companion guide to the ASME Boiler & Pressure Vessel Coded, sixth edition, vol 2., ISBN: 978-0-0791802700. ASME, New York, NY
Rao KR (ed) (2009b) Companion guide to the ASME boiler & pressure vessel code: criteria and commentary on select aspects of the boiler & pressure vessel and piping codes, third edition, volume 3, ISBN: 978–0791802694. ASME, New York, NY
Rao KR (ed) (2023) Companion guide to the ASME boiler & pressure vessel coded, sixth edition, vol 1., ISBN: 978-0791886519. ASME, New York, NY
Riccardella PC, Pitcairn DR, Giannuzzi AJ, Gerber TL (1988) Weld overlay repairs from conception to long-term qualification. Int J Press Vessel Pip 34:59–82

Rosenberg SJ, Darr JH (1948) "Stabilization of Austenitic Stainless Steel," Research Paper RP1878. J Res Natl Bur Stand 40:321–338

Sallaberry CJ, Brust FW, Kurth-Twombly E, Kurth RE, Lydell BOY (2025) Reexamination of NUREG-1829 loss-of-coolant accident (LOCA) frequency estimates, technical letter report TLR-RES/DE/REB-2025-03. USNRC, Washington, DC

Scheaffer EF (ed) (1955) Design of piping systems, 2nd revised edition. John Wiley & Sons, New York, NY

Schröpfer D, Witte J, Kromm A, Kannengiesser T (2024) Stresses in repair welding of high-strength steels—part 2: heat control and stress optimization. Weld World 68:2537–2551

Schulz H (1995) The evolution of the break preclusion concept for nuclear power plants in Germany. In: Proceedings of seminar on leak before break in reactor piping and vessels, NUREG/CP-0155. USNRC, Washington, DC, pp 31–53

Shadley JR, Merah N, Rybicki EF (1982) The effect of induction heating stress remedies on existing flaws in pipes. J Press Vess Technol 104(3):193–197

Shridhar et al (2025), Mechanical Properties and Residual Stresses in Multi-Pass AISI 304 to AISI 1018 Steel Welds by SMAW and GTAW, Scientific Reports, 15:27052

Shimizu T, Enomoto K, Araki S, Ikegami T (1980) Induction heating stress improvement for welded pipe and its effectiveness. In: Proceedings seminar on countermeasures for pipe cracking in BWRs, EPRI-WS--79-174-Vol.1. EPRI, Palo Alto, CA

Simonen FA (1995) Nondestructive examination reliability. In: Sundararajan C (ed) Probabilistic structural mechanics handbook. Springer, Boston, MA. https://doi.org/10.1007/978-1-4615-1771-9_11

Sinnes K (ed) (2018) Welding handbook, 10th edition, ISBN: 978-1-64322-014-7. American Welding Society, Miami, FL

SKI (1994) Regulations and general advice on mechanical devices in certain nuclear installations, SKIFS 1994:1. Swedish Nuclear Power Inspectorate, Stockholm, Sweden

Soyama H (2004) Introduction of compressive residual stress using a Cavitating jet in air. J Eng Mater Technol 126:123–128

Soyama H (2017) Key factors and applications of cavitation peening. Int J Peening Sci Technol 1: 3–60

Soyama H (2020) Cavitation peening: a review. Metallurgy J 10(2):270–297

Soyama H, Korsunsky AM (2022) A critical comparative review of cavitation peening and other surface peening methods. J Mater Process Technol 305:117586. https://doi.org/10.1016/j.jmatprotec.2022.117586

USNRC (1979) Seismic analysis for as-built safety-related piping systems, IE bulletin 79–14, Washington, DC

USNRC (1984a) Codes and standards for nuclear power plants. Federal Register 49(52):9711 (49 FR 9711). Washington, DC, March 15, 1984

USNRC (1984b) Report of the U.S. Nuclear Regulatory Commission piping review committee: evaluation of potential for pipe breaks, NUREG-1061, vol 3. USNRC, Washington, DC

USNRC (1988a) Technical report on material selection and processing guidelines for BWR coolant pressure boundary piping, NUREG-0313 revision 2. USNRC, Washington, DC

USNRC (1988b) NRC position on IGSCC in BWR austenitic stainless steel piping, generic letter 1988-001. USNRC, Washington, DC

USNRC (1990) Guidance for performing temporary non-code repair of ASME code class 1, 2, and 3 piping. Generic Letter 90-05. USNRC, Washington, DC

USNRC (1991) Information to licensees regarding two NRC inspection manual sections on resolution of degraded and nonconforming conditions and on operability, generic letter 91–18. USNRC, Washington, DC

USNRC (2003) Transcript of workshop on generic letter (GL) 91-18, guidance on operability, degraded and nonconformance issues, work order no. NRC-1041. NRC-ADAMS Accession No. ML032450060, Washington, DC

USNRC (2007) Standard review plant, branch technical position 3–4: postulated rupture locations in fluid system piping inside and outside containment, NUREG-800. USNRC, Washington (DC)

USNRC (2008) Estimating loss of coolant accident (LOCA) frequencies through the elicitation process, NUREG-1829. USNRC, Washington (DC)

USNRC (2020) Guidance for a technology-inclusive risk-informed, and performance-based methodology to inform the licensing basis and content of applications for licensees, certifications, and approvals for non-light water reactors, regulatory guide 1.233. USNRC, Washington, DC

USNRC (2021) Extremely low probability of rupture version 2 probabilistic fracture mechanics code, NUREG-2247. USNRC, Washington, DC

USNRC (2022) Acceptability of ASME code, section XI, division 2. In: "Requirements for Reliability and Integrity Management (RIM) programs for nuclear power plants," for non-light water reactors, regulatory guide 1.247, revision 0. USNRC, Washington, DC

USNRC (2024) Plant-specific applicability of the transition break size, DG-1428. NRC ADAMS Accession No. ML24341A159, Washington, DC

USNRC (2025) Acceptability of ASME code, Section III, Division 5, "High temperature reactors," regulatory guide 1.87 revision 3. USNRC, Washington, DC

van Erp JB (ed) (1981) Executive seminar on the future role of risk assessment and reliability engineering in nuclear regulation, NUREG/CP-0017. USNRC, Washington, DC

Vo TV, Gore BF, Eschbach EJ, Simonen FA (1989) Probabilistic risk assessment based guidance for piping in-service inspection. Nucl Technol 88:13–20

Vo TV, Smith BW, Simonen FA, Doctor SR (1990) Development of generic in-service inspection priorities for pressure boundary component. Nucl Technol 92:291–299

Vo T, Gore B, Simonen F, Doctor S (1994) A pilot application of risk based methods to establish in-service inspection priorities for nuclear components at Surry unit 1 nuclear Power Station. NUREG/CR-6181, Washington, DC

Wang M, Zhang B, Xu J, Chen S (2025) A pilot application study on risk-informed in-service inspection methods for pipelines in HPR1000 nuclear power plants: a case study of the RCV system. Energies 18:4753. https://doi.org/10.3390/en18174753

Wilkowski GM, Chao K-S (eds) (1990) Leak-before-break: further developments in regulatory policies and supporting research, NUREG/CP-0109. U.S. Nuclear Regulatory Commission, Washington (DC)

Yoon B, Kim Y, Lee J (2014) Steam generator small bore piping socket weld inspection using the phased Array ultrasonic technique. J Nucl Sci Technol 51(2):231–239

Yoshimura S, Kanto Y (2020) Probabilistic fracture mechanics for risk-informed activities. In: Fundamentals and applications, ISBN 978–4–907200–07-7, probabilistic fracture mechanics subcommittee, atomic energy research committee. The Japan Welding Engineering Society, Tokyo

Chapter 6
Analytical Contexts and Guidelines

6.1 Basic Concepts

Probabilistic piping reliability analysis relies on fundamentals, operating experience, and modeling. The quality of the analysis depends on a thorough understanding of piping principles, covered in Chap. 2 (*Piping Reliability and Nuclear Safety*), Chap. 3 (*Historical Perspectives on Piping Reliability*), Chap. 4 (*Material Degradation: Learning from Operating Experience*), and Chap. 5 (*Pressure Boundary Integrity*). Operating experience data and experimental data are crucial for developing appropriate model inputs.

Techniques used to determine probabilistic failure metrics for piping are designed to accurately reflect the physics of failure. This is achieved by validating both input parameters and results with experimental findings and real-world operational data, making it a multidisciplinary analytical task. A comparison of various methods for evaluating probabilistic failure within a clearly defined boundary is documented in IAEA-TECDOC-1988 (IAEA 2021). The report provides statements regarding how technical approaches are chosen for probabilistic piping reliability analysis.

1. All implementations required extensive input data preparation, and some form of computer code implementation to facilitate the quantification processes, including the uncertainty analyses.
2. Implementations of the methods tended to be computationally intense, necessitating pre-planning of the computations and post-processing of results.
3. Regardless of the selected methodology, a successful implementation depended on the experiences that had been obtained by the analysts from previously performed practical applications.
4. The value and importance of operating experience data cannot be understated. All methods benefited from and required information on material performance and pipe failures.

B. Lydell, *Knowledge Preservation in Nuclear Safety*,
https://doi.org/10.1007/978-3-032-25463-4_6

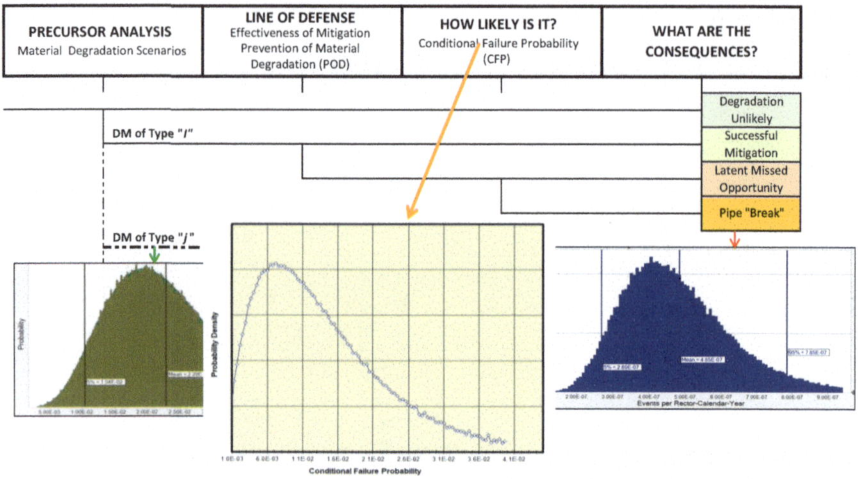

Fig. 6.1 The structural integrity risk triplet (This adaptation was originally made by the author in 2018 as part of a research project funded by the Canadian Nuclear Safety Commission (Project ID 87055-15-0214). The figure also appears in IAEA Nuclear Energy Series Publication No. NR-T-2.16 (2023))

Consideration of uncertainty is essential in probabilistic piping reliability analysis. An event tree, Fig. 6.1, is used as a representation of a piping integrity "risk triplet." It addresses what can go wrong, how likely it is to happen, and what the consequences are of a structural failure. The "risk triplet" is a fundamental concept in risk analysis, describing risk as a combination of three key elements: the scenario or event that can go wrong, the probability or likelihood of that scenario occurring, and the consequences or impact of that scenario if it happens. Each element in the risk triplet is represented by a probability density function as described by Kaplan[1] and Garrick[2] (Kaplan and Garrick 1979; Kaplan and Garrick 1980). The risk triplet representation is illustrated in Fig. 6.1 and described as follows:

[1] S. Kaplan (1931–2011), a leading contributor to the risk sciences and nuclear reactor physics. Among his specific contributions to risk assessment were the discrete probability distribution method for probabilistic calculations and the two-stage Bayesian technique for data analysis. He was elected to the National Academy of Engineering in 1999, and he was a Fellow of the Society for Risk Analysis (SRA).

[2] B.J. Garrick (1930–2020), founder of the B. John Garrick Institute for the Risk Sciences, University of California at Los Angeles (UCLA). An international authority on quantitative risk assessment, and a Distinguished Adjunct Professor of Materials Science and Engineering at UCLA. "John Garrick was one of the founders and a global leader of the probabilistic risk assessment discipline, and a highly regarded force in advancing the field in theory and practice" (this citation is from https://www.risksciences.ucla.edu/). Garrick and his co-workers were instrumental in developing the "Sundesert Nuclear Power Plant Reliability Engineering Program." Reliability and risk analysis insights were inherently part of the design framework for San Diego Gas & Electric's (SDG&E) proposed Sundesert Nuclear Power Plant (1972–1978). The design of the proposed two-unit PWR plant was heavily rooted in the defense-in-depth philosophy.

$$\langle S_i, P_i, C_i \rangle$$

An analysis of structural integrity begins by identifying possible material degradation scenarios (S_i) including the initiation of material degradation and the potential for growth of a pipe flaw in the through-wall direction. This step produces a frequency of a degraded state. Next, the analysis investigates the "line of defense" against a degraded pressure boundary integrity, i.e., the probability of detecting a degraded state before going through-wall producing a detectable leakage. This is followed by an assessment of the likelihood that a material degradation scenario propagates into loss of pressure boundary integrity (P_i). Finally, the consequence (C_i) of a structural failure is represented by the product of S_i and P_i. The notion of "probability" is intertwined with the definition of structural failure of piping.

In "*what can go wrong*," an analysis considers the incubation period and crack/flaw growth for different combinations of materials and degradation mechanisms (DMs). An incubation period refers to the total time that a certain piping pressure boundary goes through crack nucleation until the crack progresses to a critical degree such that there is an active crack growth in the through-pipe-wall direction. The quantitative assessment produces an estimate of the frequency of failure of a pressure boundary. Failure can be a crack or leak that results in repair without immediate effect on operations.

In "*line of defense*," an analysis considers the different strategies for early detection of a degraded piping pressure boundary so that a significant structural failure is prevented. Early detection encompasses reliability of a leak detection system and probability of detection (POD) of a crack before it penetrates a piping pressure boundary. Renewal theory is used to analyze the effect on the frequency of failure by, for example, leak detection and nondestructive examination.

In "*how likely it is to happen*," the analysis considers how an undetected pipe flaw or a minor through-pipe-wall leakage propagates to form a significant structural failure. This is represented by a conditional failure probability representing the margin to a pressure boundary failure.

Probabilistic piping reliability analysis estimates the frequency of material degradation such as a weld with a through-wall flaw, the conditional probability that the flaw will grow to a critical flaw size and corresponding major loss of structural integrity, and the frequency of having a loss of structural integrity of a certain magnitude. To paraphrase Apostolakis (1981), probabilistic piping reliability analysis is essentially the quantification and propagation of the uncertainties associated with random quantities embedded in the structural integrity risk triplet (Table 6.1).

Ideally, the operating experience is such that sufficient data exists so that probabilistic failure metrics are obtained through direct statistical estimation. This is a classical "data-driven" approach, and to obtain a pipe failure precursor frequency, the analyst obtains the number of event occurrences over time and assumes that the frequency follows a homogeneous or nonhomogeneous Poisson distribution. Homogeneity implies that a piping system and its operating environment are the same for all plants from which the failure data has been obtained. Changes in homogeneity

Table 6.1 Basis for how to characterize the risk triplet

Element in the structural integrity risk triplet	Parameter	Formula	Description
C_i	ρ	Combination of probability density functions for λ and *CFP*, respectively, using Monte Carlo multiplication	Frequency of a major structural integrity failure as a function of an equivalent break size (EBS) or through-wall mass or volumetric leak rate. The exact definition depends on the context of an analysis
S_i	λ	$\lambda_{ik} = \frac{n_{ik}}{\tau_{ik}} = \frac{n_{ik}}{f_{ik} N_i T_i}$	"Precursor rate" or "failure rate" lambda, which is the rate of being in a certain state, e.g., crack of certain dimension (through-wall depth as percentage of wall thickness, crack opening area, crack opening displacement, or through-wall leak). The failure rate is calculated as the number of observed failures over an observation period
P_i	*CFP*	$CFP = \frac{\alpha}{\alpha+\beta}$	The formula is symbolic, parameter α is the number of significant structural failures, and parameter β represents the number of challenges to the pressure boundary integrity

occur because of the use of different materials, different reliability and integrity management, and varying times to failure.

Parameter λ in Table 6.1 is referred to as the pipe failure rate, or precursor rate, and is estimated from the number of "failures" that have been observed in a given period. Parameter τ is an exposure term, consisting of reactor operating years or equivalent full power years of operation and number of piping components that produced the relevant operating experience. The precursor rate is a simple maximum likelihood estimator with a single parameter lambda, which is the mean number of events over time. It is a representation of being in a certain degraded state of structural integrity, e.g., cracked, thinned, or leaking, and can be statistically described as a Poisson point process. The CFP represents the conditional probability that a preexisting flaw in the piping pressure boundary transitions from a stable to an unstable state.

Depending on the objective of an analysis, the term failure has different definitions, oftentimes referred to as a "breach," "break," "leakage," or "rupture." However, any of these terms are rather meaningless unless further qualifications are provided. Of essence is the estimation of the frequency of a degraded structural integrity state as a function of its consequence in terms of the mass or volumetric through-wall flow rates or the dimensions of the area of a compromised pipe wall. Results of probabilistic piping reliability analyses are usually presented in the form of a cumulative failure frequency (y-axis) versus consequence (x-axis). The consequence can be the through-wall mass or volumetric flow rate or it can be in terms of equivalent, or effective, break size (EBS). A higher frequency is normally associated with smaller-size piping or smaller EBS and less uncertainty. There is not a universally accepted presentation format, however. As we shall see, past assessments of piping structural integrity have in general been inconsistent in the use of terminology. Oftentimes, the term probability has inadvertently been used in lieu of frequency of failure. The underlying statistics of "major" pipe failures can be described as Markovian in the sense that these events depend on what happened in the past with respect to missed inspections, inadequate repairs, less-than-adequate material selections, and so on.

6.2 Terminology

When evaluating and comparing outcomes from various piping reliability models, the definition of "failure" is crucial. Ambiguity often arises regarding pipe failure mode terminology. Failure indicates a compromise in the integrity of the piping pressure boundary. The manner in which pipe material deteriorates and the resulting consequences of such degradation significantly influence the modeling approach.

The term "failure" describes occurrences of degraded pressure boundaries that have resulted in repair or replacement action causing plant downtimes lasting from days to many months. Oftentimes, these repair and replacement activities have been accompanied by experimental and theoretical work to establish the causal factors of material degradation and to develop countermeasures. "Failure" also includes through-pipe-wall leakages of various magnitudes that require immediate attention to meet operational rules, good engineering practices, and compliance with regulations. A degraded (cracked or thinned) or leaking piping pressure boundary represents a precursor to a possible significant structural failure (SSF), and the margin between "precursor" to "significant" can be of great importance. A sudden, energetic loss of pressure boundary integrity invariably results in spatial effects such as flooding or spraying on surrounding equipment and structures and automatic shutdown of the reactor. Querying a pipe failure database for information on specific attributes invariably results in small failure populations. Hence, this operating experience points to low failure frequencies per plant system and operating year. In other words, a pipe failure of any kind is not common, and a significant structural failure is quite rare indeed.

Codes and standards for piping integrity give details on pipe failure mode determination and the methodology to be applied when assessing the fitness for continued operation, given a degraded piping pressure boundary. Crack growth is characterized as either subcritical/stable crack growth or critical/unstable crack

Table 6.2 An example of pipe failure mode definitions

"Rupture" mode	Equivalent break size, EBS (mm)	Through-wall flow rate, Q (kg/s)
Pressure boundary failure	$15 < \text{EBS} \leq 50$	$0.5 < Q \leq 5$
Small breach	$50 < \text{EBS} \leq 100$	$5 < Q \leq 20$
Breach	$100 < \text{EBS} \leq 250$	$20 < Q \leq 100$
Large breach	$250 < \text{EBS} \leq 500$	$100 < Q \leq 400$
Major breach	$\text{EBS} > 500$	$Q > 400$

growth. A subcritical crack is a crack whose stress intensity factor is below the critical value. Typically, it is assumed that fatigue cracks and stress corrosion cracks are related to the stress intensity factor, *K*, unless under extreme loading scenarios such as water hammer or beyond-design-basis accidents where large plastic deformation occurs. The stress intensity factor is used in fracture mechanics to predict the stress intensity near the tip of a crack caused by a remote load or residual stresses. If the energy available for an incremental extension of a rapidly propagating crack falls below the material fracture toughness, the crack arrests.

In probabilistic fracture mechanics (PFM), a "leak" is a wall-penetrating defect which is stable. A wall-penetrating defect which fails due to the applied loads is considered as a "rupture" with an unstable crack growth rate. In PSA, the definition of the consequence of a pipe failure is important. It is not meaningful to seek answers to questions like "What is the frequency of a pipe leak?" or "What is the frequency of pipe rupture?" In probabilistic terms, the definition of failure versus its consequence affects the modeling approach, including the modeling of uncertainty. The consequence of a pipe failure can be characterized in terms of the through-wall mass or volumetric flow rate, e.g., kg/s or m^3/s. The size of a through-wall pipe flaw can also be used and oftentimes is expressed in terms of an equivalent break size (EBS) or crack opening area (COA). An example of pipe failure mode definitions is given in Table 6.2. Considerations of the consequence of a potential pipe failure determine the technical approach to modeling and uncertainty characterization.

6.3 Analytical Contexts

The effectiveness of probabilistic piping reliability analysis depends on the specific objectives of the study and the application domain for its outcomes. Each analytical domain imposes distinct usability requirements. The selection of methods follows a three-tiered process that considers the overall context of the analysis, end-user expectations, and validation criteria. Table 6.3 presents examples from seven different analytical contexts:

1. PSA applications for existing and new reactor designs: US regulations stipulate the development and/or update of a PSA for a Design Certification (DC) or

Table 6.3 Analytical contexts of probabilistic piping reliability analysis

Objective(s)	Analysis objective	Description
PSA and PSA applications	LOCA initiating event frequencies	Quantification of location-specific LOCA frequencies as a function of EBS or through-wall flow rate
	High-energy line break initiating event frequencies	Quantification of location-specific high-energy line break frequencies inside or outside the containment
	Internal flooding initiating event frequencies	Pipe failure frequencies account for all potential flood sources. Includes consideration of high-energy and moderate-energy piping systems outside reactor containment structures
	Support system failure initiating event frequencies	Examples include loss of cooling, loss of instrument air, and loss of electrohydraulic control system
	Development of age-dependent reliability characteristics	For plants in long-term operation, provide risk-informed assessments of the effectiveness of different aging management programs
	Fire PSA	Pipe failure frequencies associated with firewater systems, and piping systems transmitting flammables, e.g., hydrogen, oils
RIM per ASME XI division 2	Evaluation of applicable degradation mechanisms	Review of applicable degradation mechanisms and definition of target OE datasets for different combinations of materials and pipe sizes
	RIM program development for a new reactor	Quantification of reliability targets for different ISI locations to determine the type of RIM needed
	Impact of leak detection and nondestructive examination on piping reliability	Evaluation of how reliability characteristics are influenced by various surveillance programs such as leak detection, leak and pressure tests, and in-service inspections
	RIM program updates	Updates to account for new OE, piping design changes, and/or results of PSA updates
ASME XI non-mandatory appendix R	RI-ISI program development for an existing gen-II or gen-III light water reactor	Degradation mechanism analysis (the qualitative element of RI-ISI), pipe failure potential analysis in which different methods are applied to obtain probabilistic failure metrics
	RI-ISI program update	Updates to account for new OE, piping design changes, and/or results of PSA updates
	Risk-informed HELB analysis	Pipe failure potential analysis

(continued)

Table 6.3 (continued)

Objective(s)	Analysis objective	Description
Operability determination	Risk-informed "significance determination"	Application of plant-specific PSA to determine the risk significance of a degraded or failed piping component
	Structural evaluation of flawed piping component	Determination whether a flawed component is suitable for continued operation, e.g., until next planned outage of sufficient length to accommodate a repair or replacement
PFM applications	Frequency of crack initiation	PFM model validation and development of conditional failure probability models, including uncertainty distributions
	Crack opening area or crack opening displacement versus through-wall leak rate	
	Aspect ratios based on operating experience	
	Frequency of significant structural failures	Probability of "rupture" analysis to produce hazard curves as a function of different RIM strategies
Probabilistic benchmarking	Intercomparison of different PFM codes	To understand differences in probabilistic fracture mechanics (PFM) software design, and improved understanding of the effectiveness of leak detection and nondestructive examination
	Intra-comparison of different methodologies	Improve the understanding of assumption about crack initiation and crack growth
	Model validation	Compare results of probabilistic piping reliability analysis with operating experience data
Licensing and regulatory oversight	Aging management program review, periodic safety review, changes to licensing basis, accident precursor analysis	Applications that involve consideration of piping OE and probabilistic piping reliability analysis within the scope of regulatory oversight processes

Combined Operating License (COL) application under 10 CFR Part 52 (USNRC 2016, 2024a).

2. Reliability and integrity management (RIM) program development: Probability of a pressure-boundary failure is compared against a reliability target in defining an appropriate strategy for inspection. Consideration is given to probabilistic rate of flaw occurrence, flaw temporal progression, frequency of periodic inspection, probability of detecting a flaw, and time to make corrective action.
3. ASME BPVC Section XI, Nonmandatory Appendix R, is concerned with risk-informed in-service inspection (RI-ISI) program development, implementation, and updating to account for new operating experience insights.

4. PFM applications: Validation of input parameters against relevant operating experience data. Comparing model predictions with failures observed in the operation of nuclear power plants.
5. Probabilistic benchmarking and validation of the results of probabilistic piping reliability analysis.
6. Operability determination: Applying risk insights to determine the safety significance of degraded or failed piping.
7. Reactor licensing: Development and application of transition break size (TBS) criteria, leak-before-break analyses, aging management program review, periodic safety review, license renewal and subsequent license renewal review, changes to licensing basis, accident precursor analysis, and advanced reactor piping integrity.

Expectations and technical requirements that are associated with the different analytical contexts are summarized in Table 6.3. The IAEA Nuclear Energy Series Report NES NR-T-2.16 (IAEA 2023) is a "good practice" guideline for the calculation of pipe failure rates. The guideline is in the form of a seven-step framework that evolved from practical analyses. With respect to the performance of high-energy line break analysis within the scope of internal flooding PSA and the development of internal flooding initiating event frequencies caused by raw water cooling and fire protection system pipe failure, the Electric Power Research Institute (EPRI) has developed analysis guidelines and supporting pipe failure rates and rupture frequencies for a broad set of piping systems. For analyses that utilize probabilistic fracture mechanics (PFM), the Canadian Nuclear Safety Commission (CNSC) and the United States Nuclear Regulatory Commission have supported work to develop best practice guidelines for how to implement PFM.

The USNRC issued a Regulatory Guide 1.178 in 1998, which describes an approach that is acceptable to the staff for developing RI-ISI programs. Section 2.1.5 of the current revision of this regulatory guide is concerned with "assessment of pipe failure potential" (USNRC 1998):

> *The failure potential used in or in support of the analysis should be appropriate for the specific environmental conditions, degradation mechanisms, and failure modes for each piping location. When data are analyzed to develop a categorization process relating degradation mechanisms to failure potential, the data should be appropriate and publicly available. When an elicitation of expert opinion is used in conjunction with, or in lieu of, probabilistic fracture mechanics analysis or operating data, a systematic process should be developed for conducting such an elicitation. In such cases, a suitable team of experts should be selected and trained (see NUREG/CR-5424, "Eliciting and Analyzing Expert* Judgment*—A Practical Guide," issued January 1990, and NUREG-1563, "Branch Technical Position on the Use of Expert Elicitation in the High-Level Radioactive Waste Program," issued November 1996).*

> *The methodology, process, and rationale used to determine the likelihood of failure of piping segments should be independently reviewed during the final classification of the risk significance of each segment. Referencing applicable generic topical reports approved by the NRC is one acceptable means to standardize the process. When new computer codes are used to develop quantitative estimates, the techniques should be verified and validated against established industry codes and available data. When data are used to evaluate the*

likelihood of piping failures, the data should be submitted to the NRC or referenced by an NRC-approved topical report. As stated in RG 1.174, "data, methods, and assessment criteria used to support regulatory decision-making should be well documented and available for public review." It is the responsibility of the licensee to provide the data, methods, and justification to support its estimation of the failure potential of piping segments.

The domains of piping reliability analysis refer to the intended utilization of analysis results. There are three types of end users of results in support of (1) regulations, (2) operations, and (3) reactor design. In a regulatory domain, a piping reliability analysis may be performed in response to new regulatory requirements, to make changes to an existing licensing basis, or to assess the risk significance of pipe failure. In an operational domain, a piping reliability analysis may be performed to optimize ISI and aging management programs (e.g., adding or removing inspection locations) or to assess the risk significance of a leaking piping component. Finally, in a new reactor design domain, a piping reliability analysis would support the evaluation of different piping system design concepts, development of an RIM program, and/or leak-before-break analysis. Examples of requirements for probabilistic piping reliability analysis are given in Table 6.4.

6.4 Analysis Guidelines

Six analysis guidelines are summarized in this section. Four of the guidelines deal with probabilistic fracture mechanics applications. A fifth guideline prepared by the IAEA has a broader scope in that it recognizes that different methodologies can be applied to a common problem. A sixth guideline deals with probabilistic piping reliability analysis in the context of internal flooding PSA. There are commonalities between the guidelines, especially with respect to how to organize computational processes and the flow of information between the different tasks, from input data definition to result presentation.

6.4.1 CNSC Guideline for the Performance of PFM

The Canadian Nuclear Safety Commission (CNSC) has sponsored work to develop a best practice guideline document that summarizes the results of a literature review of current practices for PFM assessments (Modarres and Kirk 2021). Based on the results, a guideline document for best practices for PFM analyses and evaluations was developed to assist in future PFM applications and regulatory considerations. The guideline adopted a set of working principles to perform probabilistic analysis that includes PFM applications. It also summarized some past examples of PFM guidelines and applications. It then identified seven critical steps for practical PFM analysis. The guidance includes a summary of the essential elements of each step. It

Table 6.4 Examples of analysis requirements

PSA task	Description	Modeling requirements
Initiating event frequency	LOCA due to primary coolant pipe failure	The exact definition of LOCA is a function of plant design, including emergency core cooling system make-up capability. Engineering calculations support the plant-specific classification of the different sets of LOCA frequencies that are needed. The LOCA frequency calculation accounts for degradation mechanisms and all the locations within the primary pressure boundary that potentially could produce a LOCA of certain magnitude. The PSA model input can be in the form of a cumulative pipe failure frequency versus the through-wall mass flow rate or equivalent break size. The analysis acknowledges locations that were stress relieved during construction or at later dates to mitigate possible SCC effects. Consideration of RIM (e.g., leak detection and ISI) is an explicit analysis requirement
	Main steam line break inside and outside the containment	Assessment of location-specific pipe failure rates. Engineering analyses support the definition of calculation cases followed by detailed reviews of piping arrangement drawings, heat balance diagrams, piping and instrumentation diagrams, and isometric drawings. The analysis addresses dynamic pipe break effects and spatial impact of consequential actuation of fire protection sprinklers that results in full flow from fire protection system pumps into the turbine building with localized or more widespread flooding
	Feedwater line break inside and outside the containment	
Internal flooding initiating event frequency	Assessment of all possible sources of internal flooding caused by pipe failure	Internal flooding PSA is concerned with pipe failure locations inside the reactor building, auxiliary building, service building, and turbine building. Pipe failure rates are developed for multiple systems, pipe sizes, locations within a plant, and operating environments. Flood frequencies are provided not only in terms of the basic three flood modes of spray (up to 5 kg/s), flood (5 to 125 kg/s), and major flood (>125 kg/s), but also in more finely defined flood rate intervals to support refinement of the modeling of flood-induced accident sequences

(continued)

Table 6.4 (continued)

PSA task	Description	Modeling requirements
		in an IF-PSA. More refined flood rate ranges may be required. An analyst may also use cumulative pipe to allow for more specific applications

provides ample references from a literature review to support any technical statement and to provide resources for further reading. The seven steps are as follows:

1. Specifying how the PFM model will be used: This involves understanding metric (s) such as crack size and leak rate to quantify the extent of the degradation mechanism of interest and how this degradation evolves with time. Tracking these metrics is important because the PFM code needs to relate the metric to a final structural failure probability.
2. Definition, documentation, and representation of the component parts of the overall model: This task provides guidance on the definitions of the various forms of input data, multiple models, path of information flow between these, and documentation. The guidance emphasizes a graphical representation of the PFM process to aid with communicating the details of the model between different team members, decision-makers, and external stakeholders.
3. Definition of uncertainty characterization and quantification: A critical element of PFM is the characterization and quantification of uncertainties in the model itself and inputs and outputs from models. The guidance document elaborates on the need to identify models and inputs that should be designated as uncertain, nature of these uncertainties (aleatory or epistemic, and whether they be explicitly quantified and represented separately), and models and inputs that should be represented as constants or conservative values. The guidance discusses the approach to formulating the PFM model and its "connectivity," which describes how model and input uncertainties propagate through the entire PFM analysis.
4. Software development, documentation, and tracking: PFM analyses are complex and rely on detailed computer software to implement the calculations specified in the first three steps. The guidance discusses good practice methods for developing the associated software requirement document and an associated software quality assurance program, which tracks and controls software development. Additionally, the guidance discusses the need for a verification and validation program to ensure that the various parts of the PFM model appropriately represent the underlying physical phenomena (validation) and the computer code meets the specification of the software requirement document (verification).
5. PFM model understanding and refinement: The development of a PFM code of any complexity is rarely a single-step process. Various revisions and each version of the code often reveal the need for improvements and/or refinements. The guidance includes discussions on the use of sensitivity studies as a vital tool to highlight the need for modifications and code improvements.

6. Review and approval of PFM model: The complexity and novelty of PFM models can lead to extensive review processes (e.g., involving subject matter experts, regulatory staff, and outside stakeholders) before a PFM model is deemed acceptable for use. The elements of a good practice review and approval process are provided in the guidance document.
7. Regulatory decision-making: Once the PFM analysis is complete, thoroughly reviewed, and all documentation submitted, it remains for the regulator to decide if the safety case made with PFM is acceptable and demonstrates an appropriate level of safety. Unless a risk-based or risk-informed decision-making process is used formally against agreed-upon risk acceptance criteria, which is typically not the case, factors other than the outcomes of the PFM analysis are considered. The guidance document provides a general discussion of the best practices in developing acceptance criteria.

6.4.2 USNRC Guidelines for the Performance of PFM

NUREG/CR-7278 (Lund et al. 2022), on probabilistic fracture mechanics (PFM), is a companion document to Regulatory Guide (RG) 1.245, Revision 0, "*Preparing Probabilistic Fracture Mechanics (PFM) Submittals*" (USNRC 2022). The NUREG report describes a graded approach to developing PFM submittal documentation, provides a generalized technical basis for conducting PFM analyses, and constitutes the technical basis for RG-1.245. This guidance outlines a procedure that includes this suggested graded approach for PFM analyses and submittals. This RG discusses acceptable ways to present PFM analyses in regulatory submittals to the USNRC and can be used to demonstrate compliance with a wide range of regulations.

NUREG/CR-7278 describes a hypothetical process for conducting PFM analysis. This process is aligned with the position on documentation elements given previously in a technical letter report (TLR), "*Important Aspects of Probabilistic Fracture Mechanics Analyses*" (Raynaud et al. 2018). Section 3 of the NUREG defines five steps in PFM analysis (Table 6.5).

RG 1.245 uses the term "quantity of interest" (QOI) as a quantity that is directly tied to a regulatory decision. QOI is related to a model output or to a function of outputs; for a PFM model to be useful, understanding the relationship between the model output and the QOI is critical. For example, suppose the QOI is the probability of rupture by year for a single pipe. For each set of inputs, the PFM model may output the year in which rupture occurs. The QOI is then estimated by calculating the frequency of rupture by year across many realizations of this single pipe's performance. In the xLPR probabilistic fracture mechanics code, a "realization" refers to a specific, single run of the probabilistic model (Homiack et al 2021) (USNRC 2021). Each realization represents one possible scenario of crack initiation, growth, and propagation based on the input parameters, uncertainties, and model assumptions.

Table 6.5 Analytical steps in PFM per RG 1.245 and NUREG/CR-7278

Step	Objective
1	Translating regulatory requirements into an analysis plan
1.1	Define how PFM analyses will be used as a technical basis for regulatory action, including the criteria to be used to support proposed regulatory action
1.2	Define the quantity of interest and how it relates to the model output and acceptance criteria
1.3	Determine the suitability of the PFM code for the specific application. What are the material degradation mechanisms for which PFM codes are developed?
1.4	Identify key elements of the PFM application that will determine how to conduct the analysis. Simplifying assumptions and approximations may be necessary based on the complexity of the problem or, conversely, may be justified because the problem at hand is inherently not complex
2	Model input uncertainty characterization
2.1	Identify uncertain model inputs: The purpose of this step is to determine which model inputs are treated with uncertainty and, if relevant, the aleatory or epistemic uncertainty for each input
2.2	Specify probability distributions on uncertain inputs
3	Estimation of quantity of interest and associated uncertainty
3.1	Select a method for propagating uncertainty in the model inputs through the model to estimate the quantity of interest and the associated uncertainty
3.2	Assess the statistical convergence of quantity of interest estimates from model outputs given a sampling scheme
3.3	Conduct sensitivity analyses to determine input uncertainty importance
3.4	Provide a final estimate, with associated uncertainty, of the quantity of interest and to visualize results
4	Sensitivity studies to assess the credibility of modeling assumptions
4.1	Determine a set of sensitivity studies: The purpose of this action is to identify important assumptions that merit further scrutiny to understand what might happen if these assumptions were changed. For example, in the study of a plant, the distribution of a specific input could have been calibrated using information from a global set of similar but different plants
4.2	Conduct sensitivity studies and present results
5	Draw conclusions from analysis results
5.1	Interpret analysis results: The purpose of this action is to synthesize the information gathered in steps 1–4 and draw conclusions from this information
5.2	Iterate on the analysis process to refine model results: The purpose of this action is to determine whether additional analyses are required to draw informative conclusions from the modeling

6.4.3 UK Working Principles

Formed in 2018 by leading structural integrity specialists from industry and academia in the United Kingdom, the *Nuclear Structural Integrity Probabilistic Working Group* developed a document titled the *Nuclear Structural Integrity Probabilistic Working Principles* (NSIPWG 2019). These working principles are separated into three categories: firstly, the benefits that are enabled by probabilistic approaches; secondly, guidance on application; and thirdly, the validation requirements and

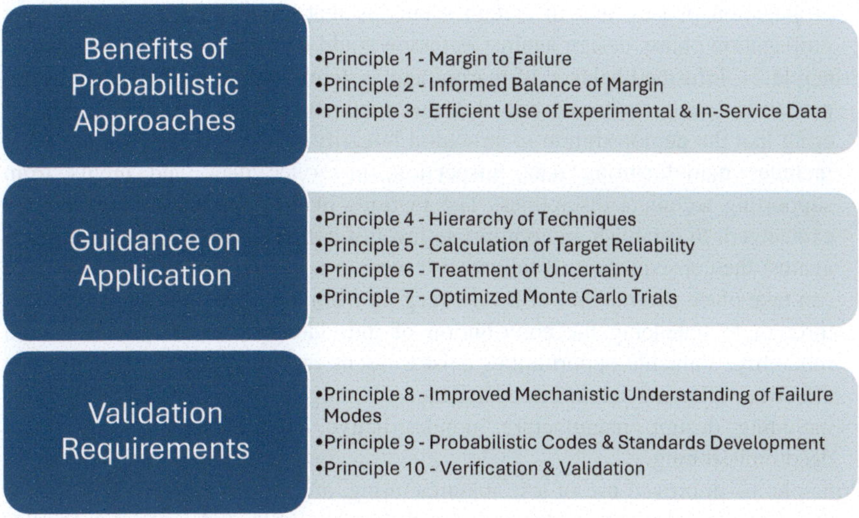

Fig. 6.2 Working principle categories

future work needed to enable routine use of the probabilistic approaches (Fig. 6.2). The working principles are defined as follows:

Principle 1—margins to failure are quantified probabilistically: The probability of failure is compared with a target reliability for the particular structural region of interest. The target reliability can be derived from the system-level probabilistic safety assessment. It is noted that a traditional deterministic margin to failure, usage factor (or utilization ratio), will have an associated probability of failure that is not explicitly expressed. For example, satisfying the strength, shakedown, and fatigue criteria in the ASME III Boiler and Pressure Vessel Code is often associated with a 10^{-5} probability of failure. However, this can be misleading as the inference for well-designed and built pressure vessels was for catastrophic failure only. The probability of pressure boundary leakage, based on operating experience, is clearly greater than this. Reference is made to the development of reliability-based load and resistance factor design (LRFD) methods for piping (ASME 2007).[3,4] Leakage due to fatigue will require incubation, initiation, and

[3] Load factors are applied to nominal loads to account for variations and uncertainties in their magnitude and distribution. For example, live loads often have higher load factors than dead loads because they are more variable. Resistance factors are applied to the nominal resistance of the material to account for uncertainties in material strength and other factors. Resistance factors are typically less than 1.0. LRFD is concerned with the reliability of ASME III Class 2 and 3 piping.

[4] ASME CRTD-86 provides results based on the following tasks: (1) a state-of-the-art assessment and selection of reliability theories, (2) review and evaluation of existing strength models for piping, (3) selection of strength models and equations that deemed suitable for LRFD development, (4) preliminary analysis of basic random variables to characterize their uncertainties, and (5) development of LRFD guidelines and criteria.

propagation phases, and in certain scenarios that are tolerant of a significant propagation phase, design against initiation could be excessively conservative.

Principle 2—informed balance of margin and systems approach: The power of the probabilistic approaches described in this document can be unlocked by considering that the demonstration of structural integrity is part of a larger system that includes manufacturing data, inspection, in-service data, and inputs from supporting technical disciplines. The margins in the individual inputs can be exchanged, to establish an appropriate level of conservatism for a predicted life against the consequences of failure, if it were to occur. The exchange of margin can take place at a number of levels, ranging from inputs to individual analyses, through to balancing the contribution of individual failure modes to overall reliability. Valuable opportunities exist to focus resources where they are most effective, allowing an informed balance of margin throughout the life cycle, including design, manufacture, nondestructive examination, operation, and decommissioning.

Principle 3—improved use of test, manufacturing, and in-service data: Test data is expensive and difficult to obtain, particularly when irradiated material properties are required. Similarly, all manufacturing data and in-service data have associated infrastructure, measurement, and processing costs. Probabilistic approaches enable the entire dataset to be used in the assessment of the proximity to failure in contrast to deterministic approaches that typically discard the majority of the information to use a select few data points. Operating experience from plant sensors or condition monitoring systems can be used to actively update the reliability assessment as more data is obtained. There are standard techniques for doing this such as Bayesian analysis that are well established in other industries. This type of analysis can also be useful to inform infrastructure decisions on the application of sensors and condition monitoring technologies as the worth of the sensor data can be evaluated in comparison to other system data.

Principle 4—application using hierarchy of techniques: An analytical failure model must provide a good representation of the actual failure mode of the component or system and consider potential interaction between failure modes. The treatment of failure modes should be consistent, for example, consistent assessment of through-wall leakage in contrast to initiation, and the time period and time dependence associated with the failure modes should be understood and accounted for if necessary.

Principle 5—calculation of target reliability: The application of probabilistic structural integrity techniques is inextricably linked to the calculation of an appropriate target reliability figure to which structural failure probabilities are compared. One approach to the calculation of target reliability figures is to use published and accepted damage frequency data in conjunction with the system-level PSA to "reverse engineer" the target reliability. Reference is made to a technical paper published in the Journal of Power and Energy Systems (Kurisaka et al. 2011). In the calculation of the target reliability, consideration should be given to the potential interaction of different failure modes in the structure, the aggregation of component failures, and the availability, or otherwise, of protection systems. Reference is also made to the ASME/RIM programs for high-temperature gas-cooled reactors and liquid-metal reactors.

Principle 6—selection of appropriate distributions, correlations, and sampling: The distributions applied to input data should be as accurate as possible and applicable to the data type; for example, normal distributions are generally not appropriate for datasets that can only contain values greater than zero. Lognormal distributions are generally applicable to material property distributions. It is expected that quantification of uncertainty will not be possible for all data inputs, and in such cases, it is necessary to make a conservative judgment. If sufficient data is available, direct sampling from the data histogram may also be a viable alternative. The form of the distribution may or may not be important and depends on whether the trials that dominate the failure probability are within the available dataset. If the region of interest is within the tails of the distribution, sensitivity studies should be undertaken to understand the influence of distribution shape. The relative importance of the tails can be reduced by sampling from more variables, if available.

Principle 7—application of response surface/surrogate models: Response surface or surrogate models can be used to improve the speed of individual trials within a Monte Carlo analysis. Sensitivity studies should be undertaken to establish if the response surface or surrogate model is appropriate. If a "design of experiments" approach has been used to generate a response surface or surrogate model, a selection of independent trials should be undertaken to demonstrate the quality of fit.[5]

Principle 8—improved mechanistic understanding of failure modes: Developing an improved mechanistic understanding of the failure mode is key to releasing margin from the structural analysis.

Principle 9—development of probabilistic design codes and standards: A limited set of the international suite of structural integrity codes and standards includes probabilistic guidance. Involvement in international codes and standards bodies such as the ASME committees is required to demonstrate the benefits of probabilistic approaches and lead the development of probabilistic content. There is also a valuable opportunity to develop codes and standards for future high-temperature designs, fusion, and small modular reactors.

Principle 10—verification and validation: Well-established techniques are available for the application of Monte Carlo and FORM/SORM approaches, including numerous commercial off-the-shelf products and in-house codes, although there is no universally accepted tool. As such, it is recommended that independent

[5]The design of experiment approach is a systematic, statistical method for testing the effects of multiple input factors on output responses simultaneously. It involves a series of planned experiments that make purposeful changes to variables to identify the cause-and-effect relationships, often with the goal of optimizing a process or product with minimal resources. The approach relies on core principles like replication, randomization, and statistical analysis to ensure valid and reliable conclusions, unlike traditional methods that test one factor at a time.

analysis is undertaken to provide verification. This can be achieved by using an independent calculation; verification of Monte Carlo analysis can be demonstrated using a separate Monte Carlo analysis or FORM/SORM analysis (Du 2005), if valid. Running independent Monte Carlo analyses may introduce an excessive computational burden. Validation of the overall probabilistic approach based on the observation of in-service failures may be appropriate for cases with a large number of repeated structures, such as AGR superheater tubes, but is not appropriate for the small number of highest-level reliability components. In these cases, there is a validation precedent from other high-integrity industries such as rail and aerospace where target reliability-based acceptance criteria are used in structural integrity assessment.

6.4.4 NURBIM Probabilistic Fracture Mechanics Guidelines

The Nuclear Risk-Based Inspection Methodology (NURBIM) was a project funded by the European Union under the fifth Framework Programme (2001–2004) (Schulz et al. 2004). The objective of NURBIM was to advance the recommendations of the European Network of Risk-Informed In-service Inspection (EURIS) and subsequent work of the European Network for Inspection and Qualification (ENIQ) to develop improved procedures to identify where the highest probability of failure is located in passive systems, structures, and components and to provide quantitative measures of the associated risk. NURBIM produced two documents associated with implementations of PFM: (1) NURBIM report D2 "*Definition of a Set of Criteria That Should be Met by a Suitable Structural Reliability Model*" and (2) NURBIM report D4 "*WP-4, Review and Benchmarking of SRMs and Associated Software*." NURBIM noted that the implementation of PFM takes on many different forms (Brickstad et al. 2004).

With respect to verification and validation of PFM software, NURBIM Report D2 stated that the following criteria should be met (Chapman 2004):

1. The basic programming can be shown to have suitable quality assurance documentation.
2. That the scope, analytical assumptions, and limitations of the modeling capability are well defined.
3. That the analytical assumptions in (2) are well grounded and based on the theory that is accepted as representative of the situations considered by the given structural reliability model (SRM).
4. That the model is capable of reproducing the data on which its analytical assumptions are based, and examples are provided that can demonstrate its general agreement with the available experimental data.
5. Attempts have been made to show how the model compares with the world or field data, accepting the inherent limitation of this data.

6. That the model has been benchmarked against other SRM models within the same field or scope and that any differences are adequately explained.

With respect to item 5, comparisons with operating experience, NURBIM Report D2 noted:

Since the objective of any SRM is to provide a realistic estimate for structural failure rates within industry, it would seem very logical to argue that the historical data from the industry on such failures should be fundamental to the V&V of any SRM! Unfortunately, there are inherent difficulties in using this data as a means of V&V for an SRM. The first difficulty with this type of data is that it is by definition a simple statistic. Using the frequentist approach to the definition of failure probability from this type of data gives the following:

$$\textit{Probability of failure per year} = \frac{N_f}{N}$$

where N_f is the sum of all the observed failures for a particular family of components or welds and N is the total operating years' experience for that family of components or welds.

It can be seen that this equation gives just the one value for the total family i.e., a simple statistic. This simple statistic is, in reality, a point estimate of the mean probability of failure for the family of components or welds. It tells us nothing about the difference in the failure probability within the family, that is to say, the distribution of failure over the family. An SRM, on the other hand, attempts to assign this concept of failure probability associated with every member of the family as a function of the construction, inspection, operational history, etc.

The only way in which an SRM can give a mean value is to run the SRM for all the components or welds within the family. In order to do this, the input data for the SRM would need to be known, i.e., the cyclic stresses, the number of cycles, and the environment. If this were available, the SRM would give a distribution associated with that family, which in turn would provide a mean or expected number of failures for that family. This value could then be compared with the single estimate from the plant experience.

In this way, the industrial experience could be seen as a very large experimental program. Unfortunately, it is unlikely to have been a very well-controlled experiment because the data required to run the SRM would probably not have been recorded!

This leaves us with the difficult problem of identifying a statistic, based on a number of failures, from a set of components, such as a set of pipe welds, which can be considered as a family. For such a set to be considered as a family, it would be necessary to show that:

(a) Their size and thickness were similar.
(b) The welding procedures/inspection standards were similar.
(c) There was sufficient knowledge about transient history such that again these could be seen as being similar.

Such a precise set of data may prove somewhat elusive.

Finally, there is the question of the absolute levels of some of the failure probabilities that need to be assessed. Within the nuclear industry, there is now something of the order of 10,000 years of operating experience. If there are several examples of specific components or sites within each plant, then this volume of experience should provide some confidence, albeit limited, in failure probability rates down to 10^{-4} or even 10^{-5} per year if there are sufficient examples per plant. While many of the failure probabilities may be within this range or greater, it can be seen that the probability of validating failure rates of 10^{-5} or less from this source is not a viable option for the near future.

> However, as the number of SRM analyses grows, the results they produce will build into a global picture of the expected outcome in terms of component failures. This global assessment of the failure probability can then be compared with the actual global data. At a more specific level, if an SRM is used in a safety campaign for a given utility, then the SRM-generated global data can be compared with the utility-specific data and the world data in order to provide a validation/sanity test of the complete analysis.

With respect to PFM software, NURBIM Report D4 proposed the following requirements and recommendations:

Requirements

1. The PFM theory and technical basis should be published and independently reviewed.
2. The PFM model and the associated software should address the relevant degradation mechanisms under consideration.
3. The PFM model and the associated software should be able to evaluate failure frequencies for leak events and ruptures.
4. A sensitivity study using the PFM model and the associated software should be presented, addressing the relevant degradation mechanism under consideration. In the sensitivity study, failure probabilities for "small leaks" to "ruptures" should be evaluated for variations of input parameters and shown to be consistent with expectations and the given PFM theory assumptions.
5. Sample calculations should be presented where the assigned input parameters should be described, and sources of the data assignments should be given. The probability distributions and internally assigned parameters in the PFM software should be documented. Also, the limitations of the PFM software should be clearly identified.

Recommendations

1. The PFM software should be benchmarked against at least one other open-source PFM software for the relevant material degradation mechanism under consideration. The report on such a benchmark should be published and independently reviewed.
2. The PFM software should be benchmarked against operating experience data using actual plant failure frequencies. For damage mechanisms where no ruptures have occurred, leak frequencies may be used for the comparison.
3. Hardwired formulations should be avoided as far as possible, except to avoid the risk of misuse of the software. Formulations of input data (type of probability distribution and its random properties) should, as far as possible, be decided by the user. This also involves the possibility of using a deterministic input for the variables.
4. The PFM software should enable the user to extract control variables from the results. This is important for the user to be able to check the solutions and understand why a certain result is achieved. Examples of control variables (for a cracking damage mechanism) are:

(a) Initial crack sizes, crack shapes during the subcritical growth, and critical crack sizes at leak and rupture
(b) Time to leak and time to rupture
(c) Stress intensity factors and J-integrals[6]
(d) Crack opening areas and leak flow rates for through-wall cracks

5. The influence of inspections should be included in the SRM and the associated software in order to quantify risk reductions from repeated inspections.
6. When rupture probabilities are evaluated, it is important to model leak-before-break events. In this context, an adequate model is important with respect to crack opening areas, leak flow rates, and leak flow rate detection.
7. The software used should be clearly identified. It is desired that new information or better modeling assumptions should be continuously incorporated into the SRM and the associated software so that the generated results may reflect the best current knowledge.

6.4.5 IAEA Guidelines for the Estimation of Pipe Failure Rates

IAEA Nuclear Energy Series No. NR-T-2.16 (Winkler 1968) (IAEA 2023) presents a review of good practices for the assessment of piping reliability parameters for advanced water-cooled reactors. Good practices are those processes and analytical tasks that would be expected in piping reliability analysis for the results to be realistic representations of piping structural integrity. Section 3 of NR-T-2.16 presents a framework on how to organize a piping reliability analysis task. The framework has evolved over a period of multiple decades, and it has benefited from technical insights gained from practical applications performed in different contexts and by different analysts. The framework consists of seven steps (or activities). The objective of the analysis framework is to promote consistency in how an analysis is organized, executed, and documented. The seven steps are as follows:

1. Define the evaluation boundary. An evaluation boundary defines the piping system and its components (e.g., bends or elbows, pipes, tees, welds) for which reliability parameters are to be derived. This step requires numerous sets of information, such as piping and instrumentation diagrams, isometric drawings, and in-service inspection program plans.
2. Identify the potential and observed degradation mechanisms. Degradation mechanism assessment is a formal, systematic process for identifying the possible

[6]The J-integral represents a way to calculate the strain energy release rate, or work (energy) per unit fracture surface area, in a material. Experimental methods were developed using the integral that allowed the measurement of critical fracture properties in sample sizes that are too small for linear elastic fracture mechanics to be valid.

degradation mechanisms, and it is based on many decades of accumulated knowledge about the conjoint requirements for pipe degradations.

3. Extract and screen the operating experience (OE) data that correspond to the evaluation boundary. The pipe failure OE data to be extracted from a pipe failure database corresponds to the evaluation boundary for which the reliability parameters are to be determined. However, regardless of the selected quantification scheme, the analyst is expected to examine the existing, relevant OE from water-cooled reactors to determine if there is a basis for reconciling model output with real data. Annex III provides supporting information with reference to possible sources of piping OE data and piping reliability parameters.
4. Select the strategy for performing an analysis. Based on a specific evaluation boundary, in this step, the calculation case(s) and the specific reliability metrics are defined. Supporting engineering calculations are identified for the purpose of calculation case definition, such as specific pipe failure modes to be considered and with reference to potential spatial impacts as characterized by a zone of influence. Examples of engineering calculations include transient thermal-hydraulic analysis and simulation. Leak rate calculations are done on the basis of equivalent break size (EBS), crack opening displacement (COD), or crack opening area (COA). Section 7 of the IAEA publication includes a discussion on methodological selection criteria that are based on insights from practical applications as well as insights obtained from benchmark studies.
5. Perform the analysis by applying one or more methods and method implementations. This step is implemented by:
   (a) Defining input parameters
   (b) Documenting assumptions
   (c) Determining the types of sensitivity analyses to be performed
   (d) Performing an integrated uncertainty analysis and validating the model output
6. Synthesize the insights and results. Certain follow-up (or sensitivity) studies may have to be performed once a base-case set of probabilistic failure metrics has been obtained.
7. Document analysis: The documentation consists of assumptions, input parameters, model outputs, and result interpretation. End-user requirements in an analysis affect how the results are processed and documented.

6.4.6 EPRI Guidelines for Performance of Internal Flooding PSA

EPRI's "*Guidelines for Performance of Internal Flooding Probabilistic Safety Assessment*" include step-by-step guidance for the performance of an internal flood probabilistic safety assessment (IF-PSA) (Fleming et al. 2009). The scope of this guide includes the treatment of piping failures in low-energy systems and selected high-energy systems, i.e., systems containing water and steam with pressure

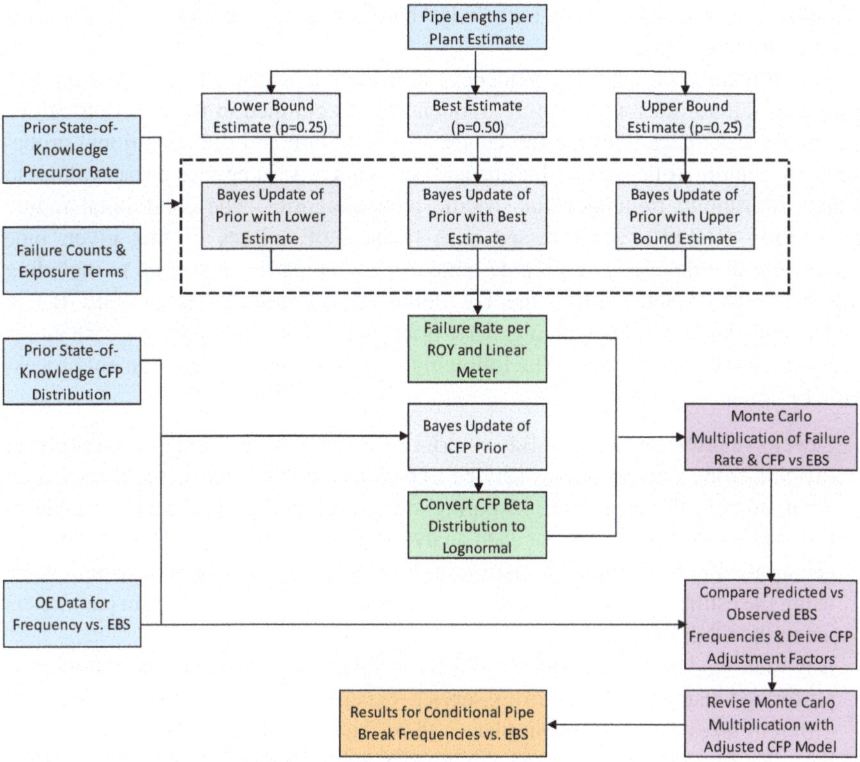

Fig. 6.3 EPRI pipe rupture frequency calculation process

and temperature above saturated conditions. The inclusion of high-energy systems is necessary because a high-energy line break (HELB) can lead to flooding and spatial dependencies. All fluid sources outside of the containment structure that have potential to cause flooding and/or HELB impacts are to be considered in IF-PSA. Section 5 of the guidelines offers guidance for flood-induced initiating event frequency quantification, with technical details presented in a companion report.

The companion report includes user instructions in the form of graphical representations of the calculation processes and for how to apply the probabilistic failure metrics in IF-PSA (Fig. 6.3) (Fleming et al. 2023, 2025). Also included are extensive tabulations of piping system failure rates and flood frequencies. A software tool, FRANX, has been developed to assist in developing and quantifying flood scenarios. All flood frequencies and adjustment factors are provided in Excel tables and a .csv exportable input file for FRANX Version 11 in Appendices A–E of the companion report. Training videos have been developed for use of the new FRANX files, including additional videos or knowledge base articles on the use of FRANX for internal flooding PSA.[7]

To estimate pipe rupture frequency, a modeling approach is employed that correlates failure rates with rupture frequencies. As outlined in the equation below, this model calculates both the frequency of pipe failure and the conditional probability of rupture following a failure and subsequently integrates these values to derive the rupture frequency for specific pipe categories. The conditional failure probability, $P_i(R_x | F)$, is expressed as a fraction of failures of that given pipe component that develops into a pipe break for each pipe break size up to including complete offset rupture of the pipe, i.e., double-ended guillotine break (DEGB).

The calculation begins with a generic prior distribution for each pipe case and is updated with the service data. The following steps describe the development process for CFP models:

1. Develop a prior distribution: Bayes' prior distribution is developed to incorporate insights from generic piping service experience and results of previous expert elicitations in the estimation of piping system reliability. This yields a table of CFP values vs. break sizes for each analyzed pipe size category. This is fitted to a constrained non-informative distribution (CNID) using the beta distribution, by using the estimates as mean values and estimating the beta distribution parameters using the CNID method.
2. Bayesian update (Stage 1): A Bayesian update of the prior distribution is performed using evidence from service experience on frequency of leaks of different sizes for systems similar to those being analyzed. This yields a new beta distribution and reflects the combined evidence expressed in the prior and the industry service data.
3. Bayesian update (Stage 2): A second-stage Bayes' update is applied to calibrate the CFP model to synchronize the predictions made by the pipe failure model to the observed industry data on the frequency of leaks of different magnitudes. This step was added by observing that predictions made by the EPRI pipe failure model used in previous revisions to the failure report were somewhat overestimating the frequency of leaks of different sizes observed in the industry service data.
4. Uncertainty treatment: Adjustments are made to the uncertainty distribution of the CFP model to account for modeling uncertainty. This includes the basic

[7] FRANX is a software tool for analyzing PSA internal and external event risk, such as fire, internal flooding, seismic events, or other scenario-based hazards. FRANX performs the following functions: 1) manages the list of scenarios that need risk calculations, 2) manages the mapping of the plant components to the elements of the PSA model that are affected by the event, 3) calculates the risk of each scenario using the PSA model, 4) manages the process of screening the scenarios starting with conservative assumptions and then refining the assumptions as necessary, 5) provides analysis and diagnostic information that allows determination of the primary contributors to the risk, and 6) creates a single-top configuration risk model that can be used by tools such as the EPRI EOOS Configuration Risk Management software. These results enable further refinement of the scenario assumptions and identification of new scenarios.

assumption in the pipe failure model that pipe failures can be regarded as precursors to pipe ruptures of various sizes. This is done by converting the means of the beta distribution from the Bayesian update of the CFP priors to means of lognormal distributions in which the lognormal range factor expresses a larger degree of uncertainty than is reflected in the Bayes' updated beta distribution. The lognormal distribution range factors were set based on the following engineering judgments:

(a) A range factor (RF) of 2 was assigned to a CFP value of 1E-2.
(b) An RF of 30 was assigned to a CFP value of 3E-5.
(c) For specific CFP values other than the above, an RF was calculated based on log-log linear interpolation/extrapolation between these points.

The failure rate distributions and CFP distributions for different pipe sizes are combined using a Monte Carlo multiplication procedure to produce results in the form of cumulative frequencies for each equivalent break size and pipe size.

6.5 Expert Judgment and Expert Elicitation

Expert elicitation (EE), also known as expert judgment elicitation (EJE), involves synthesizing the opinions of three or more subject matter experts (SMEs) in situations where uncertainty arises due to incomplete or insufficient data on a physical phenomenon. Expert judgments are well-founded opinions based on analysis, experiments, knowledge, and experience, provided by experts to address technical problems. These judgments reflect the expert's state of knowledge at the time they respond to a technical inquiry (Apostolakis and Mosleh 1979; Winkler 1968). Expert judgment plays an important role in assessing and improving piping reliability, especially when data is limited or unavailable. Experts can provide valuable insights based on their experience and knowledge, aiding in failure prediction, risk assessment, and development of in-service inspection strategies.

Expert elicitation is a scientific consensus methodology that may involve analytical work within a clearly defined context. It is frequently employed in the study of rare events, such as significant structural failures of piping systems. This approach enables the "parametrization" of the topic under investigation and generally quantifies uncertainty by offering upper and lower bound estimates. Expert elicitation remains a valuable source of data in probabilistic piping reliability analysis, especially when one or more of the following situations exist:

- No other data for answering the technical problem or issue is available.
- High variability characterizes the data.
- Subject matter experts question the applicability or validity of operating experience data.

- Existing operating experience data needs to be supplemented, interpreted, or incorporated with a new model or updated calculations of probabilistic failure metrics.
- Analysts need to determine the state of knowledge about what is currently known, what is not known, and what is worth learning.

Expert elicitation has sometimes been a principal component of the technical basis for important regulatory decisions. The following are examples of formal expert elicitations:

1. NUREG-1150, *Severe Accident Risks: An Assessment for Five U.S. Nuclear Power Plants*, utilized expert judgment to assess core damage frequency, failure frequencies, failure modes, recovery actions, accident progression, and source term behavior, among other phenomena to characterize the risk associated with severe accidents in operating reactors (USNRC 1989) (USNRC 1990). Details on aggregation of results of an expert elicitation to assess core damage frequency attributed to pipe failure were documented in NUREG/CR-4550, Volume 2 (Wheeler et al. 1989).
2. Expert elicitation has been used for probabilistic seismic hazard assessment (PSHA) for siting and design of nuclear facilities. Due to large uncertainties in the geoscience data and in their modeling, multiple model interpretations are often possible. This has led to disagreements on the selection of ground motion for design at a given site. A "Senior Seismic Hazard Analysis Committee" (SSHAC) was formed in 1993 under the sponsorship of the United States Department of Energy, the USNRC, and EPRI. This committee concluded that the differences in PSHA results were due to procedural rather than technical differences and proposed a formal method for utilizing expert judgment in PSHA.
3. Expert judgment has been required to assess the performance of a high-level waste repository, specifically, to predict future climates, characterize waste degradation, and transport if the waste package is breached and to perform volcanic hazard analysis. According to Shrader-Frechette (1992): "*Some types of expert judgments are obviously unavoidable in both science and risk assessment. Hence, the problem is not expert judgments, as such, but the frequent absence of any ideas about the limits of error or uncertainty associated with them in many Yucca Mountain assessments and the tendency of experts to be overconfident about the effects of their judgments.*"
4. Human reliability analysis relies on expert judgment and sometimes on formal expert elicitation input (Reiman 1996; Peng et al. 2020).
5. Expert elicitation has been used to estimate loss-of-coolant accident (LOCA) frequencies. These frequencies provided the basis for selecting the transition break size proposed in the risk-informed revision of the emergency core cooling system acceptance criteria (USNRC 2008a).
6. The Phenomena Identification and Ranking Table (PIRT) process is wherein an expert panel is convened to rank potential material degradation scenarios according to their judgment of susceptibility and current state of knowledge. Two examples are:

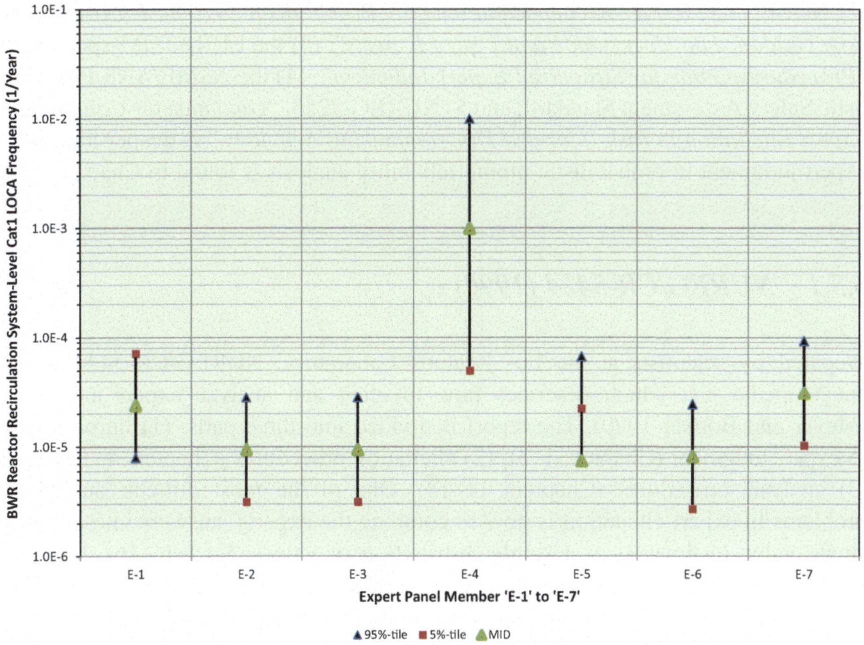

Fig. 6.4 Expert elicitation input to LOCA frequency assessment

(a) Expert Panel Report on Proactive Materials Degradation Assessment (PMDA), NUREG/CR-6923 (Andresen et al. 2007)
(b) Expanded Materials Degradation Assessment (EMDA), NUREG/CR-7153 (Busby 2014)

Arguments against expert elicitations include a possible lack of consistency and transparency in how an elicitation is conducted and in how results are combined, or aggregated, into a single estimate of, say, a frequency of occurrence of a rare event. An example is presented on potential complications that might arise when aggregating results from eight expert elicitation panel members (Fig. 6.4), one of which provided a significant outlier. Before pursuing aggregation of results, what were the underlying assumptions and what risk metric was considered in the analyses performed by the fictitious expert panel member "G." In this example, a small (Cat1) loss-of-coolant accident corresponds to a through-wall mass flow rate of >6 kg/s. The most significant event to date has produced a mass flow rate of about 0.2 kg/s![8] The question then becomes how to reconcile expert elicitation inputs.

Five examples of guidance for the conduct of expert elicitation are (1) NUREG/CR-5424, "*Eliciting and Analyzing Expert Judgment: A Practical Guide*";

[8]Duane Arnold on June 14, 1974.

(2) NUREG/CR-6372, "*Recommendations for Probabilistic Seismic Hazard Analysis: Guidance on Uncertainty and Use of Experts*"; (3) the EURATOM-sponsored "*Procedures Guide for Structured Expert Judgment*"; (4) the ASME/ANS Probabilistic Safety Assessment Standard; and 5) NUREG-2255, "*Guidance for Conducting Expert Elicitation in Risk-Informed Decisionmaking Activities.*" A deeper look into expert judgment in probabilistic piping reliability analysis is found in Chap. 7.

6.5.1 NUREG/CR-5424 (1990)

Prepared by scientists at the Los Alamos Laboratory, NUREG/CR-5424 is an encyclopedic text which describes how to elicit and analyze expert judgment (Meyer and Booker 1990). The report is divided into three parts: (1) introduction to expert judgment (Chapters 1–3), (2) elicitation procedures (Chapters 4–10), and (3) analysis procedures (Chapters 11–18). One of the most difficult analytical problems in expert elicitation is how to combine the experts' answers into either a single value (estimator) or a single distribution of values. Sixteen (16) different methods of mathematical aggregation are described and compared in the report. The authors make the point that there is no shortage of techniques available for mathematical aggregation; however, many of these techniques impose restrictions on the data, the experts, the analyst, and the interpretations of results.

6.5.2 NUREG/CR-6372 (1997)

Volume 2 of NUREG/CR-6372 consists of an Appendix J, "*Guidance on Technical Facilitator/Integrator Principles and Procedures,*" which documents details on how to integrate diverse expert interpretations of information needed in performing probabilistic seismic hazard analysis (PSHA). Compared to earlier and later "guidelines," NUREG/CR-6372 gives in-depth information on mathematical aggregation. The guidance on technical facilitator/integrator (TFI) processes and procedures covers the following topics (Budnitz et al. 1997):

1. Aggregation of the elicited information using mathematical schemes and consensus schemes. On mathematical aggregation, the authors of the report make the following statement:

   > Mathematical aggregation has several advantages. The logic is transparent and completely checkable. Combination formulas can isolate and separate specific assessments of dependence, expertise, and overlap, so that sensitivity studies are straightforward. Unfortunately, given the current state of the art, there are several substantial disadvantages to mathematical aggregation, at least if applied in a cookbook fashion. Mathematical models are not advanced enough to include all the factors that are important. A survey of the literature on this subject reveals that each model makes assumptions that may be only partially relevant to the problem at hand; worse, important

considerations, such as the degree of interdependence of the expert inputs, either are not handled at all by most models or are addressed in a very limited way. This view is consistent with Winkler's statement (Statistical Science, 1968).[9] "My own feeling is that different combining rules are suitable for different situations, and any search for a single, all-purpose, 'objective' combining procedure is futile." Similarly, Kaplan (1992)[10] states: "Needless to say, the expert opinion 'problem', so called, is never going to be 'solved' by any single mechanical or cookbook-type procedure. The expert information approach should be viewed as one more tool in the toolbox. A skillful user will create the right mix to fit each situation and his or her own style.

2. Guidance on expert elicitation facilitation: A technical facilitator integrator (TFI) must act as a scientist, as an information elicitor, and as a facilitator for group interactions. The substantive expert in the TFI team is essential in clarifying and facilitating scientific interchange and in summarizing points of agreement and disagreement. It is preferable to have the TFI substantive expert lead most of the technical discussions, with the TFI normative expert taking a supporting role in the face-to-face group discussions.
3. Guidance on integration: issues and models: This section documents important aggregation issues that are commonly encountered in practice:
   (a) Different degrees of expertise.
   (b) Outliers: If the issue is dealt with directly, outlier interpretations can be dealt with explicitly by forming judgments about their value conditional on the logic and empirical support that is supplied. For example, if a proponent supplies an outlier estimate, supplies a credible rationale, and provides explicit supporting data and interpretations, then that estimate ought to be given significant weight. However, if the proponent cannot technically support an outlier position, then the position should be given less, if any, weight.
4. Mathematical aggregation model for independent experts: This section of the guidance presents the "Bayesian normal model" for independent experts who provide a point of a scalar quantity. In this case, the experts provide estimates of some uncertain quantity, μ. Each expert's input is in the form of a forecasting error, ϵ, which denotes the difference between an estimate and the true value. If the forecasting errors are assumed to be normally distributed, the result of a Bayesian analysis is that the TFI's aggregated distribution is normal as well and, in the case of three experts, with mean,

[9] Winkler, R. L. (1968), "The Consensus of Subjective Probability Distributions," Management Science, 15(2): B61-B75.

[10] Kaplan, S. (1992), "Expert Information vs. Expert Opinions. Another Approach. to the Problem of Eliciting/Combining/Using Expert Opinion in PRA," Reliability Engineering and System Safety, 35:61–72.

$$\mu_* = \lambda_1\mu_1 + \lambda_2\mu_2 + \lambda_3 y_3 \tag{6.1}$$

where λ_i represents the weights assigned to each expert's input and which sums to unity.

5. Nonindependent experts: Explicitly dealing with dependence among proponent experts is in many ways the toughest expert integration issue of all. Stated simply, the degree of overlap in expert databases and/or models and reasoning processes can significantly affect the ability to integrate the knowledge of a group of experts. Here is a simple example: two expert weather forecasters use different methods and data to come up with two forecasts of rain of 20% and 80%. If the experts are equally credible, you might conclude that you should average the two estimates to obtain a "best estimate" of 50%. Suppose then that a third forecaster says 80%. How does this affect your state of information? Common sense says that it matters a great deal whether the third forecaster provided an independent forecast or simply used the same data and reasoning as the other 80% expert. The generally accepted decision analysis definition of expert dependence (Morris 1974) is based on a conditional probability statement: two experts, El and E2, are considered to be dependent when the probability distribution on E2's estimate (before he or she provides it), given the true value of the variable of interest, depends also on El's estimate. In other words, E2's estimate is correlated, not only with the true value, but also with El's estimate. Morris presents a theory of expert resolution which is consistent with the Bayesian or subjectivist view of probability.
6. Dependence among experts: Aggregation schemes are provided for the case where a forecasting error is conditional on the knowledge of not only the true value pi but also the preceding experts' estimates.
7. Aggregation based on equal weights: The attraction of equal weights is that it avoids at least two extremely difficult issues: (1) no one is forced to make a judgment on who is the best expert, and (2) no one is forced to make what can be very difficult assessments as, if not equal, what is an acceptable option.

6.5.3 European Expert Elicitation Protocol (2000)

A proposed European expert elicitation protocol published in the year 2000 consists of three parts: (1) preparation for elicitation, (2) performing elicitation, and (3) integration of elicitation results (Cooke and Goossens 2000). During the years 1990–1999, the European Commission and the USNRC undertook a joint uncertainty study of accident consequence codes for nuclear power plants using structured expert judgment. The purpose was to perform an uncertainty analysis of two severe accident consequence codes, COSYMA[11] and MACCS,[12] and to develop a basis for how to conduct structured expert judgment for quantifying uncertainty. The European expert elicitation protocol consists of the following steps:

1. Preparation for elicitation:
   (a) Definition of case structure: Defining the case structure for an expert judgment study involves studying the physical/biological/environmental models for whose quantification expert judgment is to be employed. The parameters in the models have of course been assigned values, but these values may be uncertain. The activity of the analyst in this phase may be described as answering the following questions:
   (b) Identification of target variables: Once the case structure of a specific model is defined, the target variables over which uncertainty distributions are required must be identified. Target variables are those parameters of the model in question which satisfy three criteria: (1) the values of the parameters are uncertain, (2) the uncertainty cannot be quantified with historical and/or measurement data, and (3) the uncertainty is expected to have a significant impact on the uncertainty of one or more endpoints of the model.
   (c) Identification of query variables: Variables which will be presented to the experts in the elicitation and for which they will quantify their uncertainty are termed query variables. It is assumed that the target variables to be quantified via structured expert judgment have been identified. The question is how to obtain uncertainty distributions over these variables. In defining the variables about which experts will be asked, two rules apply: (1) ask for values of observable or potentially observable quantities, and (2) formulate questions in a manner consistent with the way in which an expert represents the relevant information in his knowledge base.
   (d) Identification of performance variables: Empirical control is built in the elicitation procedure by asking experts to assess calibration or seed variables. These are variables whose values are or will be known to the analyst within the frame of the exercise but not to the expert. Seed variables are important for assessing the performance of the combined experts' assessments. Seed variables also form an important part of the feedback to experts, helping them

[11]COSYMA (Code System from MARIA) is a program package for assessing the offsite consequences of accidental releases of radioactive material to atmosphere, developed as part of the MARIA program (Methods for Assessing the Radiological Impact of Accidents). Mainframe and PC versions of COSYMA are distributed to interested users by arrangement with the European Commission.

[12]The MELCOR Accident Consequence Code System (MACCS) is a fully integrated, engineering-level computer code developed at Sandia National Laboratories for the USNRC. MACCS simulates the impact of severe accidents on the surrounding environment. It uses a site's weather data to determine hypothetical land contamination levels, doses to individuals, health effects and risks on populations based on protective action recommendations, and economic losses resulting from an accident, https://maccs.sandia.gov/maccs.aspx.

to gauge their subjective sense of uncertainty against quantitative measures of performance

(e) Identification of experts.
(f) Selection of experts.
(g) Definition of elicitation format document: When the case structure is finalized, the target and query variables are identified, the potential dependencies are identified, and the seed variables are defined, the elicitation document can be prepared.
(h) Dry-run exercise: The dry-run exercise aims at finding out whether the case structure document and the elicitation format document are unambiguously outlined and whether they capture all relevant information and questions. One or two people experienced in the field of interest should be asked to provide comments on both documents.
(i) Expert training session: The experts are requested to provide subjective assessments on the query variables. They will represent their subjective assessments in terms of quantile points (e.g., 5%, 50%, and 95%). Most experts are unfamiliar with quantifying their degree of belief in terms of quantiles. For that reason, a training session is recommended with all experts attending the session. Such a meeting is also useful to discuss the case structures or evaluation boundaries.

2. Perform elicitation:

   (a) Expert elicitation sessions
   (b) Presentation of results

3. Integration of elicitation results:

   (a) Combination of expert assessments
   (b) Discrepancy and robustness analysis
   (c) Feedback
   (d) Post-processing analyses
   (e) Documentation

6.5.4 ASME/ANS Probabilistic Risk Assessment Standard

Section 14.3 of the ASME/ANS Probabilistic Risk Assessment Standard (ASME 2022), titled "*Use of Expert Judgment*," outlines the requirements for utilizing expert judgment outside of the PSA analysis team to address specific technical issues. The standard states that the guidance from NUREG/CR-6372 and NUREG-1563

(USNRC 1996) may be applied to meet these requirements. The technical requirements for expert judgment are specified as follows[13]:

1. Explicitly and clearly define the objective and intended use of the information that is being sought through the use of outside expert judgment.

   (a) Determine the need for outside expert judgment. Outside experts should be used for any of the following or related reasons:

      Complete experimental data exists that the PSA analysts know has been interpreted differently by different outside experts.
      More than one conceptual model exists for interpreting the technical issue, and judgment is needed as to the applicability of the different models.
      Judgments are required to assess whether bounding assumptions or calculations are appropriately conservative.
      Uncertainties are large and significant, and judgments of outside technical experts are useful in illuminating the specific issue.

   (b) Identify the expert judgment process. The PSA analysis team shall determine:

      The degree of importance and the level of complexity of the issue.
      Whether the process will use a single entity (individual, team, company, etc.) that will act as an evaluator and integrator and will be responsible for developing the community distribution or will use a panel of experts and a facilitator/integrator. The facilitator/integrator shall be responsible for aggregating the judgments and community distributions of the expert panel so as to develop the composite distribution of the informed community.

   (c) Identify and select experts. The PSA analysis team shall identify one or more experts capable of evaluating the relative credibility of multiple alternative hypotheses to explain the available information. These experts shall evaluate all available hypotheses and bases of inputs from the literature and from proponents and resource experts and shall provide their own input and their representation of the community distribution.
   (d) Assign responsibility for the expert judgment. The PSA analysis team shall assign responsibility for the resulting judgments, either to an integrator or shared with experts. Each individual expert shall accept responsibility for the individual judgments and interpretations.

The term "community distribution" refers to the range of opinions or beliefs held by a group of experts on a particular topic, essentially representing the spread of their individual judgments across a spectrum, often visualized as a probability density curve where the peak represents the most common viewpoint, and the tails represent the more extreme opinions within the expert community.

[13]This is paraphrased from the 2013 edition of the standard. The reader is encouraged to review the 2022 edition to identify changes, if any, to the guidance.

6.5.5 NUREG-2254

NUREG-2255, "*Guidance for Conducting Expert Elicitation in Risk-Informed Decisionmaking*" (USNRC 2024b), was prepared in response to a Staff Requirements Memorandum COMGEA-11-0001, "*Utilization of Expert Judgment in Regulatory Decision Making.*" This guidance consists of three parts:

1. A recommended expert elicitation process
2. Supplemental guidance for implementing an expert elicitation process
3. Technical basis for expert elicitation

Section 3.8 of NUREG-2255 is an overview of different mathematical aggregation methods and highlights pitfalls, including the potential for misrepresenting expert judgments, drawing misleading conclusions about scientific information, and adding biases to the conclusions. Sensitivity analysis is a necessary part of the integration step to determine the robustness of group estimates. According to the guidance, the largest sensitivity is associated with the method used to aggregate individual expert panelist estimates to obtain group estimates.

6.6 The Probabilistic Failure Metrics of Pipe Integrity

Piping pressure boundary leakage is prohibited by design codes and standards. Operational rules are in place in case a piping pressure boundary is compromised, whether a pressure boundary breach produces a minute drop leakage of process medium or more significant leakage. Therefore, applying probability theory, a measure of pipe integrity is the probability P_i for exceeding some limit state i within a defined period. When this occurs, an unintentional or disallowed condition is reached which leads to a prompt operability determination followed by an action. Deriving a limit state probability must consider the time interval, and hence, a frequency of being in a degraded state must be calculated. Each element in the structural integrity risk triplet is characterized by a probability density function, which takes many different definitions, details of which are given below.

6.6.1 Material Degradation Scenarios, S

The "enabler" for probabilistic piping reliability is information on the possibilities for a preexisting condition, e.g., cracked, thinned, or leaking pipe. Depending on a specific evaluation boundary, the only information available consists of observation of degraded conditions, experimentally obtained or through operating experience. The data of interest are manifold but expressed as a "failure rate" with a precise definition, e.g., number of observed circumferential cracks connected to the inside pipe wall and of a certain length and through-wall depth and for a certain nominal pipe diameter and material. Or it can be in terms of the number of multiple circumferential cracks within a heat-affected zone and of certain crack dimensions.

Parameter S_i represents the "precursor" to pipe failure and accounts for crack initiation and the potential for growth of a pipe flaw in the through-wall direction. This step produces a frequency of a degraded state. Next, the analysis investigates the "line of defense" against a degraded pressure boundary integrity, i.e., the probability of detecting a degraded state before going through-wall producing a detectable leakage. This is followed by an assessment of the likelihood that a material degradation scenario propagates into loss of pressure boundary integrity, parameter P_i. Finally, the consequence, C_i, of a pipe is represented by the product of S_i and P_i:

$$C_i = S_i \times P_i \tag{6.2}$$

The precursor analysis considers the incubation period and crack/flaw growth for different combinations of materials and degradation mechanisms (DMs). An incubation period refers to the total time that a certain piping pressure boundary goes through crack nucleation until the time the crack progresses to a critical degree such that there is an active crack growth in the through-pipe-wall direction. The quantitative assessment produces an estimate of the frequency of failure of a pressure boundary. The definition of what constitutes a failure depends on the purpose of an analysis. Failure can be a crack or leak that results in repair without immediate effect on operations or shutdown of the reactor. The input data comes from the operating experience review. To calculate the precursor rate, λ, one can start with a broad generic prior distribution. To address uncertainty in prior distribution, a constrained non-informative distribution (CNID) method can be used in which the mean value of a gamma distribution is anchored to the mean estimate (Atwood et al. 2003). The precursor rate is expressed in units of failures per reactor operating year and the number of components that are exposed to a certain material degradation mechanism. Next, the impact of aging, A, and reliability and integrity management, I, may be assessed to account for age-dependent in-service degradation and in-service inspection, respectively.

6.6.2 *Likelihood of Failure, P*

The analysis of the likelihood of failure, P, considers how an undetected pipe flaw or a minor through-pipe-wall leakage suddenly propagates to form a significant structural failure, potentially with dynamic effects on surrounding plant areas, structures, and equipment. This is represented by a conditional failure probability model that represents the margin to a major pressure boundary failure. The goal of the analysis is to establish a set of *CFPs* vs. break size for each consequence category (Table 6.6).

Table 6.6 An example of consequence categories

Plant type	LOCA category	Through-wall mass flow rate (kg/s)	Equivalent break size (EBS) (mm)
BWR	1	>6.3	12.7
	2	>100	46.6
	3	>300	82.6
	4	>1600	176.8
	5	>6000	456.2
PWR	1	>6.3	12.7
	2	>100	41.3
	3	>300	76.2
	4	>1600	176.8
	5	>6000	355.6
	6	>32,000	786.4

The *CFP* can be represented by a cumulative failure probability versus an equivalent break size or in terms of through-wall mass flow rate in kg/s. An insight from assessments of operating experience is that the slope of the *CFP* curve shows a relatively strong decrease in probability with an increasing failure magnitude. Another way of interpreting the results is to say that the conditional rupture probability is driven by the diameter of the pipe, i.e., the smaller the pipe, the lesser the steepness.

6.6.3 *Failure Consequences, C*

The consequence of a pipe failure is represented by the combination of two probability density functions. One is for the precursor rate, and another is for the likelihood of failure producing a certain release of process medium. Monte Carlo uncertainty analysis is performed to combine the uncertainties in the precursor frequency and conditional failure probability.

An intended application determines the formulation of how the consequence results are represented and computed. Result presentation can take many forms, e.g., cumulative or instantaneous hazard curves for a given consequence category or cumulative "consequence" versus equivalent pipe break sizes or through-wall mass or volumetric flow rate (Fig. 6.5).

Assuming that a pipe failure scenario of interest is defined as consequences in the range from x_1 to x_2 in Fig. 6.5, the frequency of such a consequence interval is computed from

$$F(x_1, x_2) = F(x_1) - F(x_2) \tag{6.3}$$

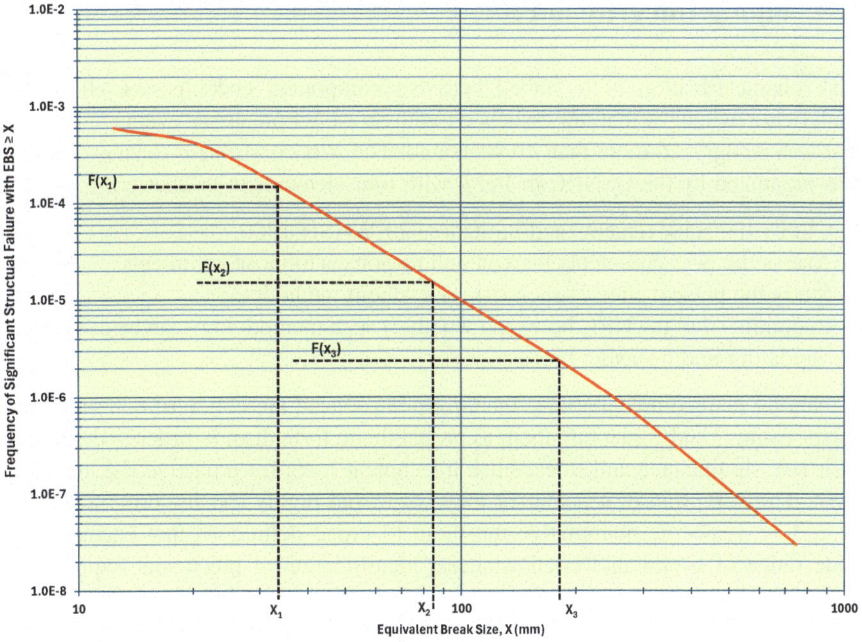

Fig. 6.5 Pipe failure frequency versus equivalent break size

The slope of the cumulative failure curve in Fig. 6.5 indicates a strong decreasing frequency versus increasing consequence. This means that the frequency of a range of equivalent break sizes tends to be dominated by the break sizes on the left-hand side of the range or side with the smaller equivalent break. If the consequences of the pipe failure scenario are different from the left-hand side to the right-hand side of the interval, the interval should be subdivided to avoid a conservative combination of high frequency based on the left-hand side of the curve and the high consequences from the right-hand side. Linear interpolation on a log-log curve of cumulative frequency versus break size determines the cumulative break frequency for a break size intermediate between two known points, e.g., x_1 and x_2.

The cumulative frequency for an equivalent break size between x_1 and x_2 is calculated from the following relations:

$$F(x_1, x_2) = 10^{log\,F(x_1) + m(log\,y - log\,x_1)} \tag{6.4}$$

$$m = \frac{log\,F(x_2) - log\,F(x_1)}{log\,x_2 - log\,x_1} \tag{6.5}$$

6.7 Piping Integrity in PSA

Risk characterization of a "failed" passive component encompasses all of the analytical challenges that are associated with so-called precursor events, a concept that was recognized more than 40 years ago. The *Risk Assessment Review Group*[14] was organized by the USNRC in 1977, with four elements to its charter:

1. Clarify the achievements and limitations of WASH-1400.
2. Assess the peer comments thereon and responses to those comments.
3. Study the present state of such risk assessment methodology.
4. Recommend to the NRC how (and whether) such methodology can be used in the regulatory and licensing process.

Among its recommendations, the Group introduced the notion of defining "precursor events" and subjecting them to probabilistic formalism in order to determine their risk significance and to establish a technical basis for proactive risk management. The definition of a precursor may vary from highly specific criteria such as exceeding a specific quantitative threshold to broad definitions that encompass a wide range of events and circumstances. Identification of precursors requires the review of operational events for instances in which plant functions that provide defense in depth against core damage have been challenged or compromised. Operational events involving piping pressure boundary failure can be associated with (1) reactor trip caused by a loss of feedwater due to pipe failure, (2) small-break LOCA, or (3) spatial impacts caused by pipe failure within the auxiliary building or turbine building (Fig. 6.6).

In response to recommendations of the Risk Assessment Review Group, the USNRC initiated the Accident Sequence Precursor (ASP) Program in 1979. The ASP Program is concerned with the identification and documentation of operational events that have involved portions of core damage sequences and with the estimation of associated frequencies and probabilities (USNRC 2008b). The term risk impact assessment is sometimes used in lieu of ASP analysis. There are many reasons for pursuing ASP-type evaluations, for example, as an integral aspect of regulatory oversight processes to determine risk significance of operational events such as a primary pressure boundary pipe leak. As another example, a precursor analysis may be performed in order to determine whether a certain event is reportable.

[14] Also referred to as the "Lewis Committee" after the Chairman of the Risk Assessment Review Group, Professor Harold W. Lewis (1923–2011), Department of Physics, University of California at Santa Barbara.

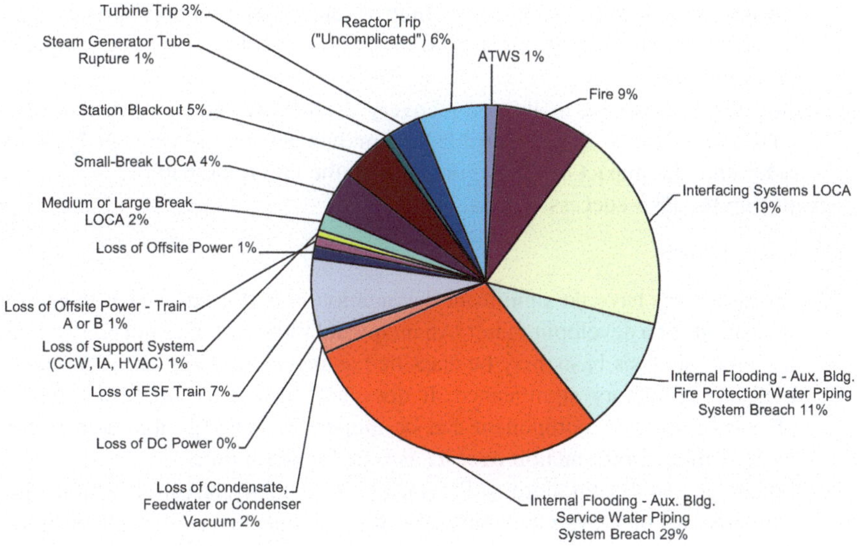

Fig. 6.6 Example of PSA results organized by initiating event category

6.7.1 Relevant PSA Relationships

The failures being considered involve both significant structural breakdowns and various types of early warning events. For instance, a major structural failure like a double-ended guillotine break usually leads to additional consequences, such as equipment being sprayed or areas of the plant getting flooded. The way these events are handled in PSA can be described as follows:

1. Significant structural failures
   (a) These result in reactor trip/turbine trip and require some form of replacement action on the component in order to restore the component to operability.
   (b) Significant structural failures are modeled by setting the basic event to TRUE (basic event probability of 1.0) and setting its non-recovery probability, if applicable, to TRUE. Consideration for spatial impacts complicates modeling.
2. Degraded failures

(a) Degraded failures can prevent a system or train from meeting the success criteria modeled in the PSA model.
(b) Degraded structures may fail from a more severe external event or fail at a condition outside its rated specifications.
(c) Degraded failures are generally modeled by one of the following applications:

Adjusting the failure probability to a higher value, based on appropriate engineering analysis, to reflect increased likelihood of failure (e.g., due to aging, crack growth).

Setting the basic event to its non-recovery probability (based on a recovery analysis) when it is not feasible to conduct an engineering analysis to determine the impact of the degradation on the failure probability.

Adjusting the PSA success criteria.

(d) Incipient failures

Incipient failures have no significant degradation in performance, but there are indications of a developing fault. An incipient failure that does not conform to its safety analysis basis may be classified as inoperable. The term "inoperable" has regulatory significance. It does not necessarily imply a state of physical failure. A component can be inoperable and still able to perform its PSA mitigation function over its assumed mission time.

Although an incipient failure will typically lead to a corrective action, the corrective action may or may not make the component unavailable to perform its function.

The figure of merit for accident precursor analyses is the conditional core damage probability (CCDP) for initiating events and the increase in core damage probability (*ΔCDP*) or importance for conditions. The importance is the measure of the incremental increase between the CCDP for the period in which the condition existed and the nominal core damage probability (CDP) for the same period. Two thresholds are noted in the Accident Sequence Precursor Program in the order of increasing severity:

1. An initiating event with a CCDP or a condition with a ΔCDP greater than or equal to 1×10^{-6} is called a precursor. The CCDP represents the probability that a core would have gone to a damaged state conditional on a specific initiating event and the actual plant equipment, and operator responses are accounted for. CCDP represents the remaining probabilistic "margin" or defense in depth relative to core damage at a precise moment in time, that of the event itself.
2. An initiating event with a CCDP or a condition with a ΔCDP greater than or equal to 1×10^{-3} is called a "significant precursor."

The ASP estimates the CCDP given a degraded condition for the time this degradation existed. The nominal CDP, which accounts for normal maintenance during this time, is subtracted from the CCDP to obtain the change in CDP due to the degraded condition alone (without consideration of any specific maintenance configuration that might have existed). This numerical result is then normalized to arrive at a delta CDF in units of "per year":

$$\Delta CDP = CCDP - CDP \tag{6.6}$$

Table 6.7 Relationship between INES scale and $ICCCP_{Event}$

INES scale	$ICCCP_{Event}$
3	$1 > ICCDP_{Event} \geq 1 \times 10^{-2}$
2	$1 \times 10^{-2} > ICCDP_{Event} \geq 1 \times 10^{-4}$
1	$1 \times 10^{-4} > ICCDP_{Event} \geq 1 \times 10^{-6}$
0	$1 \times 10^{-6} > ICCDP_{Event} \geq 1 \times 10^{-8}$

The Swiss Federal Nuclear Safety Inspectorate (ENSI) uses PSA in assessing the risk significance of an operational event. According to regulatory guideline ENSI-A06/e, the following probabilistic safety indicators shall be determined and assessed (Table 6.7) (ENSI 2016):

1. The maximum annual conditional core damage frequency, $CCDF_{i,\ max}$.
2. The incremental cumulative core damage probability, *ICumCDP*
3. Assessment of the trend of the above risk metrics
4. Assessment of the trend of these safety indicators
5. Reporting of the contributions to *ICumCDP* in terms of contributions from maintenance, repair, test, and reactor trip:

$$ICumCDP = \sum_{i=1}^{n} ICCDP_i \tag{6.7}$$

$$\sum_{i=1}^{n} ICCDP_i = \sum_{i=1}^{n} \left(CCDF_i - CDF_{Baseline}\right) \times \frac{\Delta t}{8760\ (hours\ per\ year)} \tag{6.8}$$

6.7.2 Acceptance Criteria

Probabilistic piping reliability analyses typically yield very low failure frequencies, often less than 1×10^{-6} per piping component per year, with significant uncertainties. Validation against various modeling approaches is required to assess the realism of these results. Meaningful comparisons require the definition of specific "figures of merit" to be calculated. Within the context of probabilistic safety assessment and risk-informed applications, piping pressure boundary integrity is generally incorporated into the initiating event frequency modeling element.

The relevant PSA parameter can be represented as a cumulative pipe break frequency relative to a set of consequence threshold values, which are expressed by an equivalent break size (EBS) in terms of mass flow rate or the corresponding hole diameter in the pipe wall that would generate a given flow rate. The "failure mode" of interest may range from minor through-wall leaks to major structural failures such as double-ended guillotine breaks. An analysis encompasses multiple sets of input parameters that consider the impact of various degradation mechanisms on structural integrity, effectiveness of in-service inspection and leak detection,

Table 6.8 Proposed acceptance criteria for PFM applications

Proposed application	Intent of application	Numerical acceptance criteria	Definition of acceptance limit
EPRI MRP-362: Basis for ASME Section XI Code Case N838—Flaw Tolerance Evaluation of Cast Austenitic Stainless Steel (CASS) Piping Components (Griesbach et al. 2016)	Evaluate design-intent probability of rupture for ASME class 1 CASS piping	1.0×10^{-6}	Annual probability of rupture per crack
NUREG-1874: Recommended Screening Limits for Pressurized Thermal Shock (USNRC 2010)	Demonstrate that the probability of failure of a class 1 reactor pressure vessel due to PTS is extremely low	1.0×10^{-6}	Annual frequency of occurrence of through-wall cracks cumulative for all welds in the vessel
xLPR Version 2.0 Technical Basis Document Acceptance Criteria (Hardies et al. 2016)	Demonstrate that the probability of failure of class 1 piping with dissimilar metal welds is extremely low	1.0×10^{-6}	Annual frequency of rupture cumulative for all susceptible welds

degradation mitigation techniques, and chemical and mechanical properties of the materials involved. Probabilistic acceptance criteria for pipe failure frequencies have been suggested, as indicated in Table 6.8.

When establishing acceptance criteria from PFM applications, it becomes important to ensure that the definition of failure is clearly defined, especially if results of different assessments will be combined or compared for risk-informed decision-making. For example, consider the information in Table 6.8. While all three applications propose the use of the same numerical value, 1×10^{-6}, the definitions are different in each case. Three observations arise from the comparison:

- The EPRI MRP-362 (Griesbach et al. 2016) criteria use an annual probability, while the other two examples use annual frequencies. Annual probability and frequency are not equivalent reliability parameters by definition. These parameters are only approximately equivalent for special cases established using annual timescale and when the numerical values are small.
- The EPRI MRP-362 (Griesbach et al. 2016) criteria apply to an evaluation of a single crack subject to a single service-level load. For the NUREG-1874 (USNRC 2010) and the proposed xLPR applications, the acceptance criteria would be compared to a summation of frequencies for all individual flaws in the system, considering all loading scenarios.
- The definition of failure for the NUREG-1874 criteria is the propagation of a crack through the wall thickness of a pipe. The evaluation does not consider the stability of the through-wall cracking. The criteria for the EPRI MRP-362 and

proposed applications of the xLPR computer code allow for the consideration of through-wall cracking stability.

As a result of the differences associated with these three observations, the acceptance limits for these three examples would generate different reliability targets. Creating probabilistic acceptance criteria for pipe failure frequencies and conditional probabilities is complex and time-consuming. It involves both system-level targets (top-down) and component-specific processes (bottom-up). The criteria and methodology must align with regulatory frameworks, international and national standards, and guidelines for fitness for service, operability, and PSA. These criteria inform decisions on matters such as:

- Acceptability of continued operation for another fuel cycle, given a preexisting, non-leaking or leaking pipe flaw.
- Acceptability of changing an existing in-service inspection program.
- Acceptability of the risk posed by a degraded piping component.

Acceptance criteria aim to evaluate probabilistic results. For instance, if the pipe failure frequency is shown to be below a specific value, it could justify a favorable decision. However, uncertainty in reliability metrics complicates assessments.

6.7.3 Safety Assessment Principles for Nuclear Facilities

The United Kingdom's Office for Nuclear Regulation (ONR) has developed a set of *safety assessment principles* (SAPs), together with supporting *technical assessment guides* (TAGs), to guide regulatory judgments and recommendations when undertaking technical assessments of nuclear site licensees' safety submissions. The primary purpose of the SAPs is to provide inspectors with a framework for making consistent regulatory judgments on the safety of activities. With respect to "*Engineering Principles: Pressure System*," SAP-284 and SAP-287 state (ONR 2020):

SAP-284: *The general lack of adequate reliability data for the disruptive failure of metal components and structures leads to assessments being based primarily on established engineering practice. As a result, although the radiological consequences of the failure of some components or structures may be significant (into the range where there are societal risks), it is not possible to calculate a plausible failure frequency for inclusion in a fault analysis. At best, it might be possible to adopt a representative failure rate that would allow the effects of the component or structure failure to be included in a fault analysis in a nominal way or as a sensitivity study. If the safety case is sensitive to the failure frequency, then the estimate will need substantial support from engineering analyses and engineering judgment. At the least, an engineering judgment would be needed to confirm that the component or structure in question has characteristics similar to those in the database used to determine reliability values. If lines of protection exist to cope with the effects of the initiating component or structure failure, the overall case may not demand high confidence in the structural integrity claim.*

SAP-286: *A rule of thumb, generally accepted in the United Kingdom for many years, is that it is difficult to substantiate a claim of much less than about 1×10^{-7} per vessel-year for the* *gross failure of a reasonably sized pressure vessel. Therefore, a claim that gross failure of a pressure vessel can be discounted is not plausible for failure rates much better than 1×10^{-7} to 1×10^{-8} per vessel-year. There is no generally accepted lowest plausible failure frequency for individual welds, for instance, individual pipe welds. As a general guide, claims for pipework weld failure rates for gross failure (e.g., guillotine failure) much better than 1×10^{-8} to 1×10^{-9} per weld-year should not be considered plausible. This implies that a facility safety case should not rely on claims that gross failure can be discounted for large numbers of pipework and similar welds.*

6.7.4 Risk Determinations

Significance determination (SD) is used in the United States when applying risk insights to determine the safety significance of regulatory inspection findings (USNRC 2018). The SD process (SDP) is part of the reactor oversight process (ROP), which uses a three-phased approach to determine the significance of inspection findings in the initiating events, mitigating systems, and barrier integrity. It estimates the increase in annualized CDF risk due to identified deficiencies in licensee performance that led to unavailability of equipment or safety functions or to the increase in initiating event frequencies. This increase is measured from the normal annualized CDF that results from routine plant operation. The additional risk contributions caused by deficient licensee performance (as characterized by the SDP) are assumed to be additive to this normal annualized CDF, which already includes the risk contribution due to the probabilities of equipment failures expected occasionally for industrial facilities of this size and complexity. With a common thread, two examples of SD analyses are given below:

6.7.4.1 Assessing Risk Significance of High-Cycle Fatigue Failure

On November 8, 2013, the Oconee Nuclear Station (ONS) Unit 1 control room received indications of potential reactor coolant system leakage within the containment (USNRC 2013; USNRC 2014a). On November 10, 2013, following initial indication of reactor coolant system (RCS) leakage in the vicinity of a high-pressure injection (HPI) isolation valve, operations personnel initiated a power reduction to 20% full power. On November 11, 2013, a containment entry determined that the leak was from a circumferential crack in the Code Class 1 safe end-to-pipe butt weld between the HPI injection nozzle and valve.

The regulatory inspectors determined that the failure to ensure that a station nondestructive examination procedure was adequate to achieve acceptable coverage for the ultrasonic examination of the affected weld was a performance deficiency. The procedure did not contain necessary steps to achieve acceptable examination coverage when limitations were encountered, nor did it contain guidance on actions to take when acceptable examination coverage was unattainable. The inspectors determined that the finding was more than minor because it affected the design control attribute of the initiating event cornerstone and adversely affected the cornerstone objective in that an undetected crack resulted in reactor coolant system pressure boundary leakage and a forced shutdown of Unit 1. The finding was determined to require a detailed risk analysis because the finding could result in a leak which exceeded the RCS leak rate for a small loss-of-coolant accident (LOCA). There was no immediate safety concern because the crack was repaired. The inspectors determined that the licensee did not create and maintain complete, accurate, and up-to-date documentation to ensure acceptable coverage for ultrasonic examinations. As a result of the significance determination, a "White Finding" was issued:

> Low-to-moderate safety significance, which is quantitatively greater than 10^{-6} and less than or equal to 10^{-5} ΔCDF or greater than 10^{-7} and less than or equal to 10^{-6} ΔLERF. Qualitatively, a White significance indicates an acceptable level of performance by the licensee, but outside the nominal risk range. Cornerstone objectives are met with minimal reduction in safety margin.

Based on different assumptions regarding the conditional failure probability given the through-wall flaw, the USNRC risk analysts determined the ΔCDF to be in the range of 1.3×10^{-6} to 2.8×10^{-6}. The significance determination process resulted in an increased regulatory oversight. In response, an independent assessment of the event was performed. In this assessment, two different models of piping reliability were applied, probabilistic fracture mechanics and a probabilistic decomposition approach.

In the probabilistic decomposition approach, precursor rates were derived from the operating experience and the conditional failure probability uncertainty distributions derived using a state-of-knowledge resource on LOCA frequency (USNRC 2008a). Based on estimates of the plant-specific initiating event frequencies for HPI line-induced small and medium LOCA initiating event frequencies, the change in CDF due to occurrence of the leak event was estimated to be about 5×10^{-7} per reactor calendar year. A sensitivity analysis was performed using a CFP model that was based on probabilistic fracture mechanics, and the results are indicated in Fig. 6.7.

6.7.4.2 Assessing Risk Significance of Thermal Fatigue

On December 22, 2014, the North Anna Unit 1 reactor power was reduced from 100% to 30% to allow containment entry into the RCS loop rooms to investigate an unidentified leak rate of 3.4×10^{-3} kg/s. During a containment walkdown, the licensee discovered steam coming from underneath the lagging on the "B" RCS intermediate loop. The licensee's investigation identified a pressure boundary leak on the "B" loop drain piping between the loop connection and 1-RC-68, the "B" loop cold leg drain isolation valve. At that time, the limiting condition for operation action statement of TS 3.4.13.b, RCS 6.3×10^{-2} kg/s unidentified leakage, was entered, which required placing the unit in Mode 3 within 6 h and Mode 5 within 36 h. Due to

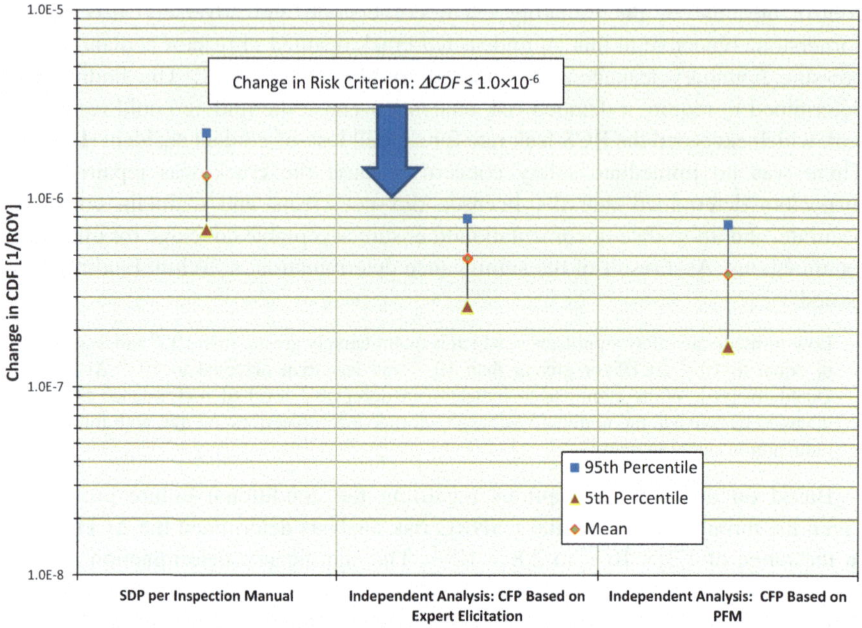

Fig. 6.7 Change in CDF due to pipe failure

the pressure boundary leakage, the event was reported at 0057 h on December 23, 2014, in accordance with 10 CFR 50.72(b)(2) for "initiation of plant shutdown required by technical specifications" and 10 CFR 50.72(b)(3)(ii)(A) for "any event or condition that results in the condition of nuclear power plant including its principle safety barriers being seriously degraded." On December 23, 2014, the unit was placed in Mode 3 (hot standby), and at 1629 h on December 23, 2014, the unit was placed in Mode 5 (cold shutdown) (USNRC 2014b).

The licensee determined that the cause of the through-wall leak in the "B" loop drain line elbow was thermal fatigue of the elbow. Limitations in the industry generic model used to predict swirl penetration thermal fatigue in stagnant RCS branch lines allowed the nondestructive examination (NDE) test frequency of the "B" loop drain line elbow to be set non-conservatively, which resulted in the thermal fatigue cracking to go unmonitored. The inspectors reviewed the licensee's root cause evaluation and the Electric Power Research Industry (EPRI) Material Reliability Program (MRP) 146, Management of Thermal Fatigue in Normally Stagnant Non-Isolable Reactor Coolant System Branch Lines (Deardorff et al. 2005).

The NRC inspectors determined that the licensee had appropriately evaluated the DN50 "B" loop drain line elbow and implemented the NDE test frequency requirements in accordance with the licensee's commitments on the current industry EPRI MRP 146 standard guidelines. No violations of ASME Code Section XI requirements were identified. The licensee is working with EPRI and industry peers to develop a new model and/or new guidance to better predict the impact of the thermal fatigue and other sources of thermal loading. Since the cause was not reasonably within the licensee's ability to foresee and correct, the inspectors concluded that there was no performance deficiency associated with this issue. The NRC inspectors utilized available risk-informed tools to assess the safety significance of this RCS pressure boundary leakage issue, and a detailed risk assessment for this issue was performed by a regional senior reactor analyst in accordance with NRC Inspection Manual Chapter 0609 Appendix A (USNRC 2015). The major analysis assumptions included a conservative 1-year exposure interval, the 50 mm break that was considered a small loss-of-cooling accident (SLOCA), conditional core damage probability for SLOCA from the licensee's full scope North Anna PSA model, and conditional rupture probability of 5.56×10^{-2} from an EPRI report (Fleming et al. 1999). The result of the analysis was a risk increase for the condition of 1.4×10^{-6} per year, representing low-to-moderate safety significance, also known as a White Finding.

Nonetheless, the inspectors determined that a violation of TS occurred because it constituted pressure boundary leakage. North Anna Technical Specifications Limiting Condition for Operation (LCO) 3.4.13.a requires, in part, that RCS leakage be limited to no pressure boundary leakage when in Modes 1, 2, 3, and 4. The associated action statement requires, in part, that with any (RCS) pressure boundary leakage, the unit be in hot standby within 6 hours and in cold shutdown within the following 36 h. Contrary to the above, on December 23, 2014, it was discovered that RCS pressure boundary leakage did exist while the unit was in Modes 1, 2, 3, and 4 and that the unit was not placed in hot standby within 6 hours and in cold shutdown within the following 36 hours. Although a violation of the TS occurred, the violation was not attributable to an equipment failure that was avoidable by reasonable licensee quality assurance measures or management controls. Therefore, the violation of TS 3.4.13.a was not associated with a licensee performance deficiency. Corrective actions resulting from the significance determination were as follows:

1. Performed an extent of condition NDE of all three cold leg loop drain lines for Unit 1. Results indicated low-level craze cracking and a circumferential defect in the similar elbow on "C" drain loop piping. No issues were found in the "A" drain loop piping.
2. Replaced the cracked elbow from the "B" loop drain piping.
3. Provided the failed "B" loop elbow/piping for materials' failure analysis. The results of the evaluation were used to confirm the direct cause of thermal fatigue and provide more insight on the failure mechanism.
4. Performed an evaluation in accordance with ASME Section XI Sub-article IWB-3640, Evaluation Procedures and Acceptance Criteria for Austenitic Piping, of the Unit 1 "C" loop drain elbow. The results determined that it was acceptable to operate the unit until the next refueling outage, during which, the repairs were made.
5. Discontinued sampling of RCS cold leg chemistry.
6. Informed the EPRI Materials Reliability Program of this failure and continued to work with the industry on this and similar issues.

The direct cause of the through-wall leak in the "B" loop drain line elbow was a thermal fatigue failure of the elbow. Limitations in the industry generic model used to predict swirl penetration thermal fatigue in stagnant RCS branch lines allowed the nondestructive examination (NDE) test frequency of the "B" loop drain line elbow to be set non-conservatively, causing thermal fatigue cracking to go unmonitored. The root cause was a legacy issue from 2009 due to the generic model not accurately predicting fatigue damage due to swirl penetration. A contributing cause was the fact that Unit 1 chemistry samples of the RCS have been collected from the "B" cold leg loop drain line since 2004. Chemistry sampling added additional thermal stresses, which accelerated the fatigue failure in the "B" loop drain line elbow. The screening calculation performed in 2010 to identify lines susceptible to thermal fatigue was performed by site and corporate engineering. The screeners did not have a strong questioning attitude to identify chemistry sampling as a thermal fatigue contributor. A questioning attitude would have prompted a discussion with chemistry department personnel for additional information on sample duration and frequency.

6.7.4.3 Observations

Both examples of significance determination assessments responded to operational events with less-than-adequate in-service inspection processes as their root cause. It is noteworthy that the regulatory significance determinations yielded similar results using very different assumptions regarding the conditional failure probability, 6.0×10^{-4} in the case of high-cycle fatigue failure and 5.6×10^{-2} in the case of thermal fatigue failure. The independent analysis of the high-cycle fatigue failure was performed to highlight the many technical considerations and assumptions that go into the detailed quantitative significance determination process.

A robust probabilistic piping reliability analysis requires a meticulous approach to integrating varied sources of input data. At each stage, statistical rigor is paired with engineering judgment to ensure that model assumptions and results remain grounded in physical reality. Calibration of parameters often involves iterative refinement: analysts may return to the original data or consult subject matter experts to resolve ambiguities and reduce epistemic uncertainties. Furthermore, as new data emerges from continuing operations or updated experiments, models are updated to reflect the evolving understanding of system behavior and degradation mechanisms. Once the foundational data is gathered and harmonized, models are constructed to represent both random and systematic influences on pipe failure. This includes accounting for known degradation mechanisms, inspection and maintenance histories, and external influences such as environmental conditions or material variations. Bayesian frameworks are particularly effective here, providing a means to blend prior knowledge with new evidence and allowing for explicit quantification of uncertainty in reliability parameters. Chapter 7 expands on Bayesian analysis frameworks.

References

Andresen PL, Ford FP, Gott K, Jones RL, Scott PM, Shoji T, Staehle RW, Tapping RL (2007) Expert panel report on proactive materials degradation assessment, NUREG/CR-6923. USNRC, Washington

Apostolakis G (1981) Bayesian methods in risk assessment. In: Lewins J, Becker M (eds) Advances in nuclear science and technology, vol 5. Springer, Boston. https://doi.org/10.1007/978-1-4613-9919-3_5

Apostolakis G, Mosleh A (1979) Expert opinion and statistical evidence: an application to reactor Core melt frequency. Nucl Sci Eng 70:135–149

ASME (2007) Development of reliability-based load and resistance factor design (LRFD) methods for piping, Research and Development, report CRTD-86, ASME Special Working Group on probabilistic methods in design. ASME, New York

ASME (2022) Standard for level 1/large early release frequency probabilistic risk assessment for nuclear power plant applications, ANS/ASME RA-S-1.1-2022. ASME, New York, NY

Atwood CL, LaChance JL, Martz HF, Anderson DL, Englehardt M, Whitehead D, Wheeler T (2003) Handbook of parameter estimation for probabilistic risk assessment, NUREG/CR-6823. USNRC, Washington

Brickstad B, Chapman OJV, Schimpfke T, Schulz H, Muhammed A (2004) Nuclear risk-based inspection methodology for passive components (NURBIM), WP-4, review and benchmarking of SRMs and associated software, NURBIM report D4, European atomic energy community (EURATOM), Brussels, Belgium

Budnitz RJ, Apostolakis G, Boore DM, Cluff LS, Coppersmith KJ, Cornell CA, Morris PA (1997) Recommendations for probabilistic seismic Hazard analysis: guidance on uncertainty and use of experts, NUREG/CR-6372. USNRC, Washington, DC

Busby, J.Y. (2014). Expanded materials degradation assessment (EMDA) volume 1: Executive summary of EMDA process and results, NUREG/CR-7153, USNRC, Washington

Chapman OJV (2004) Nuclear risk-based inspection methodology for passive components (NURBIM), definition of a set of criteria that should be met by a suitable structural reliability model, NURBIM report D2, European atomic energy community (EURATOM), Brussels, Belgium.

Cooke RM, Goossens LJH (2000) Procedures guide for structured expert judgment, EUR 18820 EN, Commission of European Communities, directorate-general XI (environment and nuclear safety) directorate D (DG 11 CCJH), Luxembourg/Brussels

Deardorff A, Keller J, Spain L (2005) Materials reliability program: management of thermal fatigue in normally stagnant non-isolable reactor coolant system branch lines (MRP-146), 1011955. EPRI, Palo Alto

Du X (2005) Probabilistic engineering design, chapter 7, first order and second reliability methods. University of Missouri, Rolla

ENSI (2016) Probabilistic safety analysis (PSA): applications, guideline for Swiss nuclear installations, ENSI-A06/e. Eidgenössisches Nuklearsicherheitsinspektorat, Brugg

Fleming K, Mikschl T, Bidwell D, Read J, Wiedemeier B (1999) Piping system failure rates and rupture frequencies for use in risk informed in-service inspection applications, TR-111880. EPRI, Palo Alto

Fleming K, Lydell B, Bullington L, Dimitrijevic V, Sarrack A, Wall I, Weston R, Schneider R (2009) Guidelines for performance of internal flooding probabilistic risk assessment, 1019194. EPRI, Palo Alto, CA

Fleming KN, Lydell BOY, Presley M, Mosleh A, Chalgham W (2023) Pipe rupture frequencies for internal flooding probabilistic risk assessments: revision 5, 3002024904. EPRI, Palo Alto

Fleming KN, Lydell BOY, Presley M, Mosleh A, Chalgham W (2025) Initiating event frequencies for internal flooding and high energy line break PRAs. J Nucl Eng. https://doi.org/10.3390/jne6030037

Griesbach T, Cofie N, Harris D, Dedhia D, Alleshwaram A (2016) Materials reliability program: technical basis for ASME section XI code case N-838 - flaw tolerance evaluation of cast austenitic stainless steel (CASS) piping components (MRP-362 rev. 1), 3002007383. EPRI, Palo Alto

Hardies R, Rudland D, Lupold T, Alley D, Csontos A, Dinsmore S, Dyle R, Harrington C (2016) xLPR version 2.0 technical basis document: acceptance criteria, USNRC and EPRI, NRC_ADAMS accession no. ML16271A436

Homiack M, Facco G, Benson M, Erickson M, Harrington C (2021) Extremely low probability of rupture version 2 probabilistic fracture mechanics code, NUREG-2247. USNRC, Washington

IAEA (2021) Technical insights from benchmarking different methods for predicting pipe failure rates in water cooled reactors. IAEA-TECDOC-1988, Vienna

IAEA (2023) Methodologies for assessing pipe failure rates in advanced water cooled reactors, nuclear energy series no. NR-T-2.16, Vienna, Austria

Kaplan S (1992) Expert information vs. expert opinions. Another approach. To the problem of eliciting/combining/using expert opinion in PRA. Reliab Eng Syst Saf 35:61–72

Kaplan S, Garrick BJ (1979) On the use of a Bayesian reasoning in safety and reliability decisions-three examples. Nucl Technol 44:231–245

Kaplan SJ, Garrick BJ (1980) Try probabilistic thinking to improve powerplant reliability. Power Magaz 3(March):56–61

Kurisaka K, Nakai R, Asayama T, Takaya S (2011) Development of system based code (1) reliability target derivation of structures and components. J Power Energy Syst 5(1):19–32

Lund L et al (2022) Technical basis for the use of probabilistic fracture mechanics in regulatory applications, NUREG/CR-7278. USNRC, Washington

Meyer MA, Booker JM (1990) Eliciting and analyzing expert judgment: a practical guide, NUREG/CR-5424. USNRC, Washington

Modarres M, Kirk M (2021) Review of current open literature for best practices probabilistic fracture mechanics assessment procedures including development of draft guidelines, RSP-690.1, prepared for the Canadian Nuclear Safety Commission. DML International, Ottawa

Morris PA (1974) Decision analysis expert use. Manag Sci 20(9):1233–1319. https://doi.org/10.1287/mnsc.20.9.1233

NSIPWG (2019) Nuclear structural integrity probabilistic working principles, nuclear structural integrity probabilistic working group. The UK Forum for Engineering Structural Integrity, Hutton

ONR (2020) Safety assessment principles for nuclear facilities, 2014 edition, revision 1. Office for Nuclear Regulation, Bootle. http://www.onr.org.uk/saps/index.htm

Peng L, Yongping Q, Juntao H, Jiejuan T, Jun Z, Zhizhong L (2020) Expert judgments for performance shaping factors' multiplier design in human reliability analysis. Reliabil Eng Syst Saf. https://doi.org/10.1016/j.ress.2018.12.022

Raynaud P, Kirk M, Benson M, Homiack M (2018) Important aspects of probabilistic fracture mechanics analyses, TLR-RES/DE/CIB-2018-01. USNRC, Washington

Reiman L (1996) Expert judgment in human reliability analysis. In: Cacciabue PC, Papazoglou IA (eds) Probabilistic safety assessment and management '96. Springer, London. https://doi.org/10.1007/978-1-4471-3409-1_104

Schulz H, Schimpfke T, Brickstad B, Chapman OJV, Shepherd B, Kelly S, Olsson S, Wintle J, Muhammed A, Simola K (2004) Nuclear risk-based inspection methodology for passive components (NURBIM), Final report. European Atomic Energy Community (EURATOM), Brussels

Shrader-Frechette K (1992) Risk estimation and expert judgment: the case of Yucca Mountain. RISK 3:4

USNRC (1989) Severe accident risks: an assessment for five U.S. nuclear power plants, NUREG-1150 volumes 1, 2 and 3. USNRC, Washington

USNRC (1990) Special committee review of the nuclear regulatory commission's severe accident risks report (NUREG-1150), NUREG-1420. USNRC, Washington

USNRC (1996) Branch technical position on the use of expert elicitation in the high-level radioactive waste program, NUREG-1563. USNRC, Washington

USNRC (1998) An approach for plant-specific risk-informed Decisionmaking for Inservice inspection of piping, RG 1.178 Revision 0. USNRC, Washington

USNRC (2008a) Seismic considerations for the transition break size, NUREG-1903. USNRC, Washington

USNRC (2008b) United States nuclear regulatory commission accident sequence precursor (ASP) program summary description, Washington, DC. NRC-ADAMS accession no. ML1392A106.

USNRC (2010) Recommended screening limits for pressurized thermal shock (PTS), NUREG-1874. USNRC, Washington

USNRC (2013) LER 13-004-00 for Oconee, unit 1, regarding high cycle fatigue resulted in reactor coolant leak and unit shutdown, Washington, (DC). NRC-ADAMS Accession No. ML14015A141

USNRC (2014a) Oconee Nuclear Station—NRC inspection report number 05000269/2014–011, preliminary greater than green finding and related apparent violation, Washington, DC. NRC-ADAMS Accession No. ML14177A333

USNRC (2014b) LER 14–003-00 for North Anna unit, regarding technical specification required shutdown due to reactor coolant system pressure boundary leakage, Washington, DC. NRC-ADAMS Accession No. ML15058A035

USNRC (2015) North Anna Power Station—NRC integrated inspection report 05000338/2015003 and 05000339/2015003 and exercise of enforcement discretion, Washington, DC. NRC-ADAMS Accession No. ML15316A34

USNRC (2016) Design certification/combined license: assessing the technical adequacy of the advanced light-water reactor probabilistic risk assessment for the design certification application and combined license application, DC/COL-ISG-028. Interim Staff Guidance, Washington

USNRC (2018) An approach for using probabilistic risk assessment in risk-informed decisions on plant-specific changes to the licensing basis, regulatory guide 1.174, revision 3, Washington, DC.

USNRC (2021) Extremely low probability of rupture version 2 probabilistic fracture mechanics code, NUREG-2247. USNRC, Washington

USNRC (2022) Preparing probabilistic fracture mechanics submittals, regulatory guide 1.245 revision 0, Washington

USNRC (2024a) Content of risk assessment and severe accident information in light-water power reactor construction permit applications, DRA-ISG-2024-XX. Interim Staff Guidance, Washington

USNRC (2024b) Guidance for conducting expert elicitation in risk-informed Decisionmaking activities, NUREG-2254. USNRC, Washington

Wheeler TA, Hora SC, Cramond WR, Unwin SD (1989) Analysis of Core damage frequency from internal events: expert judgment elicitation, NUREG/CR-4550, vol 2. USNRC, Washington

Winkler RL (1968) The consensus of subjective probability distributions. Manag Sci 15(2):B61–B75

Chapter 7
Principles of Probabilistic Piping Reliability Analysis

7.1 Fundamentals

The underlying theme in this book is the use of piping operating experience data to inform probabilistic assessments of pressure boundary integrity of piping systems within existing and new nuclear power plants. Different models of piping reliability have been proposed, addressing gaps due to a lack of relevant operational experience insights necessary for statistical evaluations of significant failure piping pressure boundaries. Additionally, some models have been developed to incorporate the physics of failure aspects associated with crack growth and propagation in thick-walled piping. A simple road map for probabilistic pipe failure model development is illustrated in Fig. 7.1.

An analysis begins with studies of material degradation scenarios and frequency of failure, corresponding to the "*precursor rate*" heading in Fig. 7.1. Next, the likelihood of failure is assessed given an undetected degraded piping pressure boundary, corresponding to the "*CFP model development*" heading in Fig. 7.1. A final step is to apply a convolution operation to the two probability distributions to obtain an uncertainty distribution for a significant structural failure having a specified consequence of certain magnitude. Evaluating and statistically characterizing operating experience data and converting these interpretations into probability density functions are the main topics of this chapter. "Data specialization" is an important part of probabilistic piping reliability analysis. Such specialization involves updating generic, industry-wide data parameters with plant-specific data. The data updating is accomplished using a Bayesian parameter estimation approach in which well-qualified generic data is represented by a prior distribution. Examples of data specialization include the following:

- Update of existing, referenceable piping reliability parameter estimates by using new operating experience data. We refer to this as "routine" or ordinary data specialization in which a new set of operating experience data is incorporated into an existing analysis.

B. Lydell, *Knowledge Preservation in Nuclear Safety*,
https://doi.org/10.1007/978-3-032-25463-4_7

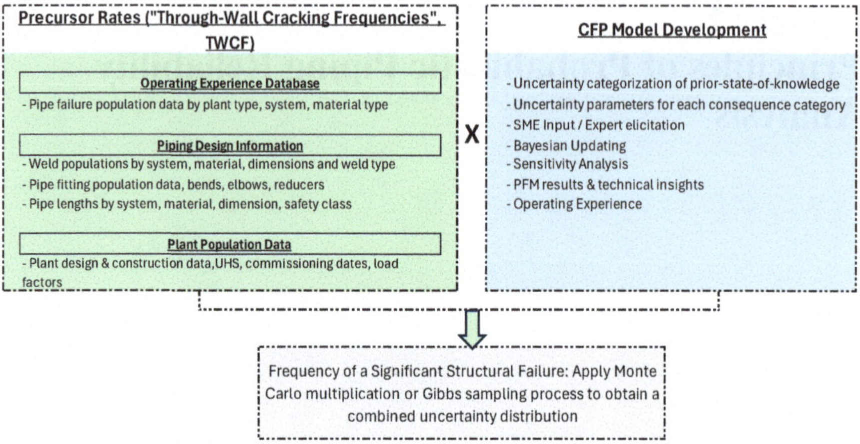

Fig. 7.1 Road map for data-driven probabilistic pipe failure model development

- Modifying generic piping reliability parameter estimates to account for impact on the structural reliability by changes to an inspection program or material degradation mitigation through mechanical stress improvement or using material resistant to degradation.
- Derivation of material degradation mechanism-centric pipe failure rates and conditional failure probabilities.
- Derivation of piping reliability parameters for advanced reactors on the basis of existing industry-wide operating experience data to a new piping design for which there is no prior operating experience. This type of data specialization involves applications of the full knowledge base that is embedded in a comprehensive pipe failure database.
- Probabilistic safety assessment applications involve developing pipe rupture frequencies for different through-wall flow rate categories, for example, "spray events" ($\leq$5 kg/s), "widespread spatial impact" (between 5 kg/s and 100 kg/s), and "major, widespread spatial impact flooding" ($>$100 kg/s). To remove conservatism, refined treatment of flow rate ranges is done by parsing pipe rupture frequency for flow rate ranges of varying sizes warranted.

7.2 Data-Informed Probabilistic Pipe Failure Model

"Data-informed" implies that a model of piping reliability as much as possible builds on interpretations and evaluations of operating experience data. The failure or precursor rate λ and the conditional failure probability (CFP) term are assessed using a Bayesian reliability analysis. Depending on the context of an analysis, the frequency of a pipe failure having a specific consequence is estimated at a piping component level or at a piping system level. Subject to epistemic uncertainty, the frequency of failure is derived from

$$F(IE_x) = \sum_i m_i \times \rho_{ix} \tag{7.1}$$

F(IE_x) is the frequency of a structural failure as a function of its consequence in terms of an equivalent break size or through-wall volumetric flow rate. In Eq. (7), parameter m_i is the number of welds (or fittings or inspection locations) of type *i*; each type is determined by pipe size, weld type, material degradation mechanisms, and inspection (e.g., leak monitoring and nondestructive examination). The frequency of structural failure of component type *i* with an equivalent break size *x* or through-wall mass flow rate *i* is derived from

$$\rho_{ix} = \sum_k \lambda_{ik} \times CFP_{xk} \times I_{ik} \tag{7.2}$$

where

λ_{ik} Pipe component or weld precursor rate. Expressed in units of failures per reactor operating year and the number of components exposed to a certain degradation mechanism.

CPF_xk Conditional probability of pipe failure with consequence *x* in terms of an equivalent break size or through-wall flow rate, given a degraded pipe material condition (*DC*) for weld of type *i* susceptible to degradation mechanism *k*.

I_{ik} Ratio of hazard rates for weld type *i* and material degradation mechanism *k*, subject to epistemic uncertainty. This factor is derived from a Markov model of piping reliability. This factor is a function of the reliability of leak detection and nondestructive examination.

Finally, the precursor rate is derived from

$$\lambda_{ik} = \frac{n_{ik}}{\tau_{ik}} = \frac{n_{ik}}{f_{ik} N_i T_i} \tag{7.3}$$

where

n_{ik} Number of failures of pipe component *i* due to degradation mechanism *k*, with little epistemic uncertainty.

τ_{ik} Component exposure population for welds of type i susceptible to degradation mechanism *k*.

f_{ik} Estimate of the fraction of the component exposure population for weld type *i* that is susceptible to degradation mechanism *k*.

N_i

Estimate of the average number of pipe welds of type i per reactor in the reactor operating year exposure for the operating experience data query used to determine n_{ik}, subject to epistemic uncertainty.

T_i Total exposure in reactor operating years for the data collection for component type i, with little or no uncertainty.

All the inputs to the three basic equations are obtained from an operating experience database of sufficient depth and quality. The conditional failure probability requires input from expert elicitation or probabilistic fracture mechanics (PFM). This method consists of two basic elements: (1) the precursor rate (λ) that is calculated using a Bayesian analysis technique to update the probability distribution representing a prior state of knowledge with the evidence from observations of pipe failure events and (2) the conditional failure probability (CFP) of a certain magnitude. The model used for relating precursor rates and conditional failure probabilities is straightforward. The reason for this approach is that in many cases, there is insufficient data available to estimate rupture frequencies directly from the service data. There is adequate data to estimate precursor rates. Conditional failure probabilities are estimated from a combination of operating experience data, engineering judgment, expert elicitation, and PFM.

Bayesian uncertainty distributions for the above reliability parameters and prior distributions are developed for the parameters λ_{ik} and CFP_{xk}. These distributions are updated using evidence from the failure and exposure term data. The failure rate exposure terms are uncertain because of plant-to-plant variability in, for example, pipe lengths and weld populations. The ROY of service experience has a relatively small amount of uncertainty; therefore, this uncertainty is dominated by the uncertainty in the estimated pipe population, i.e., number of elbows, tees, welds, or length of piping. The uncertainty in the piping component exposure term is represented by a discrete probability distribution that characterizes different hypotheses about the values of the exposure terms, which requires three Bayesian updates for each precursor rate. The resulting posterior distributions for each parameter are then combined using a sampling procedure to obtain uncertainty distributions for the pipe failure frequencies, e.g., Monte Carlo simulation or Gibbs sampling.[1]

An implementation of the Bayesian approach begins with the development of a prior distribution, either through empirical Bayes or standard Bayes methods. In empirical Bayes, the prior distribution is developed from the operating experience data. In standard Bayes applications, the prior distribution is fixed before any failures are observed. Engineering judgment can be used to characterize the uncertainty in pipe failure frequency estimates. A model of the conditional failure probability is used to extrapolate estimates of failure frequency of minor leaks to major structural failures. The probability can be represented as a cumulative failure probability

[1] Gibbs sampling is a Markov chain Monte Carlo (MCMC) algorithm used to generate samples from a joint probability distribution when the joint distribution is difficult to sample from directly. It works by iteratively sampling each variable from its conditional distribution, given the current values of all other variables.

versus an equivalent break size (EBS) expressed in terms of through-wall mass flow rate in kg/s or size of the hole in the pressure boundary converted to an equivalent diameter in mm (i.e. an EBS). An insight based on operating experience is that the slope of the CFP curve shows a relatively strong decrease in probability with increasing failure magnitude.

The terms "structural reliability model" (SRM) and probabilistic fracture mechanics (PFM) model have been used interchangeably. However, SRM refers to a broad framework for assessing a structure's overall safety by considering uncertainties in loads, materials, and dimensions, while a PFM model is a specialized type of structural reliability analysis that focuses specifically on components with existing cracks. PFM uses fracture mechanics to evaluate the probability of an axial or circumferential crack growing to a critical size under probabilistic loads, materials, and crack sizes, making it a more focused approach than general structural reliability analysis.

7.3 Models of Failure

The factors governing the structural integrity of piping systems are numerous and varied. Synthesizing these parameters into a model of piping reliability and defining the necessary input parameters can be done based on fracture mechanics, experiments, and risk science. What is the meaning of "risk science"? Aven (2020) gives the following perspective:

> Simplified, the risk science can be viewed as comprising two main components. The first one is to support scientific knowledge generation for specific activities, for example when studying risk related to climate change. The scientific work is typically multidisciplinary or interdisciplinary, building on competences from various fields and sciences, including risk science. ... The second risk science component is the generic, fundamental risk science, referred to as type B. It covers scientific knowledge related to the development of concepts, principles, approaches, methods, and models for understanding, assessing, characterizing, managing, and governing risk.

The answer to a question like to what extent piping operating experience can support the derivation of required model inputs lies in an analyst's ability to utilize probability theory and applied risk and reliability methodologies while recognizing the "pipe fundamentals." Ideally, the operating experience is such that sufficient data exists so that probabilistic failure metrics are obtained through direct statistical estimation. This is a classical data-driven approach, and, to obtain a pipe failure frequency, the analyst obtains the number of event occurrences over time and assumes that the frequency follows a Poisson distribution since it is based on the number of event occurrences over time. Data homogeneity implies that a piping system and its operating environment are the same for all plants from which the failure data has been obtained. Changes in homogeneity occur because of the use of different materials, different reliability and integrity management, varying time to failure, etc.

Brief descriptions are presented of two useful distributions in the modeling of pipe failure, the binomial distribution, and the Poisson distribution. The former is used to model on-demand type of failures, e.g., significant structural failure given a hydraulic pressure pulse. The latter models the probability of a certain number of pipe failures occurring over a specific time interval, assuming a constant average rate of occurrence. Poisson distribution deals with discrete events that can be counted (e.g., number of failures, number of defects) rather than continuous values (e.g., time to failure). A challenge in applying these distributions is to apply a sufficient amount of engineering insights into the estimation processes.

7.3.1 Binomial Distribution

The binomial distribution is used when an experiment can have only two outcomes (which are, naturally, mutually exclusive and exhaustive, like success-failure); the probability of, say, success is p and of failure 1-p. The binomial distribution has two parameters, the probability of "success" p and the number of trials n. The experiment is repeated n times, and p is assumed constant throughout; then p(r) gives the probability of exactly r successes in n trials and $P(x)$ the probability of at most x successes:

$$p(r) = \binom{n}{r} p^r (1-p)^{n-r} = \frac{n!}{r((n-r)!} p^r (1-p)^{n-r} \tag{7.4}$$

where $r = 0, 1, \ldots\ldots n$ and $0 \leq p \leq 1$

$$P(x) = \sum\nolimits_{r=0}^{n} \binom{n}{r} p^r (1-p)^{n-r} \tag{7.5}$$

7.3.2 Poisson Distribution

In reliability engineering, the Poisson distribution is used to model the probability of a certain number of events occurring over a specific time interval, based on a constant average rate of occurrence. The Poisson distribution pertains to discrete events that can be counted, such as the number of failures or defects, rather than continuous values like time to failure:

$$p(r) = e^{-\lambda t} \frac{(\lambda t)^r}{r!} \tag{7.6}$$

$$F(x) = \sum_{r=0}^{n} p(r) \tag{7.7}$$

Since the Poisson distribution describes the probability of a certain number of events occurring in a fixed interval of time, it is suitable for modeling random events.

7.4 Failure Rates and Hazard Rates

The failure rate is defined for non-repairable component populations as the instantaneous rate of failure for the survivors to time t during the next instant of time. The next instant the failure rate may change and the units that have already failed play no further role since only "survivors" count. The hazard rate $h(t)$ is a measure of the rate at which failures occur, accounting for the size of the population exposed to a hazard at a given time, i.e., those which are still functioning at time t. It is the ratio of the number of failures in a given time period to the number of components exposed to failure. The hazard rate is expressed as (Kapur and Lamberson 1977)

$$h(t) = \frac{f(t)}{1 - F(t)} = \frac{f(t)}{R(t)} \tag{7.8}$$

The hazard rate is related to the first failure of a non-repairable component. The probability that a system will fail at some time $T < t + \Delta t$ given that it has not failed at $T = t$ is:

$$h(t)\Delta t = P[T < t + \Delta t | T > t] \tag{7.9}$$

$$h(t)\Delta t = \frac{P[T > t \cap (T < t + \Delta t)]}{P[T > t]} = \frac{P[t < T < t + \Delta t]}{P[T > t]} \tag{7.10}$$

$$h(t)\Delta t = \frac{f(t)\Delta t}{R(t)} \tag{7.11}$$

Reliability, $R(t)$, is also related with the hazard rate as

$$R(t) = e^{-\int_0^t h(t)dt} \tag{7.12}$$

The exponential distribution is the simplest of all lifetime models. The probability density function (PDF) of the exponential distribution is

$$f(t) = \lambda e^{-\lambda t} \tag{7.13}$$

The cumulative distribution function is

$$F(t) = 1 - e^{-\lambda t} \tag{7.14}$$

The reliability function is

$$R(t) = e^{\lambda t} \tag{7.15}$$

The exponential failure rate is therefore equal to

$$h(t) = \frac{f(t)}{R(t)} = \frac{\lambda e^{-\lambda t}}{e^{-\lambda t}} = \lambda \tag{7.16}$$

The exponential hazard rate is therefore constant and equal to the scale parameter λ, which is quantified as the number of failures over time.

7.5 Renewal Processes

An implicit consequence of relying on operating experience data is that the derived reliability parameters account for different types of in-service inspection program implementation processes, e.g., leak detection technologies and in-service inspection. The influence of in-service inspection and leak detection on pipe failure rates is accounted for by a structural integrity management factor I_{ik}. Markov models are parametric models useful in analyzing probabilistic behavior of piping systems over time. The Markov model is illustrated in Fig. 7.2.

This four-state model is used to quantify parameter I_{ik}. The modeling starts with a representation of the integrity of a piping component in a set of discrete and mutually exclusive states. At any instant in time, the piping component is permitted to change state in accordance with competing processes appropriate for that state. A Markov model state refers to flaws, leaks, or ruptures. The processes that can create a state change are degradation mechanisms acting on the pipe and the processes for inspecting or detecting flaws and leaks and repair of damage before progressing to a complete structural failure. The degradation mechanisms that act on a piping component are represented by failure rates.

All failure processes of this model can be evaluated using operating experience data. A piping component can be in four mutually exclusive states: *S* (success or free of flaws), *F* (flawed or cracked), *L* (leaking, non-active leakage, or active leakage with leak rate within technical specification limit), or *R* (significant structural failure). The time-dependent probability that a piping component is in each state S, C, F, or L is described by a differential equation. Under the assumption

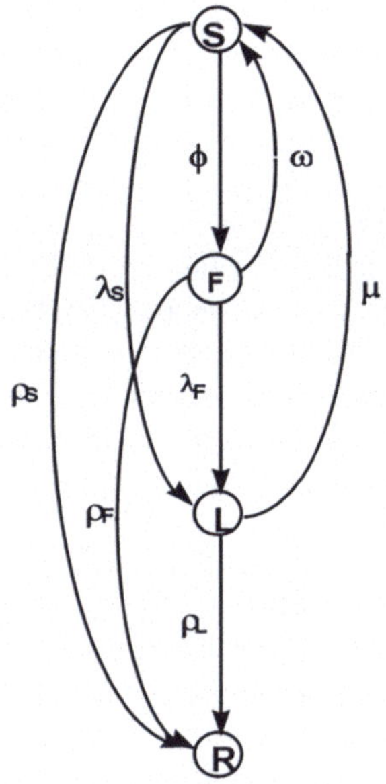

PIPING INTEGRITY STATES	
S = Success (or undamaged state); F = Crack (non-through wall flaw); L = Leaking through-wall flaw (leak rate is within Technical Specification limit); R = Rupture or major structural failure	
State Transitions:	
ϕ	Occurrence of non-through wall flaw
λ_S	Occurrence of a leak (L) given no flaw – from success state to leak
λ_F	Occurrence of leak given a pipe flaw (‘*F*’)
ρ_S	Occurrence of rupture given no flaw
ρ_F	Occurrence of rupture given a flaw
ρ_L	Occurrence of rupture given a leak
μ	Detect through in-service inspection and repair leak
ω	Inspect and repair a non-through wall flaw

Fig. 7.2 Four-state Markov model of piping reliability

that all the state transition rates are constant, the Markov model equations will consist of a set of coupled linear differential equations with constant coefficients. The reliability term needed to represent a piping pressure boundary failure is the hazard rate *h{t}* which is time dependent.

The hazard rate is defined as

$$h(t) = \left(\frac{1}{1 - R(t)} \times \frac{dR(t)}{dt} \right. \tag{7.17}$$

$$1 - R(t) = S(t) + F(t) + L(t) \tag{7.18}$$

The hazard rate is a function of time and the parameters of the Markov model. The hazard rate is the time-dependent frequency of pipe rupture. For the four-state Markov model, *h{t}* is expressed as a function of the six parameters: an occurrence rate for detectable flaws (ϕ); a failure rate for leaks given the existence of a flaw (λ_F); two rupture frequencies, including one from the initial state of a flaw (ρ_F) (break-before-leak, BBL) and another from the initial state of a leak (ρ_L) (leak-before-break, LBB); a repair rate for detectable flaws (ω); and a repair rate for leaks (μ). The latter two parameters dealing with repair are further developed by the following simple models:

$$\omega = \frac{P_{FI} \times P_{FD}}{(T_{FI} + T_R)} \tag{7.19}$$

Parameter P_{FI} is the probability that a piping element with a flaw will be inspected per inspection interval. This parameter has a value of 0 if it is not in the inspection program and 1 if it is in the inspection program. For inspected elements, a value of 1 is used for any ISI inspection case and 0 for the case of no ISI. The element may be selected for inspection directly by being included in the sections sampled for ISI inspection or indirectly by having a rule such that if degradation is detected anywhere in the system, the search for degraded locations will be expanded. The parameter P_{FD} is the probability that a flaw will be detected given this element is inspected. This is the reliability of the inspection program and is equivalent to the probability of detection (POD). Chapter 8 expands on how to assess the POD.

The parameter T_{FI} is the mean time between inspections for flaws, i.e., the inspection interval. The parameter T_R is the mean time to repair once a flaw has been detected. Depending on the location of the weld to be repaired, the actual weld repair could take on the order of several days to much more than a week. Accounting for time to prepare for repair, NDE, root cause evaluation, etc., the total outage time attributed to the repair of a Class 1 weld is on the order of 1 month or more. Since the repair time is always combined with T_{FI}, and T_{FI} could be 10 years, in practice, the results are insensitive to assumptions regarding T_R. An estimate of the repair rate for leaks can be determined from

$$\mu = \frac{P_{LD}}{(T_{LI} + T_R)} \tag{7.20}$$

$$\mu = \frac{P_{LD}}{(T_{LI} + T_R)} \tag{7.21}$$

Parameter P_{LD} is the probability that the leak in the element will be detected per leak inspection or detection period. The parameter T_{LI} is the mean time between inspections for leaks. For primary system piping, the time interval between leaks can be essentially instantaneous if the leak is picked up by radiation alarms to the period between leak tests performed on the system. The parameter T_R is the mean time to repair and should be the minimum of the actual repair time and the time associated with cooldown to enable repair and any waiting time for replacement piping.

Opportunities for leak detection depend on the system in which the leak occurs as well as the specific location and size of the leak. For example, a significant leak in the reactor coolant would create a containment radiation alarm in the control room. In these cases, the time to inspect and repair is limited by technical specifications on primary system leakage and time to shut down the plant and begin the process of repair. Other leaks may not cause an alarm but would be subject to possible detection during operator walk visual inspections every shift or other opportunity for leak detection. There are some leaks that may only be detected upon periodic leak testing, which may occur less often as required to meet codes and standards for different classes of piping. Chapter 8 expands on leak detection principles.

An observation about the leak repair term μ, in comparison to the flaw repair term ω, is that for most leaks, the detection possibilities are not normally limited to some predetermined population of welds that are inspected. However, leak testing often provides an opportunity to inspect all locations system wide. Hence, given a leak of significant magnitude anywhere in the system, the probability of leak detection tends to be quite high. By contrast, most of the locations that could produce a non-leaking flaw are never inspected according to ASME rules, in which case the flaw repair rate term ω is zero. Also, the time between successive inspections for leaks tends to be much shorter than for volumetric examination of welds, with virtually instantaneous detection in cases when the leak triggers an alarm in the control room. Hence, the Markov model provides the capability to account for the leak-before-break principle. The extent to which this principle contributes to reducing the probability of a rupture is only a function of the relative values of the Markov model transition rates. The four-state Markov model can be represented by the following system state transition matrix (or the transition matrix of the Markov chain):

$$T = \begin{bmatrix} -\varphi & \phi & 0 & 0 \\ \omega & -(\omega + \lambda_F + \rho_F) & \lambda_F & \rho_F \\ \mu & 0 & -(\mu + \rho_L) & \rho_L \\ 0 & 0 & 0 & 0 \end{bmatrix}$$

Let $S(t)$, $F(t)$, $L(t)$, $R(t)$ represent the time-dependent probabilities of being in the states S, F, L, and R, respectively. These state probabilities can be obtained by solving the system of differential equations shown below, with the initial condition $S(0) = 1$ and $F(0) = L(0) = R(0) = 0$ and subject to the condition $S(t) + F(t) + L(t) + R(t) = 1$. The initial condition corresponds to a flaw-less pipe:

$$\frac{dS(t)}{dt} = \omega F(t) + \mu L(t) - \phi S(t) \tag{7.22}$$

$$\frac{dF(t)}{dt} = \phi S(t) - (\omega + \lambda_F + \rho_F) F(t) \tag{7.23}$$

$$\frac{dL(t)}{dt} = \lambda_F - (\mu + \rho_L) L(t) \tag{7.24}$$

$$\frac{dR(t)}{dt} = \rho_F F(t) + \rho_L L(t) \tag{7.25}$$

Solving the above equations determines the time-dependent probabilities of occupying each state. Assuming that all the transition rates are constant, the Markov model equations consist of a set of coupled linear differential equations with constant coefficients. These equations can be solved analytically or numerically (Fleming

Table 7.1 Parameters of a four-state Markov model

Model/ parameter	Description
$\omega = \frac{P_I P_{FD}}{(T_{FI}+T_R)}$	Rate of inspection and repair non-through-wall defect
$\mu = \frac{P_{LD}}{(T_{LI}+T_R)}$	Rate of detection and repair of a leak
P_I	Probability that a piping element with a flaw will be inspected per inspection interval
P_{FD}	Probability of detecting a flaw in a pipe segment
P_{LD}	Probability that a through-wall flaw will be detected during a walkdown inspection
T_{FI}	Mean time between inspections for flaws (inspection interval)
T_R	Mean time to repair when detected. If the detection is made with reactor @ power, the repair time should be the minimum of the actual repair time and the time associated with any technical specification limiting condition for operation if the leak rate exceeds technical specification requirements
T_{LI}	Mean time between inspections for leaks
$I_{ix} = \frac{h_{ix}(t)}{h_i(t)}$	Integrity management factor for component *i* for RIM strategy *x* at piping system *t*. as shown in this equation, this factor can be expressed as a product of separate integrity management and piping system factors. The RIM strategy includes a specification of how often leak tests and inspections and nondestructive examination are performed on the piping system component and how effective they are
$h_{ix}(t)$	Hazard rate for component *i* at piping system *t* and RIM strategy *x*. this is determined from the solution of the ordinary differential equations that describe the Markov model
$h_i(t)$	Hazard rate for component *i* at piping system of *t* and RIM strategy corresponding to an "average" component in the service data. For example, for ASME III class 1 components, the average component is determined by 25% of the components being subjected to nondestructive examination exams every 10 years, the remaining 75% not being subjected to nondestructive examination, and 100% of the components being subjected to a system leak test once every refueling outage

2004; Fleming et al. 2010). The parameters of the Markov model are described in Table 7.1.

7.6 Semi-Markov Models of Piping Reliability

The Markov model formulation assumes constant transition rates. Asymptotically, this model shows a constant hazard rate and thus does not account for a potentially increasing or decreasing hazard rate. A semi-Markov process (SMP) model (Grabski 2015), on the other hand, assumes user-defined distributions for time spent during state transitions and hence allows for a physics-of-failure-oriented piping reliability analysis. Typically, a statistical distribution (lifetime distribution such as a Weibull or lognormal distribution) is used to represent crack initiation, and the growth can be represented by a Paris-Erdogan crack growth law. Motivations for an SMP model formulation are usually based on an assumption of inadequate OE data but may also be based on a desire for more in-depth assessment of material aging phenomena.

Stochastic differential equations (SDEs) may be used to resolve the SMP transitions. An SDE is a differential equation in which one or more of the terms are a stochastic process, resulting in a solution which is also a stochastic process. Numerical solution of stochastic differential equations is an evolving field, and especially stochastic partial differential equation is a young field relatively speaking. Almost all algorithms that are used for the solution of ordinary differential equations will work very poorly for SDEs, having poor numerical convergence. A standard textbook on SDE, and its theory and application, is authored by Øksendal (2003).

Semi-Markov process models have been applied to different piping reliability analysis problems. Di Maio et al. summarize SMP, an analysis of thermal fatigue of a mixing tee in a PWR operating environment (2016). The Monte Carlo simulation method is used to model the time-dependent transition rates. According to Di Maio et al., the integration of physical models of fatigue (e.g., initiation and propagation) allows for a more realistic degradation process modeling. Included in the paper by Di Maio et al. are results from a benchmark study where Markov and SMP models were applied to the same thermal fatigue problem. Both approaches indicated that pipe rupture due to thermal fatigue is a "non-credible event" for the first 10–15 years of plant operation. In the long term (beyond 15 years of operation), the probability of pipe rupture given thermal fatigue is shown to be an order of magnitude greater using an SMP as opposed to a Markov modeling approach. Thermal fatigue damage is determined by factors such as mode of operation, leak tightness of isolation valves, mass flow rates in "run" pipes, and branch pipes connected to a mixing tee. Ample operating experience data does exist including examples of significant pipe failures occurring within a very short period of plant operation, less than 1 year.

Collins (2010, 2011) has applied an SMP model to the problem of primary water stress corrosion cracking (PWSCC) of Alloy82/182 in a PWR environment. A Weibull distribution was selected for the characterization of crack initiation, and MRP-135 (Rashid et al. 2010) was selected for crack propagation. The results indicate that when using a Weibull scale parameter of 4 years, a maximum hazard rate of about 1×10^{-5} per year occurs at about 10 years and then appears to decrease towards an asymptote of about 1×10^{-7} per year.

Veeramany (2011, 2012) explores different assumptions about the pipe flaw initiation time on the basis of the coefficient of variation. The Weibull distribution was used to model flaw initiation, and the distributions for the time spent between various states are represented in a matrix form called the "kernel matrix" of the process. The kernel matrix and the initial state occupied by the process completely define the stochastic behavior of the semi-Markov process. Given these as the input, the statistical time behavior of the process is described by a system of linear integral equations.

7.7 Discrete Probability Distributions

Discrete probability distributions (DPDs) have been suggested for use in risk analysis calculations to simplify the numerical computations which must be performed to determine failure probabilities (Kurth and Cox 1985). DPDs are used to model events with a finite number of possible outcomes, where each outcome has a specific probability used to quantify the likelihood of events and potential risks. Three-point DPDs have found use in pipe failure rate estimation to account for the variability, degradation mechanism susceptibility, and plant-to-plant weld population variability, which becomes an important input to the exposure term definitions. Ten- to fifteen-point DPDs have found use in probabilistic fracture mechanics to model the initial flaw and stress distributions.

Assume that we are interested in estimating a weld precursor failure rate for locations within reactor coolant systems susceptible to thermal fatigue. The estimation process involves querying a database for relevant failure data and the number of plant-years that produced the failures. Multiplying the number of susceptible welds and the number of plant-years would yield one estimate that could be used to perform a Bayesian update. This product would characterize neither the uncertainty in the estimates of the number of welds in a system of interest nor the fraction of these welds susceptible to thermal fatigue. To address uncertainty in precursor rate calculations, three Bayesian updates are performed using lower bound, best estimate, and upper bound weld population data. The lower bound for susceptible welds is calculated as the number of plant-years times a lower-bound degradation susceptibility fraction, multiplied by half the weld population. Figure 7.3 shows the three-bin distributions.[2]

7.8 Properties of the Lognormal Distribution

Very often in applied reliability and risk analysis, the primary variables are assumed to be lognormally distributed. Justifications for using the lognormal distribution

come from the central limit theorem,[3] which states that the product of

[2]For an engineering perspective on the use of DPDs, see Kaplan, S.J., Garrick, B.J. (1980), "Try Probabilistic Thinking to Improve Power plant Reliability," Power Magazine, March, pp. 56–61.

[3]The central limit theorem (CLT) states that, under appropriate conditions, the distribution of a normalized version of the sample mean converges to a standard normal distribution. This holds even if the original variables themselves are not normally distributed. There are several versions of the CLT, each applying in the context of different conditions. The theorem is a key concept in probability theory; it implies that probabilistic and statistical methods that work for normal distributions can be applicable to many problems involving other types of distributions: https://en.wikipedia.org/wiki/Central_limit_theorem.

Material Degradation Susceptibility Fraction Distribution	
Fraction	Probability
0.05	0.25
0.15	0.50
0.80	0.25
The above data would be based on formal degradation mechanism analysis & engineering judgment	

Number of Welds Distribution	
Number of Welds	Probability
N × 0.5	0.25
N	0.50
N × 1.5	0.25
The above data would be based on reviews of a representative set of in-service inspection programs	

Fig. 7.3 An example of three-bin distributions for precursor rate calculations

n independent random variables is a lognormally distributed random variable for large n. Some useful properties of this distribution are summarized below. The lognormal density function is given in the following equation:

$$p(x)\frac{1}{x\sigma\sqrt{2\pi}}\,exp\left[-\frac{(lnx-\mu)^2}{2\sigma^2}-\right] \tag{7.26}$$

The lognormal distribution has considerable applications in PSA and works well for active and passive components like piping. This form of the density function has two parameters, μ and σ. Several characteristic values of the distribution are as follows:

$$\text{Mean}: \alpha = \exp\left(\mu + \frac{\sigma^2}{2}\right)^2 \tag{7.27}$$

$$\text{Variance}: \beta^2 = e^{2\mu} + \sigma^2\left[e^{\sigma^2}-1\right] = \alpha^2\left[e^{\sigma^2}-1\right] \tag{7.28}$$

$$\text{Mode}: x_m = \exp(\mu - \sigma^2) \tag{7.29}$$

The two parameters of a lognormal distribution are referred to as the location parameter and scale parameter, respectively. When lognormal data is transformed using the natural logarithm, μ can be viewed as the mean of the transformed data and σ as the standard deviation (of the transformed data). Inverting Eq. (7.27) and (7.28), one gets the parameters μ and σ as functions of the mean and the variance, i.e.,

$$\mu = ln\alpha - \frac{\sigma^2}{2} \tag{7.30}$$

$$\sigma^2 = \ln\left[\frac{\beta^2}{\alpha^2}+1\right] \tag{7.31}$$

Useful percentiles of the lognormal distribution are

$$\text{5th percentile}: x_{05} = \exp(\mu - 1.645\sigma) \tag{7.32}$$

$$\text{50th percentile (median)}: x_{50} = e^{\mu} = \sqrt{x_{05} \times x_{95}} \tag{7.33}$$

$$\text{95th percentile}: x_{95} = \exp(\mu + 1.645\sigma) \tag{7.34}$$

$$\text{Range factor (RF)}: RF = \sqrt{\frac{x_{95}}{x_5}} = e^{1.645\sigma} \tag{7.35}$$

Since ln(x) is normally distributed, the most convenient way to view a lognormal distribution is to write

$$x = me^{\sigma z} \tag{7.36}$$

When z in Eq. (7.36) is a standard normal variate, the x is lognormally distributed. Parameter m is the median value, and σ is the lognormal standard deviation or the "multiplicative" standard deviation. The reason for the latter term is seen in Eq. (7.36) from the fact that z is increased by the additive amount, 1.0, and then x is increased by the multiplicative factor e^{σ}. The multiplier e^{σ} plays the same role in a lognormal curve as the additive quantity σ plays in a normal curve. That is, when x changes by the factor e^{σ}, the cumulative probability changes by one standard deviation's worth. An important property of the distribution relates to the product of variables. Assuming that two variables, X and Y, are independent lognormally distributed variables, and

$$V = X \times Y \tag{7.37}$$

then

$$V = \Lambda\left(\mu_x + \mu_y, \sqrt{\sigma_x^2 + \sigma_y^2}\right) \tag{7.38}$$

Using Eq. (7.36), we get

$$V = m_x m_y \exp\left(\sigma_x z_x + \sigma_y z_y\right) \tag{7.39}$$

Since the sum of two normal curves is a normal curve, we have

$$\sigma_x z_x + \sigma_y z_y = \sigma_v z_v \tag{7.40}$$

where

$$\sigma_v = \sqrt{\sigma_x^2 + \sigma_y^2} \tag{7.41}$$

From Eq. (7.39), it follows that V is a lognormal variate with median $m_x m_y$ and lognormal standard deviation σ_v.

7.9 Properties of the Gamma Distribution

The gamma distribution is an extension of the exponential distribution. It is flexible to deal with decreasing, constant, and increasing failure rates. The probability density function is given by

$$f(t) = \frac{\beta^\alpha}{\Gamma(\alpha)} t^{\alpha - 1} \exp(-t\beta) \tag{7.42}$$

where $\Gamma(\alpha)$, a standard mathematical function, denotes the gamma function evaluated at α. The parameters α and β are referred to as the shape and scale parameter, respectively. The gamma distribution is the maximum entropy probability distribution (both with respect to a uniform base measure and with respect to a $1/x$ base measure) for a random variable X for which $E[X] = \theta = \alpha/\beta$ is fixed and greater than zero. The principle of maximum entropy states that the probability distribution which best represents the current state of knowledge about a system is the one with the largest entropy, in the context of precisely stated prior data. Entropy is a measure of uncertainty. To provide a link to operating experience data and to maximize the uncertainty, a common approach is to use the constrained non-informative distribution method. In probabilistic safety assessment, this method has been used for example in estimating initiating event frequencies for internal flooding events. The constrained non-informative distribution method, which permits an informed estimate of the mean of an uncertainty distribution but otherwise is non-informed to maximize uncertainty, is comprised of the following steps:

- The mean of the gamma distribution is set to the point estimate of the failure rate obtained from the service data, i.e., number of occurrences (n) divided by number of reactor-years (T).
- The alpha (α) and beta (β) parameters are set to apply the constrained non-informative distribution method ($\alpha = 0.5$); method for the gamma distribution is defined as follows:

$$Mean = \frac{\alpha}{\beta} = \frac{n}{T} \tag{7.43}$$

where $\alpha = 0.5$ and $\beta = 0.5/\left(\frac{n}{T}\right)$. If the gamma distribution is used as a prior distribution in a Bayes' update, and there is plant-specific data on the number of failures m and the plant's exposure t, the Poisson likelihood function is selected, and the posterior distribution is also a gamma distribution with parameters α' and β' calculated as follows:

$$\alpha' = \alpha + m \tag{7.44}$$

$$\beta' = \beta + t \tag{7.45}$$

The problem of defining a prior distribution is now down to defining parameters n (i.e., the industry-wide experience) and T (i.e., the observation period during which n was observed). The prior state of knowledge on which the analyst bases the parameter selection comes from the operating experience and engineering analyses.

7.10 Properties of the Beta Distribution

The beta distribution is a family of continuous probability distributions defined on the interval [0, 1] parameterized by two positive shape parameters, denoted by alpha (α) and beta (β), that appear as exponents of the random variable and control the shape of the distribution. The generalization to multiple variables is called a Dirichlet distribution.[4] The beta distribution has been applied to model the behavior of random variables limited to intervals of finite length in a wide variety of disciplines. In Bayesian estimation, the beta distribution is the conjugate prior probability distribution for the binomial distribution. Beta distribution is extensively used in reliability engineering and probabilistic safety assessment. Its probability density function is given by

$$f(p; \alpha, \beta) = \frac{1}{\beta(\alpha,\ \beta)}\, p^{\alpha-1}\,(1-p)^{\beta-1} \tag{7.46}$$

$$\beta(\alpha, \beta) = \int_0^1 z^{\alpha-1}\,(1-z)^{\beta-1} dz,\quad 0 < \mathrm{p} < 1, \alpha > 0, \text{and } \beta > 0 \tag{7.47}$$

The beta distribution is widely used in Bayesian statistics as it has a conjugate prior when associated with a binomial likelihood distribution. When a beta distribution of parameters α and β is used as a prior distribution and is combined with a binomial distribution of parameters f and s as a likelihood, the posterior distribution is a beta distribution with parameters $\alpha' = \alpha + f\, and\, \beta' = \beta + s$, where f is the number of failures and s is the number of successes. Although the conjugate prior mechanism is straightforward, the elicitation of a beta distribution (i.e., estimating parameters of the beta distribution) is not as easy. Nevertheless, methods exist to determine the parameters of a beta distribution (Table 7.2). Two methods are described in more detail: the "method of moments" and the "Pearson-Tukey" method.

---

[4]Named after the German mathematician Peter Gustav Lejeune Dirichlet, the Dirichlet distribution is a multivariate probability distribution used to model the probabilities of events in a finite set. It is a generalization of the beta distribution and is often used as a prior distribution in Bayesian statistics.

Table 7.2 Methods to determine the parameters of beta distribution

Method	Required input	Outputs
Maximum likelihood estimators	–	No closed-ended solution
Method of moments	M, σ^2	Refer to Sect. 7.10.1
Program evaluation and review technique (PERT)	m: The most likely value	– If m < 0.13, $\mu = \frac{2}{2+\frac{1}{m}}$ and $\sigma = \left[\frac{m^2(1-m)}{1+m}\right]^{\frac{1}{2}}$ – If 0.13 ≤ m ≤ 0.87, $\mu = \frac{4m+1}{6}$ and $\sigma = \frac{1}{6}$ – If m > 0.87, $\mu = \frac{1}{3-2m}$ and $\sigma = \left[\frac{m(1-m)^2}{2-m}\right]^{\frac{1}{2}}$ $\alpha = \left[\frac{(1-\mu)\mu}{\sigma^2} - 1\right]\mu$ $\beta = \left[\frac{(1-\mu)\mu}{\sigma^2} - 1\right](1-\mu)$
Method based on skewness	Positive or negative skewness (m) High, moderate, or little skewness	– If distribution is positively skewed: m is in [0.1; 0.4] If skewness is high/moderate/little: $\alpha = 2.5/3.5/4.5$ $\beta = \frac{\alpha-1}{m} - (\alpha - 2)$ – If distribution is negatively skewed: m is in [0.6; 0.9] If skewness is high/moderate/little: $\beta = 2.5/3.5/4.5$ $\alpha = \frac{1+m(\beta-2)}{1-m}$
Method based on the degree of confidence	*Min* *Max* Positive or negative skewness (m) k: Degree of confidence in m	$\mu = \frac{Max+Min+km}{k+2}$ $\sigma^2 = \frac{Max-Min}{c}$ $\alpha = \left[\frac{(1-\mu)\mu}{\sigma^2} - 1\right]\mu$ $\beta = \left[\frac{(1-\mu)\mu}{\sigma^2} - 1\right](1-\mu)$
Pearson-Tukey	5th (optimistic), 50th (most likely), and 95th (pessimistic) percentiles	Closed-ended solution, Sect. 7.10.2

7.10.1 Method of Moment

The method of moments is an accepted method of estimation of parameters of a distribution that equates the moments of distribution with the sample mean and variance. The nth moment of a probability density function, with probability function f, is defined as

$$E(p^n) = \int_{-}^{+} p^n df(p). \tag{7.48}$$

For the beta distribution, the moment-generating function is

$$E(p^t) = \int_0^1 \frac{\Gamma(\alpha+\beta)}{\Gamma(\alpha)\Gamma(\beta)} p^{\alpha-1}(1-p)^{\beta-1} p^t dp. \tag{7.49}$$

An advantage of the method of moments is that it provides a simple expression of the first and second moments for a beta distribution in function of the distribution parameters, α and β:

$$E(p) = \frac{\alpha}{\alpha+\beta} \tag{7.50}$$

$$E(p^2) = \frac{(\alpha+1)\alpha}{(\alpha+\beta+1)(\alpha+\beta)}. \tag{7.51}$$

Therefore, the mean and the variance of the beta distribution are

$$\begin{cases} \mu = \dfrac{\alpha}{\alpha+\beta} \\ \sigma^2 = E(p^2) - (E(p))^2 = \dfrac{\alpha\beta}{(\alpha+\beta)^2(\alpha+\beta+1)} \end{cases} \tag{7.52}$$

In the method of moments, the assessor provides his estimates of the mean μ and the variance σ^2 of the beta distribution. From these estimates, the parameters of the beta distribution can be determined by using the following formulas:

$$\begin{cases} \alpha = \left(\dfrac{\mu(1-\mu)}{\sigma^2} - 1\right)\mu \\ \beta = \left(\dfrac{\mu(1-\mu)}{\sigma^2} - 1\right)(1-\mu) \end{cases} \tag{7.53}$$

7.10.2 Pearson-Tukey Method

Three-point approximations are widely used when assessing parameters of distributions because there is usually a good sense of what the median and two bounds are, for instance, the 5th, 50th, and 95th percentiles. Extensive work has been done to determine the best approximations of the mean and standard deviation of a beta distribution, based on three percentiles. Keefer and Bodily (1983) claimed that the approach by Pearson and Tukey (1965) is the best of all three-point estimate approximations.

In the Pearson-Tukey approximation, an "expert" is asked to provide the 5th, 50th, and 95th percentiles, denoted as $C_{.05}$, $C_{0.50}$, and $C_{0.95}$ in the equations below. The "expert" can be an expert elicitation panel or a PFM analyst. According to the approach proposed by Pearson[5] and Tukey,[6] the estimated mean μ and variance σ^2 are obtained as follows:

$$\mu = 0.185\ (C_{0.95} + C_{0.05}) + 0.63\ C_{0.50} \tag{7.54}$$

$$\sigma_0^2 = \left(\frac{C_{0.95} - C_{0.05}}{3.25}\right)^2 \tag{7.55}$$

$$\sigma^2 = \left[\frac{C_{0.95} - C_{0.05}}{3.29 - \frac{0.1(C_{0.95}+C_{0.05} - 2C_{0.50})^2}{\sigma_0^2}}\right]^2 \tag{7.56}$$

$$\begin{cases} \alpha = \left[\dfrac{(1-\mu)\,\mu}{\sigma^2} - 1\right]\mu \\ \beta = \left[\dfrac{(1-\mu)\,\mu}{\sigma^2} - 1\right](1-\mu) \end{cases} \tag{7.57}$$

The values of 0.181 and 0.63 are the upper/lower and midpoint probabilities, respectively, as assigned by an expert, while the coefficients 3.25 and 3.29 were established using a procedure defined by Pearson and Tukey.

7.11 Properties of the Weibull Distribution

The Weibull distribution has its origin in the study of component reliability in terms of time to failure and has found widespread use in practical reliability engineering analysis. The underlying work behind the Weibull distribution and subsequent applications became an important role in the evolution of the reliability engineering discipline.[7] The hazard rate for the Weibull distribution is time dependent. It is a three-parameter distribution which covers a variety of time-to-failure shapes, e.g., decreasing, constant, and increasing. The three parameters are α, β, θ, or the scale, shape, and location parameter, respectively.

---

[5]E.S. Pearson (1885–1980), an accomplished statistician and appointed a full professorship at University College London in 1936. His career was focused on industrial applications of statistical methodology. In 1965, together with J.W. Tukey, he published the paper "Approximate Means and Standard Deviations Based on Distances Between Percentage Points of Frequency Curves." He was elected a Fellow of the Royal Society in 1966.

[6]J.W. Tukey (1915–2000), an American mathematician and statistician, best known for the development of the "fast Fourier transform (FFT) algorithm and boxplot," the "Tukey range test," the "Tukey lambda distribution," and the "Pearson-Tukey" three-point beta distribution approximation. He joined the Princeton University mathematics department in 1945 where he remained until his retirement in 1985.

[7]The development of the Weibull distribution stems from the study of fatigue life problems in the 1940s.A large number of texts have been published on its application to reliability engineering problems.

Applications of the Weibull distribution include time to failure given IGSCC in austenitic stainless steel piping, aging trend analysis of piping susceptible to corrosion mechanisms or flow-assisted degradation, and assessment of fatigue crack initiation frequencies. Both types of applications have proven useful because of the extensive pipe failure data. The probability density function is given by

$$f(t) = \frac{\beta t^{(\beta-1)}}{\alpha^{\beta}} \times \exp\left[-\left({}^{t}/_{\alpha}\right)^{\beta}\right], \quad t, \alpha, \beta > 0 \tag{7.58}$$

Using

$$h(t) = \frac{dF(t)/dt}{R(t)} = \frac{f(t)}{R(t)} \tag{7.59}$$

The hazard function then becomes

$$h(t) = \left(\frac{\beta}{\alpha}\right)\left(\frac{t}{a}\right)x^{\beta-1}, \qquad \beta, \alpha > 0, 0 < t < \tag{7.60}$$

In the above equation, if $0 < \beta < 1$, then the Weibull distribution characterizes early-life failures, assumed to have a decreasing hazard rate. For $\beta = 1$, the Weibull distribution corresponds to an exponential distribution, and if $\beta > 1$, the Weibull distribution characterizes the wear-out characteristics as in material aging. Further details on practical application to piping reliability are found in Sect. 7.17, "*Structural Integrity Adjustment Factors*."

7.12 Cohort Effects

Different analytical approaches are utilized when deriving probabilistic failure metrics based on the age of a piping component or on temporal variations in the operating experience. The term "temporal change" refers to the dependence of descriptive failure statistics (such as the mean, median, and confidence bounds) on the manner in which failure data is analyzed. Temporal trends may stem from factors such as material aging, alterations in physical properties of piping, or more fundamental elements like changes in design and construction practices, as well as the codes, standards, and regulations applicable at plant commissioning.

Additionally, the term "cohort effect" is used in statistics to describe the underlying causes of changes observed within an event population. A cohort is any event population that shares common, well-defined features that relate to piping design, operating environment, and susceptibility to material degradation. A cohort effect is a change which characterizes an event population of a certain period, which is independent of the process of aging. Intergranular stress corrosion cracking

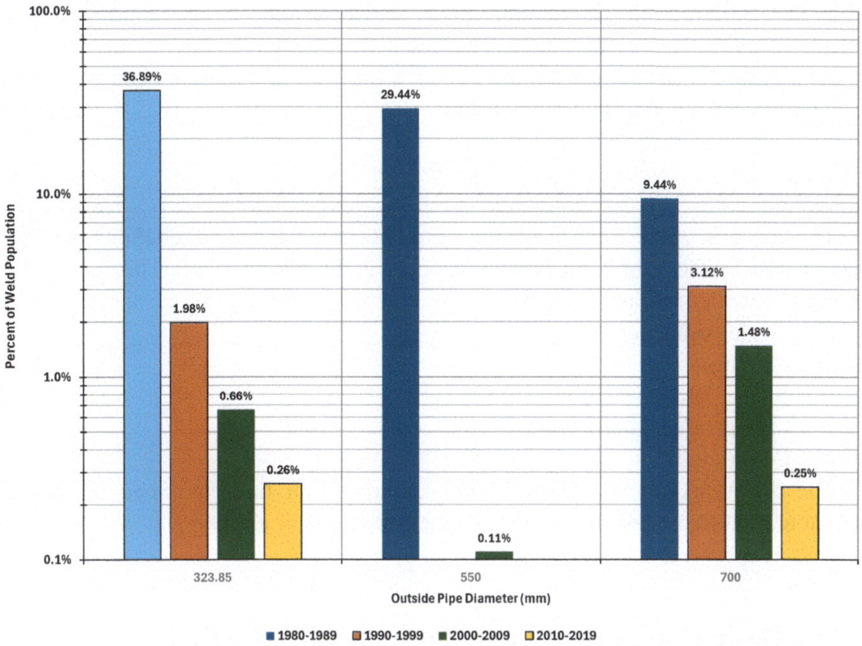

Fig. 7.4 Percentage of IGSCC flaws as a function of pipe size and time period

(IGSCC) in BWR primary system piping is used to exemplify the temporal evolution of this degradation mechanism. The percentage of cracked welds in operating plants is illustrated in Fig. 7.4.

This operating experience is for early design generation BWR plants in operation in Japan, the United States, Switzerland, and Taiwan and three classes of large-diameter reactor recirculation piping. Specifically, General Electric BWR/3 and BWR/4 reactors were commissioned in the period 1970–1982. The operating experience is normalized against the total weld population within each pipe size. In the time period 1980–1989, about a third of the weld population for piping with outside diameters of 325 mm and 550 mm experienced cracking.

The degree of cracking susceptibility may change in any given alloy/environment system. Relatively small changes in the details of the "material," "stress," or "environmental" components can mitigate the problem, and, in many cases, a "defense-in-depth" mitigation approach of positively modifying two of the system conditions has been adopted by many utilities (e.g., the use of Type 316NG stainless steel for replacement piping, plus the implementation of hydrogen water chemistry). The pre-1990 operating experience is significantly different from the post-1989 operating experience, which is indicative of a strong cohort effect.

Correlating the different populations of data with corresponding exposure terms (plant population that generated the failures, operating time, and weld populations) provides a basis for establishing "a priori" failure rate distributions for each category of IGSCC failure data. Oversimplifying a statistical analysis by ignoring cohort effects results in incorrect conclusions about the structural integrity of a piping system. A model of the pipe failure rate as a function of aging and cohort effects is expressed by a power-law relationship:

$$\lambda(t) = \theta t^{\alpha} \times e^{\gamma\tau} \tag{7.61}$$

The first term on the right-hand side represents time-dependent reliability, and the second term on the right-hand term corresponds to the cohort effects. The parameters in Eq. (1) are defined as

θ = Scale parameter
t = Plant age
α = Shape parameter
γ = Cohort effect coefficient
τ = Cohort age

The three unknown parameters, Θ, β, and γ, can be determined using the maximum likelihood estimation (MLE) method (Yuan et al. 2008).

7.13 Basic Concepts in Bayesian Reliability Analysis

Bayesian reliability analysis employs Bayesian inference to predict and evaluate the reliability of systems or components. Bayesian inference is a comprehensive statistical framework that utilizes Bayes[8] theorem to update the failure rate or probability for a hypothesis as more evidence or information becomes available (Fig. 7.5). The process of Bayesian reliability analysis can be idealized by dividing it into the following three steps:

1. Setting up a full probability model, joint probability distribution for all observable and unobservable quantities in a problem. This is the prior state of knowledge and should be consistent with the knowledge about the underlying scientific problem and the data collection process. Bringing engineering insights into the process of defining the prior distribution parameters is an essential part of Bayesian analysis.
2. Conditioning on observed data: calculating and interpreting the appropriate posterior distribution, i.e., the conditional probability distribution of the unobserved quantities of ultimate interest, given the observed data.
3. Evaluating the fit of the model and the implications of the resulting posterior distribution: How well does the model fit the data? Are the conclusions reasonable? How sensitive are the results to the prior state-of-knowledge modeling

[8] T. Bayes (1702–1761), an ordained minister and statistician. Bayes set out his theory of probability in "Essay Towards Solving a Problem in the Doctrine of Chances," *which was published posthumously in the Philosophical Transaction of the Royal Society of London in 1764.*

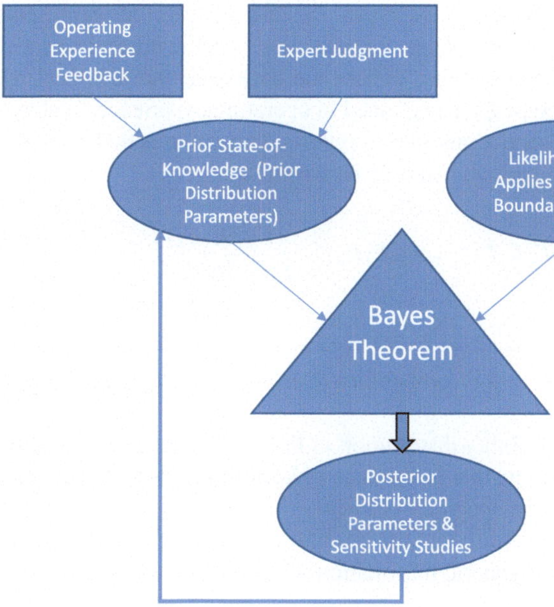

Fig. 7.5 Bayes theorem in probabilistic piping reliability analysis

assumptions? In response, one can alter or expand the model and repeat the three steps.

The first step used to be seen as a major stumbling block in Bayesian analysis. Just where do the models come from? How do we go about constructing appropriate probability specifications? What if an incorrect prior probability distribution is used? Can a generic prior distribution be used when calculating plant-specific piping reliability parameters? In the early days, some classical statisticians equated Bayesian analysis with hocus-pocus; those days are long gone, however. Guidance on issues relating to the development of priors is given in this book. Answers to the question "Where do the prior state-of-knowledge models come from?" are embedded in operating experience data, experimental data, and practical engineering insights. The dependence of conclusions on subjective prior distributions can be examined and explored through operating experience data analysis, assuming that the data is sufficiently comprehensive and vetted. A primary motivation for applying a Bayesian approach is that it facilitates a common-sense interpretation of statistical conclusions. There are three types of information available for parameter S in the structural integrity risk triplet:

- E_1: General engineering knowledge of piping design, welding technology, and material degradation mechanisms
- E_2: Historical performance in other plants similar to the one in questions
- E_3: The past experience in the specific application being studied

The information (or evidence) of types E_1 and E_2 together represents the "generic" information and E_3 the plant- or application-specific information. Generic information is contained in operating experience databases and technical reports. Similarly, information of type E_3 is contained in operating experience databases and plant records such as in-service inspection reports, corrective action requests, and work orders. The Bayes' theorem can now be written as

$$p(\lambda|E_1E_2E_3) = p(\lambda|E_1E_2) \times \frac{p(E_3|\lambda E_1E_2)}{p(E_3|E_1E_2)} \tag{7.62}$$

where

$p(\lambda| E_1E_2E_3) =$ The "posterior" probability distribution in light of the evidence $E_1E_2E_3$

$p(\lambda| E_1E_2) =$ The probability distribution in light of the generic information E_1 and E_2 only, that is, the state of knowledge "prior" to the plant- or application-specific evidence

$p(E_3| E_1E_2) =$ The probability that the specific evidence E_3 would be observed, given the generic information

For purposes of numerical calculation, one can discretize the λ-axis, establishing a set of possible values, λ_j. The denominator of the above equation can then be written as

$$p(E_3|E_1E_2) = \sum_j p(\lambda_j|E_1E_2)p(E_3|\lambda_jE_1E_2) \tag{7.63}$$

$$p(\lambda|E_1E_2E_3) = \frac{p(\lambda|E_1E_2)p(E_3|\lambda_jE_1E_2)}{\sum_j p(\lambda_j|E_1E_2)p(E_3|\lambda_jE_1E_2)} \tag{7.64}$$

The denominator term is the sum of the numerator terms so that the posterior probabilities automatically add to 1.0 as they should. It remains to describe how the numerator terms, the likelihood, and the prior are calculated. The evidence can be in the form of k failures out of n trials or k failures in T hours. Respective likelihood is written as

$$p(E_3|E_1E_2) = \frac{n!}{k!(n-k)!}\lambda_j^k(1-\lambda_j)^{n-k} \tag{7.65}$$

$$p(E_3|E_1E_2) = \frac{(\lambda_jT)^k}{k!}e^{-\lambda_jT} \tag{7.66}$$

Since the factorial of zero is equal to one, evidence of zero failures ($k = 0$) therefore is handled within the Bayesian framework, just like any other evidence.

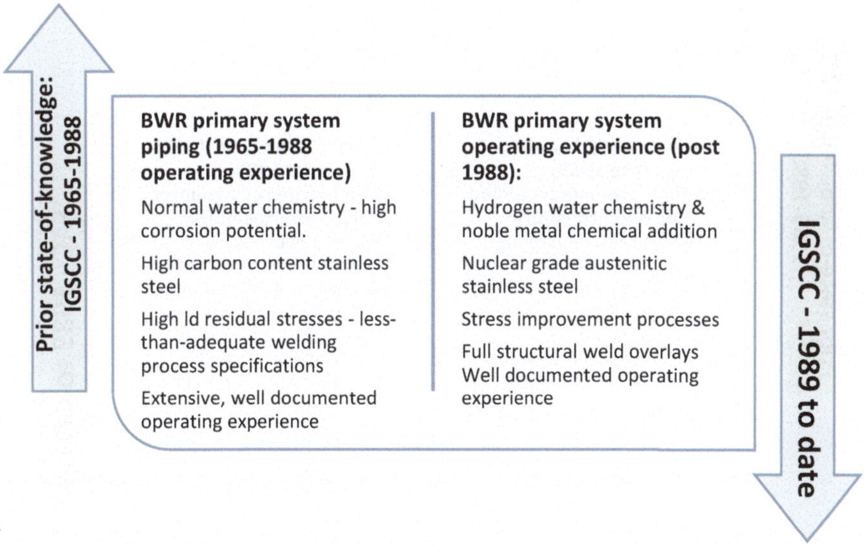

Fig. 7.6 IGSCC prior and posterior state of knowledge

7.13.1 Prior Pipe Failure Rate Distributions

Developing prior distributions has been seen as a very controversial aspect of Bayesian analysis. Typical advice is that considerable care should be taken when selecting priors, and the process by which priors are selected must be documented carefully. Two examples are offered to answer the question on what the conditions for developing empirical prior distributions are: 1) intergranular stress corrosion cracking (IGSCC) and 2) flow-accelerated corrosion (FAC). Both material degradation mechanisms are well understood and managed through specific reliability and integrity management processes. The boiling water reactor IGSCC experience is used to illustrate how the related knowledge base has evolved (Fig. 7.6).

For the prior state of knowledge, the operating experience is extensive. The demarcation between the two domains 1965–1988 and post-1988 is logical in that it is strongly connected with reactor regulations and industry initiatives that were implemented to address and mitigate intergranular SCC (USNRC 1988a, b). The available operating experience lends itself to the development of empirical Bayes priors that are input to an assessment of the IGSCC mitigation factor of improvement (Lydell et al. 2022a, b).

Similar to IGSCC, the FAC failure experience can be divided into two groups, before and after implementing effective reliability and integrity management processes to monitor and mitigate pipe wall thinning (OECD/NEA 2014). This is illustrated in Fig. 7.7.

Plant operators have monitored FAC since early plant life and have performed repairs and replacements as warranted by pipe wall wear rates and pipe failures. FAC

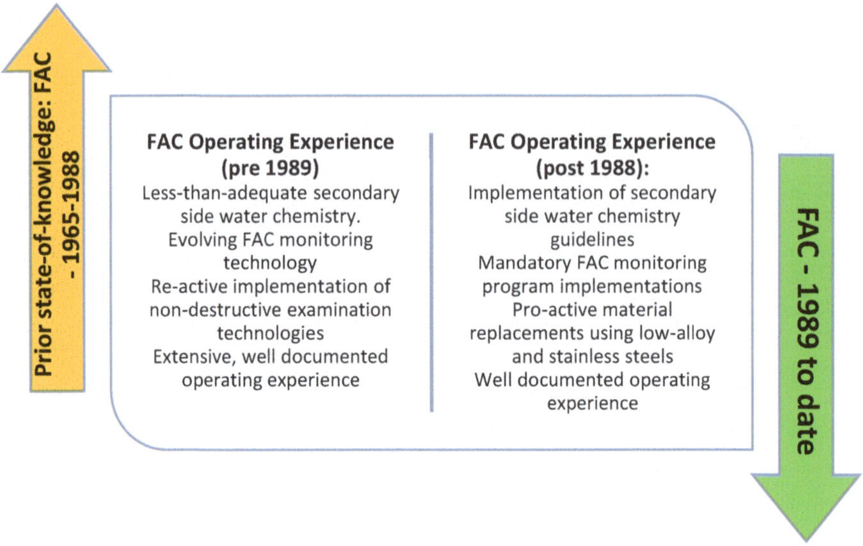

Fig. 7.7 FAC prior and posterior state of knowledge

was considered to be mainly a problem in two-phase flow (e.g., wet steam) systems. The first case of single-phase FAC-induced pipe failure was reported in 1985. A second case of single-phase FAC-induced pipe failure was reported in 1986 when a DN450 suction line to a main feedwater pump failed catastrophically. In July 1987, the United States Nuclear Regulatory Commission (NRC) issued Bulletin No. 87–01, "*Thinning of Pipe Walls in Nuclear Power Plants.*" This Bulletin requested all licensees for nuclear power plants holding an operating license or construction permit to submit information concerning their programs for monitoring pipe wall thickness. Later, in 1989, Generic Letter 89–08 (*Erosion/Corrosion-Induced Pipe Wall Thinning*) was issued, requesting all licensees to implement a long-term FAC detection program to prevent pipe failures in high-energy (single- and two-phase) carbon steel piping systems. The programs developed by each utility used plant-specific conditions, industry-wide operating experience, engineering judgment, NDE techniques, and specialized computer programs to predict the rate of FAC on a component-by-component basis.

7.13.2 Informative Priors

An "informative prior distribution" in Bayesian analysis is a prior probability distribution that incorporates specific knowledge or beliefs about a parameter before observing any data, effectively expressing a strong opinion about where the true value of the parameter is likely to lie, often based on previous studies, expert opinions, or theoretical considerations, in contrast to a non-informative prior which expresses minimal or no prior knowledge about the parameter.

Unlike a non-informative prior, an informative prior significantly influences the resulting posterior distribution, pulling it towards the region of higher probability density within the prior distribution. When substantial prior information is available, like results from previous studies or expert opinions, using an informative prior can enhance the accuracy of Bayesian inference by incorporating this knowledge. Prior information's strength dictates its impact on the posterior distribution. When using informative priors, it is crucial to perform sensitivity analysis to assess how much the posterior results depend on the chosen prior distribution. The prior distribution must come from sources other than the current data. Ideally, generic data provides the basis for prior belief.

7.13.3 Non-informative Priors

The Jeffreys non-informative prior is intended to convey little prior belief or information, thus allowing the data to speak for themselves. This is useful when no informed consensus exists about the true value of the unknown parameter. It is also useful when the prior distribution may be challenged by people with various agendas. Some authors use the term reference prior instead of "non-informative prior," suggesting that the prior is a standard default, a prior that allows consistency and comparability from one study to another. With Poisson data, the Jeffreys non-informative prior is obtained if the shape parameter of a gamma distribution is taken to be $\alpha = 1/2$ and the parameter β is taken to be zero. Jeffreys prior (also called *Jeffreys-rule prior*),[9] is used in Bayesian parameter estimation. It is an uninformative prior, which means that it gives vague information about probabilities. It is usually used when a suitable prior distribution is not available. However, an uninformative prior is used to make sure that it has little impact on the posterior distribution because it makes minimal assumptions about the model:

$$CFP_{Posterior} = \frac{(2 \times SSF + 1)}{(2(SSF + F) + 2)} \tag{7.67}$$

$$F = \frac{(2 \times (SSF + F) + 1)}{2 \times T} \tag{7.68}$$

[9] Sir Harold Jeffreys (1891–1989) as a British mathematician, statistician, geophysicist, and astronomer. His book, Theory of Probability (1939), played an important role in the revival of the objective Bayesian view of probability. He taught mathematics at the University of Cambridge, and he was knighted in 1953. One of his major contributions was on the Bayesian approach to probability.

7.13.4 Constrained Non-informative Prior

This prior is a compromise between an informative prior and Jeffreys non-informative prior. The mean of the constrained non-informative prior uses prior belief, but the dispersion is defined to correspond to little information. These non-informative prior distributions are described by Atwood (1994). For Poisson data, the constrained non-informative prior is a gamma distribution, with the mean given by prior belief and the shape parameter $\alpha = 0.5 = 1/2$. That is,

$$Mean = \frac{\alpha}{\theta} = \frac{n}{T} \tag{7.69}$$

$$\theta = \frac{0.5}{\frac{n}{T}} \tag{7.70}$$

7.13.5 Empirical Bayes

In a "standard Bayes" application, the prior distribution is fixed before any data are observed or when very limited data are available. In contrast to this approach, empirical Bayes methods are procedures for statistical inference in which the prior probability distribution is estimated from the operating experience data. The development of "location-specific" prior weld failure rate distributions is used as an example of how empirical Bayes can be applied. In empirical Bayes, the prior is often estimated from the data itself, rather than being explicitly specified. The selected example of interest is concerned with the estimation of weld failure rates attributed to IGSCC. A representative in-service inspection isometric drawing of one loop of this system is shown in Fig. 7.8.

The main recirculation system consists of multiple welds in DN300, DN400, and DN600 stainless steel piping. There are ASME Section XI Category B-F and B-J welds in the system. The former category consists of dissimilar metal welds between the piping and the reactor pressure vessel inlet and outlet nozzles. The latter category consists of similar metal butt welds in multiple locations within the system, such as pipe-to-bend, pipe-to-pipe, and pipe-to-reducer welds. Assume that an analyst is developing a loss-of-coolant accident (LOCA) initiating frequency model. In defining the prior state of knowledge, the following information should be acknowledged:

- IGSCC was first reported in the mid-1960s, and by the end of the 1970s, relatively large instances of IGSCC failures had been reported at BWR plants.
- By the mid-1980s, all BWR plants in the United States had reported a very large number of IGSCC failures. This led to industry and regulatory efforts to better understand the underlying contributing factors to IGSCC initiation and propagation.

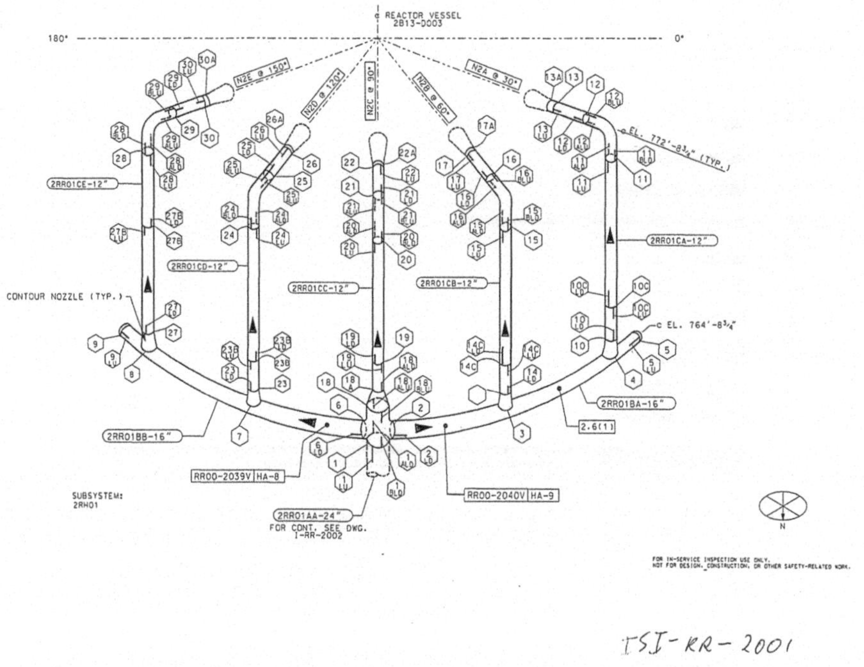

Fig. 7.8 BWR reactor recirculation system ISI isometric drawing. (Reproduced from Holman, G. S., Chou, C.K. (1988), Probability of Failure in BWR Reactor Coolant Piping, NUREG/CR-4792, U.S. Nuclear Regulatory Commission, Washington, DC)

Weld precursor failure rates correlate with weld location in the piping system, likely due to piping layout, welding complexity, and operational or residual stresses. Figure 7.9 shows weld configuration versus failure fraction. How can one acknowledge potential location-based trends in weld failures through data specialization? Figure 7.10 shows the results of combining average weld failure rates with piping layout data. For a weld location of type "*i*" and size "*j*," the failure rate is expressed as follows:

$$\lambda_{ij} = \frac{F_{ij}}{W_{ij} \times T} \tag{7.71}$$

$$S_{ij} = \frac{F_{ij}}{F_j} \tag{7.72}$$

$$A_{ij} = \frac{W_j}{W_{ij}} \tag{7.73}$$

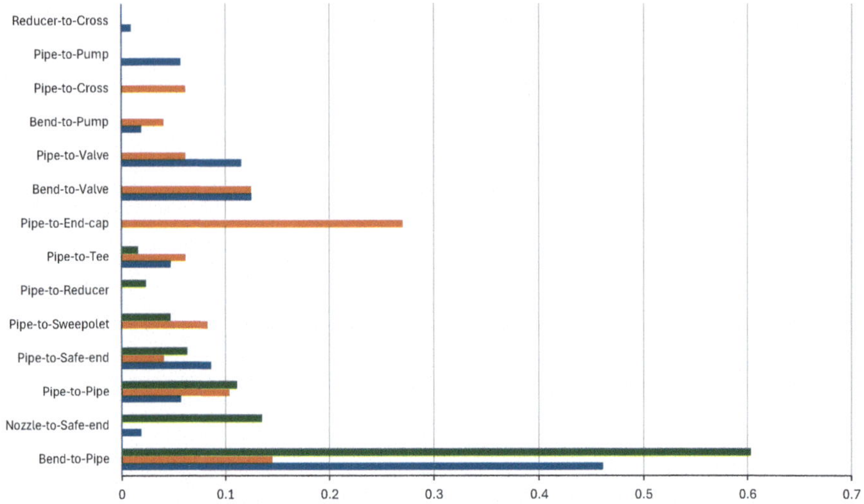

	Bend-to-Pipe	Nozzle-to-Safe-end	Pipe-to-Pipe	Pipe-to-Safe-end	Pipe-to-Sweepolet	Pipe-to-Reducer	Pipe-to-Tee	Pipe-to-End-cap	Bend-to-Valve	Pipe-to-Valve	Bend-to-Pump	Pipe-to-Cross	Pipe-to-Pump	Reducer-to-Cross
DN300	0.6032	0.1349	0.1111	0.0635	0.0476	0.0238	0.0159							
300 < DN ≤ 550	0.1458		0.1042	0.0417	0.0833		0.0625	0.2708	0.125	0.0625	0.0417	0.0625		
DN > 550	0.4615	0.0192	0.0577	0.0865			0.0481		0.125	0.1154	0.0193		0.0577	0.0096

Fraction of IGSCC Failures

Fig. 7.9 Fraction of IGSCC failures by location in BWR main recirculation piping

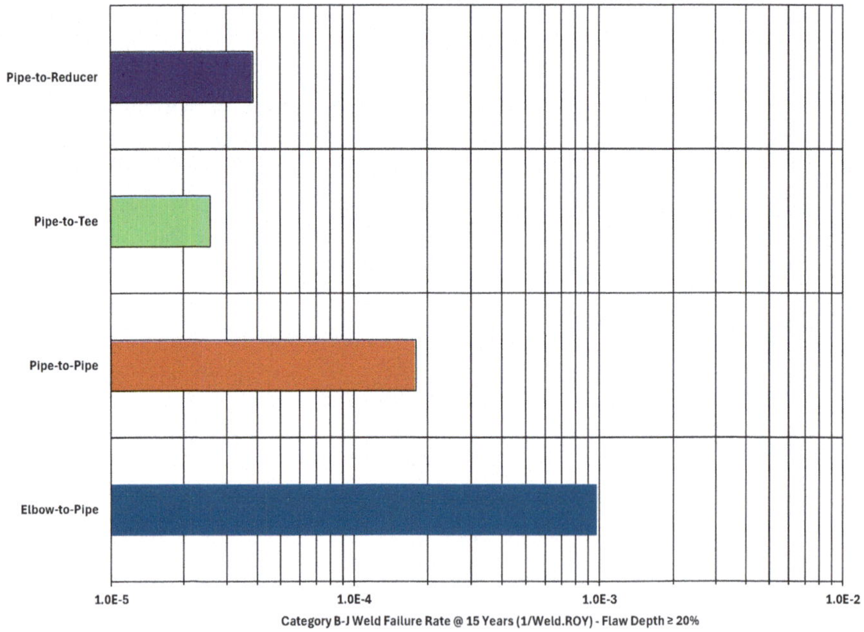

Fig. 7.10 Prior weld failure rates with adjustment for location dependency

$$\lambda_{ij} = \frac{F_j \times S_{ij}}{W_{ij} \times T} \tag{7.74}$$

$$\lambda_{ij} = A_{ij} \times S_{ij} \times \lambda_j \tag{7.75}$$

where

λ_j Failure rate for weld in a piping system of nominal diameter *ji* susceptible to IGSCC.

λ_{ij} Failure rate of weld in location *j* within the piping system.

A_{ij} An attribute factor which defined as the ratio of the total number of welds of a certain size *j* to the number of welds of type *i*. A_{ij} is a correction factor and accounts for piping layouts (or routing) such that there are certain number of welds, say, between 90-degree elbows and pipes or between two sections of straight pipe.

F_{ij} Number of "*Ij*" weld failures.

S_{ij} A susceptibility factor accounts for stressors that are imposed on welds in different locations within a piping system. For a given pipe size and system, the susceptibility is expressed as the fraction of welds of type *Ij* that failed due to a certain degradation mechanism. The stressors include weld residual stresses, fit-up stresses, normal operating stresses, and stresses imposed by unanticipated pressure transients.

W_{ij} Number of "*Ij*" welds.

W_i Number of "*i*" welds in the piping system.

7.14 Updating Generic Pipe Failure Rates with Plant-Specific Data

Tabulated pipe failure rates in referenceable, peer-reviewed reports can in principle be used as "generic" or "industry-wide" data that represent an informed prior distribution. However, it is important to understand the underlying technical basis for the selected generic prior, and this section provides some technical insights into this matter. The Bayesian approach has been widely utilized to analyze significant risks of internal flooding accident scenarios in probabilistic safety assessments. The first step in the Bayesian process involves collecting plant-specific failure and exposure term data, followed by the careful selection of the relevant prior distribution parameters. In the context of probabilistic piping reliability, the extent of implementations of Bayesian updates to generic priors is not well documented.

An exception is found in the full-scope site Level 3 probabilistic safety analysis (PRA) project (L3PSA project) for a two-unit pressurized water reactor reference plant (Vogtle Unit 1 and 2).[10] This work was sponsored by the USNRC. Volume 3b of the PSA documents an analysis of internal flooding initiating event frequencies that are attributed to service water system pipe failure (USNRC 2022a, b).[11] Plant-specific pipe failure rates were obtained through Bayesian updates using plant-specific operating experience and generic pipe failure rates taken from EPRI 3002000079, *Pipe Rupture Frequencies for Internal Flooding Probabilistic Risk Assessments, Revision 3* (Fleming and Lydell 2013). The plant-specific data considered for the USNRC internal flooding PRA covered the period January 1, 1990, to December 31, 2012. The plant-specific data were taken from the following sources:

- Plant-specific operating experience submitted to the CODAP international database
- Search of plant-specific Licensee Event Reports (LERs) for events including "leak," "leakage," or "flood" in the title through December 2012
- NRC Reactor Operating Experience Data (NROD) website,[12] which includes plant-specific EPIX/ICES records

For Safety Class 3 service water system piping, the data search found zero (0) events involving piping having a nominal diameter of greater than DN250 and zero failures in 42 critical reactor-years for the two reference plant reactor units. The data search also found six events involving piping having a nominal diameter of greater than DN50 but less than or equal to DN100. In contrast, querying this book's pipe failure database of reference yields results shown in Fig. 7.11.

EPRI's methodology combines Bayesian updating with data specialization and periodic updates to create a system for refining pipe failure rate predictions and adapting to new information. This approach allows the piping reliability models to learn incrementally, moving from a general prior belief (as represented in the first edition of EPRI's pipe failure report) to a more specific posterior belief as new, specialized data has been incorporated in subsequent editions. Periodic updates have been performed to ensure that the models used to assess piping reliability have remained current and relevant. In order to perform a technically sound Bayes' update of pipe failure rates, the following questions arise:

---

[10] Vogtle Units 1 and 2 are nuclear power plant units and are Westinghouse four-loop 1200 Mwe pressurized water reactors that have been in commercial operation since 1987 and 1989, respectively.

[11] The identity of reference plant is not stated in the L3PRA documentation, but based on the information in Volume 2 of the documentation, it is possible to infer which plant it is, i.e., Vogtle Units 1 and 2.

[12] The NROD website allows NRC and nuclear industry personnel to search and view IDCCS coded data directly. In addition to providing a user interface for searching the coded data, the website also provides various reports, links, and documents related to PRA data. NROD also links to the online version of the reliability and availability data system (RADS). RADS provides NRC and industry a source of unit-specific component-level data on reliability. New reports or features may be added to the website to enhance utility to the NRC staff. Help documents shall also be included on the website and shall also be maintained as changes are made.

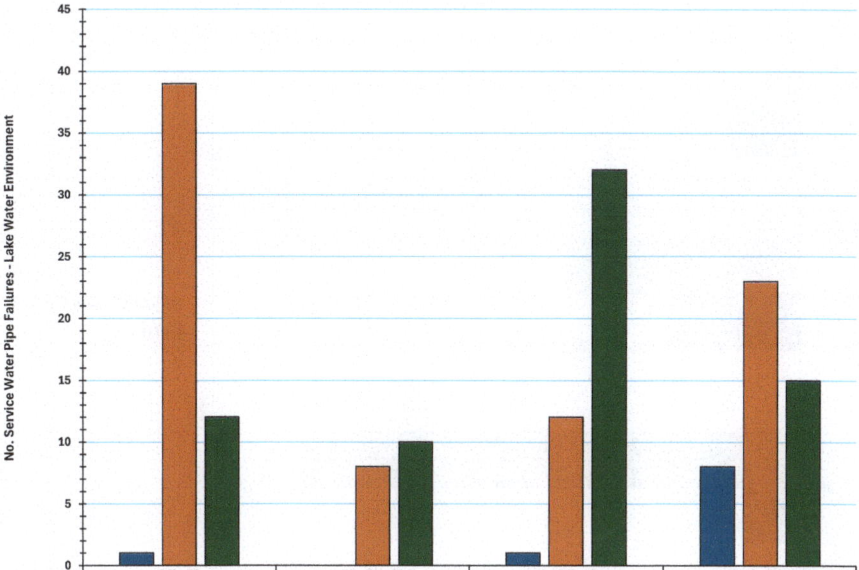

	DN ≤ 50	50 < DN 100	100 < DN ≤ 250	DN > 250
1980 -1989	1		1	8
1990-2012	39	8	12	23
2013-2024	12	10	32	15

Fig. 7.11 Service water pipe failure data applicable to four-loop PWR plants

- Is the plant-specific data for pipe failures and linear meters of pipe of exposure being collected and analyzed in a manner that is consistent with the treatment of generic data in the generic estimates?
- Is there a significant plant-to-plant or site-to-site variability in the failure rate data that is reflected in the generic distributions?
- There is a question about whether plant-specific data should be removed from the generic data to avoid overcounting the same evidence in two places. This is a generic issue in Bayes' updating with plant-specific data, but it is usually ignored under the assumption that the contribution to the generic distributions from any specific plant is small.
- There is strong evidence of aging in successive updates of the pipe failure rate estimates in four editions of the EPRI report. The evidence is a progressive trend upwards in the calculated average failure rate as new evidence is applied.

Absent precise details on how the L3PRA Bayesian update was performed, a simple example is used to demonstrate a process for updating generic pipe failure rates. This demonstration uses the operating experience with Safety Class 3 service water piping of DN750. Restricted to the brackish-water/seawater environment, Fig. 7.12 illustrates the SW pipe operating experience for plant "X" for which Bayesian updating is intended, versus the overall SW operating experience less the event population from plant "X."

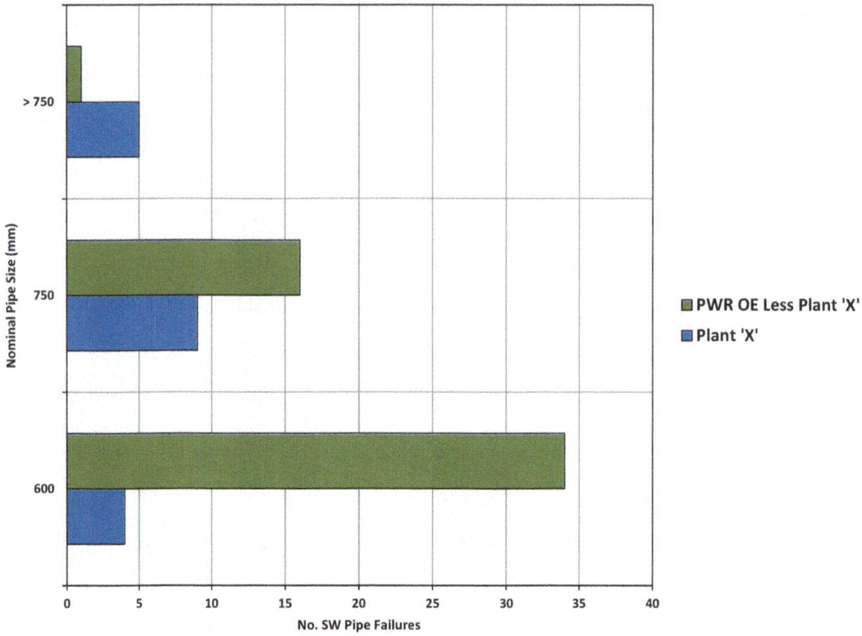

Fig. 7.12 Service water pipe failure data

The plant-to-plant operating experience variability is illustrated in Fig. 7.13. This comparison of the operating experience data raises the following question with respect to how to best perform a Bayesian update. Given the relative weight of the evidence provided by the prior and plant-specific data, what would be an appropriate strategy for performing a Bayesian update? Should a selected prior be taken straight from a referenceable data source or should a new, informed prior distribution be developed? A reason for re-baselining a generic prior pipe failure rate distribution is to avoid double-counting the plant-specific operating experience, assuming that the technical basis for a generic failure rate is fully documented. In this simple demonstration, the input to the prior state of knowledge is known, thus enabling the establishment of a "neutral" prior.

Table 7.3 shows the results of a strategy for developing plant-specific pipe failure rates. In a first step, a generic pipe failure rate for SW piping of diameter DN750 is calculated using the applicable event population and exposure term data, with the plant-specific data subtracted. Next, the new results are updated using the plant-specific input data for the same time period covered by a referenceable data source. This yields a plant-centric prior distribution, which is used to perform an update with new data. The "referenceable data source" in this example, as in the L3PRA study, is EPRI 3002000079, *Pipe Rupture Frequencies for Internal Flooding Probabilistic Risk Assessments, Revision 3.*

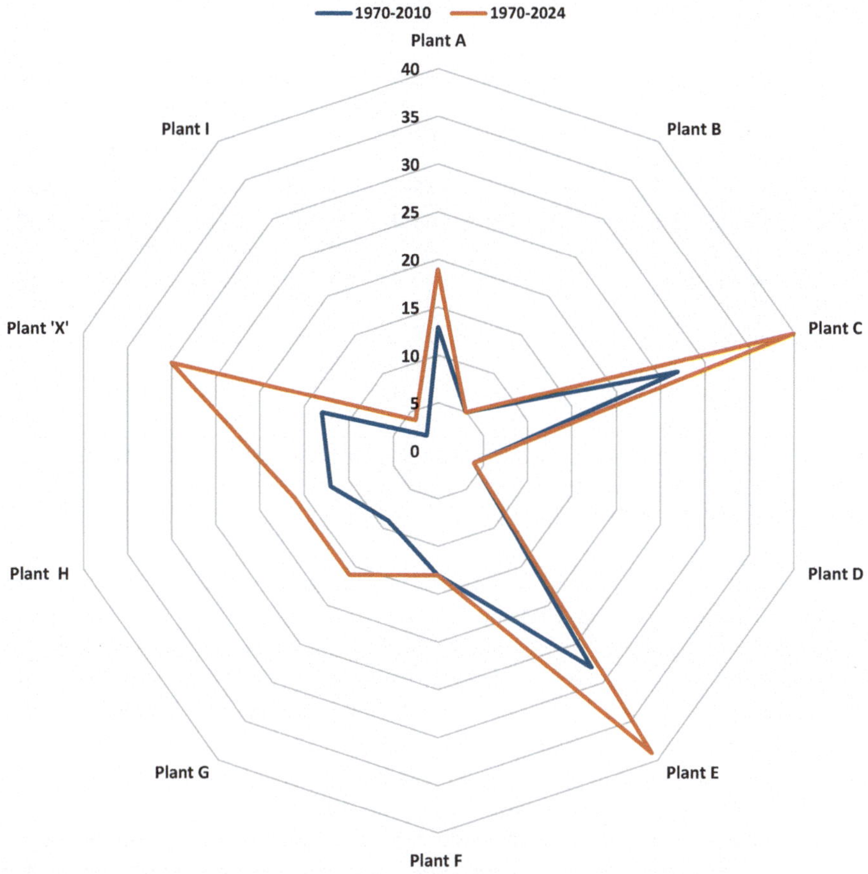

Fig. 7.13 Plant-to-plant variability in service water pipe failure experience

7.15 Models of Conditional Pipe Failure Probability

Models of the conditional failure probability (CFP) are utilized to extrapolate estimates of failure frequency for precursor events, such as minor through-wall leaks progressing to major structural failures. CFP can be depicted by a cumulative failure probability versus an equivalent break size (EBS), expressed in terms of through-wall mass flow rate in kg/s or size of the pressure boundary hole converted to an equivalent diameter. An insight from assessments of operating experience is that the slope of the CFP curve demonstrates a significant decrease in probability with increasing failure magnitude. Another way to interpret the results is that the conditional rupture probability is influenced by the pipe diameter; specifically, smaller pipes exhibit a lesser steepness. However, this assumption is not consistently supported by probabilistic fracture mechanics analyses, which suggest that the CFP

Table 7.3 Results for different Bayesian update strategies

Assumption	Prior pipe failure rate uncertainty distribution		Plant "X" posterior pipe failure rate uncertainty distribution—DN750 piping(1/ROY-linear meters)				
	α	θ	Mean	5%-tile	50%-tile	95%-tile	EF
Generic prior: >DN250	0.5	6779.8	6.72E-04	3.04E-04	6.38E-04	1.16E-03	1.82
Plant "X" neutral prior: >DN250	0.5	5998.9	7.31E-04	3.32E-04	6.94E-04	1.26E-03	1.82
Plant-specific prior: DN750	0.5	2732.4	1.33E-03	6.45E-04	1.27E-03	2.22E-03	1.75
Plant-specific prior: >DN250	0.5	7369.2	7.31E-04	3.54E-04	6.98E-04	1.22E-03	1.75

"Neutral prior" means that the plant-specific data embedded in the generic prior has been removed to establish a new, "corrected generic" prior

The error factor (EF) is defined as $95\%-tile/_{Median}$

is driven by the percent cracked circumference more than the diameter of a pipe and for pipe diameters ≥DN100 (Rudland 2019).

It is useful to explore piping operating experience by organizing the data by safety class, number of observed failures, and corresponding through-wall mass flow rates. An example is illustrated in Fig. 7.14, which shows the number of failures on the y-axis and the corresponding equivalent break size (EBS) on the x-axis. To date, all known significant Safety Class 1 pipe failures have involved small-diameter piping, with mass flow rates less than or equal to about 10 kg/s.

The operating experience with safety-related service water system piping is extensive. Accounting for the in-plant piping and belowground piping within the owner-controlled area, the operating experience includes the full spectrum of pressure boundary failures, from small leaks to massive flood events. The latter are limited to failures in belowground piping, however.

7.15.1 Expanded View on Empirical CFP Models

An expanded viewpoint on the feasibility of developing empirical CFP models is investigated in this section. "CFP model" means the collection of uncertainty distributions as a function of a full spectrum of equivalent break sizes, from minor to very large. A first step in this investigation is to obtain insights from the piping operating experience. This is done by organizing the operating experience according to safety classification and type of failure. The latter is characterized as "degraded" (DEG), "through-wall leak" (TWL), or a "significant structural failure" (SSF). The latter type oftentimes is referred to as either a "pipe break" or a "pipe rupture." The TWL category covers minor leaks to significant leaks but less than 10 kg/s pipe

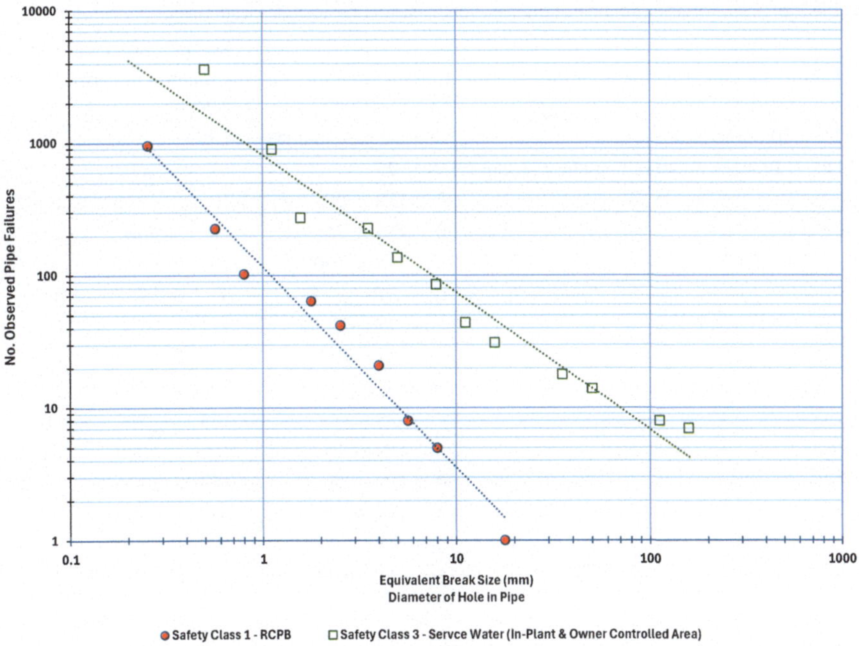

Fig. 7.14 Exploring operating experience for possible CFP vs. EBS correlations

Table 7.4 Material degradation mechanisms versus structural integrity category

	No. of failures by structural integrity category		
Material degradation mechanism	DEG	TWL	SSF
Corrosion	706	2633	131
Erosion-corrosion	219	478	20
Erosion-cavitation	105	207	2
FAC	1598	770	151
High-cycle fatigue	208	2174	362
Low-cycle fatigue	28	295	18
IGSCC	1607	420	0
PWSCC	78	90	0
TGSCC	218	297	3
Thermal fatigue	179	108	7

through-wall mass flow rate. The term "degraded" implies that a repair or replacement program has been implemented to ensure that the structural integrity of the piping is maintained. Categories DEG (or non-through-wall) and TWL, if left undetected, are to be seen as precursors to SSF.

Two tables are generated in this first step of the investigation. Table 7.4 responds to the question: Which degradation mechanism has produced SSFs? This table clearly shows that it is primarily various corrosion mechanisms, flow-accelerated

Table 7.5 Safety class and pipe diameters versus structural integrity category

ASME III Safety Class	Nominal pipe diameter	No. of failures by structural integrity category DEG	TWL	SSF
Class 1: Primary pressure boundary piping	DN ≤ 25	61	603	17
	25 < DN ≤ 50	59	130	2
	50 < DN ≤ 75	48	46	2
	75 < DN ≤ 100	107	54	2
	100 < DN ≤ 150	92	21	1
	150 < DN ≤ 250	213	19	0
	250 < DN ≤ 550	706	73	0
	DN > 550	395	16	0
Class 2	DN ≤ 25	145	797	90
	25 < DN ≤ 50	108	289	11
	50 < DN ≤ 75	153	126	3
	75 < DN ≤ 100	63	90	4
	100 < DN ≤ 150	147	119	4
	150 < DN ≤ 250	145	97	5
	250 < DN ≤ 550	274	57	5
	DN > 550	53	15	1
Class 3	DN ≤ 25	48	871	113
	25 < DN ≤ 50	119	557	18
	50 < DN ≤ 75	78	231	10
	75 < DN ≤ 100	111	334	10
	100 < DN ≤ 150	71	342	8
	150 < DN ≤ 250	141	344	14
	250 < DN ≤ 550	236	443	20
	DN > 550	126	304	24
Non-safety: Balance of plant piping	DN ≤ 25	60	541	195
	25 < DN ≤ 50	139	271	58
	50 < DN ≤ 75	64	74	24
	75 < DN ≤ 100	98	168	31
	100 < DN ≤ 150	104	156	30
	150 < DN ≤ 250	400	145	48
	250 < DN ≤ 550	495	140	53
	DN > 550	83	75	27

corrosion (FAC), and high-cycle fatigue that have been responsible for significant structural failures. Hence, in principle, the corresponding operating experience can be used to develop empirical models of the CFP.

Table 7.5 responds to the question: Which combination of piping safety class and nominal pipe diameter has produced SSFs? The table provides additional insights; IGSCC and PWSCC have not produced any SSFs. Another important insight is, of

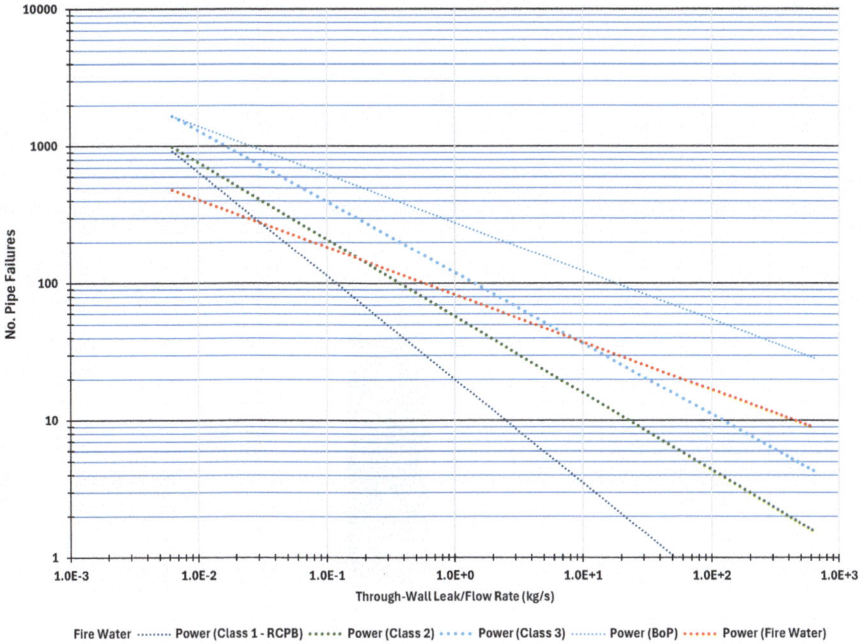

Fig. 7.15 Pipe failures as a function of observed through-wall flow rate

course, that small-diameter piping is more susceptible to failure than larger-diameter piping.

A second step in this investigation is to plot the number of pipe failures versus the observed through-wall flow rate (Fig. 7.15). Converting the y-axis to a frequency scale yields additional insights into the CFP modeling endeavor. Developing the associated uncertainty distributions is a relatively straightforward process in principle for Safety Class 2 and 3 systems, non-safety systems, and firewater system piping. Insights into the CFP of Safety Class 1 piping are obtained by estimating the frequency of operational leakage as a function of the equivalent break size. The results are shown in Fig. 7.16.

7.15.2 CFP Model Development Absent Real-Life Data on SSF

What are the prospects for using operating experience as a basis for CFP model development? Using Bayesian concepts, there are two ways of responding to this question: (1) for small-diameter piping ≤DN75, sufficient operating experience exists not only to determine the parameter of a prior probability distribution but also for Bayesian updates, and (2) for piping having a nominal diameter greater than

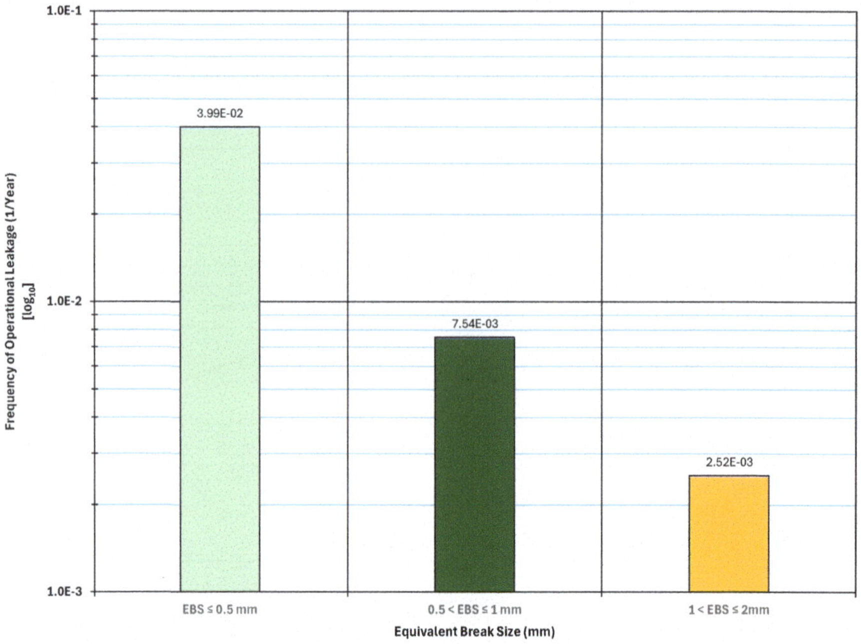

Fig. 7.16 Frequency of operational leakage from Class 1 piping

DN75, the setting of the prior probability distribution parameters requires subject matter expert (SME) input combined with operating experience input. Absent operating experience data on pipe rupture to support direct estimation of the conditional failure probability, an assessment can utilize insights from experimental work, expert elicitation, and/or probabilistic fracture mechanics. Three methods for Bayesian priors and Bayesian updates are as follows:

Method 1: Under the assumption that the beta distribution reasonably well represents the conditional pipe failure probability, the prior state-of-knowledge parameters of the distribution are estimated by assuming a non-informative $\alpha = 0.5$ and with β derived from operating experience.

Method 2: Under the assumption that a beta distribution reasonably well represents the conditional pipe failure probability, the prior state-of-knowledge parameters of the distribution are estimated using the Pearson-Tukey approach, followed by a Bayesian update. Here, the expert opinion is in the form of the 5th, 50th, and 95th percentiles ($C_{.05}$, $C_{0.50}$, $C_{0.95}$), which may come from expert elicitation or probabilistic fracture mechanics calculations.

Method 3: A consensus-based approach to defining a prior state-of-knowledge probability distribution in a Bayesian framework. The distribution parameters are obtained using reverse engineering.

7.15.2.1 Method 1

A practical approach to calculating conditional pipe failure probabilities is to use a Bayesian reliability analysis method in which the prior CFP uncertainty distribution is expressed by a beta distribution. The beta distribution takes on values between 0 and 1 and is defined by two parameters *A* and *B*. The mean of the beta distribution is $CFP_{Mean} = A/(A + B)$. If the abstract parameters $A = B = 1$, the beta distribution takes on a flat distribution between 0 and 1. If $A = B = ½$, the distribution is referred to as Jeffreys non-informative prior and is a *U*-shaped distribution with peaks at 0 and 1. Expert opinion can be incorporated by selecting *A* and *B* to correspond to that expert's estimate of the mean failure probability.

The two-parameter beta distribution has been well accepted to model conditional probabilities because of its versatility and ease of use in a Bayesian framework. However, assessing the parameters *A* and *B* is not trivial since an infinite number of combinations of *A* and *B* can produce the correct value of *p*. Insights from operating experience are an entry point for developing engineering-oriented bounding values for the CFP, that is, our Bayesian prior state of knowledge of the CFP for different combinations of systems, operating environments, and loading conditions. To illustrate a meaning of this, assume that an expert estimate of the CFP is 1.0×10^{-2}; then, the two parameters could be selected as $A = 1$ and $B = 99$ as per $CFP_{Mean} = A/(A + B)$. These abstract parameters can be associated with the number of failures and the number of successes in examining operating experience data to obtain an estimate of the conditional pipe failure probability.

The beta distribution has some convenient and useful properties for use in Bayes' updating. The prior state-of-knowledge distribution, representing an analyst's understanding of piping performance, can be represented as a prior distribution by defining a set of initial *A* and *B* values, denoted as A_{Prior} and B_{Prior}. These prior parameters are selected in such a way that they are representative of engineering estimates (e.g., PFM results) prior to the collection of pipe failure data. Then, when looking at the relevant service experience, if there are *N* failures and *M* successes, the Bayes' updated, or posterior, distribution is also a beta distribution with $A_{Posterior} = A_{Prior} + N$ and $B_{Posterior} = B_{Prior} + M$. Here, *N* corresponds to the number of structural failures of some well-defined magnitude and in some specialized combination of pipe size and material, and *M* corresponds to the total number of failures that do not result in a structural failure in the corresponding pipe size/material combination.

Since many different parameter combinations will produce the same mean value, selecting well-justified *A* and *B* parameters is not a trivial task. For leak-before-break piping, it can be concluded that parameter *A* has to be a relatively small number, <1, and parameter *B* has to be a large number. Expert elicitation combined with the Pearson-Tukey method can be used to determine the two parameters.

7.15.2.2 Method 2

Three-point approximations are widely used when assessing parameters of distributions because there is usually a good sense of what the median and two bounds are (for instance, the 5th, 50th, and 95th percentiles). Extensive work has been performed to determine the best approximations of the mean and standard deviation of a beta distribution based on three percentiles. Subject matter experts (SMEs) are asked to provide the 5th, 50th, and 95th. The information provided by an expert may be based on results obtained by advanced structural reliability models, e.g., PFM models. Next, the parameters of the corresponding beta distribution are obtained by using the Pearson-Tukey method. The calculation process is illustrated in Fig. 7.17.

An interesting observation is that practical applications, in which the expert opinion is in the form of PFM input, yield α values near 0.5 but with large to very large β values.

7.15.2.3 Method 3

Method 3 uses expert opinions published in NUREG-1829 (USNRC 2008). However, NUREG-1829 was not intended for developing CFP uncertainty distributions, which may be seen as a limitation of this method. The individual expert elicitation input data was not part of the NUREG documentation, however. The process involves a six-step reverse engineering approach.

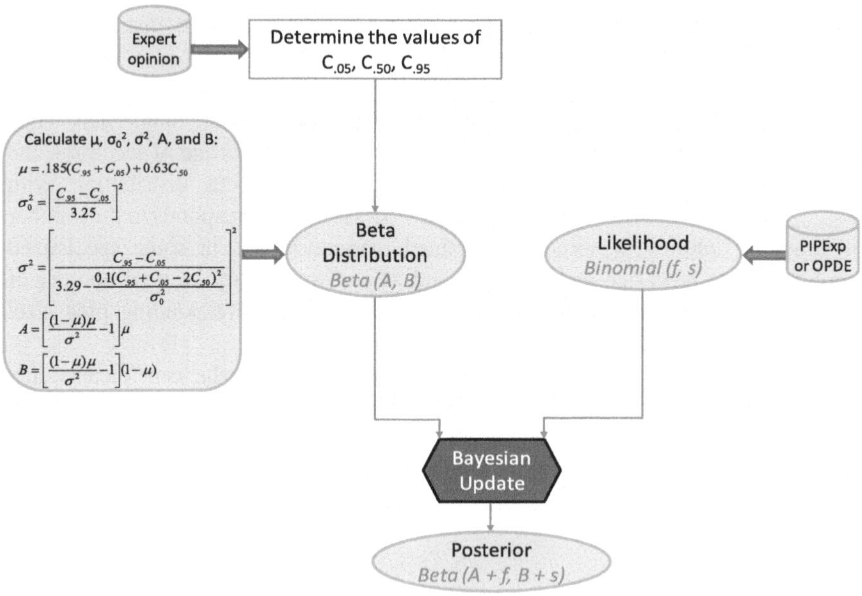

Fig. 7.17 Pearson-Tukey approach to beta parameter estimation

1. Extract the subject matter expert (SME) input to the expert elicitation. Each SME documented the quantitative inputs on Microsoft® Excel worksheets.
2. Select an evaluation boundary to define the CFP model categories, i.e., system and pipe sizes.
3. Develop a decomposition model for how to derive a conditional failure probability from LOCA frequency and precursor rate. The decomposition model views LOCA as a two-stage process involving the stochastic occurrence of pipe flaws and the stochastic rupture events given a degraded condition. It is a first-order approximation of a physical process model. An advantage of this approach is the ability to make use of existing operating experience in various forms, including information on observed degraded states and early stages of pipe and weld failure, as indicators of far less frequent rupture events.
4. Develop composite CFP uncertainty distributions. In this step, the SME distributions are combined into a single composite distribution.
5. Select a "target LOCA frequency (TLF)" for each evaluation boundary.
6. Develop CFPs from TLFs. This step uses the target LOCA frequencies and information on the underlying precursor rates for each evaluation boundary to derive a CFP model that, when linked with the precursor rate analyses, will reproduce the target LOCA frequencies.

The NUREG-1829 expert elicitation process required each member of an expert panel to qualitatively and quantitatively assess important LOCA contributing factors and quantify their uncertainty. The quantitative responses were combined to develop BWR and PWR total LOCA frequency estimates for each contributing panelist. The distributions for six LOCA size categories and three distinct time periods evaluated were represented by four parameters (mean, median, 5th and 95th percentiles). Finally, the individual estimates were aggregated to obtain group estimates, along with measures of panel diversity.

Method 3 uses results of the expert elicitation as input to the development of CFP uncertainty distributions as a function of equivalent break size. The expert elicitation results come from Expert Elicitation Response Sheets.[13] In a first step, a lognormal geometric mean of the expert elicitation participant results is obtained for each component case. Next, a single lognormal distribution for each analysis case and LOCA category is obtained by taking the geometric mean of the medians of the experts' lognormal distributions as the composite distribution median and the geometric mean of the range factors of the experts' lognormal distributions as the composite distribution range factor. Next, the precursor failure frequency results of respective analysis case were converted to fit a lognormal distribution, the *precursor* frequency. Finally, Eqs. (7.77), (7.78), and (7.79) are applied to calculate the CFP distributions to be used as prior distribution for each combination of precursor

[13] "BWR Piping Raw Data" (NRC-ADSAMS Accession No. ML080560008) and "PWR Piping Raw Data" (NRC-ADAMS Accession No. ML080560011). Note that the Expert Elicitation Response Sheets are not included in NUREG-1829, but available separately from the USNRC Public Document Room.

frequency and CFP. Since the use of lognormal distributions enables the LOCA frequency to be expressed as the product of a lognormally distributed *precursor* frequency and a lognormally distributed *CFP*, the parameters of the *CFP* distribution may be calculated using the following equations:

$$Median_{CFP(i)} = \frac{Median_{TLF(i)}}{Median_{Precursor}} \tag{7.77}$$

$$RF_{CFP(i)} = e^{1.645 \times \sigma_{CFP(i)}} \tag{7.78}$$

$$\sigma_{CFP_i} = \sqrt{\left(\frac{ln(RF_{TLF_i})}{1.645}\right)^2 + \left(\frac{ln(RF_{Precursor})}{1.645}\right)^2} \tag{7.79}$$

where

$Median_{CFP(i)}$ is the median of the lognormal distribution for the CFP in LOCA category *i* given a precursor failure.

$Median_{TLF(i)}$ is the median of the target LOCA frequency for category *i*.

$Median_{Precursor}$ is the median of the lognormal distribution for precursor frequency associated with a base-case analysis.

$RF_{CFP(i)}$ is the range factor of the lognormal distribution for the conditional probability of pipe rupture in LOCA category *i* given pipe failure, equal to $SQRT(95^{th}/5^{th})$ of the lognormal distribution.

σ_{CFP_i} is the logarithmic standard deviation for the lognormal distribution for the conditional failure probability of pipe rupture in LOCA category *i* given pipe failure.

RF_{TLF_i} is the range factor of the lognormal distribution for target LOCA frequency for category *i*.

$RF_{Precursor}$ is the range factor for the lognormal distribution for the *precursor* frequency.

Medians and range factors of the CFP distributions are calculated from the medians and range factors of the target LOCA frequency distributions using Eqs. (7.78) and (7.79). Using the properties of the lognormal distribution, the remaining parameters of the distributions are calculated for each evaluation boundary (Table 7.6).

NUREG-1829 developed base cases so that the panelists did not have to provide absolute frequencies during the elicitation (Table 7.7). Instead, they chose appropriate base cases and provided a relative ratio to express the difference between the base cases and the important contributing variables that they identified. This technique was based on the premise that relative ratios are easier to assess and therefore more accurate than absolute numbers. The decision to use and the application of these base cases were made by each individual panelist. Some panelists utilized them extensively; others considered the relative trends expressed by the base-case estimates; and others chose not to utilize them at all and performed their own assessments. In general, the panelists using the base-case estimates chose to anchor their responses to

Table 7.6 LOCA frequency calculation cases

BWR LOCA-sensitive piping		Pipe inside diameter (mm)—typical of BWR/4	DEGB (mm)
System	Pipe segment(s)		
Reactor recirculation system (RR)	RR discharge	711.2	1005.8
	RR suction	273.1	386.1
	RR manifold	558.6	789.9
Main steam system (MS)	MS—Inside containment	455.6	645.2
Feedwater system (FW)	FW—Inside containment	387.4	548.6
	FW—Inside containment	273.1	386.1
Shutdown cooling (SDC)/low-pressure coolant injection (LPCI)	SDC influent lines from RR loops to containment penetration and LPCI injection line	363.5 317.5	513.1 449.6
High-pressure coolant injection (HPCI)	HPCI discharge piping—BWR/5 and BWR/6	242.9	342.9
Low-pressure core spray (LPCS)	LPCS discharge piping	242.9	342.9
Reactor water cleanup (RWCU) system	RWCU from one RR	193.7	274.3
PWR LOCA-sensitive piping		Pipe inside diameter (mm)—Typical of WE 4-loop	DEGB (mm)
System	Pipe segment(s)		
Reactor coolant system (RCS)	RCS cold leg	698.5	987.6
	RCS hot leg	736.6	1041.4
	RCS crossover leg	787.4	1112.5
	Pressurizer surge line	325.4	459.7
Residual heat removal (RHR) system	RHR suction line—Off RCS hot leg	257.2	363.2
	RHR discharge to cold leg	215.9	304.8
	RHR discharge to cold leg	173.1	243.8
	RCS hot leg recirculation	173.1	243.8
Safety injection system (SIS)	Accumulator discharge to cold leg	257.2	363.2
	RCS hot leg recirculation	173.1	243.8

base-case results obtained using PPR results instead of PFM results. The BWR base cases assumed a BWR/4 plant type. The PWR hot leg and surge line base cases were modeled based on a three-loop Westinghouse plant, while the HPI make-up line was representative of a Babcock & Wilcox (B&W) plant.

NUREG-1829 provides estimates of the frequencies for LOCAs based on a set of LOCA categories selected to span the break sizes and leak rates that are typically modeled in PSA. The approach to using information in NUREG-1829 to develop estimates of the CFP is based on the following observations and information presented in that document.

– Base-case results were presented for well-defined evaluation boundaries for BWR and PWR plants, which comprise part of the primary pressure boundary. For each

Table 7.7 NUREG-1829 base-case definitions

Plant type	Base case	Relevant operating experience
BWR	BWR1	U.S. IGSCC experience involving reactor recirculation DN700 DN300 dissimilar weld cracks, respectively. BWR/4 operating experience 1970–2004
	BWR2	
	BWR3	Safety class 1 feedwater system dissimilar metal weld operating experience
PWR	PWR1	PWSCC failure at V.C. On October 7, 2000. This event is described in Chap. 4, Sects. 4.9.3 and 4.9.6
	PWR2.1	Pressurizer surge line dissimilar metal weld at pressurizer vessel. This base case was defined to address the possibility of PWSCC at this location
	PWR2.2	Pressurizer surge line at reactor coolant system hot leg. This base case was defined to address the possibility of thermal fatigue cracking at this location
	PWR3	Oconee 2 thermal fatigue failure on April 21, 1997. An un-isolable leak in the high-pressure injection (HPI) line from a through-wall crack in the weld connecting the HPI pipe and the safe end of the reactor coolant loop. This event is described in Chap. 4, Sects. 4.7.4 and 4.7.6

component, four independent estimates were provided for each applicable LOCA category: two estimates based on a statistical analysis of operating experience and simple CFP models, and two estimates based on probabilistic fracture mechanics analyses. These base-case results were provided as input to the experts, and some experts chose to use them as anchors for their respective inputs.

– As part of the elicitation, most of the experts provided input to the estimation of LOCA frequencies for respective evaluation boundary including those that were evaluated in the base-case studies.
– The CFP models used in the base-case studies assumed that each pipe failure is a precursor to a LOCA. Each of these models starts with an estimate of the pipe failure rate (precursor failure rate). The model integrates the failure rates, which are estimated using service data, with the more significant pipe failures that produce LOCAs. This is accomplished by defining the conditional probability of a break of a given size given a pipe failure. Another way to look at this model is that pipe failures are assumed to represent challenges to the system and that upon each challenge, there is a probability of experiencing a break of a given size. By considering the full range of different break sizes, all the LOCA frequency categories can be quantified.
– The previously described "data-informed model" of piping reliability provides the capability to estimate failure rates at each location within a piping system pressure boundary. Conversion of the LOCA frequency inputs in NUREG-1829 from a LOCA frequency basis to a conditional probability requires establishing target LOCA frequencies that when combined with the failure rate model will produce the same target LOCA frequencies.

Based on the above information and insights, the information from NUREG-1829 can be used to convert LOCA frequencies versus LOCA category to conditional probabilities versus an equivalent break size. The model used to convert information on unconditional rupture frequencies to conditional failure probabilities makes use of the NUREG-1829 base-case results. Each term in this model is subject to epistemic uncertainty, which is to be estimated. Therefore, this model and the base-case analysis of the failure rates are used to derive epistemic uncertainties for the CFPs in each LOCA category. This produces a set of target LOCA (TLF) frequency distribution parameters that have been selected to incorporate the epistemic uncertainties developed in NUREG-1829. This approach makes use of there being a technical basis for the precursor failure rate estimates from service data and a well-reviewed and extensively applied Bayes' uncertainty analysis method. These estimates were part of the information that was available to each NUREG-1829 expert. Since there have been no LOCA category 1, 2, 3, 4, 5, or 6, the expert elicitation results of all the experts constitute an extrapolation from the existing service data. Therefore, Method 3 simply assumes that the variability in the expert elicitation inputs for LOCA frequency represents the epistemic uncertainty in the LOCA frequency for each evaluation boundary.

The expert elicitation included a request for estimates of LOCA frequencies for specific piping components. One set of numbers provided by the experts was LOCA frequencies by LOCA category in terms of a mid-value (Mid), an upper bound (UB), and a lower bound (LB), with the understanding that those would be interpreted as medians, 95%-tiles, and 5%-tiles of a lognormal uncertainty distribution. For symmetric inputs (i.e., when UB/Mid = Mid/LB), which were provided in most cases, these distributions were assumed to be lognormal distributions. For asymmetric inputs provided by the experts, a specific split lognormal distribution was assumed. The expert distributions were combined into a single composite distribution. NUREG-1829 discussed two approaches for developing expert composite distributions: the mixture distribution method and the geometric mean method:

- **Mixture Distribution Method**

  A single mixture distribution was developed for each combination of component and LOCA category by combining the LOCA frequency distributions provided by each expert. The mixture distribution approach assumes that the expert panel is a random sample from the population of all experts and that the goal is to obtain an unbiased estimate of the aggregate distribution function of the LOCA frequencies averaged over the population of all experts.[14] In cases where experts did not provide inputs for each LOCA category, the mixture distribution was developed only for those experts providing inputs for that category. In all cases, a minimum of seven experts provided input, and the majority of cases had nine.
- **Geometric Mean Method**

  When this method was used in NUREG-1829, it was oriented towards the calculation of the total LOCA frequency rather than the LOCA frequency for

[14] Wheeler, T.A., Hora, S.C., Cramond, W.R., Unwin, S.D. (1989), Analysis of Core Damage Frequency from Internal Events: Expert Judgment Elicitation, NUREG/CR-4550 Volume 2, USNRC, Washington, DC.

multiple locations. Another contrast was the use in NUREG-1829 of split lognormal distributions, whereas this book uses lognormal fitting based on preserving medians and 95%-tiles. A single lognormal distribution for each component and each LOCA category is defined by taking the geometric mean of the medians of the experts' lognormal distributions as the composite distribution median and the geometric mean of the range factors of the experts' lognormal distributions for the LOCA frequencies as the composite distribution range factor. As with the mixture method, in this study, the input lognormal distributions provided by the experts were fit to lognormal distribution by matching the 50th and 95th percentiles. A summary of the composite distribution parameters derived is provided in Table 7.8.

The composite distributions generated by the mixture distribution method produce much broader ranges of uncertainty than those obtained by the geometric mean method. As discussed in NUREG-1829, the upper and lower bounds of the mixture distributions are heavily influenced by the experts' extreme high-side and low-side inputs, respectively, whereas the distribution percentiles from the geometric mean method more fairly represent the experts' inputs.

The CFP distribution parameters were obtained by first developing the LOCA frequencies and then calculating the CFP distribution parameters using formulas for calculating the parameters for the product of two lognormal distributions and precursor rates from Appendix D of NUREG-1829. The CFP uncertainty distributions in Table 7.9 were derived using Eqs. (7.1), (7.2), and (7.3). The medians and range factors of the CFP distributions were calculated from the medians and range factors of the target LOCA frequency distributions. Then, using the properties of the lognormal distribution, the remaining parameters of the distributions can be calculated.

7.16 Expert Elicitation and Probabilistic Piping Reliability

Three examples of the use of expert elicitation in the field of probabilistic piping reliability analysis are the component cooling water system (CCWS) pipe rupture frequency study that supported NUREG-1150, a risk-based method for establishing in-service inspection priorities, and NUREG-1829 on loss-of-coolant accident (LOCA) frequency estimation. The objective of NUREG-1150 (*Severe Accident Risks: An Assessment for Five U.S. Nuclear Power Plants*) was to provide an update of WASH-1400 estimates of core damage frequency. The purpose of NUREG-1829 was to establish a risk-informed revision of the emergency core cooling regulatory requirements, which considers estimates of LOCA frequencies as a function of pipe break size.

Table 7.8 Composite distributions for NUREG-1829 results based on geometric mean method

Evaluation boundary	Equivalent break size (mm)	LOCA frequency (1/Year)			
		Mean	5%-tile	Median	95%-tile
Reactor coolant system hot leg	≥0.5	4.08E-07	9.32E-09	1.21E-07	1.57E-06
	≥ 1.5	1.28E-07	2.25E-09	3.34E-08	4.95E-07
	≥ 3	6.51E-08	1.01E-09	1.59E-08	2.52E-07
	≥6.75	2.59E-08	2.49E-10	4.96E-09	9.88E-08
	≥14	1.50E-08	6.70E-11	1.90E-09	5.37E-08
	≥31.5	3.16E-09	4.84E-12	2.18E-10	9.78E-09
Reactor coolant system cold leg	≥0.5	1.47E-07	3.27E-09	4.30E-08	5.66E-07
	≥1.5	5.20E-08	9.07E-10	1.35E-08	2.01E-07
	≥3	2.19E-08	3.33E-10	5.31E-09	8.48E-08
	≥6.75	7.85E-09	7.41E-11	1.49E-09	2.99E-08
	≥14	4.54E-09	1.94E-11	5.60E-10	1.62E-08
	≥31.5	1.10E-09	1.56E-12	7.23E-11	3.36E-09
Pressurizer surge line	≥0.5	3.60E-07	1.33E-08	1.34E-07	1.35E-06
	≥1.5	1.26E-07	3.46E-09	4.09E-08	4.83E-07
	≥3	6.45E-08	1.29E-09	1.79E-08	2.49E-07
	≥6.75	1.92E-08	2.47E-10	4.28E-09	7.41E-08
	≥14	2.72E-09	4.22E-11	6.66E-10	1.05E-08

Table 7.9 CFP distribution priors derived from target LOCA frequencies

		Conditional failure probability distribution parameters			
Evaluation boundary	EBS (mm)	Mean	5%-tile	50%-tile	95%-tile
RCS hot leg	≥12.5	1.46E-03	1.84E-04	9.10E-04	4.50E-03
	≥37.5	3.31E-04	1.35E-05	1.29E-04	1.23E-03
	≥75	1.65E-04	5.01E-06	5.61E-05	6.28E-04
	≥171.5	5.74E-05	1.49E-06	1.81E-05	2.20E-04
	≥350	2.49E-05	4.54E-07	6.62E-06	9.65E-05
	≥800	5.84E-06	1.06E-07	1.55E-06	2.26E-05
RCS cold leg	≥12.5	1.20E-03	1.50E-04	7.48E-04	3.72E-03
	≥37.5	2.74E-04	1.31E-05	1.15E-04	1.00E-03
	≥75	1.13E-04	4.92E-06	4.54E-05	4.18E-04
	≥171.5	3.58E-05	1.49E-06	1.41E-05	1.33E-04
	≥350	1.59E-05	4.25E-07	5.09E-06	6.10E-05
	≥800	4.48E-06	9.17E-08	1.26E-06	1.73E-05

7.16.1 Component Cooling Water System Pipe Rupture Frequency

The Zion Nuclear Power Plant[15] was one of the reference plants in NUREG-1150 (Ortiz et al., 1991). The design of the component cooling water system (CCWS) at Zion was such that a potential existed for a pipe rupture in certain parts of the CCWS that would fail the entire system. Such an event would constitute an initiating event that could lead to core damage. Emergency core cooling capability would be lost, and the reactor coolant pump seals would be without cooling as well. This could result in a potential LOCA through the pump seals with no capability to replace the lost inventory. The issue of concern was the frequency with which, under normal operating conditions, the CCWS serving the Zion units would fail due to a pipe rupture event. It was selected as an expert elicitation issue due to the potentially significant impact of the CCWS pipe rupture rate upon the overall core damage frequency at Zion Unit 1. This issue had been raised in an earlier review of the Zion Probabilistic Safety Study (ZPSS).[16] At the time of the review, no data on low-temperature/low-pressure pipe rupture had been deemed to not be available, so the CCWS pipe rupture sequence in ZPSS was quantified using pipe rupture frequencies associated with primary system piping.

[15]The two Zion nuclear four-loop Westinghouse PWRs were permanently shut down in February 1998.

[16]The Zion Probabilistic Safety Study (ZPSS) was undertaken in early 1980 and completed in August the following year. The study responded to a request by USNRC staff in 1979 to evaluate the following: (1) measures to reduce the likelihood of core melt events, (2) measures to mitigate the effects of core melt events irrespective of the likelihood of such events, and (3) interim measures to improve safety while the above two items were under consideration.

Table 7.10 Results of CCWS pipe integrity expert elicitation

Quantile	Aggregated pipe rupture frequency (Year^{-1})
5th	$7. \times 10^{-8}$
50th	3.0×10^{-6}
Mean	1.1×10^{-4}
95th	3.5×10^{-4}

The owner/operator of Zion agreed about the potential severity of the CCWS pipe rupture scenario but felt that the design and construction of the piping were of a "superior nature" and that the application of high-energy pipe rupture frequencies to moderate-energy piping was inaccurate. Therefore, the NUREG-1150 study team formed an expert panel to assess the Zion CCWS pipe rupture frequency. There was no suggested decomposition or specific structure given to the panel. Each expert developed a model to describe the uncertainty in the estimates of the CCWS pipe rupture frequency. The three expert panel members developed curves to express the uncertainty of the estimate for pipe rupture frequency. Next, the three curves were aggregated by taking the arithmetic average. The results are shown in Table 7.10.

An important insight from this expert elicitation is that employing a straightforward arithmetic mean for combining results tends to disproportionately emphasize higher outliers. This well-recognized issue can be addressed by reconvening the expert panel members to discuss assumptions and reach an industry consensus on moderate-energy piping integrity matters.

7.16.2 Expert Elicitation and Risk-Informed In-Service Inspection

An assessment of the risks from severe accidents in five commercial nuclear power plants in the United States was performed in the mid- to late-1980s. It was a reassessment of the accident risks that had been studied in the Reactor Safety Study. This reassessment was published as NUREG-1150. To support the analysis of certain key issues in the accident progression analysis, expert panels were convened. Fourteen accident progression, containment loadings, and structural response issues were considered by four panels. These panels considered a wide range of information available from experiments and computer calculations. Using expert elicitation methods, probability distributions were developed based on the experts' interpretations of these issues (Wheeler et al. 1989).

At the time of the NUREG-1150 expert elicitation, the Pacific Northwest National Laboratory (PNNL) had conducted a multi-year program for the USNRC regarding nondestructive evaluation reliability for in-service inspection (ISI) of light water reactors (LWRs). The goals of this program were to determine the reliability of the then-current ISI of pressure boundary systems and components and to develop recommendations that can ensure a suitably high inspection reliability. The long-term objective was to develop recommendations for improved inspection plans that

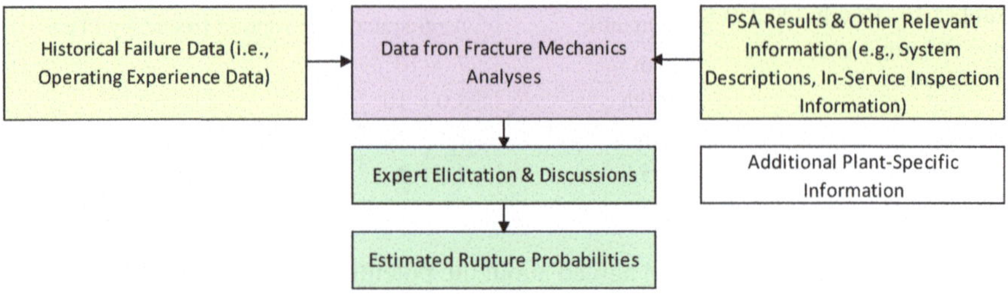

Fig. 7.18 Information provided to an expert panel

address inspection samples, frequencies, and methods. To meet the program objectives, PNNL developed a risk-based technique as a guide for establishing inspection priorities. The results of PSA were known to be useful for estimating the safety consequences resulting from the failure of components of a system. However, the methodology also required estimates of the probabilities of failure for pressure boundary components, and PNNL encountered a significant lack of failure probability data at the required level of detail. Probabilistic fracture mechanics models had addressed a few components of interest, but it was also recognized that those calculations had a high degree of uncertainty. To overcome this data deficiency, PNNL conducted an expert judgment elicitation to address the issue of failure probabilities. The goal was to obtain numerical estimates for probabilities of catastrophic or disruptive failures in pressure boundary systems and components in a PWR plant (Voo et al., 1989) (Vo et al., 1991). Component "rupture" was defined as a break or leak that is greater than make-up capability and/or that can disable the systems' intended function. In his study, small leaks that may cause system degradation were not included. Figure 7.18 shows the information that was used to obtain the desired estimates from the experts.

Prior to the expert judgment elicitation workshop, PNNL sent reference materials to the experts, including data sources, reports, probabilistic models, and results of probabilistic safety assessment studies. Panel members were asked to study these materials and to make initial estimates of failure probabilities. At the meeting, a formal presentation was provided for each system addressed. Presentations covered technical descriptions, historical component failure mechanisms, elicitation statements, suggested approaches, questionnaire forms, and any materials that supported the issue descriptions. The presentations were followed by discussions. The experts provided their knowledge regarding plant design and operation, failure history, material degradation mechanisms, and methods for recomposition and aggregation of the data.

Each expert then completed questionnaire forms that addressed location-specific rupture probabilities for the systems of interest. Following the elicitation meeting, the information provided by the experts was recomposed and aggregated. A written analysis of each issue including the initial recomposition, additional plant-specific data, and other relevant information was then returned to each expert for review. The purposes of this review were to provide experts the opportunity to revise their earlier

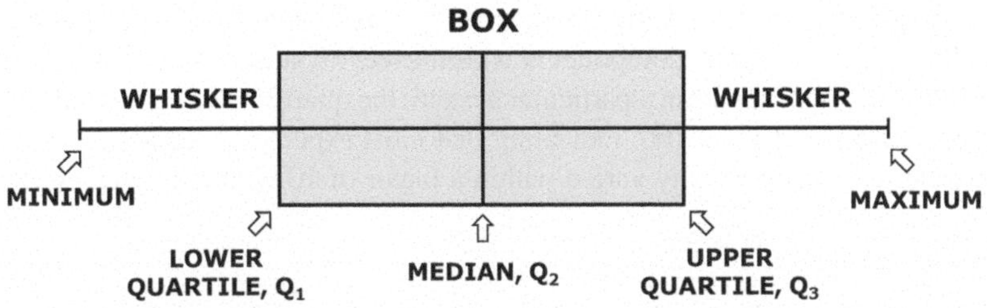

Fig. 7.19 Box-and-whisker plot. (Reproduced from https://www150.statcan.gc.ca/n1/edu/power-pouvoir/ch12/5214889-eng.htm)

assessments and to ensure that potential misunderstandings were identified and resolved and that the documentation correctly reflected the judgment of the experts involved. The experts were asked to use this information to revise their original estimates as needed. Most importantly, the experts were asked to state the rationale for their assessments. The revised individual judgments were recomposed and aggregated again to provide a single final composite judgment for each issue.

Following the review and revision of the results provided by the experts, the results were compiled into distributions. The distributions were then used to determine a best estimate for every component present in the dataset. The geometric mean was used because of the small sample size and because the dataset was not unimodal or symmetric. This estimator is influenced by the central values of the dataset and is not so influenced by the extreme values. And the population quartile was chosen for uncertainty estimates. Next, the best estimates obtained from the experts were summarized in a series of box-and-whisker plots (or boxplots) (Fig. 7.19). These plots represent the distribution of estimated failure (rupture) probabilities provided by various experts. An individual plot graphically displays the following features of the distribution:

- The whiskers identify the extreme upper and lower values of the distribution.
- The box itself locates the 25% and 75% quantiles (i.e., quartiles) of the distribution. In other words, 50% of the data points will lie within the box.
- The line within the box intersecting with the circle or dot is the median of the distribution.

The 25% and 75% quartiles were chosen for uncertainty estimates to eliminate the two most extreme values of the distribution. In descriptive statistics, a boxplot is a method for graphically demonstrating the locality, spread, and skewness of groups of numerical data through their quartiles. The experts provided a broad range of responses regarding failure probabilities, and this trend is entirely consistent with the large uncertainties associated with the performance of the components being addressed. In interpreting the range of responses, there was no attempt to seek consensus from the experts. Rather, the responses were evaluated in a statistical sense, with a central tendency being established as a single best estimate probability for use in risk-based studies. For all systems selected for the study, plots showed that failure frequency estimates varied between 1.0×10^{-9} and 1.0×10^{-3} failure/yr., depending on the systems, components within systems, and component locations. For a given component within a particular system, the quartile range generally varied between a factor of 10 and 100, indicating that most experts' estimates agreed. The component medians generally varied within a factor of 10 within a given system.

7.16.3 Expert Elicitation and LOCA Frequency

The LOCA frequency expert elicitation reported in NUREG-1829 (USNRC 2008) was performed between 2003 and 2004 to develop boiling water reactor (BWR) and pressurized water reactor (PWR) LOCA frequencies as a function of pipe break size for three distinct time periods: (1) present-day (25-year fleet average at the time of elicitation), (2) end-of-plant license (40-year fleet average), and (3) end-of-plant first license renewal (60-year fleet average). These estimates were based on the responses from an expert panel, with the aim of obtaining estimates that represent a type of group consensus. Additionally, another objective was to reflect both the uncertainty in each panelist's estimates and the diversity among the individual estimates. The method used to aggregate the results of the expert elicitation is illustrated in Fig. 7.20 and summarized as follows:

- The expert panel members provided a mid-value (MV), an upper bound (UB), and a lower bound (LB) for each elicitation question. These numbers were assumed to correspond to the median, 95th percentile, and 5th percentile, respectively, of the panelist's subjective uncertainty distribution. The goal was to estimate LOCA frequencies for four plant-type combinations (BWR piping, BWR non-piping, PWR piping, and PWR non-piping) and combine these to estimate total BWR and PWR passive component LOCA frequencies.
- The responses for each panelist were propagated to obtain individual total LOCA frequency estimates and then combine them to obtain group estimates of the bottom-line parameters.
- The individual responses for the anchoring frequencies and adjustment ratios were assumed to be independent split lognormal distributions and were multiplied to determine the piping system frequencies. These system frequencies were then summed for all contributing piping systems to obtain the un-normalized piping contribution. The un-normalized piping contribution was then multiplied by the piping contribution factor (Block 1.6) to determine the piping contribution (Block 1.7).
- The non-piping calculations were analogous except for an additional step required to determine the non-piping component from the subcomponent frequencies. The individual responses for the anchoring frequencies (Block 1.9) and adjustment ratios (Block 1.10) were assumed to be independent split lognormal distributions and multiplied to determine the non-piping subcomponent frequencies (Block 1.11). These subcomponent frequencies were then summed for all

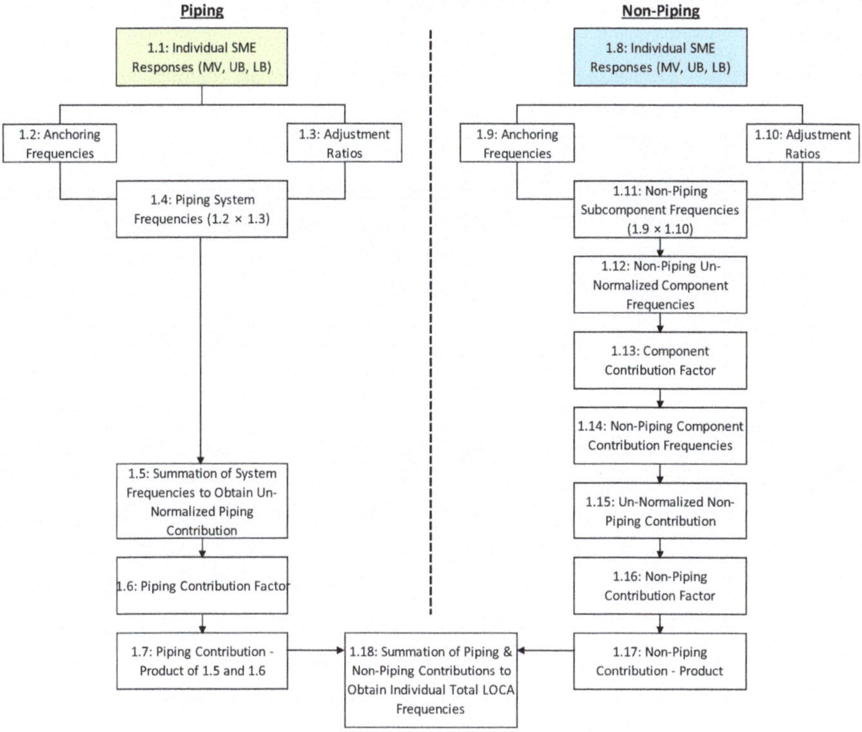

Fig. 7.20 NUREG-1829 method for aggregating expert elicitation results

contributing non-piping subcomponents to obtain the un-normalized non-piping component frequencies (Block 1.12). These frequencies were then multiplied by the component contribution factor (Block 1.13) to determine the non-piping component frequencies (Block 1.14). The non-piping component frequencies were summed to calculate the un-normalized non-piping contribution (Block 1.15), which was then multiplied by the non-piping contribution factor (Block 1.16). This resulted in the non-piping contribution (Block 1.17). The piping and non-piping contributions were then summed to develop individual estimates of the bottom-line parameters (mean, median, 95th, and 5th percentiles) of the total LOCA frequencies (Block 1.18). Finally, the individual estimates of the bottom-line parameters were combined to develop the group estimates and associated confidence intervals (Block 1.19 and Table 7.11). Plots of the individual expert inputs for LOCA frequencies are shown in Figs. 7.21, 7.22, 7.23, 7.24, 7.25, 7.26 and 7.27.

When the results were aggregated using either the geometric (GM) or the arithmetic mean (AM) method, the group as a whole supported the notion that larger breaks are much less likely than smaller breaks in the same pipe. Whether using the GM or AM method, only changes were noted in the shape of the decreasing

Table 7.11 NUREG-1829 total BWR and PWR LOCA frequencies

LOCA category	EBS (mm)	("Non-pipe" to "pipe") ratio	Total Mean	Non-pipe Mean	Pipe-only contribution Mean	5%-tile	50%-tile	95%-tile
BWR @ 25 years								
1	12.7	0.24	6.50E-04	1.26E-04	5.24E-04	1.77E-06	2.05E-04	1.73E-03
2	47.6	0.4	1.30E-04	3.71E-05	9.29E-05	3.09E-07	3.58E-05	3.02E-04
3	82.3	0.11	2.90E-05	2.87E-06	2.61E-05	8.51E-08	9.85E-06	8.32E-05
4	177.8	0.06	7.30E-06	4.13E-07	6.89E-06	2.80E-08	2.66E-06	2.25E-05
5	457.2	0.07	1.50E-06	9.81E-08	1.40E-06	4.05E-09	4.68E-07	3.95E-06
PWR @ 25 years								
1	12.7	32	7.30E-03	7.08E-03	2.21E-04	8.70E-07	1.01E-04	8.50E-04
2	47.65	6.7	6.00E-04	5.22E-04	7.79E-05	3.06E-07	3.54E-05	2.99E-04
3	82.3	0.36	1.60E-05	4.24E-06	1.18E-05	4.63E-08	5.35E-06	4.52E-05
4	177.8	1.1	1.60E-06	8.38E-07	7.62E-07	3.00E-09	3.47E-07	2.93E-06
5	457.2	0.5	2.00E-07	6.67E-08	1.33E-07	5.24E-11	6.07E-09	5.12E-08
6	787.4	1.4	2.90E-08	1.69E-08	1.21E-08	4.75E-11	5.49E-09	4.64E-08

frequency with increasing break size curve. In each case, the consensus of the experts was that the frequency of a complete break of a pipe is two to three orders of magnitude less than that of a 12 mm opening in the same pipe.

7.17 Structural Integrity Adjustment Factors

Generic component failure rates are often adjusted to account for specific reliability attributes and factors not originally considered in an initial assessment. This process, known as data specialization, goes beyond typical Bayesian analysis by tailoring generalized cases to specific applications. Examples of relevant adjustment factors include the assessment of the factor of improvement (FOI) achieved through material degradation mitigation processes and techniques, assessment of the effect of leak detection and in-service inspection on piping reliability, and assessment of potential aging phenomena on piping reliability.

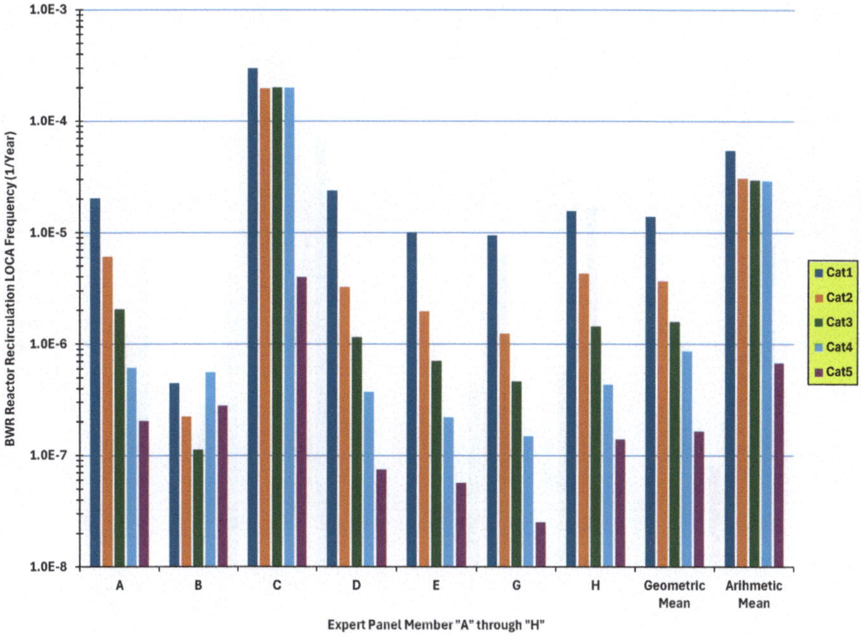

Fig. 7.21 Expert inputs to GE BWR reactor recirculation system LOCA frequency

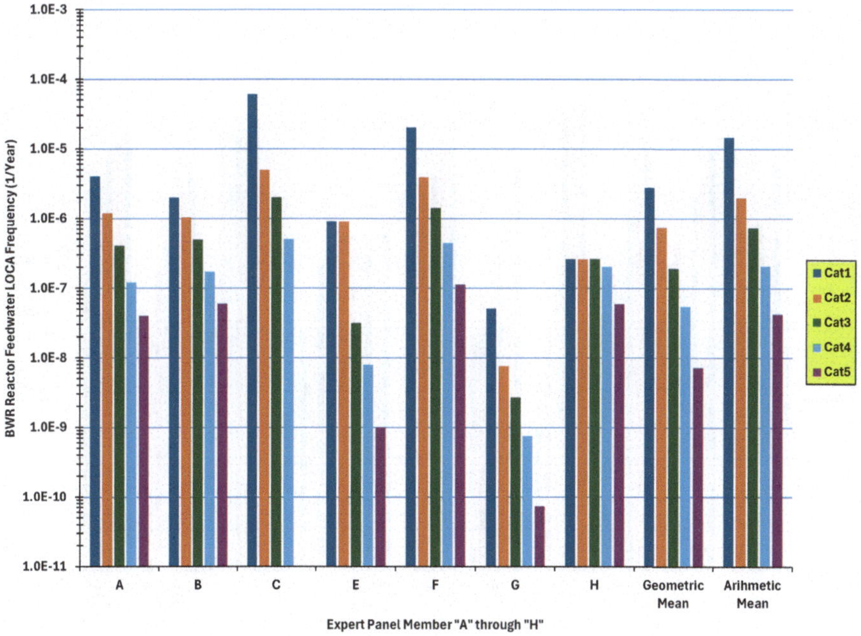

Fig. 7.22 Expert input to General Electric BWR feedwater system LOCA

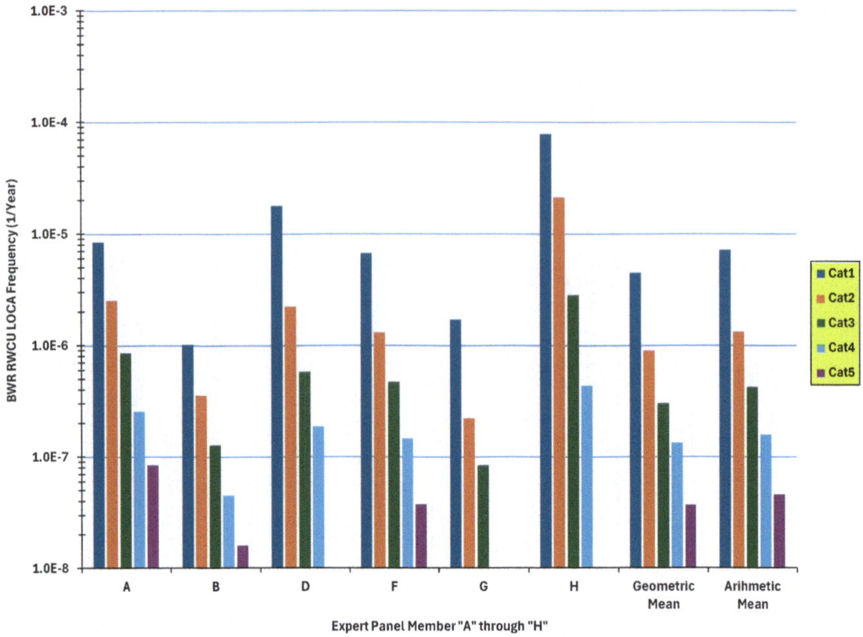

Fig. 7.23 Expert input for General Electric BWR reactor water cleanup system LOCA

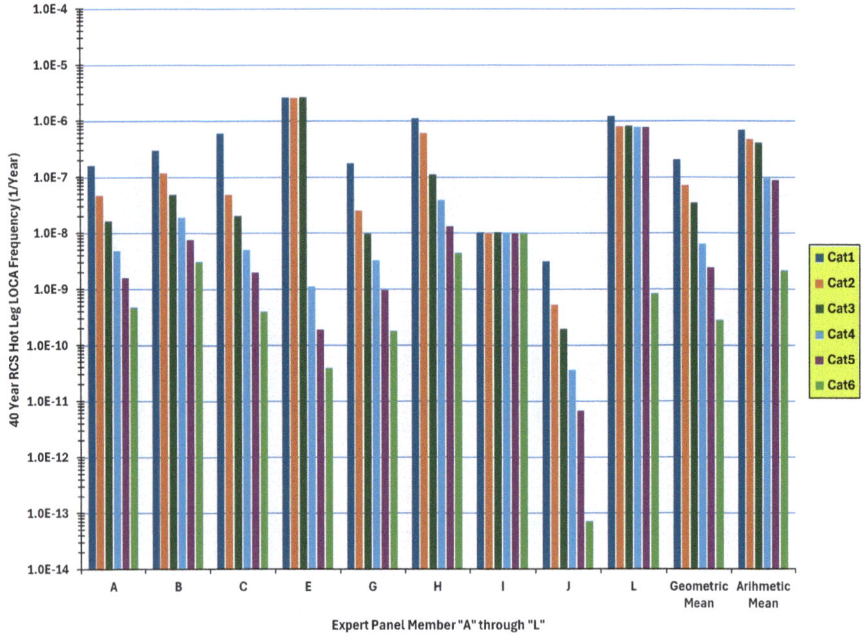

Fig. 7.24 Expert inputs for RCS hot leg LOCA frequencies

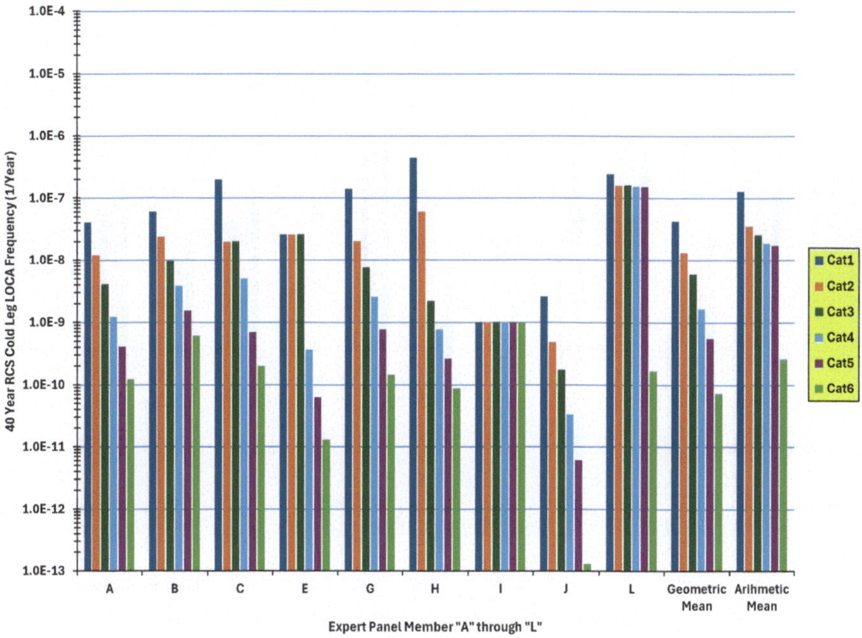

Fig. 7.25 Expert inputs for RCS cold leg LOCA frequencies

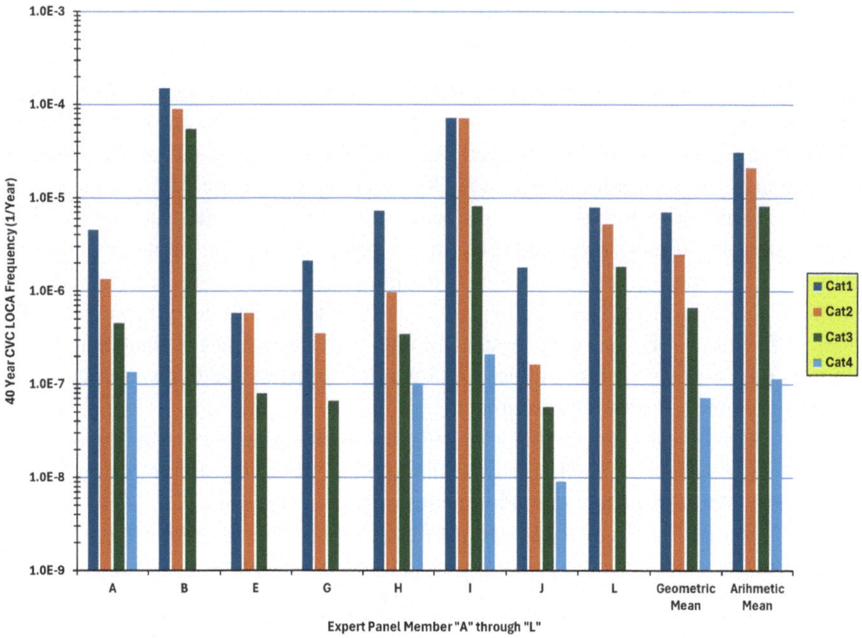

Fig. 7.26 Expert panel inputs for PWR chemical and volume control system LOCA frequency

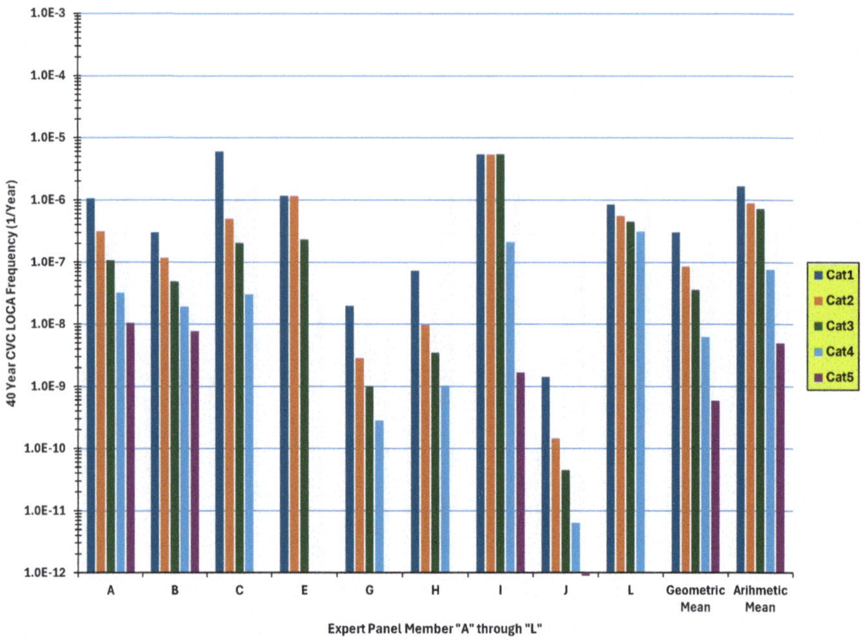

Fig. 7.27 Expert panel inputs for PWR residual heat removal system LOCA frequency

7.17.1 *Aging Factor Assessment*

Age-dependent pipe failure rate estimation can be conducted using various analysis strategies to determine piping reliability parameters based on the age of the affected pipe section at the time of its observed failure, the age of the plant, or the temporal changes in piping operation experience. Aging encompasses the cumulative impact of time-dependent factors that may influence the pipe failure rate, such as physical degradation due to corrosion and erosion mechanisms, and the implementation of inspection and maintenance programs. Trends in pipe failure rates are impacted by physical alterations to structural integrity caused by aging and design modifications (e.g., substituting original carbon steel piping with corrosion-resistant materials). Various data pooling methodologies, data window truncation studies, and curve-fit models have been thoroughly examined.

One way to investigate whether a failure rate λ is changing over time is to perform separate calculations for each year in a study period, obtaining a point estimate and confidence interval for each year. Plot them side by side and see if they seem to show a trend. A more formal way to investigate for the presence of a trend is to construct a model for λ that includes the possibility of a trend, for example a log-linear model:

$$\lambda = \exp(a + bt) \tag{7.80}$$

$$\log(\lambda) = a + bt \tag{7.81}$$

where t denotes calendar time, e.g., calendar year. If b is negative, the trend is decreasing, and a plot of λ versus time is an exponentially decreasing curve. For small datasets, the analysis is more complicated than simply fitting a straight line to the values of $log(n_i/t_i)$, where n_i and t_i are the event count and exposure time in year i, respectively. One fundamental reason is that small datasets include counts of zero, and log(0) is undefined. An alternative approach builds on technical insights from many decades of studying ASME III Safety Class 3 service water piping reliability (Lydell and Fleming 2006, 2010, 2013, 2018; Lydell et al. 2018; Fleming et al. 2023). The service water pipe failure population was fairly evenly spread across five decades, 1971–2020, so the failure population was placed in ten 5-year bins. For each analysis case and time interval:

- Determine the average age of the plants that produced the pipe failures.
- Using the CNID method with a prior failure rate of $2.0\text{x}10^{-5}$, calculate the temporal failure rate.
- Using Microsoft® Excel, plot a line of best fit through the scatterplots of failure rate vs. plant age resulting in a log-linear power model with scale and shape factors. The age factor A_{it} is determined from

$$AF_{it} = \frac{\alpha_{it} t^{\beta_i}}{\lambda_i} \tag{7.82}$$

where

$AF_{it} =$ Aging factor at pipe age t for component i defined by system, design, or code class and pipe size
$\alpha_{it} =$ Scale parameter for component i
$\beta_i =$ Shape parameter for component i
$\lambda_i =$ Average pipe failure rate for component i

Different data pooling strategies were used to assess their impact on R-squared, which indicates how well a regression line fits the data points. The different data pooling strategies are shown in Table 7.12. The aging factor (AF) analysis was performed for different pipe diameters and types of ultimate heat sink (UHS). As a sensitivity case, the calculations were repeated after placing the data in five 10-year bins ("Method 2") (Fig. 7.28).

Table 7.12 Data pooling strategies

Data pooling strategy	Variables			No. of analysis cases	Impact on R^2	
	Plant type	Pipe size	UHS type			
Base case	☒	☒	☒	24	0.2–0.9	
1	☒	☐	"Blank" = pooled data	12	BWR	0.78–0.95
2					PWR	0.45–0.93
					Case 2: BWR and PWR	0.71–0.93
3	☒		☒	6	BWR	0.79–0.93
					PWR	0.48–0.85
4	☒	☒		9	BWR	0.19–0.93
					PWR	0.70–0.93
					BWR and PWR: Small diameter	0.87
Color code	Method 1: 10-bin analysis					
	Method 2: 5-bin analysis					

7.17.2 Factor-of-Improvement Assessment

The factor of improvement (FOI) is a measure of the effectiveness of a certain material degradation mitigation process. It is a quantitative measure that can be obtained from probabilistic fracture mechanics studies, experimental data, or field experience data. An early example of the former type of FOI analysis was published by L.M. Arnett (1976). This study was based on a probabilistic model that related the probability of pressure vessel failure to the effectiveness of pre-service inspection and in-service inspection. A spectrum of failure probabilities was calculated, based on weld flaw characteristics, linear elastic fracture mechanics parameters such as stress intensity factor, fracture toughness, critical flaw size, and maximum allowable crack size. The FOI was defined as

$$FOI = \exp(\alpha \times NI) \tag{7.83}$$

where α is a constant that is a function of the linear elastic fracture mechanics parameters and the effectiveness of nondestructive examination. NI is the number of inspection, including the pre-service inspection and potential weld repairs prior to

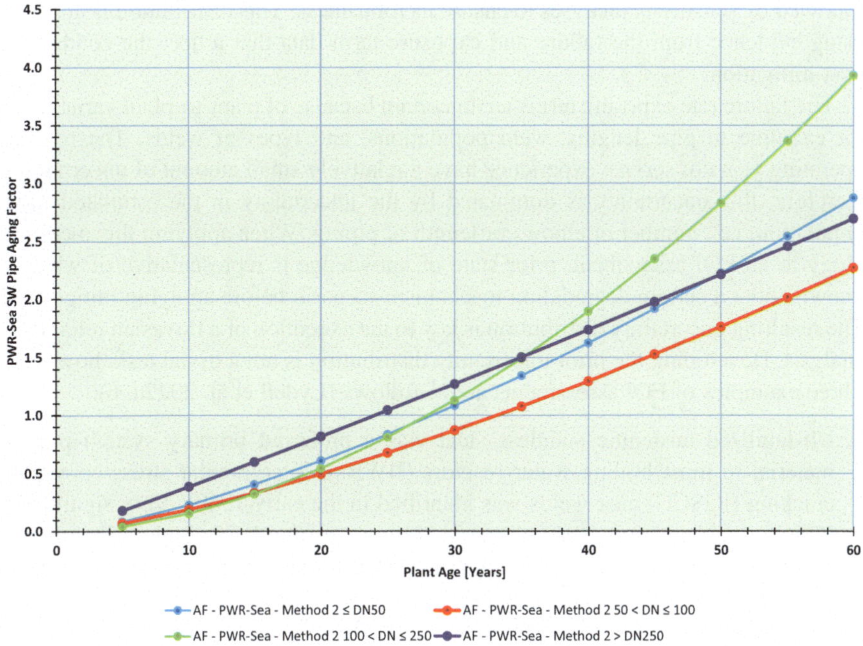

Fig. 7.28 Aging factor assessment results

placing a pressure vessel in service.[17] Where experimental data is used, the time to crack initiation is used, and in the latter case, a probabilistic assessment is performed by calculating the before-and-after-mitigation piping pressure boundary failure rate, respectively.

A general representation of the FOI is given by

$$FOI = \frac{\lambda_{Before}}{\lambda_{After}} \tag{7.84}$$

The pipe failure rate λ is calculated using a Bayesian reliability analysis technique. In the development of a Bayesian uncertainty distribution for the λ_{Before} parameter, a prior distribution is developed representing the state of knowledge before mitigation accounting for the applicable operating experience data. The technical basis for the prior distribution is obtained through statistical analysis

[17] Arnett summarized his work as follows: "When very conservative assumptions are made for all factors that affect the exponential relation, factors of improvement of approximately 8000 are expected for three in-service inspections, optimally spaced during the service life of the vessel. For more reasonably assumed values, the factor of improvement is much larger, even in excess of 10^{13}. These large factors assume 100% inspection of the vessel and are attained only when the schedule of the in-service inspections is optimized. The optimum schedules show large deviations from those in which the inspections are equally spaced throughout the service life of the vessel."

followed by sensitivity analyses to ensure its robustness. This distribution is updated using evidence from the failure and exposure term data that reflect the conditions post-mitigation.

The failure rate exposure terms are uncertain because of plant-to-plant variability, for example in pipe lengths, weld populations, and types of welds. The reactor operating years of service experience have a relatively small amount of uncertainty; therefore, this uncertainty is dominated by the uncertainty in the estimated pipe population, i.e., number of elbows or length of piping. When applying this methodology to an FOI assessment, prior state of knowledge is representative of what is known about a certain degradation mechanism scenario before applying mitigation. The resulting uncertainty distribution is key to the execution of a Bayesian reliability analysis. Developing the prior uncertainty distribution is not a trivial task, however. Three examples of FOI assessments are as follows (Lydell et al. 2022a, b):

1. Unstabilized austenitic stainless steel was a preferred primary system piping material in most boiling water reactors (BWRs). Intergranular stress corrosion cracking (IGSCC) near welds was identified in the early 1960s. Very significant IGSCC incident rates were observed throughout the 1970s and well into the 1980s. The approaches to resolving this IGSCC problem were to implement strict primary water chemistry control, to apply stress improvement measures using the mechanical stress improvement process (MSIP) or the structural weld overlay (SWOL) process, and/or to replace the original material with more resistant material. The combined effect of these measures has proven quite effective. One approach would be to analyze the operating experience before 1988 and to use it as a prior state of knowledge in a Bayesian reliability analysis. An underlying assumption is that post-1987, most if not all plants implemented steps to mitigate IGSCC.
2. Alloy 600 is used extensively in PWR plants. Already in the 1960s, the materials science community determined that this alloy was susceptible to SCC in certain high-temperature environments. The first piping primary-water SCC events in commercial PWR plants were reported in the mid-1980s. Alloy 690 and its weld metals, Alloy 52 and Alloy 152, were determined to be preferential replacement materials. Like IGSCC, the PWSCC operating experience and the mitigation implementations are well documented. An evaluation of this operating experience determines the boundary conditions for the prior state of knowledge, i.e., the justifications for selecting the before/after mitigation implementation dates. Two sensitivity analyses were performed: a) use of Alloy 690 from the outset and under an assumption of "perfect" SCC resistance and b) developing a prior state of knowledge using the existing operating experience through 2008 and updating the prior uncertainty distribution with the post-2008 operating experience with dissimilar metal welds subjected to mitigation.
3. Carbon steels, commonly used in raw water-cooling systems like the safety-related service water system (SWS), are prone to corrosion and flow-assisted degradation (FAD). Despite attempts to replace carbon steel with austenitic stainless steel (e.g., Type 304 or Type 316) to combat microbiologically

influenced corrosion (MIC), issues persisted. High-alloy austenitic stainless steels (e.g., UNS N08367 and S31254), introduced in the 1970s and used in nuclear power plants from the early 1980s, were more resistant to MIC due to their high chrome, nickel, molybdenum, and nitrogen content. Developing prior state-of-knowledge uncertainty distributions is straightforward, but posterior distributions are more complex. Expert elicitation can help formulate hypotheses on resistance to environmental degradation, which can then be tested against screened operating experience data.[18]

7.17.3 Piping Design Code and Pressure Boundary Integrity

Examples of piping design codes include the ASME B31.1 *Power Piping Code* and ASME Section III *Rules for Construction of Nuclear Facility Components.* First published in 1942, ASME B31 has its roots in the broader ASME B31 Code for Pressure Piping, which began in 1926 with the aim of standardizing piping design. Initially, the B31 Code was a single document, but it later evolved into separate sections addressing specific industries, with B31.1 focusing on power plant piping and related systems. Over the decades, ASME B31.1 has been updated and revised to reflect advancements in engineering, materials, and regulatory requirements, impacting power plant and industrial piping design practices. ASME III, or the Nuclear Code, was developed to establish standards for the design, fabrication, and inspection of nuclear power plant components. Its history is rooted in the need for safety regulations as nuclear energy technology advanced, building upon existing ASME codes for pressure vessels. The first edition was published in 1963 for reactor pressure vessels and subsequently expanded in 1971 to cover the entire primary pressure boundary inside the containment. Eventually, the ASME III 1974 edition was expanded to also cover Safety Class 2 and 3 piping in subsections NC and ND, respectively.

Section 1 of the 1955 edition of B31.1 encompassed the minimum requirements for design, manufacture, test, and installation of power piping systems, as defined for steam-generating plants, central heating plants, and industrial plants. Section 6 of the code provided minimum standards for fabrication but also included approved methods for providing thermal expansion and flexibility in piping systems. Specific requirements that governed the piping design were as follows:

1. Paragraph 122—Thickness Pipe: The equation, identical to that used in current codes, was described.
2. Paragraph 607—Allowable Stresses: This paragraph permitted the allowable stress to be increased by 20% for short-time overloading conditions.

[18]Lydell, B., Fleming, K. (2014), Piping System Failure Rates for Corrosion Resistant Service Water Piping, 3002002787, EPRI, Palo Alto, CA.

3. Paragraph 622—Stresses and Reactions: This paragraph introduced methods for computing the allowable expansion stress and stress range reduction factors. The paragraph also stated that the effects of pressure, weight, and other sustained loadings shall not exceed the basic allowable stress and that if these stresses are less than the allowable, then the difference can be added to the allowable thermal expansion stress range. Guidance was given for calculating the longitudinal pressure stress based on the inside area, but there were no specific formulas provided for computing the axial stresses due to sustained and occasional moments.
4. Paragraph 623—Supports: This paragraph required that "*design and spacing of supports shall be checked to assure that the sum of the longitudinal stresses due to weight, pressure, and other sustained external loading does not exceed the basic allowable stress*." No formula was provided for computing stress.

Unlike ASME B31.1, which uses the maximum stress theory, the design requirements for ASME Section III Class 1 piping are based on the maximum shear stress theory. Maximum normal stress theory predicts failure in brittle materials when the maximum principal stress reaches the ultimate strength, whereas maximum shear stress theory predicts failure in ductile materials when the maximum shear stress reaches the shear yield strength. Per ASME III rules, piping designs must satisfy a series of equations for different loadings to be considered acceptable. The allowable stresses for B31.1 and ASME Classes 2 and 3 generally follow those from ASME Section I, *Power Boilers*. Before 1973, B31.1 did not rigorously address fatigue.

A number of code cases were issued in the 1960s that provided additional guidelines for the design of nuclear power plant piping. The first edition of ASME B31.7, titled "*Nuclear Power Piping*," was approved and published in August 1969. It was a section within the broader ASME B31 Code for Pressure Piping, specifically dedicated to the design, materials, fabrication, and installation of piping systems in nuclear power plants. It was an interim document that was eventually incorporated into Section III of the ASME Boiler and Pressure Vessel Code. In an addenda to B31.1 published in June 1973, changes were made which modified the power piping code to be much more like it appears in the early versions of ASME Section III Class 2 and 3 and in the current version of ANSI B31.1.

Would piping designed to B31.1, as opposed to ASME III, exhibit less structural integrity over time? As we learn in Chap. 5, a great many variables determine piping pressure boundary integrity performance over time. Pipe dimensions, material selections, welding process specifications, methods and techniques for reliability and integrity management, and associated variables are manifold. Factors to be considered are found in Tables 7.13 and 7.14. Some insights about design code of reference versus the achieved level of pressure boundary integrity may be inferred from operating experience data. Such an analysis involves a significant amount of data screening and data reduction. To evaluate the impact of weld quality on piping pressure boundary integrity, an analysis starts by correlating operating experience data with factors affecting failure likelihood. An example is used to demonstrate how

Table 7.13 Pressure boundary integrity management and defense in depth

Rules for design and fabrication of pressure boundary components		Defense in depth			
Ensuring "no pressure boundary failure"					
ASME B31.1	ASME III	ASME XI Base		ASME XI appendix R	Owner-defined ISI
Prevent design and construction errors		Pre-service inspection (PSI)	In-service inspection (ISI)	Risk-informed ISI	
Limited-scope fatigue damage evaluation	Expanded fatigue damage evaluations, consideration of cumulative usage factors (CUFs)	Identify defects, if any, before "N" symbol stamping	Identify flaws and flaw growth	Optimized selection of inspection locations	Plant-specific ISI program plan, beyond ASME XI requirements, responding to plant-specific operating experience and license renewal or subsequent license renewal obligations
No consideration of pipe anchor movements prior to 1973. Post-1973 editions include specific requirements for flexibility and support	Requires that effects of anchor movements and thermal expansion/contraction be considered for piping integrity	Perform additional PWHT and hydrostatic pressure test if needed. Flaws have been found in field PSI required by ASME Section XI of completed welds already accepted in accordance with Section III. ASME B31.1 Chapter VI provides testing requirements, e.g., not less than 5% of circumferential butt and miter groove welds shall be examined fully by random radiography	Ensure pressure boundary integrity through systematic, periodic inspections. Prescriptive selection of locations to be subjected to NDE	Moves from generic rules to plant-specific risk insights, targeting high-safety significant areas and reducing unnecessary checks on less critical systems	
Qualification of welding procedures and of the performance of welders and welding operators shall conform to the requirements of the BPVC Section IX, "Welding, Brazing, and Fusing Qualifications"	Qualified welding procedures, certified welders, preheat/interpass temperature control, and rigorous NDE and pressure testing.				

Table 7.14 Factors influencing piping pressure boundary integrity

Design codes (e.g., ASME B31.1 and ASME III)	Pre-service inspection (PSI)/in-service inspection (ISI)	Operational considerations and industry initiatives	Regulatory requirements
Dimensions	Basic PSI/ISI	Water chemistry control guidelines	Leak detection
• Diameter • Wall thickness/mechanical strength	Enhanced ISI—Risk-informed ISI	Pipe stress improvement, e.g., peening, IHSI, weld overlay	ISI requirements including inspection programs with focus on specific material degradation mechanisms
	Reliability and integrity management		
Weld joint quality	*Pressure testing requirements*	*NDE qualification (methods and personnel)—Probability of detection (POD)*	*Aging management*
• Basic quality factors • Increased quality factors	Hydrostatic pressure testing requirements		*Regulatory guides* (a few examples): • RG 1.246: Acceptability of ASME code, Section XI, Division 2, requirements for RIM programs for NPPs, for non-LWRs • RG 1.201: Guidelines for categorizing structures, systems, and components in nuclear power plants according to their safety significance
	Flaw acceptance criteria	Qualification of ISI methods for specific degradation mechanisms and geometries	
Stresses	*• Code cases specific to high-energy and moderate-energy piping integrity management • Code cases associated with flow-assisted material degradation mechanisms*	*Qualification of new materials through experimental work and testing*	
• Internal pressure stresses • External pressure stresses • Longitudinal pressure stresses • Allowable displacement stress range		RI-ISI methodology development	
		RIM methodology development	
Materials	*NDE qualification requirements*		*Generic letters (a few examples)* *• GL 88-01: NRC position on intergranular stress corrosion cracking (IGSCC)* *• GL 89-08: FAC-induced Pipe Wall thinning* *• GL 90-05: Guidance for performing temporary non-code repair of ASME code class 1, 2, and 3 piping*
• Listed material specifications; ferritic, austenitic, nickel base. • Temperature limitations. • Impact testing methods—Charpy V.	NDE certification requirements		
	Code repair requirements		
Materials—"precautionary considerations"			
• Degradation of materials in service. • Dynamic effects—Contraction/expansion. • Thermal fatigue at mixing points.			

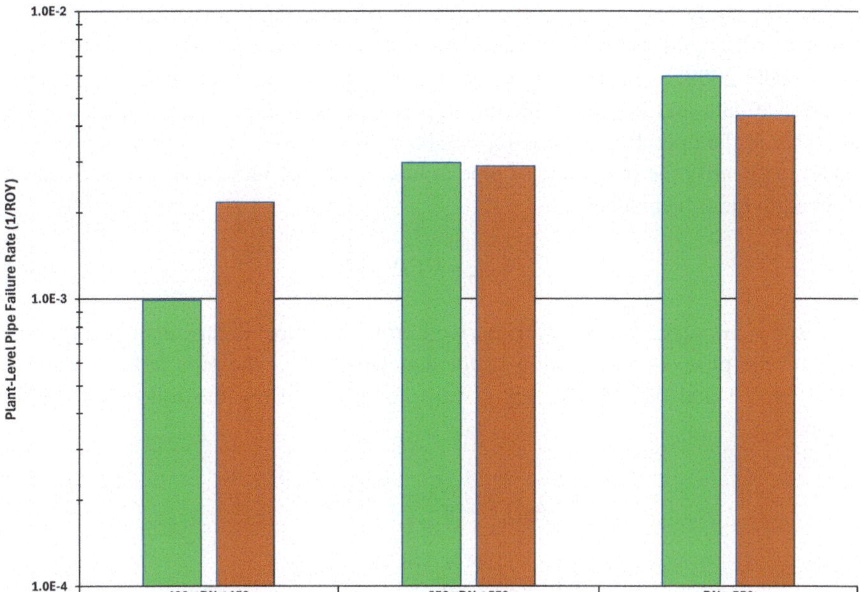

	100 < DN ≤ 150	250 <DN ≤ 550	DN > 550
B31.1	9.93E-04	2.98E-03	5.96E-03
ASME III	2.17E-03	2.90E-03	4.34E-03

Fig. 7.29 PWR Safety Class 1 piping—failures attributed to high WRS and excessive heat input

this can be done. "Raw" datasets for Safety Class 1 piping are categorized according to nominal pipe size and the underlying causes of failure listed below:

- High weld residual stresses (WRS): Internal, locked-in stresses in a material that remain after a welding process, caused by nonuniform heating and cooling cycle that creates differential volumetric expansion and contraction.
- Excessive heat input: Reduces toughness and can create brittle area.
- Lack of fusion: Weld does not properly fuse with the base metal.
- Slag inclusion: Hidden, welding defect where nonmetallic, solid by-products (slag) become trapped inside the weld metal or between weld passes.
- Repair weld: A weld subjected to high residual stresses as a result of extensive repairs, e.g., during original construction.

Next, the analysis determines the reactor operating years that produced the observed failure. With the failure data organized by nominal pipe size, the analysis includes a step to estimate point estimates of the plan-level precursor rates. Results of the analysis are shown in Fig. 7.29.[19]

[19] Details of the analysis are as follows: the US-only operating experience for two distinct periods: 1960–1989 and 1990–2025. This distinction was made to address fundamental differences in reliability and integrity management. For the latter period, operating experience of direct relevance was produced during 1381 (ROYs) for ASME III plants and 1006.8 ROYs for B31.1 plants.

Another study expanded on the treatment of uncertainty in assessing possible differences in piping pressure boundary integrity due to piping design code (Fleming and Lydell 2008). The assessment was based on light water reactor operating experience data and began by identifying the piping design code of reference for all plants in the operating experience database. The analysis was based on a simple model to quantify the frequency of pipe failure ρ_i for a certain consequence in terms of an equivalent break size:

$$\rho_i = \lambda_i \times CFP_i \times C_z \tag{7.85}$$

A series of scale factors were derived from the pipe failure data in order to estimate the pipe code correction factor denoted by C_z. The pipe failure rates for B31.1 piping and ASME III piping were developed using the following simple formulas:

$$\lambda_{B31.1} = \frac{N_{31.1}}{T_{31.1}} = 0.24 \tag{7.86}$$

$$\lambda_{ASME\ III} = \frac{N_{ASME\ III}}{T_{ASME\ III}} = 0.16 \tag{7.87}$$

$$\lambda_{Avg} = \frac{T_{31.1} \times \lambda_{31.1} + T_{ASME\ III} \times \lambda_{ASME\ III}}{T_{31.1} + T_{ASME\ III}} = 0.20 \tag{7.88}$$

Assuming that an estimate exists of λ_{Avg} for a specific safety-related piping, estimates of the pipe design code correction factors can be developed from the following:

$$C_{31.1} = \frac{\lambda_{31.1}}{\lambda_{Avg}} = 1.20 \tag{7.89}$$

$$C_{ASME\ III} = \frac{\lambda_{ASME\ III}}{\lambda_{Avg}} = 0.80 \tag{7.90}$$

Therefore, if the pipe failure rates are specialized to the design code, the failure rates for B31.1 would be expected to be about 20% higher than those estimated previously and about 19% lower if they were specialized for ASME Section III. It can be shown that there is no statistically significant difference between piping designed to ASME III Safety Class 1 and ASME III Safety Class 2. Even in the case of B31.1 piping, variations on the order of 20% higher or lower are insignificant in comparison to the underlying uncertainties, which span several orders of magnitude. Uncertainty distributions for pipe failure were revised by applying an uncertainty distribution for the design code adjustment factors as defined above. For $C_{31.1}$, a discrete probability distribution was assigned to represent the uncertainty in this factor based on engineering judgment:

$C_{31.1}$	Probability
1.0	0.05
1.1	0.15
1.2	0.60
1.5	0.15
2.0	0.05

It is unreasonable to expect this factor to be less than 1.0 or greater than 2.0 based on the large amount of data that was analyzed to obtain the point estimates. The uncertainty in $C_{ASMEIII}$ was assumed to be fully correlated to that for $C_{31.1}$, an assumption made to preserve the mean failure rate. The discrete probability distribution fixes the mode at the value of the point estimate obtained above of 1.2 and due to the skewed right property of the distribution; the mean of the uncertainty distribution is 1.26. This is accomplished by randomly sampling from the discrete probability distribution for $C_{31.1}$ and then calculating a value of $C_{ASME\ III}$ using the above equations. The formula for this calculation is given by the following equation:

$$C_{ASME\ III} = \frac{(N_{31.1}(1 - C_{31.1}) + N_{ASME\ III}}{N_{ASME\ III}} \tag{7.91}$$

In addition to the uncertainty in the design code correction factor, the sources of uncertainty include (1) uncertainty in the assumed Bayes' prior distributions, (2) uncertainty in the pipe weld counts and component years of experience, and (3) uncertainty due to the binning of pipe rupture sizes from a continuous curve into discrete rupture sizes which reflects the variability in pipe rupture size across a discrete range. Analysis results for DN250 carbon steel (CS) and stainless steel (SS) are shown in Fig. 7.30. Rather small variations in pipe failure rate due to ASME design code class were found, with small variations in relation to the underlying uncertainties in the failure rate estimates.

7.17.4 Pipe Failure Probability at Low Pressure and Temperature

This section examines pipe failures during cold shutdown. Four ASME III Class 1 and 2 piping systems of interest are as follows: in BWR plants, cooling is handled by the RHRS and water quality by the RWCU; in PWR plants, the RHRS provides cooling, while the CVCS manages inventory and chemistry. A piping pressure boundary failure would lead to a non-isolable or isolable loss-of-coolant accident.

During cold shutdown of a BWR plant, the RWCU system is operated on a reduced or intermittent basis using the following methods:

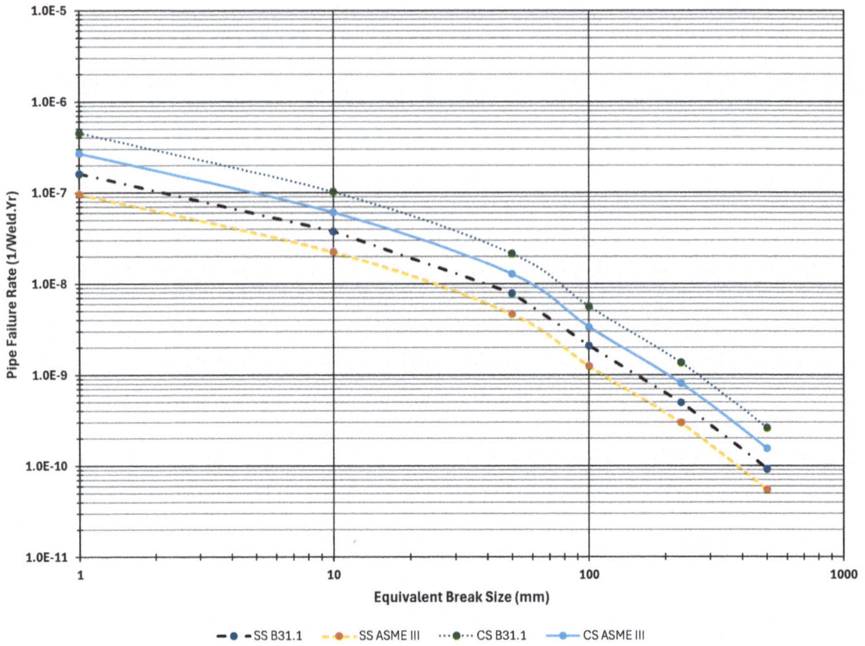

Fig. 7.30 Design code adjusted failure rates for DN250 piping

- Recirculation: Water is taken from the reactor vessel (from the suction of the reactor recirculation pumps and the bottom head) and circulated through the RWCU's filter/demineralizers before being returned to the vessel.
- Flow control: Operators control flow with valves to adjust the volume and direction of water processed by the system.
- Demineralizer bypass: During low-pressure conditions in shutdown, a bypass valve can be used to route water around the demineralizers. This allows the system to continue circulating and cleaning water for vessel level and pressure control.
- Blowdown for volume control: The system also has the ability to "blow down" or remove water from the reactor vessel. This is necessary during certain shutdown phases when other systems, such as the control rod drive (CRD) system, are adding water.

During cold shutdown of a PWR plant, the CVCS is used alongside the RHRS to manage reactor coolant inventory, pressure, and chemistry control. Since the reactor coolant pumps (RCPs) are off, the CVCS is aligned with the RHR system to continue critical functions that support maintenance and refueling. Key functions of the CVCS during cold shutdown are as follows:

- Inventory control: The CVCS works with the RHRS to maintain the correct water level in the reactor coolant system (RCS). This is crucial because as the reactor cools, the volume of the coolant decreases significantly.
- Pressure control: During cold shutdown, the RCS is typically at atmospheric pressure. For some operations, it may be kept at a low pressure, and the CVCS provides the control needed for this "solid plant" operation, where the system is filled with water.
- Decay heat removal: While decay heat is much lower during cold shutdown than during power operation, it still must be removed. The RHR system is the primary method for this, but the CVCS plays a supporting role by ensuring that there is sufficient coolant available for recirculation through the RHR heat exchangers.
- Boron concentration management: Cold shutdown requires sufficiently high boron concentration in the coolant to ensure that the reactor remains subcritical and cannot restart. The CVCS is used for controlled boration or de-boration to adjust the boron concentration as needed. For instance, de-boration is necessary to reduce concentration after a period of prolonged shutdown, which can be done by using the system's ion exchange capabilities.
- Degasification: Before opening the reactor coolant pressure boundary for maintenance or refueling, the CVCS degasification system removes dissolved radioactive gases and hydrogen from the coolant. This minimizes the release of radioactivity to the atmosphere. CVCS is also used to remove air that has dissolved into the coolant when the system is open to the atmosphere.
- Reactor coolant pump seal injection: Even when the RCPs are secured, seal water may still be supplied for maintenance purposes.

Low-power and shutdown PSA (LPSD) examines accident scenarios involving loss of primary coolant inventory as well as accident scenarios involving the spatial impact of pipe failures within containment, reactor building, or service building structures. Figures 7.31 and 7.32 illustrate the PWR-specific operating experience associated with the aforementioned systems in terms of through-wall pipe leaks (TWL) and more significant structural failures (SSF). This information signifies a starting point for additional data processing to obtain the required probabilistic failure metrics organized by pipe size, equivalent break size, and/or mass flow rate.

Piping precursor frequencies are estimated utilizing a Bayesian approach informed by operating experience data. The prior state of knowledge is established using a constrained non-informative distribution (CNID). For PWR low-power and shutdown PSA, the pre-2000 operating experience is used to inform the prior distribution, while post-1999 data provides for the Bayesian update. The rationale for this timeframe relates to the Civaux Unit 1 pipe failure on 12 May 1998, where a DN250 Schedule 40 seam-welded elbow in the RHR system failed significantly during an intermediate shutdown phase, with a through-wall flow rate of approximately 8 kg/s. Thermal fatigue was identified as the cause, and this event prompted a

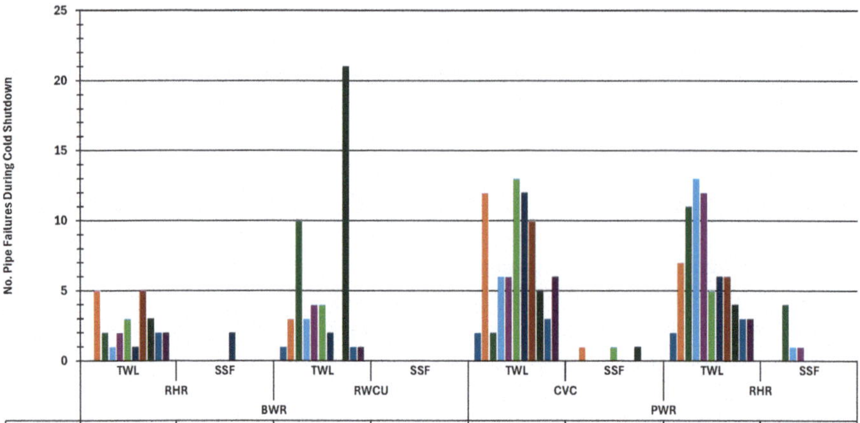

	BWR RHR TWL	BWR RHR SSF	BWR RWCU TWL	BWR RWCU SSF	PWR CVC TWL	PWR CVC SSF	PWR RHR TWL	PWR RHR SSF
1971-1975			1		2		2	
1976-1980	5		3		12	1	7	
1981-1985	2		10		2		11	4
1986-1990	1		3		6		13	1
1991-1995	2		4		6		12	1
1996-2000	3		4		13	1	5	
2001-2005	1	2	2		12		6	
2006-2010	5				10		6	
2011-2015	3		21		5	1	4	
2016-2020	2		1		3		3	
2021-2025	2		1		6		3	

Fig. 7.31 Pipe failures during periods of cold shutdown

redesign of the RHR system for the N4 class of reactors.[20] Results from the precursor frequency analysis are given in Table 7.15.

A proposed conditional pipe failure probability model for shutdown modes builds on technical insights offered by two studies referred to as NUREG-1829 and NUREG/CR-6144. The objective is to utilize those insights to modify a consensus-based set of full-power LOCA frequencies to obtain LOCA frequencies for plant operational states at low pressure and temperature.

During 1989, the USNRC initiated an extensive program to examine the potential risks during low-power and shutdown operations. Surry Power Station (PWR) was selected as one of the reference plants. Completed in 1994, the study was published as NUREG/CR-6144; the reader is referred to Chap. 3 of this book for more details. The pipe failure-induced LOCA frequency analysis assessed the potential loss-of-coolant accident scenarios that are unique to shutdown states. Full-power LOCA frequencies were taken from NUREG-1150 (USNRC 1989) and adjusted to account for the unique low-power/shutdown system configurations. For plant operational states (POS) 1 (reactor shutdown and low power) and 15 (reactor startup), full-power frequencies were used. For states 2 (cooldown to 175 °C) and 14 (startup with steam

[20]The N4 class of reactors is four-loop PWRs that were commissioned in the mid-1990s. An important design feature was the use of Alloy 690 rather than Alloy 600 steam generator tubing material.

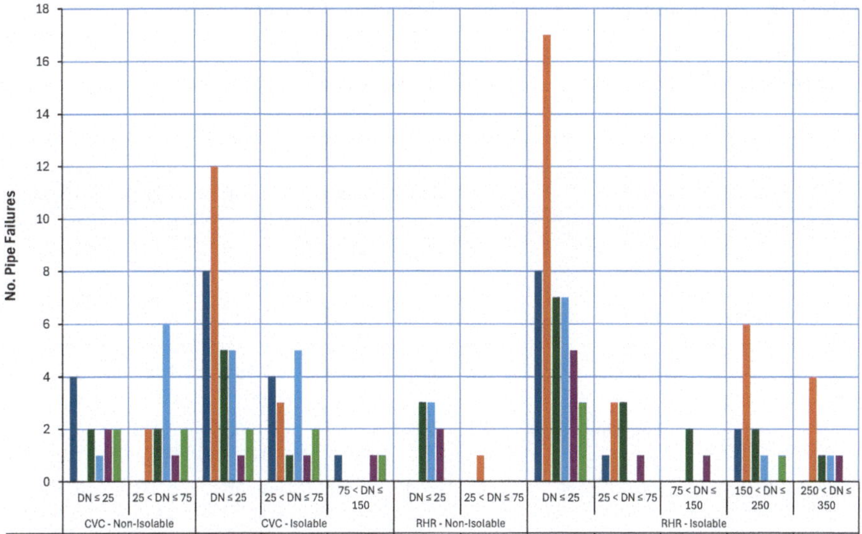

	CVC - Non-Isolable DN ≤ 25	CVC - Non-Isolable 25 < DN ≤ 75	CVC - Isolable DN ≤ 25	CVC - Isolable 25 < DN ≤ 75	CVC - Isolable 75 < DN ≤ 150	RHR - Non-Isolable DN ≤ 25	RHR - Non-Isolable 25 < DN ≤ 75	RHR - Isolable DN ≤ 25	RHR - Isolable 25 < DN ≤ 75	RHR - Isolable 75 < DN ≤ 150	RHR - Isolable 150 < DN ≤ 250	RHR - Isolable 250 < DN ≤ 350
1970-79	4		8	4	1			8	1		2	
1980-89		2	12	3			1	17	3		6	4
1990-99	2	2	5	1		3		7	3	2	2	1
2000-09	1	6	5	5		3		7			1	1
2010-19	2	1	1	1	1	2		5	1	1		1
2020-25	2	2	2	2	1			3			1	

Fig. 7.32 Expanded LPSD pipe failure experience summary

generators), the full-power frequencies were divided by a factor of 20.[21] For states 3 and 4 (cooldown to ambient temperature) and 12 and 13 (preparing for startup and heat up to 175 °C), the full-power frequencies were divided by a factor of 30. As a justification, these ratios were derived from "*statistical considerations of structural mechanics in the case of the pressurizer surge line (large LOCA).*" NUREG/CR-6144 did not provide specific details on how these ratios were obtained, however.

The quotation is from page 4–76 of NUREG/CR-6144. The technical basis is likely to have been based on NUREG/CR-4469, "*Nondestructive Examination (NDE) Reliability for Inservice Inspection of Light Water Reactors*," in which probabilistic fracture mechanics analyses were performed to establish pipe rupture frequencies. Additional relevant information is found in the study of Vo et al., (1991).

Absent operating experience data for medium and large LOCA events, which would allow for direct calculation of conditional failure probability, assessments instead rely on findings from experiments, expert judgments, and probabilistic fracture mechanics. A reverse engineering process can be used to create a consensus-driven uncertainty distribution for prior state of knowledge regarding

[21] Comprehensive definitions of low-power and shutdown plant operating states are found in Bley, D.C., Stetkar, J.W. (1984), Zion Nuclear Power Plant Residual Heat Removal PRA, NSAC-84, Nuclear Safety Analysis Center, Palo Alto, CA.

Table 7.15 System-level pipe failure precursor frequencies for LPSD modes

System-level precursor frequency (1/ESDY)—through-wall leak							
Analysis case	Diameter	CNID prior		Posterior—2000–2025 operating experience			
		α	β	Mean	5%-tile	50%-tile	95%-tile
CVC—Non-isolable	DN ≤ 25	0.5	44.7	6.98E-03	2.90E-03	6.56E-03	1.25E-02
	25 < DN ≤ 75	0.5	67.1	1.17E-02	6.24E-03	1.13E-02	1.86E-02
CVC—Isolable	DN ≤ 25	0.5	10.7	1.13E-02	5.75E-03	1.08E-02	1.83E-02
	25 < DN ≤ 75	0.5	33.5	1.09E-02	5.58E-03	1.05E-02	1.78E-02
	75 < DN ≤ 150	0.5	268.3	2.47E-03	5.66E-04	2.15E-03	5.47E-03
RHR—Non-isolable	DN ≤ 25	0.5	89.4	6.60E-03	2.75E-03	6.21E-03	1.18E-02
	25 < DN ≤ 75	0.5	268.3	4.94E-04	1.94E-06	2.25E-04	1.90E-03
RHR—Isolable	DN ≤ 25	0.5	8.4	2.06E-02	1.28E-02	2.02E-02	2.99E-02
	25 < DN ≤ 75	0.5	38.3	1.92E-03	2.25E-01	1.51E-03	5.00E-03
	75 < DN ≤ 150	0.5	134.2	1.71E-03	2.01E-04	1.35E-03	4.45E-03
	150 < DN ≤ 250	0.5	26.8	3.25E-03	7.44E-04	2.83E-03	7.19E-03
	250 < DN ≤ 350	0.5	53.7	3.14E-03	7.19E-04	2.73E-03	6.94E-03

conditional pipe failure. The NUREG-1829 study serves as the basis for how this approach is applied. Experts estimated LOCA frequencies based on both system type and effective break size. An aggregation of elicited frequencies produced system-specific LOCA frequency information which, alongside precursor rates, allows for determining conditional failure probabilities for each system:

$$Frequency\ of\ LOCA = LF = \lambda_{Precursor} \times CFP \tag{7.92}$$

$$CFP = \frac{LF}{\lambda_{Precursor}} \tag{7.93}$$

To adjust full-power conditional failure probabilities, the initial step involves determining CVC and RHR system LOCA frequencies as they relate to equivalent break size. By utilizing the respective system-level precursor rates, the associated CFP lognormal uncertainty distributions can then be established. Subsequently, two

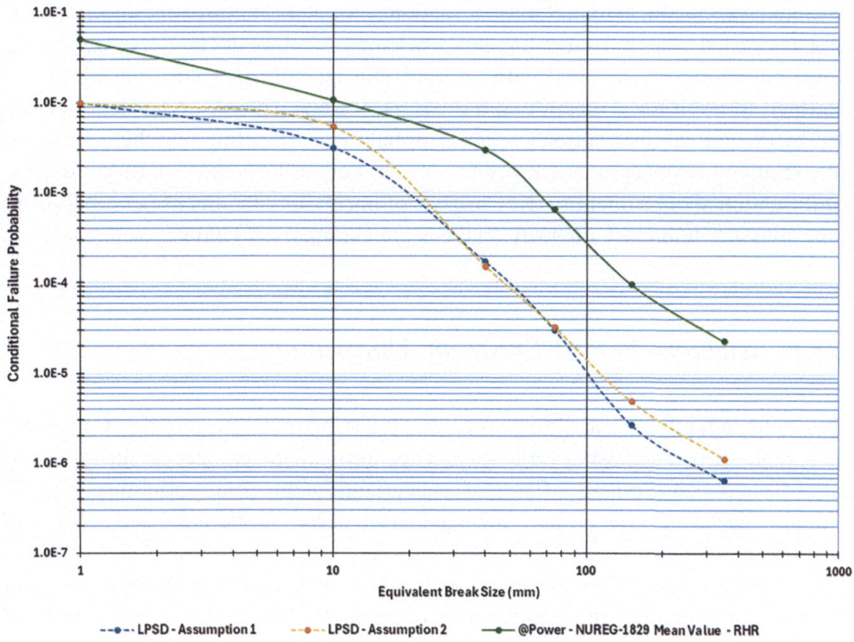

Fig. 7.33 Conditional failure probability distributions for LPSD PSA application

assumptions are made regarding the modification of full-power CFP under conditions of reduced pressure and temperature:

1. Assume that the lower-bound NUREG-1829 full-power CFP is representative of the upper-bound CFP distribution under shutdown conditions and calculate the mean and the 5th and 50th percentiles of the new CFP.
2. Use the multipliers from NUREG/CR-6144 to derive a CFP under low pressure and temperature.

Figure 7.33 demonstrates that both approaches to adjusting full-power CFP parameters produce comparable outcomes, which merits attention. Additional research should be conducted, providing opportunities for future nuclear engineers and scientists to pursue this line of inquiry.

7.18 Uncertainty Analysis

Uncertainty analysis focuses on identifying and quantifying the sources of uncertainty in the results of an analysis. Ultimately, this is an engineering activity, meaning that the sources of uncertainty that are relevant to structural reliability analysis can be classified into two categories: (1) aleatory uncertainties, being associated with physical uncertainty or randomness, and (2) epistemic uncertainties, being associated with understanding or knowledge. Aleatory uncertainty refers to the natural randomness associated with an uncertain quantity. Aleatory uncertainty is quantified through the collection and analysis of data. The observed data may be fitted by theoretical distributions, and the probabilistic modeling may be interpreted in the relative frequency sense. Epistemic uncertainty reflects the lack of knowledge or information about a quantity. Key aspects of uncertainty treatment in probabilistic piping pressure boundary integrity analysis are summarized below.

7.18.1 Aleatory Versus Epistemic Uncertainty

Aleatory uncertainty, or variability, is associated with randomness and is sometimes called random or stochastic variability. This form of uncertainty is intrinsic and cannot be reduced. Sources of aleatory uncertainty stem from variability in weld residual stresses, chemical and mechanical material properties, and operating environment.

Epistemic uncertainty is attributed to the lack of knowledge about events and processes. In principle, this uncertainty is reducible because it is a function of information. Epistemic uncertainty is sometimes called subjective or internal uncertainty and divides into two major subcategories: model uncertainty and parameter uncertainty. Epistemic uncertainty can, in principle, be narrowed or eliminated through acquisition by the analyst of additional information.

An IAEA Nuclear Energy Series publication (IAEA 2023) offers a process for translating abstract uncertainty concepts into practical guidelines for probabilistic piping reliability analysis. It begins with fundamentals, as deep knowledge of piping design principles, material degradation, and empirical data associated with piping in different operating environments. The treatment of uncertainty differs according to the type of piping system. An example from the IAEA publication:

1. Primary system piping is connected to the reactor pressure vessel. Predominant materials are stainless steels, carbon steel clad with stainless steel, and nickel-base alloys. The conjoint requirements for material degradation are well known. Nevertheless, pipe flaws sometimes develop in locations believed to be immune to degradation.

   (a) Parametric uncertainty relates to the uncertainty in the frequency of crack initiation and propagation. This uncertainty is influenced by welding processes, material selection, NDE techniques, and stress improvement processes.
   (b) Modeling uncertainty means uncertainty in the conditional failure probability. Different models yield different answers, potentially with significant impact on the 5th, 50th, and 95th percentiles of the pipe failure frequency.
   (c) Completeness uncertainty addresses whether an analysis accounts for all material degradation scenarios.

When combined with the aleatory uncertainty, the epistemic uncertainty parameters provide a basis for quantitatively estimating the uncertainty bounds for a specific reliability parameter. The combined uncertainty parameters represent both the irreducible and event-driven issues that are associated with any given analysis case. This uncertainty contribution comes from the knowledge of the analyst integrating the information about material degradation and determining the likelihood of major pipe failure challenging the safety of a plan. As information about the pipe failure context and knowledge about possible influences of RIM increase, the epistemic uncertainty decreases.

7.18.2 Sources of Uncertainty

There are manifold sources of uncertainty to consider: completeness of the operating experience, including variability in material degradation susceptibility, and plant-to-plant variability in piping system designs. Table 7.16 outlines the approach to managing uncertainty throughout each phase of the analysis.

Table 7.16 Treatment of uncertainty in probabilistic piping reliability analysis

Analysis step	Uncertainty treatment
Prior distribution for precursor rate	A constrained non-informative distribution (CNID) can be used to characterize uncertainty in the prior state of knowledge
Collect failure data for each analysis case and pipe size. Develop estimates of pipe lengths and weld populations organized by nominal pipe size and system. Calculate reactor operating years of experience relevant to respective analysis case. Characterize uncertainty in the estimated pipe lengths or weld population per plant	Perform three Bayes' updates of the CNID prior distribution for the precursor rate: One for the best estimate of pipe length or weld population, one assuming that pipe length/weld population is X% higher, and one assuming that pipe length/weld population is X% lower
	Using engineering judgment, assign an X% probability that the best estimate of pipe length is the correct value, Y% probability that the lower estimate is correct, and Z% that the upper estimate is correct
	Use Monte Carlo simulation to calculate a mixture distribution combining the above distribution using the stated probability weights
Characterize uncertainty in weld residual stresses	Develop discrete probability distributions that reflects what is known about welding processes and welding quality
Develop prior distribution for the conditional failure probability (CFP) of a significant structural failure	Use engineering judgment based on insights from analysis of service data, results of probabilistic fracture mechanics analyses, and expert elicitation to estimate the probabilities for discrete levels of break size. Use the CNID method to characterize the prior distributions

(continued)

Table 7.16 (continued)

Analysis step	Uncertainty treatment
Collect operating experience on pipe failures as a function of through-wall leak raters and convert them to equivalent break sizes	Perform Bayes update of the prior distributions to obtain Bayes' posterior distributions for each conditional probability of rupture size
Apply adjustment factors to each value of CFP to ensure that frequencies of SSF agree with the observed industry data. This step should be considered when analyzing reliability of safety class 2 and 3, balance-of-plant piping, and small-diameter piping	Adjustment factors are developed by comparing the predicted SSF frequencies against the observed frequencies from the operating experience data. Adjustment factors are calculated to make the predictions consistent with the industry data
Develop overall uncertainty in the frequency of SSF for each analysis case	Perform Monte Carlo uncertainty analysis to combine the uncertainties in precursor rate and CFP. Fit the resulting distributions to lognormal distributions by matching the means and estimating the range factor (RF) from the Monte Carlo distribution percentiles

References

Arnett LM (1976) Optimization of in-service inspection of pressure vessels, DP-1428. Savannah River Laboratory, Aiken

Atwood CL (1994) Constrained noninformative priors, INEL-94/0074. Idaho National Laboratory, Idaho Falls

Aven T (2020) Risk science contributions: three illustrating examples. Risk Anal 40(10): 1889–1899. https://doi.org/10.1111/risa.13549

Collins DH et al (2010) Treatment of passive component reliability in risk-informed safety margin characterization, INL/EXT-10-20013. Idaho National Laboratory, Idaho Falls

Collins DH (2011) Reliability modeling of nuclear power plant subsystems using statistical flowgraphs., LA-UR-11-02211. A presentation prepared for the 2011 joint statistical meeting of the American Statistical Association. Los Alamos National Laboratory, Los Alamos

Fleming KN (2004) Markov models for evaluating risk-informed in-service inspection strategies for nuclear power plant piping systems. Reliab Eng Syst Saf 83(1):27–45

Fleming KN et al (2010) Treatment of passive component reliability in risk-informed safety margin characterization, FY 2010 report, INL/EXT-10-20013. Idaho National Laboratory, Idaho Falls

Fleming K, Lydell B (2006) Pipe rupture frequencies for internal flooding PRAs, Revision 1, 1013141. EPRI, Palo Alto

Fleming K, Lydell B (2008) Assessment of PBMR passive component reliability: evaluation of pipe design code selection on helium inventory control system piping reliability, Proprietary Technical Report prepared for PBMR Pty, Centurion.

Fleming K, Lydell B (2010) Pipe rupture frequencies for internal flooding PRAs, Revision 3, 1021086. EPRI, Palo Alto

Fleming K, Lydell B (2013) Pipe rupture frequencies for internal flooding PRAs, Revision 3, 3002000079. EPRI, Palo Alto

Fleming K, Lydell B (2018) Pipe rupture frequencies for internal flooding PRAs, Revision 4, 3002012997. EPRI, Palo Alto

Fleming K, Lydell B, Mosleh A, Chalgham W (2023) Pipe rupture frequencies for internal flooding probabilistic risk assessments, Revision 5, 3002024904. EPRI, Palo Alto

Grabski F (2015) Semi-Markov processes: applications in system reliability and maintenance., ISBN 978-0-12-800518-7. Elsevier, Waltham

IAEA (2023) Methodologies for assessing pipe failure rates in advanced water cooled reactors. Nuclear energy series no. NR-T-2.16, Vienna. ISBN: 978-92-0-150422-7

Kapur KC, Lamberson LR (1977) Reliability in engineering design., ISBN: 0-471-51191-9. Wiley, New York, pp 12–15

Keefer DL, Bodily SE (1983) Three-point approximations for continuous random variables. Manag Sci 29(5):595–609

Kurth RE, Cox DC (1985) Discrete probability distributions for probabilistic fracture mechanics. Risk Anal 5(3):235–240

Lydell B, Fleming K (2014) Piping system failure rates for corrosion resistant service water piping, 3002002787. EPRI, Palo Alto

Lydell B, Fleming L, Roy J-F (2018) Analysis of possible aging trends in the estimation of piping system failure rates for internal flooding PRA. In: W14–376, Proc. PSAM 14 probabilistic safety assessment and management, 16–21 September. UCLA, Los Angeles

Lydell B, Chitose K, Riznic J (2022a) Material degradation mitigation & new degradation scenarios: lessons learned from the analysis of operating experience data. In: Proc FONTEVRAUD 10 international symposium: contribution of materials investigations and operating experience to LWRs' safety, performance and reliability, 19–22 September, Avignon, France

Lydell B, Chitose K, Focht E (2022b) Material degradation mitigation & new degradation scenarios: lessons learned from the analysis of operating experience data. In: Proc. FONTEVRAUD-10, international symposium on the contribution of materials investigations and operating experience to LWR safety, performance and reliability, Avignon, France

Maio FD, Colli D, Zio E, Tao L, Tong J (2016) A multi-state physics modeling approach for the reliability assessment of nuclear power plant piping systems, HAL Id: hal-01265883, https://hal.archives-ouvertes.fr/hal-01265883

OECD/NEA (2014) CODAP Topical Report on Flow Accelerated Corrosion (FAC) of Carbon Steel and Low Alloy Steel Piping in Commercial Nuclear Power Plants, NEA/CSNI/R(2014)6, Boulogne-Billancourt, France.

Ortiz NR, Wheeler TA, Breeding RJ, Hora S, Meyer MA, Keeney RL (1991) Use of expert judgment in NUREG-1150. Nucl Eng Des 126:313–331

Pearson ES, Tukey JW (1965) Approximate means and standard deviations based on distances between percentage points of frequency curves. Biometrika 52(3/4):533–546

Rashid J, Dunham R, Wong F (2010) Material reliability program: development of material constitutive model for irradiated austenitic stainless steels (MRP-135 Revision 1), 1020958. Electric Power Research Institute, Palo Alto

Rudland D (2019) Estimation of the Impact of Pipe Diameter on Rupture Using xLPR. In: 3rd International Seminar on Probabilistic Methodologies for Nuclear Applications, Rockville, MD

USNRC (1988a) NRC position on IGSCC in BWR austenitic stainless steel piping, Generic Letter No. 88-01, Washington, DC.

USNRC (1988b) Technical report on material selection and processing guidelines for BWR coolant pressure boundary piping, NUREG-0313 revision 2. USNRC, Washington, DC

USNRC (2008) Estimating Loss-of-Coolant Accident (LOCA) frequencies through the elicitation process, NUREG-1829, vol Volume 2. USNRC, Washington, DC

USNRC (2022a), Probabilistic Risk Assessment (PRA) project: volume 2: background, site and plant description, and technical approach, Washington, DC. NRC-ADAMS accession no. ML22067A232.

USNRC (2022b) Level 3 probabilistic risk assessment (PRA) project: volume 3b: reactor, at-power, level 1 PRA for internal flooding. NRC-ADAMS accession no. ML22067A213, Washington, DC

Yuan X-X, Pandey M, Riznic J (2008) A point process model for piping failure frequency analysis using OPDE data. In: ICONE16–48078, proc. 16th international conference on nuclear engineering ICONE16, ASME, New York

Veeramany, A. and Pandy, M. (2010), "Reliability analysis of nuclear piping systems using semi-Markov process model," Ann Nucl Energy, 38:1133–1139, 2011. https://doi.org/https://doi.org/10.1016/j.anucene.2010.12.012
Veeramany A (2012) Multi-state reliability analysis of nuclear power plant systems., PhD Thesis. University of Waterloo, Waterloo
Vo TV, Gore BF, Eschbach EJ, Simonen FA (1989) Probabilistic risk assessment based guidance for piping in-service inspection. Nucl Technol 88(1):13–20. https://doi.org/10.13182/NT89-A34332
Vo TV, Heasler PG, Doctor SR, Simonen FA, Gore BF (1991) Estimates of rupture probabilities for nuclear power plant components: expert judgment elicitation. Nucl Technol 96(3):259–271
Wheeler TA, Hora SC, Cramond WR, Unwin SD (1989) Analysis of core damage frequency from internal events: expert judgment elicitation, NUREG/CR-4550, vol Volume 2. USNRC, Washington, DC
Øksendal B (2003) Stochastic differential equations. An introduction with applications., ISBN 978-3-642-14394-6. Springer, New York

Chapter 8
Reliability of Leak and Flaw Detection

8.1 Leak Detection

Various systems are employed to promptly notify operations personnel of any leakage within a piping system, as illustrated in Fig. 8.1. A timely response enables the identification and isolation of leaks to ensure that their consequences remain within acceptable parameters. Multiple, diverse, and redundant leak detection methods are implemented to maintain a high level of detection reliability.

General Design Criterion 30, "*Quality of Reactor Coolant Pressure Boundary*," of Appendix A to 10 CFR Part 50, "*General Design Criteria for Nuclear Power Plants*," requires that means be provided for detecting and, to the extent practical, identifying the location of the source of reactor coolant leakage.[1] Regulatory Guide 1.45, "*Reactor Coolant Pressure Boundary Leakage Detection Systems*," was issued in 1973 (USNRC 2008). It described acceptable methods of implementing the guide with regard to the selection of leakage detection systems for the reactor coolant pressure boundary.

Regulatory Guide 1.45 recommends the use of at least three different detection methods in reactors to detect leakage from the primary pressure boundary. Monitoring of both containment sump flow and airborne particulate radioactivity is recommended. A third method can involve either monitoring of condensate flow rate from air coolers or monitoring of airborne gaseous radioactivity. Regulatory Guide 1.45 also recommends that leak rates from identified and unidentified sources be monitored separately to an accuracy of 3785 cm^3/min and that indicators and alarms for leak detection be provided in the main control room.

[1] https://www.nrc.gov/reading-rm/doc-collections/cfr/part050/part050-appa

B. Lydell, *Knowledge Preservation in Nuclear Safety*,
https://doi.org/10.1007/978-3-032-25463-4_8

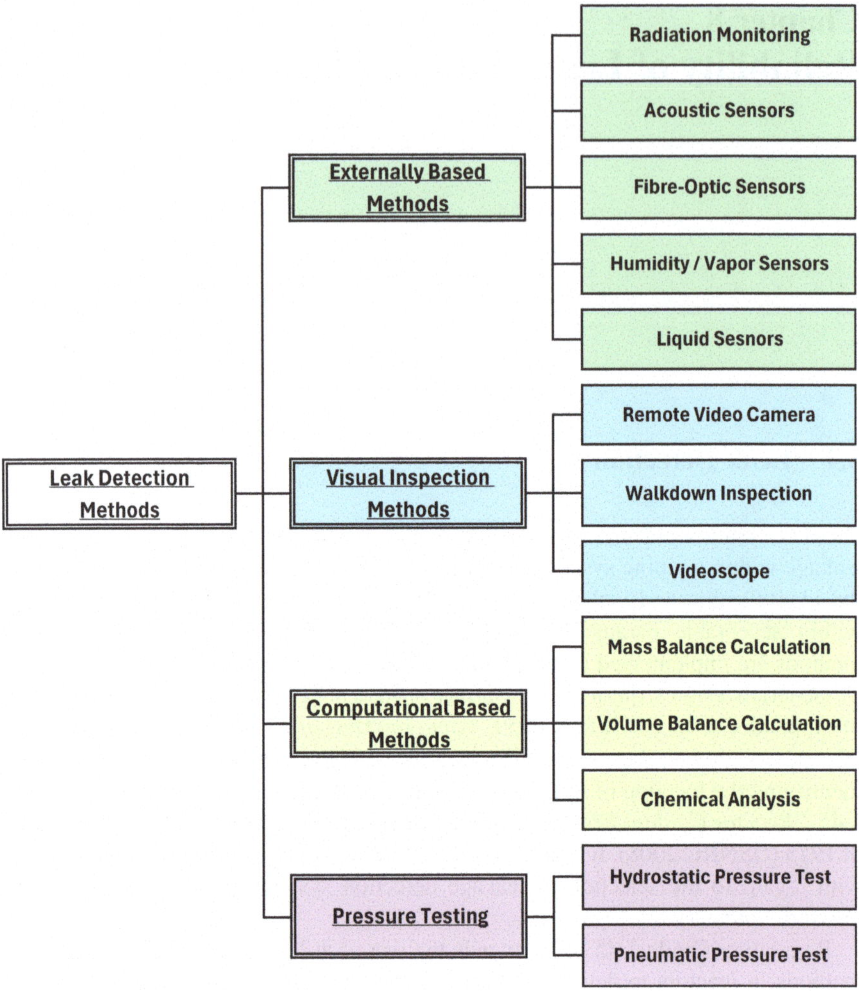

Fig. 8.1 Leak detection methods

8.1.1 BWR Primary System Leak Detection

The drywell floor drain sump measurement monitors the normal design leakage collected in the floor drain sump. This leakage consists of leakage from the control rod drive system, valve flange leakage, closed cooling water, air cooler drains, and any leakage not connected to the equipment drain sump. Leakage from the reactor coolant pressure boundary may be detected by any one of a variety of independent monitored variables, such as drywell pressure and temperature, sump-level changes, and containment gaseous and particulate radioactivity levels. While some of these systems are not redundant in themselves, it is not postulated that any one event could render all means of leak detection inoperable. The normal operational method for monitoring leakage in the drywell is with the drywell floor drain sump leak detection system (LDS).

Collection in excess of background leakage would indicate reactor coolant leakage. Equipment sump drain temperature is also monitored and thus will indicate leakage. Leakage from various equipment located in the drywell is piped directly to the drywell equipment drain sump. The sump itself is covered, and all of the various drain lines are open only to the equipment they serve, thereby forming a closed system which receives leakage only from identified sources.

The drywell floor drain and equipment drain sumps are provided with redundant-level transmitters that supply remote-level indications. The level indicators are equipped with switches that start and stop the pumps as required and provide signals to the LDS. The LDS has two timers that are energized and de-energized by the sump-level switches to determine if the leakage rate into the sump is in excess of the expected design value. One timer is started each time the sump is pumped down to the low-level set point, at which time the sump pumps are automatically stopped. Should the sump fill up to the high-level set point prior to the expiration of time on the timer, an alarm is sounded in the main control room indicating a leakage rate into the sump in excess of the design limit.

8.1.2 PWR Primary System Leak Detection

The leakage detection system consists of three diverse systems: (1) containment sump-level alarm system, (2) containment atmosphere particulate radioactivity monitoring system, and (3) containment atmosphere gaseous radioactivity monitoring system. The containment normal sump-level detection and alarm system consists of instrumentation that senses and indicates sump level and provides an alarm. A sensor in the sump sends a signal to the level indicator/alarm in the control room. Should sump level exceed a predetermined value, the indicator/alarm provides an alarm to alert operators of the condition. The water collected in the sump is then processed by the miscellaneous waste processing system. The containment atmosphere particulate and gaseous radioactivity monitors consist of offline detectors, sample pumps, pump controls and signal conditioning, recording, indication, and alarm functions. The offline sample pumps take suction from the containment ventilation system—this provides a representative sample of the containment atmosphere—and deliver the sample to the monitors. The particulate monitor is a fixed filter, scintillation-type detector. The detector continuously monitors the radiation level of the filter and provides a signal which is conditioned so it can be read by the indicating and recording electronics in the control room. The signal is sent to the recorders for trending purposes. Should the samples collected by the filter exceed predetermined limits, the radiation monitoring electronics also provide an alarm to alert operators of the condition. After the particulates are collected by the filters, the cleaned

Table 8.1 ASME XI Article IWB 5230 test pressures

Test temperature (°C)	Test pressure
40 or less	$1.10 \times p_0$
95	$1.08 \times p_0$
150	$1.06 \times p_0$
200	$1.04 \times p_0$
260 or greater	$1.02 \times p_0$

p_0= nominal operating pressure corresponding to 100% rated reactor power. Linear interpolation at intermediate temperatures is permissible

atmosphere sample is sent to the scintillation detector gaseous monitor where the radiation level is continuously monitored.

8.1.3 ASME XI Pressure Testing Requirements for Repair/Replacement Activities

Unless exempted by ASME XI Article IWA-4540(b), repair/replacement activities performed by welding on a pressure-retaining boundary shall include a hydrostatic or system leakage test in accordance with Article IWA-5000 prior to or as part of returning to service (Table 8.1). Only welds made in the course of a repair/replacement activity require pressurization and a VT-2 visual examination for the location of any leaks during the test. A VT-2 examiner visually scans components (sometimes with mirrors, borescopes, or video scopes) for signs of leakage, looking for discontinuities like cracks or corrosion that could cause failure.

8.1.4 ASME XI Hydrostatic Pressure Testing

ASME Boiler and Pressure Vessel Code Section XI, *Rules for Inservice Inspection of Nuclear Power Plant Components*, requires a 10-year hydrostatic pressure testing at or near the end of the 10-year in-service inspection interval.[2] Article IWC-5000 of ASME XI specifies that the hydrostatic test pressure shall be at least 1.10 times the system pressure P_{sv} for systems with design temperature of 95 °C or less and at least 1.25 times the system pressure P_{SV} for systems with design temperature above 95 °C. The system pressure P_{sv} shall be the lowest pressure setting among the number of safety or relief valves provided for overpressure protection within the boundary of

[2]The standard ASME Section XI in-service inspection (ISI) interval for Class 1 and 2 components is 10 years. However, ASME has published Code Case N-921 as a voluntary alternative, allowing a 12-year inspection interval. The NRC conditionally accepts this code case for those implementing the 2017 or later editions of the ASME BPV Code.

the system to be tested. For systems (or portions of systems) not provided with safety or relief valves, the system design pressure P_d can be substituted for P_0, normal operating pressure. However, in such cases, alternative testing must be pre-approved by the regulatory body.

Hydrostatic pressure tests are time-consuming. In some cases, improper test procedures can lead to over-pressurization, deformation, leaks, or catastrophic failures. Industry experience further demonstrates that leaks are seldom discovered as a result of hydrostatic test pressure driving a preexisting flaw through-wall; rather, they are found in most cases at normal operating conditions.

8.1.5 ASME Code Case N-498-1

ASME Section XI hydrostatic testing verifies the tightness and integrity of pressure vessels and piping systems, especially in nuclear power plants, by applying pressure to a liquid-filled system to reveal potential leaks or flaws. While effective, its practicality can face challenges in accessibility and time, and, if not properly dried after testing, water left inside a pressure boundary can lead to corrosion. Possible practical challenges include the following:

- Time consumption: Hydrostatic tests can be time-consuming, especially for large and complex systems.
- Accessibility issues: Access to all areas of the system being tested for leak detection may be difficult or restricted, especially in older or complex installations.
- System design limitations: Some system designs might make it impractical to perform hydrostatic testing, such as open-ended systems or those with unique valve arrangements. A good example is branch lines off a PWR reactor coolant system hot leg but that cannot satisfy the code test requirement for nominal operating pressure associated with 100% rated power.

As an alternative to the ASME XI 10-year system hydrostatic test requirement, Code Case N-498-1, "*Alternative Rules for 10-year System Hydrostatic Testing for Class 1, 2, and 3 Systems*," was approved in 1994. The code case provided increased testing flexibility and could, in principle, considerably reduce or eliminate relief requests associated with welded repairs or installation of replacement items by welding. This code case states that the following rules shall be used:

- A system leakage test shall be conducted at or near the end of each inspection interval, prior to reactor startup.
- The boundary subject to test pressurization during the system leakage test shall extend to all Class 1 pressure retaining components within the system boundary.
- Prior to performing the VT-2 visual examination, the system should be pressurized to nominal operating pressure for at least 4 hours for insulated systems and

10 minutes for non-insulated systems. The system shall be maintained at nominal operating pressure during the performance of the VT-2 visual examination.
- Test temperatures and pressures shall not exceed limiting conditions for the hydrostatic test curve as contained in the plant technical specifications.
- The VT-2 visual examination shall include all components within the boundary identified in (a)(2) above.
- Test instrumentation requirements of ASME XI IWA-5260 are not applicable.

Technical specifications for operating plants specify the reactor coolant system (RCS) surveillance requirements. RCS pressure and temperature limits for heat-up, cooldown, low-temperature operation, criticality, and hydrostatic testing as well as heat-up and cooldown rates shall be established and documented in a Pressure and Temperature Limit Report (PTLR)[3] for the following: "RCS pressure and temperature (P/T) limits" and "low-temperature overpressure protection (LTOP)."

8.2 History of NDE Reliability

Conducted under a USNRC research program on the "*Integration of NDE Reliability and Fracture Mechanics*," Dr. Spencer Bush of the Pacific Northwest National Laboratory prepared a three-volume study on the reliability of nondestructive examination techniques.[4] This study provided detailed technical documentation of the evolution of NDE reliability in the 1960s and 1970s. The program started in 1977, with results published in 1983 as NUREG/CR-3110, consisting of a series of white papers summarizing the state of knowledge based on literature surveys and the author's extensive understanding of the subject matter.

- NUREG/CR-3110 Volume 1, Chapters 1 to 6, 594 pages (Bush 1983a): This volume includes a review of national and international codes and standards for NDE, a review of early programs to develop NDE qualification criteria, and methods of flaw detection reliability.
- NUREG/CR-3110 Volume 2, Chapters 7 to 11, 507 pages (Bush 1983b): This volume includes a review of the capabilities and limitations of ultrasonic examination techniques, a theoretical background to flaw detection and sizing, a review of reliability and limitations of various NDE techniques, metho, an assessment of the probability of failure of piping systems on the basis of available statistic, and a review of models of probability of detection.

[3]The PTLR is a plant-specific document that outlines the permissible reactor vessel pressure and temperature limits, heat-up/cooldown rates, and low-temperature overpressure protection (LTOP) system limits.

[4]The NRC began funding a multi-year program at Pacific Northwest Laboratory (later PNNL) in 1977. Its initial aim was to assess the reliability of in-service inspections in nuclear power plants and use fracture mechanics to understand safety implications of NDE unreliability.

– NUREG/CR-3110 Volume 3, Chapters 12 to 15, 520 pages (Bush 1983c): Chapter 12 evaluates factors that are important in characterizing flaws for meaningful input into fracture mechanics equations. Obvious factors are flaw size and type, orientation, and locations. Chapter 13 concentrates on (a) factors influencing the reliability of piping containing austenitic stainless steel weldments, (b) actual data pertinent to reliability flaw detection, and (c) probability models suggested for flaw detection and sizing as well as mathematical analyses of factors influencing reliability of detection. Chapter 14 is a review of ASME B&PV Code Sections Section V ("*Nondestructive Examination*") and Section XI ("*Rules for Inservice Inspection*"). Chapter 15, "*Conclusions and Recommendations*," is a distillation of the previous 14 chapters.

Volume 1 begins with a summary of the conclusions of an IAEA-sponsored technical committee meeting of the International Working Group on Reliability of Reactor Pressure Components (IWG-RRPC)[5] held in Kobe, Japan, in April 1977. The topic of the meeting was "*Use of Non-destructive Testing Techniques for In-Service Inspection of Reactor Pressure Components.*" Some conclusions are reproduced below:

> *The question of the reliability of ultrasonic techniques in detecting rejectable defects was mentioned several times, and the important programs of 11 round-robin ultrasonic testing (UT) work being done currently in* the *United States and Europe were mentioned. It is hoped that the results be made available to the IWG-RRPC at an early stage for them to discuss the implications and need for further work.*
>
> The present information indicates that conventional UT techniques, especially as practiced with manual operation, have limitations in the sizing and detection of flaws.
>
> Ultrasonic examinations of welds in austenitic stainless steel components using conventional ultrasonic equipment and techniques are generally less effective, as compared to those used for welds in carbon steel components, owing to the increased acoustic attenuation exhibited by these materials. Furthermore, large-amplitude ultrasonic indications may occur in weld regions that are free of rejectable defects. The acoustic energy attenuation problems combined with the false indications encountered during the examination of austenitic stainless steel welds have emphasized the need for development of more effective ultrasonic methods to supplement the examination processes presently used during the fabrication of stainless steel components. Improvements should be sought with emphasis on those aspects which will improve the probability of detecting any important flaws and taking appropriate action.
>
> Less information existed on the reliability of sizing and location of any defects reported. It was noted that the work on B-scan, focused probe, acoustic holography, and phased-array and signal analysis devices all showed considerable evidence of some improvements, and it is desirable that a coordinated international program to assess the accuracy, reliability, and reproducibility of the various possible sizing methods be established and the IWG-RRPC be invited to consider this, possibly in collaboration with other international and national

[5]The International Working Group on the Reliability of Reactor Pressure Components (IWG-RRPC) was established by the International Atomic Energy Agency (IAEA) to foster information exchange and research regarding the reliability of pressure boundary components in both power and research reactors. The IWG-RRPC held its first meeting in 1966. During the first couple of years, the activities were mainly concerned with reactor pressure vessel integrity.

organizations. In particular, additional NDE including standard and advanced UT, acoustic holography, radiographic testing (RT), and penetrant testing (PT) would be a desirable extension of the Pressure Vessel Research Committee (PVRC) work on 201, 202, and 251J plates and of the PVRC European Plate Inspection Steering Committee (PISC) program to better understand the limitations of geometry, weldment design, and material differences on the detectability and precise sizing of flaws. A compilation of past and ongoing work in various countries would be a valuable first step. Possibly, the Electric Power Research Institute (EPRI) could provide the service.[6]

The meeting noted that there was a real problem arising from ensuring that NDE personnel can maintain their effective ability for sustained periods under ISI conditions and also that there was a need to minimize radiation exposure of highly qualified personnel. Any improvements in these aspects that could be obtained by mechanization of equipment, automation of the collection, recording, and interpretation of data would be a major gain and should be actively encouraged. Furthermore, all participating countries are invited to provide the technical secretariat with the data available concerning personnel radiation exposure, both total per plant and subdivided according to details of specific examinations and locations, incurred as a result of in-service inspection. The technical secretariat is encouraged to file and periodically publish such data.

Another related 5-year program was initiated in 1979 and entitled "*Integration of Nondestructive Evaluation Reliability and Fracture Mechanics*." Objectives of this program were as follows (Becker et al. 1981):

- Determine the reliability of ultrasonic ISI performed on commercial LWR primary piping systems.
- Using fracture mechanics analysis, determine the impact of (NDOE) unreliability on system safety and determine the level of inspection reliability required to assure a suitably low probability of piping failure.
- Evaluate the degree of reliability improvement that could be achieved using improved and advanced NDE techniques.
- Based on material, service, and NDE uncertainties, formulate recommended revisions to ASME Section XI and regulatory requirements needed to assure a suitably low probability of system failure.

The first phase of the project has the principal objectives of scoping and identifying the limits of NDE reliability, estimating current levels of reliability, and making preliminary fracture mechanics analysis. The second phase of the program had two primary objectives: (1) measurement of NDE reliability through round-robin inspections and (2) fracture mechanics analysis for the purpose of establishing the required level of NDE reliability. Phase three included evaluations of improved and advanced inspection techniques. Phase four established a unified set of recommendations for modifications of existing code and regulatory requirements necessary to achieve a suitably high level of inspection reliability.

[6]The plate designations 201, 202, and 251 J refer to specific test specimens or mock-ups used in nondestructive evaluation (NDE) reliability studies, particularly those related to the nuclear power industry. These plates are part of research programs designed to assess the effectiveness and reliability of NDE techniques for detecting flaws in critical components.

8.3 European-American Workshops on NDE Reliability

The "*European-American Workshop (EAW) on Reliability of NDE*" was a series of seven international workshops focused on the reliability of NDE methods, bringing together experts from academia and industry to discuss advancements and challenges in the field. The first EAW was held in Berlin, Germany, in 1997, with a focus on the then-perceived discrepancy[7] in how to guarantee the quality of NDE using EPRI's Performance Demonstration Initiative (PDI) and the European approach by the European Network for Inspection Qualification (ENIQ) (DGfZP 1997). The conference was organized jointly by the Federal Institute for Materials Research and Testing (BAM) (Germany) and the National Institute of Standards and Technology (NIST) (USA).

- PDI, based on practical performance demonstration under field conditions.
- ENIQ, based on definition of the physical parameters and their tolerances, e.g., physical reasoning.

The main achievement of the first workshop was the conceptual modular reliability model in terms of the reliability formula, which simply says that the total reliability of an NDE system is composed of the intrinsic capability (IC) (which stands for the physical principle behind the defect indication and its technical realization as an upper bound), application parameters (APs) (describing the realistic circumstances like UT coupling, limited access, noise, heat, radiation), and human factors (HFs) present in each application. As a continuation of EAW, the International Society for Optics and Photonics (SPIE) in 2023 organized the eighth International Workshop on Reliability of NDT/NDE (Kanzler and Meyendorf 2023).

8.4 NDE Qualification Methods

NDE reliability evolved from the work of the Pressure Vessel Research Council (PVRC), which was formed in 1945 by the Welding Research Council in the United States.[8] In 1965, the PVRC initiated a program to investigate the reliability of

[7]The main discrepancy was that ENIQ focused on a holistic, risk-informed, evidence-based approach emphasizing technical justification and demonstrated performance within a broader safety context for in-service inspection (ISI), while EPRI leaned towards more traditional, quantitative probability of detection (POD) studies, leading to differing philosophies on how to prove that an NDE system was reliable enough, with ENIQ stressing documented expert reasoning alongside trials, not just raw POD numbers.

[8]The Welding Research Council, a 501(c)(3) not-for-profit scientific research corporation, brings to bear in the solution of problems in welding and pressure vessel technology the talents of diverse science and engineering specialists. They exchange knowledge, share perspectives, and develop and guide R&D activities. As needed, the Council organizes and manages cooperative programs to advance codes and standards. It also incorporates the Pressure Vessel Research Council (PVRC)

ultrasonic NDE for nuclear reactor pressure vessels (RPVs) using thick-section test specimens containing various defects.[9] Marked differences in performance between teams using the same nominal procedure were revealed, and even lower detection performance was observed when the inspections were performed through RPV cladding. This led to a second PVRC test program commencing in 1974 to test inspection procedures based on the 1974 edition of ASME Section XI.

Three specimens from the second PVRC program were offered to the Committee on the Safety of Nuclear Installations (CSNI) Working Group of the OECD/NEA for a European NDE exercise performed under the guidance of the Steering Committee of the Program for Inspection of Steel Components (PISC).[10] It was envisioned that this exercise would augment the PVRC program and provide additional statistical data to quantify the probability of detection as a function of location and sizing of flaws using ultrasonic techniques. This resulted in three PISC exercises and a very extensive volume of technical information including databases of test results.

8.4.1 Program for Inspection of Steel Components (PISC)

The scope of the OECD/NEA and the European Commission (EC) joint PISC exercises was the detection and sizing of flaws and defects in large-scale assemblies of welded components typical of those used in light water reactors, largely by the application of ultrasonic testing (UT) techniques. The first phase of PISC studied the capability of a particular UT procedure based on the minimum requirements of ASME Section XI procedure, Boiler and Pressure Vessel Code (ASME) 1974/1977 procedure. The second phase consisted of further round-robin tests (RRTs)[11] including work on reactor pressure vessel nozzle assemblies and using several UT procedures and techniques. The third phase focused on NDE performance demonstration using real geometry assemblies containing realistic defects and service-induced defects. In addition to reactor pressure vessel specimens, the work

---

and Materials Properties Council (MPC). The WRC publishes Technical Bulletins, with a recent example being WRC 503 (2023), "Compendium of Temperature-Dependent Physical Properties for Pressure Vessels, Piping and Tankage Materials." The technical bulletins are available for purchase from https://accuristech.com/about-us/

[9] Details on the fabrication of NDE qualification specimens are found in references /9.9/ /9.28/. An example of commercially available test specimens is offered by Sonaspection International Ltd., https://sonaspection.com/

[10] OECD/NEA provides the secretariat of the PISC managing board, which consisted of representatives of 14 countries.

[11] The term "round-robin test" refers to a method used in quality control to evaluate the consistency and accuracy of test results across multiple laboratories, instruments, or teams of inspectors. It involves sending the same sample to different labs or instruments and comparing the results to identify any discrepancies or variations. For more details, see for example ASTM E691 *Standard Practice for Conducting an Interlaboratory Study to Determine the Precision of a Test Method*; https://store.astm.org/e0691-21.html

covered other evaluation boundaries within the primary reactor coolant pressure boundary. This included steam generator tubes using eddy current technology, as well as RRT on austenitic steel welds and dissimilar metal weld assemblies. Further details of the three PISC program phases are given below:

1. **PISC-I Program (1974–1978)**: The program investigated the capability of a modified PVRC version of the 1974 ASME XI procedure (subsequently referred to as the "PISC procedure") to detect, locate, and size defects (Nichols et al. 1987). A second aim was to compare the performance of the PISC procedure with alternative procedures already used in Europe. Of the 34 teams participating in this program, 28 used the PISC procedure. The program experienced problems associated with the presence of many unintended manufacturing flaws in the butt welds and base materials. There was also no independent oversight of the inspections and a lack of a well-defined reporting format for the results. The results of PISC I were published in 1978 and revealed serious inadequacies in the PISC procedure regarding the detection and sizing of significant through-wall defects, ca. 50% defect detection probability for a 25 mm through-wall crack-like defect. Alternative procedures tested in PISC I gave a much improved performance, ca. 80% defect detection probability for a 25 mm crack-like defect. A second-round test program (PISC II) was therefore initiated to test these alternative procedures more rigorously and perform supporting parametric studies.
2. **PISC II Program (1979–1985)**: The major results of PISC II were presented at the SMIRT 9 Seminar No. 3 in Varese, Italy, in 1985, and full results which had been published in the official green report of CEC/OECD were presented and discussed by the participants at Ispra, Italy, in October 1986 (PISC 1986). PISC-II round-robin test (RRT) results showed that it was possible to improve on the industrial NDE practice; consequently, some changes to the 1974/1977 editions of the ASME Section XI were proposed. The PISC II parametric studies on the effect of defect characteristics showed the importance of the type of defect (planar or volumetric), the crack tip aspect, the depth, the angular position, and the surface roughness. These results in fact mirrored the trends shown by the PISC II RRTs on welded assemblies. It was thus concluded that the blocks used for the PISC parametric studies, which contained cheap "realistic" artificial defects, were good enough for the performance evaluation of NDE techniques. In 1986, in light of the PISC results, the ASME Section XI was modified to reduce the recording level to 20% DAC and to require longitudinal wave probes.[12]
3. **PISC-III Program (1986–1994)**: This program conducted tests on austenitic casting and pipe welds and dissimilar metal welds. Unlike PISC-I and II, which used artificial specimens, PISC-III used real specimens containing service-induced defects. The program was closed in 1994 and comprised of eight actions

[12]Distance amplitude correction (DAC) provides a means of establishing a graphic "reference-level sensitivity" as a function of sweep distance on the A-scan display. An A-scan is a one-dimensional graph where ultrasonic echo amplitude is plotted as a function of time.

that produced a large database of test results for statistical evaluations (Nichols and Crutzen 1993; Borloo and Lemaitre 1994):

- Action 1, "Real Contaminated Structures," collected results from specific investigations on service-induced defects in materials and structures of the primary circuit of light water reactors. An objective of Action 1 was the study of real defects from an NDT response point of view with the determination of a correlation between real and artificial realistic defects to be used for test assemblies.
- Action 2, "Full Scale Vessel Tests," to validate results obtained by vessel inspection procedures in the PISC-II exercise but using more realistic inspection conditions. It consisted of a BWR full-scale reactor pressure vessel to which were added modular full-scale PWR components. Nine teams participated from 1988 to 1990 in Phase 1 concerning the sizing of selected defects in the full-scale vessel components. The results demonstrated sizing capability on simple planar flaws such as fatigue cracks but difficulties in classifying complex flaws with consequent unacceptable sizing errors. Eight organizations participated in Phase 2. This phase was aimed at the validation of ASME-type procedures by an international team using an ISI automatic scanner. Destructive examination was performed by trepanation of flawed areas of the vessel.
- Action No. 3, "Nozzles and Dissimilar Metal Welds," was initiated in 1986 and completed in 1991. It included RRT of four safe ends representing a difficult technical aspect of ISI. In this work, nozzle/safe-end assemblies from Japan, Italy, Spain, and the United States were inspected by teams in 13 countries. This work was considered essential for the design of performance demonstration procedures. It was concluded that the ability of NDE in detecting and sizing flaws was "*somewhat disappointing overall*" and did not reach the level of effectiveness that was usually called for in actual performance demonstration work.
- Action No. 4, "Austenitic Steel Testing," applied the PISC-II methodology to the primary circuit piping of light water reactors. RRTs for capability assessment and parametric studies were performed in the 1990–1993 timeframe. Twenty-five teams participated in one or more phases. The wrought-to-wrought pipe section RRT was the first one to be completed. Parametric study blocks were made for the study of the influence of the metallurgical structure and of the flaw characteristics. The inspection of the assemblies made of wrought steel provided good detection results of IGSCC, fatigue cracks, and notches.
- Action No. 5, "Steam Generator Tube Testing," involved RRTs of individual tubes of steam generators containing service-induced and artificial defects. Thirty teams participated in the RRTs. The RRTs were completed in 1993.
- Action No. 6, "Mathematical Modeling of NDE," had the objective of validating experimental mathematical models and performing parametric studies to assess the importance of defect characteristics.

- Action No. 7, "Human Reliability Studies," sought to evaluate the influence of human interpretation of inspection results, equipment malfunctions, and human interaction on the overall inspection procedure. The experimental studies involved six inspectors and were performed under laboratory conditions. They were completed by studies in an environmental laboratory to investigate the effect of working conditions and duration of shift/program. The assessment of the results showed that these types of test could identify, illustrate, and demonstrate various aspects of unreliability such as the importance of the calibration stage, personality of the operator, duration of his involvement in a particular series of trials, and ambient conditions.
- Action No. 8, "Support to Code and Standard (C&S) Organizations," to provide direct support through the PISC group of experts to C&S organizations. This support involved three aspects: (1) informing the relevant codes and standards technical committees of PISC results; (2) preparation of technical reports by PISC members related to codes and standards problems for the benefit of national technical groups, the International Standardization Organization (ISO) and the International Institute of Welding (IIW) which elaborate standards; and (3) supporting work done in view of performance demonstration.

8.4.2 NDE Qualification Requirements

Field experience with nondestructive evaluation (NDE), including instances where NDE failed to detect flaws, has led to the development of enhanced procedures and processes aimed at improving NDE reliability. NDE qualification addresses the effectiveness of NDE technologies, encompassing both hardware and software, as well as the competency of NDE personnel. It is essential for individuals performing NDE to demonstrate proficiency in equipment calibration and in detecting and characterizing flaws.

NDE technology includes manual and automated flaw detection techniques. In order to determine the ability of NDE to reliably detect subsurface or surface flaws, systems have been devised to ensure that NDE personnel have the proper training, have passed written and practical examinations, and have enough experience to properly perform NDE tasks using the applicable test method or technique. Personnel that have met all three of these requirements may then be certified, which is defined in several different ways under the various NDE systems. ASME XI Article IWA-2300 defines the NDE personnel qualification requirements.

There are two main types of NDE certification systems, "utility-based" and "central" certification systems:

- Utility-based certification system in which the nuclear power plant owner/operator is responsible for the administration of the training and the examinations of their own NDE personnel, as well as the documentation of the required training, examinations, and experience in accordance with an in-house standard or recommended practice. Most utility-based systems do allow the owner/operator to accept training and examination services provided by outside agencies provided that they are properly documented, and the employer has determined that the content of those services meet their own company requirements as described in the owner's/operator's NDE procedures.

– Upon proof of certification, the owner/operator may issue a certificate, which can be a formal certificate or in letter format and can authorize their personnel to perform NDE tasks. In all utility-based systems, the employer is responsible for authorizing their personnel to perform such work. Because utility-based certification is usually tailored to plant-specific needs, the resulting certifications expire when an employee leaves the company that issued the certification.

Central certification systems are systems in which the examinations are administered by an independent third-party certification body based on a central certification standard. To be eligible to sit for these examinations, prospective candidates must provide acceptable documentation of their training and experience to the certification body. Upon successful completion of the third-party examinations, the certification body will issue a certificate attesting to the fact that the named certificate holder has met the requirements and passed the examinations described in the third-party certification system. The employer can then choose to accept the third-party certificate(s) as proof of certification. As with employer-based systems, the employer has the ultimate responsibility to certify (authorize) the certificate holder to perform NDE tasks. Most certification programs have three levels of NDE qualification: Level I, Level II, and Level III. The American Society for Nondestructive Testing (ASNT) has developed standards and recommended practices for qualification and certification of NDE personnel:

– Certification standard: National or international documents describing the requirements for the qualification and certification of NDE personnel.
– Recommended practice: A formal document that provides nationally or internationally recognized guidelines and describes the qualification and certification process for NDE personnel. If mandated by governing codes, standards, specifications, or contract documents, these guidelines become requirements for the specified project.
– Certification system: The combination of the standard or recommended practice governing the certification requirements, third-party certification program (if applicable) or employer's written practice, and additional employer documents used in the administration of their certification program.
– Certification program: The documented employer's or certification body's procedures and processes based on a standard or recommended practice, which defines the requirements of that specific program.

The 2024 edition of the ASNT Recommended Practice No. SNT-TC-1A defines the NDE personnel qualification levels as follows (ASTM 2024):

- NDE Level I personnel should be qualified to properly perform specific calibrations, specific NDE, and specific evaluations for acceptance or rejection determinations according to written instructions and to record results. The NDE Level I should receive the necessary instruction and supervision from a certified NDE Level II or III individual.
- NDE Level II personnel should be qualified to set up and calibrate equipment and to interpret and evaluate results with respect to applicable codes, standards, and specifications. The NDE Level II should be thoroughly familiar with the scope and limitations of the methods for which he is qualified and should exercise assigned responsibility for on-the-job training and guidance of trainees and NDE Level I personnel. The NDE Level II should be able to organize and report the results of NDE tests.
- NDE Level III personnel should be capable of developing, qualifying, and approving procedures; establishing and approving techniques; interpreting codes, standards, specifications, and procedures; as well as designating the particular NDE methods, techniques, and procedures to be used. The NDE Level III should be responsible for the NDE operations for which he is qualified and assigned and should be capable of interpreting and evaluating results in terms of existing codes, standards, and specifications. NDE Level III should have sufficient practical background in applicable materials, fabrication, and product technology to establish techniques and to assist in establishing acceptance criteria when none are otherwise available. The NDE Level III should have general familiarity with other appropriate NDE methods, as demonstrated by an ASNT Level III basic examination or other means. NDE Level III, in the methods in which he is certified, should be capable of training and examining NDE Level I and II personnel for certification in those methods.
- NDE Instructor: The term "NDE Instructor" is used in the ANSI/ASNT Standard CP-189, "*Qualification and Certification of Nondestructive Testing Personnel*" (ASTM 2016), to describe an individual with the skills and knowledge to plan, organize, and present classroom, laboratory, demonstration, and/or on-the-job NDE instruction, training, and/or education programs.

8.4.2.1 ASME Rules

Appendix VIII of ASME XI requires qualification of the procedures, personnel, and equipment used to detect and size flaws in piping, bolting, and reactor pressure vessel. Each organization (e.g., owner or vendor) will be required to have a written program to ensure compliance with the requirements. The general requirements for qualification of nondestructive examination personnel contained in Section XI of the B&PV Code, IWA-2300, are amended by Section XI, Appendix VIII. Appendix VIII describes the additional requirements for performance demonstration of ultrasonic examination systems that integrate personnel, equipment, and procedures into a single entity. ASME Code Section XI, Appendix VIII, also includes supplements that contain specific instructions for the conduct of performance demonstrations, including test specimen requirements, conduct of performance demonstrations, and acceptance criteria. The Appendix VIII requirements are organized into the following five articles:

1. Article 1000: Scope of Applicability of Appendix VIII.
2. Article 2000: General Procedure and Personnel Examination System Requirements.
3. Article 3000: Qualification Requirements for Detection and Sizing Tests, Essential Variables, and Requalification.
4. Article 4000: Essential Variable Tolerance Requirements.
5. Article 5000: Qualification and Requalification Record Requirements.

Additional mandatory requirements are provided in supplements to Appendix VIII of ASME B&PV Code Section XI. These supplements address additional requirements associated with specific material, component welds, examination processes, and other issues. For piping welds, qualification supplements are included for wrought stainless, ferritic, dissimilar metal, and weld overlay component types. In general, NDE examination procedures, equipment, and personnel must qualify for both the detection and sizing acceptance criteria in Appendix VIII. Typically, length-sizing tests are conducted in conjunction with the detection test; however, a separate depth-sizing test is required. For piping welds, qualification is generally based on the following criteria:

- Demonstrate at least 80% detection with minimal false calls.
- The root mean square (RMS) error between measured and true flaw length must be less than or equal to 19.1 mm.
- The RMS error between measured and true flaw depth must be less than or equal to 3.2 mm.

The adoption of performance-based requirements for ultrasonic inspections changed the ASME code requirements and the role of the code in procedure development. In doing so, the ASME code shifted the responsibility of reliable procedure development to NDE engineers. The primary charter of ASME code committees also changed. Rather than specifying procedure requirements, the code committees must focus now on developing guidance for performance demonstration testing requirements that are realistic and cost-effective and will ensure that reliable ultrasonic examination procedures are implemented in the field. Finally, these performance demonstration criteria now permit quantitative descriptions of the reliability of the inspector population.

8.4.2.2 CORDEL

In 2007, the World Nuclear Association established the *Cooperation in Reactor Design Evaluation and Licensing Working Group* (CORDEL WG) with the aim of stimulating a dialogue between the nuclear industry (including reactor vendors, operators, and utilities) and regulators on the benefits and means of achieving a worldwide convergence of reactor safety standards for reactor designs. The Codes and Standards Task Force (CSTF) of the CORDEL WG was set up in 2010 to collaborate with the Standards Developing Organizations (SDOs) and the Multinational Design Evaluation Program (MDEP) Codes and Standards Working Group (CSWG) on the international convergence of mechanical codes and standards related to the design and quality of nuclear power plant components that are important to safety.

According to CORDEL, there are three stages that NDE personnel have to undergo before they are qualified to perform a test on Class 1 components (not required for Class 2 and 3 components) in most countries. These are certification, training, and qualification. The first stage, certification, is required by nuclear design codes and national regulators. For example, in the United States, this includes an American Society for Nondestructive Testing (ASNT) SNT-TC-1A-type certification under ASME Sect. III, while in France, it includes EN ISO 9712 in the RCC-M code for the qualification and certification of NDE personnel. The certification required by each of the design codes investigated is summarized in the MDEP-CSWG code comparison report (MDE 2013).[13]

An independent third party is required to validate the qualification of the certificate holder. This qualification is normally specific to the control of Class 1 components. The ENIQ framework and Section XI requirements are the two major sets of requirements for certification of NDE personnel for ISI of nuclear-grade components internationally. It is important to understand the differences between the personnel certification schemes described by EN ISO 9712:2012 (ISO 2012) and ASNT SNT-TC-1A and qualification of an NDE personnel inspection system in accordance with the ENIQ Methodology for Qualification (EMQ). ASME Appendix VIII describes the requirements for qualification of procedures, equipment, and personnel.

- The certification processes (ISO 9712 and SNT-TC-1A) explain how to qualify the workforce for a basic assessment of the basic knowledge of a method or technique. These certification processes provide confidence that an operator who is successfully certified in accordance with the process has broad knowledge of the principles, application, and capability of a particular NDE method for a range of situations.

[13]The Multinational Design Evaluation Programme (MDEP) was established in 2006 as a multinational initiative taken by national safety authorities to develop innovative approaches to leverage the resources and knowledge of the national regulatory authorities involved in the review of new reactor nuclear power designs. MDEP incorporates a broad range of activities including (a) enhancing multilateral cooperation within existing regulatory frameworks, with a view to harmonization of regulatory requirements and practices; (b) facilitating multinational convergence of international and intergovernmental safety standards and safety goals through interactions with other multinational organizations and industry; and (c) implementing the MDEP products to facilitate licensing of new reactors as well as oversight of their construction and commissioning activities: https://www.oecd-nea.org/mdep/

– The qualification of the personnel or a procedure according to ENIQ's qualification methodology or EPRI Appendix VIII programs define performance-based requirements. A qualification of an NDE system in accordance with the ENIQ or Appendix VIII methodologies is a more stringent process that demonstrates that the combination of inspection system and personnel with an acceptable knowledge, obtained via a training dossier, is capable of achieving very specific defect detection and sizing criteria in a particular situation.

8.4.2.3 European NDE Qualification Methodology

The European Network for Inspection and Qualification (ENIQ) is a network dealing with the reliability and effectiveness of nondestructive testing (NDT) for nuclear power plants. ENIQ is supported by European nuclear utilities and is working in the areas of qualification of NDT systems and risk-informed in-service inspection (RI-ISI). ENIQ Recommended Practices are itemized in Table 8.2. ENIQ guidance documents can be downloaded from the NUGENIA website, via http://nugenia.org/eniq-reports.

8.4.2.4 Performance Demonstration Initiative (PDI)

The Performance Demonstration Initiative (PDI) was formed by utilities in the United States in 1991. Its primary focus is to provide efficient and technically sound implementation of ASME Section XI Appendix VIII performance demonstration requirements. The EPRI NDE Center is the Performance Demonstration Administrator (PDA) for the program.[14] Using common protocols and sample sets, the need for site-specific performance demonstrations is minimized. Piping, bolting, and reactor pressure vessel (RPV) shell weld performance demonstration testing began in 1994.

With the publication of the revised 10 CFR 50.55a rule on September 22, 1999, the NRC required expedited implementation of all Appendix VIII requirements.[15] As a result, PDI initiated activities for Phase II of the program. Phase I of the PDI program addressed important problems that could have significant impacts on the cost and effectiveness of the performance demonstration approach. These included the nozzle inner-radius examination from the outside surface, nozzle-to-shell weld examinations, and dissimilar metal welds. The PDI program addresses the sample fabrication and operational requirements for a unified performance demonstration

[14]The PDA provides guidelines and other program-related information to licensed users of the PDI program through a dedicated website http://www.epriq.com

[15]§ 50.55a codes and standards (https://www.nrc.gov/reading-rm/doc-collections/cfr/part050/part050-0055a.html) list documents approved for incorporation by reference.

Table 8.2 ENIQ recommended practices and scope of NDE qualification

RP No.	Title	Year	Scope of NDE qualification
1	Influential/essential parameters (ENIQ Report No. 6)	1998	RP1 should assist those involved in inspection qualification in how to use and implement the concept of influential/essential parameters in agreement with the spirit of the European methodology. The main objectives of this recommended practice are to: • Explain the concept of influential/essential parameters. Indicate how the concept can be used in inspection qualification according to the European methodology. Give advice concerning the classification of influential parameters. Give examples of parameters which can be influential as a function of the specific inspection to be qualified for two cases: An ultrasonic inspection of welds and an eddy current inspection of steam generator tubes
2	Strategy and recommended contents for technical justifications (ENIQ Report No. 39)	2010	RP2 defines a list of recommended contents for writing technical justifications. It should assist those producing technical justifications to identify the material that might be included. It should also assist in producing technical justifications in a uniform format throughout Europe
3	Strategy document for technical justifications (now part of RP2)	1998	RP3 describes a strategy on how to use and implement the concept of technical justification, which is an important element of the ENIQ European methodology for qualification of NDT. The main objectives are to: • Explain the different purposes of technical justifications. • Indicate how the specific purpose or application of the technical justification may affect its contents. Give guidance on the relative weight to be given to test piece trials and technical justification considering a number of factors such as level, available evidence, and specific application
4	Recommended contents for the qualification dossier (ENIQ Report No 13)	2018	RP4 should assist those doing qualifications to identify the material which might be included in the qualification dossier, which is defined as an assembly of all the information relevant to the

Table 8.2 (continued)

RP No.	Title	Year	Scope of NDE qualification
			definition and execution of the qualification. It should also assist in producing qualification dossiers in a uniform format throughout Europe, an essential element in providing a general framework for a scheme of recognition of qualifications performed in the EU. Note that the concept of dossier is not that of a single document or report but rather that of a file in which key documents of the qualification are inserted
5	Guidelines for the Design of Test Pieces and Conduct of Test Piece Trials (ENIQ Report No. 42)	2011	RP5 provides guidelines for the design of test pieces and the conduct of test piece trials, once they have been decided (for example, as a result of the analysis done in the technical justification) that they are required. It refers especially to those test piece trials (open or blind) that are supervised by the qualification body
6	The use of modelling in inspection qualification (ENIQ Report No. 45)	2011	RP6 deals with the use of mathematical modeling in inspection qualification. Mathematical models have been developed by several organizations for various inspection situations and, where applicable, can provide valuable evidence on inspection capability for inclusion in a technical justification. Authors of technical justifications may therefore be considering the use of models. This RP provides advice on: • Types and range of mathematical models which are available. • How the models can be used to generate evidence for technical justification. Important considerations and constraints in using models
7	Recommended general requirements for a body operating qualification of non-destructive tests (ENIQ Report No. 58)	2018	RP7 provides guidance on the minimum criteria that a body operating qualification of nondestructive testing should follow if it is to be recognized as impartial, independent of operational pressures, competent, and reliable. Three types of qualification body are considered within the RP: • Type 1: A qualification body which is

Table 8.2 (continued)

RP No.	Title	Year	Scope of NDE qualification
			an independent third-party organization. • Type 2: A qualification body which is an independent part of the utility's organization set up on a permanent or long-term basis. • Type 3: An ad hoc qualification body set up for a specific qualification. The RP is mainly intended to provide guidance on the requirements for qualification bodies of types 1 and 2. An ad hoc qualification body, type 3, being more temporary and inspection specific in nature will generally be established in a less formal way than qualification bodies of types 1 and 2. However, many parts of the RP should still provide useful guidance for setting up an ad hoc qualification body. The RP should assist those who want to establish a qualification body and those who have to audit the competence of a qualification body. It should also assist in providing a general framework for a scheme of recognition of qualifications performed in the European Union
8	Qualification levels and approaches (ENIQ Report No. 59)	2018	RP8 provides guidance on the setting of qualification level and on determining the qualification approach based partly on this choice of level. The qualification level required reflects the assurance required that the inspection would attain its objectives in demonstrating structural integrity and may depend on, e.g., the safety significance of the component and the role of the inspection in assuring structural integrity. In practice, qualification can be done with varying degrees of complexity and cost. The way such work is carried out is defined in this document as the "qualification approach" and needs to consider both the structural integrity significance and the difficulty of each specific inspection. The qualification approach determines to what extent the various aspects of qualification, i.e., technical justification, open trials, and blind trials, are included in a particular case

Table 8.2 (continued)

RP No.	Title	Year	Scope of NDE qualification
9	Verification and validation of structural reliability models and associated software to be used in risk-informed in-service inspection programs (ENIQ Report No. 30)	2007	RP9 summarizes the verification and validation (V&V) requirements that a probabilistic fracture mechanics (PFM) model and associated software should satisfy in order to be suitable for such purpose. These requirements are mainly based on the work performed within the NURBIM project
10	Personnel qualification (ENIQ Report No. 38)	2010	RP10 assists those involved in the qualification of inspection personnel to meet the principal objective of personnel qualification, i.e., to ensure that those carrying out an inspection are appropriately trained, experienced, and examined to ensure that it is applied correctly and effectively. Detailed guidance on how to conduct personnel qualification, handle qualifications to be renewed, and define the role and responsibilities of the parties involved in the personnel qualification process is provided
11	Guidance on expert panels in risk-informed in-service inspection (ENIQ Report No. 53)	2017	RP11 is indirectly coupled to NDE qualification. It is concerned with the composition of an expert panel consisting of a decision-maker, RI-ISI project leader, a panel leader, and panel members. The decision-maker is responsible for approving the final decisions of the panel. The project leader has to ensure that all the relevant information is made available to the expert panels. The expert panel should be guided by a panel leader as in facilitator. This should be an individual with thorough knowledge and experience of the whole technology process of a nuclear power plant. The expert panel should contain several panel members with expertise in all the relevant technical areas related to the RI-ISI process. The panel members should be experts in their specialist field and, additionally, have a good general understanding of in-service inspection processes and RI-ISI methodology. The following expertise should be covered by the members of the expert panel: • In-service inspections and NDE.

Table 8.2 (continued)

RP No.	Title	Year	Scope of NDE qualification
			• Materials engineering and piping failure operating experience. • Structural integrity and piping design and stress analysis. • Probabilistic safety assessment. • Plant operations and maintenance practice.
12	Strategy and recommended contents for inspection procedures (ENIQ Report No. 63)	2019	RP12 has as its main objective to support licensees, qualification bodies, and inspection vendors to produce and assess an inspection procedure. RP12 includes what and how the contents in an inspection procedure should be described. An appendix gives examples on the content of different types of inspection procedures
13	Qualification of non-destructive testing systems that make use of machine learning (ENIQ Report No. 65)	2021	RP13 has as its main objective to support licensees, qualification bodies, and inspection vendors to produce and assess inspection procedures that use machine learning (ML) for automated data analysis. For the most part, the qualification of nondestructive testing (NDT) systems that utilize ML is similar to qualifying more traditional NDT systems. RP13 provides guidance on the specific considerations related to the use of ML in the qualification process

program operated by EPRI. The program contains established requirements for the demonstration test blocks as well as the rules and protocol for the program operation.

8.5 Models of Probability of Detection

A foundation of probabilistic piping reliability rests on the structural integrity risk triplet, regardless of methods, techniques, or computational algorithms. NDE reliability is defined as "*the probability of detecting a crack in a given size group under the inspection conditions and procedures specified.*" It is a line of defense in preventing a pipe flaw from growing into a through-wall pipe leak with an operational impact. How is this line of defense characterized probabilistically, and what are the technical justifications for a certain NDE reliability probability density function? Probability of detection (POD) is a measure of how well an inspection procedure can detect a specific type of flaw.[16] The exact meaning of the resulting POD must be clearly defined. For example, is the estimate a mean value or some lower bound considering uncertainty? POD is a conditional probability that depends on certain conditions, and those conditions must be defined and explained. The probability of detecting a flaw, whether connected to the inside diameter of the pipe or embedded in the base metal or weld heat-affected zone, can be expressed by the following equation:

$$POD = \frac{TP}{TP + FN} \tag{8.1}$$

where

$$TP = \text{"true positive" test data}$$

$$FN = \text{"false negative"} \ \textit{test data}$$

A "false-negative" result occurs when a flaw or defect is present in an item, but the NDE method fails to detect it, potentially leading to significant structural failures if the fault goes undetected. A true-positive result means the NDE method correctly identifies a flaw or defect that actually exists in the material or component being tested. POD cannot be measured directly, but can be estimated by the following:

- Statistical inference from experimental data such as data obtained from NDE qualification programs. MIL-HDBK-1823A, a United States Department of Defense handbook, provides guidance on probability of detection (POD) curve estimation and is a widely used resource in the aerospace industry (USDOD 2009). The handbook is frequently cited in POD research reports. Appendix G of the handbook describes statistical methods for analyzing NDE data and producing POD curves. The work that led to the handbook began in the mid-1970s.
- Expert elicitation to establish curves of POD as a function of flaw size, representing differing levels of NDE reliability in ultrasonic examination of welds.
- Calculations with theoretical models, e.g., an approach referred to as the European American Workshop (EAW) reliability formula or modular reliability model. It is a theoretical model of NDE reliability which accounts for the intrinsic capability (IC) of the devices that used to detect flaws, factors of application (AP), and human factors (HFs) involved in NDE. The objective of this approach for measuring NDE reliability is to provide a validated testing system.
- Statistical inference from operating experience data. How can this be done? It requires access to a very large volume of failure data, associated in-service inspection records, and results from root cause analyses, necessitating deep

[16] An excellent tutorial is available at https://www.nde-ed.org/NDEEngineering/POD/index.xhtml

database interrogation to extract qualitative and quantitative information about NDE reliability.

NDE qualification tests produce information on measured data, such as flaw depth $\widehat{a}$ (or "a-hat") versus actual flaw depth a. The results of a large number of qualification tests can be represented by so-called $\widehat{a}$ versus a plots. These plots can be converted to POD plots by using statistical models as referred to in the following book sections.

8.5.1 Definitions

The probability of crack (or flaw) detection is defined as the probability that a crack with certain specified physical characteristics (e.g., size) is found during an inspection and correctly classified as a crack. Probability of crack detection is measured directly by summing the number of times a crack of a given category is successfully detected and then dividing by the total number of times the crack is subject to inspection. The POD related to pipe flaw dimensions can be assessed through tests that evaluate the performance of various NDE inspection teams in identifying flaws within a specified specimen. ASME Section XI specifies the rules for conducting nondestructive in-service examinations.

Prior to 1989, the ultrasonic examination requirements were very prescriptive. Section XI attempted to identify and define all essential variables necessary to ensure that reliable UT practices would be used at nuclear power plants. In 1989, Section XI discontinued defining the essential variables and adopted a performance-based philosophy that relied on the qualification of UT procedures, equipment, and personnel used to detect and size flaws in piping. These new requirements were incorporated into Section XI in mandatory Appendix VIII, titled "Performance Demonstration for Ultrasonic Examination Systems." These requirements apply to all personnel who detect, record, or interpret indications or size flaws in welds or components. Each plant owner or operator must have a written program that ensures compliance with these code requirements. Appendix VIII is organized into the following five articles:

1. Article VIII-1000: Scope of Applicability. Performance demonstration requirements apply to personnel who detect, record, or interpret indications or size flaws in welds or components.
2. Article VIII-2000: General Examination System Requirements. The examination procedure shall contain a statement of scope that specifically defines the limits of procedure applicability (e.g., materials, thickness, diameter, product form).
3. Article VIII-3000: Qualification Requirements for Detection and Sizing Tests, Essential Variables, and Requalification.
4. Article VIII-4000: Essential Variable Tolerances.
5. Article VIII-5000: Record of Qualification. The organization's performance demonstration program shall specify the documentation that shall be maintained

as qualification records. Documentation shall include identification of personnel, NDE procedures, and equipment and specimens used during qualification and results of the performance demonstration.

Additional mandatory requirements are provided in supplements to Appendix VIII. These supplements address requirements associated with specific material, component welds, examination processes, and other issues. For piping welds, qualification supplements are included for wrought stainless, ferritic, dissimilar metal, and weld overlay component types. In general, NDE examination procedures, equipment, and personnel must qualify for both the detection and sizing acceptance criteria in Appendix VIII. For piping welds, qualification is generally based on the following criteria:

- Demonstrate at least 80% detection with minimal false calls.
- The root mean square (RMS) error between measured and true flaw length must be less than or equal to 19.1 mm.
- The RMS error between measured and true flaw depth must be less than or equal to 3.2 mm, with RMS error calculated as follows:

$$RMS = \sqrt{\left[\frac{\sum_{i=1}^{n}(m_i - t_i)^2}{n}\right]} \tag{8.2}$$

where

$$m_i = measured\ flaw\ size$$

$$t_i = true\ flaw\ size$$

$$n = number\ of\ flaws\ measured$$

The *Performance Demonstration Initiative* (PDI) was formed in 1991, (Ammirato et al 1993). It is designed to implement a unified industry approach to the performance demonstration requirements of ASME Section XI, Appendix VIII. Using common protocols and sample sets, the need for site-specific performance demonstrations is minimized. Performance demonstration testing began in 1994. With the publication of a revised 10 CFR 50.55a rule in 1999, the USNRC required an expedited implementation of all Appendix VIII requirements.[17]

[17] NRC-ADAMS Accession No. ML993280386.

8.5.2 Early Work

Based on round-robin tests performed in the early 1980s, Simonen and Woo (1984) reported on the detection capability of ultrasonic inspection of carbon steel and stainless steel piping. A "round-robin" ultrasonic inspection is a blind comparative study where multiple teams independently inspect the same piping specimens using ultrasonic testing (UT) to assess reliability and effectiveness of methods and procedures. The round-robin test used three types of piping, including 813 mm diameter thick-walled ferritic steel pipe with stainless steel cladding on the inside surface, 838 mm thick-walled centrifugally cast stainless pipe, and DN250 Schedule 80 wrought stainless steel pipe. The S-shaped normal cumulative distribution function was used to fit the round-robin test data:

$$POD = \Phi = \frac{1}{\sqrt{2\pi}} \int_{-}^{x} exp\{t^2/2\}dt \tag{8.3}$$

$$x = U + B \times ln\frac{a}{t} \tag{8.4}$$

where U and B are constants related to pipe materials, ultrasonic inspection procedures, and inspection access. Parameters a and t are crack depth and pipe wall thickness, respectively. Based on the round-robin tests, three POD curves were defined as follows (Figs. 8.2 and 8.3):

- "Poor team": This POD curve represents a lower bound on performance among round-robin teams. This team did follow the minimum ASME code requirements of 1981:

$$POD = \Phi[0.432 + 0.162\ln(a/t] - ferritic\ steel \tag{8.5}$$

$$POD = \Phi[0.24 + 1.485\ln(a/t] - stainless\ steel \tag{8.6}$$

- "Good team": This POD curve represented the better teams in the round-robin tests:

$$POD = \Phi[1.75 + 0.583\ln(a/t] - ferritic\ steel \tag{8.7}$$

$$POD = \Phi[1.526 + 0.533\ln(a/t] - stainless\ steel \tag{8.8}$$

- "Advanced team": This POD curve represented the performance that may be achieved with improved procedures and existing technology. It assumed a 0.9999 flaw detection probability for a through-wall flaw and a probability of detection of about 90% for a flaw with a depth of 10% of pipe wall thickness:

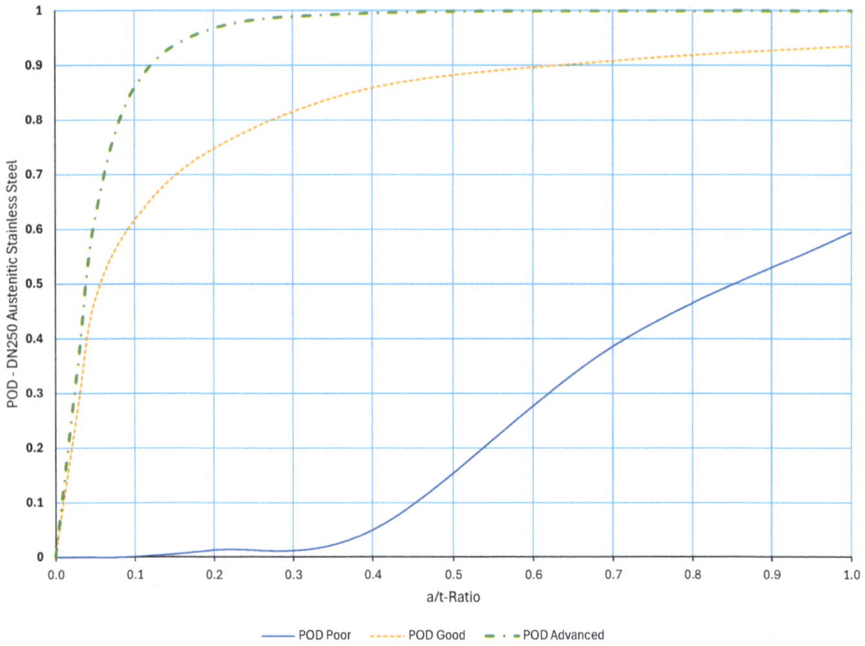

Fig. 8.2 Ferritic steel POD curves

$$POD = \Phi[3.63 + 1.106 \ln(a/t)] - ferritic \& stainles\ steel \tag{8.9}$$

8.5.3 *POD Estimates Based on Statistical Inference*

Probabilistic piping reliability uses real-life data on flaw occurrence rates and NDE/leak detection performance to assess piping reliability accurately. POD curves play a role by showing variability in projected NDE performance for specific analysis cases, accounting for uncertainty in different NDE techniques and processes. Typically, POD is plotted as a single variable (flaw size "α") in an S-shaped curve that increases with "α," denoted as $POD(\alpha)$. Typically, the $POD(\alpha)$ for a particular application is determined through the collection of a large amount of empirical data. Guidelines for performing an NDE reliability assessment study based on POD are described in MIL-HDBK-1823A.

Probabilities of detection (POD) curves are used to quantify the reliability of NDE systems. An NDE system refers to the complete set of equipment, procedures, and personnel used to inspect piping components for flaws without causing damage. It encompasses various techniques and technologies that allow for the detection and evaluation of crack-like flaws in base metal, welds, or weld heat-affected zones.

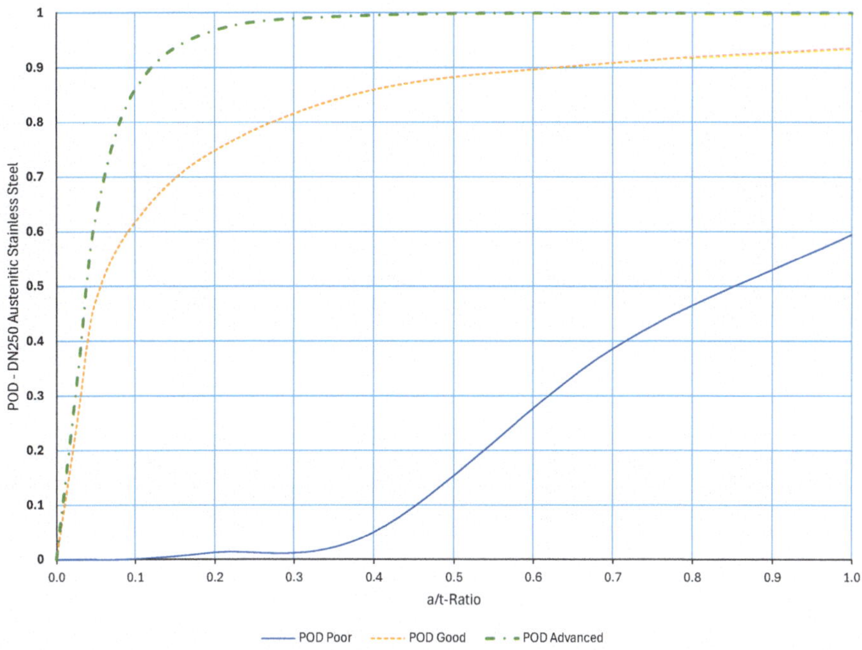

Fig. 8.3 Stainless steel POD curves

These POD curves are developed from performance demonstration data (Murgatroyd 1988; Doctor et al. 1989).

Beginning around 1980, several studies were conducted to determine the POD in an effort to quantify its reliability and effectiveness of nondestructive examination and to understand how it could be improved (Hedden et al. 2002). Such studies are typically performed by fabricating representative test blocks and having them inspected by several teams. The number of blocks and inspections required is dictated by the amount needed to achieve statistically relevant results. Many publications allude to the "90/95" criterion for reliability, which is the flaw size at which POD is 90% with a confidence of 95% or greater. A 95% confidence limit was assumed to provide the required degree of conservatism, but without specific justification for such a choice. In order to achieve 95% confidence, a significant number of inspections would be required.

Berens and Hovey of the University of Dayton Research Institute (UDRI) developed a probabilistic framework for POD (Berens and Hovey 1981, 1982; Berens 1989). In this framework, a model of POD assumes that there is a distribution of detection probabilities because of differences in inspector skills, use of manual or automated inspection technologies, etc. The two researchers identified seven different $POD(\alpha)$ functions to be applied when developing POD curves from NDE qualification data, i.e., the "hit/miss data." An acceptability test was performed, with acceptability criteria based on (1) goodness of fit, (2) normality of deviations

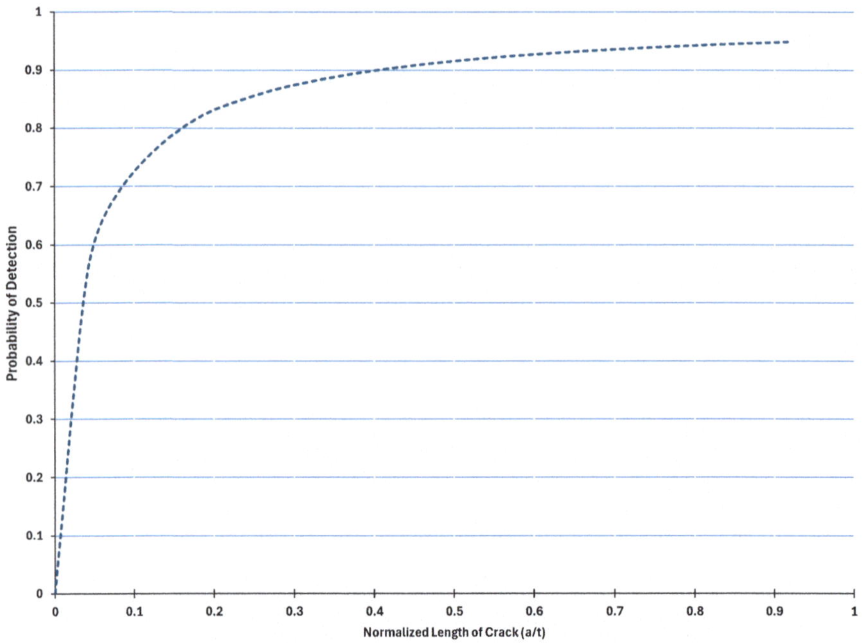

Fig. 8.4 Probability of detecting a flaw in DN50 XXS double-extra-strong piping

from fit, and (3) equality of variance of deviations from fit for all cracks. The latter criteria refer to the assumption that different hit/miss samples have similar levels of variability around their means. The test indicated that the log-logistic distribution was the most acceptable and with the $POD(\alpha)$ function written as

$$POD(a) = \frac{\exp(\beta_1 + \beta_2 lna)}{1 + \exp(\beta_1 + \beta_2 lna)} \tag{8.10}$$

The flaw size α is expressed as the length of crack. The two parameters, β_1 and β_2, are determined from curve fitting with empirical data. $POD(\alpha)$ is the conditional probability that a flaw of size "α" if present and within the qualified procedure scope will be detected. This probability is inferred from the experimental qualification results only, without consideration of field application variables including human performance being attempted. $POD(a)$ does not include the effect of flaws that have been detected but improperly dismissed by gross under-sizing. This log-logistic formulation was originally developed for nondestructive evaluation of airframe structures. Applying this formulation to infer information on the expected $POD\left(\frac{a}{t}\right)$ versus crack depth in a weld heat-affected zone, where a is crack depth and t is wall thickness, would yield results as indicated in Fig. 8.4.

Consider results from NDE qualification tests with y observations versus x "targets." Recognizing that there is some randomness in the tests, the underlying relationship between the dependent and independent values can be written as

$$y_i = \beta_1 + \beta_2 x_i + \varepsilon_i \tag{8.11}$$

The literature on statistical analysis of NDE qualification test data is extensive. Under the assumption that a model of $POD\left(\frac{a}{t}\right)$ is linear, the NDE test data can be analyzed using ordinary least square regression. ENIQ Report No. 41 describes and recommends best statistical practices for producing POD vs. flaw size curves from hit/miss data (Doctor et al. 1989). In many cases, NDE qualification data are analyzed using logistic regression on detection scores using the following mode (Gandossi and Annis 2010):

$$POD\left(\frac{a}{t}\right) = logistic\left(\beta_1 + \beta_2 \times \frac{a}{t}\right) \tag{8.12}$$

Two types of POD curves are used: (1) POD related to a specific NDE technique and (2) POD related to a specific NDE procedure. Due to variability in the inspection system and other system-related properties, an inspection system can only detect a certain range of defects. The POD curve for the inspection procedure shows the probability of detection when using certain inspection techniques combined with the inspection procedure. In general, the detection probability for the inspection procedure will be lower than for the inspection technique, since the POD curve for the inspection assumes an ideal defect resolution. Various types of human factor influences are incorporated in the inspection procedure-related POD curve.

Uncertainty in *POD*(a) is characterized by a confidence interval, which is a range of POD with the meaning that there is a certain specified probability, say 95%, that the true POD will be within that interval. An example statement can be that 95% confidence interval for POD is [0.88–0.95]. In this example, this means that there is a 5% probability that the true POD is less than 0.88 or greater than 0.95. Usually, only the lower value of the confidence interval is considered to be relevant in conservative assessment of inspection capability. Confidence intervals can also be constructed as one-sided, where the upper limit is 1.0. The confidence interval is strongly influenced by the size of the available datasets used to assess POD. Small numbers of data points typically result in wide confidence intervals even if the mean (or true) value of POD is high.

8.5.4 POD Curves Based on Expert Elicitation

Khaleel and Simonen (2000) developed a numerical approach to predict the probability that a fabrication flaw in a reactor pressure vessel will extend by fatigue crack growth mechanisms and become a through-wall flaw. A probabilistic fracture mechanics (PFM) model was formulated for the growth of surface and buried cracks and was implemented into the PFM computer code developed for reactor pressure vessel structural integrity analysis. Characteristic POD curves for the detection of fatigue cracks based on expert judgment were input into the PFM model. The approach was to establish three POD curves that represented widely differing levels of NDE performance in ultrasonic examination of welds.

An informal expert judgment elicitation was performed using staff at PNNL with specialized knowledge of NDE performance data from piping inspection round-robins and from recent industry efforts in the area of NDE performance demonstrations. It was recognized that a population of inspection teams operating under field conditions can exhibit a considerable range of POD performance, even though all such teams meet minimum code requirements for inspecting nuclear piping systems. The basic premise in estimating POD curves for the present calculations was that the better teams had passed the ASME Sect. XI, Appendix VIII, performance demonstration. ASME Sect. XI requires a minimum of 10 flaws and 20 blanks to be examined and requires a failing grade for either of the following: (1) more than two false calls or (2) the team fails to detect at least 8 of the 10 flaws. The informal expert judgment also considered information and trends observed in the PNNL round-robin studies on UT inspection of wrought stainless steel piping with wall thicknesses of 15.2–25.4 mm. The NDE experts at PNNL were asked to define POD curves by estimating parameters for the POD function used in the pc-PRAISE computer code:[18]

$$P_{ND} = 1 - POD = \varepsilon + \frac{1}{2}(1 - \epsilon) = erfc[\nu(A/_{A^*}) \tag{8.13}$$

where

$$A = area\ of\ the\ crack$$

$$A^* = area\ of\ crack\ for\ 50\% P_{ND}$$

$$A^* = \frac{\pi}{4} D_B a^*$$

$$D_B = beam\ diameter\ of\ the\ ultrasonic\ probe$$

$$a^* = crack\ depth\ for\ 50\% probability\ of\ P_{ND}$$

$$\varepsilon = smallest\ possible\ P_{ND}\ for\ very\ large\ cracks$$

$$\nu = slope\ of\ the\ P_{ND} curve$$

[18] pc-PRAISE is a probabilistic fracture mechanics computer code, adapted from the PRAISE code, specifically focusing on crack initiation and growth in various materials.

Table 8.3 POD curve parameters

NDE performance	A^* ($\%^{a}/_{t}$)	ε	ν
Marginal	40	0.1	1.6
Very good	15	0.02	1.6
Outstanding	5	0	1.6

erfc = complementary error function

A range of estimates for A^* (crack area for 50% POD) was provided by the NDE experts, and a value of 1.6 was assigned to the slope parameter which was held at constant value. The slope parameter was implicitly correlated to the detection threshold parameter A^*. The value of ε was assigned to ensure that smaller values of s also correspond to smaller values of A^*. Three POD curves were selected:

1. Marginal performance: A team with a POD performance that is described by this curve would have only a small chance of passing an Appendix VIII performance demonstration, and its performance would not be representative of current-day capabilities.
2. Very good performance: This curve corresponds to the very best teams. Such teams significantly exceed the minimum level of performance needed to pass an Appendix VIII performance demonstration test.
3. Outstanding performance: This curve corresponds to a team that has a level of performance significantly better than expected from most teams that have passed an Appendix VIII type of performance demonstration. Such a team would need to apply advanced technologies and/or improved procedures that might be available in the future.

The data used as input to pc-PRAISE computer code to describe the three POD curves are given in Table 8.3, with results shown in Fig. 8.5. The same POD curves with the POD function expressed in terms of the normalized flaw depth ($^{a}/_{t}$) were used for all pipe sizes. The parameter a* is the crack depth for 50% probability of non-detection. The slope parameter was assumed to be constant for the selected range of POD curves. The smallest possible P_{ND} for large cracks is the largest for the marginal POD (i.e., 10%) and was the smallest for the outstanding POD curve (i.e., 0.5%).

The curves were intended to bound the performance expected from inspection teams operating in the field. Results had shown that POD capability appeared to be the most limiting factor with regard to the overall capability of in-service inspections to reduce leak probabilities. The effects of flaw sizing errors are relatively small when calculations are based on inputs for flaw sizing capabilities and acceptance standards that are representative of current NDE capabilities and code requirements. However, the calculations show that gross errors in flaw sizing or significant departures from current flaw acceptance standards could negate the expected benefits of inspection methods that exhibit outstanding performance in the area of flaw detection.

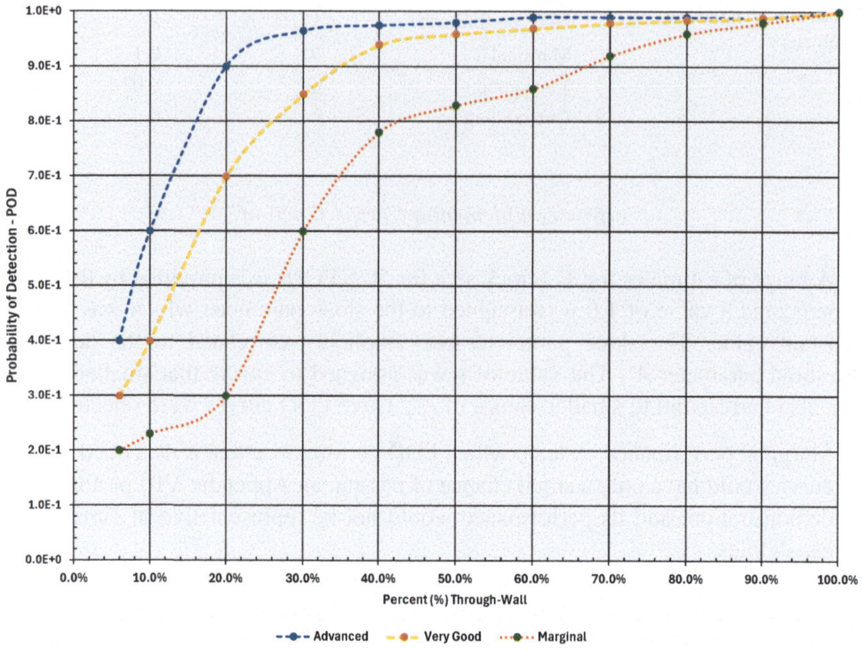

Fig. 8.5 POD curves for three levels of NDE qualification

8.5.5 *POD Curves Based on a Combination of Qualification and MTO Tests*

Swedish research has combined insights from NDE qualification tests (QTs) and MTO tests to produce POD curves for IGSCC and fatigue flaws (Enkvist et al. 2001; Jelenik et al. 2005). Man-technology-organization (MTO) test is a term used to analyze and promote safety within complex industrial systems by examining the dynamic interactions between three core elements: the human (man), the technology, and the organization. This type of test is performed to assess the effects of time pressure and noise on NDE. Based on the results from both types of tests using a set of NDE test blocks with IGSCC and fatigue flaws, the following POD model has been proposed:

$$POD = \Phi(0.6503 + \ln(\alpha) - 0.5285 \times QT_{IGSCC} - 0.5015 \times MTO) \quad (8.14)$$

The underlying logic is that if qualification tests only consider flaws attributed to IGSCC, then $QT_{IGSCC} = 1$ and $MTO = 0$ in the above equation. If only fatigue cracks are considered in MTO tests, then $QT = 0$ and $MTO = 1$. Finally, if only fatigue cracks are considered in qualification tests, then $QT_{IGSCC} = 0$ and $MTO = 0$. This indicates that flaws attributed to IGSCC are more difficult to detect than fatigue flaws.

8.5.6 POD Calculations Based on Experimental Models

Using the modular reliability model, the three different main influencing elements, i.e., intrinsic capability (IC), application parameters (APs), and human factors (HFs), are, in the first instance, investigated separately. The intrinsic capability stands for the pure physical-technological process of the signal detection caused by the waves or the rays from a material defect in the presence of noise (driven by the material and the devices). This intrinsic capability is the upper bound of possible reliability. Already, when measuring this intrinsic capability for thick-walled components, the original one-parameter POD must be extended to a multiparameter POD, where, in addition to the defect size, a number of additional physical parameters, such as the grain size distribution (or attenuation), defect depth, and angle or surface roughness, must be considered.

Real life cycle assessments must evaluate signal responses from actual defects. Industrial factors such as coupling conditions, limited accessibility, heat, and environmental vibrations affect reliability. If these conditions are controlled, the reduction in reliability can be quantified; otherwise, expect fluctuations in field reliability. Human factors, including inspectors' performance, workplace design, procedures, teamwork quality, system interactions, organization, and company relationships, also play a critical role. Comparing "ideal" and "real" inspections highlights how current practices and standards support reliable testing and identifies gaps. It is essential to demonstrate how various factors affecting NDE reliability impact the acceptance or rejection of safety-critical parts.

The modular approach allows for the integration of the European American Workshop Reliability Formula (Kanzler and Meyendorf (2023) from 1997), which defines total reliability (R) as a combination of intrinsic capability (IC), application parameters (APs), and human factors (HFs). In the first example, we determine the POD of radiographic crack testing using IC modeling and AP and HF statistical tests. The second example examines the impact of flaw shape on the POD of automated ultrasonic testing. The third example involves measuring the wall thickness of insulated tubes through manual and automated evaluation of radiographic films. Key parameters are optimized using modeling and experimental data, with the overall system assessed by statistical analysis of measurement deviations:

$$R = f(IC) - g(AP) - h(HF) \tag{8.15}$$

where

f(*IC*) is the intrinsic capability (IC) of the method or system, essentially establishing an upper bound for reliability based on the physics and inherent capabilities of the devices.

g(*AP*) accounts for factors related to the industrial application (AP), such as restricted access in the field or environmental conditions, which can negatively impact reliability.

h(*HF*) captures the influence of human factors (HFs) on reliability, recognizing that human actions and procedures can significantly impact the overall reliability of a system.

8.5.7 POD Calculations Based on Operating Experience Data

Reviews of the piping operating experience show that NDE failures occur when NDE personnel rationalize away detected flaws (i.e., the flaws are not seen as recordable), unqualified NDE procedures are used, a qualified NDE procedure is poorly implemented, a qualified NDE procedure misses a flaw, or NDE is not used where known degradation mechanisms are active. Based on the operating experience data:

- No inspection procedure can succeed if a detected indication is rationalized away.
- When ASME Appendix VIII qualified procedures are used appropriately, ultrasonic inspections have shown to have a high reliability.
- The reliability of inspection procedures declines sharply when the procedures deviate from the Appendix VIII qualified procedures.
- Appendix VIII qualified procedures can miss flaws.

Operating experience highlights ongoing NDE reliability challenges. New requirements focus on visually inaccessible materials like cast austenitic stainless steel using newly qualified NDE technology. While UT technology can detect minor discontinuities, interpreting these flaws remains difficult. The NDE qualification process defines clear limits, but assessing out-of-scope indications may overestimate smaller flaws.

From a perspective of assessing POD from service data, a starting point would be to organize the data by type of structural failure and plant operating mode at the time of detection. A structural failure can be self-revealing, i.e., through a detectable leakage and direct impact on plant operations, or hidden in the sense that the extent of material degradation can only be determined through nondestructive examination. Hidden failure means a failure that is not immediately evident to operations or maintenance personnel and has the potential for loss of piping pressure boundary integrity. A first step in database interrogation to extract quantitative POD insights is to define a query that produces failure by method of detection versus piping safety classification and plant operating mode at the time of detection (Fig. 8.6). In this figure, "part through-wall" is the same as "hidden."

Safety Class 1 and 2 piping is subjected to mandatory periodic in-service inspections that involve visual examinations, nondestructive examination, and pressure testing as prescribed by codes and standards. For Safety Class 3 piping, visual examinations and leak tests are performed periodically and as prescribed by codes and standards. The balance-of-plant (BOP) category includes the piping systems that are required for energy conversion, e.g., steam, condensate, and feedwater piping.

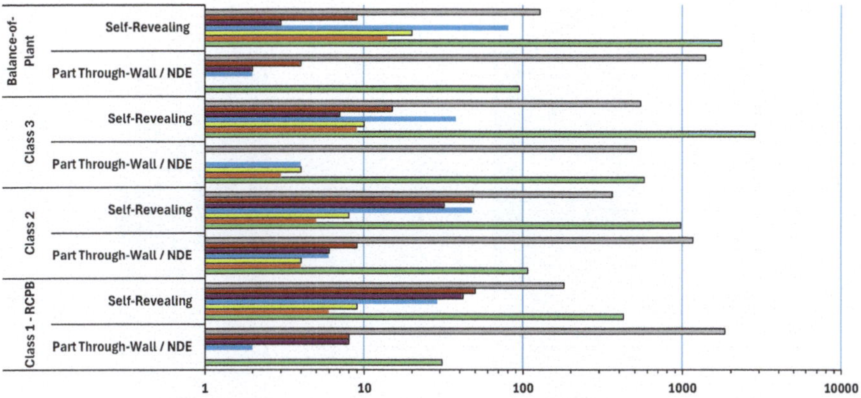

	Class 1 - RCPB		Class 2		Class 3		Balance-of-Plant	
	Part Through-Wall / NDE	Self-Revealing	Part Through-Wall / NDE	Self-Revealing	Part Through-Wall / NDE	Self-Revealing	Part Through-Wall / NDE	Self-Revealing
CSD - Cold Shutdown	1828	181	1163	361	509	545	1390	128
HSD - Hot Shutdown	8	50	9	49		15	4	9
HSB - Hot Standby	8	42	6	32		7	2	3
Starting Up	2	29	6	48	4	38	2	81
Shutting Down	1	9	4	8	4	10	1	20
Low Power Operation		6	4	5	3	9		14
Power Operation	31	423	107	971	574	2808	95	1745

No. Pipe Failures

Fig. 8.6 Basis for statistically inferred probability of detection

The piping in the Safety Class 3 and BOP categories is susceptible to corrosion and flow-assisted degradation.

"Owner-defined" in-service inspection programs are commonplace, which means that NDE technology is used to detect "hidden failures" before progressing to leakage. An owner-defined in-service inspection program for nuclear power plants is a mandatory program, established and implemented by the plant owner, to examine structures, systems, and components for potential deterioration, ensuring adequate safety measures and promoting maintenance requirements during operation and outages.

Many analytical steps are needed to go from Fig. 8.6 to empirical POD estimates. Focusing on Class 1 and Class 2 piping, the NDE requirements, past and current, are well defined. The part through-wall flaws have all been discovered through NDE. A database query for Class 1 and Class 2 piping greater than DN100 returns a population of "hidden" failures that were successfully found through ultrasonic testing. Next, a review of the in-service inspection histories reveals instances where NDE failed to detect a subsurface flaw, ultimately resulting in through-wall leakage. Follow-up analyses could include further refinements of the data analysis to single out systems, weld categories, materials, pipe wall thickness, etc. Based on a preliminary analysis of this operating experience returns, the results are illustrated in Fig. 8.7.

What about POD estimation for Safety Class 3 and balance-of-plant piping? Such analyses start with reviews of the owner-defined and mandatory in-service

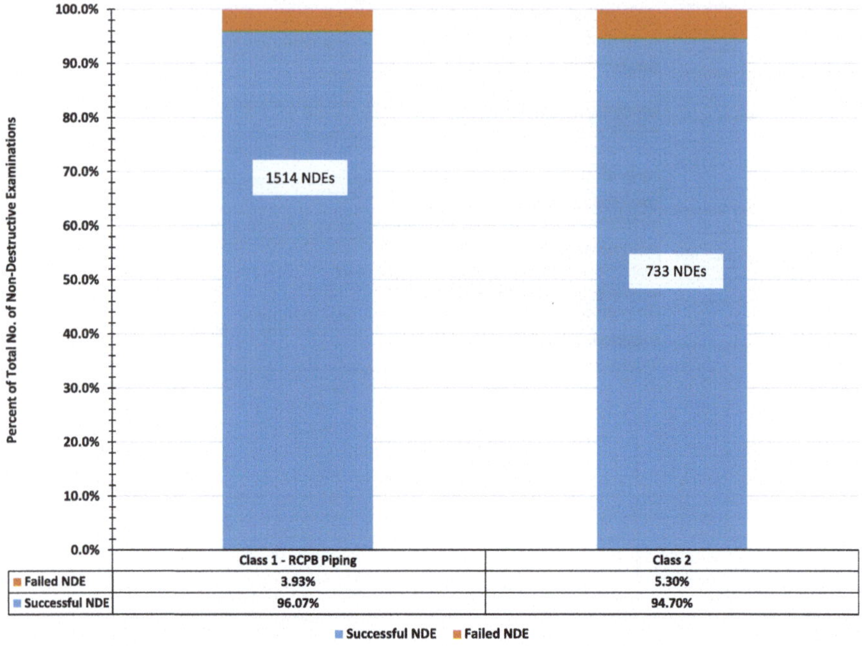

Fig. 8.7 Empirical POD estimates for ASME III Class 1 and 2 piping

inspection requirements. An important Class 3 system is the service water system for which piping integrity management involves the use of internal flooding PSA insights and plant-specific operating experience to define inspection location sampling strategies. The most common volumetric inspection methods are built on ultrasonic testing and radiography. For buried and belowground service water piping regulatory requirements and industry guidelines, mandate specific programs for piping integrity management.

Generic Letter 89-08 in the United States asked all licensees to create long-term programs to detect FAC in high-energy carbon steel piping systems. Each utility develops its program based on specific plant conditions, industry experience, engineering judgment, NDE techniques, and computer analysis. A long-term program should include a clear definition, documented results, analysis of susceptible systems, inspection and repair or replacement of damaged components, trending of inspection data, and ongoing analysis based on findings.

As stated in NSAC-202L (EPRI 2013a, b), "… *Accurate inspections are the foundation of an effective FAC program.*" NSAC-202 is an EPRI report with recommendations for an effective flow-accelerated corrosion program. Wall thickness measurements will establish the extent of wear in a given component, provide data to help evaluate FAC trends, and provide data to refine the predictive model. Thorough inspections are key to fulfilling these needs. A piping failure database includes information on ISI histories and ISI programmatic errors or weaknesses applicable to specific events. The objective of any FAC program is to predict wear rates and prevent through-wall flaws or break-before-leak vents from occurring. Despite the dedicated efforts to comprehensively manage FAC, pipe failures do occur due to less-than-adequate implementation of inspection plans. Included in the database are numerous examples of programmatic deficiencies. Listed below are examples of FAC failures that were attributed to programmatic deficiencies:

- Through-wall leak develops prior to next scheduled inspection. This could be indicative of higher-than-predicted wear rates. It also entails instances where pipe routing has been modified without corresponding change to FAC monitoring program. An example of the latter type of deficiency is the failed extraction steam drain line in a Japanese BWR plant in 2010. The turbine rotors had been replaced in 2009 to achieve higher thermal efficiency. The design change involved rerouting the drain piping, which in turn resulted in an increased drain flow and higher-than-anticipated wear rate.
- Failure occurs in a location identified by predictive program as susceptible. However, this information may not have been transferred to the inspection plan. An example of this type of deficiency is the ruptured first-stage extraction steam line at a PWR plant in the United States. The event occurred during the first quarter of 2009. The failed line was listed in the FAC program documentation as susceptible but was not monitored at the failed location.
- Failure occurs due to FAC software model input errors. During the third quarter of 2006, a DN150 moisture separator reheater drain line failed at a PWR plant in the United States. The FAC monitoring program included inaccurate wear prediction due to a computer software modeling error that was due to "failure of the organization to establish a proper level of verification." It was also determined that the preparers, reviewers, and approvers of the FAC procedure did not recognize that a formal second level of verification was needed to ensure a quality software model to ensure the safety of plant personnel and plant power generation reliability.
- Less-than-adequate FAC program implementation: In 1999, a PWR plant experienced a rupture of a DN150 MSR drain line. The failure occurred in a straight section of pipe immediately downstream a 45-degree elbow. A causal factor associated with this event is the manner in which the FAC inspection program provided guidance for inspecting downstream piping when performing fitting inspections. The program did not consider downstream piping inspections as mandatory, although it is a common practice. However, this downstream piping was not inspected when the upstream 45-degree elbow was inspected during previous refueling outages. Had this piping been previously inspected, it is probable that an abnormal wear rate would have been detected. A second causal factor associated with this event was the perception that the upstream 45-degree elbow would be the most susceptible component for failure due to FAC. This causal factor contributed to the decisions made during previous refueling outages to not inspect the downstream piping associated with this elbow.

8.6 Human Reliability Considerations in NDE

Results from the PISC-II Program suggested that differences existed among manual inspectors in terms of their skills, knowledge, and working practices, which could exert a significant influence on the reliability of an inspection. Therefore, a program of work on human reliability studies was initiated in the PISC-III Program as Action 7, with the objectives of studying and identifying causes of variability in inspection activities and identifying some of the factors influencing the reliability of inspection in industrial conditions (Murgatroyd and Crutzen 1994). It was foreseen that the information from Action 7 would aid in the development of methods for reducing the incidence of human error in inspection activities.

The underlying philosophy of the studies was to observe professionally qualified manual inspectors at work in typical inspection tasks and to note and record the standard of performance achieved and any errors made. To this end, special facilities and techniques were developed which would enable comprehensive monitoring and recording of an inspector's activities to be performed in as unobtrusive a manner as possible. The two main test facilities were the "reliability studio" and the "transportable environmental laboratory" (TEL). The former was used for tests in an environment representative of laboratory conditions, and the latter for tests in simulated industrial conditions. The inspection conditions and environment in these facilities could be controlled according to the program specification. Broadly, this specified normal 8-h day working for a period of 5 days in the laboratory and 12-h shift working over 11 days in the TEL with ambient temperatures of up to 27 ° C, humidity in the range of 40–50%, and a background noise level of 80 dB in the working area.

The monitoring equipment included video cameras and recorders, a computerized system for monitoring physiological condition, and a computer-based inspection simulator (the simulator), which recorded the details of each inspection, including the coverage achieved during scanning and the effectiveness of ultrasonic coupling. Other equipment included ferritic steel test plates and items for testing calibration repeatability.

Special tests were devised for the studies, and these included evaluation of calibration repeatability, inspections on the steel plates and simulator, and assessment of the accuracy of plotting and reporting manual inspection data. The main parameter measured in the inspections was flaw detection frequency (FDF), and this was used to assess and compare the capability of the inspectors and to investigate how performances varied between the laboratory-type conditions and the simulated industrial environment. However, the precise value of this parameter has no significance outside the conditions applicable to this test program.

A psychological monitoring program was applied throughout the studies to gather information on the inspectors' attitude to the work, and their physical and mental condition, with the aim of correlating the information with their inspection performance over the test period. The program specified that the inspectors taking part in the studies should be professionally qualified to ASNT Level 2 or their national equivalent and have recent experience of working in nuclear inspection tasks. Six were hired from major inspection organizations in Europe, and they and their companies were guaranteed complete anonymity in the reporting of the results. The test program was managed by invigilators who also operated the monitoring equipment and controlled the documentation.

LTA-RIM means that a preexisting flaw has not been identified before exceeding acceptance criteria. This would be equivalent to a missed opportunity to prevent through-wall leakage. When RIM fails to detect flaws, one or more of the following factors are often present (Aldrin et al. 2006; Bertović 2016; Sanquist et al. 2018):

- Accepting a rejectable flaw indication for continued operation: This could be due to misinterpretation of NDE results.
- Misinterpreting flaw signals by qualified personnel due to complexity of the examination (for example, when several embedded flaws are stacked together and appear to be connected).
- Rationalizing away detected defects.
- Using an improperly qualified NDE technique.
- Poorly implemented owner-defined inspection program.
- Missing a flaw with a qualified procedure.
- Allowing human factors to influence results.

  - Time pressure: Outage schedule constraints and last-minute changes increase time pressure. Concerns about radiation dose limits increase time pressure. Analysis of indications may require more time than planned.
  - Physical environment and access limitations: Nondestructive examination (NDE) access limitations refer to situations where the geometry or location of a component restricts the ability to apply certain NDE techniques, meaning certain areas within the component may not be fully inspected due to limited access points or complex shapes, often requiring specialized equipment or techniques to reach hard-to-reach areas. Poor accessibility may require NDE personnel to arrange themselves in uncomfortable positions for extended periods of time and limit the ability to see the component being inspected or the equipment being used. High levels of heat and noise can increase stress, affect the ability to see or hear equipment signals, and make communication with team members more difficult. Poor lighting can be particularly challenging for visual examinations, impacting the ability to distinguish cracks from surface scratches.

8.7 How to Use POD in Probabilistic Piping Reliability Analysis

The frequency of failure of piping component type i with an equivalent break size x (size of hole in the pipe) or through-wall mass flow rate, subject to epistemic uncertainty, is derived from

$$\rho_{ixt} = \lambda_i P_i(R_x|F) \times I_{ix} \tag{8.16}$$

where

I_{ix} is the integrity management factor. It accounts for leak detection and in-service inspection.

Markov analysis can be used to assess the integrity management factor for different piping systems. Two Markov model state transition rates of interest are the inspection-and-repair transition ω and the leak-detection-and-repair transition μ. For the former transition rate, the following model can be used:

$$\omega = \frac{P_I \times POD}{(T_{FI} + T_R)} \tag{8.17}$$

where

$P_I =$ Probability that a piping element with a flaw will be inspected. When inspection locations are inspected at random, this parameter is related to the fraction of the pipe segment that is inspected during each interval and the ability of the inspection strategy to pinpoint the location of possible flaws in the pipe. When locations for the inspection are fixed, this term is either 0 or 1, depending on whether the pipe segment is inspected. This parameter corresponds to the fraction of the locations in the population that are subjected to NDE.

$POD =$ The conditional probability given that the location being inspected has a flaw that meets the criteria for repair according to codes and standards.

$T_{FI} =$ Mean time between inspections for flaws, i.e., the inspection interval.

$T_R =$ Mean time to repair when detected. An underlying assumption is that any significant flaw that is detected will be repaired. Because T_{FI}, which is measured in years, is normally much greater than T_R, which can multiple to hours or to months depending on the complexity of the repair. Markov model results tend not to be sensitive to assumptions regarding the repair time.

Similarly, the repair rate for leaks can be estimated using the following model:

$$\mu = \frac{P_{LD}}{(T_{LD} + T_R)} \tag{8.18}$$

where

$P_{LD} =$ Probability that a leak will be detected.

$T_{LD} =$ Mean time between inspections for leaks.

$T_R =$ As defined above, but for full-power applications. This time should be the minimum of the actual repair time, and the time associated with any technical specification limiting condition for operation (LCO) if the leak rate exceeds technical specification requirements.

Opportunities for leak detection depend on the system in which the leak occurs as well as the specific location and size of the leak. For example, in the reactor coolant system (RCS) of an LWR, minor leaks would create an immediate alarm in the control room. In these cases, the time to inspect and repair is limited by technical specifications on RCS leakage and the time to shut down the plant and begin the repair process. Other leaks might not cause an immediate alarm but would be subject to possible detection during operator walkdown visual inspections every shift or any other opportunity for leak detection. Some leaks might be detected only upon periodic leak testing, which might occur less often as required to meet ASME rules for different classes of pipe according to ASME Section XI and other requirements for leak testing.

A Markov model offers the capability to account for the leak-before-break principle. The extent to which this principle contributes to reducing the probability of a rupture is a function only of the relative values of the Markov model transition rates. The integrity management factor is derived from the following equation:

$$I_{ix} = \frac{h_{ix}(t)}{h_{iBASE}(t)} \tag{8.19}$$

where

I_{ix} = Integrity management factor for component i for RIM strategy x. This factor can be expressed as a product of separate integrity management and piping system factors. RIM strategy (Table 8.4) refers to how often leak tests and inspections and NDE are performed on the piping system component and how effective they are (Fleming et al. 2023).

$h_{ix}(t)$ = Hazard rate (time-dependent rupture mode frequency) for component i and RIM strategy x. this is determined from the solution of the ordinary differential equations that describe the Markov model as explained more fully below. The piping system is set at the average piping system in the plant population responsible for the failure rate and conditional rupture frequency estimates.

$h_{iBASE}(t)$ = Hazard rate for component i and RIM strategy corresponding to an "average" component in the service data.

8.8 Operating Experience Insights

Instances of less-than-adequate RIM have involved deficiencies in an inspection process to detect preexisting flaws before exceeding acceptance criteria. An operating experience database records failure information in a tiered way. First, the most fundamental information about an event is recorded. This includes a free-format

Table 8.4 RIM strategy definitions and their relationships to the Markov model of piping reliability

RIM case	Description	Analysis strategy
RIM-0	Base case	Maintain current procedures and practice for leak inspections and nondestructive examinations (NDE).
RIM-1	Enhanced leak testing	A written procedure for quarterly leak inspection that identifies all pipe sections for which this strategy is to be implemented. Implementation of the procedures should be capable of performing visual examination of applicable pipe sections that are effective in identifying the presence of leaks. At least 50% of the applicable pipe sections should be accessible for visual exams to verify leak tight integrity over the sections of piping system that the integrity management factor is being applied. If there is water on the floor, source of leakage should be determined. If the applicable pipe section is determined to be leaking, the leak should be repaired or replaced. Assume quarterly leak testing at a 50% probability of finding and repairing a detectable leak and no NDE.
RIM-2	Robust leak testing	Develop written procedures for performance of quarterly leak inspection that identifies all pipe sections that are applicable to the RIM-2 strategy. Implementation of the procedures should be capable of performing white glove examination, or equivalent, of applicable pipe sections that are effective in identifying the presence of leaks. At least 90% of the applicable pipe sections should be accessible for the visual exams to verify leak-tight integrity over the sections of piping system that the integrity management factor is being applied[1]. If there is water on the floor, source of leakage should be determined. If the applicable pipe section is determined to be leaking, the leak shall be addressed by appropriate repair and replacement. Assume quarterly leak testing at a 90% probability of finding and repairing a detectable leak and no NDE.
RIM-3	Robust leak testing and enhanced NDE	Same leak testing as in RIM-2 with additional 10-year NDE that has the capability to detect a flaw in the applicable pipe sections that exceeds the NDE criteria for repair with a 50% POD of finding a detectable flaw. If the applicable pipe section is determined to be leaking in the quarterly leak inspection or if a flaw exceeding the repair criteria is found in the 10-year NDE, the leak or flaw shall be addressed by appropriate repair and replacement. Assume quarterly leak testing at a 90% probability of finding and repairing a detectable leak combined with a 10-year NDE with a 50% POD of finding and repairing a detectable flaw.
RIM-4	Robust leak testing and robust NDE	Same leak testing as in RIM-3 but with a 10-year NDE that has the capability to detect a flaw in the applicable pipe sections that exceeds the NDE criteria for repair with a 90% POD of finding a detectable flaw. If the applicable pipe section is determined to be leaking in the quarterly leak

Table 8.4 (continued)

RIM case	Description	Analysis strategy
		inspection or if a flaw exceeding the repair criteria is found in the 10-year NDE, the leak or flaw shall be addressed by appropriate repair and replacement. The locations covered by this strategy should be incorporated into the plant's existing augmented ISI programs. Assume quarterly leak testing at a 90% probability of finding and repairing a detectable leak combined with a 10-year NDE with a 90% probability of finding and repairing a detectable flaw.

Before applying RIM-3 to ASME Safety Class 3 piping, it is recommended to do the following: Determine if there are any applicable license renewal or subsequent license renewal aging management commitments for the pipe section in question. Where applicable, review the augmented ISI program(s) for the pipe section(s) and the inspection history, (a) inspection interval, (b) NDE technology, and (c) results—If through-wall leaks detected post-NDE implementation, what were the steps taken to make NDE program improvements?
If "a" or "b" does not apply, assess the feasibility of performing NDE

event narrative that describes the sequence of events, including plant response, consequence, in-plant location of the failed piping component, dimensional data, and component type. This is followed by recording the known ISI history, including the date when the failed component was last inspected, method of NDE qualification if a qualified method had been used, and any NDE performance deficiencies or failures. This section includes summaries about selected RIM failures, including failure to select the proper NDE locations and examination volumes, inadequate implementation of NDE procedures, and failure to calibrate and use an NDE tool.

8.8.1 High-Level Overview of RIM Failures

The ASME Section XI, "Rules for Inservice Inspection of Nuclear Power Plant Components," and equivalent codes provide requirements for examination, testing, and inspection of components and systems and repair/replacement activities in a nuclear power plant. The ultrasonic inspection techniques that were in use in the 1970s had been adapted from the manufacturing industry and modified using lessons learned from fatigue cracking service experience. Applicable ultrasonic examination rules at the time were prescriptive and not developed for specific material degradation mechanisms. The effectiveness of nondestructive examinations was not quantified via testing. The early field experience showed that improvements in inspection requirements were needed. Therefore, the USNRC sponsored R&D to assess NDE reliability. The results indicated that a performance-based testing approach would be the most effective means for achieving the needed improvements in reliability.

LTA-RIM means that a preexisting flaw has not been identified before exceeding acceptance criteria. This would be equivalent to a missed opportunity to prevent

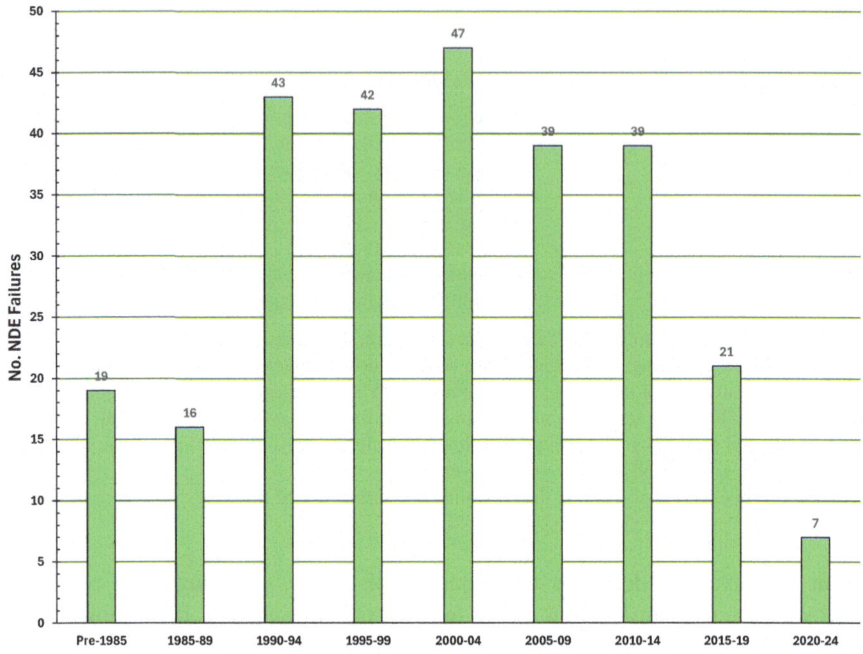

Fig. 8.8 Operating experience associated with less-than-adequate NDE

through-wall leakage. When RIM fails to detect flaws, one or more of the following factors are often present:

- Accepting a rejectable flaw indication due to misinterpretation of NDE results.
- Misinterpreting flaw signals by qualified personnel due to complexity of the examination.
- Rationalizing away detected defects.
- Using an improperly qualified NDE technique.
- Missing a flaw with a qualified procedure.
- Allowing human factors to influence results.

A series of charts, Figs. 8.8, 8.9, 8.10, 8.11, 8.12, present operating experience insights regarding LTA-RIM events. The total number of significant LTA-RIM occurrences has fluctuated but currently demonstrates a declining trend. Notably, this reduction coincides with a substantial increase in the overall frequency of inspections during the same timeframe. In Fig. 8.9, the LTA-RIM data is organized by material degradation mechanism and mode of failure. NDE process failures by plant type and piping safety class are summarized in Fig. 8.10, while Fig. 8.11 is a summary of pipe failures that have been discovered through application of pressure test processes. The through-wall flaws in Code Class 1 components that have been detected through a pressure test are manifestations of latent failures that were not previously discovered through liquid penetrant testing (More 2016) or UT

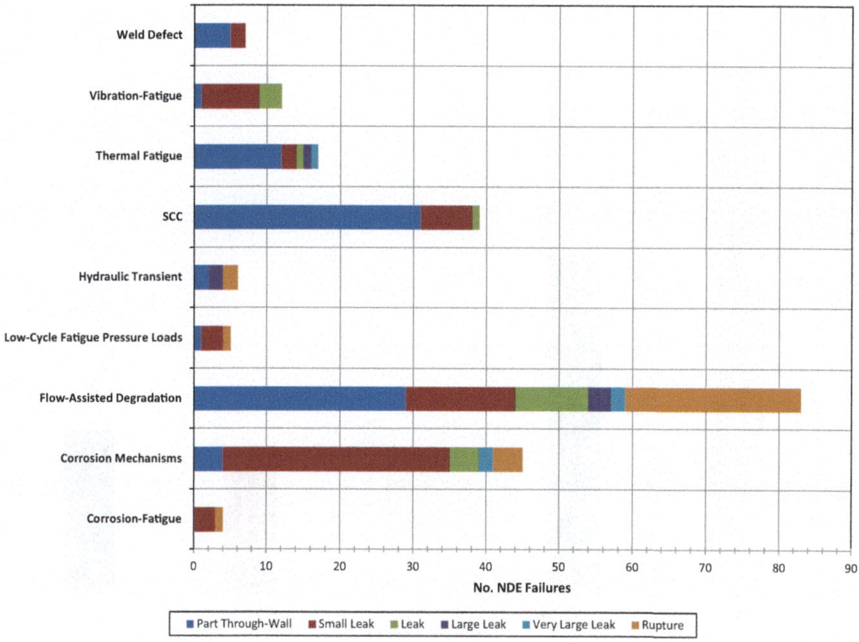

Fig. 8.9 LTA-RIM operating experience by degradation mechanisms and failure mode

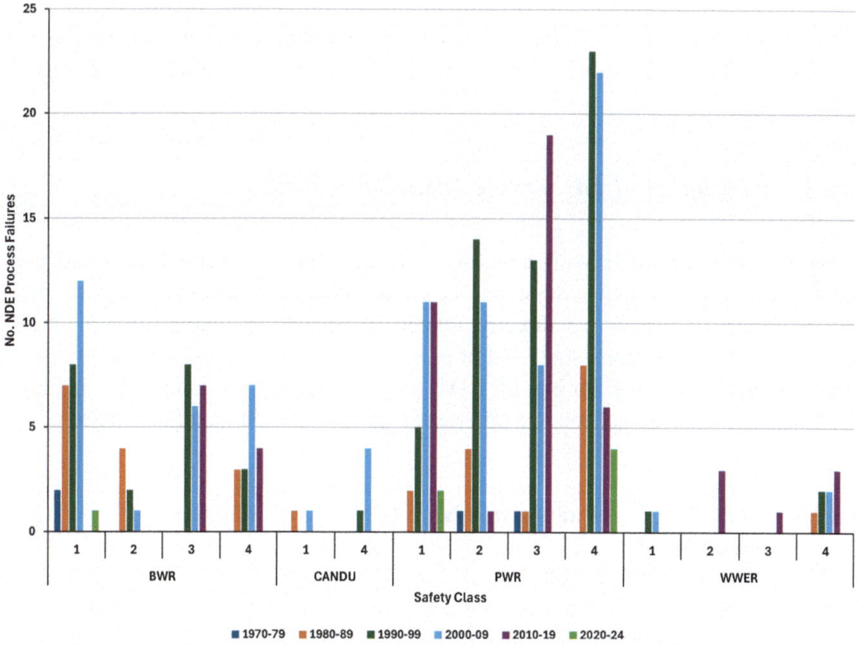

Fig. 8.10 NDE process failures by plant type and piping safety class

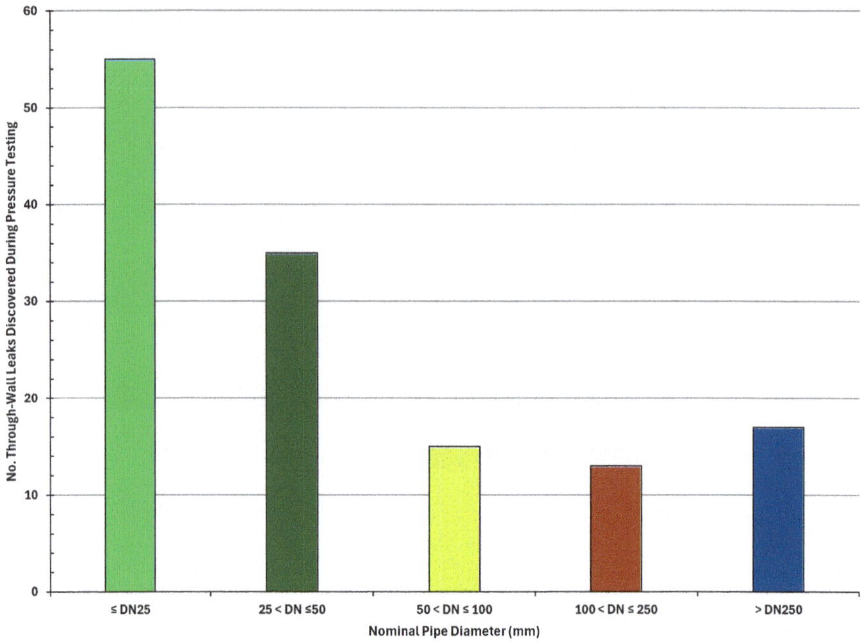

Fig. 8.11 Through-wall flaws discovered during hydrotesting

examination (Workman and Kishoni 2007). Operational impacts of hydrotesting are illustrated in Fig. 8.12. Table 8.5 lists selected FAC failures attributed to LTA-RIM.

8.8.2 Examples of In-Service Inspection Failures

Progress has been made with nondestructive examination technologies and processes; the detection and assessment of crack indications continue to pose challenges, some of which are directly related to a lack of qualification of NDE techniques, human factor issues, and/or programmatic issues. Many of the reviewed events that involve the failure of RIM to prevent passive component material degradation to exceed acceptance criteria are attributed to human factor issues.

8.8.2.1 NDE Program Failure Involving IGSCC of Dissimilar Metal Weld

An axial defect was found during a refueling outage in 2023 in a recirculation inlet nozzle safe end-to-nozzle dissimilar metal weld. The defect was characterized by phased-array ultrasonic examination and was found to be an axially oriented flaw

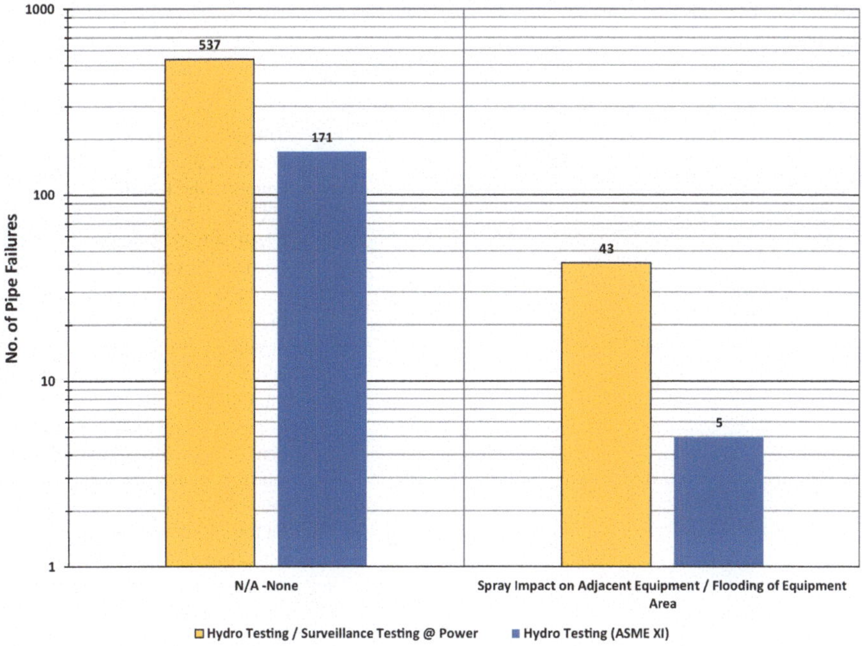

Fig. 8.12 Hydrotesting operating experience

approximately 83% through-wall. The flaw was repaired with a weld overlay during the refueling outage.

The causal evaluation identified a missed opportunity to identify the flaw due to improper setup of the UT examination system in 2007. A second missed opportunity to identify the flaw occurred during a manual phased-array ultrasonic testing exam in 2013 that documented no recordable indications. The causal evaluation noted that manual exams do not record data but only overall results, and therefore there was no capability for the plant owner to review the data to determine specifically why the exam did not identify the flaw.

An additional contributing factor was that between 1988 and 2005, the licensee mischaracterized the welds under their response to GL 88-01, "*NRC Position on IGSCC in BWR Austenitic Stainless Steel Piping*."[19] During safe-end replacement in 1982, a thin layer of the original Alloy 182 weld material was left on the reactor pressure vessel nozzles to alleviate the need to post-weld heat treat the nozzle. Subsequently, the reactor recirculation system nozzle-to-safe-end welds were categorized as Category A (resistant materials) instead of Category D (nonresistant materials, no stress improvement). This resulted in an improper examination scheduling requirement and the loss of three additional opportunities to identify the flaw.

[19] https://www.nrc.gov/reading-rm/doc-collections/gen-comm/gen-letters/1988/gl88001

Table 8.5 Examples of FAC failures attributed to less-than-adequate reliability and integrity management

Plant	Event date	System	Nominal diameter	Root cause
Tihange 2	7/1/2020	Extraction steam (EXT-STEAM)	250	The damaged pipe was located downstream of a tee which connects the fourth EXT-STEAM line to a header. CHECWORKS Steam/Feedwater Application (SFA)[a] predicted a higher wear rate in the fourth tee than in the downstream pipe. However, it is the downstream pipe where the leak occurred. For a variety of reasons, the piping component downstream of the fourth tee had never been included in the FAC inspection program. It had been assumed that the actual wall thickness downstream of the fourth nozzle was 15 mm instead of 10 mm as indicated in the isometric drawing.
Davis Besse	5/9/2015	Moisture separator reheater (MSR)	100	The failure was attributed to a data entry error in the CHECWORKS™ model for the restricting orifice, immediately upstream of the failed elbow. The input error occurred during the original computer model development in the late 1980s and resulted in a nonconservative wear rate prediction from the inception of the FAC program. Corrective actions from this event included verifying all critical design inputs to the current CHECWORKS™ models, enhancing documentation requirements for changes to the program models, and updating procedures for evaluating known accelerated wear rate configurations.

Table 8.5 (continued)

Plant	Event date	System	Nominal diameter	Root cause
Mihama 3	8/9/ 2004	Condensate (from the fourth low-pressure feedwater heater to the deaerator)	500	The failed section of pipe had not been inspected since 1976 and was not checked again because it was not expected to corrode so quickly, and it had not been thoroughly checked since 1976, as the company said.
Callaway	8/11/ 1999	MSR	150	The FAC program did not consider downstream piping inspections as mandatory, although it is a common practice. However, this downstream piping was not inspected when the upstream 45-degree elbow was inspected during previous refueling outages. A second causal factor associated was the industry experience-based perception that the upstream 45-degree elbow would be the most susceptible component for failure due to FAC.
Fort Calhoun	4/21/ 1997	EXT-STEAM	300	The inconsistencies between predicted and actual wear were due to two factors: (1) the "line correction factor" calculated by CHECWORKS for the fourth-stage extraction steam line was not within acceptable range specified in the CHECWORKS user manual and (2) the line correction factor was biased and thus under-predicted the wear rates. The owner had assumed that the affected elbow had been in service since initial operation of the plant in 1973; however, the elbow had been replaced in 1985. Therefore, the actual wear occurred over 2 years rather than the presumed 14 years.
Indian point 2	12/9/ 1993	Main steam (MS)	50	The elbow had developed a P/H leak. As part of troubleshooting, a valve operation was

Table 8.5 (continued)

Plant	Event date	System	Nominal diameter	Root cause
				performed such that a pressure pulse caused the thinned elbow to rupture. The valve operation was not in accordance with the station's operating procedures.
Sequoyah 2	3/1/ 1993	EXT-STEAM	250	At the time of the incident, the FAC program did not include the affected piping section. Subsequent inspections by the licensee revealed significant degradations in other non-safety-related steam, feedwater, and condensate piping.
Maine Yankee	7/22/ 1992	MSR	50	The FAC program did not include component upstream of the MSR flow diversion valve.
Millstone 3	12/31/ 1990	MSR	150	The affected discharge piping of the moisture separator drain pump had not been included in FAC inspection program. Investigations revealed that the FAC program had serious programmatic deficiencies. As a result of this finding, EPRI and the CHECWORKS Users Group (CHUG) sponsored plant visits to ascertain the state of a sampling of utility programs.
Surry 1	3/23/ 1990	Heater-drain	100	The affected pipe section had not previously been inspected. It was not included in the FAC inspection program.
Indian point 2	9/26/ 1989	MS	100	Caused by over-pressurization during air leakage testing. The affected elbow had been subjected to FAC-induced wall thinning.
ANO 2 (Arkansas 2)	4/18/ 1989	EXT-STEAM	150	During the 1988 refueling outage, an inspection was made of the portion of piping that ruptured; however, the vertical run of piping from the extraction nozzle to the first elbow was not inspected. The selection of

Table 8.5 (continued)

Plant	Event date	System	Nominal diameter	Root cause
				previous inspection locations had been based on engineering judgment, and the vertical run of piping was not considered to be a highly susceptible area for pipe wall thinning.

[a]CHECWORKS Steam/Feedwater Application (SFA) predicts flow-accelerated corrosion in power plants and manages associated inspection data. It helps users identify the most susceptible piping locations for flow-accelerated corrosion and erosion. It has the capability to evaluate the effects of changes to water chemistry, system operation, and configuration

8.8.2.2 Incorrectly Calculated FAC Wear Rate

Unexpected wall thinning due to flow-accelerated corrosion (FAC): Control room operators heard the sounds of rushing air or steam in the turbine building and began an investigation. The operators observed steam on the turbine deck and began a controlled power decrease. At approximately 30% reactor power, the operators initiated a manual reactor trip in accordance with plant procedures. Once the plant was shut down, workers isolated the condenser and cooled the reactor using the atmospheric relief valves. The cause of steam release into the turbine building was a ruptured elbow in DN100 pipeline. The pipe rupture caused the fire suppression system to actuate. The failed elbow in the DN100 piping was about 250 mm downstream of a restricting orifice. This orifice, which restricts the amount of scavenging steam drawn from the second-stage reheater tube bundle outlet header, was modeled in the FAC software as having an opening 75 mm in diameter when vendor drawings showed an orifice opening of 21.8 mm.

The incorrect orifice opening size was entered into the FAC software in the 1987–1989 timeframe during initial FAC software database development and caused the FAC software to incorrectly calculate the wear rate of the elbow. Because of this incorrectly calculated wear rate, no wall thickness measurements were performed directly on the failed elbow.

8.8.2.3 LTA Implementation of an Augmented Examination Program

Undetected pipe flaw: While performing planned inspections of reactor coolant system pipe welds, two flaw indications were identified on a DN50 high-pressure safety injection line connection to the reactor coolant system piping (cold leg). After confirmatory inspections and evaluation, one of the two flaw indications was determined to not meet the acceptance criteria specified in ASME Section XI. This flaw was missed during the previous refueling outage as a result of a probable skill-based human performance error. The flaw was detected during normal inspections required by the NDE Augmented Examination program, which is driven by EPRI MRP-146 ("*Thermal Fatigue in Normally Stagnant Non-Isolable Reactor Coolant System Branch Lines*") (EPRI 2005). Non-isolable branch lines connected to the reactor coolant system are susceptible to high-cycle thermal fatigue if exposed to specific operational conditions and configurations. Examples of susceptible locations are horizontal lines where in-leakage past a valve is present and lines that see turbulent swirl penetration from adjacent piping flow. Initial MRP-146 inspections were performed in 2008 and did not include the area where the flaw was discovered.

8.8.2.4 Failure of UT Procedure to Detect a Crack-like Indication

Following a controlled shutdown in response to an RCPB leak in 2013, visual inspection confirmed that the leak was located on a high-pressure injection line. The root cause evaluation determined that a crack-like indication in the failed location had existed since 2011. Furthermore, the root cause evaluation concluded that inadequate procedural guidance existed for the conduct of augmented examinations and appropriate disposition of UT examination results where conditions limited the weld volume that could be examined. A UT limitation is defined as any obstruction or condition that limits the extent of angle beam scanning or limits the extent of required coverage using straight beam scanning. When adequate weld volumes could not be examined on the nozzle safe-end-to-pipe butt weld, no procedural guidance provided weld volume acceptance criteria or directed these limitations to be entered into the corrective action program for evaluation. During the root cause investigation, metallurgical analysis documented that the crack propagated over several operating cycles. Historical NDE data revealed that the crack was visible in existing radiographs. Had the failed weld volume been adequately interrogated, the crack would have been identified before propagating through wall. The licensee reviewed the results of the previous UT examination performed in 2012 using an NDE procedure and found no reportable indications. However, in 2011, the licensee performed a radiographic examination specifically to check the condition and position of a thermal sleeve. The focus of the review was limited to that area; however, the safe-end area containing cracked weld was incidentally visible on the film. Following the current event, the licensee re-reviewed the 2011 radiographic film, and a crack-like indication was identified in the sidewall image of the weld at approximately the same location as corresponding to the current crack location. From the re-review of the film, this crack-like indication appeared to be approximately 50% through-wall.

8.8.2.5 Failure to Detect Axial Flaws in a Dissimilar Metal Weld

During an ISI of a dissimilar metal weld in an inlet-to-steam generator nozzle, several axially oriented flaws went undetected by the licensee's manual conventional

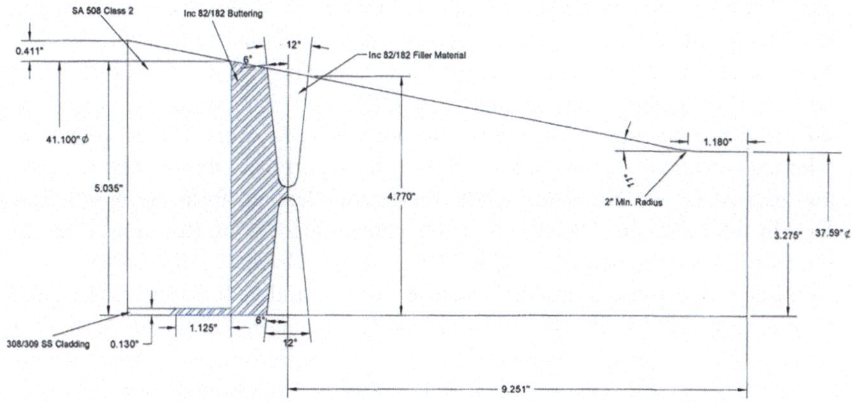

Fig. 8.13 Profile of steam generator inlet dissimilar metal weld. (Reproduced from PNNL-21546, NRC-ADAMS Accession No. ML12200A216)

ultrasonic testing (UT) technique (Anderson et al. 2012). The flaws were subsequently detected as a result of outside-diameter (OD) surface machining in preparation for a full structural weld overlay. The machining operation uncovered the existence of two through-wall flaws based on the observation of primary coolant water leaking from the weld. Further ultrasonic tests were then performed, and a total of five axially oriented flaws, classified as PWSCC, were detected in varied locations around the weld circumference. The particular hot leg safe-end-to-nozzle weld configuration has an approximate 11-degree OD taper from the thinner austenitic piping side up to the thicker carbon steel nozzle and is typical of a DMW created during steam generator replacement at Westinghouse-designed nuclear power plants (Fig. 8.13).

The level of OD taper exhibited by this particular design was not included as a blind performance demonstration mock-up used by the PDI. In response to this event, the US industry established the "NDE Improvement Focus Group" (NIFG) (Ashwin et al. 2013). Following the release of the root cause evaluation, the NIFG reviewed the root and contributing causes and concluded that these conditions might have occurred at other locations within the US fleet or commercial nuclear power plants. Therefore, the team set about establishing an extent-of-condition evaluation as an input to understanding the improvements necessary to avoid a similar occurrence in the future. The work of the NIFG resulted in several EPRI documents being published, including the following:

- Nondestructive Evaluation: NDE Improvement Focus Group Extent of Condition Actions in Response to North Anna Dissimilar Metal Weld Operating Experience. This report provided a process by which plants in the United States could categorize DM welds in terms of their risk of containing an undetected SCC flaw. The risk levels were determined by reviewing two key aspects: (1) whether the weld materials were considered susceptible to SCC (and had not been mitigated) and (2) whether the weld had ever been examined by ASME Section XI, Appendix VIII qualified, encoded UT procedures, personnel, and equipment.

- Nondestructive Evaluation: Guideline for Conducting Examinations of Dissimilar Metal Welds, Revision 3 (Lindberg et al. 2018). This report contains guidance on the planning, preparation, and execution of DM weld UT examinations. Although originally published in 2009, this report was thoroughly reviewed and revised by the NIFG to include the lessons learned from the North Anna OE. In addition, the NEI 03-08 implementation level of Revision 1 of this guideline was increased from "Good Practice" to "Needed" (NEI 2020).
- Nondestructive Evaluation: Performance Demonstration Initiative (PDI) Guidance for Improved Reliability in Ultrasonic Examinations (EPRI 2013a, b). This report contains revised versions of two preexisting Performance Demonstration Initiative (PDI) documents, *Site-Specific Mockup Requirements for Dissimilar Metal Welds, Revision C*, and *Guideline for Hands-On Practice (PDI-GL-001), Revision B*, each of which was reviewed and revised by the NIFG, based on the operating experience from North Anna Power Station and in an effort to improve the overall examination reliability associated with these products.
- Nondestructive Evaluation: 2013 Team Scanning Assessment Conducted on Behalf of the NDE Integration Committee's NDE Improvement Focus Group (Weeks et al. 2013). A scanning technique commonly called team scanning was used during the North Anna event; however, this technique lacked any recognized guidance for implementation. Team scanning is performed in areas with limited access, where one examiner is positioned at the weld and is responsible for manipulating the search unit. A separate examiner is located 2 or 3 meters away and is responsible for controlling and observing the ultrasonic instrument. In response, the NIFG developed a new set of guidelines on the use of team scanning (Appendix B of Guideline for Conducting Ultrasonic Examinations of Dissimilar Metal Welds, Revision 1). In an effort to test the reliability of team scanning, when employing these new guidelines, the NIFG conducted a blind testing experiment with three examiner teams using team scanning and following the new guidance. This report documents this experiment and the use of team scanning.

References

Aldrin JC et al (2006) Probabilistic risk assessment: impact of human factors on nondestructive evaluation and sensor degradation on structural health monitoring, AFRL-ML-WP-TP-2006-493, materials and manufacturing directorate, Air Force Research Laboratory. Air Force Material Command, Wright-Patterson Air Force Base, OH

Ammirato, F. et al (1993), Performance Demonstration Initiative: U.S. Implementation of ASME B&PV Code Section XI, Appendix XIII, GS16/5, Trans. SMiRT-12, pp 279–284.

Anderson MT, Diaz AA, Doctor SR (2012) Evaluation of manual ultrasonic examinations applied to detect flaws in primary system dissimilar metal welds at North Anna Power Station, PNNL-21546. Pacific Northwest National Laboratory, Richland

Ashwin P, Dyle R, Wirtz C (2013) Nondestructive evaluation improvement focus group extent of condition actions in response to North Anna dissimilar metal weld operating experience: Revision 1, 3002000041. EPRI, Palo Alto, CA

ASTM (2016) Standard for qualification and certification of nondestructive testing personnel (ANSI/ASNT CP-189-2016). American Society for Nondestructive Testing, Columbus

ASTM (2024) Personnel qualification and certification in nondestructive testing, ASNT recommended practice no. SNT-TC-1A. American Society for Nondestructive Testing, Columbus

Becker FL, Doctor SR, Heasler PG, Morris CJ, Pitman SG, Selby PG, Simonen FA (1981) Integration of NDE reliability and fracture mechanics, NUREG/CR-1696. USNRC, Washington, DC

Berens AP, Hovey PW (1981) Evaluation of NDE reliability characterization, AFWAL-T9-81-416C, vol 1. Materials Laboratory, Air Force Wright Aeronautical Laboratories, Wright-Patterson Airforce Base, OH

Berens AP, Hovey PW (1982) Characterization of NDE reliability. In: Review of progress in quantitative nondestructive evaluation. Plenum Press, New York

Berens AP (1989) NDE Reliability Data Analysis. In: ASM handbook, volume 17: nondestructive and quality control. ASM International, Materials Park, pp 689–701

Bertović M (2016) Human factors in Non-Destructive Testing (NDT): risks and challenges in Mechanized NDT, BAM-Dissertations Reihe Band 145. Bundesanstalt für Materialforschung- und Prüfung (BAM), Berlin

Borloo E, Lemaitre P (eds) (1994) Non-destructive examination practice and results: state of the art and PISC-III results, proceedings of the Joint CEC OECD IAEA specialists meeting held at Petten on 8–10 March 1994, EUR 15906 EN. European Commission, JRC, Institute for Advanced Materials, Petten

Bush SH (1983a) Reliability of nondestructive examination, chapters 1–6, NUREG/CR-3110, vol Volume 1. USNRC, Washington, DC

Bush SH (1983b) Reliability of nondestructive examination, chapters 7–11, NUREG/CR-3110, vol 2. USNRC, Washington, DC

Bush SH (1983c) Reliability of nondestructive examination, chapters 12–15, NUREG/CR-3110, vol 3. USNRC, Washington, DC

DGfZP (1997) Proceedings of European-American workshop determination of reliability and validation methods on NDE. ISBN 3931381-18-8. Deutsche Gesellschaft für Zerstörungsfreie Prüfung e.V, Berlin

Doctor SR, Deffenbaugh JD, Geod MS, Green ER, Heasler PG, Simonen FA, Spanner JC, Taylor TT (1989) Nondestructive examination (NDE) reliability for Inservice inspection of light water reactors, NUREG/CR-4469, vol 8. USNRC, Washington, DC

Enkvist J, Edland A, Svenson O (2001) Effects of time pressure and noise on non-destructive testing, SKI Report 01:48. Swedish Nuclear Power Inspectorate, Stockholm

EPRI (2005) Management of Thermal Fatigue in normally stagnant non-isolable reactor coolant system branch lines (MRP-146), 1011955, Palo Alto

EPRI (2013a) Nondestructive evaluation: performance Demonstration Initiative (PDI) guidance for improved reliability in ultrasonic examinations, 3002000204, Palo Alto

EPRI (2013b) Recommendations for an effective flow-accelerated corrosion program (NSAC-202L-R4) 3002000563, Palo Alto

Fleming K, Lydell B, Presley M, Mosleh A, Chalgham W (2023) Pipe rupture frequencies for internal flooding probabilistic risk assessments revision 5, 3002024904. EPRI, Palo Alto

Gandossi L, Annis C (2010) Probability of detection curves: statistical best-practices, EUR 24429 EN and ENIQ Report No. 41. European Commission, Joint Research Centre, Institute for Energy, Petten

Hedden O, Cowfer D, Batey J, Spanner J, Becker L (2002) Overview of the impact of ultrasonic examination performance demonstration on the ASME boiler and pressure vessel code. Trans ASME 124(8):254–260

ISO (2012) Non-destructive testing—qualification and certification of NDT personnel, ISO 9712 (EN). International Organization for Standardization, Geneva

Jelenik T, Tidström L, Brickstad B (2005) Probability of detection for the ultrasonic technique according to the UT-01 procedure, SKI report 2005:03. Swedish Nuclear Power Inspectorate, Stockholm

Kanzler D, Meyendorf NG (eds) (2023) 8th international workshop on reliability of NDT/NDE, proceedings, vol 12491. SPIE Digital Library. https://doi.org/10.1117/12.2684125

Khaleel MA, Simonen FA (2000) A model for predicting vessel failure probabilities including the effects of service inspection and flaw sizing errors. Nucl Eng Des 200:353–369

Lindberg J, Cinson T, Selby G, Latiolais C, Swain R (2018) Nondestructive evaluation: guideline for conducting ultrasonic examinations of dissimilar metal welds, revision 3, 3002012244. EPRI, Palo Alto

MDEP (2013) Technical report on lessons learnt on achieving harmonization of codes and standards for pressure boundary components in nuclear power plants, technical report TR-CSWG-02, Multinational Design Evaluation Programme. OECD Nuclear Energy Agency, Boulogne-Billancourt

More DG (ed) (2016) Nondestructive testing handbook, Vol. 1: Liquid Penetrant Testing (PT), 4th edn, ISBN: 978-1-57117-374-4. The American Society for Nondestructive Testing, Columbus

Murgatroyd RA (1988) Reliability of application of inspection procedures. In: OCDE/GD(97)25, paper presented at the CSNI workshop on the complementary roles of fracture mechanics and NDE in safety assessments of components, Würenlingen, Switzerland

Murgatroyd RA, Crutzen S (1994) Human reliability in inspection final report on action 7 in the PISC Programme. In: Proc. Joint CEC OECD IAEA Specialists Meeting on Non-Destructive Examination—Practice and Results, NEA-CSNI-R--1994-23. OECD Nuclear Energy Agency, Boulogne-Billancourt, France

NEI (2020) Guideline for the Management of Materials Issues, NEI 03-08 Revision 4. Nuclear Energy Institute, Washington, DC

Nichols R, Dau G, Crutzen S (1987) NDE in relation to Structural Integrity SMIRT 9 Post Conference Seminar, 26–27 August 1987, Proceedings. Int J Pressure Vessels Piping 35(1–4)

Nichols RW, Crutzen S (1993) Impact of PISC results on codes, standards and regulatory activities, PISC Report No. 28, EUR 15104 EN. Commission of the European Communities Joint Research Centre, Ispra

PISC (1986) A summary of the PISC-II project, PISC-II Report No. 1, CSNI No. 117, Programme for the Inspection of Steel Components. OECD Nuclear Energy Agency, Boulogne-Billancourt

Sanquist TF, Hughes NM, Morrow SL, Franklin CM, D'Agostino AL (2018) Human factors in nondestructive examination: manual ultrasonic testing task analysis and field research, PNNL-27441. Pacific Northwest National Laboratory, Richland

Simonen FA, Woo HH (1984) Analyses of the impact of inservice inspection using a piping reliability model, NUREG/CR-3869. USNRC, Washington, DC

USDOD (2009) Nondestructive evaluation system reliability assessment, MIL-HDBK-1823A. United States Department of Defense, Wright-Patterson Air Force Base, OH

USNRC (2008) Guidance on monitoring and responding to reactor coolant system leakage, Regulatory Guide 1.45 Revision 1, Washington, DC

Weeks P, Langevin J, Swain R (2013) Nondestructive evaluation: 2013 team scanning assessment conducted on behalf of the NDE Integration Committee's NDE Improvement Focus Group, 3002002048. EPRI, Palo Alto

Workman GL, Kishoni D (eds) (2007) Nondestructive testing handbook, Vol. 7: Ultrasonic Testing (UT), 3rd edn, ISBN: 978-1-57117-283-9. The American Society for Nondestructive Testing, Columbus

Chapter 9
Knowledge Preservation Through Data Collection

9.1 Model Input Parameters

How is operating experience data recorded and transferred into metrics of piping pressure boundary integrity? It is useful to begin this chapter with a discussion on what constitutes a model of pipe failure and how quantitative assessments can be performed (Fig. 9.1). Extensions to thc three basic pipe failure models include the Markov chain model (MCM), the multistate physics model (MSPM) (Aldemir et al. 2014; Di Maio et al. 2015), and probabilistic physics of failure (PPF) model (Beal et al. 2020). All models and their extensions rely on data on material degradation and failure.

Developing a database of suitable structure and technical content is an essential first step in formulating a technical approach that statistically profiles the failure information, supports uncertainty characterization, and provides the required input to probabilistic or physical models of pipe failure. Such a database development endeavor involves defining the scope of the data collection process and the intended applications, data quality assurance routines, data integrity and preservation, data security, and intended users. Next comes the selection of a programming language to be used and the database functionalities that need to be implemented. Intended applications, whatever they may be, determine the types of information to be extracted and converted into pipe failure model input parameters. This "model" of piping reliability can be seen as a decomposition model:

$$\rho = \lambda \times CFP \tag{9.1}$$

$$\lambda = \frac{n}{\tau} \tag{9.2}$$

These relationships seem almost self-evident. As a model for physical pipe failure, the decomposition approach considers pipe failure as happening in two stages: first, random occurrences of pipe defects, followed by random rupture events

B. Lydell, *Knowledge Preservation in Nuclear Safety*,
https://doi.org/10.1007/978-3-032-25463-4_9

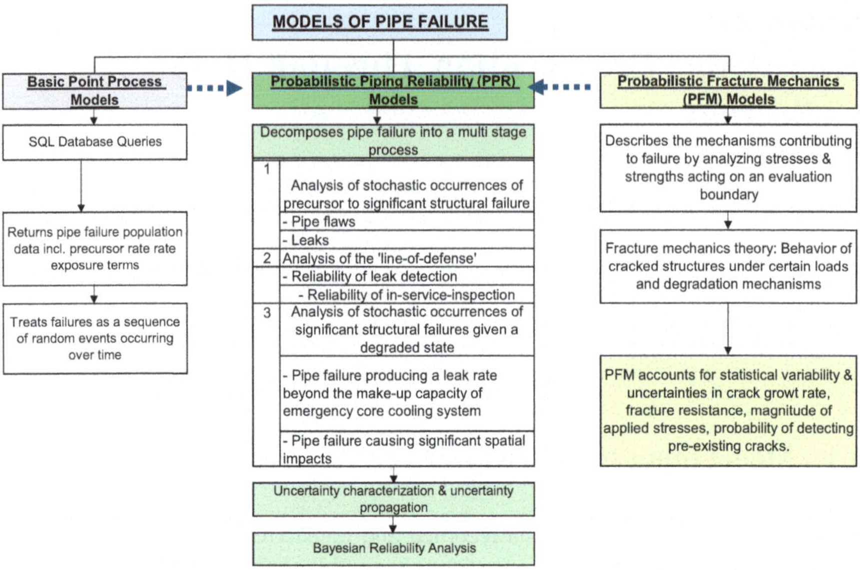

Fig. 9.1 The different models of pipe failure

in the presence of degraded conditions like surface cracks or small leaks through the pipe wall. This method provides a rough yet practical approximation of the actual failure process. One benefit of this model is its capacity to utilize existing operational data in several ways, for example, by analyzing observed instances of deterioration and initial signs of pipe or weld failure, which can serve as warnings for the much rarer major structural failures.

Deriving meaningful quantitative assessments requires a detailed breakdown of the physical factors that underlie a potential piping pressure boundary degradation or failure. Parameter ρ corresponds to the consequence of a pipe failure and corresponds to C_i in the structural integrity risk triplet. It is the frequency of a structural failure of a certain magnitude "*i*." The "magnitude" corresponds to the mass or volumetric through-pipe-wall flow rate or it can be in terms of the dimensions of the opening in the pressure boundary. An overview of how these parameters are estimated is found in Table 9.1.

Parameter λ is referred to as pipe failure rate, or "precursor" rate, and is estimated from the number of "failures" that have been observed in a given period. Parameter τ is an exposure term, consisting of reactor operating years or equivalent full-power years of operation and number of surveyed, or "exposed," piping components during an observation period. The failure rate is a simple maximum likelihood estimator described by a Poisson distribution with a single parameter lambda, which is the mean number of events over time. It is a representation of being in a certain degraded state of structural integrity, e.g., cracked, thinned, or leaking. Parameter *CFP* represents the conditional probability that a preexisting flaw in the pipe pressure boundary transitions from a stable to an unstable state in a short period of time. The

Table 9.1 Piping reliability model input parameters

Parameter	Element in the risk triplet	Formula	Description	How operating experience insights are retrieved to meet application requirements
ρ	C_i	Combination of probability density functions for λ and *CFP*, respectively, using Monte Carlo multiplication	Frequency of a pipe failure as a function of an equivalent break size (EBS) or through-wall mass or volumetric leak rate. The exact definition depends on the context of an analysis	All database applications begin by querying an SQL database for specific information. It is the requirements of an application that provide for query specifications. A query is a request for certain event populations, for action on data, or for both. A query may be defined to answer a simple question (e.g., "how many socket weld failures occurred in PWR charging systems during calendar years 2005 through 2018?"), to perform calculations, to combine data from different tables, to add, change, or delete data, or to create a new table that addresses a specific degradation mechanism. Queries that are used to retrieve data from a table or to make calculations are referred to as "select queries." queries that add, change, or delete data are called "action queries." database queries facilitate the filtering and processing of event data
λ	S_i	$\lambda_{ik} = \frac{n_{ik}}{\tau_{ik}} = \frac{n_{ik}}{f_{ik} N_i\ T_i}$	Usually referred to as "pipe failure rate" and represents the rate of	In order to study the likelihood of the occurrence of pipe

(continued)

Table 9.1 (continued)

Parameter	Element in the risk triplet	Formula	Description	How operating experience insights are retrieved to meet application requirements
			being in a certain state, e.g., crack of certain dimension (through-wall depth as the percentage of wall thickness, crack opening area, crack opening displacement) or through-wall leak. The failure rate is calculated as the number of observed failures over an observation period	failures, the objective of a statistical analysis is to examine the evolution of their rate over time. We define λ_{ik}, the rate of occurrence of events per observation period as given in the "formula column." given that events are rare and are sparsely spread in time, we take the simplest assumption that their temporal distribution follows a Poisson process; in other words, there is no correlation between successive events, and the distribution of waiting times between one event and the next one is exponential with rate λ_{ik}. Engineering analyses are needed to correlate the number of pipe failures with the piping component population that produced the failures
CFP	P_i	$CFP = \frac{R}{R+F}$	The formula is symbolic. "*R*" is the number of significant structural failures (SSFs), and "*F*" is the number of preexisting pipe flaws	The *CFP* is associated with a specific evaluation boundary in terms of piping system, precise location of the evaluation boundary within a system, and hydraulic pipe diameter. The database queries are formulated accordingly to obtain counts of *R* and *F*

physics of the conditional pipe failure probability given tight cracks in thick-walled piping is delineated in fracture mechanics. Developing probability density functions of the *CFP* is not a straightforward analysis. In some rare instances, this parameter can be estimated directly from operating experience, however.

To calculate the pipe failure precursor frequency, ρ, one can start with a broad generic prior distribution. To address uncertainty in the prior distribution, a constrained non-informative distribution (CNID) method can be used in which the mean value of a gamma distribution is anchored to the mean estimate (Atwood et al. 2003). The pipe failure rate is expressed in units of failures per reactor operating year and the number of components that are exposed to a certain material degradation mechanism. The generic prior is then updated using a Bayesian approach with data for the numerator (number of failures) and denominator (reactor operating yeas and number of components). For the pipe failure rate, the failure count (numerator) is taken from operating experience. The failure rate calculation is estimated based on the average performance of the component over the industry service life from which the data is collected. The denominator is a distribution that combines reactor operating year (ROY) data with piping population data in terms of pipe fitting counts, pipe lengths, or weld counts that are susceptible to a certain degradation mechanism.

Beyond these introductory statements, a pipe failure database may also be used for many other types of applications, including statistical analysis of censored data using hazard plotting techniques and Weibull analysis to estimate the statistical parameter of a Weibull distribution (Fig. 9.2). Such analyses may be performed using time to failure or age of a failed piping component at the time of failure. Age-dependent pipe failure rate estimation can be performed using different analysis strategies to obtain statistical characterizations of piping reliability parameters as a function of the age of an affected pipe section at the time of observed failure, age of the plant, or temporal changes in the piping operating experience.

High-level insights into trends are derived by normalizing the failure population within each period to the plant population responsible for those failures in respective operating regimes. This approach serves as an initial screening tool for identifying potential patterns within the failure data and represents an approximation. Variability among operating regimes and safety classes arises from numerous factors, such as data collection methodologies and differences in material degradation susceptibility between plants. When comparing operational experience with probabilistic fracture mechanics predictions, additional types of information are required to ensure accurate and meaningful analysis.

9.2 Database Guidelines and Industry Standards

Guidelines and industry standards exist for reliability database development and data collection (Stevens 1986; IAEA 1992; Cooke 1996a, b; Moosemiller and Weber 1998; Pyy 1999; Jurmain 2007; ISO 2016). These guidelines and standards are

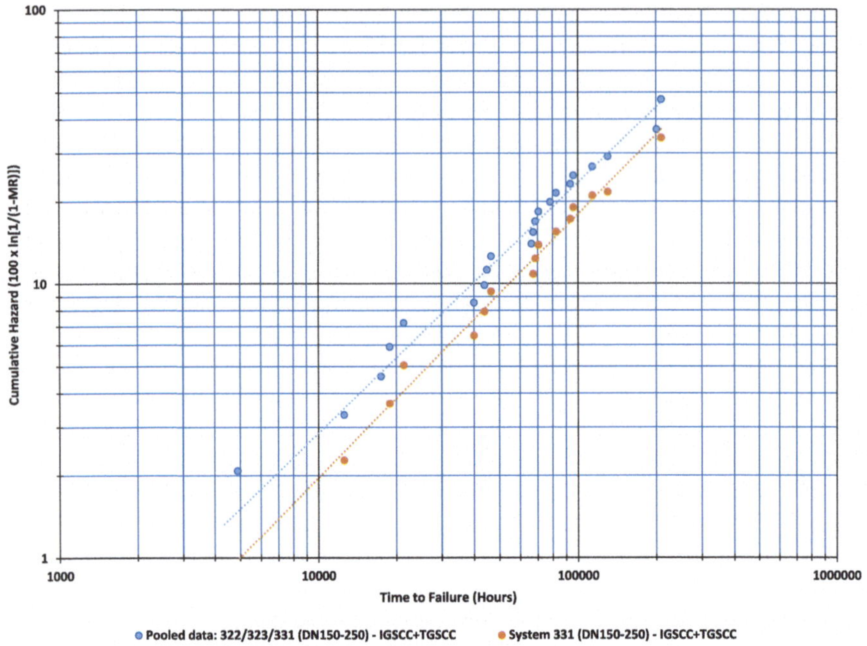

Fig. 9.2 Hazard plots and least square fits to IGSCC data. (From Appendix G of SKI Report 98-30 (1998). For details on hazard plotting techniques, see, for example, Chapter 11 in Kapur, K.C., Lamberson, L.R. (1977), Reliability in Engineering Design, ISBN:0-471-51191-9, John Wiley and Sons Inc. https://www.stralsakerhetsmyndigheten.se/publikationer/rapporter/sakerhet-vid-karnkraftverken/1998/9830/)

primarily concerned with the reliability of active components but contain the essential elements of database development, data quality, and database maintenance processes. A common theme of these guidelines and standards is twofold: (1) data quality management and (2) processes that are applied to ensure that a user has confidence in the database content and its completeness with respect to practical applications. A Nordic Nuclear Safety Research program produced the following statement on the design and use of a reliability database (Paulsen et al. 1996):

> Reliability data is not simply “there” waiting to be gathered. A failure rate is not an intrinsic property of a component like mass or charge. Rather, reliability parameters are characteristics of populations which emerge from a complex interaction of components, operating environments, and maintenance regimes. Properly defining the populations from which reliability data is to be gathered requires first and foremost a clear perception of the purpose for which the data is to be used. We may broadly distinguish three types of users of reliability data: the risk/reliability analyst interested in system performance, the maintenance engineer interested in maintenance optimization, and the component designer interested in optimal design. Among the questions which should be answered in defining relevant populations: Are there trends in the data? When can data from different sockets be pooled? How shall we analyze competing failure modes/mechanisms? Which distributions adequately describe the data? Progress has been made in recent years with regard to these questions, but this progress has not yet been translated into on-line tools for potential users.

> The principal difficulty is this: the above mentioned questions seldom receive simple answers. Rather, the proper definition of relevant populations requires intelligent interpretation on the part of the analyst. There is seldom a unique "right" interpretation; the analyst must select an interpretation which is suitable for the purpose at hand.

Observed pipe failures must be correlated with the piping component population that produced the failures and the corresponding piping component population at risk of degrading and potentially failing. Building a pipe failure database involves both failure observation and details on evaluation boundaries, system operation, piping system design, and piping component populations for different systems and nuclear power plant designs. This includes information from isometric drawings, in-service inspection program documentation, aging management programs for flow-accelerated corrosion (FAC), microbiologically influenced corrosion (MIC) and stress corrosion cracking (SCC), safety analysis reports, system operating procedures, and system training manuals.

9.2.1 ISO Standard for Reliability Data Collection Practices

ISO 14224:2016 (ISO 2016) is an international standard that outlines the systematic collection of reliability and maintenance data for equipment throughout all facilities and operations within the petroleum, natural gas, and petrochemical industries during the operational life cycle. The standard establishes principles for data collection as well as associated terminology, forming a comprehensive "reliability language" to enhance effective communication of operational experience. It also details best practices for data quality control and assurance to guide users in maintaining high standards. Reference to this standard is included due to its significance regarding data quality considerations. ISO 14224:2016 is particularly relevant to the establishment and management of databases pertaining to nuclear power plant piping operating experience. Section 7, *Quality of Data*, specifies key procedures for designing and managing a reliability database.

– ISO 14224, Section 7.1: Obtaining Quality Data.

  - Definition of data quality and completeness.
  - Data quality planning measures.
  - Verification of data quality.
  - Limitations and problems.

– ISO 14224, Section 7.2: Data Collection Process.

  - Sources of data.
  - Data collection measures.
  - Organization and training.

9.2.2 OECD/NEA Basic Principles of Collecting Pipe Failure Data

Since 2002, the OECD Nuclear Energy Agency (NEA) has operated an operating experience event database project that collects information on passive component material degradation and failure. The project has published a topical report on "*Basic Principles of Collecting and Evaluating Operating Experience Data on Metallic Passive Components*" (OECD/NEA 2018). The report responded to the following frequently asked questions:

- How is the quality of the data controlled and monitored? The response to this question is to develop and implement a QA plan.
- What is the level of database completeness and comprehensiveness? The response to this question is to implement a process for an equitable exchange of operating experience, a process for which there must be a shared commitment by all participating organizations. A tall order!
- What are the principles of how the database is populated with pipe failure event information? Those principles are documented in a coding guideline.
- Does the database support applications? Normally, it is the intended applications that determine the structure and content of a database. When a database cannot support an application, the typical response is to expand the event data classification structure by adding new or revised supporting tables containing the necessary keywords. Alternatively, the response is to implement new user facilities and to expand the scope of the database.

To achieve the objectives of the CODAP Database Project (OECD/NEA 2004, 2020), a coding format was developed and documented in the coding guideline. It builds on recognized pipe failure data analysis practices and routines that acknowledge the unique aspects of passive component reliability in different reactor operating environments, i.e., the influences of material properties, water chemistry, temperature, and pressure. For an event to be considered for inclusion in the database, it first undergoes a screening for eligibility. The objective of this initial screening is to go beyond the abstracts of event reports to ensure that only events according to the work scope definition are included in the database.

Data quality is affected from the moment the field experience data is recorded at a nuclear power plant, interpreted, and finally entered into a database system. The field experience data is recorded in different types of information systems ranging from action requests, work order systems, and ASME XI relief requests, via ISI databases and outage summary reports, to licensee event reports or reportable occurrence reports. Consequently, the details of a degradation event or failure tend to be documented to various levels of technical detail in these different information systems. Building an event database record containing the full event history often involves retrieval of information from multiple sources. Data quality is an attribute of the processes that have been implemented to ensure that any given database record (including all of its constituent elements or database fields) can be traced to the source information. The term also encompasses "fitness for use," that is, the database records should contain sufficient technical detail to support database applications. Coding guidelines define data quality as a perception or an assessment of a certain dataset's fitness to serve its purpose in a given context. Aspects of data quality include the following:

- Accuracy relative to a coding guideline and the database field definitions.
- Completeness, i.e., ensuring that a given event population captures all relevant OE and that the process of transferring operating experience data from the source to the database is done in a consistent way.
- Update and validation status.
- Relevance with respect to the work scope definition.
- Relevance with respect to aging management processes.
- Consistency across the full set of data records so that any given database query captures all relevant data. Empty database fields in a structured query language (SQL) database indicate that the information in the field is "unknown" or "pending" and represented by NULL, and an SQL query will not return records with empty fields.
- Robustness with respect to data classification: Knowledge of safety classification boundaries is needed to determine whether a pipe failure is within a safety- or non-safety-related piping pressure boundary.
- Accessibility for users and preservation of confidentiality clauses that apply to event information.

9.3 An Approach to Building a Pipe Failure Database

Different data resources are required to establish the database functionalities that are needed to populate the database and to support database interrogation by an analyst. These resources are identified and explained below. An analyst conducting probabilistic piping reliability analysis requires validated operating experience data. A database that can be queried and manipulated in various ways is essential to support the analysis requirements and specifications. Primary database activities are (1) collect data, (2) store data, and (3) retrieve data for applications. One of the most important steps in database development is to define the necessary relationships between the many different tables that are required. A pipe failure database must include the following sub-databases that are linked together:

1. Event data: This consists of event reports with extensive narratives that are classified in a certain way so that all key reliability attributes and influence factors are identified. The following are a few examples of tables that are needed to classify event reports.

    (a) Event types.
    (b) Methods of detection.

(c) Spatial impacts.
(d) Impacts on operation.
(e) Material specifications.
(f) Fabrication details.
(g) Nuclear power plant systems.
(h) Piping components.
(i) Degradation mechanisms.

2. Plant population data: This consists of key nuclear power plant information, including plant name, type of reactor, nuclear steam supply system design, design code, ultimate heat sink, date of first reactor criticality, and date of commercial operation.
3. Piping component population database: This consists of transcribed isometric drawing information. For a specific plant and piping system, the database includes all welds, fittings (bends, elbows, reducers), weld types (e.g., B-F, B-J, C-F-2), pipe spools, pipe lengths, and dimensional data (e.g., nominal diameter, hydraulic diameter).

Ideally, each record in the database should consist of a complete and unambiguous description of a specific pipe failure. Not only shall a selected database record withstand an independent review for technical accuracy and completeness, but it shall also be fit for application, by either direct or indirect database application. A direct database application implies that query results can be directly inputted to a calculation. In contrast, an indirect application implies that additional data processing is required to prepare input as specified by a stand-alone application program. A database software development specifies the types of data to be included, types of data manipulations to be implemented, the flow of data, compatibilities with other computer programs, and data security protocols.

Conducting a quantitative analysis requires considering many different parameters. Because there are so many attributes and factors influencing pipe failures, the database needs numerous fields to store this information. The data classification system should account for relevant national and international Codes & Standards, piping classifications, materials, and sizes. Typically, such a database is structured as an SQL relational database—a collection of smaller sub-databases, each dedicated to specific technical data for recording and analyzing pipe failure events. Examples of sub-databases are as follows:

– Nuclear plant information library, similar to IAEA-PRIS,[1] but expanded to include the piping design code of reference, type of ultimate heat sink, and

[1] The Power Reactor Information System (PRIS), developed and maintained by the IAEA for over five decades, is a comprehensive database focusing on nuclear power plants worldwide. PRIS contains information on power reactors in operation, under construction, or those being decommissioned. The database covers (a) reactor specification data (status, location, operator, owner, suppliers, milestone dates) and technical design characteristics and (b) performance data including energy production and energy loss data, outage, and operational event information.

cooling water system design (Bley et al. 1990). The scope of this sub-database must possess sufficient technical details to ensure that derived exposure terms do not over- or underestimate the "true" plant population that is used to estimate pipe failure rates.

- Piping system component population data in the form of transcribed isometric drawings from a representative set of nuclear power plants of different designs and design generations. It is a database in which the piping system population data is organized by systems, welds, pipe fittings, pipe lengths, dimensional data, and material designations.
- Material database, types, and designations, and methods of fabrication and welding.
- Operational impacts and repair methods, e.g., power reduction, manual reactor shutdown, reactor trip, spraying or flooding of equipment areas, repair technique, and time to repair.
- Plant system identifications for all plant types so that the systems, structures, and components of event descriptions can be correlated accurately. System identifications are NSSS specific as well as country specific. Examples of different national plant identification systems are found in Sect. 9.3.4.
- Material degradation and damage mechanisms. Results from root cause analyses.

9.3.1 Motivations and Objectives

Several attempts have been made to systematically collect data on pipe failures in nuclear power plants. Mostly, the scope of these data collections has been limited to significant, selected representative events to support emerging issues in piping structural integrity assessments. The intended applications determine the scope and process for a data collection effort. Meaningful parameter estimation requires detailed event analysis to identify pertinent reliability attributes and influence factors. Figure 9.3 offers perspectives on database users and implications on database development and maintenance processes.

Past reviews of nuclear power plant operating experience conducted in recent years have shown that existing records of baseline data (design, manufacturer, equipment qualification, and commissioning), operational histories, and maintenance histories of plant components are often not sufficiently comprehensive or readily retrievable to allow for:

- Detailed event time histories, trend analysis, and prediction of component performance and remaining service life.
- Identification and evaluation of degradation, failures, and malfunctions of piping components caused by aging effects.
- Evaluations of effectiveness of stress improvement processes.

APPLICATIONS			DATABASE INFRASTRUCTURE		
Database User	Type of Usage	Extent of Usage	Demands on Database Content	Demands on Database Maintenance Effort	Data Security
Regulatory R&D - Materials & Metallurgy, Structural Integrity, Aging Management	Exchange of Information; derive qualitative insights; R&D planning; PMDA/EMDA Updates; GALL Updates (e.g., buried piping)	Sporadic	All "significant" events of general interest. Essential to capture time-dependent degradation phenomena (DM incubation times). Effectiveness of mitigation practices over time of importance.	Driven by extent of submissions by Member Organizations (PRG)	
Regulator - Inspection & Enforcement: Safety Assessment (SDP), License Renewal/Plant Life Extension, SPAR (equiv.) apps	Resolution of Emerging Issues	Sporadic and as governed by emerging issues	Variable	Unverified assumption of "sufficient" completeness	Data integrity, including the protection of "sensitive" (proprietary & confidential) information is essential. Require well defined data access rules. Long-term database management must be ensured.
Research - Technical Support Organizations	Special needs - methods development, advanced applications; reviews of licensee submittals	Sporadic and as governed by emerging issues	Variable		
Research - Academic	Special needs - methods development	Sporadic	Variable		
Utility - ISI / RI-ISI	Degradation mechanism (DM) analysis, pipe failure potential assessment, RI-ISI program update	Driven by ISI Codes & Standards	Strive-for-Completeness; mitigation effectiveness must be captured	Sustained & continuous; data "mining" well beyond mandated reporting (LER, AO, PNO, RO, etc.). The routines for collecting information on pipe degradation & failure must be based on deep knowledge of subject matter. Data validation of critical importance.	
Utility - PSA	Resolution of Generic Issues, support to Operations & Maintenance, respond to regulatory requirements	Routine - as defined by PSA maintenance program, SDP-response, and other emerging issues	Completeness Essential - Account for plant-to-plant variability		
Utility Alliances, EPRI, VGB, INPO, NPSAG, BWROG, COG, PWROG, etc.	Support to materials reliability R&D, resolution of generic issues, guidelines & data handbooks	Determined by needs of utility members & regulatory initiatives and requirements	Completeness Essential		
PSA Practitioner	Data specializations, application of advanced methods & techniques	Routine	Completeness Essential		

Fig. 9.3 Database users and implications on database development and maintenance processes. (PMDA = Proactive Materials Degradation Assessment (NUREG/CR-6923), EMDA = Expanded Materials Degradation Assessment (NUREG/CR-7153))

9.3.1.1 Knowledge Preservation

Since the 1950s, operating experience has accumulated from more than 20,000 commercial reactor operating years. On an annual basis, each commercial reactor unit has experienced multiple instances of pipe degradation and failure of varying severity and operational impact. Michel and Sayar (2010) report on German activities in the area of knowledge preservation with the aim to ensure easy access to operating experience data and up-to-date information on material performance in different operating environments. Because of design changes and improvement in reliability and integrity management, older operating experience may no longer be applicable. However, older data establishes a basis for assessing the improvements achieved in material performance. Why is it crucial to have a comprehensive database on pipe failures spanning almost seven decades? All failure information is relevant, though its importance varies with the intended application. Section 11 of IAEA Nuclear Energy Series No. NR-T-2.16 offers more details on the subject (IAEA 2023a, b). A conceptual model of prior state of knowledge includes the following elements:

- **Material degradation insights**: The reactors prior in operation were designed and commissioned in the 1960–1980 timeframe. An extensive body of pipe failure data was "generated" during this period. The root cause evaluations, including nondestructive and destructive examinations, produced a deep understanding of the conjoint requirements for material degradation. Subsequent advances in materials science, leak detection, and nondestructive examination technologies aided the development of strategies for how to mitigate piping material degradation in certain operating environments.
- **Proactive material degradation mitigation**: By the late 1980s, the plant operators had implemented major engineering changes to enhance the structural integrity of piping systems. These changes addressed the conjoint requirements for material degradation, improved water chemistry control, use of better materials, and application of stress improvement techniques. For many water-cooled reactor piping systems, OE for the period 1990–2020 differs substantially from the period 1965–1990. These differences are well understood.
- **Learning process**: Bayesian analysis enables the derivation updated (posterior) pipe failure rates that acknowledge the material degradation "learning effects." For example, the extensive pre-1990 operating experience can be used to develop prior pipe failure rate uncertainty distributions that are updated using the post-1990 operating experience. Conceptually, it is simple to do but nevertheless should be formalized by applying the piping reliability analysis framework.
- **Advanced reactor prior pipe failure rates**: It is technically feasible to develop informed advanced reactor pipe failure rates on the basis of the existing reactor operating experience.
- **Advanced reactor posterior pipe failure rates**: As the advanced reactor operating experience becomes available, it would be technically feasible to update the

advanced reactor prior failure rates or to develop new prior pipe failure rate uncertainty distributions.

– **Advanced reactor pipe failure frequencies**: Implementing the above yields the input parameter for the development of advanced reactor pipe failure frequencies. Additional details on how to address the assessment of advanced reactor pipe failure rates are found in IAEA-TECDOC-1988 (IAEA 2021). This report identified two different analytical challenges that an analyst could face: (1) an analysis boundary consisting of material or combinations of materials for which extensive operating experience exists including experience with different types of RIM strategies and (2) an analysis boundary for which there is no known failure history. According to the TECDOC, the first of these challenges can be addressed using any of the three methodologies, given that these methods explicitly account for the effect of a RIM strategy that has been approved for advanced reactor use. The second challenge can also be addressed using any of the different methodologies. The effort needed to develop necessary input parameters is significant, however. The integrated modeling of the many uncertain input parameters requires carefully crafted implementation of uncertainty analysis.

9.3.1.2 Codes and Standards Development

Codes and standards development and implementations are proactive as well as reactive endeavors that respond to technological advances in nondestructive examination, feedback from plant operators and regulators, new advanced reactor designs, and operating experience. As new operating experience accumulates, efforts are made to make improvements in the nondestructive examination techniques to ensure early detection of material degradation. G. Park made the following statement on lessons learned based on operating experience (Park 2014):

> *The first edition of ASME Section XI was titled "Inservice Inspection of Nuclear Reactor Coolant Systems" and provided inspection requirements for the reactor coolant system only. In the foreword of the 1970 edition, it states, "In the development of this Section of the Code an attempt was made to identify those components essential for the protection of the public health and safety." As the Code was revised over the years, inspection requirements grew as operating experience warranted more detailed and comprehensive inspections. As one could imagine this would lead to higher costs and radiation exposure. We now need to review the past and look forward to the future and make the right decisions learning from past experience.*

9.3.1.3 Regulations

Regulatory authorities routinely monitor trends and patterns in material degradation mechanisms acting on structural components (Ader 2003; USNRC 2019, 2021). Technical insights from these activities influence new regulations on aging management and in-service inspection that go beyond the codes and standards that are in effect at the time of a regulatory "communiqué." Such "communiqués" may be in the form of a request for licensee actions and/or information to address issues regarding emergent or routine matters of safety or require a written response.

Gott (2002, 2011) describes a Swedish database on pipe failures in domestic nuclear power plants, a database that was operated by the Swedish Nuclear Power Inspectorate. It was set up in the early 1990s, operated until mid-2008, and included operating experience data from 1972. Originally, the database, called "STRYK," only covered components under the jurisdiction of the regulations concerning mechanical components in nuclear facilities. It was later expanded to include events of metallic parts of concrete structures (liners, rebars), primarily the containment building and associated systems.

The United States Nuclear Regulatory Commission (USNRC) has implemented an "office instruction," LIC-504 (Weerakkody 2023), to present risk-informed options to disposition emergent safety issues and document the bases of those decisions.[2] Safety issues emerge as a result of operating experiences. LIC-504 Appendix B describes a "*Simplified Approach to Risk-Informed Decision Making*." As an example, this approach has been applied to analyze possible safety implications of intergranular stress corrosion cracking observed at several locations in low-pressure safety injection and residual heat removal piping in French PWR plants between 2021 and 2023 (Rudland 2023).

9.3.1.4 PSA Applications

Progress has been made in the development of dedicated pipe failure databases and associated computational tools. Since an event database includes information on historical events, the completeness of the event population in the database always is an important factor in determining its "fitness for use." This needs to be placed in the perspective of present-day knowledge about incubation times of pipe flaws, short versus long. Five types of metrics are considered in probabilistic piping reliability analysis in support of PSA: (1) failure rate, (2) conditional failure probability, (3) inspection effectiveness, (4) effectiveness of material degradation mitigation processes, and (5) aging factors (Fleming and Lydell 2004; Lydell 2014). A piping operating experience database cannot support component-level failure rate estimation, unless the database also includes extensive piping system design information that yields information on the total piping component population that has produced the failure observations, i.e., exposure term data. System-level precursor rates as well as relative measures of piping reliability such as conditional failure probabilities can be generated by querying a database, however. The statistical robustness of such relative measures is correlated with the completeness of an event population.

[2]LIC-504 is generally used when (1) new information reveals an unforeseen hazard or a greater potential for a known hazard to occur, such as identification of an issue that may increase risk, or (2) no other NRC process exists for addressing the issue.

Based on the practical experience from database development, the inherent latency in structured data collection efforts is on the order of at least 5 years. This means that about 5 years could elapse before achieving reasonably high confidence in data completeness. In other words, around 2025, the data mining for the previous decade (2010–2019) would be expected to approach saturation, i.e., high confidence in completeness. Could "cliff edge effects" (e.g., small change in input parameter resulting in large result variation) affect an analysis due to database infrastructure factors? It depends on the maturity of inspection programs and our state of knowledge concerning certain degradation mechanisms. Considerations about the use of up-to-date failure data are intrinsically assumed to be factored into the analyses.

9.3.1.5 Aging Management and Long-Term Operation

Research into the aging of systems, structures, and components in nuclear power plants has spanned several decades, with the review of operating experience (OE) data playing a pivotal role. For instance, in 1975, the United States Nuclear Regulatory Commission established the first Pipe Crack Study Group (PCSG), tasked with evaluating the impact of stress corrosion cracking (SCC) in both boiling water and pressurized water reactors. OE data reviews were integral to the PCSG's analysis. Additionally, significant failures of condensate and feedwater piping, such as those at Trojan in 1985 and Surry in 1986, resulting from flow-accelerated corrosion (FAC), prompted similar efforts to utilize service experience and develop strategies aimed at mitigating and preventing FAC-related pipe failures.

The Generic Aging Lessons Learned (GALL) and the Generic Aging Lessons Learned for Subsequent License Renewal (GALL-SLR) programs in the United States (USNRC 2010, 2023) and the International Generic Aging Lessons Learned (I-GALL) (IAEA 2023a, b) program make extensive use of piping operating experience. An aging management review (AMR) process includes identification of aging effects and degradation mechanisms and identification of a need to develop new programs and/or enhance existing programs that manage these aging effects so that each in-scope structure, component, or commodity group will perform its intended function. The effectiveness of an aging management program (AMP) is determined based on operating experience. For example, did reliability and integrity management identify ongoing material degradation and thus prevent through-wall leakage? To determine the aging effects requiring management, the plant considers and addresses the materials, environments, and stressors that are associated with each structure, component, or commodity group under review.

An update of the AMR is performed typically every 5–10 years or as required by national regulations. The objective of the AMR is to demonstrate the suitability of the aging management process and the associated documentation. The AMR verifies that effective aging management is implemented to ensure that the safety functions of SSCs are fulfilled over the entire lifetime of the plant. On the basis of the results of the AMR for long-term operation, the existing plant programs are updated.

NEI 03–08 (Revision 4), *Industry Guideline for the Management of Materials Issues*, documents an initiative to integrate and coordinate the US industry resources devoted to the material issues (NEA 2020). A formal process for material issue identification and prioritization shall be defined. According to NEI 03-08, the process shall consider relevant international and domestic operating experience. Reviews of operating experience on relevant material issues:

> *A key feature of a robust aging management program is the evaluation and appropriate incorporation of new knowledge related to material degradation. Fleet inspection data is particularly valuable in assessing performance trends. Where component locations previously inspected by an aging management program are removed from* the *future inspection scope without identification of reasonable surrogate locations remaining in the population of components inspected (whether in individual plants or within the fleet at large), it is reasonable to obtain NRC approval prior to implementation.*

9.3.1.6 Experimental Data

The objective of experimental work is to solve knowledge base gaps in materials science and fracture mechanics. Experimental work includes laboratory tests, computational fluid dynamics and structural simulation, and large-scale tests of mock-ups subjected to different environments (pressure, temperature, chemistry parameters) and loading conditions. Results from experimental work can be essential in the characterization of uncertainties in input parameters. This becomes especially important when analyzing situations for which there is no OE data available on which to base the input parameter selections. The experimental data can also be used to estimate the structural integrity factor of improvement when replacing an existing pipe material with a new material grade. Results of the extensive experimental piping integrity work have been stored in databases.

Scott et al. (1994) describe a German-US collaboration to develop a database of pipe fracture experiments, the "CIRCUMCK" pipe fracture database. The database, which was in a spreadsheet format, included details of test parameters, experimental results from the pipe experiments, and material property data for test specimen materials.

Section 8.4 of IAEA Nuclear Energy Series No. NR-T-2.16 deals with "*Experimental Work and Model Validation*" and highlights some of the outcomes from the extensive research to determine the susceptibility of certain nickel-base alloys to SCC in a high-temperature environment. Alloy 690 and its weld metals (Alloy 52 and 152) were developed to resist SCC, and it has almost universally been the preferred replacement for Alloy 600/82/182. As of this writing, there are no known failures after several decades of field experience with Alloy 690. Without OE data, how can the resistance of Alloy 690 to primary water SCC be characterized? Crack initiation data are obtained from laboratory tests that are based, for example, on the constant extension rate tensile technique. The applied strain rates depend on the test objectives and other experimental variables. There are different approaches to determine the factor of improvement for this alloy relative to Alloy 600:

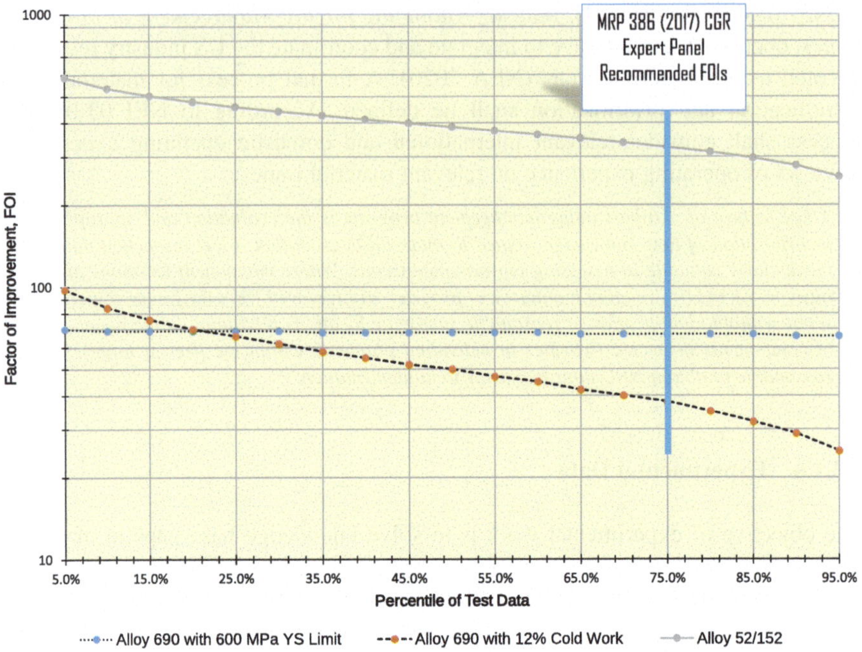

Fig. 9.4 Alloy 690 vs. Alloy 600 factor of improvement according to MRP-386

- Weibull analysis of Alloy 600 cumulative fraction of cracking by end of a test period and Bayesian-Weibull analysis of Alloy 690 specimens that did not develop cracking by the end of a test period. In this analysis, assumptions were made about the slope parameter based on engineering judgment, e.g., the slope parameter for Alloy 690 may be assumed to be the same as for Alloy 600.
- With insufficient experimental data, the factor of improvement is calculated as the fraction of the Alloy 690 test time to the time of first cracking in an Alloy 600 specimen.

In a laboratory setting, experimental tests are intended to replicate the environment that the materials are exposed to in an operating reactor. Because the incubation period of an environmental degradation mechanism may last for several years under typical operating conditions, laboratory tests are accelerated. An expert panel compiled a database of over 500 Alloy 690 crack growth rate data points and over 130 Alloy 52/152 crack growth rate (CGR) data points. These data points were evaluated and scored for data quality and assessed to determine the effects of parameters such as temperature, crack tip stress intensity factor, yield strength, and crack orientation. According to this expert panel (Hickling 2008; Moss and Was 2017; Jenks et al. 2018a, b), the recommended conservative factors of improvement were 38 for Alloy 690 versus Alloy 600 and 324 for Alloys 52/152 versus Alloys 82/182 (Fig. 9.4).

9.3.2 Philosophy

The piping systems in nuclear power plants are designed to high standards, and major failures are rare events. The rare failures have a low frequency of occurrence. Not only are the major failures rare events when viewed against a frequency scale, but they are also rare when viewed against a passive component "population scale." Each nuclear power plant contains a very large volume of piping components, e.g., many thousands of welds and several linear kilometers of piping. Therefore, the ratio of major failures by the total piping component population is a small number. Most piping failure incidents are incipient or degraded failures with minor or no immediate impact on plant operation and safety. The incipient or degraded failures have a relatively high frequency of occurrence, e.g., equal to or greater than one event per plant and year.

While the volume of technical information on operating experience with piping systems is considerable, the quality of this information varies. Some reports present detailed root cause analysis insights and results, while the majority of the reports contain cursory (and sometimes conflicting) information on the causes and consequences. The determination of root cause involves interpretation of results from visual examinations and, sometimes, detailed metallurgical evaluations of damaged or fractured piping components. In general, failure and reliability analysis of incidents involving piping systems are complex and uncertain.

9.3.3 Basic Database Structure

Piping reliability models and application requirements shape the database structure definition and development process. To assess a mature database's applicability and completeness, consider it through the lens of an expert panel defining the structure based on future needs. Continuous collection and evaluation of operating data are essential.

A pipe failure database should provide clear descriptions of how degraded conditions lead to repair or replacement decisions or affect plant operations. Defining a taxonomy based on principles and failure physics is the starting point, followed by designing the database structure and data relationships. Event classification must be concise and technically accurate. Records should be suitable for direct application (query results used immediately) or indirect use (additional processing required). Users must trust event interpretations and classifications and easily reinterpret or reclassify data as needed for new applications. Database normalization ensures consistent formatting of all entered information.

Databases are based on the taxonomy and methodology of data collection processes, from pipe failure reports to database input. The "success" of a specific database depends on those aspects. It is also important to keep a discussion of casualty coding and taxonomies from a wider perspective. Coded information has both strengths and weaknesses compared to other known techniques. Data or coded information is precise and lends itself to computer-based analysis. However, this format is perhaps not ideal for conveying the overall picture of the casualty. Coded information is necessary as an alternative approach to capturing and storing the events and causal factors of the casualties. One of the main problems with coding information is the bias of results. Different investigators can use different codes for the same events. From this perspective, descriptive information on the accident (local, people involved, consequences, etc.) and a free-text area for the description of accidents are fundamental. An event-recording system for a database should comply with industry standards and regulatory reporting guidelines and should allow for a variety of data "interrogation" exercises. It must be able to use data from other databases and have clear objectives, with possible database analysis to be made. Basically, one should have three principal levels of a pipe failure database:

1. Coded structured data which refers to information organized in a predefined format with specific codes, making it easily searchable and understandable by computers.
2. Semi-structured information such as text files and PDF files.
3. Source material: This can be linked to original information sources such as root cause evaluations and event reports.

The coded part depends on the methodology or taxonomy used. Semi-structured text and various kinds of graphical information may also be incorporated into the database. The source material will also find application in certain analysis situations and should therefore be seen as an element of the database. In practice, this may mean that the database may need to have certain references and pointers to the source material. With the current pace of development in information technology, one may also foresee techniques for computer-based reading or viewing stored material such as scanned text, photographs, and video.

9.3.4 Database Development Process

With overall structure defined, the next step is to develop a comprehensive set of drop-down menus with a standardized list of key words for event classification. The role of these drop-down menus and key words is to support database queries and data applications. Table 9.2 lists a minimum set of required drop-down menus, and Table 9.3 lists a minimum set of database field categories.

9.3.5 Piping System Component Population Data

An essential aspect of pipe failure database development is the creation of a comprehensive piping system component population database. Without it,

Table 9.2 Examples of required database drop-down menus

Drop-down menu	Description
Apparent cause	List of material degradation mechanisms
Code class	Piping safety classification, per ASME Section III, RCC-M, etc.
Collateral damage	Spatial impact, e.g., spraying, flooding, jet stream impact, and pipe whip
Corrective action	Repair method, e.g., in-kind replacement (like-for-like), temporary repair, weld repair, and weld overlay
Crack morphology	Two- or three-dimensional characterization of crack shape and orientation, e.g., intergranular and transgranular
Direction of crack propagation	Axial or circumferential crack orientation or the crack orientation can be "diffuse"
Event category	A characterization of the pipe failure, e.g., primary pressure boundary failure, loss of an affected system train in a system with redundant flow paths
Event type	A description of the failure, e.g., part through-wall crack, minor leakage, pinhole size leak, severed pipe
Impact on operation	Includes a broad range, from no impact to reactor trip/turbine trip
Location of flaw initiation	Differentiates between base metal, weld heat-affected zone, or weld metal
Material at crack initiation	Chapter 5, Sect. 5.1
Material designations	
Material types	Class of material, e.g., carbon steel, low-alloy steel, and stainless steel
Method of fabrication	Chapter 5, Sect. 5.10
Method of flaw detection	Chapter 8, Sects. 8.1 and 8.2 through 8.8
Piping component type	Chapter 5, Sect. 5.8
Plant location	The area within the plant building structures where a failure occurred, e.g., containment, drywell, auxiliary building, and turbine building
Plant operational states	Full power, low power, cold shutdown, etc.
Plant population data	Linked to a database within the database which includes plant names, type of plant, plant design type, date of initial reactor criticality, date of commercial operation, etc.
Plant system	A unified set of plant system identifiers; Tables 9.4 and 9.5
Process medium	Can be primary water, raw water, dry steam, wet steam, etc.
Repair time class	A database filter that bins actual time to repair (TTR) in classes
Pipe size class	Differentiates between small-diameter and large-diameter piping
System group	Can be RCPB, reactor auxiliary system, auxiliary cooling system, etc.
Technique of flaw detection	Chapter 8
Weld location	The location of a weld in a pipe spool, e.g., pipe to bend and pipe to elbow
Welding technique	Chapter 5, Sect. 5.10

Table 9.3 A minimum set of database fields

Database field name	Database field type
Event date	Date
Plant name (incl. Type and design generation)	Drop-down menu with link to a "nuclear plant database"
Piping design code	Drop-down menu
Plant operational state	Drop-down menu
Reference(s)	Text
Event type	Drop-down menu
Isolable primary coolant leak	Yes/no
Isolable pipe leak	Yes/no
Event category	Drop-down menu
Collateral damage	Drop-down menu
Corrective action	Drop-down menu
Impact on plant operation	Drop-down menu
Repair time	Number
Event narrative	Free text, no size limit
Quantity released	Number
Duration of leakage	Number
Through-pipe-wall mass flow rate	Number
System	Drop-down menu
Equivalent break size	Number
Supported function	Drop-down menu
Instrument line	Yes/no
Impulse line	Yes/no
Pressure relief line	Yes/no
Piping component	Drop-down menu
Location of weld in piping	Drop-down menu
Safety class	Drop-down menu
Operating regime	Early life, long-term operation, etc.
Age of component @ time of failure	Number
Age of plant at time of failure	Auto-generated number
Pipe diameter—Metric system	Number
Pipe diameter—Imperial system	Number
Wall thickness—Metric system	Number
Pipe schedule number	Number
Material type	Drop-down menu
Material designation	Drop-down menu
Process medium	Drop-down menu
Operating temperature	Number
Design temperature	Number
Operating pressure	Number
Design pressure	Number
SCC morphology	Drop-down menu
Apparent cause	Drop-down menu

Table 9.3 (continued)

Database field name	Database field type
Contributing factor(s)	Drop-down menu
Repair weld[a]	Yes/no
ISI history	Free text, no size limit
Is pipe failure attributed to less-than-adequate reliability and integrity management (RIM)?	Yes/no
Results of root cause analysis	Free text, no size limit
Availability of laboratory test results	Yes/no
Location of pipe flaw/leakage point	Free text, no size limit
Plant location where failure occurred	Drop-down menu
Method of flaw detection	Drop-down menu
Method of fabrication/welding	Drop-down menu
Flaw description	Free text, no size limit
Location of crack initiation	Drop-down menu
Direction of flaw propagation	Drop-down menu
Number of flaws in examination volume	Number
Flaw depth	Number
Flaw length	Number

Repair welding is a key contributor to stress corrosion cracking failures. For details on the structural significance of repair welds, see Slater, G. (1983), The Effect of Repair Welds on Service Performance, Report LD 22827, The Welding Institute, Cambridge, United Kingdom

bottom-up precursor rates cannot be generated. The basic structure of a pipe failure database consists of four basic elements: (1) pipe failure event information, (2) nuclear power plant information database, (3) piping system design database, and (4) database user interface. The third element is a database that consists of transcribed piping and instrumentation diagram (P&ID) and isometric drawing information that is organized by type of reactor and by plant system. Such information is necessary in correlating event descriptions with piping design information to support event classifications. While a tedious and time-consuming activity, creating a "piping component population database" has been made easier through an increased use of CAD drawings and computerized in-service inspection programs from which weld counts, etc. can be extracted.

For each plant system, each weld and pipe fitting are identified by material, pipe diameter and wall thickness, location in the piping system relative to heat exchangers, piping support pump, valves, and vessels. This database should include a representative set of nuclear power plant design generations and cover all systems of interest, i.e., ASME III Code Class 1, 2 and 3 piping as well as non-safety-related piping of interest. As an example of an existing piping component population database, Shin and Park (2010) describe the development of a nuclear materials and degradation database.

The work to correlate a certain pipe failure population with the population of plants and susceptible locations within piping systems that produced the failures can be time-consuming. It requires knowledge of plant-to-plant variability in piping

Table 9.4 Example of a nuclear plant system identification cross-reference table

System ID — common	Description	EIIS	Czech Republic	France	Germany		Sweden	
					AKS	KKS	BWR	PWR
ADS	BWR primary depressurization system (BWR)		N/A	N/A	TK, RA		314	N/A
AFW	Auxiliary feedwater system	BA		ASG	RQ		327	327
CC	Component COOLING WATER SYSTEM	CC	TF	RRI	TF	LA	435	435
COND	Condensate system	SD			RM, RN	LC	414	414
CRD	Control rod drive (insert/removal/crud removal)	AA					354	354
CS	Containment spray system	BE	TQ	EAS			322	322
CVC	Chemical and volume control system (PWR)	CB		RCV	TA, TC, TD	KB	N/A	334
CW	Circulating water system	BS					443	443
EHC	Electrohydraulic control system						442	442
EXT	Steam extraction system	SE					419	419
FPS	Fire protection system	KP	C-52			SG	861	861
FW	Main feedwater system	SJ		ARE	RL	LA	312 / 415	415
HPCS	High-pressure core spray (BWR)	BG	N/A	N/A	TJ		323	N/A
HPSI	High-pressure safety injection (PWR)	BQ	TJ	RIS	TH	JN	N/A	323
IA	Instrument air system	LD	US				751	751
LPCS	Low-pressure core spray (BWR)	BM	N/A	N/A	TK, TM		323	323
LPSI	Low-pressure safety injection (PWR)		TH	RIS	TH	JN	323	323
MS	Main steam system	SB		VVP	RA	LB	311 / 411	411
MSR	Moisture separator reheater system	SB			RB	LB	418	418
RCS	Reactor coolant system (PWR)	AB		RCP	YA, YB	JA, JE	N/A	313
RHR	Residual heat removal system	BO / BP		RRA	TH	JN	321	321

Table 9.4 (continued)

System ID — common	Description	EIIS	Czech Republic	France	Germany		Sweden	
					AKS	KKS	BWR	PWR
RR	Reactor recirculation system (BWR)	AD	N/A	N/A			313	N/A
RPV-HC	RPV head cooling system (BWR)		N/A	N/A	TC		326	N/A
RVLIS	Reactor vessel level indication system (BWR)		N/A	N/A			211	N/A
RWCU	Reactor water cleanup system (BWR)	CE	N/A	N/A	TC	KB	331	N/A
SFC	Spent fuel pool cooling system	DA	TG	PTR	TG	FA	324	324
S/G blowdown	Steam generator blowdown system (PWR)	WI		APG	RS	LA	N/A	337
SLC	Standby liquid control system (BWR)	BR	N/A	N/A			351	N/A
SW	Service water system	BI / KG	VF	SEC	VE	PE	761	761

This table is included for general information only. It is an incomplete set of system identifiers
EIIS energy industry identification system

system designs (i.e., material selections, method of fabrication, number and types of weldments, number of pipe fittings such as elbows, reducers and tees, welding technology, piping codes and standards). The plant-to-plant variability in piping system design (e.g., length of piping, number of welds) can be extensive. Access to a sufficiently complete piping system design database is required. If not readily available, it is important to obtain the needed information to support the definition of exposure terms and to characterize the uncertainties. The process of developing a piping system component population database consists of the following steps:

1. Obtain P&IDs for systems of interest, and identify all pipeline identifications, usually alphanumeric designators, with pipe diameters and line descriptions, and system designator.
2. For each pipeline ("spool piece"), locate the corresponding ISI and fabrication isometric drawings.
3. For each spool piece, add diameter, wall thickness (schedule number), and material grade.
4. From a CAD drawing or ISI Program Plan, obtain a weld list for the system of interest, and import it to an Excel spreadsheet. Review the weld list against corresponding isometric drawings, and add columns in the spreadsheet so that information on materials, dimensions, and locations within the system and relative to floor/wall penetrations is recorded. Use a P&ID as a guide for how the information is entered, which should be in the direction of the process flow.

Table 9.5 Limitations and problems in database development

Issue	Challenges
The source of data	The data source can lack the required data, and the source information can be spread over several different source documents. It is recommended to carefully evaluate this aspect in the planning measures to assess data quality and collection method
Interpretation and classification	Commonly, data are compiled from the source into a standardized format (database). In this process, the source data can be interpreted differently by various individuals. Correct definitions, training, and quality checks can reduce this problem. Ideally, the original source of data should be retained within the database as free text
Data format	The use of coded fields is fundamental to ensure data collection efficiency and consistency of data entered (e.g., correct coding of a manufacturer). Free text should, however, be included to ensure traceability in the data classification process
Data collection method	Most data needed for this category of data collection are today stored in computerized systems. By using state-of-the-art conversion algorithms and software, it is possible to transfer data among different computer platforms and programs
Competence & motivation	Data collection in the "normal" manual way can become a repetitive and tedious exercise. Therefore, take care to employ people with sufficient know-how to do the jobs, avoid using personnel with low competence/experience as data quality can suffer, and find measures to stimulate the reliability and maintenance data collection staff, e.g., by training, doing plant visits, and involving them in data analyses and application of results. Other examples include feedback on data collection results and involvement in QA processes

5. Verify weld list against fabrication isometrics; add details on branch connections, drain lines, vent lines, and instrument lines. Identify and record the presence of socket welds. For large-diameter butt welds, differentiate field welds (FWs) from shop welds (SWs).
6. For each weld, add ISI inspection category, i.e., category B-F to indicate Code Class 1 pressure-retaining dissimilar metal weld and category B-J to indicate Code Class 1 pressure-retaining similar metal weld, C-F-1 to indicate Code Class 2 welds in austenitic or high-alloy steel piping, and C-F-2 to indicate Code Class 2 welds in carbon steel or low-alloy steel piping.[3]
7. To support high-energy line break analyses and internal flooding PSA, add pipe lengths, pipe supports, floor elevations, location within the auxiliary building, service building, and turbine building.

[3] ASME Section III and Section XI define weld categories.

9.4 Data Quality, Data Curation, and Completeness

Data quality and data uncertainty are closely related. Different forms of uncertainty exist in operating experience data and in the reliability data parameters derived from it. Reliability data represents an estimate of the average performance expected from similar components, rather than a precise prediction of future component outcomes. This type of uncertainty is inherent in probabilistic analyses. Uncertainty may also be caused by incomplete information. For example, failure events recorded might only represent a sample of actual field performance, or some failures may not be reported.

The first source of uncertainty is insignificant compared to the other sources of uncertainty, however. The failure records on which the reliability parameters are based may be incomplete records of component performance. Multiple information sources coupled with detailed classification and interpretation are used to address the completeness issue. Limitations of available data and also limited detail about particular failure events mean that it will be challenging to establish data which is exactly appropriate to a component evaluation boundary under consideration. For example, applying light water reactor operating experience to advanced non-light water reactor primary pressure boundary reliability studies is challenging.

Data quality is affected from the moment the field experience data is recorded at a nuclear power plant, interpreted, and finally entered into a database system. Data quality is an attribute of the processes that are implemented to ensure that any given database record can be traced to the source information and that assumptions made in the data evaluations are clearly stated. According to ISO 14224:2016, data quality is controlled by the process that is in place for obtaining operating experience data and the data collection process itself. The process for obtaining data consists of four elements that define the level of achieved quality: (1) definition of data quality, (2) "planning measures," (3) verification of quality, and (4) limitations and problems in obtaining quality data. ISO 14224:2016 (ISO 2016) defines data quality as follows:

- Confidence in the collected failure data, and hence any analysis, is strongly dependent on the quality of the data collected. High-quality data are characterized by the following:

  (a) Completeness of data in relation to specification.
  (b) Compliance with definitions of reliability parameters, data types, and formats.
  (c) Accurate input, transfer, handling, and storage of data (manually or electronic).
  (d) Sufficient population and adequate observation period to give statistical confidence.
  (e) Relevance of data to the needs of the users.

Possible problems and limitations to be aware of when obtaining quality data are summarized in Table 9.5. On verification of quality, Section 7.1.3 of the ISO standard states that during and after the data collection exercise, analyze the data to verify consistency, reasonable distributions, proper classification, and correct interpretations in accordance with a coding guideline. This verification-of-quality process shall be documented and may vary depending on whether the data collection is for a single plant or involves several company or industry facilities. When merging individual databases, it is imperative that each data record has a unique identification. Assess the quality of data being collected as early as feasible in the data collection process in accordance with the planning measures. A suitable procedure is an assessment by the data collector, who shall be provided with guidelines for what quality measures should be focused on in accordance with the planning measures. The main objective of this early assessment is to look for any problems that require the planning measures to be immediately revised to avoid unacceptable data being collected.

The international ISO 8000 standard series defines data quality as the degree to which a set of inherent data characteristics fulfill requirements, ISO 8000-2:2022 (ISO 2015; Kanga et al. 2024). This standard identifies three categories to measure data quality:

1. Syntactic quality: the degree to which data conforms to its specified syntax.
2. Semantic quality: degree to which data corresponds to what it represents.
3. Pragmatic quality: the degree to which data is found suitable and worthwhile for a particular purpose.

Schäbe and Shubinsky (2024) offer insights into the completeness and sufficiency of statistics. A key consideration is whether the completeness of a database adequately supports the generation of inputs for models assessing piping reliability (Fig. 9.5). At the core of this concept is the idea that sufficient statistics contain all information present in a data sample that aligns with a parametric model, much like results from database queries. Typically, these statistics include the model's parameters, which may relate to materials, dimensions, operating conditions, relevant codes and standards, degradation mechanisms, and other factors. The authors illustrate this using a sample drawn from a normally distributed population, where the mean and standard deviation encapsulate all necessary information.

Furthermore, the existence of a defined model provides supplementary information beneficial for data processing. This underscores the necessity for databases concerning pipe failures to encompass comprehensive and pertinent details about reliability attributes and influencing factors. Employing artificial intelligence (AI) technologies can facilitate the extraction of deeper insights, thereby increasing the efficiency of database querying processes. AI can be tasked with identifying an appropriate parametric model for a dataset, rather than simply describing the existing data that fits such a model. It is essential, however, to validate AI-generated outputs against contextual information regarding the origin of the data and assess the plausibility of recommendations produced by AI tools.

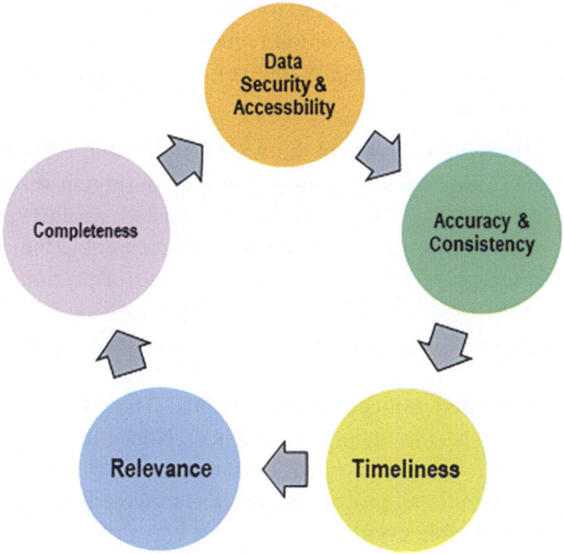

Fig. 9.5 The five dimensions of data quality

9.5 Reporting Pipe Degradation and Failure

Accessing technical information pertaining to pipe failure events is a two-step process, which begins by gathering information on reporting routines. Where is the data of interest located? These routines differ. Next step is to determine the information access rules. Fundamentally, all reporting is plant centric, meaning that the failures are recorded in condition reports and incident reports and by definition the access to those sources is restricted to plant personnel and other authorized parties. There are clear distinctions between a "recorded," "reported," and "reportable" event.

A large portion of reported incipient (precursor) failures within the reactor coolant pressure boundary (RCPB) are detected by in-service inspection during annual refueling and maintenance outages. Relaxations in plant technical specifications and nonconforming findings during outages may result in discretionary reporting, however. This means that while formal licensee event reports (LERs) would not be filed based on the NDE/ISI findings, other means of reporting could be prepared as part of summaries of the performance of outage activities (i.e., outage inspection reports). If a "significant" ISI finding by one licensee is believed to have potential generic, industry-wide implications, then that finding would be reported and result in formal dispositioning. Not only would the "discovering" licensee provide a report, but also the other licensees, which are affected by the original ISI results.

Piping systems, welds, and fittings are subject to ongoing surveillance through walkdown inspections, visual checks, leak monitoring, chemistry sampling, and nondestructive examinations, both during reactor operation and shutdown phases.

Events discovered during routine power operations are documented through a condition report or an action request, followed by a work order, formal operability determination, and finally a repair or replacement plan. The various levels include the following:

1. **Recordable flaws**: In the ASME Boiler and Pressure Vessel Code, a recordable flaw indication is a flaw detected during in-service inspection that must be documented and potentially evaluated against acceptance criteria. Specific requirements for what constitutes a recordable indication vary depending on the relevant ASME code section and the inspection method used. IWB-3514 of ASME Section XI provides acceptance standards for Category B-F pressure-retaining dissimilar metal welds and Category B-J pressure-retaining welds in piping.
2. **Recorded nonconforming events**: These are events that are recorded in the plant-specific reporting routines. Nonconforming events in nuclear reactors refer to occurrences that deviate from established safety standards, regulations, or operational procedures, such as pinhole leaks.
3. **Reportable events**: Reportable events in nuclear reactors are incidents or conditions that require notification to regulatory authorities or the International Atomic Energy Agency (IAEA). These events can range from minor operational deviations to serious accidents and are categorized based on their potential safety significance and impact. These events are classified using the International Nuclear and Radiological Event Scale (INES) (Table 9.6).

The NDE-based reporting criteria are interpreted and implemented on a case-by-case basis, and a lack of functional considerations could impose restrictions on the dissemination of reports within and outside an organization. Examples of reporting practices and requirements include the following:

– Significant incipient or degraded failures discovered during refueling or extended maintenance outages normally are reported to regulatory agencies.
– Some degraded failures during routine power operation are reported, especially those with assumed generic implications.
– Most degraded failures within the reactor coolant pressure boundary are reported, especially where there is an external leakage which is detected by the leak detection system(s) during normal plant operation.

There are exceptions to the above practices, however. As an example, to effect repairs, an RCPB leak could result in a planned shutdown of the unit. While progressing with the manual shutdown, an equipment failure occurs which is unrelated to the leak but possibly triggered by the change of plant status and causes an automatic reactor trip, say, from 50% power. In this case, an LER may be filed for the equipment failure which caused the trip directly but not filed for the piping failure as such (USNRC 2013). The pipe repair would be noted in a mandated ASME XI Owner's Activity Report, however.

Table 9.6 International nuclear and radiological event scale

Level	Classification	Description	Examples
7	Major accident	Impact on people and environment: Major release of radioactive material with widespread health and environmental effects requiring implementation of planned and extended countermeasures	There have been two Level 7 accidents: Chernobyl accident on 26 April 1986 and Fukushima accident involving a series of events beginning on 11 March 2011
6	Serious accident	Impact on people and environment: Significant release of radioactive material likely to require implementation of planned countermeasures	There have been no level 6 accidents in commercial nuclear power plants
5	Accident with wider consequences	Limited release of radioactive material likely to require implementation of some planned countermeasures. Severe core damage	Three Mile Island Unit 2 accident on 28 March 1979. The first accident that attracted worldwide attention
4	Accident with local consequences	Impact on people and environment: Minor release of radioactive material unlikely to result in implementation of planned countermeasures other than local food controls. Fuel melt or damage to fuel resulting in more than 0.1% release of core inventory	Saint-Laurent A1 (GCR), 1 7 October 1969: An error during fuel loading prevented proper circulation of the carbon dioxide, which greatly reduced the cooling of the fuel elements in a channel of the reactor core. This caused a meltdown of 50 kg of uranium. The rise in radioactivity in the reactor chamber caused an auto matic reactor trip
3	Serious incident	Impact on defense in depth: Significant failures in multiple safety provisions. No external release of radioactivity	Vandellòs I (GCR): On 19 October 1989, a lubrication oil line ruptured as a result of turbine vibration. A lack of bearing oil led, in turn, to a hydrogen explosion and oil fire. Emergency cooling and heat exchange systems were disabled
2	Incident	Impact on defense in depth: Significant failures in safety provisions	Gundremmingen-A: On 13 January 1977, severe weather induced loss of the 220 kV grid during routine power operation. As a result of overfilling the RPV, the water loading damaged the piping to one of the safety valves resulting in loss of coolant and blowdown from 8 MPa to 0.2 MPa within 20 minutes.[a]
1	Anomaly	Minor degradation of pressure boundary integrity, with significant defense in depth remaining	Penly 2: On 5 April 2012, a fire started in the containment as a result of a reactor coolant pump (RCP) lubrication oil system leak.

(continued)

Table 9.6 (continued)

Level	Classification	Description	Examples
			Later in the day, an RCP seal leakage was observed.[b]
0	Deviation	No safety significance. Operational limits have not been exceeded	$\Delta CDP \ll or \leq 10^{-6}$

[a]A more recent event is the 2004 pipe failure at the Vandellòs II nuclear power plant which was classified as an INES Level 2 event by the Spanish Nuclear Safety Council (CSN). This incident involved a failure in the essential service water (ESW) system piping. The CSN determined that there was widespread corrosion in the ESW system that the licensee had known about since 1998 but had not corrected or reported to the CSN. The incident was classified as an INES Level 2 due to the degradation of the defense in depth and weaknesses in the licensee's safety culture. INES Level 2, or an "incident," indicates a significant deficiency in safety provisions, but where safety is still maintained despite a potential additional failure

[b]It is unclear if high RCP vibrations caused a lubricating oil system fatigue failure or if the lubricating oil system failed due to other reasons. The interested reader can find additional perspectives on the issue by consulting a report from 1984 entitled "An Analysis of Postulated Fires in Reactor Coolant Pump Bays." The report is available in the NRC-ADAMS, Accession No. ML20083L740

The Canadian Nuclear Safety Commission (CNSC) and the Swiss Federal Nuclear Safety Inspectorate (ENSI) reporting requirements with respect to piping pressure boundary degradation and failure are as follows:

- CNSC (2024) reporting requirements: Each plant owner is required to submit a quarterly report on nuclear power plant pressure boundaries. The report shall include all CANDU Class 1–6 pressure boundary systems in accordance with CSA N285.0, "*General Requirements for Pressure-Retaining Systems and Components in CANDU Nuclear Power Plants/Material Standards for Reactor Components for CANDU Nuclear Power Plants*," for the purposes of reporting on pressure boundaries and pressure boundary degradation.[4]
- ENSI-B03 (2023): Paragraph 5.1.1.2.c concerns reporting of degraded reactor pressure vessel safety-related piping pressure boundaries that are discovered during nondestructive examinations. Paragraph 5.1.1.2.d concerns pipe leaks and breaks in safety-related and non-safety-related piping systems.

National codes and standards impose additional levels of reporting. Figure 9.5 illustrates the different levels of reporting. Figure 9.6 shows the distribution of sources of information on pipe failures in this book's database of reference (Fig. 9.7). Based on practices in the United States, the different levels of reporting are outlined in Table 9.7.

[4]https://www.csagroup.org/store/product/CSA_N285.0:23-CSA_N285.6_SERIES:23/

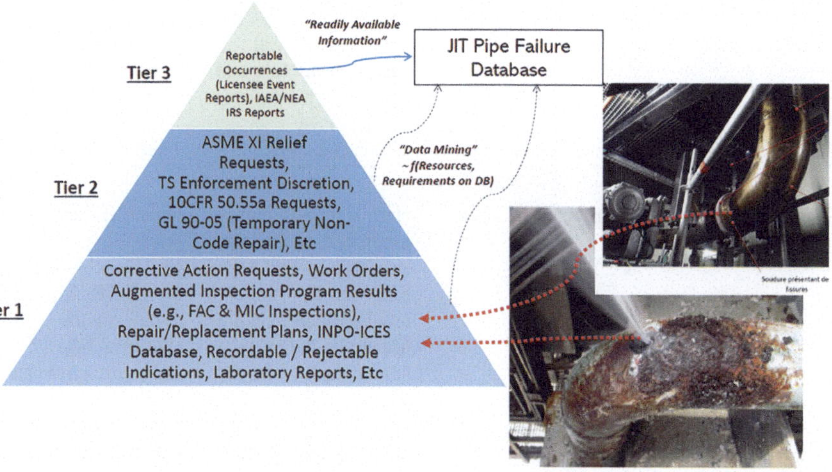

Fig. 9.6 The different levels of reporting pipe failures

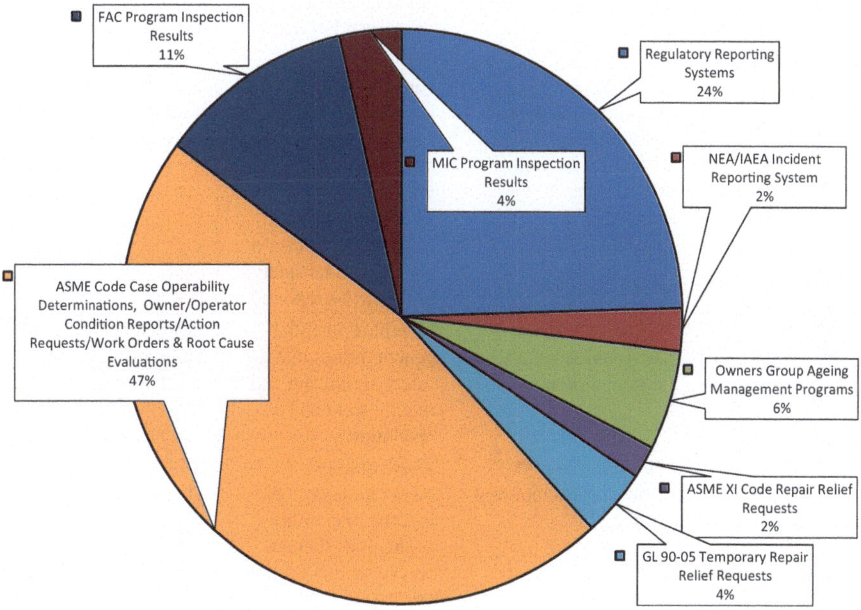

Fig. 9.7 The different sources of pipe failure data

Table 9.7 Recording and reporting of pipe failures in the United States

Reporting level	Report/database type	Level of technical content from the point of view of materials science, PSA applications, etc.	Third-party access
United States Nuclear Regulatory Commission			
Regulatory requirements and guidance	Licensee event report	High level—Requires interpretation and classification by an SME. Since the 1990s, only a few LERs on pipe failure are issued on an annual basis; <10 per year	https://lersearch.inl.gov/LERSearchCriteria.aspx. LERs from 1960 to date are also available from https://adams.nrc.gov/wba/
	Event notification (EN) system	High-level event tracking system for incoming notifications of significant nuclear events with an actual or potential effect on plant safety. An EN may result in an LER	Public domain: All ENs from January 1, 1999, onwards are available at https://www.nrc.gov/reading-rm/doc-collections/event-status/event/en.html
	10 CFR part 21—Reporting of defects and noncompliance	Part 21 implements Section 206 of the Energy Reorganization Act of 1974 (42 U.S.C. § 5846, et seq.). Requires defects in basic components posing substantial safety hazard to be reported	https://www.nrc.gov/reading-rm/doc-collections/gen-comm/
	Generic communications—Generic letters, information notices	Identifies potential generic issues that are safety significant and require technical resolution	https://www.nrc.gov/reading-rm/doc-collections/gen-comm/
	Generic letter 90-05 request for temporary repair	Applicable to moderate-energy piping, the operability evaluations that are part of the 90–05 relief request are comprehensive	Public domain; https://adams.nrc.gov/wba/
	10 CFR 50.55a(z)(2)—Application of ASME code cases	The operability evaluations that are part of the ASME code case relief request are comprehensive	Any code case relief request that is submitted to the NRC for approval is also available for review by a third party

Table 9.7 (continued)

Reporting level	Report/database type	Level of technical content from the point of view of materials science, PSA applications, etc.	Third-party access
	Biannual problem identification and resolution inspection report	Regulatory review of a licensee's corrective action program and the licensee's implementation of the program to evaluate its effectiveness in identifying, prioritizing, evaluating, and correcting problems and to confirm that the licensee is complying with NRC regulations and licensee standards for corrective action programs	Public domain; https://adams.nrc.gov/wba/
	Quarterly integrated inspection reports	A significant repository of information on passive component failures. Requires interpretation, classification, and follow-up by a subject matter expert (SME)	Public domain; https://adams.nrc.gov/wba/
	ASME XI owner's activity report	Cursory information, see below	Public domain; https://adams.nrc.gov/wba/
United States Nuclear Industry Level			
Institute of Nuclear Power Operations (INPO)	INPO Consolidated Events Database (ICES). Participating plants submit ARs, CRs, WORs, etc.—Mainly free-format narrative information—To a searchable database	Varying—Data extraction, interpretation, and classification can be time-consuming	Access restricted to ICES participating organizations
Nuclear Energy Institute (NEI)	NEI 07-07: Industry Ground Water Protection Initiative—Final Guidance Document, 2007	Reporting of below-ground/buried pipe failures causing >0.38 m^3 unintended release	Annual radioactive effluent release report submitted by licensees to the USNRC
Electric Power Research Institute (EPRI)	EPRI-CHUG: Presentation material from the biannual CHUG meetings	Varying—Data extraction, interpretation, and classification can be time-consuming	Access restricted to funding organization. Nonproprietary versions of EPRI-MRP

(continued)

Table 9.7 (continued)

Reporting level	Report/database type	Level of technical content from the point of view of materials science, PSA applications, etc.	Third-party access
			reports can be downloaded from EPRI and NRC websites
	EPRI-MRP, multiple programs—Reactor internals, thermal fatigue, aging management	NRC receives biennial summaries of the BWR and PWR reactor internal inspection results	
	Plant engineering/long-term operations/risk and safety management programs	Extensive source of material degradation and failure information organized by system, degradation mechanism, etc.	Access restricted to funding organization
NRC/EPRI-MRP Annual Technical Information Exchange meetings	Presentation material summarizing R&D status, results of reactor internal inspections, thermal fatigue inspections, etc.	High level	Available from NRC-ADAMS
GE Service Information Letter (SIL) and equivalent bulletins from the PWROG	Information on material degradation issues in response to recent operating experience	Chronological lists of events with some information on the method of discovery, extent of degradation, flaw size data, etc. plant identities not revealed	The SILs can be downloaded from the NRC public document room; see for example https://www.nrc.gov/docs/ML0501/ML050120032.pdf
Plant level			
During routine operation—Scheduled walkdown inspection	Action request (AR), condition report (CR), corrective action request (CAR), incident report (IR), nonconformance report (NCR), problem investigation process (PIP), repair/replacement plant (RRP), work order request (WOR), operability determination, and structural evaluation in	Details on the location of flaw/leak with reference to P&ID and isometric drawing—Oftentimes accompanied by photographic records, root cause analysis results, NDE results, metallographic data, flaw size data, leak/flow rates, spatial effects, risk significance, and safety significance. Piping fabrication data,	Restricted/proprietary

Table 9.7 (continued)

Reporting level	Report/database type	Level of technical content from the point of view of materials science, PSA applications, etc.	Third-party access
	accordance with applicable ASME code case	operational parameters, piping design information (material, dimensional data)	
Leak detection—Containment/drywell	Once every 8/12-h shift, RCS inventory mass balance calculation, chem. Lab trending of inflows to aux. Building/reactor building sump and holding tank	Logs, databases (various types)	Restricted/proprietary
	Containment/drywell leak detection alarm (≥6.3E-3 kg/s)	Control room logs/computer printouts, excel spreadsheets—Holding tank inflow, details of through-wall leak rates as a function of time and location, etc.	Restricted/proprietary
In-service inspection (ISI) per ASME XI—Mandatory inspections	List of "recordable indications," operability determinations/fitness-for-service (FFS) evaluations, ASME XI relief requests (for more details, see "regulatory" above). Within 120 days of the conclusion of a refueling outage, the plant owner is required to submit an ISI summary report ("owner's activity report") to the USNRC[a]	The ASME relief requests submitted to the USNRC for approval include relevant information. The 120-day owner activity reports include "abstract of repair/replacement activities required for continued service" and "items with flaws or relevant conditions that required evaluation for continued service"	ASME relief requests submitted to the USNRC are available via the NRC public document room; https://adams.nrc.gov/wba/
Owner-defined ISI programs	FAC, erosion-corrosion, MIC, and thermal fatigue programs	NDE reports, "system health reports"	Restricted/proprietary
ASME	Numerous code cases have been developed for the operability determination of degraded piping	Details on location of flaw/leak with reference to P&ID and isometric drawing—Oftentimes	When a code case has been approved for plant-specific use, an associated operability determination is kept

(continued)

Table 9.7 (continued)

Reporting level	Report/database type	Level of technical content from the point of view of materials science, PSA applications, etc.	Third-party access
		accompanied by photographic records, root cause analysis results, NDE results, metallographic data, flaw size data, leak/flow rates, spatial effects, risk significance, and safety significance. Structural integrity evaluation results are part of the required documentation	on file on-site and available for review by NRC resident inspector

[a]Through the 2017 edition of ASME BPV Code, Section XI, owners were required to prepare summary reports or owner activity reports of pre-service examination, in-service examinations, and repair replacement activities within 90 calendar days of the completion of each refueling outage. In the 2019 edition of Section XI, this timeframe was extended to 120 days. Per the Federal Register Vol. 87, No. 207 dated Thursday, October 27, 2022, the USNRC relaxed the requirement for all licensees. Licensees using Section XI, editions and addenda prior to the 2010 edition, may utilize Code Case N-778, "Alternative Requirements for Preparation and Submittal of Inservice Inspection Plans, Schedules, and Preservice and Inservice Inspection Summary Reports, Section XI, Division 1," to obtain the 120-day submittal relaxation

9.6 Pipe Failure Data Resources

A "data resource" is a structured set of data held in a computer with a view to its utilization by subject matter experts in specific types of applications. With reference to piping operating experience, the different forms of data resources are identified in this section.

9.6.1 Sources of Operating Experience Data

Data resources of interest range from "just in case" (JIC) to "just in time" (JIT). JIC databases, which contain general operating experience data stored proactively to support future risk and reliability studies, are shown in Fig. 9.8 and Tables 9.8, 9.9, and 9.10. JIT, or engineering, databases include operating experience data processed to meet specific user requirements (Table 9.11). A third type of database consists of tabulations of pipe failure rates and rupture frequencies that have resulted from research projects.

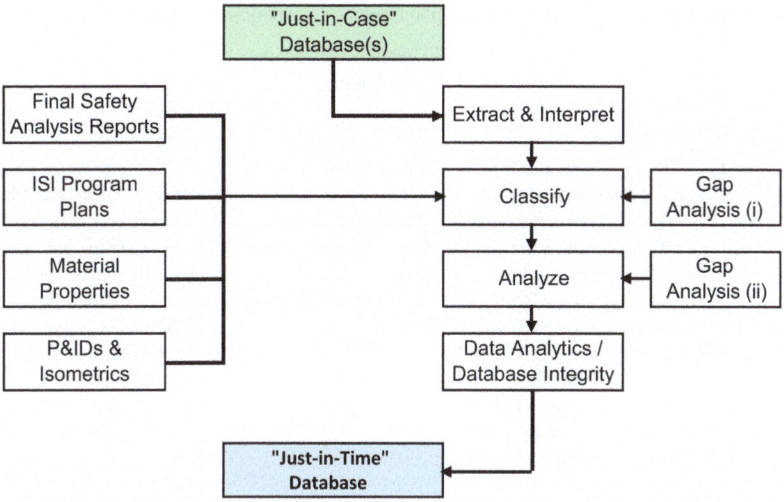

Fig. 9.8 The JIC-to-JIT "transition states"

National and international JIC repositories for nuclear power plant operating experience contain extensive information regarding failures of systems, structures, and components, as well as data on human performance factors that have adversely affected plant safety and reliability. At the most comprehensive level, these substantial collections consist of licensee event reports or abnormal occurrence reports, prepared in accordance with regulatory reporting requirements and other types of incident reports. While it is feasible to extract and analyze pipe failure data from these repositories, the process is quite time-intensive. Certain repositories are publicly accessible; however, many are restricted or proprietary in nature. For instance, the USNRC Agencywide Documents Access and Management System (ADAMS) serves as a public-domain resource, offering official recordkeeping and access to publicly available documents. Additional examples of JIC repositories include the following:

- NRC reactor operating experience data (NROD): This is a website maintained by the USNRC and the Idaho National Laboratory (INL) to provide access to coded event data for NRC staff and industry personnel. It serves as a resource for searching and viewing data related to nuclear reactor operating experience, including links to the reliability and availability data system (RADS). RADS is a database and analysis code, and the data contained in RADS is kept up-to-date by loading most of the current quarter's INPO-ICES data and by yearly loads of initiating event data from licensee event reports (Rasmuson et al. 2005).
- IAEA incident reporting system (IRS): A collaborative effort with the Nuclear Energy Agency of the Organisation for Economic Cooperation and Development (OECD/NEA) to enhance safety by sharing information on safety-related events

Table 9.8 Examples of JIC nuclear power plant operating experience databases (Part I)

Database characteristic	IAEA/NEA incident reporting system	United States Nuclear Regulatory Commission (USNRC) LER database	USNRC ADAMS public documents
Accessibility	The IRS is for use within each member state by organizations professionally involved in the nuclear power industry. Organizations wishing to nominate their staff to become generic users of the web-based IRS (WBIRS) need to contact their respective IRS national coordinator with the request	Open-access licensee event report (LER) database accessible via https://lersearch.inl.gov/LERSearchCriteria.aspx	The Agencywide Documents Access and Management System (ADAMS) is the official recordkeeping system, through which the USNRC provides access to the Publicly Available Records System (PARS) Library and the Public Legacy Library
Owner/ operator	An international system jointly managed by the IAEA and the OECD/NEA. The IRS enables participating countries to exchange operational experience to improve the safety of nuclear power plants. Established in 1980	The LER database is operated by the Idaho National Laboratory on behalf of the USNRC	The official recordkeeping system of the USNRC
Database scope	General purpose OE database	General-purpose OE database	Information of interest includes inspection reports, LERs, ASME XI relief requests, information on rejectable, recordable, and reportable pressure boundary flaws and failures
Database content (no. of data records of interest)	Includes more than 600 entries on pipe failures	>ca. 2000	Extensive, searchable source of pipe failure information dating from the early 1960s
Data submission protocol	IRS is based on the principle that each participant will provide timely information on its NPPs' OE so that it is available to all other participants	NUREG-1022 R3 (Event Report Guidelines, 10 CFR 50.72 and 50.73)	The official recordkeeping system of the USNRC
Taxonomy	Documented in the Incident Reporting System Guidelines (IAEA 2022)	Described in NUREG-1022	https://www.nrc.gov/reading-rm/adams.html
	Web-based SQL	Web-based SQL	Free format

Table 9.8 (continued)

Database characteristic	IAEA/NEA incident reporting system	United States Nuclear Regulatory Commission (USNRC) LER database	USNRC ADAMS public documents
Database type			
No. of database fields	Searchable narrative information. Each report consists of five parts: 1) nuclear power plant information, 2) narrative description of the event, 3) safety assessment, 4) cause analysis, and 5) lessons learned	Narrative information that is accessible through search functionality	Narrative information that is accessible through search functionality

at nuclear power plants and research reactors. It focuses on exchanging operating experience, lessons learned, and corrective actions to prevent similar incidents.[5]

Examples of JIT data resources, some of which include piping operating experience, include the following:

– Engineering JIT databases especially developed for probabilistic piping reliability analysis: Such database captures information on pipe failures, including the underlying cause(s) of failure. Each event is classified according to a taxonomy that acknowledges chemical and mechanical properties of materials, environmental conditions, fabrication techniques, ISI technologies, piping design parameters, and material degradation mechanisms. Querying such a database produces the input to piping reliability models. Examples of JIT databases for PSA include the following:
  - CODAP, the Component Operational Experience, Degradation and Ageing Programme project, is a collaborative effort among Nuclear Energy Agency (NEA) member countries to collect and analyze pipe failure data (OECD/NEA 2024).

Of relevance to this book are proceedings of international serial conferences dedicated to material degradation mechanisms, structural integrity issues, and related operating experience. Three examples of such proceedings are as follows:

– International Conference on Environmental Degradation (ENVDEG): Held every 3 years. ENVDEG fosters an exchange between plant operators, industry, and academic researchers concerning material degradation issues and solutions in the

[5] The International Atomic Energy Agency (IAEA) and the Nuclear Energy Agency/NEA of the OECD established the IRS in 1980 to draw conclusions from experience for improving the operation of nuclear power plants.

Table 9.9 Examples of JIC nuclear power plant operating experience databases (Part II)

Database characteristics	VERA(Vertiefte Auswertung meldepflichtiger Ereignisse)	SAPHIR	SAPIDE
Accessibility	Proprietary (GRS Intranet)	Proprietary (EDF intranet)	Proprietary (ASNR intranet)
Owner/ operator	GRS, Germany	EDF, France	Autorité de sûreté nucléaire et de radio-protection, Montrouge, France
Database scope	Processed German & international "significant events"	Central web-based repository for OE data from all EDF plants[a,b]	OE data from all EDF plants[c]
Database content (no. of data records of interest)	Assumed to be >1000	Assumed to be a "significant" number of records on pipe failures. Adopted from an IAEA-OSART report (2003)[d]: At EDF, the group Concertation Inter-Domaines (CID) selects events of interest and makes information available to the plants. In 2003, more than 15,000 events/abnormalities had been reported into the SAPHIR. The group CID is also responsible for the analysis and selection of international events. In 2003, ca. 300 events were selected from the WANO database	In 2017, this database contained >20,000 safety significant events. The number of recorded pipe failure events is unknown, assumed to be "significant"
Data submission protocol	N/A	All maintenance and in-service inspection records	Includes event reports submitted to the French nuclear regulatory authority (L'Autorité de sûreté nucléaire et de radioprotection (ASNR)) and the IRS database
Taxonomy	This database has been used to establish crack distributions based on operational experience for input to PFM models to calculate leak and break probabilities. Mean crack depth in proportion to the wall	Basic information on equipment details, event description, impact on operations	Basic information on equipment details, event description, impact on operations

Table 9.9 (continued)

Database characteristics	VERA(Vertiefte Auswertung meldepflichtiger Ereignisse)	SAPHIR	SAPIDE
	thickness as well as the mean ratio of crack depth to half of crack length was calculated		
Database type	Web-based application of an Oracle APEX database	Web-based information management system for unexpected events in the EDF operated plants	Web-based information management system
No. of database fields	Free format	Free format	78 database fields

[a]SAPHIR database is a centralized repository used to manage and track information related to safety events, abnormalities, and lessons learned across EDF's fleet of nuclear power plants. Système d'Analyse de la Physique des Incidents dans les Réacteurs (SAPHIR) serves as a tool for the operational experience feedback, collecting data from all French nuclear power plants
[b] IAEA (2003), Draft Report of the PROSPER (Peer Review of Operational Safety Performance Experience) Mission to EDF Nuclear Operations Corporate Organization, IAEA-NSNI/PROSPER/2003, Vienna, Austria
[c] "Système d'Analyse des Problèmes d'Intégrité et des Défaillances Événementielles" (SAPIDE) serves as the regulator's operating experience database for collecting and analyzing data on operational events, abnormal situations, and confinement breaches in French nuclear facilities
[d]The IAEA Operational Safety Review Team (OSART) program was established in 1982. The information about the SAPHIR database is taken from a 2003 OSART report prepared for the Penly nuclear power plant

nuclear power plants. The proceedings include case studies, https://envdeg.ampp.org/.
- The International Conference "Fontevraud": Organized by Société Française d'Énergie Nucléaire (SFEN), this conference series is dedicated to the "Contribution of Materials Investigations and Operating Experience to LWRs' Safety, Performance and Reliability." Held every 3 years, and since Fontevraud 10 (19–22 September 2022), it is organized jointly with the flow-accelerated corrosion (FAC) conference series,[6] https://sfen.org/.
- Structural Mechanics in Reactor Technology (SMiRT) (Boley 1989): The transactions of these conferences are made available to the public as part of the outreach activities of the International Association for Structural Mechanics in Reactor Technology (IASMiRT). The collections are held at https://repository.lib.ncsu.edu/handle/1840.20/40305.

[6]The Fontevraud conference is a major international symposium on materials science and engineering related to the safety and performance of nuclear reactors. It was established in 1985 to share operational experiences and findings from materials failure analyses, within the French and the international nuclear industry.

Table 9.10 Examples of JIC nuclear power plant operating experience databases (Part III)

Database characteristic	INPO Consolidated Events Database (ICES) Formerly the "Equipment Performance Exchange System" (EPIX)	Nuclear Information Archives (NUCIA)	ETH-Zürich curated events database
Accessibility	Proprietary	Public	Open-source database of nuclear events
Owner/ operator	Institute of Nuclear Power Operations (INPO), Atlanta, GA	Japan Nuclear Technology Institute (JANTI), Tokyo, Japan (Hibino and Niwa 2008; Hwang et al. 2010)	ETH Risk Center, ETH Zürich, Switzerland (Ayoub et al. 2021)
DB scope	Central web-based repository for operating experience data from all US plants	Central web-based repository for domestic and international operating experience	All events that are of safety relevance including those having an INES = 1 or 2
DB content (no. of data records)	Assumed to be a "significant" number of records on pipe degradation and failure	As of 12/2023, the public version had close to 7500 events in the database, of which ca. 650 (9%) relate to pipe degradation (cracks, pipe wall thinning) and failure (leaks and breaks)	Approximately 66% of the events are based on USNRC licensee event reports, 10% on reports by JRC Clearinghouse, 6% on reports by thc IAEA, and 5% on reports by the Federal Office for the Safety of Nuclear Waste Management (BASE) in Germany. The remaining 13% of the events are based on official reports by various regulatory bodies, news articles, publications from researchers, and others. As of August 2021, the database contained 1561 event records of which ca. 5% are pipe failures
Data submission protocol	The amount of data provided varies by utility. The raw failure records in ICES can be inconsistent	OE reports submitted by the Japanese plant operators. With restricted access, NUCIA also includes information from IRS, WANO, and INPO	N/A
Taxonomy	https://www.eiseverywhere.com/file_uploads/aef761d3464553c9bb2cafadf428d0e0_1	https://doi.org/10.1080/18811248.2008.9711446	Documented in a user manual

Table 9.10 (continued)

Database characteristic	INPO Consolidated Events Database (ICES) Formerly the "Equipment Performance Exchange System" (EPIX)	Nuclear Information Archives (NUCIA)	ETH-Zürich curated events database
	8NSCSL_ICES_ Final_Barnes.pdf		
DB type	Free-format web-based database, mostly in narrative form	Web based	First, the user will extract the files in the "ETHZ_Database. Zip" file. The "ETHZ_Database_access_tool. xlsm" contains the access tool with intuitive graphic user interface (GUI)
No. of database fields	At least 9 and with provisions to upload extensive documentation including actions taken to restore functionality	33	15

Additional information regarding pipe failures is available in topical reports that address particular material degradation mechanisms affecting piping systems. These documents are authored by subject matter experts to support ongoing research initiatives or to respond to notable new operational experiences. The endnotes provide a comprehensive selection of pertinent reference materials.

9.6.2 Handbooks of Probabilistic Failure Metrics

Two examples of "parameter handbooks" are included in Table 9.12. The term "handbook" implies that probabilistic failure metrics have been developed for a particular kind of user, e.g., a PSA analyst or a developer of risk-informed in-service inspection of reliability and integrity management programs, and that the underlying methodology, computational framework (methods and techniques), and results have been subjected to peer reviews. These handbooks contain extensive tabulations of pipe failure rates and rupture frequencies that are organized by nuclear plant system, material, nominal pipe size, operating environment, and material degradation mechanism. Access to these handbooks is restricted to users within respective funding organizations.

Table 9.11 Comparison of two JIT pipe failure databases

Database characteristic	OECD/NEA CODAP event database	PIPExp (pipe operating experience database)
Accessibility	Restricted, access via a secure server	Proprietary
Owner/operator	https://www.oecd-nea.org/jcms/pl_24823/component-operational-experience-degradation-and-ageing-programme-codap	Sigma-phase Inc., USA
Database scope	A 1998 version of what became PIPExp became the basis for CODAP (Lydell et al. 2005). Safety-related and non-safety-related piping plus selected non-piping passive components (e.g., reactor internals). Focus on "selected representative event"	Safety-related and non-safety-related piping and water hammer event database. Also includes a piping system design database with weld population data and pipe length data. Scope includes tier 1, tier 2, and tier data. A 1998 version of the database served as the base on which the OECD/NEA OPDE and CODAP database projects evolved
Supporting organizations	Regulatory authorities and technical support organizations, 14 NEA member countries participated in the 2024–2026 program	N/A
Database content (no. of data records)	About 4600 as of end of 2025. The database also includes ca. 500 records on non-piping passive component failures	>14,800 as of end of 2025 Includes a separate water hammer event database (an additional ca. 750 events). This database consists of descriptions of significant water hammer events, their root causes, and impact on structures and components
Data submission protocol	Governed by an annual work plan, QA/QC procedure, and coding guideline. CODAP is based on the principle that each participating organization shares OE data of interest to the materials science community	Since 1993, continuously updated & maintained. Database structure and QA/QC processes are documented in a proprietary coding guideline
Taxonomy	Detailed taxonomy, which is defined in a coding guideline	Detailed taxonomy with more than 700 key words relating to failure mode, root cause, materials, process medium, piping component, method of discovery, materials
Database type	Web-based SQL	SQL—Stored in MS-ACCESS
No. of database fields	89	Section 9.3 of this chapter

Table 9.12 Selected nuclear industry pipe failure rate "handbooks"

R-Book	EPRI 3002024904
Reliability data handbook for piping components in Nordic nuclear power plants, 1st edition	Pipe rupture frequencies for internal flooding probabilistic risk assessments, revision 5
Availability: Restricted to the project funding organizations. A "light version" is available by contacting Nordic PSA Group	Availability: Licensed material. Access restricted to EPRI membership. https://www.epri.com/research/products/3002009993
Prepared by LR energy Sweden AB (now the VYSUS Group)	Prepared by EPRI
Funded by Forsmark AB, OKG AB, Ringhals AB, and Swedish Radiation Safety Authority	Funded by EPRI member utilities
Content: Piping reliability parameters (failure rates and rupture frequencies) for the following systems: • ASME III Class 1 and 2 BWR systems. • ASME III Class 1 and 2 PWR systems.	Content: ASME III Class 3 and 4 piping reliability parameters (precursor rates, CFP parameters, and failure frequencies) for the following systems located outside reactor containment structures: • • Circulating water. • • Component cooling. • • Firewater. • • Balance of plant. • • Emergency core cooling systems. • • Service water system (safety class 3 and non-safety).
Date of publication: January 2011	Date of publication: August 2023
The Nordic PSA Group developed the piping reliability parameter handbook to be used in PSA. The Handbook includes pipe leak and rupture frequencies and conditional rupture probabilities based on the OPDE (2007) pipe failure event database. A first version of the R-Book was released in 2011, and it was made available to the Nordic PSA Group member organizations. https://www.npsag.org/	This report updates a 2018 EPRI report (3002012997) on piping system failure rates for use in internal flooding PSA: • pipe system failure rate and flood frequency estimates reflect service experience through December 31, 2020 • aging factors and integrity management factors have been developed for some pipe types to allow for plant-specific adjustment of the mean flood frequencies • estimation of failure rates and rupture frequencies for high-energy piping outside the containment has been improved to better isolate those piping systems that are susceptible to flow-accelerated corrosion (FAC) • includes excel tables and .Csv exportable files for import into FRANX version 11
	This report also includes failure rate estimates of corrosion-resistant service water piping for use in internal flooding probabilistic risk assessments

References

Ader C (Task Force Leader) (2003) Reactor operating experience task force report. USNRC, Washington, DC

Aldemir T, Dening R, Unwin S (2014) Methodology development for passive component reliability modeling in a multi-physics simulation environment, Project No. 11-3030. Nuclear Energy University Programs, U.S. Department of Energy, Washington, DC

Atwood CL, LaChance JL, Martz HF, Anderson DJ, Englehardt M, Whitehead D, Wheeler T (2003) Handbook of parameter estimation for probabilistic risk assessment, NUREG/CR-6823. USNRC, Washington, DC

Ayoub A, Stankowski A, Kröger W, Sornette D (2021) The ETH Zurich curated nuclear events database: layout, event classification, and analysis of contributing factors. Reliab Eng Syst Saf. https://doi.org/10.1016/j.ress.2021.107781

Beal J, Sakurahara T, Reihani S, Kee E, Mohaghegh Z (2020) An algorithm for risk-informed analysis of advanced nuclear reactors with a case study of pipe failure rate estimation. In: Proceedings of the 30th European Safety and Reliability Conference and the 15th Probabilistic Safety Assessment and Management Conference (ESREL 2020 & PSAM 15), Venice, Italy, November 2020

Bley DC, Bier VM, Johnson DH, Stetkar JW (1990) Service water systems and nuclear plant safety, NSAC-148. The Nuclear Safety Analysis Center of the Electric Power Research Institute, Palo Alto

Boley BA (1989) A short history of SMiRT: a personal view. In: 10th international conference, structural mechanics in reactor technology, August 14–19, Anaheim, CA

CNSC (2024) Reporting requirements for nuclear power plants, REGDOC-3.1.1, Ottawa

Cooke RM (1996a) The design of reliability data bases, part I: review of standard design concepts. Reliab Eng Safety Analysis 51(2):137–146

Cooke RM (1996b) The design of reliability data bases, part II: competing risk and data compression. Reliab Eng Safety Analysis 51(2):209–223

Di Maio F, Colli D, Zio E, Tao L, Tong J (2015) A multi-state physics modeling approach for the reliability assessment of nuclear power plants piping systems. Ann Nucl Energy 80:151–165

ENSI (2023) Nuclear power plant event notifications: guidelines for the Swiss Nuclear Power Plants, ENSI-B03 (in German), Brugg

Fleming KN, Lydell BOY (2004) Database development and uncertainty treatment for estimating pipe failure rates and rupture frequencies. Reliab Eng Syst Saf 86:227–246

Gott K (2002) Failures in mechanical devices of Swedish nuclear power plants 1972–2000, SKI report 2002:50 (in Swedish). Swedish Nuclear Power Inspectorate, Stockholm

Gott K (2011) Databases of operationally induced damage. In: Busby JT, Ilevbare G, Andresen PL (eds) Proceedings of the 15th international conference on environmental degradation of materials in nuclear power systems - water reactors. Springer, Cham. https://doi.org/10.1007/978-3-319-48760-1_109

Hibino A, Niwa Y (2008) Graphical representation of nuclear incidents/accidents by associating network in nuclear technical communication. Nuclear Sci Technol 45:369–377

Hickling J (2008) Materials Reliability Program: resistance of Alloys 690, 52 and 152 to Primary Water Stress Corrosion Cracking (MRP-237, Rev. 1). Summary of findings from completed and ongoing test programs since 2004, 1018130. Electric Power Research Institute, Palo Alto

Hwang S-W, Oh J-Y, Jae M (2010) Development of web-based reliability data analysis algorithm model and it's application. Ann Nucl Energy 37:248–255

IAEA (1992) Data collection and record keeping for the Management of Nuclear Power Plant Aging, Safety Series No. 50-P-3, Vienna, Austria

IAEA (2021) Technical insights from benchmarking different methods for predicting pipe failure rates in water cooled reactors. Final Report of a Coordinated Research Project, IAEA-TECDOC-1988, Vienna, Austria

IAEA (2022) IRS guidelines 2022 edition: Joint IAEA and OECD/NEA International Reporting System for operating experience (IRS), IAEA services Series19 (rev.1), ISSN: 1816-9309, Vienna, Austria

IAEA (2023a) Methodologies for assessing pipe failure rates in advanced water-cooled reactors. Nuclear Energy Series Publication. NR-T-2.16, ISSN 1995-7807, Vienna, Austria

IAEA (2023b) Ageing management for nuclear power plants: International Generic Ageing Lessons Learned (IGALL), safety report series no. 82 (rev. 2), Vienna, Austria

ISO (2015) Data quality—part 8: information and data quality: concepts and measuring, ISO 8000-8:2015. International Organization for Standardization, Geneva

ISO (2016) Petroleum, petrochemical and natural gas industries: collection and exchange of reliability and maintenance data for equipment, ISO 14224:2016. International Organization for Standardization, Geneva

Jenks A, White G, Burkardt M, Young G (2018a) Materials reliability program: recommended factors of improvement for evaluating Primary Water Stress Corrosion Cracking (PWSCC) growth rates of Thick-Walled Alloy 690 Materials and Alloy 52, 152 and variants welds (MRP-386), 3002010756. Electric Power Research Institute, Palo Alto

Jenks A, White G, Burkardt M (2018b) Materials reliability program: crack growth rates for evaluating Primary Water Stress Corrosion Cracking (PWSCC) of Thick-Wall Alloy 600 Materials and Alloy 82, 182, and 132 welds (MRP-420, revision 1), 3002014244. Electric Power Research Institute, Palo Alto

Jurmain M (2007) Guidelines for obtaining and using operating experience at fossil power plants, 1012783. EPRI, Palo Alto

Kanga J, Al Masrya Z, Varniera C, Mosallamb A, Zerhounia N (2024) A data quality management framework for equipment failure risk estimation: application to the oil and gas industry. Eng Appl Artif Intell 136(Part A). https://doi.org/10.1016/j.engappai.2024.108834

Lydell BOY (2014) Risk-informed structural integrity management: development of SMR-centric piping reliability models. In: SMR2014-3328, Proc. ASME 2014 Small Modular Reactor Symposium SMR2014. ASME, New York

Lydell BOY, Tomic B, Hegedus D, Nyman R (2005) Reliability of piping system components: framework for estimating failure parameters from service data, SKI report 1997:26, 3rd edn. Swedish Nuclear Power Inspectorate, Stockholm

Michel F, Sayar D (2010) Further development of the knowledge base for pressurized components in nuclear power plants, GRS-A-3539 (in German). Gesellschaft für Anlagen und Reaktorsicherheit (GRS) mbH, Cologne

Moosemiller M, Weber B (1998) Guidelines for improving plant reliability through data collection and analysis., ISBN: 0-8169-0751-X. Center for Chemical Process Safety, American Institute of Chemical Engineers, New York

Moss T, Was GS (2017) Factor of improvement in resistance of in stress corrosion cracking initiation of Alloy 690 over Alloy 600. In: Proc. 17th international conference on environmental degradation of materials in nuclear power systems—Water reactors, August 9–13, 2015, Ottawa, Ontario, Canada

Nuclear Energy Institute (2020) Guideline for the management of materials issues, NEI 03-08 [Rev 4], Washington, DC

OECD/NEA (2004) OECD-NEA piping failure data exchange project (OPDE): workshop on database applications, OPDE/SEC/2004)4, Boulogne-Billancourt

OECD/NEA (2018) Basic principles of collecting and evaluating operating experience data on metallic passive components, NEA/CSNI/R(2018)12, Boulogne-Billancourt

OECD/NEA (2020) Use of NEA database projects operating experience data for probabilistic safety assessment, NEA/CSNI/R(2020)12, Boulogne-Billancourt

OECD/NEA (2024) Summary of Phase IV of the Component Operational Experience, Degradation and Ageing Programme (CODAP), NEA/CSNI/R(2024)6, Boulogne-Billancourt

Park G (2014) Lessons learned based on operating experience, Section XI, Chapter 25. In: Rao KR (ed) Continuing and changing priorities of the ASME boiler & pressure vessel codes and standards., ISBN: 9780791860199. ASME, New York

Paulsen J, Dorrepaal J, Cooke R, Hokstadt P (1996) The Design and Use of Reliability Data Base with Analysis Tool. Risø-R-896(EN) (NKS/RAK-I(96)R6). Risø National Laboratory, Roskilde

Pyy P (ed) (1999) Handbook on quality of reliability data, Statistical Series No. 4, An EsReDA Working Group Report., ISBN 82-515-02535. Det Norske Veritas, Høvik

Rasmuson DM, Wierman TE, Kvarfordt KJ (2005) An overview of the Reliability and Availability Data System (RADS), INL/CON-05-00069. Idaho National Laboratory, Idaho Falls

Rudland D (2023) Risk-informed assessment of French stress corrosion cracking operational experience relative to US Fleet, NRC-ADAMS Accession No. ML23236A080. USNRC, Washington, DC

Schäbe H, Shubinsky IB (2024) How to properly apply systems of artificial intelligence. Reliab: Theory Appl 4(80):31–35

Scott P, Wilkowski G, Sturm D, Stoppler W (1994) Development of a database of pipe fracture experiments. Nucl Eng Des 151(2–3):359–371

Shin H-S, Park HB (2010) Development of nuclear materials and degradation database. Transactions of the Korean Nuclear Society Autumn Meeting, Jeju, pp 1083–1084

Stevens B (ed) (1986) Guide to reliability data collection and management, EuReDatA Project Report No. 3. Commission of the European Communities, Joint Research Centre, Ispra

USNRC (2010) Generic Aging Lessons Learned (GALL) report, NUREG-1801, revision 2, Washington, DC

USNRC (2013) Event report guidelines: 10 CFR 50.72 and 50.73, NUREG-1022 revision 3, Washington, DC

USNRC (2019) LIC-401, revision 4 "NRR reactor operating experience program," Washington, DC. NRC-ADAMS Accession No. ML19155A036

USNRC (2021) IMC 2523 NRC applications of the reactor operating experience program in NRC oversight processes, Washington, DC NRC-ADAMS Accession No. ML20321A083

USNRC (2023) Generic Aging Lessons Learned for Subsequent License Renewal (GALL-SLR), Draft Report for Comment, NUREG-2191, Volume 1, Revision 1, Washington, DC

Weerakkody SD (2023) Integrated risk-informed decision-making process for emergent issues, LIC-504 revision 6. Office of Nuclear Reactor Regulation, USNRC, Washington, DC

Chapter 10
Operating Experience Data Analytics

10.1 Data Completeness

Currently, the database referenced in this book contains records of over 14,800 pipe failures, serving as this book's core source of operating experience (OE). Although it is incomplete and continually growing, it likely includes most major structural failures identified during its creation and use. Giving a definitive assessment of its completeness is not possible. Moglia et al. (2008) offer an interesting perspective on "data completeness," which the interested reader is encouraged to explore.

Pipe failures are predictable and preventable. Using an OE database for assessing and predicting pipe failure precursor rates relies on the overall quality, accuracy, and completeness of the data collected. These factors include data sourcing, data collection methods, and data quality control measures. The source of operating experience plays an important role in determining data quality and fitness for use. Depending on the source, the data may have different levels of accuracy and trustworthiness. For example, datasets collected from tier 1, tier 2, or tier 3 sources have varying levels of technical detail that require interpretation and verification. The way data is collected and entered into an SQL database affects its accuracy and completeness. Rigorous and standardized data collection processes that rely on user guidelines produce more reliable results compared to ad hoc or poorly structured data collection methods.

Data quality control measures are important for supporting data completeness and accuracy. These measures include techniques that aim to reduce errors, inconsistencies, and biases during data collection. Applying data validation checks, following systematic data cleaning steps, and using quality control protocols can improve data accuracy. The expertise and experience of data collectors or data miners, technological infrastructure, and data storage methods also affect data completeness and accuracy. Data collectors with relevant experience tend to follow established procedures more consistently. Appropriate technological infrastructure and secure storage systems help prevent data loss, corruption, or unauthorized access, contributing to data integrity.

B. Lydell, *Knowledge Preservation in Nuclear Safety*,
https://doi.org/10.1007/978-3-032-25463-4_10

10.2 Data Mining

Data mining (Zoulikha et al. 2024), also referred to as data analytics, involves identifying the sources of pipe failure information, extracting the information in such a way that the reasons why material degradation occurred and developed into a failure are clearly explained, and transforming this information into an understandable structure for future use by a diverse group of analysts and researchers (Fig. 10.1). Any information that is not useful or incomplete is discarded. It is a dynamic, continuous learning process changing with time as a result of evolving reliability and integrity management processes (ASN 2023). As an illustration of the data mining process, and how it affects epistemic uncertainty in pipe failure rate distribution parameters, Fig. 10.2 illustrates how the content of a pipe failure database evolved between 2010 and 2025. Of interest is the number of pipe failures recorded in the period 2000–2009. At the end of 2010, 1954 failures were recorded for the period of interest. At the end of 2025, information on an additional 1000 failures had been accessed and analyzed for the same period. Data mining is the "never-ending process" of documenting operating experience and how it affects data completeness and the importance of having a sustained data collection process in

- Data mining isn't a new invention that came with the digital age. The concept has been around for over a century but came into greater public focus in the 1930s. One of the first instances of data mining occurred in 1936, when Alan Turing introduced the idea of a universal machine that could perform computations similar to those of modern-day computers.
- Organizations are now harnessing data mining and machine learning to improve everything from their sales processes to interpreting financials for investment purposes. As a result, data scientists have become vital to organizations.
- **Data mining is the process of analyzing massive volumes of data to discover important material degradation scenarios that helps solve problems, mitigate risks, and seize new opportunities.** This branch of data science derives its name from the similarities between searching for valuable information in a large database and mining a mountain for ore. Both processes require **sifting through tremendous amounts of material to find hidden value.**
- Data mining can answer questions that traditionally were too time consuming to resolve manually. Using a range of statistical techniques to analyze data in different ways, **users can identify patterns, trends and relationships** they might otherwise miss.

Fig. 10.1 The data mining process. (Adapted from https://www.talend.com/resources/what-is-data-mining/. Alan Mathison Turing (23 June 1912–7 June 1954) was an English mathematician, computer scientist, logician, cryptanalyst, philosopher, and theoretical biologist. He was highly influential in the development of theoretical computer science, providing a formalization of the concepts of algorithm and computation with the Turing machine, which can be considered a model of a general-purpose computer. Turing is widely considered to be the father of theoretical computer science)

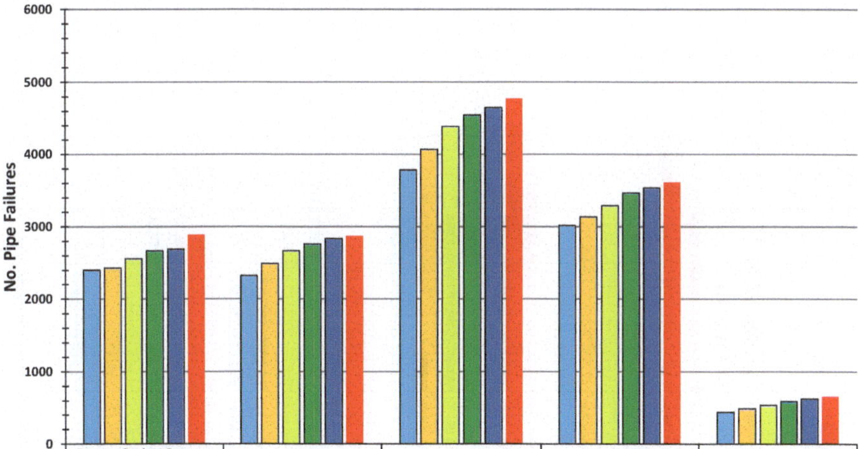

	Reactor Coolant Pressure Boundary - ASME III Class 1	ASME III Class 2	ASME III Class 3	Non-Safety - Balance of Plant Piping	Fire Water Piping System
2020	2401	2319	3781	3017	437
2021	2429	2490	4067	3135	486
2022	2555	2666	4384	3290	535
2023	2666	2760	4544	3465	585
2024	2692	2836	4647	3536	622
2025	2895	2877	4776	3619	661

Pipe Failure Population at the End of Indicated Calendar Year

Fig. 10.2 Evolution of a pipe failure database. (Database status as of 15 July 2025)

Table 10.1 Four stages of data mining

Stage	Data mining steps	Description
I	Data gathering	Extracting failure records (the raw data) from multiple and diverse sources
II	Data preparation	Classify data using a coding guideline
III	Exploring the data	Querying the database, identifying and correcting inconsistencies, if any
IV	Analyzing the collected data and making interpretations	Perform statistical characterization of the data. Looking for trends and patterns

place (Table 10.1). The term "never-ending process" implies that data mining is a dynamic process. An item of data valid for a problem at one point in time may not be valid for the same problem at some later time because of design changes, replacements using different materials, or heavier wall thickness.

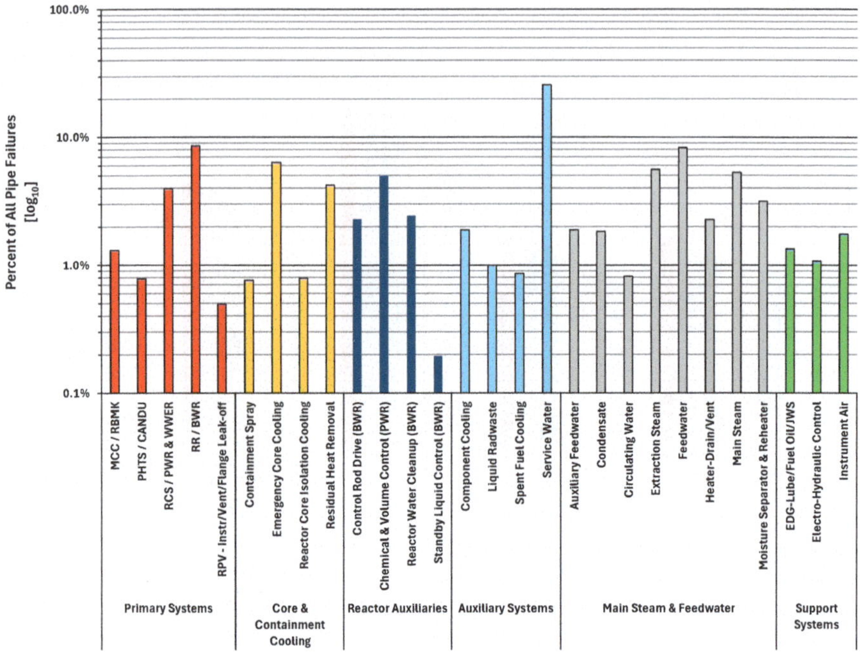

Fig. 10.3 Data collection process

10.3 Operating Experience Insights

As an introduction to deeper probes into the piping operating experience, Fig. 10.3 illustrates the data collection process with respect to nuclear power plant system groups. Effective data collection is a continuous effort, capturing both recent events and those from previous months or years; it is a dynamic and continuous process.[1] The OE data in this chart is organized by system groups and plant systems within each system group.

Piping systems are classified by safety importance: Safety Class 1 includes those connected to the reactor pressure vessel; Class 2 covers systems critical for emergency core cooling, post-accident containment heat removal, or fission product removal; and Class 3 includes systems like cooling water and auxiliary feedwater that support these functions, including residual and spent fuel heat removal. "Non-safety" systems, mainly in the turbine building for energy conversion, do not have designated safety roles, but their failure can still trigger safety system responses and reactor trips, making the term "non-safety" somewhat misleading outside of code and standard definitions.

[1] At the EUROSAFE Forum, Berlin, 2004, Learning from Experience—A Cornerstone of Nuclear Safety, 8–9 November 2004, the following statement was made by Yves Benoist: "Operating experience feedback is a permanently running activity with very significant workloads; it is an endless task, but it is an essential way of improving safety."

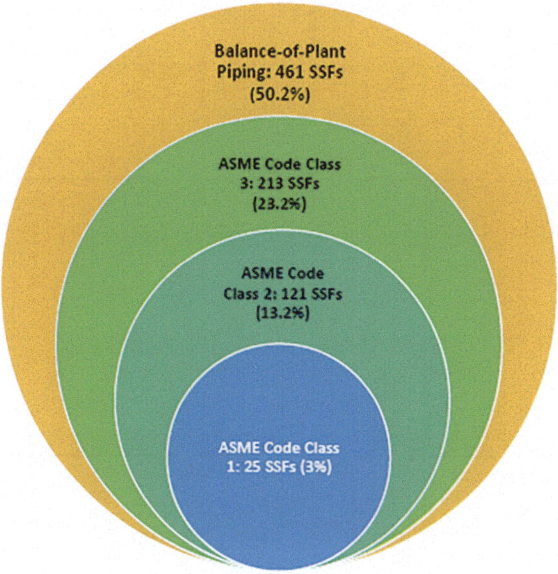

Fig. 10.4 Significant structural failures in perspective

10.4 Significant Structural Failures

Significant structural failures (SSFs) are considered rare events within piping systems, as shown in Fig. 10.4. An SSF is defined by a sudden and substantial loss of structural integrity, typically resulting in a large, rapid release of process medium, characterized by a pipe wall mass flow rate of 10 kg/s or more, along with spatial impacts caused by the event. "Noteworthy" RCPB pipe failures are given in Fig. 10.5.

SSF events have occurred in specific types of piping systems. Of the total population of SSF events, only a small proportion has involved primary system piping that cannot be isolated from the reactor pressure vessel. Notably, all such failures were limited to small-diameter piping systems located within the primary pressure boundary.

10.5 Data Reduction Schemes

Before making further inferences regarding correlations between a high-level summary of piping operating experience and specific piping integrity attributes and influence factors, additional data reductions must be performed to obtain meaningful statistical profiles. Such data reductions require knowledge of nuclear plant systems (Garland 2014; Cummins and Matzie 2018; Fennern 2018; Takeda and Inagaki 2021; Pioro 2022; Kok 2017; USDOE 1987). IEEE Std 805-1984 (IEEE 1984) is a

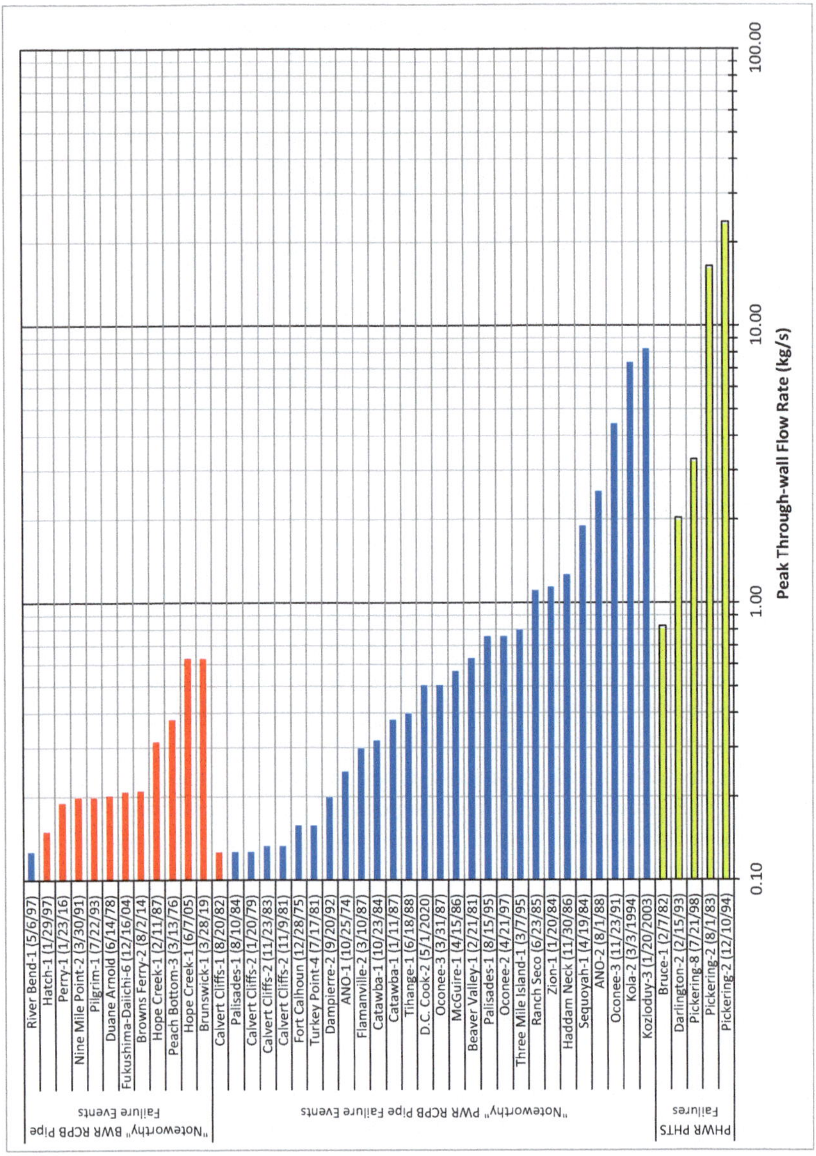

Fig. 10.5 "Noteworthy" primary system pipe failures

recommended practice[2] which provides a single source of nuclear power plant system descriptions that, along with related recommended practices concerning unique identification principles and definitions, and component function identifiers, provide a basis for uniquely identifying systems, structures, and components of light water nuclear power plant projects.

With each successive step in data reduction, the resulting event population becomes more targeted and refined. Table 10.2 shows that OE varies among different plant designs, which stems from both major differences in piping design and incomplete data. Several factors contribute to this variability, such as gaps in available data, unique aspects of certain plant designs with limited reactor operating years, and specific choices in piping materials during design. As an example, Babcock-Brown Boveri Reaktor GmbH (BBR) designed and built a single reactor unit, the German Mülheim-Kärlich nuclear power plant (Schneider and Wirtz 1991) which operated briefly before being permanently shut down due to local politics.[3] Another example involves material choices made decades ago: whereas US designs typically use carbon steel for balance-of-plant piping, German practice favored low-alloy steels known for their resistance to flow-accelerated corrosion.

The OE data in Table 10.2 is explored further in Figs. 10.6 and 10.7. The former illustrates the distribution of pipe flaws and failures in ASME III Code Class 1 piping of nominal diameter (DN) 100 mm or greater. Figure 10.8 shows the OE associated with ASME III Code Class 1 piping of nominal diameter less than DN100. This OE is organized by the time period in which a failure was reported and system category:

- FW: BWR feedwater system, within the drywell
- MS: BWR main steam system, within the drywell
- RAS: reactor auxiliary systems used for chemistry and inventory control
- RCS: PWR reactor coolant system
- RR: BWR reactor recirculation system
- SIR: Safety injection and recirculation in response to a loss-of-coolant accident and need for decay heat removal

[2] This recommended practice provided a single source of nuclear power plant system descriptions, along with related recommended practices concerning unique identification principles and definitions, component function identifiers, and implementation instructions. It gave a basis for uniquely identifying systems, structures, and components of light water nuclear power plant projects. This standard was part of a series of recommended practices, entitled the Energy Industry Identification System (EIIS), the purpose of which was to present a common language of communication that permitted a user to correlate a system, structure, or component with that of another organization for the purposes of reporting, comparison, or general communication.

[3] BBR was a joint venture between Babcock & Wilcox Company of Lynchburg, VA (USA), Brown Boveri of Mannheim (Germany), and Brown Boveri of Baden (Switzerland). Three sites were originally envisaged for the construction of BBR-PWR units in West Germany. Only the single Kernkraftwerk Mülheim-Kärlich (KMK) was constructed, however. The original B&W pressurized water reactor design was significantly modified in order to comply with the German codes and standards.

Table 10.2 Piping OE by reactor type and period of discovery

Reactor type	NSSS design	ASME Class 1		ASME Class 2	
		1960–1999	2000–2025	1960–1999	2000–2025
BWR	AEG	13		10	1
BWR	Asea-Atom	138	18	235	22
BWR	GE/HITACHI	4	36	3	9
BWR	GE/TOSHIBA	9	100	2	20
BWR	General Electric	975	168	477	159
BWR	Siemens/KWU	89	10	52	14
FBR	"Various"			24	1
GCR	"Various"	2	2	10	6
PHWR	AB Atomenergi	3			
PHWR	AECL	26	86	8	9
PHWR	CGE	3	0	1	0
PHWR	DAE	2	0		
PHWR	NPCIL	0	4	0	3
PHWR	Siemens/KWU	5	1		
PWR	Babcock & Wilcox	46	43	69	37
PWR	BBR/ABB Reaktor	1	0		
PWR	CNNC	0	6		2
PWR	Combustion engineering	110	67	158	94
PWR	FRAMACEC	3	0	1	0
PWR	Framatome	20	262	142	46
PWR	KEPCO	0	9	0	4
PWR	Siemens/KWU	23	28	48	64
PWR	WE	131	161	666	319
PWR	WE-MHI	6	32	12	30
RBMK	NIKIET	131	43	13	2
VVER	AEE	41	19	26	18
VVER	NPCIL	0	1		
VVER	AEE/SKODA	3	13	8	30
Total number of pipe flaws/failures:		1784	1109	1965	890
BWR	Boiling water reactor				
FBR	Fast breeder reactor				
GCR	Gas-cooled reactor; includes MAGNOX, advanced gas-cooled reactor (AGR), and high-temperature reactor (HTR)				
HWR	Heavy water reactor including CANDU reactors				
PWR	Pressurized water reactor				
RBMK	Reaktor Bolshoy Moshchnosty Kanalny ("high-power channel" reactor)				
VVER	Vodo-Vodyannoy Energeticheskiy Reaktor; water-water energy reactor (WWER, water-moderated and water-cooled pressurized water reactor)				

An additional perspective on the pipe failure data is offered in Fig. 10.8, in which the failure population is organized by safety class, nominal pipe size, failure mode, and period of discovery. Figure 10.9 illustrates the SSF subpopulation associated

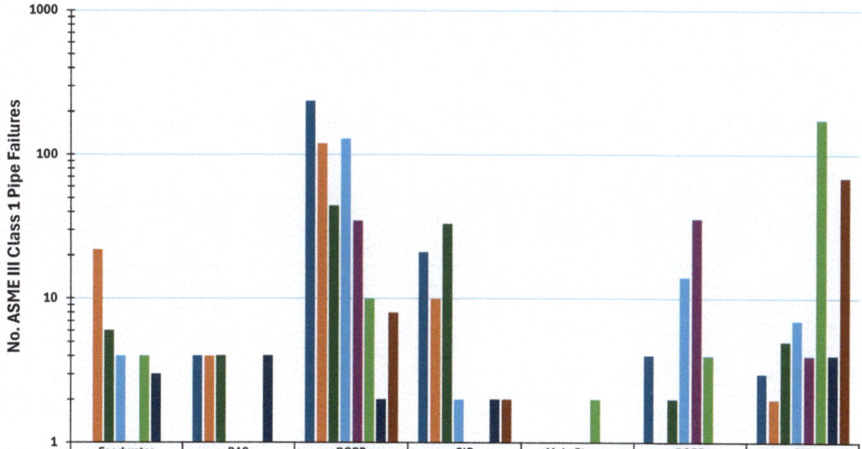

	BWR					PWR	
	Feedwater	RAS	RCPB	SIR	Main Steam	RCPB	SIR
1985-89	1	4	236	21		4	3
1990-94	22	4	120	10	1		2
1995-99	6	4	44	33		2	5
2000-04	4	1	129	2		14	7
2005-09	1	1	35	1	1	36	4
2010-14	4	1	10	1	2	4	175
2015-19	3	4	2	2	1	1	4
2020-24		1	8	2		1	69

Fig. 10.6 ASME III Code Class 1 >DN100 pipe failures

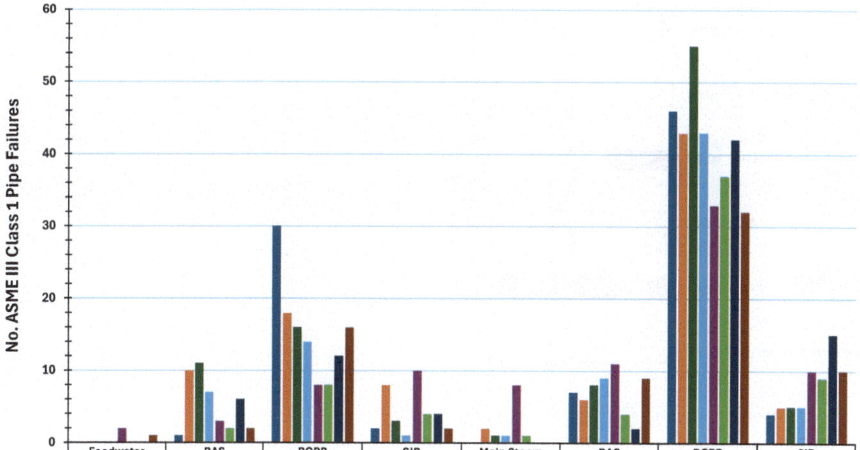

	BWR					PWR		
	Feedwater	RAS	RCPB	SIR	Main Steam	RAS	RCPB	SIR
1985-89		1	30	2		7	46	4
1990-94		10	18	8	2	6	43	5
1995-99		11	16	3	1	8	55	5
2000-04		7	14	1	1	9	43	5
2005-09	2	3	8	10	8	11	33	10
2010-14		2	8	4	1	4	37	9
2015-19		6	12	4		2	42	15
2020-24	1	2	16	2		9	32	10

Fig. 10.7 ASME III Code Class 1 ≤DN100 pipe failures

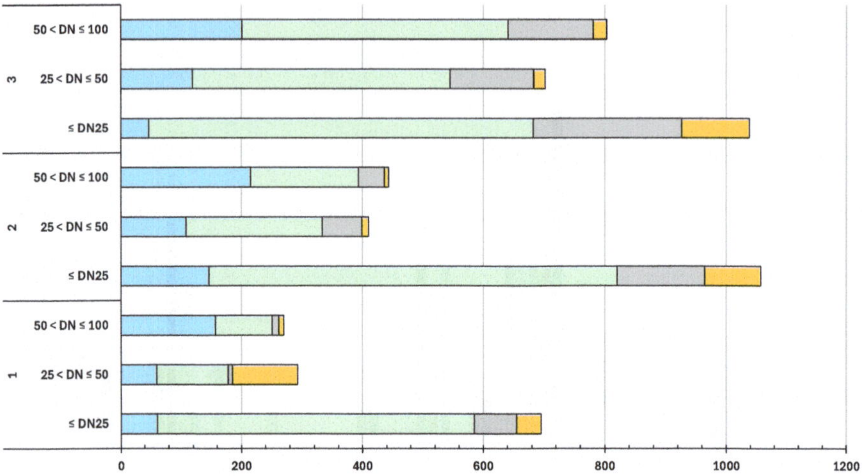

	1			2			3		
	≤ DN25	25 < DN ≤ 50	50 < DN ≤ 100	≤ DN25	25 < DN ≤ 50	50 < DN ≤ 100	≤ DN25	25 < DN ≤ 50	50 < DN ≤ 100
Crack-Part	61	60	157	146	108	215	47	119	201
Minor Leak	524	118	94	674	225	178	635	426	440
Leak	70	7	11	145	66	43	244	138	140
Large Leak	40	108	8	92	11	7	112	19	22

No. Pipe Failures by Safety Class & Nominal Diameter (mm)

Fig. 10.8 Small- and medium-diameter piping operating experience by safety classification and failure mode

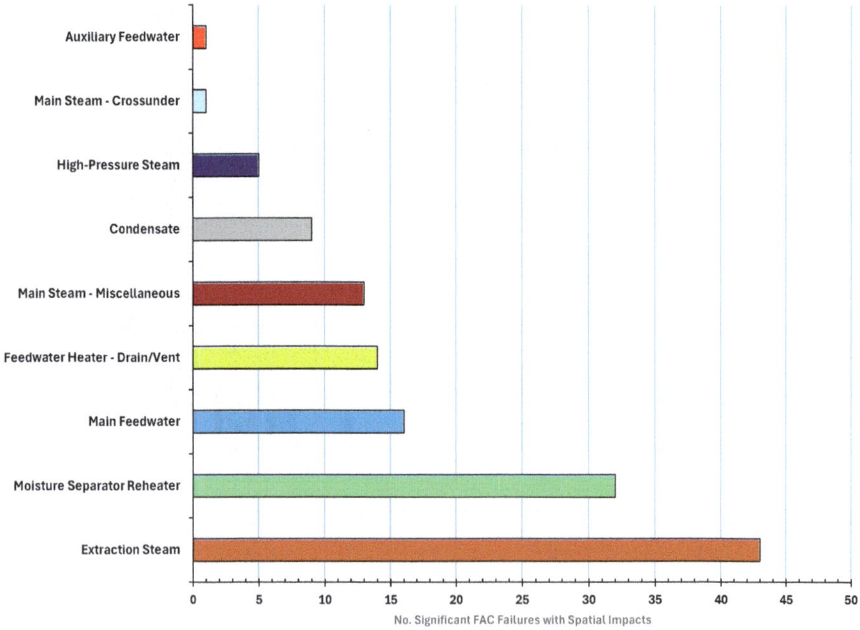

Fig. 10.9 Significant structural failures attributed to flow-accelerated corrosion

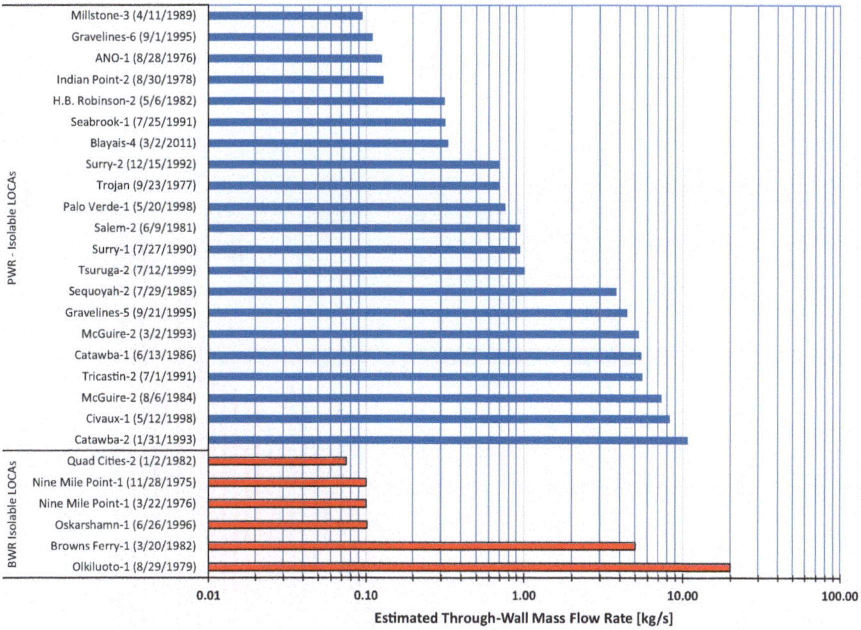

Fig. 10.10 Significant isolable primary coolant leaks caused by pipe failures

with flow-accelerated corrosion (FAC), presenting operational data categorized by the systems that encountered notable FAC failures. Figures 10.10 and 10.11 are summaries of major events involving primary coolant leaks. Figure 10.10 covers piping that is connected to the RCPB but can be isolated and also carries primary coolant. Figure 10.11 highlights significant primary coolant leak events within the Safety Class 1 boundary of PWR plants, including major failures of steam generator tubes and reactor coolant pump seals.

10.6 Effect of Pipe Wall Thickness on Pressure Boundary Integrity

Thomas provided statistical correlations among piping reliability attributes and influence factors. According to the Thomas method, the probability of pipe failure is proportional to length L, diameter D, and wall thickness. Under the assumption of equal nominal pipe size, what insights can be obtained from operating experience? The relationship between precursor rate and wall thickness can be expressed as follows:

$$\lambda_{SCH160} = \lambda_{SCH80} \times t^{-1} \tag{10.1}$$

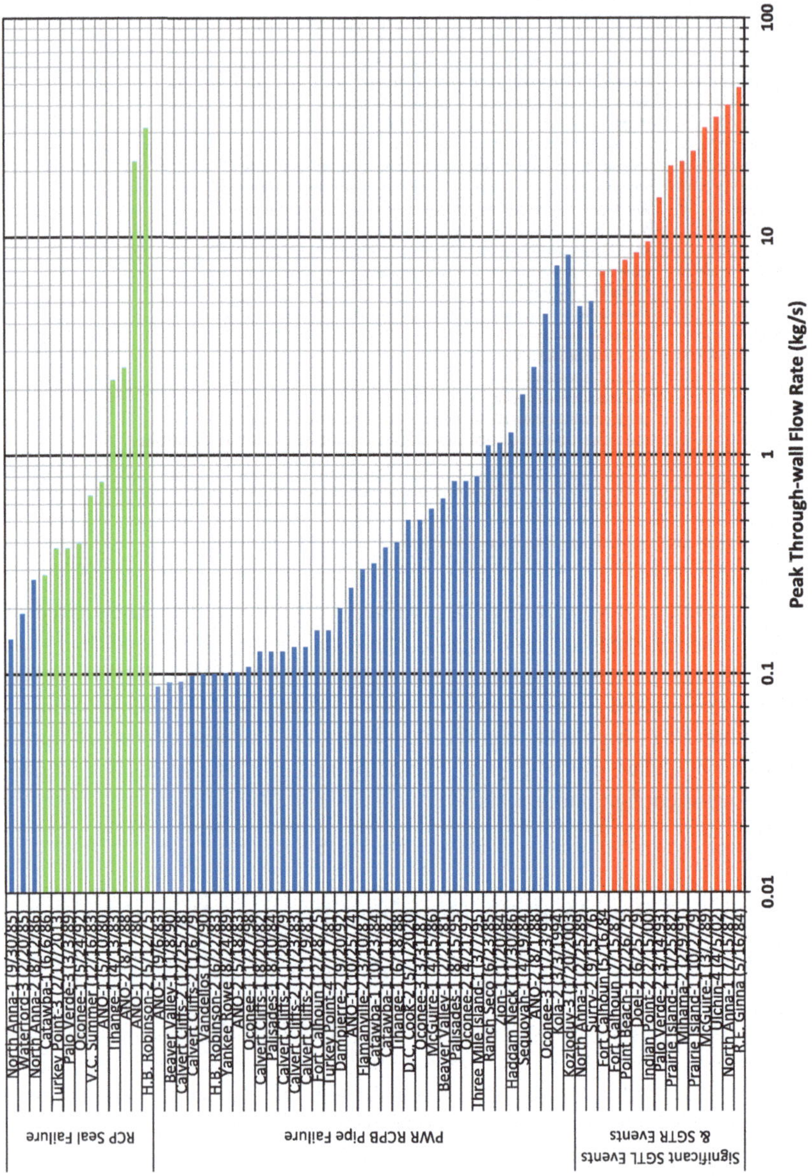

Fig. 10.11 Summary of significant RCPB failure in PWR plants

The factor t^{-1} represents the effect of wall thickness on the pipe failure propensity, with the assumption being that the thicker the pipe tube wall, the lesser the susceptibility to material loss through fretting caused by flow-induced vibration. This factor can be obtained from operating experience. Using pipe failure data for ASME III Code Class 1 and 2 small-diameter piping, it can be shown that

$$0.28 \leq t^{-1} \leq 0.55$$

The lower value is for nominal pipe size (NPS) ¾-inch (OD = 21.3 mm) and the upper value is for NPS2 (OD = 60.325 mm) stainless steel piping. The analysis is based on 900 pipe failures, and the t^{-1} factor was obtained as the ratio of number of failures in Schedule 160 piping to Schedule 80 piping of the same nominal diameter. "Failure" was equal to a small, detectable through-wall leak. A data screening approach was used to capture only those pipe failures that were attributed to high-cycle fatigue.

10.7 Early-Life Failures and Failures During Periods of Extended Operation

The experience with pipe failures during the early stages of plant operation is broad, encompassing issues discovered throughout both the commissioning phase and the initial years of commercial operation. A significant portion of these failures has been linked to various metal fatigue mechanisms. These mechanisms were often intensified by the implementation of unvalidated system operating procedures and by shortcomings in the practices used for piping fabrication and installation. Such factors contributed to a higher incidence of failures in newly operating systems. Failures occurring during periods of extended operation, specifically those beyond 40 years, have also been substantial. These events have undergone comprehensive analysis by numerous organizations, aiming to develop a technical foundation for estimating pipe failure rates that are dependent on both time and age. Such analyses have been instrumental in understanding long-term reliability and risks associated with aging piping infrastructure.

Estimation of pipe failure precursor rates as they relate to age can be approached using different analytical strategies. These strategies focus on determining piping reliability parameters, which are considered as functions of either the age of the failed pipe section, the age of the plant, or the overall temporal progression of piping operating experience. The concept of aging encompasses the cumulative effects of time-dependent factors that may influence pipe failure rates. These factors include the physical degradation of pipe materials resulting from corrosion and erosion mechanisms, as well as the influence of inspection and maintenance programs. Trends in pipe failure rates are shaped by a variety of influences. Physical changes in structural integrity, such as pipe wall thinning due to aging, play a significant role.

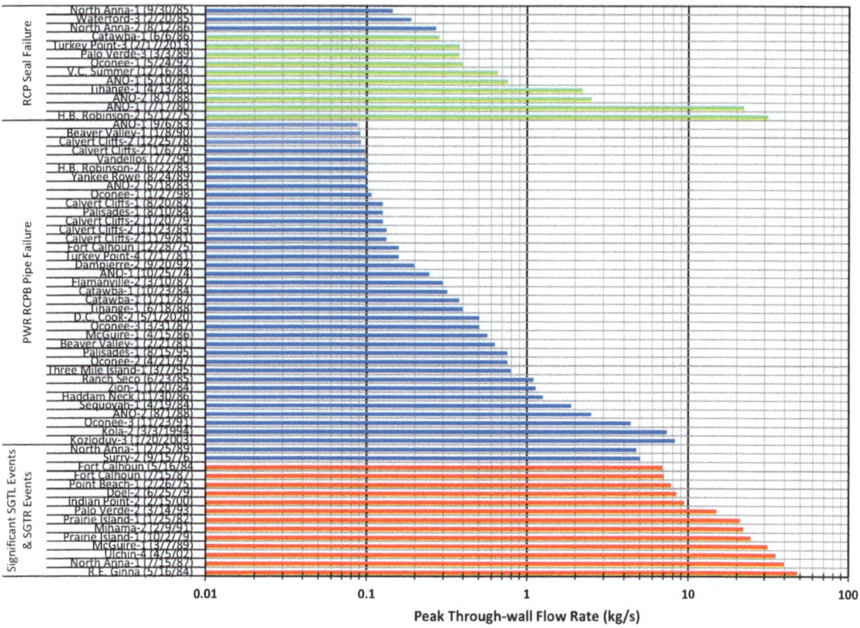

Fig. 10.12 Worldwide piping operating experience by operating regime

Additional contributing factors include modifications in reporting routines, design changes such as the replacement of original piping with materials that offer greater resistance to time-dependent degradation, and advancements in reliability and integrity management. To estimate age-dependent pipe failure rates, several methods are employed, including data pooling strategies, data window truncation techniques, and curve-fitting models.

Additional insights into the age dependency of pipe failures can be obtained as follows. Subdivide the entire failure population into five "operating regimes" (O-R) based on the age of the plant when a failure was discovered and as indicated in Figs. 10.12, 10.13, 10.14, and 10.15. Next, the failure population in a respective period is normalized against the plant population that generated the failures in a respective operating regime, an approximation, which should be viewed as an initial screening for possible trends in the failure data.

With reference to these figures, an uptick in Safety Class 3 pipe failures is primarily driven by raw water cooling pipe degradation. Underlying lack of trends (i.e., lack of up/downticks) for Safety Class 1 and 2 piping is attributed to effective piping pressure boundary integrity management processes, i.e., in-service inspection, stress improvement, water chemistry control, and use of piping materials that are resistant to stress corrosion cracking. Somewhat arbitrarily selected, "early life" covers the period from the initial reactor criticality through the second full year of commercial operation, that is, 2 years of commercial operation, plus the final commissioning phase that can take anywhere from a few months to more than a

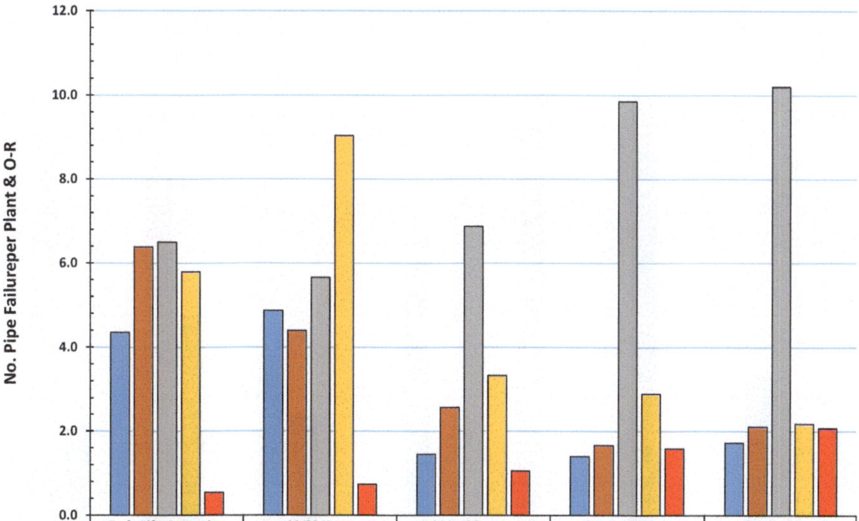

	Early Life Operation	11-20 Years	21 to 30 years	31 to 40 Years	Long Term Operation
RCPB	4.35	4.86	1.45	1.39	1.72
Class 2	6.38	4.39	2.56	1.66	2.10
Class 3	6.50	5.64	6.86	9.85	10.19
Non-Safety	5.79	9.03	3.32	2.89	2.17
Fire Water	0.54	0.74	1.05	1.59	2.07

Fig. 10.13 US piping operating experience by operating regime

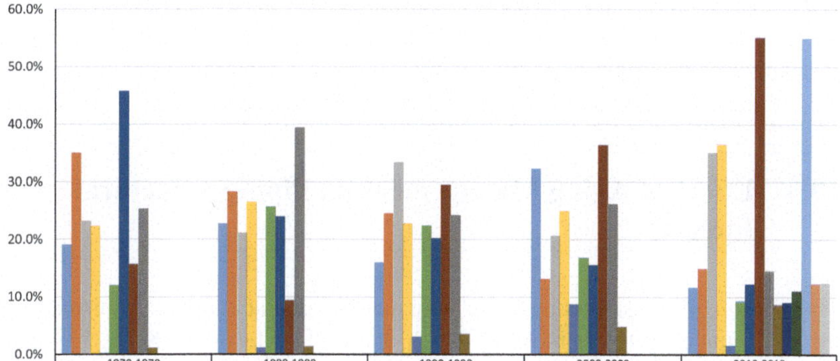

	1970-1979	1980-1989	1990-1999	2000-2009	2010-2019
ELO RCPB	19.2%	22.7%	16.0%	32.4%	11.7%
ELO Class 2	35.1%	28.3%	24.6%	13.2%	15.0%
ELO Class 3	23.1%	21.1%	33.4%	20.6%	35.0%
ELO Non-Safety	22.4%	26.6%	22.9%	25.0%	36.7%
ELO Fire Water	0.3%	1.3%	3.2%	8.8%	1.7%
MTO RCPB	12.0%	25.7%	22.4%	16.9%	9.4%
MTO Class 1	45.8%	24.0%	20.3%	15.6%	12.3%
MTO Class 3	15.7%	9.4%	29.5%	36.5%	55.2%
MTO Non-Safety	25.3%	39.4%	24.1%	26.2%	14.5%
MTO Fire Water	1.2%	1.4%	3.6%	4.9%	8.7%
LTO RCPB					9.1%
LTO Class 2					11.0%
LTO Class 3					55.1%
LTO Non-Safety					12.4%
LTO Fire Water					12.4%

Fig. 10.14 Distribution of pipe failures by operating regimes

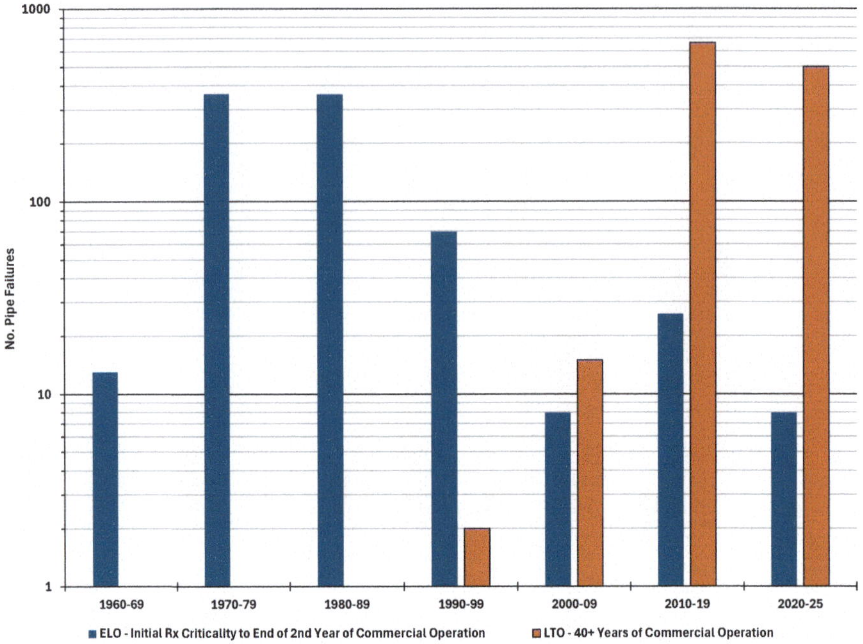

Fig. 10.15 Pipe failure experience during early life and long-term operation

year. Also, the pipe failure experience during periods of extended operation, beyond 40 years of commercial operation, is included in Fig. 10.15. Much of this operating experience is attributed to corrosion and flow-assisted degradation.

10.8 Effect of Aging Management on Piping Structural Integrity

This section briefly summarizes the main material degradation mechanisms found in metallic piping systems, most of which are well understood. Some have caused significant failures. Root cause analysis and materials research have led to detailed aging management to prevent or limit recurring issues. Evaluations of operating data, as shown in Tables 10.3, 10.4, and 10.5 and Fig. 10.16, provide insight into the effectiveness of mitigation and monitoring programs.

Results of data reduction of pipe failure data for three safety-related systems—component cooling (CC), instrument air (IA), and service water (SW)—are presented in Fig. 10.16. In this example, the data was normalized against operating experience for the, somewhat, arbitrarily selected period 1985–1999. A background to the uptick in the SW failure data is embedded in Chaps. 4 and 5 and the related reference material (OECD/NEA 2024).

Table 10.3 Aging management of material degradation (Part 1)[a]

Degradation mechanism (DM)		Material	Operating environment	Steps taken to manage/mitigate degradation (post-1980s)	Aging management effect on OE (plant age > 40 years)
Stress corrosion cracking	ECSCC	Stainless steel (SS)—Multiple grades	Multiple	Improved fabrication practices, cleanliness, pipe surface decontamination	Inconclusive. This DM is influenced by localized environmental conditions that can change with time
	TGSCC				
	IGSCC	Unstabilized austenitic SS	BWR primary water (HWC, NMCA)[b]	Improved water chemistry control, application of stress improvement process (e.g., MSIP, peening), use of SCC-resistant material, application of FSWOL, minimize the number of welds)	Significant reduction noted in SCC incident rates
		Stabilized austenitic SS			
		Alloy 82/182			
	PWSCC	Alloy 600/82/182	High-temperature PWR primary water	Use of SCC-resistant material (e.g., alloy 690/52/152), application of stress improvement process, application of FSWOL	
Fatigue	HCF (small-diameter lines)	Multiple	Multiple	Elimination of socket welds, fatigue monitoring, improved piping configuration, installation of vibration dampers	Inconclusive. An OECD/NEA report summarizes the international OE with vibration fatigue failures (OECD/NEA 2026)
	LCF	Multiple	Multiple	Improved piping configuration (e.g., added supports), enhanced welding technology	Inconclusive
	Thermal fatigue	Multiple	Hot/cold fluid mixing	Fatigue monitoring, improved NDE technology, improved system operating procedures, piping system reconfiguration	"Somewhat" inconclusive and to be further monitored and investigated (OECD/NEA 2022)
	Corrosion fatigue/	Multiple		Reconfiguration of piping (e.g., PWR feedwater nozzle	Inconclusive. Limited OE, most of which are from the 1970s to early

(continued)

Table 10.3 (continued)

Degradation mechanism (DM)		Material	Operating environment	Steps taken to manage/mitigate degradation (post-1980s)	Aging management effect on OE (plant age > 40 years)
	environmental degradation		High-temperature/high-pressure primary/secondary side system environment	configuration). Fatigue monitoring, improved plant operating procedures	1980s. Relationship between field experience and experimental and theoretical work yet to be determined

[a]This table and Table 10.4 were originally prepared by the author for NEA Publication No. 7614 (2024); see below

[b]Noble metal chemical addition (NMCA), also known as NobleChem or On-Line NobleChem (OLNC), involves adding a noble metal, like platinum (Pt), to the water chemistry of a boiling water reactor (BWR). This addition helps mitigate stress corrosion cracking (SCC) by reducing the electrochemical corrosion potential and improving the efficacy of hydrogen water chemistry

Table 10.4 Aging management of material degradation (Part 2)

Degradation mechanism		Material	Operating environment	Steps taken to manage/mitigate degradation	Aging management effect on OE (plant age $>$ 40 years)
Flow-assisted degradation (FAD)	Erosion-cavitation	Multiple	Multiple	Piping system reconfiguration, improved system operating procedures	Inconclusive—a function of the effectiveness of knowledge transfer
	Erosion-corrosion	Carbon steel	Multiple	Use of E/C-resistant materials	Inconclusive
	Flow-assisted corrosion	Carbon steel, low-alloy steel (LAS)	High-energy piping, single- or two-phase flow conditions	Use of FAC-resistant material, implementation of FAC monitoring program, improved secondary-side water chemistry	Significant reduction noted for FAC-induced leaks and ruptures (OECD/NEA 2014)
	Liquid droplet impingement erosion	LAS, stainless steel		Reconfiguration of piping, improved operating procedures/operational strategies (e.g., baseload vs. load following)	Inconclusive; the relationships between extended power uprate and LDIE susceptibility to be determined
Corrosion	Corrosion—General	Al-bronze, carbon steel, stainless steel	Raw water (e.g., brackish, lake, pond, river, sea)	Chemical treatment of water, use of corrosion-resistant material including nonmetallic material, reconfiguration of piping, application of composite repair technology, installation of cathodic protection system, etc.	Inconclusive—However, high-alloy austenitic stainless steels (e.g., AL-6XN®, 254-SMO®, and 654-SMO®) and HDPE materials perform well in corrosive environments
	Crevice corrosion				
	De-alloying—Selective leaching				
	Galvanic corrosion				
	Microbial corrosion				

Table 10.5 Material degradation temporal changes

Degradation mechanism		OE summary by time period (no. of pipe failures) 1970–2004	2005–2009	2010–2014	2015–2019	2020–2024
Stress corrosion cracking	ECSCC—BWR—Piping	83	7	3	0	0
	ECSCC—PWR—Piping	102	45	19	8	17
	IGSCC—BWR—Piping	1472	47	19	20	6
	IGSCC—PWR—Piping	153	9	27	19	71
	PWSCC—PWR—Piping	96	47	7	15	4
	TGSCC—BWR—Piping	135	14	0	1	1
	TGSCC—PWR—Piping	33	2	7	4	3
Fatigue	HCF—BWR—Socket weld	166	15	15	13	8
	HCF—PWR—Socket weld	521	48	29	27	22
	HCF—All plant types and all piping systems	1937	186	177	167	151
	LCF—All plant types and all piping system	214	30	44	38	9
	TF—BWR—All systems	88	5	5	3	3
	TF—PWR—All systems	112	6	14	21	8
Corrosion	Corrosion—FP system	221	94	136	108	62
	Corrosion—BWR—SW system	101	46	29	14	27
	Corrosion—PWR—SW system	265	83	70	61	60
	MIC—All plant types and all systems	633	195	404	444	212
Flow-assisted degradation	FAC—BWR—All systems	482	74	38	28	24
	FAC—PWRall systems	1222	167	61	41	35
	Erosion-cavitation—All plant types and all systems	154	47	41	48	28
	Erosion-corrosion—All plant types and all systems	385	91	72	56	59

10.9 Synthesis

Database applications involve multiple steps of filtering, going from the general to specifics, invariable resulting in small event populations that are subjected to further assessments before being input to a model. Characterization of the underlying sources of uncertainty becomes a critical step in the analysis of the data. Three charts are used to represent a synthesis of the accumulated piping operating experience, 1960–2025 (Figs. 10.17, 10.18, and 10.19). The charts account for failures of piping in BWRs, fast breeder reactors, gas-cooled reactors, heavy water reactors, PWRs, RBMKs, and WWERs. Both charts differentiate the operating experience data by pipe diameters and mode of pipe failure. Furthermore, the two charts account

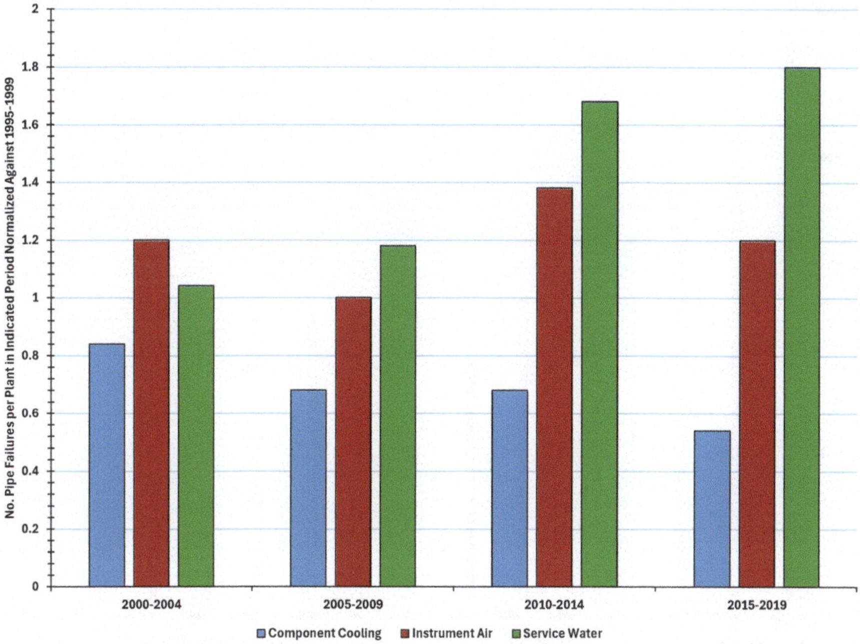

Fig. 10.16 Normalized pipe failure incident rate as a function of in-service time

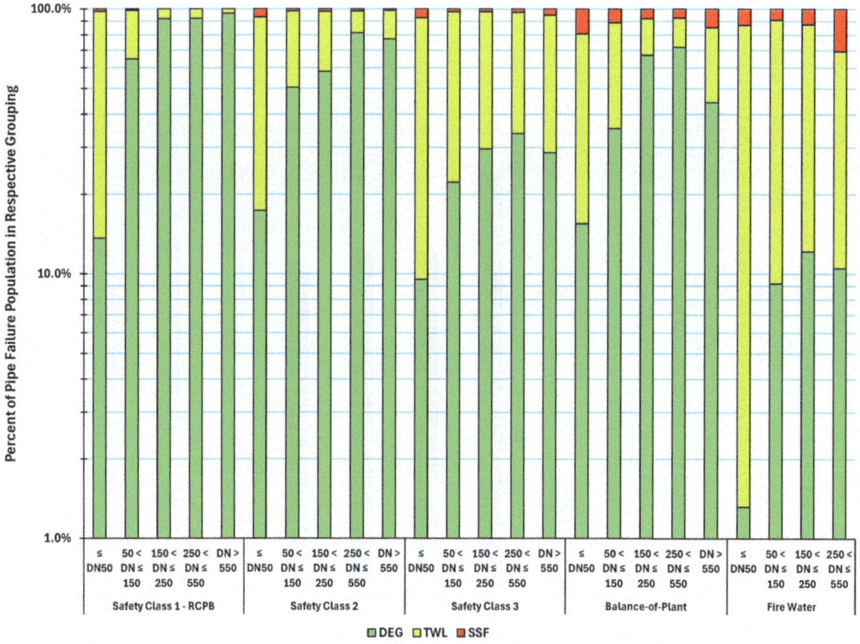

Fig. 10.17 Synthesis of the accumulated piping operating experience 1960–2025

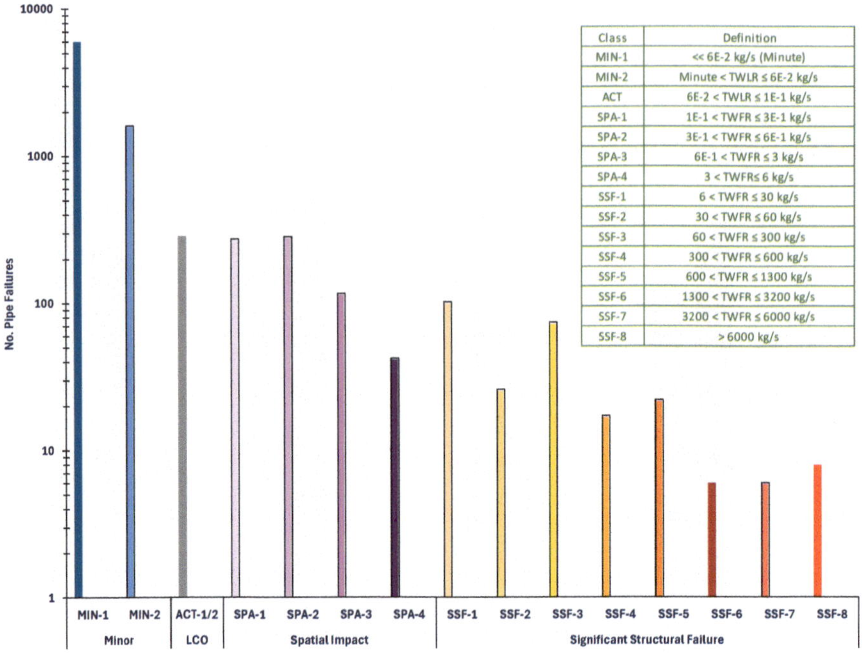

Fig. 10.18 Synthesis of the consequences of pipe failure 1960–2025

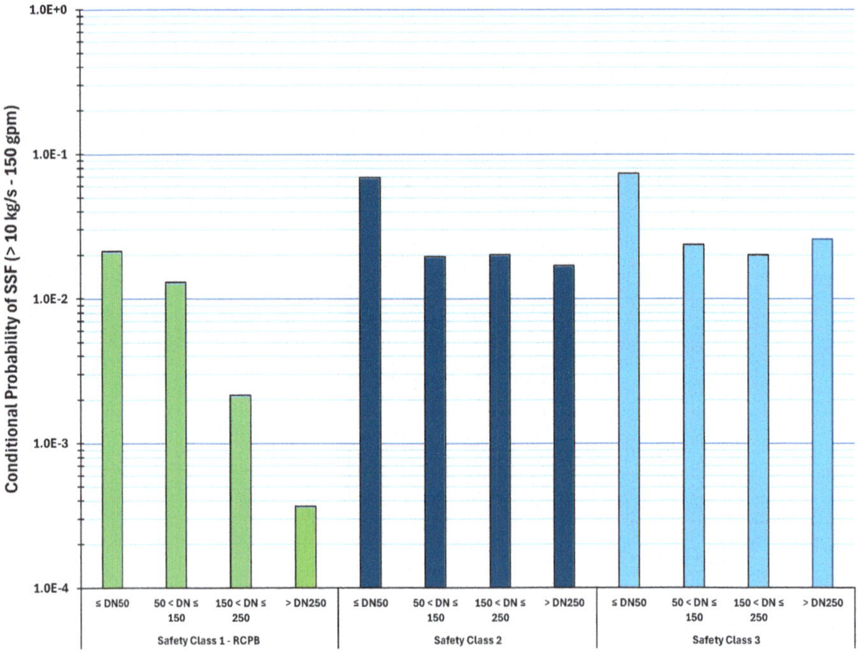

Fig. 10.19 A synthesis of the conditional probability of "significant structural failure" of safety-related piping

for in-plant piping as well as belowground piping within the owner-controlled area of a nuclear power plant. The latter category mainly applies to Safety Class 3 service water piping and firewater piping.

As noted in earlier chapters, a degraded condition (DEG) refers to any pipe flaw found by nondestructive examination that fails to satisfy the standards for continued service. A through-wall leak (TWL) is an early indicator of a potential major structural failure (SSF), which is defined as having a through-wall mass flow rate exceeding 10 kg/s.

A key takeaway from Fig. 10.17 is that in-service inspection and nondestructive examination are effective. For reactor coolant pressure boundary piping larger than DN150, over 90% of flaws are found before they penetrate the pipe wall. In contrast, ASME III Class 3 piping, which is not required to undergo regular nondestructive examinations, does not benefit from this high detection rate.

References

Autorité de Sûreté Nucléaire (2023) Nuclear accidents and developments in nuclear safety and radiation protection: how nuclear safety and radiation protection have evolved further to accidents in France and across the world. Les Cahiers Histoire de l'ASN no.1, Montrouge, France

Cummins WE, Matzie R (2018) Design evolution of PWRs: Shippingport to generation III+. Prog Nucl Energy 102:9–37

Fennern L (2018) Design evolution of BWRs: Dresden to Generation III+. Prog Nucl Energy 102: 38–57

Garland WJ (ed) (2014) The essential CANDU: a textbook on the CANDU nuclear power plant technology. UNENE, University of Network of Excellence in Nuclear Engineering, McMaster University, Hamilton. https://unene.ca/essentialcandu/

IEEE (1984) IEEE recommended practice for system identification in nuclear power plants and related facilities, IEEE Std 805-1984. Institute of Electrical and Electronics Engineers, New York

Kok KD (ed) (2017) Nuclear engineering handbook, 2nd edn ISBN: 978-1-4822-1592-2. CRC Press, Boca Raton

Moglia M, Davis P, Burn S (2008) Strong exploration of a cast iron pipe failure model. Reliab Eng Syst Saf 93:863–874

OECD/NEA (2014) CODAP topical report: Flow Accelerated Corrosion (FAC) of carbon steel & low alloy steel piping in commercial nuclear power plants. NEA/CSNI/R(2014)6, Boulogne-Billancourt

OECD/NEA (2022) CODAP topical report: a review of the "Post-1998" experience with thermal fatigue in heavy water and light water reactor piping components. NEA/CSNI/R(2022)13, Boulogne-Billancourt

OECD/NEA (2024) A review of operating experience involving passive component material degrading in periods of extended and long-term operation. NEA Publication No. 7614, Boulogne-Billancourt

OECD/NEA (2026) CODAP topical report: the international operating experience with vibration fatigue. NEA/CSNI/R(2026)NN, Boulogne-Billancourt (in press)

Pioro I (2022) Handbook of generation IV nuclear reactors, 2nd edn, ISBN: 978-0-12820588-4. Elsevier, Amsterdam

Schneider R, Wirtz P (1991) Adaptation of a pressurized water reactor of American design to the requirements of the German Standards Program and Licensing Procedure. Nucl Eng Des 127(2): 173–179

Takeda T, Inagaki Y (eds) (2021) High-temperature gas-cooled reactors., ISBN: 978-0-12-821031-2. Elsevier, Amsterdam

USDOE (1987) Overall plant design descriptions of VVER, water-cooled, water-moderated energy reactor, DOE/NE-0084, Revision 1. U.S. Department of Energy, Washington, DC

Zoulikha H, Chaima S, Leila Z-G, Wail R (2024) Data mining for next-generation reliability assessment, 3rd edn. Scientific Day of Industrial Engineering and Maintenance, National School of Advanced Technologies Ecole Nationale Supérieure des Technologies Avancées, Algiers, Algeria. https://hal.science/hal-04595149v1

Chapter 11
Conclusions: Reflecting on the Past and Planning the Future

11.1 Abundance of Data

Data on piping performance across various reactor environments comes from over 20,000 reactor operating years. Extensive knowledge is available, but what can we learn from earlier experiences in the 1950s and 1960s? The Vallecitos BWR pipe failure in September 1962 (Fig. 11.1) is a notable example. What was its significance relative to how the lessons learned shaped later nuclear power plant designs?

About 18 years after the pipe failure at the VBWR, similar—but less severe—fatigue failures occurred in BWR plants designed by Asea-Atom (Fig. 11.2). Could those failures have been prevented, and how were they mitigated? Questions expected to be answered by the readership, especially the current and pending early-career nuclear safety specialists, PSA engineers, and material scientists. The event at the VBWR was a first-in-kind (FIK) failure. FIK failures continue to occur. So, learning from operating experience remains an ongoing process. One of the lessons from the TMI-2 accident in 1979 was the need for systematic evaluation of operating experience by the nuclear industry. This need persists and remains as challenging as ever.

The evolving accumulated piping operating experience produces new insights into causes and consequences of different material degradation scenarios, including FIK failures. With this evolving operating experience come new technical approaches and implementations responding to evolving regulatory regimes, evolving codes and standards, as well as new and sometimes novel approaches to material degradation mitigation processes. Acknowledging the underlying knowledge base is important when pursuing probabilistic piping reliability analysis. History repeats itself. The abundance of piping operating experience data has not yet been fully exploited by analysts, however.

B. Lydell, *Knowledge Preservation in Nuclear Safety*,
https://doi.org/10.1007/978-3-032-25463-4_11

APED 4116
CLASS I
NOVEMBER 1962

ACCENT ON VALUE

THE CAUSE OF FAILURE
AND THE REPAIR OF
THE VBWR RECIRCULATION PIPING

By
W. R. SMITH, Sr
J. B. VIOLETTE

ATOMIC POWER EQUIPMENT DEPARTMENT
GENERAL ELECTRIC
SAN JOSE, CALIFORNIA

Fig. 11.1 Report on the 1962 VBWR pipe failure caused by thermal fatigue. (The report was prepared for the U.S. Atomic Energy Commission. The author obtained a copy of the report with assistance from the librarians of the USNRC Public Document Room and the University of Michigan Main Library. Their assistance is gratefully acknowledged)

11.2 Benefiting from Knowledge Preservation

Piping systems designed for advanced reactors are built on lessons learned from more than six decades of commercial operation of gas-cooled, heavy water, and light water reactors, including commissioning phases and extended periods of operation. Adapted from IAEA Nuclear Energy Series No. NR-T-2.16 (IAEA, 2023), building on the accumulated piping operating experience, Fig. 11.3 represents a conceptual scheme for how to estimate advanced reactor pipe failure rates and conditional failure probabilities.

11.3 The Next 20 Years

There is a constant need for reliable, high-quality data on pipe failures. As newer reactor designs are being developed and considered for deployment, information about their piping performance is missing. Previously, PSA engineers would rely on concepts like the "Thomas model" to estimate pipe failure probability. Today, there are advanced analytical techniques available that can create well-informed models for piping pressure boundary integrity and reliability, drawing from knowledge about different materials and operational conditions.

An essential foundation of probabilistic piping reliability is an effective program for developing, implementing, and maintaining a piping operating experience

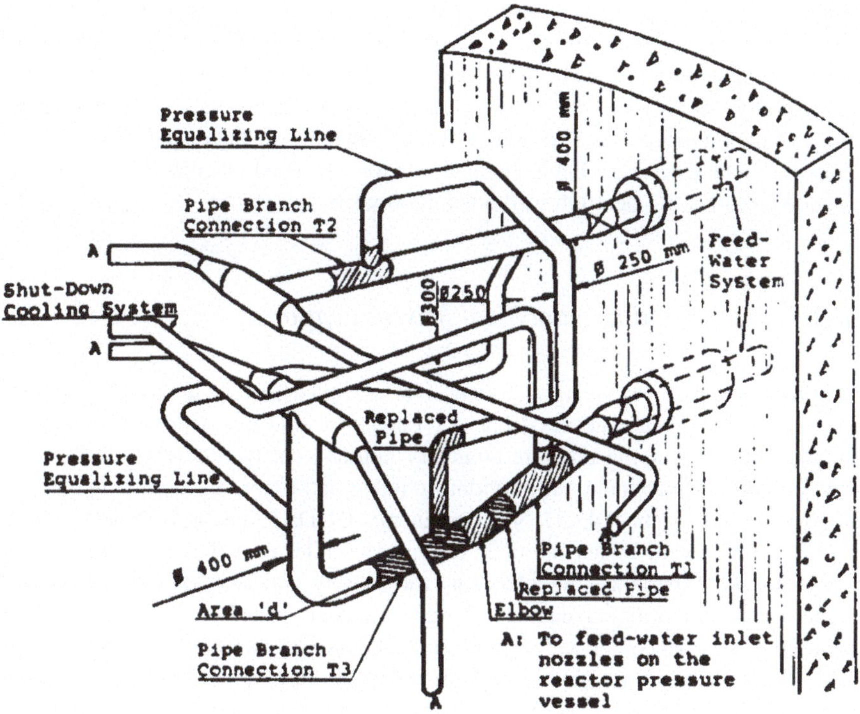

Fig. 11.2 Thermal fatigue locations in an Asea-Atom BWR. (Reproduced from Burkhart, D. (1981), Thermal Fatigue Cracking in BWR Feedwater Systems in Sweden, SWE810216-01T, AB Statens Anläggningsprovning, Stockholm, Sweden. NRC-ADAMS Accession No. ML20117L117)

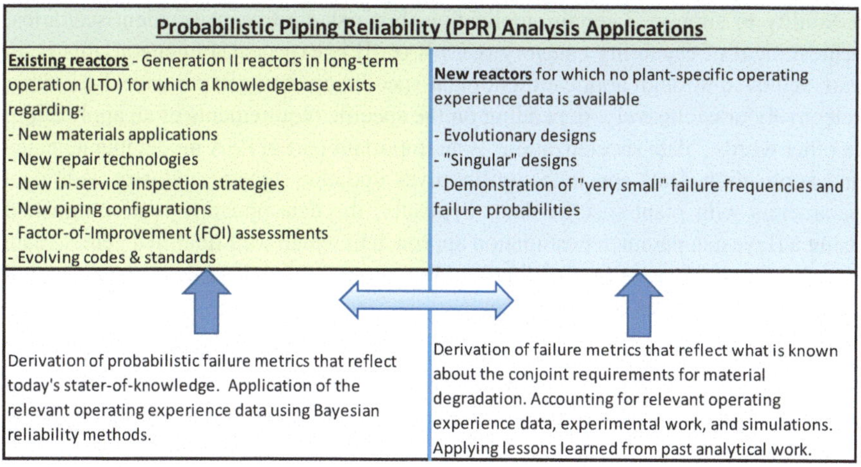

Fig. 11.3 Probabilistic piping reliability analysis for new reactors

database and curating the data therein. This is a resource-intensive activity, but with almost immeasurable benefits. An expected challenge is to ensure continued access to operating experience data. Bayesian analysis offers a powerful platform for utilizing different sources of historical data—experimental data, field experience data, and expert judgments—to form posterior distributions that respond to the needs of analysts engaged in probabilistic analyses for advanced reactors and reactors in extended periods of operation well beyond 60 years.

11.3.1 Overcoming Analytical Misconceptions

Although historical operating experience in advanced reactor probabilistic safety assessment (PSA) is sometimes debated, it remains a solid foundation for evaluating the probability of piping pressure boundary failures, but with caveats. There have been notable improvements in providing reliable data for analysts, highlighted by initiatives such as the OECD Nuclear Energy CODAP database project, which marked its 20th anniversary in 2022. This project relies on voluntary contributions of operational experiences, emphasizing the sharing and analysis of selected key events over attempting to document every incident related to piping reliability. Achieving this goal comes with significant challenges, as international collaboration depends on pledges for technical support. These, in turn, are affected by changing resources and varying levels of willingness or ability to share information. Since participation in information exchange is voluntary, cooperation can suffer if participants are not fully engaged. While this approach realistically addresses practical limitations, it might not be the best way to ensure the preservation of knowledge.

Using terminology from a PSA Standard (ASME/ANS, 2024), "*capability categories*" are assigned to the different elements of a PSA to determine its quality and its ability to support a certain application. For risk-significant accident scenarios, achievement of capability category (CC) II or III is expected. In general, this is, in part, achieved through application of plant-specific operating experience data, either selectively or exclusively, depending on the specific requirements of an application. In other words, "data specialization" is an important part of PSA model maintenance and application. Data specialization involves updating generic, industry-wide data parameters with plant-specific data. Typically, the data updating is accomplished using a Bayesian parameter estimation approach in which well-qualified generic data is represented by a prior distribution. For piping reliability, data specialization includes tasks such as the following:

- Update of existing piping reliability parameter estimates by using new operating experience data. We refer to this as "routine" or ordinary data specialization in which a new set of operating experience data is incorporated into an existing analysis.
- Modifying generic piping reliability parameter estimates to account for impact on the structural reliability by changes to an inspection program or mitigation of

material degradation application of a full structural weld overlay or mechanical stress improvement process or replacement of an existing piping system by using a material known to be resistant to environmental degradation.
- Derivation of "centered" pipe failure rates and rupture frequencies that reflect specific material degradation mechanisms. Included in this task is the development of conditional failure probability models that are conditional on the presence of a specific active or assumed inactive degradation mechanism.
- Derivation of piping reliability parameters for new reactor designs on the basis of existing industry-wide operating experience data to a new piping design for which there is no prior operating experience. This type of data specialization involves a very structured application of the full knowledge base that is associated with the lessons learned from generation I through III commercial nuclear power reactor operating experience.
- For some PSA applications, pipe rupture frequencies have been developed for different through-wall flow rate categories, for example, "spray events" ($\leq$5 kg/s), "general flooding" (between 5 kg/s and 100 kg/s), and "major flooding" ($>$100 kg/s). To remove conservatism, a refined treatment of flow rate ranges to parse the pipe rupture frequency for flow rate ranges of varying sizes may be warranted.

The quality of data specialization tasks is a function of the analyst's knowledge and experience and how a parameter estimation task is structured to adequately address a specific application requirement. Guidelines and best practices for probabilistic piping reliability analysis have been developed that address the following topics:

- Knowledge base: A fundamental basis for a qualified piping reliability analysis rests on a deep understanding of how the typically robust metallic piping systems degrade and fail or sustain damage due to different off-normal operating environments.
- Operating experience data: Under what conditions can operating experience data support quantitative piping reliability analysis? The completeness and comprehensiveness of a database are essential characteristics for a database to support the derivation of "robust" reliability parameter estimates.
- Qualitative analysis requirements: Query functions are defined to extract event population and exposure term data from a comprehensive relational database. Oftentimes, a query definition must address a complex set of reliability attributes and influence factors. The characterization of aleatory and epistemic uncertainties depends on the intrinsic qualities of a query definition.
- Quantitative analysis requirements: Pipe failure rate calculation is based on event populations that reflect different piping designs. Therefore, an established practice is to apply a Monte Carlo posterior weighting technique to synthesize the variability in weld counts and DM susceptibility. Pipe rupture frequencies are calculated for well-defined break sizes and resulting through-wall flow rates. Conditional failure probability models are needed for a predefined set of break

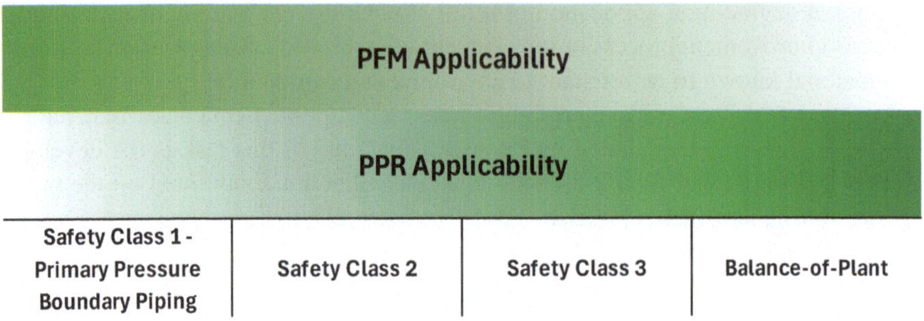

Fig. 11.4 The ranges of applicability of PFM and PPR models

size ranges. Certain follow-up studies may have to be performed once a base-case set of reliability parameters have been obtained.

The continued relevance of various models for pipe failure remains an open topic for consideration. This book identifies two modeling concepts, PFM and PPR, both having found use in quantifying reliability of piping components and systems (Fig. 11.4). PFM applies to crack-like defects, axial or circumferential, and is concerned with physics of failure. PPR relies on operating experience that involves a very broad range of material degradation scenarios and types of flaws in base or weld metal. PPR also relies on insights from PFM application to address conditional failure probabilities given degraded conditions such as minor flaws in thick-walled Safety Class 1 piping.

11.3.2 Knowledge Base

Metallic piping systems experience degradation and failure as a result of the combined influences of atypical operational and environmental factors, as well as unusual or excessive loading conditions. The triplet <***materal;environment;loading***> encapsulates the essential requirements for pipe degradation to occur. Even minor alterations to physical parameters—such as pH, corrosion potential, hydrogen content, temperature, flow rate, carbon content, or post-weld heat treatment—within this framework can significantly influence the susceptibility of piping material to degradation. Consequently, reliability analyses for piping should be grounded in a thorough understanding of how metallurgy, water chemistry, and mechanical stresses contribute to structural integrity. For instance, assessing the impact of varying material grades on piping reliability exemplifies data specialization. Similarly, quantifying reliability enhancements achieved through stress relief processes at specific weld sites offers valuable insights.

Currently available operating experience consists of two general types of pipe failures: (1) failures that are due to environmental degradation and (2) event- or stress-driven failures. The former is characterized by an incubation time for a flaw to develop followed by propagation in the through-wall direction. Flaw propagation occurs due to some driving force, e.g., weld residual stresses. These failures are time dependent and may develop over a long time period (e.g., several decades). The latter are due to upstream or downstream equipment failures or significant hydraulic transients. Under certain conditions, an event-based condition acts as an initiation site for subsequent environmental degradation.

Certain combinations of susceptibility to material degradation, type of material, and operating environments have resulted in various structural failures. In contrast, the careful application of material selection principles has led to piping systems that have not exhibited through-wall defects. Pipe degradation and failure can be mitigated by selecting an appropriate analysis strategy, which should demonstrate a comprehensive understanding of the fundamental principles of pipe degradation and failure, as well as recognition of piping design principles, codes and standards, and in-service inspection practices and requirements. There is no universal analysis strategy; for example, piping exposed to aggressive degradation mechanisms such as flow-accelerated corrosion requires different analysis techniques than cases where degradation is highly localized and evolves over extended periods.

11.4 The Future

Progress has been made regarding the development of dedicated, "just-in-time" (JIT) pipe failure databases, that is, databases that are ready to be utilized by experienced analysts. Since an event database includes information on historical events, the completeness of the event population in the database remains a consideration in determining its "fitness for use." This needs to be placed in perspective of the present-day knowledge about incubation times of pipe flaws—short versus long. Five types of metrics are considered in probabilistic piping reliability analysis: (1) precursor rate, (2) conditional failure probability, (3) inspection effectiveness, (4) degradation mitigation effectiveness, and (5) aging factors. For a pipe failure event database to support failure rate estimation, it must include extensive piping system design information that yields information on the total piping component population that has produced the failure observations. In other words, the database must include event population data as well as exposure term data. Relative measures of piping reliability such as conditional failure probabilities can be generated by querying an event database without access to exposure term data, however. The statistical robustness of such relative measures is strongly correlated with the completeness of the event population.

Completeness and comprehensiveness of an operating experience database are ensured through a sustained and systematic maintenance and update process. Completeness is an indication of whether or not all the data necessary to meet current and future analysis demands are available in the database. The comprehensiveness of a service experience database is concerned with how well its structure and content correctly capture piping reliability attributes and influence factors. A clear basis

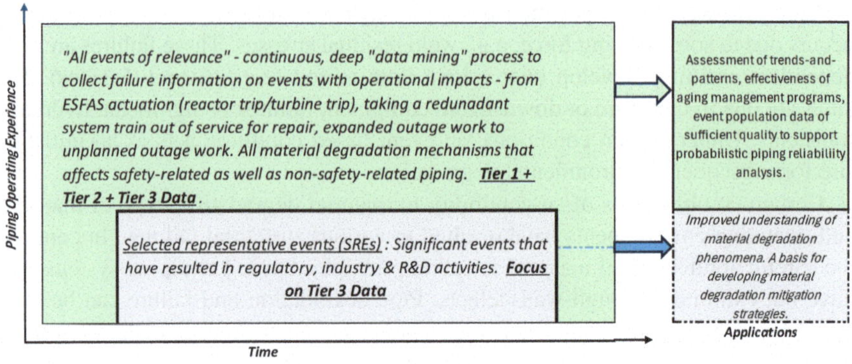

Fig. 11.5 Database comprehensiveness and fitness for use

should be included for the identification of events as failures. Based on practical experience, the inherent latency in structured data collection efforts is on the order of 5 years. This means that at least 5 years could elapse before the achievement of high confidence in data completeness. In other words, around 2035, the data mining for the period 2020–2025 would be expected to approach "saturation," meaning reasonably high confidence in the completeness of a database. Could "cliff-edge effects" affect an analysis due to database infrastructure factors? It depends on the maturity of inspection programs and our state of knowledge concerning certain degradation mechanisms. The design of and infrastructure associated with an operating experience database should be commensurate with application demands and evolving application requirements. The completeness of a relevant event population should be validated, either independently or assured through a sustained maintenance effort. The database underlying this book was developed over a period exceeding 10 person-years of effort, involving extensive analysis of technical reports, laboratory tests, plant site visits, conference attendance, and related activities (Fig. 11.5).

Codes and standards for piping design and in-service inspection undergo modernization efforts that are risk-informed and increasingly rely on probabilistic assessments of piping pressure boundary integrity. These assessments rely on high-quality data and analysis methodologies that make effective use of the knowledge base embedded in operating experience data. Reporting routines for tier 1 through tier 3 data also undergo changes that significantly affect data mining processes (Fig. 11.6).

Efforts are underway to convert the existing SQL database of reference into a secure web application. This process involves implementing protective protocols and governance measures to ensure that data remains confidential, accurate, and accessible for future users. Transitioning from an SQL platform on a local computer to a secure web-based on-premise solution to a cloud-based software solution demands multiple layers of security, including strong authentication methods, secure connection strings, service accounts with limited privileges, and managed identities

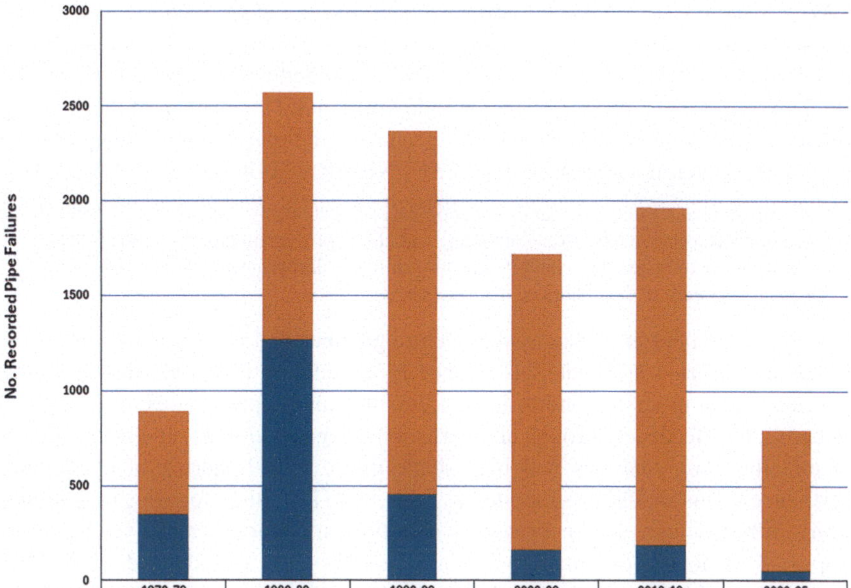

	1970-79	1980-89	1990-99	2000-09	2010-19	2020-25
Tier 1 & 2 Data	539	1307	1916	1561	1780	740
Tier 3 Data	351	1266	452	163	187	51

Fig. 11.6 Sources of piping operating experience data in the United States

appropriate for cloud environments. The process requires careful attention and multiple steps.

11.5 Recommendation for Future International Collaboration

Protecting institutional knowledge about piping pressure boundary integrity and how environmental degradation affects both metallic and nonmetallic materials is a challenging task. Preserving 60 years of operational experience ensures that future material scientists, structural engineers, and PSA engineers can easily access and learn from these valuable insights. It is important to maintain this information alongside analysts' perspectives, including the expectations and requirements imposed on analyses. As nuclear power plants remain in operation well past their original 40-year licenses and new reactors are designed and commissioned, one must consider whether full use is made of the current knowledge base. About 40 years ago, the late Dr. Spencer H. Bush made a statement on the need for "*an international body to which all failures from an agreed type of plant could be reported* ..." (Bush, 1988).

What steps have been taken to organize and update piping operating experience databases as new data emerges? While these questions may appear complex, there are answers available to address them. Before responding to these questions, the OECD/NEA CODAP project offers the following candid assessment of the progress with an international collaboration on the exchange of piping and non-piping passive component operating experience data, CSNI/NEA/R(2024)6 (OECD/NEA 2024):

> Collecting and analyzing material degradation data are technically challenging and rewarding, as well as resource intensive. Continuing with the improvement process for the CODAP event database infrastructure could be key to instilling a deeper interest in and committing adequate resources to the continued data exchange.

To advance the field, it is important for future graduate and postgraduate students in nuclear engineering and materials science to be actively involved with operating experience data processes and to have access to piping operating experience data. Gathering user feedback through an international organization about what is needed for effective data analytics will help shape future global sharing of operational experiences. This should be supported by sufficient IT funding, focusing on building strong infrastructure, secure protocols, real-time monitoring, and excellent user support. This book offers many potential research topics, including ideas for PhD theses that could drive institutional changes to promote fair and active sharing of operating experience data, ensuring enough quality and quantity to support probabilistic assessments.

Recommendations for future, sustainable international collaboration are listed below:

Organizational/Programmatic Recommendations

1. Clearly stated goals for collecting piping operating experience data and ensuring institutional support for fair but active data sharing and access: Active involvement and sufficient resources are crucial for successful international collaboration beyond routine meetings. Identify and address barriers such as complex database structures, tedious data entry, or ineffective search functions; solutions are available for these and other challenges.
2. Form collaborative relationships with academic institutions and researchers to obtain expert guidance on data completeness, quality, and peer review processes. Academics should be encouraged to actively participate in the development and ongoing maintenance of databases. Uphold engagement with organizations such as the *NEA Global Forum on Nuclear Education, Science, Technology and Policy* to support the creation of curricula that integrate operating experience data to promote nuclear safety. Develop comprehensive course materials and provide training for educators. Insights are gained from experimental research, theoretical studies, and operating experience as key elements of the course material.
3. Each collaborator should locate internal information on pipe failures and relevant accessibility standards. Using AI in data mining can help retain knowledge.

Converting JIC data to JIT data is essential, and developing strong analytic tools for database analytics is advised.

4. Implement technical oversight and peer review processes to maintain alignment with project objectives and facilitate effective engagement with industry professionals, including PFM and PSA practitioners. Conduct regular audits to assess progress.
5. Develop a formal protocol for transferring intellectual property to the designated preservation organization, ensuring that modern information technology methods are employed effectively and responsibly.

Data Completeness and Quality

1. Future collaborations should focus on practical applications to ensure and promote sustainability, leveraging technical knowledge discussed earlier. Involvement by industry is critical. Both high-quality old and new data are valuable, with established mathematical and statistical theories of reliability guiding their use. Incomplete data produces unreliable results. The accuracy of risk-informed decisions depends directly on the reliability of the underlying data. Inadequate data can lead to skewed risk estimations, causing over- or underestimation of safety or risks, while high-quality data ensures that PSA models accurately reflect the actual "as-built, as-operated" plant.[1]
2. Codes and standards (C&S) for piping pressure boundary integrity are becoming more advanced, with increased consideration of probabilistic concepts. Similarly, C&S for probabilistic safety assessment continues to evolve. Demands for high-quality data and advanced applications thereof are essential aspects of future evolutions of risk-informed decision-making and relevant C&S frameworks. Establishing a line of communication between an international collaboration on piping operating experience and the C&S community is not just helpful; it is vital for creating the aforementioned database infrastructure. Aligning data collaboration with the changing expectations and requirements of C&S is necessary.

References

ASME/ANS (2024) Level 1/large early release frequency Probabilistic Risk Assessment (PRA), ASME/ANS RA-S-1.1-2024, New York

Bush SH (1988) Statistics of pressure vessel and piping failures. J Press Vessel Technol 110(3): 225–233

IAEA (2023) Methodologies for assessing pipe failure rates in advanced water cooled reactors. Nuclear energy series No. NR-T-2.16, ISBN: 978-92-0-150422-7, Vienna, Austria

OECD/NEA (2024) Summary of Phase IV of the Component Operational Experience, Degradation and Ageing Programme (CODAP), NEA/CSNI/R(2024)6, Boulogne-Billancourt

[1]The last sentence is adapted from NEA/CSNI/R(2020)12, Use of NEA Database Projects Operating Experience Data for Probabilistic Safety Assessment. https://www.oecd-nea.org/upload/docs/application/pdf/2023-12/nea_csni_r_2020_12.pdf

Glossary

Aging Management A technical process that uses a systematic approach to identify, monitor, and control the degradation of a nuclear power plant's structures, systems, and components (SSCs) to ensure that they continue to perform their intended safety functions throughout their service life. This includes managing degradation from normal service conditions like corrosion, fatigue, and mechanical effects to prevent failures and ensure safe, continued operation, especially during extended operating periods (IAEA Safety Report Series No. 82, Revision 2).

Aleatory Uncertainty This form of uncertainty is associated with the randomness of events such as component failures, initiating events, and hazard events. This is addressed by modeling these events using probability models; for example, these are typically Poisson and binomial distributions for initiating events and component failures. These probability models represent initiating event frequencies and component failure probabilities or failure rates.

Allowed Outage Time In nuclear power plant technical specifications, it is the maximum time a safety system, component, or train can be inoperable during operation before the plant must be shut down.

Annealing (or Recover Annealing) A heat treatment process used mostly to increase the ductility and reduce the hardness of a material. This change in hardness and ductility is a result of the reduction of dislocations in the crystal structure of the material being annealed. Annealing is often performed after a material has undergone a hardening or cold working process to prevent it from brittle failure or to make it more formable for subsequent operations.

ASA/ANSI B31.1 Piping The piping systems in early commercial nuclear plants were designed in accordance with the requirements of the American Standards Association (ASA) B31.1-1955. Section 1 of B31.1-1955 was written for power piping and encompassed the "… minimum requirements for design, manufacture, test, and installation of power piping systems, as defined for steam-generating

B. Lydell, *Knowledge Preservation in Nuclear Safety*,
https://doi.org/10.1007/978-3-032-25463-4

plants, central heating plants, and industrial plants." B31.1-1955 included specific requirements for pipe wall thickness and allowable stresses.

ASME Section III Piping Division 1 of ASME Boiler and Pressure Vessel Code Section III contains requirements for piping classified as ASME Class 1, Class 2, and Class 3. ASME Section III does not delineate the criteria for classifying piping into Class 1, Class 2, or Class 3; it specifies the requirements for design, materials, fabrication, installation, examination, testing, inspection, certification, and stamping of piping systems after they have been classified as Class 1, Class 2, or Class 3 based upon the applicable design criteria and Regulatory Guide 1.26, Quality Group Classifications and Standards for Water-Steam, and Radio-Waste-Containing Components of NPPs. Subsections NB, NC, and ND of ASME III specify the construction requirements for Class 1, Class 2, and Class 3 components, including piping, respectively. Subsection NF contains construction requirements for component supports, and a newly added Subsection NH contains requirements for Class 1 Components in Elevated-Temperature Service. Subsection NCA, which is common to Divisions 1 and 2, specifies general requirements for all components within the scope of ASME Section III. The first edition was released in 1964, but the technical scope was limited to reactor pressure vessels. The 1971 edition of ASME BPVC Section III marked the transition of the code from one covering only the reactor pressure vessel to a one covering the entire primary pressure boundary in a nuclear power plant.

ASME Section XI The ASME Code Section XI (Rules for Inservice Inspection of NPP Components) provides standards for examination, in-service testing and inspection, and repair and replacement of NPP components, pressure vessels, and piping. The code details if a flaw is found within a component are acceptable for continued operation or if the component instead requires repair or replacement. The first edition of ASME Section XI was published on January 1, 1971.

ASME Section XI, Division 2 It provides the requirements for the creation of a RIM program for advanced nuclear reactor designs. RIM addresses the entire life cycle for all types of nuclear power plants; it requires a combination of monitoring, examination, tests, operation, and maintenance requirements that ensures that each system-structure-component (SSC) meets plant risk and reliability goals that are selected for the RIM program. Consideration of probabilistic safety assessment is an integral element of RIM program development.

Augmented In-Service Inspection Program An inspection program that has been developed to address observed material degradation, and an inspection that is targeted at locations where the most severe degradation is expected.

Austenitic Alloy Steel High-alloy steels with the main alloying elements being chromium (Cr) and nickel (Ni). Some high-alloy steels include niobium (Nb) to improve welding properties or titanium (Ti) to prevent intergranular corrosion and weld decay.

Backfitting According to Revision 1 of NUREG-1409 (*Backfitting Guidelines*), backfitting occurs when the USNRC (NRC) imposes new or changed regulatory requirements or staff interpretations of the regulations or requirements on nuclear power reactor licensees, certain nuclear power reactor applicants, or select nuclear material licensees. Backfitting is an integral part of the regulatory process and may be needed when the staff addresses safety or security issues. The NRC may take a backfitting action only after conducting a formal, systematic review to ensure that the action is defined and justified. This process ensures discipline, predictability, and optimal use of NRC and licensee resources. The backfitting requirements are found in 10 CFR 50.109, "Backfitting"; 70.76, "Backfitting"; 72.62, "Backfitting"; and 76.76, "Backfitting." Provisions analogous to the backfitting requirements, referred to as "issue finality provisions," are set forth in 10 CFR Part 52, "Licenses, Certifications, and Approvals for Nuclear Power Plants."

Backing Ring (BR) A backing ring is used to align two pipe sections before welding and eliminate the need for tack welding. Backing ring pins or nubs automatically set the weld gap.

Balance of Plant Piping Piping systems that are needed for power conversion, i.e., high-pressure and -temperature piping for converting thermal energy to mechanical and electric energy.

Bayesian Reliability Analysis In the Bayesian approach, a subject matter expert develops a well-informed estimate of the probability of failure distribution, the prior state of knowledge. This probability distribution is then updated as more information is collected about the structural integrity of a certain piping system component.

Boat Sampling Technique (BST) A technique developed for obtaining samples from the surface of any operating component. The technique is a nondestructive surface sampling technique as it does not cause any plastic deformation or thermal degradation of the operating component. BST can be used, remotely and in water-submerged condition, with the help of a handling mechanism. The samples are boat-shaped, having 3 mm maximum thickness, and require on the order of 3 hours for getting scooped from a location. The samples are used for metallurgical analysis to confirm the integrity of the operating component. BST incorporates mainly a sampling module, handling mechanism, and electric and pneumatic subsystems.

Boiling Water Reactor Vessel and Internals Project (BWRVIP) The project was established in 1994 by utilities in the United States in response to the discovery of intergranular stress corrosion cracking (IGSCC) in BWR core shrouds.

BONNA® Pipe A thin steel pipe embedded in reinforced concrete. It has rebar or a heavy wire mesh embedded in the outside-diameter concrete.

Branch Connection A branch connection is made when a pipe header (or run pipe) has either an equal- or a smaller-size branch line connected to it. Common branch connections are "sockolet" (or socket fitting), "weldolet," and "threadolet."

Buttering The process of adding material in welding. Weld metal is deposited on one or more surfaces to provide a metallurgically compatible weld metal for the subsequent completion of the weld.

Cavitation When the static pressure of a liquid reduces to below its vapor pressure, small vapor-filled cavities are formed in the liquid. This is called cavitation. The cavities are also called bubbles or voids. When imposed to higher pressure, the cavities collapse, thus generating shock waves (they are stronger closer to imploding cavity and weaken as it propagates away from it). Therefore, cavitation damage of significance in engineering systems is created by collapsing cavities near the metal surface causing its fatigue due to generated cyclic stress. Most often, cavitation causes most of its damage by vibration (e.g., cracked welds, broken instrument lines, loosened flanges). The erosion caused by cavitation also generates particles that contaminate the process fluid.

CFRP Systems A buried piping rehabilitation and repair technique. It consists of high-strength carbon fiber fabrics and/or glass fiber fabrics, fully saturated in a two-part 100% solid epoxy matrix. These laminates are bonded both longitudinally and circumferentially to the interior surface of the pipe forming a structural lining within the pipe. This lining can be designed to replace the degraded portions of the existing system without reliance on the degraded piping for the life of the repair, except at the terminal ends of the repair.

Charpy-V Notch Test A standardized, high-strain-rate method used to measure a material's notch toughness, which is its ability to absorb energy and resist brittle fracture under a sudden impact.

Cliff-Edge Effect In risk and reliability analysis, a cliff-edge effect refers to a situation where a very small change in an input parameter leads to a sudden shift in the outcome of a computation. It exposes underlying deficiencies in assumptions and/or calculational procedure.

Code Repair A repair on a piping system that is performed according to the standards set by the American Society of Mechanical Engineers (ASME) to maintain its structural integrity, safety, and regulatory compliance after it has been placed in service. A detailed degradation mechanism evaluation and material assessment before repairs specify accepted repair methods like welding or composite wraps, mandate rigorous qualification and testing of repair procedures and personnel, and require comprehensive documentation to ensure equipment integrity and safety for a defined operating life.

Cohort In reliability engineering, a cohort is a group of components tracked over time to analyze failure patterns based on shared traits like construction date, degradation mechanism, or weld type. Defining cohorts allows for more detailed analysis than considering all failures together, helping analysts spot hidden trends and make targeted, accurate predictions and improvements.

Cold Working Cold working refers to the process of strengthening metal by changing its shape without the use of heat. Subjecting the metal to this mechanical stress changes the mechanical properties and the environmentally assisted crack growth rate. The crack growth rate increases rapidly with the increase of the level of cold working.

Component Boundary Defines the physical boundary of a component required for system operation. A component boundary definition should be consistent with the parameter database supporting PSA model quantification. For piping components, the component boundary is established through degradation mechanism evaluations (see below).

Coriou Effect In 1959, the French materials scientist Henri Coriou and his co-workers published the results of laboratory tests which indicated that Alloy 600 in a high-temperature (350 °C), pure water environment is susceptible to primary water SCC. The so-called *Coriou effect* was met with skepticism until the early 1980s when an increasing number of primary water SCC failures were reported.

Crack Morphology The microscopic path and appearance of cracks as they propagate through a material under the combined action of tensile stress and a corrosive environment. The two principal types of SCC crack morphology are intergranular and transgranular, and the specific form depends heavily on the material's microstructure, corrosive environment, and stress conditions.

Crevice Corrosion Crevice corrosion occurs in a wetted or buried environment when a crevice or area of stagnant or low flow exists that allows a corrosive environment to develop in a component. It occurs most frequently in joints and connections or points of contact between metals and nonmetals, such as gasket surfaces, lap joints, and under-bolt heads. Carbon steel, cast iron, low-alloy steels, stainless steel, copper, and nickel-base alloys are all susceptible to crevice corrosion. Steel can be subject to crevice corrosion in some cases after lining/cladding degradation.

Critical Crack Length The length of a crack (either axial or circumferential) for which the crack will propagate unstably for a given set of loading conditions.

Cumulative Usage Factor (CUF) Refers to the design basis for Class 1 nuclear power plant components subjected to cyclic service in ASME Section III. Transient events that the plant might credibly experience are evaluated to establish a design basis for plant equipment. The fatigue analyses rely on the definition of design-basis transients that envelope the expected cyclic service and the calculation of a cumulative usage factor (CUF). The CUF is the ratio of the number of cycles experienced by a structure or component divided by the number of allowable cycles of that structure or component. In accordance with ASME Section 111, Subsection NB, the cumulative usage factor must be less than 1.0.

Dealloying (Selective Leaching) "Dealloying" or "selective leaching" refers to the selective removal of one element from an alloy by corrosion processes. A common example is the dezincification of unstabilized brass, whereby a weakened, porous copper structure is produced. The selective removal of zinc can proceed in a uniform manner or on a localized (plug-type) scale. It is difficult to rationalize dezincification in terms of preferential Zn dissolution out of the brass lattice structure. Rather, it is believed that brass dissolves with Zn remaining in solution and Cu replating out of the solution. Graphitic corrosion of gray cast iron, whereby a brittle graphite skeleton remains following preferential iron dissolution, is a further example of selective leaching. During cast iron graphitic corrosion, the porous graphite network that makes up 4–5% of the total mass of the alloy is impregnated with insoluble corrosion products. As a result, the cast iron retains its appearance and shape but is weaker structurally. Testing and identification of graphitic corrosion are accomplished by scraping through the surface with a knife to reveal the crumbling of the iron beneath.

Delayed Hydride Cracking (DHC) A subcritical crack growth mechanism occurring in zirconium alloys as well as other hydride-forming materials that requires the formation of brittle hydride phases at the tip of a crack and subsequent failure of that hydride resulting in crack extension. Hydrogen in solution in the zirconium alloy is transported to the crack tip by diffusion processes where it precipitates as a hydride phase. When the precipitate attains a critical condition, related to its size and the applied stress intensity factor, KI, fracture ensues, and crack extends through the brittle hydride and arrests in the matrix. Each step of crack propagation results in crack extension by a distance approximately the length of the hydride.

Degradation Mechanism Phenomena or processes that attack (wear, erode, crack, etc.) the pressure-retaining material over time and might result in a reduction of component pressure boundary integrity. It should be noted that damage mechanisms and degradation mechanisms could interact to cause major, catastrophic passive component failures.

Degradation Mechanism Evaluation The identification of degradation mechanisms in a pipe segment by comparing actual piping design and operating conditions to a well-defined set of material and environmental attributes. The evaluation considers plant-specific as well as generic operating experience involving cracking and leakage.

Design Certification PSA In the United States, a design certification (DC) is achieved through a regulatory rulemaking process, which addresses the various safety issues associated with the proposed NPP design, independent of a specific site. A DC application contains a final safety analysis report (FSAR) that is supported by a design-specific PSA.

Double-Ended Guillotine Break A condition for which a circumferential through-wall crack propagates around the entire circumference of the pipe such that the cracked pipe section severs into two pieces and the two ends are displaced relative to their pipe axes to allow for full flow from each end. It refers to a scenario that has been a design basis for safety regulations like 10 CFR 50.46, *Acceptance Criteria for Emergency Core Cooling Systems for Light-Water Nuclear Power Reactors*. However, this requirement is being revised to a risk-informed approach that uses a transition break size (TBS), recognizing that DEGB is extremely unlikely. This shift is supported by the leak-before-break (LBB) principle, which shows that flaws in pipes would likely be detected by a small leak before a catastrophic break could occur, and probabilistic risk assessments that show the probability of DEGB is very low.

Ductile Crack Growth With ductile fracture, a crack moves slowly and is accompanied by a large amount of plastic deformation around the crack tip. A ductile crack will usually not propagate unless an increased stress is applied and generally cease propagating when loading is removed.

Defense-in-Depth (DID) A nuclear safety principle that uses multiple layers of physical, administrative, and procedural barriers to prevent and mitigate accidents and radiation releases from nuclear facilities. This strategy ensures that if one layer of protection fails, subsequent layers are in place to compensate, relying on the redundancy and diversity of systems and measures to protect the public, workers, and the environment from harm.

Enhanced Visual Examination The enhanced visual examination-1 method is intended for the visual examination of surface breaking flaws. Any visual inspection for cracking requires a reasonable expectation that the flaw length and crack mouth opening displacement meet the resolution requirements of the observation technique. The enhanced visual examination-1 specification augments the VT-i requirements to provide more rigorous inspection standards for SCC.

Environmentally Assisted Cracking A localized deformation process accelerated by local corrosion in addition to mechanical stresses or strains. The cracking of structural materials in NPPs may proceed along grain boundaries, i.e., intergranular SCC, underlining the role of dissolution, or through the grains, i.e., transgranular SCC, underlining the role of mechanical loading.

Epistemic Uncertainty The uncertainty that comes from the lack of knowledge. This lack of knowledge comes from many sources. Inadequate understanding of the underlying processes, incomplete knowledge of the phenomena, or imprecise evaluation of the related characteristics.

Equivalent Break Size The calculated size of a hole in a pipe given a certain through-wall flow rate and for a given pressure.

Equipment Performance and Information Exchange Database (EPIX) The history of the Institute of Nuclear Power Operations (INPO) EPIX database began in 1997 as a voluntary industry initiative to collect equipment reliability and availability data. It replaced the older NPRDS database and collects data on component failures and degraded performance from approximately 100 US commercial nuclear power plants. This data was vital for programs like the Maintenance Rule, the Mitigating Systems Performance Index (MSPI), and the Reactor Oversight Process (ROP) and is used by the USNRC to update risk assessments, although specific plant data remains proprietary.

Erosion Cavitation Erosion cavitation is the process when material surface deteriorates due to creation of vapor or gas pockets inside the flow of liquid, thus causing surface material loss. Once cavitation is present, the erosion is caused by the bombardment of vapor bubbles on material surface. When these bubbles strike the surface, they collapse or implode. Although a single bubble imploding does not carry much force, over time, the small damage caused by each bubble accumulates. The repeated impact of these implosions results in the formation of pits. Also, like erosion, the presence of chemical corrosion enhances the damage and rate of material removal. Erosion-cavitation has been observed in PWR stainless steel decay heat removal and charging system piping.

Erosion-Corrosion Erosion is the destruction of metals by the abrasive action of moving fluids, usually accelerated by the presence of solid particles or matter in suspension. When corrosion occurs simultaneously, the term erosion-corrosion is used. The term "erosion-corrosion" applies to moderate energy carbon steel piping (e.g., raw water piping).

Expert Elicitation Involves synthesizing the opinions of three or more subject matter experts (SMEs) in situations where uncertainty arises due to incomplete or insufficient data on a physical phenomenon. Expert judgments are well-founded opinions based on analysis, experiments, knowledge, and experience, provided by experts to address technical problems.

Factor of Improvement Also referred to as relative improvement factor, the factor of improvement is an estimate of the crack growth rate in one type of material relative to another material type. It is an indication of the estimated improvement in material performance as a function of material chemical composition, mechanical properties, local stresses, degradation mitigation techniques, etc. The factor of improvement can be determined on the basis of testing in laboratory environments, analytical work (e.g., statistical analysis of test data), expert elicitation, and/or operating experience data.

Failure A condition for which a component or system is no longer capable of performing its design function. Depending on the context, failure can be defined as either the condition for which the piping system is no longer capable of maintaining internal pressure or when the pipe experiences a double-ended guillotine break. National codes and standards (e.g., the ASME Boiler and Pressure Vessel Code, Section XI, Rules for Inservice Inspection of NPP Components) give specific details on what constitutes a failure.

Ferrite Content Ferrite is the ferromagnetic, body-centered cubic, microstructural constituent of variable chemical composition in iron-chromium-nickel alloys. Austenitic stainless steels are essentially free of ferrite, which is magnetic. Cast products of these alloys typically have some ferrite present. These alloys also form some ferrite when they are cold worked or work hardened. In both cases, the products will show a magnetic tendency. Ferrite can be detrimental to corrosion resistance in some environments. There are also applications where magnetic characteristics interfere with the performance of the end product. The ferrite content of the cast alloy can be controlled through alloy composition. Carbon, nitrogen, nickel, and manganese are strong austenite formers, and increasing their content in the alloy will reduce the tendency for ferrite formation. There are several different methods of predicting the ferrite content, but one of the more common is the DeLong diagram. Ferrite reduces the tendency for the steel to solidification crack during cooling. It is not uncommon for 304 castings (CF8) to contain 8–20% ferrite. The cast ingot composition of wrought 304 stainless steel is also balanced to have 1–6% ferrite since this reduces the chance of cracking during forging or hot working.

Fillet Weld A type of weld used to join two pieces of pipe or a pipe and a fitting, where the pieces meet at an angle, typically a 90° angle, to form a T-joint or lap joint. The weld fills the internal corner with weld metal, forming a roughly triangular cross section that fuses the two pieces together. Common uses include welding fittings to pipes or joining branch pipes to header pipes.

Fitness for Service A multidisciplinary engineering approach that is used to determine if equipment is fit to continue operation for some desired future period. The equipment may contain flaws, have sustained damage (e.g., corrosion or creep damage), or have aged so that it cannot be evaluated by use of the original construction codes. A typical FFS assessment may involve several engineering disciplines such as stress analysis, metallurgy, nondestructive examination, and probabilistic fracture mechanics.

Flashing Flashing occurs when a high-pressure liquid flows through a valve or an orifice to a region of greatly reduced pressure. If the pressure drops below the vapor pressure, some of the liquid will be spontaneously converted to steam. The downstream velocity will be greatly increased due to a much lower average density of the two-phase mixture. The impact of the high-velocity liquid on piping or components creates flashing damage.

Flaw An imperfection or unintentional discontinuity that is detectable by nondestructive examination.

Flaw Aspect Ratio Ratio of length of deepest crack to depth of deepest crack.

Flow-Accelerated Corrosion Flow-accelerated corrosion is "a process whereby the normally protective oxide layer on carbon or low-alloy steel dissolves into a stream of flowing water or water-steam mixture." It can occur in both single-phase and two-phase regions. The cause of flow-accelerated corrosion is a specific set of water chemistry conditions (e.g., pH, level of dissolved oxygen), and there is no mechanical contribution to the dissolution of the normally protective iron oxide (magnetite) layer on the inside pipe wall.

Flow-Induced Vibration The term flow-induced vibration is used to describe outside-diameter pipe wear (or wall loss) caused by the interaction of flow-induced vibration and reflective metal insulation.

Full Structural Weld Overlay (FSWOL) A structural reinforcement and SCC mitigation technique through application of an SCC-resistant material layer around the entire circumference of the treated weldment. The minimum acceptable FSWOL thickness is 1/3 the original pipe wall thickness. The minimum length is $0.75 \times \sqrt{(R \times t)}$ on either side of the dissimilar metal weld to be treated, where R is the outer radius of the item and t is the nominal thickness of the item.

Fusion (or Heat Fusion) In the context of HDPE piping design and installation, fusion techniques are used to join pipes together. It is a welding process used to join two different pieces of a thermoplastic. This process involves heating both pieces simultaneously and pressing them together. The two pieces then cool together and form a permanent bond.

Gap Analysis Identifies discrepancies between raw data and a data classification scheme. It identifies missing or erroneous data (gaps from errors/omissions) to pinpoint areas needing improvement (e.g., more data, revised processes) to achieve completeness and quality measures. "Where are we now, where do we want to be, and what is missing to get there?" Data collection gaps refer to missing information associated with reliability attributes and influence factors.

General Corrosion An approximately uniform wastage of a surface of a component, through chemical or electrochemical action, free of deep pits or cracks.

Grayloc Region An inspection area at or near the mechanical connection between a primary heat transport system (PHTS) feeder tube and a fuel channel of the CANDU primary system. Grayloc® is a trade name of a type of mechanical connector that provides metal-to-metal seals in piping systems.

Heat-Affected Zone An area between a weld and base metal is a non-melted area that was exposed to high temperatures and therefore has undergone changes in its properties. The heat-affected zone can vary in its size and severity, dependent on the materials involved, the intensity and concentration of heat, and the process employed. Welding with high heat input (i.e., fast heating) has faster cooler rates compared to welding with low heat input (i.e., slow heating) and thus has smaller heat-affected zones. Conversely, welding with low heat input results in a larger heat-affected zone. As the speed of the welding decreases, the size of a heat-affected zone increases. Weld geometry also plays a role in the size of the heat-affected zone. The heat-affected zone problems can be mitigated by performing pre- and/or post-weld heat treatment. In high-temperature cutting processes, the depth of the heat-affected zone is associated with the cutting process itself, cutting speed, material properties, and material thickness. Similar to welding, cutting metals at high temperatures and slow speeds tends to lead to large heat-affected zones. Cutting metal at high speeds tends to reduce the width of the heat-affected zone because the zone experiences sufficient heat for a long enough time, and the layer undergoes microstructure and property changes that differ from the parent metal. These changes are usually undesirable and ultimately serve as the weakest part of the component. For example, microstructural changes can lead to high residual stresses, reduced material strength, increased brittleness, and decreased resistance to corrosion and/or cracking. As a result, many failures occur in the heat-affected zone.

Heat Balance Diagram (HBD) A schematic representation of the steam/condensate flow streams associated with the power generation cycle. The diagram contains information of steam properties (e.g., pressure, temperature, enthalpy, and steam mass flow rate) at every junction of the cycle. HBD provides input parameters to the development of predictive flow-accelerated corrosion management programs.

High-Energy Piping A piping system for which the maximum operating temperature exceeds 94 °C or the maximum operating pressure exceeds 1.9 MPa.

High-Performance Stainless Steel Standard-grade stainless steels are those having relatively low chromium and molybdenum contents, e.g., Types 304 and 316. The "high-performance" stainless steels, also referred to as "super-austenitic" stainless steels, are defined as those with substantially higher chromium, molybdenum, and nitrogen content.

Hydrogen-Induced Cracking Stepwise internal cracks that connect adjacent hydrogen blisters on different planes in the metal or to the metal surface. Externally applied stress is not needed for the formation of hydrogen-induced cracking. In steels, the development of internal cracks (sometimes referred to as blister cracks) tends to link with other cracks by a transgranular plastic shear mechanism because of internal pressure resulting from the accumulation of hydrogen. The linkup of these cracks on different planes in steels has been referred to as stepwise cracking to characterize the nature of the crack appearance. Hydrogen-induced cracking is commonly found in steels with (a) high impurity levels that have a high density of large planar inclusions and/or (b) regions of anomalous microstructure produced by segregation of impurity and alloying elements in the steel.

Hydrostatic Pressure Test A pressure test conducted during a plant or system shutdown at a pressure above nominal operating pressure or system pressure for which overpressure protection is provided.

Incident Reporting System (IRS) An international system jointly managed by the IAEA and the Nuclear Energy Agency of the Organisation for Economic Cooperation and Development. The IRS enables participating countries to exchange operational experience to improve the safety of nuclear power plants. The system was established in 1980 and has 33 participating countries. Since the IRS focuses on significant events important for the international community, it should be viewed neither as a source for statistical studies nor for component reliability studies. In addition, the system is designed for specialists of the nuclear community as a source of detailed information on analysis and lessons learned from operating experience. Access to the IRS is restricted to authorized users.

Induction Heat Stress Improvement The induction heat stress improvement process changes the normally tensile stress on the inside-diameter surface of weld heat-affected zones to compression. This process involves induction heating of the outer pipe surface of completed girth welds while simultaneously water cooling the inside with flowing water. Thermal expansion caused by the heating coil yields the outside surface in compression, while the cool inside surface yields in tension. After cooldown, contraction of the outside diameter causes the direction of stress to reverse, leaving the inside diameter in compression and the outside diameter in tension.

Institute of Nuclear Power Operations (INPO) Consolidate Events Database (ICES) The proprietary ICES database evolved from EPIX to provide a central repository for sharing nuclear power plant equipment failure and reliability data. It replaced the Nuclear Plant Reliability Data System (NPRDS) (see below) to meet industry needs for standardized, component-level information exchange.

International Nuclear Event Scale (INES) A tool for promptly and consistently communicating to the public the safety significance of events associated with sources of ionizing radiation. The INES scale applies to any event associated with the use, storage, and transport of radioactive material and radiation sources, whether or not the event occurs at a facility; this includes events involving the loss or theft of radioactive sources or packages and the discovery of orphan sources, such as sources being discovered in scrap metal. Events are rated at seven levels: levels 1–3 are "incidents" and levels 4–7 "accidents."

Kaplan-Garrick Risk Triplet Defines risk as the answer to three fundamental questions: (1) What can happen (the scenario)? (2) How likely is it to happen (the probability)? (3) If it happens, what are the consequences? Each of these is represented as a triplet of the form $<S, P, C>$, where S is the scenario, P is its probability, and C is its consequence. Risk analysis involves identifying all possible scenarios (S), determining their likelihood (P), and evaluating their potential impact (C).

Latent Failure A degraded material condition that may lie dormant for a long period before leading to a visible flaw (e.g., through-wall crack, active leakage).

Leak-Before-Break (LBB) Generally referred to as a methodology whereby it is shown that a crack can be detected by leakage under normal operating conditions and that crack would be stable at normal plus safe shutdown earthquake loads. Sometimes also referred to as a condition whereby a surface crack breaks through the pipe thickness and remains stable even if the break-through occurs at emergency or faulted loads. The leak-before-break principle does not apply to systems that are susceptible to fatigue cracking, SCC, or water hammer.

Less-Than-Adequate (LTA) In the context of a root cause analysis of RIM program deficiencies, the term "less-than-adequate" (LTA) is used to characterize a procedure that lacks something essential to successfully perform an activity.

Limiting Condition for Operation (LCO) According to the technical specifications, LCO is the lowest functional capability or performance level of a piece of equipment required for safe operation of a nuclear plant. When an LCO cannot be met, the reactor is to be shut down, or the licensee follows any remedial action permitted by technical specifications until the condition can be met.

Liquid Droplet Impingement The liquid droplet impingement is caused by the impact of high-velocity droplets or liquid jets. Normally, liquid droplet impingement occurs when a two-phase stream experiences a high-pressure drop (e.g., across an orifice on a line to the condenser). When this occurs, there is an acceleration of both phases with the liquid velocity increasing to the point that, if the liquid strikes a metallic surface, damage to the surface will occur. The main distinction between flashing and liquid droplet impingement is that in flashing, the fluid is of lower quality (mostly liquid with some steam), and with liquid droplet impingement, the fluid is of higher quality (mostly steam with some liquid).

Liquid Penetrant Examination Uses liquids to detect cracks in materials. In the mid- and late 1930s, Robert and Joseph Switzer worked with processes incorporating visible colored dyes in the penetrant to give better contrast. In 1941, they introduced processes using fluorescent dyes which, when viewed under a black light, produced contrasts superior to those obtainable with the visible dyes. The fluorescent method was quickly accepted by the military for aircraft part examination. Since then, the use of both color-contrast and fluorescent penetrants has spread to practically all fields of manufacturing, and new and improved PT products are constantly being developed.

Loss-of-Coolant Accident Those postulated accidents that result in a loss of reactor coolant at a rate in excess of the capability of the reactor make-up system from breaks in the reactor coolant pressure boundary, up to and including a break equivalent in size to the double-ended rupture of the largest pipe of the RCS.

Mechanical Stress Improvement Process (MSIP®) A patented process that was invented, developed, and first used in 1986 for mitigating SCC in nuclear plant weldments. MSIP® works by using a hydraulically operated clamp which contracts the pipe on one side of the weldment. A typical tool design consists of a specially designed hydraulic box press for bringing the clamp halves together. By contracting the pipe on one side of the weldment, the residual tensile stresses are replaced with compressive stresses.

Median Rank An estimate of the unreliability (cumulative probability of failure) for each failure data point at a 50% confidence level. It represents the median of the distribution of the ith ordered observation from a sample.

Microbiologically Influenced Corrosion (MIC) A process where corrosion is caused or accelerated by the activity of microorganisms, such as bacteria, on metal surfaces like piping. These microbes form biofilms (slimes) on the pipe's interior, which can produce corrosive byproducts like acids or sulfides. This activity leads to accelerated, severe pitting corrosion, often appearing as pinholes and significant deterioration under deposits formed by the microbes. MIC is common in low-flow or dead-leg areas: Areas with stagnant water, such as "dead legs" or low-velocity sections of piping, are prime locations for MIC. The interfaces where different materials or phases meet, such as a water/sediment or water/air interface, can harbor MIC-causing organisms.

Minimum Pie Wall Thickness Criterion The 87.5% minimum pipe wall thickness requirement in pipe design and manufacturing is primarily based on the allowable manufacturing tolerances for pipe wall thickness. For seamless pipes, the most common under-tolerance is 12.5%, meaning the actual wall thickness at any point should not be less than 87.5% of the nominal wall thickness specified in the design.

Miter Groove Weld Joins two angled pipes or tubing to form an elbow by cutting their ends at matching angles (e.g., 45 degrees each), creating a V-shaped groove for deep weld penetration.

Mixture Distribution Combining several probability distributions to represent various failure types and subpopulations, and Bayesian inference helps estimate their parameters and uncertainties, enabling more realistic reliability assessments than single-component models can achieve.

Moderate-Energy Piping A piping system for which the maximum operating temperature is less than 94°C or the maximum operating pressure is less than 1.9 MPa.

Multi-Pass Weld A welding technique that builds a strong, thick weld by applying multiple layers, or passes, of weld metal one on top of the other to completely fill a joint. This process is necessary for joining thick materials that cannot be welded in a single pass, as it allows for better control of heat input, reduces the risk of weld defects, and creates a more robust and reliable weld.

NDE Qualification In the context of nondestructive examination, qualification includes technical justification, which involves assembling all the supporting evidence for inspection capability (results of capability evaluation exercises, feedback from site experience, applicable and validated theoretical models, physical reasoning) and may include practical trials using deliberately defective test pieces.

Nominal Pipe Size A North American set of standard sizes for pipes used for high or low pressures and temperatures. "Nominal" refers to pipe in nonspecific terms and identifies the diameter of the hole with a nondimensional number (for example, a 2 inch nominal steel pipe consists of many varieties of steel pipe with the only criterion being 2.375 inch (60.3 mm) OD). Specific pipe is identified by pipe diameter and another nondimensional number for wall thickness referred to as the schedule (Sched. or SCG, for example, "2 inch diameter pipe, Schedule 40"). The European and international designation equivalent to nominal pipe size is DN (diamètre nominal/nominal diameter/Durchmesser nach Norm), in which sizes are measured in millimeters.

Non-safety Related with Special Treatment (NSRST) Components classified as NSRST relate to pressure boundary integrity by ensuring that these components, while not classified as safety related, still have a high degree of assurance (beyond normal industrial practices) that they can perform functions to prevent or mitigate events that might otherwise compromise the integrity of the reactor coolant pressure boundary (RCPB) or other critical pressure boundaries.

Notch Toughness A material's ability to absorb energy and resist fracturing when a stress concentrator (like a crack, flaw, or notch) is present, especially under rapid loading or low temperatures, indicating its capacity to transition from ductile to brittle failure. It is measured using tests like the Charpy V-notch test, where a notched sample breaks under impact, quantifying energy absorption to assess if the weld metal, heat-affected zone (HAZ), or base metal will be resilient or fail catastrophically.

Nuclear Plant Reliability Data System (NPRDS) Initiated by the American National Standards Institute Subcommittee N18-20, the NPRDS was designed to gather, store, and report on the reliability and failure statistics of nuclear power plant systems and components. It was considered the first major effort in the United States to collect data from participating utilities, with the goal of a comprehensive dataset by the mid-1970s.

Nuclear Safety Analysis Center (NSAC) The center was established soon after the accident at Unit 2 of the Three Mile Island Nuclear Power Plant near Harrisburg, Pennsylvania, in March 1979. The initial efforts of NSAC were directed towards a detailed analysis of the TMI accident and the results thereof. Subsequently, the functions were expanded to include (1) the analysis of generic nuclear safety issues, (2) serving as a repository for information resulting from the effort to restore the Three Mile Island plant, and (3) serving as a clearinghouse for technical information related to the safety and licensing of nuclear power plants. The center was dissolved in the early 1990s.

Optimized Weld Overlay (OWOL) A subset of the full structural weld overlay (FSWOL) process. It has been developed for larger geometries (e.g., RCS hot and cold leg nozzles) where FSWOL application becomes too time consuming for a typical refueling outage. The optimized weld overlay thickness is less than that of a full structural weld overlay in order to allow completion in the time available in a typical refueling outage for larger geometries.

PFM Realization A "realization" is a single run of the deterministic fracture mechanics model with one set of sampled input parameters. In probabilistic fracture mechanics, uncertainty is addressed by modeling inputs (like crack size, material properties, stresses, and inspection effectiveness) as probability distributions. Each realization samples values from these distributions, uses them in the deterministic model to simulate crack growth, and generates one outcome (e.g., time to leak or crack size). By running potentially millions of realizations, probabilities of failure are determined through statistical analysis of the collective results.

Phenomena Identification and Ranking Technique (PIRT) A systematic way of gathering information from experts on a specific subject, and ranking the importance of the information, in order to meet some decision-making objective. It has been applied to many nuclear technology issues including material degradation assessment in order to help guide research or develop regulatory requirements.

Pipe Schedule Designation The pipe schedule number (SN) is defined as SN = 1000 × (*P*/SE), where *P* is the operating pressure in lb/in^2 and SE is the allowable stress range multiplied by joint efficiency in lb/in^2 (i.e., how much stress the material can be subjected to).

Pipe Welding Positioner A mechanical device that holds and rotates pipes and other cylindrical workpieces to assist welders. It provides a stable, adjustable platform for the pipe, allowing it to be rotated for optimal welding angles, which improves weld quality, increases efficiency, and reduces welder fatigue.

Power Reactor Information System (PRIS) Developed and maintained by the International Atomic Energy Agency (IAEA), PRIS is a comprehensive database on nuclear power plants worldwide. The database covers reactor specification data (status, location, operator, owner, suppliers, milestone dates) and technical design characteristics. It also offers performance data including energy production and energy loss data, outage, and operational event information.

Precipitation Hardening (PH) Precipitation-hardened stainless steels are chromium- and nickel-containing steels that give an optimum combination of the properties of martensitic and austenitic grades. They are known for their ability to gain high strength through heat treatment, and they also have the corrosion resistance of austenitic stainless steels. The high tensile strength comes after a heat treatment process that leads to precipitation hardening, which is achieved through the addition of one or more of the elements Cu, Al, Ti, Nb, and Mo.

Probability of Detection The probability that a flaw of a certain size will be detected, and it is conditional on factors such as wall thickness, nondestructive examination personnel qualifications, and flaw orientation.

Probabilistic Physics of Failure (PPF) Model Uses mathematical tools that combine the physics-based understanding of failure mechanisms with statistical uncertainty to predict the reliability of a component or system. These models are used in reliability engineering to account for variations in material properties, manufacturing, and operating conditions and to predict a component's time to failure and remaining useful life by incorporating the probability of different failure modes like fatigue, wear, or corrosion.

Primary Water Stress Corrosion (PWSCC) Intergranular stress corrosion cracking nickel base of Ni-base alloys in PWR primary water environment.

Radiographic Examination A nondestructive testing method of inspecting materials for hidden flaws by using the ability of short-wavelength electromagnetic radiation (high-energy photons) to penetrate various materials.

Reliability and Integrity Management Those aspects of the plant design and operational phase that are applied to provide an appropriate level of reliability of SSCs and a continuing assurance over the life of the plant that such reliability is maintained. The most recent edition of the ASME Boiler and Pressure Vessel Code Section XI was issued in 2019. Division 2 of the 2019 edition provides the requirements for the creation of the RIM program for advanced nuclear reactor designs. The RIM program addresses the entire life cycle for all types of NPPs. It requires a combination of monitoring, examination, tests, operation, and maintenance requirements that ensures that each SSC meets plant risk and reliability goals that are selected for the RIM program.

Repair Weld A weld subjected to high residual stresses as a result of extensive repairs, e.g., during original construction. Such regions may be subjected to premature SCC.

Residual Stresses Those stresses that remain in an object (in particular, in a welded component) even in the absence of external loading or thermal gradients. In some cases, residual stresses result in significant plastic deformation, leading to warping and distortion of an object. In others, they affect susceptibility to fracture and fatigue.

Richardson Number (R_i) Named after Lewis Fry Richardson (1881–1953), a dimensionless parameter used in the identification and assessment of locations in a piping system potentially susceptible to stratification.

Risk-Informed In-Service Inspection Risk informed ISI methodologies are applied to determine the risk significance and failure potential of a piping component. This enables the targeting in-service examination resources to locations that are truly risk significant, providing the ability to capture or minimize risk and thereby improving plant reliability while keeping radiation doses to workers as low as reasonably achievable. The failure potential is determined through degradation mechanism analysis. The risk significance is determined through pipe failure frequency assessment.

Root Mean Square (RMS) Evaluation NDE qualifications include the use of an ultrasonic sizing procedure which should be developed and qualified for equipment, technique, and sizing examination personnel. At least ten flawed specimens should be used in the performance demonstration. RMS evaluation should be used to demonstrate adequate sizing performance. Acceptable flaw sizing performance demonstration is achieved when the RMS is 12.5% or less.

Root Weld Gap The intentional space between two pieces of metal before welding. This gap is essential for achieving full penetration, ensuring that the weld metal fuses through the entire thickness of the joint and creates a strong, solid bond. Without a proper root gap, the arc may not reach the bottom of the joint, resulting in poor penetration and a weak weld.

Round-Robin Test An analysis or experiment performed independently several times and by different teams. This can involve multiple teams performing analyses using a common set of data with the use of a variety of methods. A round-robin test may be conducted to determine the reproducibility of an analysis, from input parameters to interpretation of results.

Schaeffler Diagram A chart used by welders and metallurgists to predict the microstructure of stainless steel after it has been heated and then cooled rapidly. It plots the "chromium equivalent" (a measure of ferrite-stabilizing elements like chromium) against the "nickel equivalent" (a measure of austenite-stabilizing elements like nickel) to show how the chemical composition will determine if the final structure will be austenitic, ferritic, martensitic, or a combination, like duplex.

Selective Leaching Also referred to as dealloying, demetalification, parting, and selective corrosion, it is a corrosion type in some solid solution alloys, when in suitable conditions, a component of the alloys is preferentially leached from the material. The less noble metal is removed from the alloy by a microscopic-scale galvanic corrosion mechanism. The most susceptible alloys are the ones containing metals with high distance between each other in the galvanic series, e.g., copper and zinc in brass.

Significance Determination Process (SDP) Part of the United States Nuclear Regulatory Commission's "Reactor Oversight Process." The SDP estimates the increase in annualized CDF risk due to identified deficiencies in licensee performance that led to unavailability of equipment or safety functions or to increase in initiating event frequencies. This increase is measured from the normal annualized CDF that results from routine plant operation. The additional risk contributions caused by deficient licensee performance (as characterized by the SDP) are assumed to be additive to this normal annualized CDF, which already includes the risk contribution due to the probabilities of equipment failures expected occasionally for industrial facilities of this size and complexity.

Significant Structural Failure (SSF) Loss of piping structural integrity with direct operational impact. In this text, such a failure produces a through-wall mass flow rate greater than 10 kg/s.

S–N Curve A plot of the magnitude of an alternating stress versus the number of cycles to failure for a given material. Typically, both the stress and number of cycles are displayed on logarithmic scales. SN curves were developed by the German scientist August Wöhler during the investigation of an 1842 train crash in Versailles, France. In this crash, the axle of the train locomotive failed under the repeated "low-level" cyclic stress of everyday usage on the railroad. While investigating, Wöhler discovered that cracks formed and slowly grew on an axle surface. The cracks, after reaching a critical size, would suddenly propagate and the axle would fail. The level of these loads was less than the ultimate strength and/or yield strength of the material used to manufacture the axle.

Socket Weld A pipe attachment detail in which a pipe is inserted into a recessed area of a pipe (e.g., elbow), valve, or flange. Socket welds are mainly used for small-diameter pipes, generally, for piping whose nominal diameter is DN50 or smaller.

Sockolet A pipe fitting with a "run end" and a "branch end." The run end connects to the larger diameter pipe, and the branch end connects to the smaller diameter pipe. The connection to the run pipe is a fillet weld.

Solid Particle Erosion Damage caused by particles transported by fluid stream rather than by liquid water or collapsing bubbles. If hard, large particles are present at sufficiently high velocities, damage will occur. In contrast to liquid droplet impingement, the necessary velocities for solid particle erosion are quite low. Surfaces damaged by solid particle erosion have a very variable morphology. Manifestations of solid particle erosion in service usually include thinning of components, a macroscopic scooping appearance following the gas/particle flow field, surface roughening (ranging from polishing to severe roughening, depending on particle size and velocity), lack of the directional grooving characteristics of abrasion, and, in some but not all cases, formation of ripple patterns on metals.

Spatial Impact In nuclear safety regulations, the spatial impact of pipe failure refers to the localized physical, environmental, and structural damage caused by a pipe rupture on surrounding equipment and buildings. Regulatory bodies require that safety systems remain functional even if a pipe breaks, necessitating analyses that consider the zone of influence (ZOI) of a failure.

Stabilized Austenitic Stainless Steel Refers to steel containing "stabilizers" such as titanium and niobium, which form carbides that are more stable than chromium carbides. Research Paper RP1878 (Journal of Research of the National Bureau of Standards, Volume 40, April 1948) documents a study on the resistance of different austenitic stainless steels to IGSCC. According to the paper, IGSCC susceptibility "may be decreased or eliminated by the addition of titanium or columbium, in conjunction with stabilizing heat treatment." Type 321 (EN 1.4541) is an example of a Ti-stabilized austenitic stainless steel. Type 347 is an example of an Nb-stabilized austenitic stainless steel.

Steam Hammer In fluid dynamics, "steam hammer" refers to two related phenomena in steam systems: a violent pressure spike caused by a slug of condensed water moving at high speed, and a similar effect caused by the sudden collapse of steam bubbles against a surface. The first type, also known as differential shock, happens when steam condenses in a pipe, creating a "slug" of water that accelerates and then slams into an obstruction. Thermal shock happens when steam quickly condenses on a cold surface, forming a vacuum that draws in water or triggers condensation and collapse, creating a shock wave. Both can cause loud banging noises and severe damage to pipes, fittings, and other equipment.

Strain-Induced Corrosion Cracking Strain-induced corrosion cracking is used to refer to those corrosion situations in which the presence of localized dynamic straining is essential for crack formation (i.e., initiation and propagation) to occur, but in which cyclic loading is either absent or restricted to a very low number of infrequent events. Strain-induced corrosion cracking has been observed in particular in German NPPs made of higher-strength carbon steel and low-alloy steel.

Stress Analysis Isometric Drawing A specialized 3D-styled, single-line diagram of a piping system used to visualize, model, and calculate structural loads, including thermal expansion and pressure, to ensure system safety. These drawings highlight critical node points (supports, branches, valves) for FEA analysis and provide essential data for stress reports.

Stress Corrosion Cracking SCC is a localized non-ductile failure which occurs only under the combination of three factors: (1) tensile stress, (2) aggressive environment, and (3) susceptible material. The SCC failure mode can be intergranular SCC or transgranular SCC.

Stress-Oriented Hydrogen-Induced Cracking Arrays of cracks that are aligned nearly perpendicular to the applied stress, which is formed by the linkup of small hydrogen-induced cracking in steel. Tensile stress (residual or applied) is required to produce stress-oriented hydrogen-induced cracking. Stress-oriented hydrogen-induced cracking is commonly observed in the base metal adjacent to the heat-affected zone of a weld, oriented in the through-thickness direction. Stress-oriented hydrogen-induced cracking may also be produced in susceptible steels at other high-stress points such as from the tip of mechanical cracks and defects or from the interaction between hydrogen-induced cracking on different planes in the steel.

Structural Reliability Model Sometimes, the terms PFM and structural reliability models are used synonymously. While there are some common elements in the underlying methodologies, a structural reliability model concerned is "stress-strength" model in which it is assumed that a component fails if the applied stress exceeds its strength, a type II structural reliability model in some classification schemes.

Super-Emergency Feedwater System (SEFW) A safety-related system that is specific to WWER-type reactors. It is designed to respond to the total loss of heat-sink accident scenarios (i.e., total loss of normal and auxiliary feedwater). The system injects feedwater directly into the steam generator secondary side by separate stainless steel feedlines from demineralized water tanks.

Technical Specification (TS) Establishes requirements for safety limits, limiting safety system settings, limiting control settings, limiting conditions for operation, surveillance requirements, design features, and administrative controls.

Thermal Aging Embrittlement A time- and temperature-dependent change in the microstructure of the material which leads to a reduced ductility and deterioration of the fracture toughness and the impact properties. The material will show an increased embrittlement over time.

Thermal Stratification Hot water can flow above cold water in horizontal runs of piping when the flow (hot water into a cold pipe or cold water into a hot pipe) does not have enough velocity to flush the fluid in the pipe. The temperature profiles in the pipe where the top of the pipe is hotter than the bottom cause the pipe to bow along with the normal expansion at the average temperature.

Thermal Striping Incomplete mixing of high-temperature and low-temperature fluids near the surface of structures with subsequent fluid temperature fluctuations gives rise to thermal fatigue damage to wall structures.

Through-Wall Crack Instability A condition where a through-wall crack grows around the pipe circumference in a rapid manner.

Tinel Coupling A pipe coupling made of Tinel material (half titanium and half nickel). The coupling is machined to predetermined dimensions with the inside diameter smaller than the tube outside diameter. Passing a mandrel through the coupling inside diameter while submerged in liquid nitrogen at cryogenic temperature will expand the coupling. It will remain as expanded while submerged in liquid nitrogen. When the coupling is installed to couple two pieces of tubing together, it will automatically return to its predetermined dimensions when exposed to room temperature. The shape memory is the result of a change in the crystal structure of the alloy known as reversible austenite to martensite phase transformation.

Transgranular Stress Corrosion Cracking Transgranular SCC is caused by aggressive chemical species especially if coupled with oxygen and combined with high stresses.

Trial Phosphate Ester A fire-resistant self-extinguishing electrohydraulic control fluid especially developed for steam turbine application. According to the U.-S. National Fire Protection Association (NFPA) Recommended Practice 850 ("Fire Protection for Electric Generating Plants"), a fire-resistant fluid "... is difficult to ignite and does not sustain combustion due to its low heat of combustion."

Tritium in Nuclear Power Plants Most of the tritium produced in nuclear power plants stems from boron, absorbing neutrons from the plant's chain reaction. Nuclear reactors use boron, a good neutron absorber, to help control the chain reaction. Towards that end, boron either is added directly to the coolant water or is used in the control rods to control the chain reaction. Much smaller amounts of tritium can also be produced from the splitting of Uranium-235 in the reactor core or when other chemicals (e.g., lithium or heavy water) in the coolant water absorb neutrons. Like normal hydrogen, tritium can bond with oxygen to form water.

When this happens, the resulting "tritiated" water is radioactive. Tritiated water (not to be confused with heavy water) is chemically identical to normal water, and the tritium cannot be filtered out of the water.

Unified Numbering System (UNS) An alloy designation system in use in North America. It consists of a prefix letter and five digits designating material composition. For example, a prefix of S indicates stainless steel, and C indicates copper, brass, or bronze alloys.

Valve Chatter Chattering is the rapid opening and closing of a pressure-relief valve. The resulting vibration may cause misalignment, valve seat damage, and, if prolonged, mechanical failure of valve internals and associated piping. Some causes of chattering include excessive inlet pressure drop, excessive backpressure, an oversized relief valve, and a relief valve that must handle widely varying flow rates.

Visual Examination (or Testing) The oldest and most commonly used NDE method, which may be defined as "an examination of an object using the naked eye, alone or in conjunction with various magnifying devices, without changing, altering, or destroying the object being examined." Per ASME XI, there are three different VT methods: VT-1, VT-2, and VT-3.

VT-1 Examination As specified by ASME Section XI, a limited visual examination which is the observation of exposed surfaces of a part, component, or weld to determine its physical condition including such irregularities as cracks, wear, erosion, corrosion, or physical damage.

VT-2 Examination Per ASME XI, VT-2 is a visual surface examination to locate evidence of leakage from pressure-retaining components.

VT-3 Examination A limited visual examination specific to ASME Section XI which is the observation to determine the general mechanical and structural condition of components and their supports, such as the verification of clearances, settings, physical displacements, loose or missing parts, debris, corrosion, wear, erosion, or loss of integrity at bolted or welded connections. The VT-3 examinations include examinations for conditions that could affect operability of functional adequacy of snubbers and constant load and spring-type supports. The VT-3 examination is intended to identify individual components with significant levels of existing degradation.

Von Mises Stress A value used to determine if a given material will yield or fracture. It is mostly used for ductile materials, such as metals. The von Mises yield criterion states that if the von Mises stress of a material under load is equal or greater than the yield limit of the same material under simple tension, then the material will yield. Richard von Mises was an Austrian mathematician and aerodynamicist who developed the von Mises stress theory in 1913, which is used to predict when ductile materials will yield under complex stress states.

Water Hammer If the velocity of water or other liquid flowing in a pipe is suddenly reduced, a pressure wave results, which travels up and down the pipe system at the speed of sound in the liquid. Water hammer occurs in systems that are subject to rapid changes in fluid flow rate, including systems with rapidly actuated valves, fast-starting pumps, and check valves.

Weld Inlay A mitigation technique defined as application of primary water SCC-resistant material (Alloy 52/152) to the inside diameter of a dissimilar metal weld that isolates the primary water SCC susceptible material (Alloy 82/182) from the primary reactor coolant.

Welding Procedure Specification (WPS) A formal written document describing welding procedures, which provides direction to the welder or welding operators for making quality production welds as per the code requirements. The purpose of the document is to guide welders to the accepted procedures so that repeatable and trusted welding techniques are used. A welding procedure specification is developed for each material type and for each welding type used.

Weld Residual Stress Refers to the internal, locked-in stresses in a material that remain after a welding process, caused by nonuniform heating and cooling cycle that creates differential volumetric expansion and contraction. These stresses can be tensile and can reduce the structural integrity of a weld by increasing susceptibility to fatigue and stress corrosion cracking.

Index

B. Lydell, *Knowledge Preservation in Nuclear Safety*,
https://doi.org/10.1007/978-3-032-25463-4

The manufacturer's authorised representative in the EU is Springer Nature Customer Service Centre GmbH, Europaplatz 3, 69115 Heidelberg, Germany. If you have any concerns regarding our products, please contact ProductSafety@springernature.com

Printed and bound by CPI Group (UK) Ltd, Croydon, CR0 4YY
07/10/2026
02229351-0014